lationship

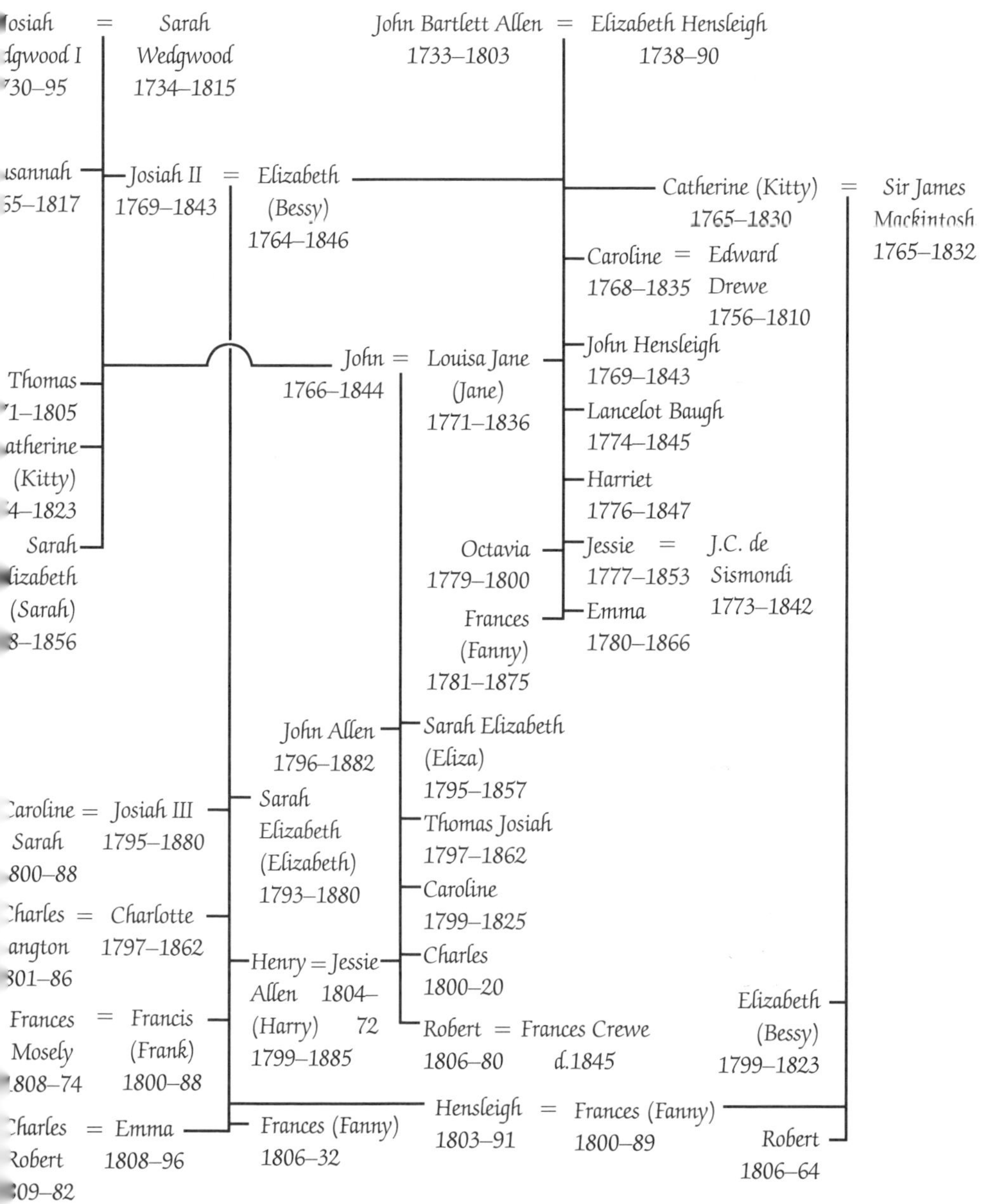

osiah
dgwood I
730–95
Sarah
Wedgwood
1734–1815
John Bartlett Allen
1733–1803
Elizabeth Hensleigh
1738–90
usannah
55–1817
Josiah II
1769–1843
Elizabeth
(Bessy)
1764–1846
Catherine (Kitty)
1765–1830
Sir James
Mackintosh
1765–1832
Caroline
1768–1835
Edward
Drewe
1756–1810
John Hensleigh
1769–1843
John
1766–1844
Louisa Jane
(Jane)
1771–1836
Thomas
71–1805
Lancelot Baugh
1774–1845
atherine
(Kitty)
4–1823
Harriet
1776–1847
Sarah
lizabeth
(Sarah)
8–1856
Octavia
1779–1800
Jessie
1777–1853
J.C. de
Sismondi
1773–1842
Frances
(Fanny)
1781–1875
Emma
1780–1866
John Allen
1796–1882
Sarah Elizabeth
(Eliza)
1795–1857
Caroline
Sarah
800–88
Josiah III
1795–1880
Sarah
Elizabeth
(Elizabeth)
1793–1880
Thomas Josiah
1797–1862
Caroline
1799–1825
Charles
angton
301–86
Charlotte
1797–1862
Henry
Allen
(Harry)
1799–1885
Jessie
1804–72
Charles
1800–20
Frances
Mosely
808–74
Francis
(Frank)
1800–88
Robert
1806–80
Frances Crewe
d.1845
Elizabeth
(Bessy)
1799–1823
Charles
Robert
309–82
Emma
1808–96
Frances (Fanny)
1806–32
Hensleigh
1803–91
Frances (Fanny)
1800–89
Robert
1806–64

THE CORRESPONDENCE OF

CHARLES DARWIN

Editors

FREDERICK BURKHARDT   DUNCAN M. PORTER

SHEILA ANN DEAN   SAMANTHA EVANS

SHELLEY INNES   ALISON M. PEARN

ANDREW SCLATER   PAUL WHITE

SARAH WILMOT

Assistant Editor

CHARLOTTE BOWMAN

Research Associates

ANNE SCHLABACH BURKHARDT

ANNA-K. MAYER

This edition of the Correspondence of Charles Darwin is sponsored by the American Council of Learned Societies. Its preparation is made possible by the co-operation of Cambridge University Library and the American Philosophical Society.

Support for editing has been received from the Alfred P. Sloan Foundation, the Andrew W. Mellon Foundation, the National Endowment for the Humanities, the National Science Foundation, the Pew Charitable Trusts, the British Academy, the British Ecological Society, the Isaac Newton Trust, the Natural Environment Research Council, the Royal Society of London, the Stifterverband für die Deutsche Wissenschaft, and the Wellcome Trust. The National Endowment for the Humanities's grants (Nos. RE-23166-75-513, RE-27067-77-1359, RE-00082-80-1628, RE-20166-82, RE-20480-85, RE-20764-89, RE-20913-91, RE-21097-93, and RZ-20393-99) were from its Program for Editions; the National Science Foundation's funding of the work was under grants Nos. SOC-75-15840, SES-7912492, SES-8517189, SBR-9020874, and SES-0135528. Any opinions, findings, conclusions, or recommendations expressed in this publication are those of the editors and do not necessarily reflect the views of the grantors.

Charles Darwin in 1865
Photograph by the London Stereoscopic & Photographic Company
Courtesy of the Wellcome Library, London

THE CORRESPONDENCE OF CHARLES DARWIN

VOLUME 13 1865

SUPPLEMENT TO THE CORRESPONDENCE

1822–1864

CAMBRIDGE
UNIVERSITY PRESS

PUBLISHED BY THE PRESS SYNDICATE OF THE UNIVERSITY OF CAMBRIDGE
The Pitt Building, Trumpington Street, Cambridge, United Kingdom

CAMBRIDGE UNIVERSITY PRESS
The Edinburgh Building, Cambridge CB2 2RU, UK
40 West 20th Street, New York, NY 10011-4211, USA
477 Williamstown Road, Port Melbourne, VIC 3207, Australia
Ruiz de Alarcón 13, 28014 Madrid, Spain
Dock House, The Waterfront, Cape Town 8001, South Africa

http://www.cambridge.org

First published 2002

Citation:
Burkhardt, Frederick *et al.*, eds. 2002. *The correspondence of Charles Darwin.*
Vol. 13. Cambridge: Cambridge University Press.

Printed in the United Kingdom at the University Press, Cambridge

Typeface Monotype Baskerville 10/12 pt. *System* EmTex

A catalogue record for this book is available from the British Library

ISBN 0 521 82413 3 hardback

In Memoriam

GEORGE PEMBER DARWIN

1928–2001

CONTENTS

ILLUSTRATIONS

CALENDAR LIST OF LETTERS

The following list is in the order of the entries in the *Calendar of the correspondence of Charles Darwin*. The first part includes all those letters that are listed in the *Calendar* for the year 1865, and those that have been redated into 1865. The second part includes the letters published in the supplement to this volume. Alongside the *Calendar* numbers are the corrected dates of each letter. A date or comment printed in italic type indicates that the letter has been omitted from this volume, except where the letter has been moved out of 1865 into the Supplement to this volume.

Letters acquired after the publication of first edition of the *Calendar*, in 1985, have been given numbers corresponding to the chronological ordering of the original *Calendar* listing with the addition of an alphabetical marker. Many of these letters are summarised in a 'Supplement' to a new edition of the *Calendar* (Cambridge University Press, 1994). The marker 'f' denotes letters acquired after the second edition of the *Calendar* went to press in 1994.

2435. 23 Mar [1865]
3493. 1 Apr 1865
3494. 4 Apr [1865]
4330. [3 Nov 1865]
4370. [Apr–May 1865?]
4381. 6 Jan [1865]
4467. 19 Apr [1865]
4603. [27 Aug – 1 Sept 1865]
4610. 7 Sept [1865]
4639. 19 Oct [1865]
4643. 24 Oct [1865]
4726. *[1863–6]*
4727. [before 1 Oct 1865?]
4728. *[before 2 Oct 1866]*
4729. [late Feb–May 1865]
4730. *[before 28 Mar 1872]*
4731. *[24 Mar–20 Apr 1866]*
4732. 1 Jan 1865
4733. 1 Jan 1865
4734. 1 Jan 1865
4735. 2 Jan 1865
4736. 3 Jan 1865
4737. 3 Jan 1865
4738. 4 Jan [1865]
4739. 5 Jan [1865]
4740. 6 Jan [1865]
4741. 6 Jan [1865]
4742. 7 Jan [1865]
4743. [8–18 Jan 1865]
4744. 9 Jan [1865?]
4744a. *10 Jan [1868]*
4744f. [before 7 Jan 1865]
4745. 15 Jan 1865
4746. 16 Jan 1865
4747. 17 Jan 1865
4747a. 18 Jan [1865]
4747f. 17 Jan [1865]
4748. 19 Jan [1865]
4749. [20 Jan 1865]
4750. 20 Jan 1865
4751. 20 Jan 1865
4752. 22 Jan [1865]
4753. 23 Jan 1865
4754. [26 Jan 1865]
4755. 27 Jan [1865]
4756. 28 Jan 1865
4756f. 28 Jan [1865]
4757. 29 Jan [1865]
4758. 30 Jan [1865]
4759. 31 Jan [1865]
4760. 1 Feb [1865]
4761. *1 Feb 1864*
4762. 2 Feb [1865]
4763. [3 Feb 1865 or 1866?]
4764. [14–18 Jan 1865]
4765. 3 Feb 1865

SUPPLEMENT, 1822–64

INTRODUCTION

In 1865, the chief work on Charles Darwin's mind was the writing of *The variation of animals and plants under domestication* (*Variation*). At the beginning of the year he was optimistic about publishing it that autumn, but a recurrence of illness forced him to abandon this plan, and he did not resume work on the book until December. However, several smaller projects came to fruition in 1865, including the publication of his long paper on climbing plants in the *Journal of the Linnean Society*, and, arising from this, the editing of excerpts from Fritz Müller's letters on climbing plants to make another paper. Darwin also submitted a manuscript of his hypothesis of pangenesis, intended for publication in *Variation*, to Thomas Henry Huxley for evaluation, and persuaded his friend Joseph Dalton Hooker to comment on a paper on *Verbascum* (mullein) by CD's protégé, John Scott, who was now working in India.

Darwin's transmutation theory continued to be discussed in both scholarly and popular publications. A lengthy discussion written by George Douglas Campbell, duke of Argyll, appeared in the religious weekly, *Good Words*. Darwin received news of an exchange of letters on his theory in a New Zealand newspaper; the letters were written by a supporter of his, Samuel Butler, and, according to Butler, the bishop of Wellington. Darwin's theory was discussed at an agricultural congress in Erfurt, Germany, and the proceedings were reported in the *Gardeners' Chronicle*. At the end of the year, Darwin was elected an honorary member of the Royal Society of Edinburgh.

The year was marked by three deaths of personal significance to Darwin: Hugh Falconer, a friend of Darwin's and prominent supporter of (though not a convert to) Darwin's theory, died at the end of January; Robert FitzRoy, captain of the *Beagle* on Darwin's 1831 to 1836 voyage, committed suicide at the end of May; and William Jackson Hooker, director of the Royal Botanic Gardens, Kew, and J. D. Hooker's father, died in August. There was also a serious dispute between two of Darwin's friends, John Lubbock and Charles Lyell. These events all inspired sombre reflections in Darwin's letters. Nevertheless, there were lighter moments. All the Darwin children were thriving, and when illness made work impossible, Darwin and Hooker read a number of novels, and discussed them in their letters.

Darwin's first letter to Hooker of 1865 suggests that the family had spent a cheerful Christmas together. 'We are all well', Darwin wrote, '& I enjoy much having all the Boys at home: they make the house jolly' (letter to J. D. Hooker, 7 January [1865]). Darwin was ready to submit his paper on climbing plants to the Linnean Society of London, and though he was clearly not perfectly well, his tone was enthusiastic and energetic. However, on 31 January, Hugh Falconer died after a sudden illness. Falconer was 56, almost the same age as Darwin himself. Falconer had seconded Darwin's nomination for the Copley Medal of the Royal Society of London in 1864, had staunchly supported his candidacy, and had tried hard to persuade Darwin to accept the award in person (see *Correspondence* vol. 12). In early January Falconer had written to Darwin's brother, Erasmus Alvey Darwin, to reassure Darwin (or at least Erasmus) that he had 'stuck up' for *On the origin of species* (*Origin*), which the Council of the Royal Society had failed to include among the grounds of the award (see letter from Hugh Falconer to Erasmus Alvey Darwin, 3 January 1865). Erasmus forwarded his letters to Darwin, and Darwin responded warmly: 'Your letter is by far the grandest eulogium which I have ever received, or shall ever receive; and if one half—or one quarter be true and not exaggerated by your great kindness, I may well rest content that I have not laboured in vain' (letter to Hugh Falconer, 6 January [1865]).

Darwin's response to the news of Falconer's death was at first calm and philosophical: 'I am much grieved; It will be a great loss to science. What a lot of knowledge of all kinds has perished with him. He was always a most kind friend to me. So the world goes.—' (letter to J. D. Hooker, 2 February [1865]). However, Hooker, at the time recovering from a bout of influenza, wrote to Darwin at some length about Falconer's life and death, concluding gloomily: 'The inconceivability of our being born for nothing better than such a paltry existence as ours' is, gives me some hope of meeting in a better world. What does it all mean.— When we think what millions upon millions of lives & intellects it has taken to work up to a knowledge of gravity & Natural selection, we really do seem a contemptible creation intellectually & when we feel the death of friends more keenly the older we grow, we do strike me as being corporeally most miserable, for we have no pleasures to compensate fully for our griefs & pains: these alone are unalloyed' (letter from J. D. Hooker, 3 February 1865). Darwin, now 'haunted' by Hooker's account of Falconer's last sufferings, responded with an even more cosmic sense of futility, and with only the remotest sense of consolation: 'I quite agree how humiliating the slow progress of man is; but everyone has his own pet horror, & this slow progress, or even personal annihilation sinks in my mind into insignificance compared with the idea, or rather I presume certainty, of the sun some day cooling & we all freezing. To think of the progress of millions of years, with every continent swarming with good & enlightened men all ending in this; & with probably no fresh start until this our own planetary system has been again converted into red-hot gas.— Sic transit gloria mundi, with a vengeance' (letter to J. D. Hooker, 9 February [1865]).

Darwin had another cause for gloom: his health was poor throughout 1865, and at the beginning of the year he despaired of finding a doctor who could ease his symptoms. He wrote to Hooker for recommendations but added, 'I know it is folly & nonsense to try anyone' (letter to J. D. Hooker, 7 January [1865]). He particularly hated being ill because it affected his ability to read and write, and this, and his experiments, were, he claimed, important for his enjoyment of life.

He wrote to Charles Lyell on 22 January [1865], 'unfortunately reading makes my head whiz more than anything else. I am able most days to work for 2 or 3 hours & this makes all the difference in my happiness'. At the end of April, Darwin's condition worsened to the extent that he felt unable to do any work other than looking over the proofs of the second French edition of *Origin* (see CD's 'Journal', Appendix II). In May, he invited a new doctor, John Chapman, to Down and began a course of Chapman's ice treatment, which involved placing bags of ice along the spine. (A detailed list of Darwin's symptoms, which Darwin wrote for Chapman, is reproduced in Appendix IV.) The treatment proved to be ineffective, and Darwin had given it up by early July (see letter to J. D. Hooker, [10 July 1865]). In July, he consulted the physician Henry Bence Jones, who put him on a strict diet involving 'eating very little of anything, & that almost exclusively bread & meat' (letter to Asa Gray, 15 August [1865]). By October, Darwin thought he might be slowly getting better, attributing the improvement to Jones's diet (see letter to T. H. Huxley, 4 October [1865]). It was not until December, however, that he announced that he was 'able to write about an hour on most days' (letter to J. D. Hooker, 22 December [1865]).

Nevertheless, the letters Darwin wrote and dictated during the year show that he achieved a good deal and kept up some level of scientific activity–even if only being read to–during all but his worst periods of illness. *Variation*, which he had begun in January 1860, and which was intended to explain his theory of descent and the operation of natural selection in greater detail than *Origin*, was his greatest concern. In the early months of the year there are a number of letters to and from the poultry expert William Bernhard Tegetmeier concerning the draft of a manuscript on poultry that Tegetmeier was reading for Darwin and the commissioning of artists to make woodcuts for *Variation*. In March Darwin wrote to his publisher, John Murray, 'Of present book I have 7 chapters ready for press & all others very forward, except the last & concluding one' (letter to John Murray, 31 March [1865]). In April he authorised Murray to advertise the book, but warned: 'my health is extremely precarious and . . . it depends upon this and upon nothing else whether my book will be ready for the press in the autumn' (letter to John Murray, 4 April [1865]). In early June, he wrote to Murray to announce that there was now 'no chance' that he would be ready to publish in the autumn, assuring him, however, 'I am never idle when I can do anything' (letter to John Murray, 2 June [1865]). It was not until 25 December that he resumed work on *Variation* where he had left off in April (CD's 'Journal', Appendix II). In the event, because of further illness and delay, the book was not published until January 1868.

As was usual with him, however, Darwin had a number of smaller projects on hand at the same time as the great one. In January he expected his manuscript on climbing plants back from the copyist, and was wondering whether to send it to the Linnean Society, or to the Royal Society of London, who he thought might be more willing to bear the expense of the woodcuts (letter to J. D. Hooker, 7 January [1865]). After sending the manuscript to the Linnean, he complained to Hooker: 'For the last ten days I have done nothing but correct refractory sentences, & I loathe the whole subject like tartar emetic' (letter to J. D. Hooker, 19 January [1865]). An abstract of the paper was read before the Linnean Society on 2 February, and in April Darwin wrote to his friend Asa Gray, a botanist in the United States, that he had started correcting proofs in preparation for its publication: 'I think it contains a good deal new & some curious points, but it is so fearfully long, that no one will ever read it. If, however, you do not *skim* through it, you will be an unnatural parent, for it is your child' (letter to Asa Gray, 19 April 1865; Darwin noted at the beginning of 'Climbing plants' that he had been led to the subject by a paper of Gray's). This work seems to have continued after the worst phase of Darwin's illness, in May, for in early June Darwin was ordering books that he needed for references, probably from the Linnean Society (letter to [Richard Kippist], 4 June [1865]). The paper was published in a double issue of the *Journal of the Linnean Society* in June, and a commercial offprint was also made, as well as a set for the author (Freeman 1978).

The publication of 'Climbing plants' led to a small collaborative exercise later in the year. One of the people to whom Darwin sent a copy of the paper was Fritz Müller, in Destêrro (now Florianópolis), Brazil. Darwin had received a copy of Müller's book, *Für Darwin*, a study of the Crustacea with reference to Darwin's theory of transmutation, in or before November 1864 (*Correspondence* vol. 12, letter to Ernst Haeckel, 21 November [1864]). Since it was, according to Darwin, in difficult German, he had it translated, and wrote to Müller in August 1865 that he had just finished hearing it read aloud (letter to Fritz Müller, 10 August [1865]). Over the next few months, Müller submitted to Darwin a wealth of observations on climbing plants in Destêrro, excerpts from which Darwin edited and submitted in October to the Linnean Society for publication in Müller's name (see letter from Fritz Müller, [12 and 31 August, and 10 October 1865]; since it is impossible to disentangle the original letters from the published version and rejoin them with other fragments that remain, the article is reprinted in this volume). Before submitting the letters to the Linnean Society, Darwin enlisted the help of Daniel Oliver, a botanist at Kew, to check one or two points that he did not clearly understand (letter to Daniel Oliver, 20 October [1865]). Darwin was particularly interested in Müller's observations of tendrils formed by the modification of branches, and made alterations to the second edition of *Climbing plants* reflecting Müller's findings. Although he wrote to Müller, 'I am sorry to say my health keeps so weak that I am not able to do any scientific work' (letter to Fritz Müller, 20 September [1865]), he clearly read Müller's letters (or heard them read) with close and active interest, raising questions and suggesting new lines of research.

A similar, though not so immediately productive, collaborative effort took place in the beginning of the year when John Scott, a protégé of Darwin's whom Darwin and Hooker had recently helped to find employment in India (see *Correspondence* vol. 12), sent Darwin the results of his experiments on *Verbascum*. Darwin had suggested to Scott in 1862, when Scott was working at the Royal Botanic Garden,

Edinburgh, that he should repeat Karl Friedrich von Gärtner's experiments on *Verbascum*, in which crosses between differently coloured varieties of the same or of different species were found to produce less seed than the parallel crosses between similarly coloured varieties (see *Correspondence* vol. 10, letter to John Scott, 19 November [1862]). Darwin had already written to Hooker of Gärtner's experiments: 'I do not think any experiment can be more important on Origin of species; for if he is correct, we certainly have what Huxley calls new physiological species arising' (*Correspondence* vol. 9, letter to J. D. Hooker, 28 September [1861]). Scott had evidently started his crossing experiments in 1863 (see *Correspondence* vol. 11, letter from John Scott, 21 September [1863]), and wrote up his results on his voyage to India in late 1864, despite suffering from sea-sickness (letter from John Scott, 21 July 1865). This may have been unwise: Thomas Thomson, a friend of Hooker's, described by him as a 'scientific arithmetician', found errors in Scott's calculations, and Hooker announced that the paper was 'not fit for publication' and 'awfully tedious to read', though he praised Scott's 'industry & ability' (letter from J. D. Hooker, [10 March 1865]). Scott took these criticisms, no doubt diplomatically conveyed by Darwin, in good part, and with some shame, and said that he would take up the work again when he had time (letter from John Scott, 21 July 1865); at the time of writing, he had recently accepted the position of curator of the botanic gardens at Calcutta. Scott eventually published his results in the *Journal of the Asiatic Society of Bengal* (Scott 1867), and Darwin summarised them in *Variation* 2: 106–7, concluding, 'it follows from Mr. Scott's experiments, that in the genus Verbascum the similarly and dissimilarly-coloured varieties of the same species behave, when crossed, like closely allied but distinct species'.

Darwin also continued to ponder some of the theoretical issues he intended to raise in *Variation*. At the end of May, a month in which his health had been particularly bad, he sent Thomas Henry Huxley a fair copy of a manuscript in which he attempted to connect the facts he had collected about 'bud—and seminal—variation, ... inheritance, reversion, effects of use & disuse &c', and which he intended to publish in *Variation*. Darwin seems to have submitted his 'Provisional hypothesis of pangenesis', as it later became, to Huxley's judgment with some trepidation. 'It is a very rash & crude hypothesis', he wrote, with characteristic self-deprecation, 'yet it has been a considerable relief to my mind, & I can hang on it a good many groups of facts.' (Letter to T. H. Huxley, 27 May [1865].)

The hypothesis of pangenesis that Darwin outlined to Huxley was briefly as follows: each individual cell throws off minute particles (gemmules) that circulate in the bodily fluids and are capable of generating new cells, though they remain dormant until required. The gemmules are drawn together by mutual attraction and into the right places in the right order to, for instance, reconstruct (in organisms that have that capability) a tail or other limb where one has been torn off. Cells throw off gemmules at all stages of development and all gemmules circulate until required: thus embryo gemmules thrown off by embryo cells come together, when conditions are right, to form what Darwin called the germ (the female element) or the male element; the germ, if fertilised (or even if not, in the case of parthenogenesis), grows into a new embryo. Gemmules could also remain dormant for several generations before being sparked into activity. The hypothesis was clearly intended to offer a comprehensive account of heredity and development: Darwin thought that pangenesis could explain both sexual and asexual reproduction, as well as reversion and the regrowth of body parts.

The importance to Darwin of this hypothesis is obvious from his letter to Huxley, despite his apparent modesty, and the letter and associated manuscript throw an interesting light on his thinking. The hypothesis of pangenesis was ambitious in scope, and shows Darwin's work as a theorist. This provides a contrast with the painstaking amassing of data, designed eventually to build overwhelming support for his theory, if not to provide scientific proof, that was to all appearances his usual method of working. He asked Huxley not for detailed criticism, but for a brief verdict on whether he should publish it or not. Huxley's answer does not survive, but some idea of what he wrote to Darwin may be gathered from Darwin's letter to him of 12 July [1865]. Huxley had evidently pointed out some similarity between Darwin's views and those of Georges Louis Leclerc, comte de Buffon, and Charles Bonnet; Darwin wrote back: 'I do not doubt your judgment is perfectly just & I will try to persuade myself not to publish.' Huxley swiftly qualified what Darwin had apparently seen as a rather negative reaction in his letter to Darwin of 16 July 1865: 'Somebody rummaging among your papers half a century hence will find Pangenesis & say "See this wonderful anticipation of our modern Theories—and that stupid ass, Huxley, prevented his publishing them"... I am not going to be made a horrid example of in that way— But all I say is publish your views—not so much in the shape of formed conclusions—: as of hypothetical developments of the only clue at present accessible—and don't give the Philistines more chances of blaspheming than you can help'. A much revised and expanded version of the 1865 manuscript was published as chapter 27 of *Variation* in 1868.

Outside Darwin's own circle, discussions of the theory of transmutation were still taking place both in learned societies and in the popular press. In December 1864, George Douglas Campbell, the duke of Argyll, had delivered an address to the Royal Society of Edinburgh criticising *Origin.* Like Charles Lyell, who wrote to Darwin on 16 January 1865 praising the address, Campbell accepted Darwin's theory of transmutation only within certain limits. Campbell distinguished between laws such as natural selection, which determined the success of existing species or modifications, and 'creational laws', which produced modifications or new forms of life. According to Campbell, Darwin had frequently forgotten this distinction in *Origin*, and had written 'of natural selection "producing" this and that modification of structure' (G. D. Campbell 1864, pp. 275–6). Campbell argued further that natural selection could not account for the existence of beauty in the natural world, since it could only explain the existence of structures that were of direct use to organisms in the struggle for existence (*ibid.*, pp. 276–81). Darwin responded to Lyell's account in some detail (see letter to Charles Lyell, 22 January [1865]), ending pointedly: 'If ever you sh[d]. speak with the Duke on the subject please say how much interested I was with his Address & tell him about Sexual Selection.' The message may not have been sent, or if it was, it made no impression on Campbell, for he published similar arguments in 1865 in the religious weekly *Good Words*.

Campbell's reservations may have confirmed Darwin's view that Scottish scientific circles were in general not favourable to his theory; certainly he was surprised to find himself in December elected an honorary member of the Royal Society of Edinburgh. 'Here is a really curious thing, considering that Brewster is President & Balfour secretary', he wrote to Hooker on 22 December [1865]. David Brewster had criticised *Origin* for 'poisoning the foundations of science, and disturbing the serenity of the Christian world' (Brewster 1862, p. 3). John Hutton Balfour, though he had sent Darwin plant specimens on occasion, and was a friendly correspondent, had objected strongly to the potential application of the theory of transmutation to humans (see *Correspondence* vol. 10, letter from J. H. Balfour, 14 January 1862). According to Hooker, Balfour's prejudice against John Scott, who had worked under Balfour at the Royal Botanic Garden, Edinburgh, was compounded by Scott's interest in Darwin's theory (*Correspondence* vol. 11, letter from J. D. Hooker, 10 June 1863). However, probably the most enthusiastic encomium of the year also came from Scotland, from James Shaw, a schoolmaster with strong literary inclinations who wrote to Darwin: 'Newton's ocean of wonder on whose shore he gathered shells seems no longer a dark unnavigable sea crested with howling breakers. Here and there through its awful bosom the strong vision with which you have supplied us enables us to see at least that there are other islands whose conditions are explicable by our own' (letter from James Shaw, 20 November 1865). Shaw had also corresponded with Campbell in defence of Darwin.

Elsewhere, there was also encouraging news. In Germany, Darwin's theory, his supporters in that country informed him, was making good progress. In May, Darwin received the first instalment of Friedrich Rolle's *Der Mensch* (Rolle 1866), a study of the development of human society in the light of *Origin* and recent discoveries in geology. In September, a general agricultural congress held in Erfurt, Germany, discussed, among other things, Darwin's theory; this was reported in the *Gardeners' Chronicle*. Samuel Butler, the author and artist, and, in the 1860s, a supporter of Darwin's theory, had recently returned from New Zealand, and sent Darwin his pamphlet, *The evidence for the resurrection of Jesus Christ* (letter to Samuel Butler, 30 September [1865]). He later sent a copy of a letter defending Darwin's theory that he had written to the *Press*, a New Zealand newspaper; this was part of an pseudonymous exchange of letters with, he strongly suspected, the bishop of Wellington (letter from Samuel Butler, 1 October 1865). Darwin continued his correspondence with the entomologist Benjamin Dann Walsh in the Midwestern United States, and Asa Gray wrote a long review of 'Climbing plants' in the *American Journal of Science and Arts*, in which he described Darwin's botanical works as showing 'a genius for biological investigation, and a power of turning common materials and ordinary observations to high scientific account' (A. Gray 1865–6, pp. 273–4). Darwin had also written to Gray on 19 April to express his happiness and relief at the end of the American Civil War.

More painful matters were also discussed in the correspondence. At the end of May, the dispute between Charles Lyell and John Lubbock over alleged plagiarism by Lyell was forced upon Darwin's notice. The affair is discussed in detail in Appendix V. In 1865, Lubbock published *Prehistoric times*, a work on human antiquity, adding a note to his preface asserting that Lyell in his *Antiquity of man*, published in 1863, had made unacknowledged use of Lubbock's earlier articles in the *Natural History Review*. He also cited a statement by Lyell in *Antiquity of man* that the pages containing his discussion of Danish kitchen-middens had been written *before* the publication of an article by Lubbock on the same subject and suggested that the statement was 'an inadvertence'. Though Lubbock had raised the matter with Lyell before publishing, this note, tantamount to an accusation of plagiarism, had not been authorised or even seen by Lyell, and when he read it in its published form, he wrote to Lubbock to protest. Not receiving a satisfactory response, he forwarded copies of his and Lubbock's letters on the subject to a number of friends, including Darwin, at the end of May. The letter sent to Darwin and its enclosures have not been found, so Lyell's letter to Hooker, which must have been substantially similar in content and had the same enclosures, is published in this volume.

Hooker, while acknowledging Lyell's fault, thought Lubbock's behaviour the more reprehensible, given Lyell's greater age and professional seniority (see letter from J. D. Hooker, [2 June 1865]); but Darwin's feelings seem to have been more evenly balanced, and he found the whole business distressing, perhaps the more so as his illness made it difficult for him to read attentively (see letter to J. D. Hooker, [4 June 1865]). The fact that both men were friends of his must also have made the crisis particularly painful. Lyell had been to some extent his mentor, but had not endorsed Darwin's theory of transmutation in print as explicitly as Darwin felt he had a right to expect from their private communications; Lubbock, a younger friend, had been encouraged and supported in his scientific work by Darwin, and supported Darwin's theory publicly. Though Hooker implicitly expected Darwin to take an active role, writing, 'I should like to join you in healing this ugly breach' (letter from J. D. Hooker, [2 June 1865]), Darwin seems to have avoided becoming involved. He wrote to Hooker, 'I doubt whether you or I or any one c^{d} do any good in healing this breach. Time alone could do it if it ever can be done' (letter to J. D. Hooker, [4 June 1865]); the hard work of reconciliation seems in the end to have been done by Huxley. In fact, Darwin's immediate anxieties seem to have been personal: perhaps because of his awareness of the collaborative nature of much of his work, he expressed a fear that he and Hooker might one day similarly fall out (letter to J. D. Hooker, 1 June 1865).

The correspondence throws light also on the personal lives of Darwin's closest friends. For Hooker in particular 1865 must have seemed on the whole a trying year. In January he had influenza (letter from F. H. Hooker, [27 January 1865]); before June his wife, Frances, had a miscarriage, from which she did not quickly recover, and Hooker took her to Teesdale, County Durham, to convalesce (letter from J. D. Hooker, [15 June 1865]). This was evidently a happy and reasonably lazy time for Hooker. He wrote, 'I am studying the moraines all day long with as much enthusiasm as I am capable of after laying in bed till 9 eating heavy breakfasts & looking forward to dinner as the summum bonum of existence' (letter from J. D. Hooker, 13 July 1865); his observations on the moraines were published in the *Reader* (Hooker 1865), so he was, in fact, still working. In August, back at Kew, many of the family, including Hooker, contracted a throat infection. Hooker's father, William Jackson Hooker, who also caught it, rapidly succumbed to it and died on 12 August. Hooker himself contracted rheumatic fever after sleeping on the floor beneath an open window in his father's sick-room, and was unable to be with his father when he died or to attend the funeral. In early September, the family left Kew again for Hooker to convalesce, first with friends in Notting Hill, London, then in Buxton, Derbyshire. Even if he was unable to write in person, Hooker's enthusiasm for science and scientific society was evidently undimmed: Frances, writing to Darwin to notify him of their movements, wrote on Hooker's behalf, 'He asks if you saw the article of M^r. Croll in the last Reader on the displacement of the Earth's axis(?) by the ice of the glacial epoch?— & also do you know that the Reader has been sold to the Anthropologicals?' (letter from F. H. Hooker, 6 September [1865]). Before he returned from his convalescence, Hooker had been offered the directorship of the Royal Botanic Gardens, Kew (letter from F. H. Hooker, 13 September [1865]; he had been assistant director while his father was director). Upon his return to work in late October, he embarked on what must have been an exhausting programme of reform and reorganisation (letter from J. D. Hooker, [3 November 1865]).

Another distressing event was the suicide of Robert FitzRoy on 30 April 1865. The letters sent to Darwin by friends contain speculation about the reasons for his action and a detailed account of the funeral. FitzRoy, who was head of the Meteorological Department at the Board of Trade, had suffered much criticism from the scientific community over his national system of weather-forecasting (see letter from J. D. Hooker, 2 May 1865 and nn. 2 and 5), and he may have suffered, or imagined, other slights. Darwin, however, seems to have seen FitzRoy's action as an outcome of mental instability he had observed in him many years before: 'I was astounded at news about FitzRoy; but I ought not to have been, for I remember once thinking it likely; poor fellow his mind was quite out of balance once during our voyage. I never knew in my life so mixed a character. Always much to love & I once loved him sincerely; but so bad a temper & so given to take offence, that I gradually quite lost my love & wished only to keep out of contact with him' (letter to J. D. Hooker, 4 May [1865]). Darwin contributed to a fund set up to support FitzRoy's children (see letter from Charles Shaw, 3 October 1865).

Darwin's own family do not seem to have given him much anxiety over the course of the year. In January he wrote to Hooker that all the family were well, including Horace, who had been sickly for some years and was unable to attend school: 'Our boy Horace has made a sudden start in power of walking & that, I think, is a very good sign of real improvement in health' (letter to J. D. Hooker, 7 January [1865]). All the children living away from home made frequent visits. William was working as a banker in Southampton. George was at Trinity College, Cambridge; in September Darwin was consulting, or encouraging George to consult, a friend, the civil engineer Edward Cresy, about his career (letter to Edward Cresy, 7 September [1865], and letter from Edward Cresy, 10 September 1865). Francis and Leonard were still at school in Clapham, south-west London, and Horace was seeing a private tutor. Henrietta took a long holiday at Hengwrt, a house in Wales rented by the Hensleigh Wedgwoods for the summer, and Elizabeth was evidently attending school, and spent some time travelling in Europe (Emma Darwin's diary (DAR 242), *Emma Darwin* (1905), 2: 207–8). Henrietta was called upon to give her opinion of the Lyell–Lubbock plagiarism affair (letter from J. D. Hooker, [2 June 1865]), and Darwin reported that she was 'much interested' (letter to J. D. Hooker, [17 June 1865]), though her only known comment on the affair, to her mother, ends, 'I wish people weren't so foolish'. In November, Darwin and Emma visited Erasmus in London (CD's 'Journal', Appendix II). Wedgwood and Darwin relatives visited Down House frequently, and Hooker also came for a short stay in March (Emma Darwin's diary, DAR 242).

While Darwin and Hooker were ill, they both read or had read to them novels and other literature, and this was evidently an activity that brought the families together. Darwin wrote to Hooker, 'My good women-kind read to me a great deal, but I dare not ask for much science & am not sure that I could stand it' (letter to J. D. Hooker, 27 [or 28 September 1865]). In June he wrote to Hooker that they had enjoyed an 'Indian novel' that Hooker had evidently recommended: it was probably Iltudus Thomas Prichard's *How to manage it*, a love-story set in the Indian Mutiny of 1857 to 1858 (letter to J. D. Hooker, [17 June 1865]). They were now interested in reading more on the Mutiny. Hooker wrote from Buxton, where he was convalescing (letter from J. D. Hooker, [26 September 1865]), that he had read Joseph Sheridan Le Fanu's *Uncle Silas*, which, from his description, was something of a thriller, and George Eliot's *The mill on the Floss*, which he was 'ravished' with: 'what a clever person the authoress is, I like it even better than Adam Bede'. He also read Hamilton Aïdé's *Carr of Carrlyon*, pronouncing it 'utter trash after G. Eliot'. Darwin replied (letter to J. D. Hooker, 27 [or 28 September 1865]): 'I did not enjoy "Mill on the Floss" so much as you, but from what you say we will read it again. Do you know Silas Marner: it is a charming little story.' The Darwins were also reading 'American novels', Miriam Coles Harris's *Rutledge*, and Laura Curtis Bullard's *Christine: or, woman's trials and triumphs.* In non-fiction, they had read or were reading William Lecky's *History of the rise and influence of the spirit of rationalism in Europe*, Edward Burnett Tylor's *Researches into the early history of mankind*, and William Gifford Palgrave's *Narrative of a year's journey through central and eastern Arabia.*

By the end of December, Darwin and Hooker's letters to each other are again full of botany and news of scientific society. Darwin finished the year well placed to continue work on *Variation*, and with correspondents now ready to make observations for him in India (John Scott) and Brazil (Fritz Müller). Although not well enough to receive virtual strangers like Samuel Butler (letter to Julius von Haast, 26 December [1865]), Darwin was looking forward to a visit from Hooker (letter to J. D. Hooker, [31 December 1865]) and promising to explain about his '*so-called* hybrids of Lythrum' when they met. The last two months of the year also saw letters from George Henslow, the son of Darwin's mentor at Cambridge, John Stevens Henslow, from John Traherne Moggridge in Mentone, sending orchid specimens, from Frederick William Farrar, writing on language, and from Roland Trimen in Cape Town. His last letter of the year is a typical example of his energy and kindness. 'As for your thinking that you do not deserve the C[opley] Medal,' he rebuked Hooker, 'that I declare is mere insanity.—'

This volume also contains a supplement of 106 letters written or conjectured to have been written before 1865 that have been discovered or redated since the publication of the volumes they should have appeared in. It also contains letters that cannot be dated more precisely than to a span of years but that were probably written before 1865. It includes a small and remarkable collection of letters written by Darwin when he was 12 years old. Over half of the letters in the Supplement are from the period before the publication of *Origin*, including a number from the period after the *Beagle* voyage and before the move to Down, Kent, when Darwin was living in London. There are letters commenting on *Origin*, including two from Charles Lyell, who had been sent the proof-sheets before publication. Letters after 1859 throw light on Darwin's continuing researches and on the response to *Origin*, and there are letters to and from Darwin's children, giving a glimpse of a lively family atmosphere. However, Darwin wrote poignantly to his son William on 30 November [1861]: 'Mamma is in bed with bad Headach.— Miss. L. is very bad with headach.— Lenny has got a slight headach.— I am not very bright— The day is raining torrents, the children are ennuied—so I have not heart to write— GoodBye | My dear old man | C. Darwin.'

ACKNOWLEDGMENTS

The editors are grateful to the late George Pember Darwin and to William Huxley Darwin for permission to publish the Darwin letters and manuscripts. They also thank the Syndics of Cambridge University Library and other owners of the manuscript letters who have generously made them available.

The work for this edition has been supported by grants from the National Endowment for the Humanities (NEH), the National Science Foundation (NSF), and the Wellcome Trust. The Alfred P. Sloan Foundation, the Pew Charitable Trusts, and the Andrew W. Mellon Foundation provided grants to match NEH funding, and the Mellon Foundation awarded grants to Cambridge University that made it possible to put the entire Darwin correspondence into machine-readable form. Research and editorial work have also been supported by grants from the Royal Society of London, the British Academy, the British Ecological Society, the Isaac Newton Trust, the Jephcott Charitable Trust, the Natural Environment Research Council, and the Wilkinson Charitable Foundation. The Royal Society has helped meet the costs of publication. The Stifterverband für die Deutsche Wissenschaft has provided funds to edit Darwin's correspondence with German naturalists.

Cambridge University Library and the American Philosophical Society (APS) have generously made working space and many services available to the editorial staff.

Since the project began in 1975, we have been fortunate in benefiting from the interest, experience, and practical help of many people in many places, and the editors hope that they have adequately expressed their thanks to them individually as the work proceeded.

Without the expert help of John L. Dawson of the Literary and Linguistic Computing Centre of Cambridge University, the initial computerisation of the correspondence would not have been possible, for the work on both the *Calendar* and the *Correspondence* required the solution of many novel technical problems. Iain Burke, Peter Dunn, Patricia Killiard, Adrian Miller, Chris Sendall, Tomasz Waldoch, and Ellis Weinberger have helped maintain and extend our access to the Library's electronic resources, and provided essential technical support. Simon Buck provided invaluable assistance in setting up automatic typesetting procedures. Gratitude is expressed to C. A. Tripp for providing the editors with an invaluable research tool, enabling them to access texts through search and retrieval software.

English Heritage has responded most generously to requests for information and for material from the collections at Down House, Downe. We are particularly grateful to Tori Reeve, curator of Down House, Nick Biddle, and the staff of the English Heritage Picture Library. Richard Darwin Keynes kindly made available

Darwin family material in his possession. Ursula Mommens has also provided letters and other materials that belonged to her grandfather, Francis Darwin.

Libraries all over the world have given indispensable help by making available photocopies of Darwin correspondence and other manuscripts in their collections. The institutions and individuals that furnished copies of letters for this volume can be found in the List of provenances. The editors are extremely indebted to them all. We are also grateful to the many people who have transmitted information regarding the whereabouts of particular letters and, in many cases, have generously provided copies of such letters for the project's use.

The editors make daily use of the incomparable facilities of Cambridge University Library and have benefited greatly from its services and from the help and expertise of its staff, particularly the staff of the Manuscripts Department. We are especially grateful to the former University Librarian, Frederick W. Ratcliffe, to his successor, Peter K. Fox, and to the Keeper of Manuscripts and Archives, Patrick Zutshi, for their generous support. Other members of the library's staff who have frequently responded to the editors' requests are: Marjolein Allen, Gerry Bye, Kathleen Cann, Colin Clarkson, Barry Eaden, Les Goodey, David J. Hall, John Hall, Tony Harper, Isabel Holowaty, Brian Jenkins, Morag Law, Elisabeth Leedham-Green, David Lowe, Peter Meadows, Anne Murray, William Noblett, Adam Perkins, Jayne Ringrose, Mark Scudder, Clive Simmonds, Nicholas Smith, Anne Taylor, Nicola Thwaite, Godfrey Waller, Cynthia Webster, and Jeremy Wong. The fetchers in the Rare Books reading room have also patiently dealt with the editors' often complex requirements, as have the staff of the Map Room.

At the American Philosophical Society Library, a splendid collection of Darwiniana and works in the history of science has been continuously available to the editorial staff since the inception of the project. Whitfield J. Bell Jr, secretary of the society until 1983, serves on the United States Advisory Committee for the project and has done his utmost to further its work. The editors have also benefited from the co-operation of Edward Carter II, Robert S. Cox, Roy C. Goodman, Carl F. Miller, Elizabeth Carroll-Horrocks, and Bertram Dodelin, all of the APS Library.

The editors would like to acknowledge the assistance of Rodney Dennis, Jennie Rathbun, and Susan Halpert of the Houghton Library, Constance Carter of the Science Division of the Library of Congress, and Judith Warnement and Jean Cargill of the Gray Herbarium of Harvard University, who have all been exceptionally helpful in providing material from the collections in their charge.

In Britain, the editors have often received assistance from Mary Sampson, archivist of the Royal Society; Rex E. R. Banks, Vicki Veness, and the late John Thackray of the Natural History Museum Library; Gina Douglas, librarian of the Linnean Society of London; and Sylvia FitzGerald, John Flanagan, and Kate Pickard of the Library of the Royal Botanic Gardens, Kew. We would also like to thank Malcolm P. Marjoram of the British Library (Manuscripts); Richard G. Williams, Mrs Felton, and Anne Barrett (college archivist) of Imperial College of Science, Technology and Medicine; Christine Fyfe, archivist of Keele University;

George R. Else, of the Department of Entomology, Natural History Museum, London, and Dr and Mrs Courtney of Christ's College, Cambridge. Invaluable assistance has been received from local studies librarians and county archivists, too numerous to name individually, although Elizabeth Silverthorne of Bromley Central Library and Donald Gibson of the Centre for Kentish Studies deserve special mention for their repeated assistance.

Virginia Murray of John Murray (publishers) Ltd has graciously provided the editors with copies of the Darwin correspondence in the Murray archive and with related information. Leonard G. Wilson kindly provided the editors with photocopies of manuscript material in the possession of the family of Sir Charles Lyell. George Beccaloni, Milton Forsyth, Quentin Keynes, Chris Kohler, Joseph Sakmyster, and John Wilson have generously provided photocopies of or information about newly found letters, and Richard Brent provided photocopies of letters to B. P. Brent together with information based on his own research.

Among the others who have advised and assisted the editors in their work are David Allen, Maria Asp, William Bynum, Eva-Marie Engels, Kanchi Gandhi, Nick Gill, Ruth Goldstone, Julius Held, Sandra Herbert, Desmond King-Hele, Gren Lucas, Jim Moore, William Newman, Muriel Palmer, Henri Quinn, Andrew Rogerson, Jim Secord, Garry J. Tee, and Trudy Walker. The editors are also pleased to acknowledge the invaluable support of the members of the British and United States Advisory Committees.

Thanks are also due to all former staff of the Darwin Correspondence Project: Sarah Benton, Heidi Bradshaw, Janet Browne, Mario Di Gregorio, Kate Fletcher, Hedy Franks, Joy Harvey, Arne Hessenbruch, Thomas Junker, David Kohn, Sarah Lavelle, William Montgomery, Perry O'Donovan, Stephen V. Pocock, Marsha L. Richmond, the late Peter Saunders, Anne Secord, the late Sydney Smith, Nora Carroll Stevenson, Jonathan R. Topham, and Sarah Wilmot.

The editors also gratefully acknowledge the valuable assistance of Philip Hardie, Kathleen Lane, and Béatrice Willis, and extend thanks to Jean Macqueen for providing the index to the volume.

In addition, we would like to acknowledge the contribution of colleagues who left the staff during the preparation of this volume. During 2001 we lost the valuable services of Jyothi Krishnan Unni, Anna-K. Mayer, and Tyler Veak, all of whom we wish well in all they undertake in the future. Also during 2001 and 2002 we enjoyed the great pleasure of the company of Gene Kritsky, who joined the project as a volunteer, assisting with transcription of manuscript material and helping with entomological queries, and also of Martin Oldfield, who assisted us in creating a publicly available electronic Calendar to the correspondence.

Finally, we mourn the loss of John D'Arms, late president of the American Council of Learned Societies and great friend of the project.

LIST OF PROVENANCES

The following list gives the locations of the original versions of the letters printed in this volume. The editors are grateful to all the institutions and individuals listed for allowing access to the letters in their care. Access to material in DAR 261 and DAR 263, formerly at Down House, Downe, Kent, England, is courtesy of English Heritage.

Alexander Turnbull Library (National Library of New Zealand), Wellington, New Zealand
Alfred Denny Museum, University of Sheffield, Sheffield, England
American Philosophical Society, Philadelphia, Pennsylvania, USA
Archives Départmentales de Maine-et-Loire, Angers, France
Autographia (dealers), San Rafael, California, USA
Bloomsbury Book Auctions, London, England
Bodleian Library, Oxford, England
Bonhams (dealers), Knightsbridge, London, England
Richard Brent (private collection)
British Library, London, England
British Museum, London, England
Julian Browning (dealer), London, England
Sherman Bull (private collection)
Cambridge University Library, Manuscripts Department, Cambridge, England
Centre for Kentish Studies, Maidstone, Kent, England
Christie's Images, Vauxhall, London, England
Christie's London (dealers), London, England
Christie's New York (dealers), New York, USA
Christ's College, Cambridge, England
Cleveland Health Sciences Library, Cleveland, Ohio, USA
Francis A. Countway Library of Medicine (Harvard Medical Library), Boston, Massachusetts, USA
CUL. *See* Cambridge University Library
DAR. *See* Cambridge University Library
Elgin Museum, Moray, Scotland
Field Museum of Natural History, Chicago, Illinois, USA
Fitzwilliam Museum, Cambridge, England
Forschungsinstitut und Natur-Museum Senckenberg, Senckenburg, Frankfurt am Main, Germany
Gerald M. Friedman (private collection)

Gallery of History (dealer), Las Vegas, Nevada, USA
Gray Herbarium of Harvard University, Cambridge, Massachusetts, USA
Ernst-Haeckel-Haus, Jena, Germany
Eilo Hildebrand (private collection)
Oliver N. Hooker (private collection)
Imperial College of Science, Technology and Medicine, London, England
Journal of the Linnean Society (publication)
Quentin Keynes (private collection)
Kinnordy MS. *See* Lady Lyell
C. C. Kohler (dealer), Dorking, Surrey, England
Tim Lewens (private collection)
Linnean Society of London, Piccadilly, London, England
LuEsther T. Mertz Library Archives. *See* New York Botanical Garden
K. M. Lyell ed. 1881 (publication)
Lady Lyell (private collection)
Magdalen College, Oxford, England
Milne Home 1891 (publication)
Möller 1915–21 (publication)
Morristown National Historical Park, Morristown, New Jersey, USA
John Murray Archive, London, England
Natural History Museum, Cromwell Road, London, England
New York Botanical Garden, Bronx, New York, USA
Oxford University Museum of Natural History, Oxford, England
Remember When Antiquities, Acton, Maine, USA
Josh B. Rosenblum (private collection)
Royal Botanic Garden Edinburgh, Edinburgh, Scotland
Royal Botanic Gardens, Kew, Richmond, Surrey, England
Royal College of Surgeons of England, London, England
Royal Horticultural Society, London, England
Royal Society, London, England
Joseph R. Sakmyster (dealer), Chicago, Illinois, USA
David Schulson Autographs Inc., New York, USA
Scott Polar Research Institute, Cambridge, England
Shrewsbury School, Shrewsbury, England
R. M. Smythe (dealer), New York, USA
Sotheby's London (dealer), London, England
Sotheby's New York (dealer), New York, USA
Staatsbibliothek zu Berlin – Preussischer Kulturbesitz, Berlin, Germany
J. A. Stargardt (dealer), Berlin, Germany
Christophe Stickel Autographs, Pacific Grove, California, USA
Stokes 1846 (publication)
Transactions of the Hawick Archæological Society (publication)
University of Akron, Akron, Ohio, USA

University of Edinburgh, Edinburgh, Scotland
University of Virginia, Charlottesville, Virginia, USA
University of Wales Bangor (Archives Department), Bangor, Gwynedd, Wales
R. Wallace ed. 1899 (publication)
Wellcome Library for the History and Understanding of Medicine, London, England
Yale University, New Haven, Connecticut, USA

A NOTE ON EDITORIAL POLICY

The first and chief objective of this edition is to provide complete and authoritative texts of Darwin's correspondence. For every letter to or from Darwin, the text that is available to the editors is always given in full. The editors have occasionally included letters that are not to or from Darwin if they are relevant to the published correspondence. Volumes of the *Correspondence* are published in chronological order. Occasional supplements will be published containing letters that have come to light or have been redated since the relevant volumes of the *Correspondence* appeared. Letters that can only be given a wide date range, in some instances spanning several decades, are printed in the supplement following the volume containing letters at the end of their date range. The first such supplement was in Volume 7 and included letters from 1828 to 1857. The second supplement is in this volume, and includes letters from 1822 to 1864.

Dating of letters and identification of correspondents

In so far as it is possible, the letters have been dated, arranged in chronological order, and the recipients or senders identified. Darwin seldom wrote the full date on his letters and, unless the addressee was well known to him, usually wrote only 'Dear Sir' or 'Dear Madam'. After the adoption of adhesive postage stamps in the 1840s, the separate covers that came into use with them were usually not preserved, and thus the dates and the names of many recipients of Darwin's letters have had to be derived from other evidence. The notes made by Francis Darwin on letters sent to him for his editions of his father's correspondence have been helpful, as have matching letters in the correspondence, but many dates and recipients have had to be deduced from the subject-matter or references in the letters themselves.

Transcription policy

Whenever possible, transcriptions have been made from manuscripts. If the manuscript was inaccessible but a photocopy or other facsimile version was available, that version has been used as the source. In many cases, the editors have had recourse to Francis Darwin's large collection of copies of letters, compiled in the 1880s. Other copies, published letters, or drafts have been transcribed when they provided texts that were otherwise unavailable.

The method of transcription employed in this edition is adapted from that described by Fredson Bowers in 'Transcription of manuscripts: the record of variants', *Studies in Bibliography* 29 (1976): 212–64. This system is based on accepted principles of modern textual editing and has been widely adopted in literary editions.

The case for using the principles and techniques of this form of textual editing for historical and non-literary documents, both in manuscript and print, has been forcefully argued by G. Thomas Tanselle in 'The editing of historical documents', *Studies in Bibliography* 31 (1978): 1–56. The editors of the *Correspondence* followed Dr Tanselle in his conclusion that a 'scholarly edition of letters or journals should not contain a text which has editorially been corrected, made consistent, or otherwise smoothed out' (p. 48), but they have not wholly subscribed to the statement made earlier in the article that: 'In the case of notebooks, diaries, letters and the like, whatever state they are in constitutes their finished form, and the question of whether the writer "intended" something else is irrelevant' (p. 47). The editors have preserved the spelling, punctuation, and grammar of the original, but they have found it impossible to set aside entirely the question of authorial intent. One obvious reason is that in reading Darwin's writing, there must necessarily be reliance upon both context and intent. Even when Darwin's general intent is clear, there are cases in which alternative readings are, or may be, possible, and therefore the transcription decided upon must to some extent be conjectural. Where the editors are uncertain of their transcription, the doubtful text has been enclosed in italic square brackets.

A major editorial decision was to adopt the so-called 'clear-text' method of transcription, which so far as possible keeps the text free of brackets recording deletions, insertions, and other alterations in the places at which they occur. Darwin's changes are, however, recorded in the back matter of the volume, under 'Manuscript alterations and comments', in notes keyed to the printed text by paragraph and line number. All lines above the first paragraph of the letter (that is, date, address, or salutation) are referred to as paragraph '0'. Separate paragraph numbers are used for subscriptions and postscripts. This practice enables the reader who wishes to do so to reconstruct the manuscript versions of Darwin's autograph letters, while furnishing printed versions that are uninterrupted by editorial interpolations. The Manuscript alterations and comments record all alterations made by Darwin in his letters and any editorial amendments made in transcription, and also where part of a letter has been written by an amanuensis; they do not record alterations made by amanuenses. No attempt has been made to record systematically all alterations to the text of copies of Darwin letters included in the correspondence, but ambiguous passages in copies are noted. The editors believe it would be impracticable to attempt to go further without reliable information about the texts of the original versions of the letters concerned. Letters to Darwin have been transcribed without recording any of the writers' alterations unless they reflect significant changes in substance or impede the sense; in such cases footnotes bring them to the reader's attention.

Misspellings have been preserved, even when it is clear that they were unintentional: for instance, 'lawer' for 'lawyer'. Such errors often indicate excitement or haste and may exhibit, over a series of letters, a habit of carelessness in writing to a particular correspondent or about a particular subject.

Capital letters have also been transcribed as they occur except in certain cases, such as 'm', 'k', and 'c', which are frequently written somewhat larger than others as initial letters of words. In these cases an attempt has been made to follow the normal practice of the writers.

In some instances that are not misspellings in a strict sense, editorial corrections have been made. In his early manuscripts and letters Darwin consistently wrote 'bl' so that it looks like 'lb' as in 'albe' for 'able', 'talbe' for 'table'. Because the form of the letters is so consistent in different words, the editors consider that this is most unlikely to be a misspelling but must be explained simply as a peculiarity of Darwin's handwriting. Consequently, the affected words have been transcribed as normally spelled and no record of any alteration is given in the textual apparatus. Elsewhere, though, there are misformed letters that the editors have recorded because they do, or could, affect the meaning of the word in which they appear. The main example is the occasional inadvertent crossing of 'l'. When the editors are satisfied that the intended letter was 'l' and not 't', as, for example, in 'stippers' or 'istand', then 'l' has been transcribed, but the actual form of the word in the manuscript has been given in the Manuscript alterations and comments.

If the only source for a letter is a copy, the editors have frequently retained corrections made to the text when it is clear that they were based upon comparison with the original. Francis Darwin's corrections of misreadings by copyists have usually been followed; corrections to the text that appear to be editorial alterations have not been retained.

Editorial interpolations in the text are in square brackets. Italic square brackets enclose conjectured readings and descriptions of illegible passages. To avoid confusion, in the few instances in which Darwin himself used square brackets, they have been altered by the editors to parentheses with the change recorded in the Manuscript alterations and comments. In letters to Darwin, square brackets have been changed to parentheses silently.

Material that is irrecoverable because the manuscript has been torn or damaged is indicated by angle brackets; any text supplied within them is obviously the responsibility of the editors. Occasionally, the editors are able to supply missing sections of text by using ultraviolet light (where text has been lost owing to damp) or by reference to transcripts or photocopies of manuscript material made before the damage occurred.

Words and passages that have been underlined for emphasis are printed in italics in accordance with conventional practice. Where the author of a letter has indicated greater emphasis by underlining a word or passage two or more times, the text is printed in bold type.

Paragraphs are often not clearly indicated in the letters. Darwin and others sometimes marked a change of subject by leaving a somewhat larger space than usual between sentences; sometimes Darwin employed a longer dash. In these cases, and when the subject is clearly changed in very long stretches of text, a new paragraph has been started by the editors without comment. The beginnings of letters, valedictions, and postscripts are also treated as new paragraphs regardless of whether they appear as new paragraphs in the manuscript. Special manuscript devices delimiting sections or paragraphs, for example, blank spaces left between sections of text and lines drawn across the page, are treated as normal paragraph indicators and are not specially marked or recorded unless their omission leaves the text unclear.

Occasionally punctuation marking the end of a clause or sentence is not present in the manuscript, but the author has made his or her intention clear by allowing, for example, extra space or a line break to function as punctuation. In such cases, the editors have inserted an extra space following the sentence or clause to set it off from the following text.

Additions to a letter that run over into the margins, or are continued at its head or foot, are transcribed at the point in the text at which the editors believe they were intended to be read. The placement of such an addition is only recorded in a footnote if it seems to the editors to have some significance or if the position at which it should be transcribed is unclear. Enclosures are transcribed following the letter.

The hand-drawn illustrations and diagrams that occur in some letters are reproduced as faithfully as possible and are usually positioned as they were in the original text. In some cases, however, it has been necessary to reduce the size of a diagram or enhance an outline for clarity; any such alterations are recorded in footnotes. The location of diagrams within a letter is sometimes changed for typesetting reasons. Tables have been reproduced as close to the original format as possible, given typesetting constraints.

Some Darwin letters and a few letters to Darwin are known only from entries in the catalogues of book and manuscript dealers or mentions in other published sources. Whatever information these sources provide about the content of such letters has been reproduced without substantial change. Any errors detected are included in footnotes.

Format of published letters

The format in which the transcriptions are printed in the *Correspondence* is as follows:

1. *Order of letters*. The letters are arranged in chronological sequence. A letter that can be dated only approximately is placed at the earliest date on which the editors believe it could have been written. The basis of a date supplied by the editors is given in a footnote unless it is derived from a postmark, watermark, or endorsement that is recorded in the physical description of the letter (see section 4, below). Letters with the same date, or with a range of dates commencing with that date, are printed in the alphabetical order of their senders or recipients unless their contents dictate a clear alternative order. Letters dated only to a year or a range of years precede letters that are dated to a particular month or range of months, and these, in turn, precede those that are dated to a particular day or range of dates commencing with a particular day.

2. *Headline*. This gives the name of the sender or recipient of the letter and its date. The date is given in a standard form, but those elements not taken directly

from the letter text are supplied in square brackets. The name of the sender or recipient is enclosed in square brackets only where the editors regard the attribution as doubtful.

3. *The letter text.* The transcribed text follows as closely as possible the layout of the source, although no attempt is made to produce a type-facsimile of the manuscript: word-spacing and line-division in the running text are not adhered to. Similarly, the typography of printed sources is not replicated. Dates and addresses given by authors are transcribed as they appear, except that if both the date and the address are at the head of the letter they are always printed on separate lines with the address first, regardless of the manuscript order. If no address is given on a letter by Darwin, the editors have supplied one, when able to do so, in square brackets at the head of the letter. Similarly, if Darwin was writing from an address different from the one given on the letter, his actual location is given in square brackets. Addresses on printed stationery are transcribed in italics. Addresses, dates, and valedictions have been run into single lines to save space, but the positions of line-breaks in the original are marked by vertical bars.

4. *Physical description.* All letters are complete and in the hand of the sender unless otherwise indicated. If a letter was written by an amanuensis, or exists only as a draft or a copy, or is incomplete, or is in some other way unusual, then the editors provide the information needed to complete the description. Postmarks, endorsements, and watermarks are recorded only when they are evidence for the date or address of the letter.

5. *Source.* The final line provides the provenance of the text. Some sources are given in abbreviated form (for example, DAR 140: 18) but are listed in full in the List of provenances unless the source is a published work. Letters in private collections are also indicated. References to published works are given in author–date or short-title form, with full titles and publication details supplied in the Bibliography at the end of the volume.

6. *Darwin's annotations.* Darwin frequently made notes in the margins of the letters he received, scored significant passages, and crossed through details that were of no further interest to him. These annotations are transcribed or described following the letter text. They are keyed to the letter text by paragraph and line numbers. Most notes are short, but occasionally they run from a paragraph to several pages, and sometimes they are written on separate sheets appended to the letter. Extended notes relating to a letter are transcribed whenever practicable following the annotations as 'CD notes'.

Quotations from Darwin manuscripts in footnotes and elsewhere, and the text of his annotations and notes on letters, are transcribed in 'descriptive' style. In this method the alterations in the text are recorded in brackets at the places where they occur. For example:

'See Daubeny ['vol. 1' *del*] for *descriptions of volcanoes in [*interl*] S.A.' *ink*

means that Darwin originally wrote in ink 'See Daubeny vol. 1 for S.A.' and then deleted 'vol. 1' and inserted 'descriptions of volcanoes in' after 'for'. The asterisk before 'descriptions' marks the beginning of the interlined phrase, which ends at the bracket. The asterisk is used when the alteration applies to more than the immediately preceding word. The final text can be read simply by skipping the material in brackets. Descriptive style is also used in the Manuscript alterations and comments.

Editorial matter

Each volume is self-contained, having its own index, bibliography, and biographical register. A chronology of Darwin's activities covering the period of each volume and translations of foreign-language letters are supplied, and additional appendixes give supplementary material where appropriate to assist the understanding of the correspondence. A cumulative index is planned once the edition is complete. References are supplied for all persons, publications, and subjects mentioned, even though some repetition of material in earlier volumes is involved.

If the name of a person mentioned in a letter is incomplete or incorrectly spelled, the full, correct form is given in a footnote. Brief biographies of persons mentioned in the letters, and dates of each correspondent's letters to and from Darwin in the current volume, are given in the Biographical register and index to correspondents. Where a personal name serves as a company name, it is listed according to the family name but retains its original order: for example, 'E. Schweizerbart'sche Verlagsbuchhandlung' is listed under 'S', not 'E'.

Short titles are used for references to Darwin's books and articles and to collections of his letters (e.g., *Descent*, 'Parallel roads of Glen Roy', *LL*). They are also used for some standard reference works and for works with no identifiable author (e.g., *Alum. Cantab.*, *Wellesley index*, *DNB*). For all other works, author–date references are used. References to the Bible are to the authorised King James version unless otherwise stated. Words not in *Chambers 20th century dictionary* are usually defined in the footnotes with a source supplied. The full titles and publication details of all the books referred to are given in the Bibliography. References to archival material, for instance that in the Darwin Archive at Cambridge University Library, are not necessarily exhaustive.

Darwin and his correspondents consistently used the term 'fertilisation' for both fertilisation (the fusion of sperm with egg) and pollination (the transfer of pollen from anther to stigma); the first recorded usage of the term 'pollination' was in 1875 (*OED*). 'Fertilisation' in Darwin's letters and publications often, but not always, refers to pollination. In the footnotes, the editors have attempted to use the modern terms in the context of Darwin's descriptions of experiments. When he is quoted directly, his original usage is never altered.

The editors use the abbreviation 'CD' for Charles Darwin throughout the footnotes. A list of all abbreviations used by the editors in this volume is given on p. xl.

The Wedgwood an

Robert Waring Darwin = Susannah Wedgwood
1766–1848 1765–1817

- Henry Parker = Marianne
  1788–1856 1798–1858
  - Robert b.1825
  - Henry 1827–92
  - Francis 1829–71
  - Charles b.1831
  - Mary Susan
    1836–93
- Caroline Sarah = Josiah III (Jos)
  1800–88 1795–1880
  - Sophy Marianne
    1838–9
  - Katherine Elizabeth
    Sophy (Sophy) 1842–1911
  - Margaret Susan
    1843–1937
  - Lucy Caroline
    1846–1919
- Susan Elizabeth
  1803–66
- Erasmus Alvey
  1804–81
- Charles Robert = Emma 1808–96
  1809–82
  - William Erasmus 1839–1914
  - Anne Elizabeth 1841–51
  - Mary Eleanor Sept.–Oct. 1842
  - Henrietta Emma (Etty)
    1843–1927
  - George Howard 1845–1912
  - Elizabeth 1847–1926
  - Francis (Frank) 1848–1925
  - Leonard 1850–1943
  - Horace 1851–1928
  - Charles Waring 1856–8
- Charles Langton = Emily Catherine
  1801–86 (Catherine) 1810–66

Darwin Families up to 1865

- Josiah Wedgwood II 1769–1843 = Elizabeth (Bessy) Allen 1764–1846
  - Sarah Elizabeth (Elizabeth) 1793–1880
  - Mary Anne 1796–8
  - Charlotte 1797–1862 = Charles Langton 1801–86
    - Edmund 1841–75
  - Henry Allen (Harry) 1799–1885 = Jessie Wedgwood 1804–72
    - Louisa Frances 1834–1903
    - Caroline b.1836
    - John Darwin 1840–70
    - Anne Jane 1841–77
    - Arthur 1843–1900
    - Rowland 1847–1921
  - Francis (Frank) 1800–88 = Frances Mosley 1808–74
    - Godfrey 1833–1905
    - Amy 1835–1910
    - Cicely Mary 1837–1917
    - Clement 1840–89
    - Lawrence 1844–1913
    - Constance Rose 1846–1903
    - Mabel Frances 1852–1930
  - Hensleigh 1803–91 = Frances (Fanny) Mackintosh 1800–89
    - Frances Julia (Snow) 1833–1913
    - James Mackintosh (Bro) 1834–64
    - Ernest Hensleigh 1838–98
    - Katherine Euphemia (Effie) 1839–1931
    - Alfred Allen 1842–92
    - Hope Elizabeth 1844–1934
  - Frances (Fanny) 1806–32

ABBREVIATIONS

AL	autograph letter
ALS	autograph letter signed
DS	document signed
LS	letter in hand of amanuensis, signed by sender
LS(A)	letter in hand of amanuensis with additions by sender
Mem	memorandum
(S)	signed with sender's name by amanuensis
CD	Charles Darwin
CUL	Cambridge University Library
DAR	Darwin Archive, Cambridge University Library
del	deleted
illeg	illegible
interl	interlined
underl	underlined

TRANSCRIPTION CONVENTIONS

[some text]	'some text' is an editorial insertion
*[*some text*]*	'some text' is the conjectured reading of an ambiguous word or passage
[some text]	'some text' is a description of a word or passage that cannot be transcribed, e.g., '*3 words illeg*'
⟨ ⟩	word(s) destroyed
⟨some text⟩	'some text' is a suggested reading for a destroyed word or passage
⟨*some text*⟩	'some text' is a description of a destroyed word or passage, e.g., '*3 lines excised*'

THE CORRESPONDENCE OF CHARLES DARWIN
1865

From Joseph Dalton Hooker 1 January 1865

Kew
Jany 1/65

Dear Darwin

I have told Mr Stainton that I have referred his letter to me to you— kindly answer it—if you can.[1]

I find plenty of Cucurbitaceæ to have sticking ends to their tendrils, & expect that it is a common enough phenomenon in the Order.[2] From what you say of the form of the cells in Hanburya I should have inferred something of the sort, as such cells are common on the undersurface of the leaves of various vines & you say they have sticking tendrils.[3] Certainly the suppressed potentiality of so many organs of plants to play so many parts, is one of the most wonderful of your discoveries. The more I think of it, the more pregnant it is: one day it will itself play a prodigious part, in the interpretation of both morphological & physiological facts.

I have read Sabines complete address (I had seen only extracts before) & am indignant & disgusted at the mutilation & emasculation of what I wrote— Especially about *Lythrum* & *Linum*, which he has made nonsense of & the use your observations will be in interpreting, no end of phenomena not yet guessed at.[4] Poor old man, he is ill still, & I am beginning to fear that my ill-natured prophecy, that the Presidentship would be the death of him, may come true.—[5]

Have you read Huxleys (I suppose) slashing leader in todays Reader.[6] it is uncommonly able &c: but as usual with him, he goes like a desert whirlwind over the ground scorching blasting & suffocating all opposing objects, & leaving nothing but dry bones on the ground. The vegetation he withers was one of vile weeds to be sure, but vile weeds are *green*, & all is *black* after him

I have done little but dissect *Cucurbitaceæ* since I wrote last.[7] I wish I had the energy when doing each Nat. Ord: to show how each prevalent & characteristic feature shades off at either end, or waxes & wanes in the series of Genera.[8] An absolute character is very rare in an order. Cucurbi. is a very curious order—especially as to stamens.[9]

I will let you know if I hear of Hoffmeisters book,[10] or Hildebrands paper.[11] & Oliver will keep you advised of Bot. Zeit. articles.[12]

Tell me when you write how your health is.

My book on Geog. Distrib. is *nowhere*—[13] I wish it were only begun.

The Reader seems to me dull & rather aimless—the articles too learned & heavy for men who work hard during the week, & want some enticement to read Science on Sunday. Huxley has taken the scientific Editorship I am told;[14] but he never stuck to any thing of the kind long, & I have no hopes of it's succeeding under him— he is far too good for such work, & has no aptitude for it—: no man can write such good articles upon Science as he can, but he is no caterer for the public, & never can be: he wants breadth of sympathy

We are all well, & the children very happy— I have no news of any kind.

Did I send you enclosed letter of Thwaites with a passage for you?[15]

Ever y^{rs} affec | J. D. Hooker.

[Enclosure]

Peradenia, Ceylon
28th Oct 1864

Dear Hooker,

The last mail brought a kind & most welcome letter from you and, in case I should omit to do so, will at once answer your question about *Nepenthes*: it certainly climbs by means of its tendrils and I enclose you the only example I happen to have of the tendrils firmly grasping any object but this will show you what they can do.[16] The plant is most common in open swampy ground but when it gets amongst bushes it takes to climbing & I have often been struck with its beauty under such circumstances. Nepenthes does not grow in this neighbourhood and I have never succeeded in keeping plants alive planted out in the Garden.[17] I will get M^{r}. W. Ferguson[18] to look at & gather a fine specimen illustrating its climbing propensities or capabilities & will send it to you another time.

I am glad to hear you have conquered Melastomads.[19] It is certainly best to make several attacks at distant intervals, in order to overcome the difficulties of marshalling such a puzzling family as the *Melastomads*, as after periods of rest differences are better estimated as to their importance, and undue prominences get softened down in the more general view one is able to take some time after working at particular genera or groups.

It gratifies me much to learn that Müller thinks well of my arrangement of the Ceylon Euphorbs, as they cost me much labour & thought[20]

I have heard nothing yet of the box of seeds you mention as sending through M^{ssr}. David Power,[21] so I suppose it will come by the next mail steamer.

In travelling here the other day I was particularly struck with the resemblance in colour the natives exhibit to different varieties of soil here. It was curious to observe, sometimes, how precisely of one colour were the native & the ground upon which he was sitting or lying. I wonder whether this colour served at one time as a protection from his enemies whether wild beasts or men—& that so it originated— ask Darwin what he thinks of this view of the matter, when you write to him.

I am glad to see that Colenso [showed] himself at the meeting of the Association at Bath and was so well received.—[22] I have been greatly delighted with the perusal of his work.[23] A true & safe reformer he appears to me to be. It is curious how long the world has been taking things on trust in matters of such immense importance & it is well that the change of opinion is coming on so gradually, or what a smash there would be!

Always my dear Hooker | Your affectionate friend | G. H. K. Thwaites

P.S. Will you kindly send this little note to Berkely when you next write to him: it contains a curious [Sphæria] upon a Fly—[24]

DAR 102: 1–3; Royal Botanic Gardens, Kew, DC vol. 162 doc. 224

[1] The letter from Henry Tibbats Stainton has not been found; however, CD's reply to Hooker of 7 January [1865] indicates that Stainton's letter concerned a translation of Karl Friedrich von Gärtner's *Versuche und Beobachtungen über die Bastarderzeugung im Pflanzenreich* (Gärtner 1849) that was to be undertaken by the Ray Society. Stainton was secretary of the Ray Society, which was founded in 1844 with the object of printing works of natural history and had produced a number of translations of foreign works (see Curle 1954). CD had recommended Gärtner 1849 for translation in his letter to the Ray Society, [before 4 November 1864] (*Correspondence* vol. 12; see also letter to J. D. Hooker, [*c.* 23 September 1864]). Hooker also supported the proposed translation; however, it was not undertaken (see *Correspondence* vol. 12, letters from J. D. Hooker, 16 September 1864 and [28 September 1864], and Curle 1954, pp. 25–6). A heavily annotated copy of Gärtner 1849 is in the Darwin Library–CUL (see *Marginalia* 1: 256–98). It is cited extensively in *Origin*, *Variation*, and *Forms of flowers* on the subject of hybrid sterility. For a discussion of the importance of Gärtner's work to CD's research, see *Correspondence* vol. 10, Appendix VI.

[2] CD included Hooker's comment in 'Climbing plants', p. 78.

[3] See *Correspondence* vol. 12, letter to J. D. Hooker, 10 December [1864] and n. 6. As part of his research on climbing plants, CD had observed the development of discs at the tips of tendrils in different families, including Cucurbitaceae. He noted that these discs had a variety of adhesive properties, enabling the plants to climb in different conditions (see *Correspondence* vol. 12, letter to Asa Gray, 28 May [1864], letter from Asa Gray, 11 July 1864, and letter to J. D. Hooker, 4 December [1864]). From his observations of *Hanburya mexicana*, a species of Cucurbitaceae, CD inferred that the plant might be an incipient form, since the adhesive discs were of no apparent use (see letter to J. D. Hooker, 10 December [1864], and 'Climbing plants', p. 78). In 'Climbing plants', p. 104, CD concluded that three genera of the Cucurbitaceae presented 'a nearly perfect gradation from a common tendril to one that forms an adherent disk at its tip'. CD's notes on this subject are in DAR 157.1: 140, DAR 157.2: 52 and 65–7, and DAR 187: 2. CD recorded in his Journal for 1864 (*Correspondence* vol. 12, Appendix II) that he had finished the paper on climbing plants on 15 September; however, he continued his observations and made small changes to the manuscript until it was sent to the Linnean Society of London on 18 January (see letter to J. D. Hooker, 19 January [1865]). The paragraphs on adherent discs in *Hanburya*, and on the development of discs in Cucurbitaceae, are inserted in CD's working copy of the manuscript in DAR 18.1: 146d–e and DAR 18.2: 199 v., 199a v.

[4] Hooker refers to the anniversary address written for the 30 November 1864 meeting of the Royal Society by its president, Edward Sabine. At the meeting, George Busk accepted the Copley Medal on CD's behalf. Sabine's address described the grounds on which the award had been given, and included a discussion of CD's 'Two forms in species of *Linum*' and 'Three forms of *Lythrum salicaria*' (see Sabine 1864, p. 510). Hooker had provided information on CD's botany to assist Sabine in composing this portion of the address (letter from Edward Sabine to J. D. Hooker, 14 November 1864, Royal Botanic Gardens, Kew, letters to J. D. Hooker, vol. 18, letter 218). Hooker had previously expressed

dissatisfaction with Sabine's account of CD's botany after reading a shortened and edited version of the address in the 3 December 1864 issue of the *Reader*, pp. 708–9 (see *Correspondence* vol. 12, letter from J. D. Hooker, [6 December 1864]). Hooker probably refers to the publication of Sabine's address in the *Proceedings of the Royal Society of London* 13 (1864): 497–517. Individual numbers of the journal were issued to fellows of the Royal Society at intervals during the year. The portion of Sabine's address on the Copley award is reproduced in *Correspondence* vol. 12, Appendix IV.

5 Sabine had contracted influenza prior to the 30 November meeting of the Royal Society (see *Correspondence* vol. 12, letter from Elizabeth Juliana Sabine, 7 December [1864]). For a discussion of Sabine's presidency, see M. B. Hall 1984, pp. 104–8.

6 The reference is to Thomas Henry Huxley's unsigned article 'Science and "Church policy"', which appeared in the 31 December 1864 issue of the *Reader*, p. 821 ([T. H. Huxley] 1864b). The attribution is based on the letter from T. H. Huxley, 15 January 1865. The article criticised 'leading statesmen' and 'ecclesiastical dignitaries' for their lack of regard for science, and addressed in particular the remarks on science made by Benjamin Disraeli in his recent speech on church policy (Disraeli 1864). For a discussion of Huxley's article, see A. Desmond 1994–7, 1: 331–2.

7 Hooker was preparing an account of the family Cucurbitaceae for *Genera plantarum* (see Bentham and Hooker 1862–83, 1: 816–41). See *Correspondence* vol. 12, letter from J. D. Hooker, [6 December 1864].

8 Hooker had expressed similar frustrations with regard to his exclusively descriptive taxonomic work for *Handbook of the New Zealand flora* (Hooker 1864–7); having accumulated a large number of facts on variable genera and species, he had intended to write a theoretical introduction to the book (see *Correspondence* vol. 12, letter from J. D. Hooker, [26 or 27 April 1864]).

9 For CD's interest in transitional forms in the family Cucurbitaceae, see n. 3, above. See also letter to J. D. Hooker, 19 January [1865]. For Hooker's description of the stamens of Cucurbitaceae, see Bentham and Hooker 1862–83, 1: 816.

10 In his letter to Hooker of 10 December [1864] (*Correspondence* vol. 12), CD reported having seen a foreign advertisement for 'Handbuch zur Physiologie' by Wilhelm Hofmeister. The advertisement has not been found; it evidently referred to *Handbuch der physiologischen Botanik*, a series of monographs to be published under the general editorship of Hofmeister (Hofmeister ed. 1865–77). Hofmeister wrote two monographs in the series, *Die Lehre von der Pflanzenzelle* (Hofmeister 1867), and *Allgemeine Morphologie der Gewächse* (Hofmeister 1868). CD's copy of Hofmeister 1867 is in the Darwin Library–Down.

11 CD had informed Hooker in his letter of 10 December [1864] (*Correspondence* vol. 12) that he expected to hear of a paper on dimorphic *Pulmonaria* by Friedrich Hildebrand. CD had corresponded with Hildebrand about dimorphism in *P. officinalis* in June 1864 (see *Correspondence* vol. 12, letter from Friedrich Hildebrand, 21 June 1864, and letter to Friedrich Hildebrand, 25 June [1864]). Hildebrand sent CD a copy of his paper 'Dimorphismus von Pulmonaria officinalis' in February 1865 (Hildebrand 1865, pp. 13–15; see letter to J. D. Hooker, 15 [February 1865]). CD's annotated copy of the paper is in the Darwin Pamphlet Collection–CUL. It is discussed in 'Illegitimate offspring of dimorphic and trimorphic plants', pp. 430–1, and *Forms of flowers*, pp. 101–3.

12 In his letter to Hooker of 10 December [1864] (*Correspondence* vol. 12), CD had asked Daniel Oliver to check *Botanische Zeitung* for references to material on dimorphism.

13 CD had asked about Hooker's proposed book on geographical distribution in his letter of 10 December [1864] (*Correspondence* vol. 12). Hooker and CD had a long-running interest in the geographical distribution of plants (see, for example, Hooker 1853, and *Correspondence* vol. 6, letter from J. D. Hooker, 9 November 1856). Hooker is cited extensively in the chapters on geographical distribution in *Origin*. Although he never wrote a general book, he continued to publish articles and addresses on the subject (see, for example, Hooker 1867 and Hooker 1881).

14 The *Reader*, a weekly review of literature, science, and art, was started in January 1863 (see Sullivan ed. 1984 and North 1997, pp. 4066–8). CD was enthusiastic about the journal (see, for example, *Correspondence* vol. 12, letter to J. D. Hooker, 26[–7] March [1864]). In 1864, editorial control passed increasingly into the hands of Huxley, Joseph Norman Lockyer, and other scientific practitioners (see Barton 1998, pp. 439–40). Near the end of 1864, the journal was purchased and reorganised by a

consortium that included CD's friend, John Lubbock. To help support the journal, CD bought shares in the Reader Limited Company in November 1864 (CD's Account book–banking account (Down House MS); see also *Correspondence* vol. 12, letter to John Lubbock, 19 November [1864]).

15 See enclosure. The reference is to George Henry Kendrick Thwaites.

16 In the spring of 1864, CD had asked Hooker on several occasions to observe the climbing habit of the pitcher plant *Nepenthes* (see *Correspondence* vol. 12, letter from J. D. Hooker, 9 [March] 1864, n. 22; see also letter to Daniel Oliver, [22 July 1864]). Hooker evidently passed CD's query to Thwaites; however, CD's description of two *Nepenthes* species in 'Climbing plants', pp. 46–7, is based largely on specimens that he obtained from the nursery firm of James Veitch (see *Correspondence* vol. 12, letters to J. D. Hooker, 10 June [1864] and [28 September 1864]). CD classified *Nepenthes* among the leaf-climbers, concluding that, at least when young, the tips of the leaves coil around a stick in order to support the pitcher with its load of secreted fluid. His notes on the genus are in DAR 157.1: 111–12 and DAR 187: 1. At some point, CD appears to have inserted the paragraph on *Nepenthes* into his working copy of the manuscript (see DAR 17.2: 90, 90a, and 90 v., and in 'Climbing plants', p. 46).

17 Thwaites refers to the Peradeniya botanic gardens, Ceylon, of which he was director (*DNB*).

18 William Ferguson.

19 Hooker had been working on the taxonomy of the Melastomaceae, a family of tropical and subtropical plants, for *Genera plantarum* (Bentham and Hooker 1862–83, 1: 725; the Melastomaceae correspond approximately to the modern family Melastomataceae). See *Correspondence* vol. 12, letter from J. D. Hooker, [26–7 April 1864].

20 Thwaites probably refers to Johann Müller and to his own description of the Euphorbiaceae in Thwaites 1858–64, 4: 268. Müller visited Hooker at the Royal Botanic Gardens, Kew, in September 1864 (see *Correspondence* vol. 12, letter from J. D. Hooker, 16 September 1864).

21 David Edward Power was a merchant with premises at 110 Fenchurch Street, London (*Post Office London directory* 1865).

22 The reference is to John William Colenso's reception at the annual meeting of the British Association for the Advancement of Science in Bath on the evening of 14 September 1864. See Cox 1888, 1: 257, for Colenso's account of the meeting, and Barton 1998, p. 437. See also *Correspondence* vol. 12, letter from J. D. Hooker, [19 September 1864].

23 Thwaites refers to the first parts of Colenso 1862–79, a work of biblical criticism that had led to religious controversy (see Guy 1983 and *Correspondence* vol. 12, letter from E. A. Darwin, 1 February [1864] and nn. 3 and 5; see also *Correspondence* vol. 12, letter from J. D. Hooker, 16 February 1864).

24 Miles Joseph Berkeley was an expert on British fungi. *Sphaeria* is a fungal parasite.

From Henrietta Anne Huxley 1 January 1865

Dear M^{r}. Darwin

Hal[1] has just brought me your note containing your slyly disparaging remarks on my beloved Tennyson—& quoting "as a gem"

'And he meant, he said
he meant,
Perhaps he meant, or partly
meant you well.'[2]

In the first place it was very mean of you to give the lines without the context shockingly Owenlike[3]

Secondly. The lines only convince me more than ever that Tennyson is quite master of his situation. Could you better render In words, the desire in the wife's mind to do justice, to—her enemy I suppose for I have not read "Sea Dreams", together with the conflicting feeling which yet possessed her of his insincerity? I am very pleased that Tennyson accredits the feminine mind with such a strong sense of justice.

I now refer to the book— I am grieved to find that a philosopher of your repute—should have damaged your reputation for accuracy so greatly as to tell me that the quotation was from "Enoch Arden" whereas it was from "Sea Dreams"— If the "facts?!" in the Origin of Species are of this sort—I agree with the Bishop of Oxford—[4]

Yours too sincerely | Henrietta Huxley
love to your dear wife & ask her for a screed.

New Year's Day | 1865.

DAR 166: 284

[1] Thomas Henry Huxley
[2] In the letter to T. H. Huxley, 5 November [1864] (*Correspondence* vol. 12), CD had quoted this passage from 'Sea Dreams', by Alfred Tennyson. CD had remarked, 'Such a gem as this is enough to make me young again & like poetry with pristine fervour'. The poem appeared in the volume *Enoch Arden, etc* (Tennyson 1864), p. 105.
[3] CD had protested to Huxley and others about the 'false & malignant' review of *Origin* by Richard Owen ([Owen] 1860): 'I never saw such an amount of misrepresentation' (see *Correspondence* vol. 8, letter to T. H. Huxley, 9 April [1860]; see also letter to Charles Lyell, 10 April [1860]).
[4] The reference is to Samuel Wilberforce, bishop of Oxford, who had criticised *Origin* in his article in the *Quarterly Review* ([Wilberforce] 1860). See also *Correspondence* vol. 8, Appendix VI.

From Thomas Henry Huxley 1 January 1865

Jany 1. 1865

My dear Darwin

I cant do better than write my first letter of the year to you—if it is only to wish you and yours, your fair share (& more than your fair share if need be) of good for the New year— The immediate cause of my writing however, was turning out my pocket & finding therein an unanswered letter of yours containing a scrap on which is a request for a photograph[1]—which I am afraid I overlooked— At least, I hope I did and then my manners wont be so bad— I inclose the latest version of myself (N.B. another will be shortly published by my wife but the likeness is not warranted to be so accurate)[2]

I wish I could follow out your suggestion about a book on Zoology[3] (By the way please to tell Miss Emma that my last *is* a book.[4] Marry come up! Does her ladyship call it a pamphlet?)

But I assure you that writing is a perfect pest to me unless I am interested— and not only a bore but a very slow process— I have some popular lectures on Physiology which have been half done for more than a twelvemonth & I hate the sight of them because the subject no longer interests me & my head is full of other matters[5]

So I have just done giving a set of Lectures to working men on the various Races of Mankind which really would make a book in Miss Emma's sense of the word & which I have had reported— But when am I to work them up?[6] Twenty four Hunterian Lectures loom between me & Easter—[7] I am dying to get out the second volume of the book that is not a book but in vain.[8]

I trust you are better though the last news I had of you from Lubbock was not so encouraging as I could have wished—[9]

With best wishes & remembrances to Mrs Darwin

Ever yours | T H Huxley

Thanks for 'für Darwin'—[10] I had it—

DAR 166: 304

[1] See *Correspondence* vol. 12, letter to T. H. Huxley, 5 November [1864].

[2] The photographs have not been found.

[3] In his letter to Huxley of 5 November [1864] (*Correspondence* vol. 12), CD encouraged him to write a 'Popular' treatise on zoology.

[4] In his letter to Huxley of 5 November [1864] (*Correspondence* vol. 12), CD included some remarks by Emma Darwin on Huxley's recently published *Lectures on the elements of comparative anatomy* (T. H. Huxley 1864a): 'I don't call that a Book . . . I want something that people can read . . .'

[5] Huxley had given a Friday evening course on physiology at the School of Mines in 1863 (see L. Huxley ed. 1900, 1: 246, and *Correspondence* vol. 11, letter from T. H. Huxley, 2 July 1863). These lectures were eventually published as *Lessons in elementary physiology* (T. H. Huxley 1866).

[6] Huxley's lectures 'On the various races of mankind' were delivered at the end of 1864 as part of a regular series of evening courses for 'working men' at the School of Mines (see L. Huxley ed. 1900, 1: 249, and Bibby 1959, pp. 97–100). The lectures were not published. No report of the lectures has been found. Huxley had also discussed human races in two lectures of his Hunterian course, 'On the structure and classification of the Mammalia', delivered at the Royal College of Surgeons at the beginning of 1864. Abstracts of these lectures appeared in the *Medical Times and Gazette*, 26 March 1864, pp. 343–4, and 2 April 1864, pp. 369–70. The lectures were also reported in the *Reader*, 27 February 1864, pp. 266–7. Huxley's work on human races is discussed in Di Gregorio 1984, pp. 160–84.

[7] Huxley was required to deliver an annual course of twenty-four lectures as Hunterian Professor at the Royal College of Surgeons (see L. Huxley ed. 1900, 1: 234–6).

[8] T. H. Huxley 1864a. A second volume was not published.

[9] CD had reported on his poor health in his last two letters to John Lubbock (see *Correspondence* vol. 12, letters to John Lubbock, 19 November [1864] and 21 December [1864]). Although CD's health had improved for some of 1864, he still suffered intermittently from sickness and weakness (see *Correspondence* vol. 12).

[10] The reference is to *Für Darwin* (Müller 1864), a developmental history of the Crustacea, presented as a validation of CD's theory of transmutation. CD had received the book from Fritz Müller in 1864 (see *Correspondence* vol. 12, letter to Ernst Haeckel, 21 November [1864]). A lightly annotated copy is in the Darwin Library–CUL (see *Marginalia* 1: 609). CD later helped to finance an English translation, which was published in 1869 (see letter to Fritz Müller, 16 March [1868] (*Calendar* no. 6014), and Möller ed. 1915–21).

From Henry Holland 2 January 1865

Brook Street.
Jany. 2^{d}. | 1865

My dear Charles,

First, let this note convey to yourself, & to all your family, my earnest good wishes for the year just begun, & for all & many years succeeding it.[1] Receive these wishes as equally cordial & sincere; but do not occupy your time in answering them. I shall understand the reciprocity of kind feeling, without troubling you to write it.

Next, let me thank you for the paper on the 3 forms of the Lythrum Salicaria & their sexual relations[2]—a most curious research; & showing further (if further proof were needful) that it is to the *primal mystery of Generation* that we must look for explanation of the phenomena of Life, & the succession of Life on the Earth.— It is hard to say how far we can ever get into this mystery; but your researches run in the right road.—[3]

I despair of being ever told, in the shape of a physical law, why the nose of the Grandson is a copy of the Grandfathers, with another configuration of nose between.

You have probably seen Cobbold's curious statement (in his book on Entozoa) of the 4 successive stages of development, & successive *habitats*, of the tape-worm species[4]

The allusion to these things carries me back in memory, to your capital monograph on the Cirripeds, & their parasitic & bi-sexual peculiarities.[5]

I write hastily, but ever, my dear Charles, your's aff^{y} | H Holland

P.S. | I gladly see that you are inducting your Son into your researches[6]

The D. of Argyll sent me a few days ago his Address as President of the R Society of Edinburgh—chiefly occupied with the question of Origin of Species[7]

DAR 166: 245

1 Holland occasionally served as a physician to CD, and had been consulted on the illnesses of CD's children, Henrietta and Horace (see, for example, *Correspondence* vol. 8, letter to J. D. Hooker, 5 June [1860], and *Correspondence* vol. 10, letter from Henry Holland, 26 March [1862]).

2 'Three forms of *Lythrum salicaria*'. Holland's name appears on CD's presentation list for the paper (see *Correspondence* vol. 12, Appendix III).

3 Holland had reviewed scientific works for the *Edinburgh* and *Quarterly Reviews* for many years. He revised his 1859 article 'Life and organisation' to include a commentary on *Origin* (see Holland 1859 and Holland 1862, pp. 98–9). CD was highly critical of Holland's reviews (see, for example, *Correspondence* vol. 4, letter to J. D. Hooker, [18 April 1847], and *Correspondence* vol. 11, letter to J. D. Hooker, 23 [June 1863]). In the letter to Charles Lyell, 25 October [1859] (*Correspondence* vol. 7), CD expressed his hope that Holland would not be chosen to review *Origin* for the *Quarterly Review*, although he later noted that Holland went an 'immense way' with CD and his supporters (*Correspondence* vol. 7, letter to Charles Lyell, [10 December 1859]). For Holland's views on *Origin*, see also *Correspondence* vol. 7, letter from Henry Holland, 10 December [1859].

4 Holland refers to the discussion of the development of the common tapeworm (*Taenia solium*) by Thomas Spencer Cobbold in his monograph on internal parasites (Cobbold 1864, pp. 220–2). Cobbold

described the life-cycle of the tapeworm as it migrated from its parent colony in the bowel of the host, through drains and ditches, until it was ingested first by pigs, and then, in the form of 'measly pork', by humans. A review of Cobbold 1864 had appeared in the 26 November 1864 issue of the *Reader*, pp. 668–70; CD was a regular subscriber to the journal (see letter from J. D. Hooker, 1 January 1864 and n. 14).

5 The reference is to CD's work on 'complemental males' in barnacles. In several species of Cirripedia, CD had observed minute males attached to the bodies of hermaphrodites, and differing greatly from them in size and structure (see *Living Cirripedia* (1851), pp. 231–2 and 281–93, *Living Cirripedia* (1854), pp. 23–30, and Newman 1993, pp. 377–81). In his letter to J. D. Hooker, 10 May 1848 (*Correspondence* vol. 4), CD remarked that the discovery was suggestive of the development of separate sexes, and thus of significance for his theory of species (see also *Correspondence* vol. 4, Appendix II, pp. 399–400). CD later suggested that complemental males may have developed in Cirripedes to enable intercrossing between two hermaphrodites (see 'On the males and complemental males of certain cirripedes, and on rudimentary structures', *Nature* 8 (1873): 431–2; *Collected papers* 2: 177–82).

6 In 'Three forms of *Lythrum salicaria*', pp. 172, 174, CD referred to observations and camera-lucida drawings made by his son William (see *Collected papers* 2: 108–9, 110; for William's work with *Lythrum*, see *Correspondence* vols. 10 and 11). William also made extensive observations for CD on dimorphism in *Pulmonaria* and other genera in 1862, 1863, and 1864 (see *Correspondence* vols. 10–12).

7 The reference is to the opening address delivered to the Royal Society of Edinburgh by its president, George Douglas Campbell, eighth duke of Argyll (G. D. Campbell 1864). See letter from Charles Lyell, 16 January 1865.

From Hugh Falconer to Erasmus Alvey Darwin 3 January 1865

21 Park Crescent N.W.
3^{d}. Jany 1865

My Dear M^{r}. Darwin

D^{r}. Sharpey called today and brought the letter about which I spoke to you.[1] I send it enclosed—and should like your Brother to see it—as he has seen the other notes—and may think from Sabines citation of my words in one of them, that I was hard upon the "Origin".[2]

You will see—taking the whole passage—that I stuck up staunchly for the "Book"—and urged it as an additional claim[3]—and I should be sorry that Charles Darwin should think otherwise.

Ask him to return the note—as it belongs to D^{r}. Sharpey—being the original, and I shall have to send it back to him.

Yours very Sinly | H. Falconer

P.S. on second thoughts—you need not trouble your Brother—with the note.[4] It will be sufficient for me that you have seen what I have said.

H F

DAR 164: 23

1 See *Correspondence* vol. 12, letter from Hugh Falconer to William Sharpey, 25 October 1864. In his letter to Sharpey, Falconer gave the grounds on which he supported CD for the Royal Society's Copley Medal. Falconer had seconded George Busk's nomination of CD for the medal at the meeting of the Royal Society Council on 23 June 1864; however, he was unable to attend the meeting at which the

discussion of the award of the medals was to begin. The Council voted to award CD the medal on 3 November 1864 (Royal Society, Council minutes).

2 The president of the Royal Society, Edward Sabine, had written an address for the 30 November 1864 meeting of the Society, based in part on information he received from Falconer (letter from Edward Sabine to William Sharpey, 29 December 1864, Royal Society, Misc. Mss. 19, no. 41). The address contained the remark that *Origin* had not been included among the grounds of the Copley award. A controversy arose over whether Sabine's address had misrepresented the views of the Council (see *Correspondence* vol. 12, Appendix IV). Hooker had written to CD in his letter of 2 December 1864 (*Correspondence* vol. 12) about the 'small breeze' at the anniversary meeting, noting that Falconer was 'grievously put out' and had written a letter to Sabine on the subject. The 'other notes' to which Falconer refers have not been identified; no letter in which Falconer's words are cited by Sabine has been found.

3 In the letter to William Sharpey, 25 October 1864 (*Correspondence* vol. 12), Falconer urged the Council of the Royal Society to consider *Origin* as 'a strong additional claim on behalf of Mr. Darwin for the Copley Medal'. In his letter of 3 November 186[4] (*Correspondence* vol. 12), Falconer wrote: 'Your friends—including myself did not fail to stand up for "the Origin of Specs"—as establishing a strong claim.'

4 Despite Falconer's concerns, reiterated in his letter to E. A. Darwin of 5 January [1865], the enclosed letter was forwarded to CD (see letter to Hugh Falconer, 6 January [1865]).

From Ludwig Rütimeyer 3 January 1865

Dear & much respected Sir

Adding to the answer of your letter from Dec. 13[1] my best thanks for the highly interesting memoir, in the meantime arrived, on Lythrum,[2] I must begin with lamenting that I have not been able till today to promote very much the proposed monograph about wild oxen,[3] to which you have so largely contributed by sending me one of the most interesting types.[4] Yes the Plates destined to that memoir are all finished since long and also a great deal of Wood-engravings and the text for the european Species. But having perhaps inconsiderately extended my purpose over the Linnean Genus Bos in general, I have much been arrested by the necessity of seeking the materials for foreign species in the different museums of the continent, and also by the unevitable consequence to compare the fossil species. Nevertheless I am as much at the work as the many incumbent other obligations permit,[5] so that I hope, that before the end of the year we just began, the labour will be done.

As to the skull from Lord Tankerville, I received from you, it is indubitable, that it is the *purest descendant* known of the fossil Bos primigenius, be it a remnant of the extinct wild race, as I presume, or a flock grown wild (Verwildert) of a formerly tamed breed;[6] certainly the flock of Lord Tankerville, the very allied *Pembroke*-breed and the predominant type in our lake-habitations of the stone age do not differ otherwise from the extinct primigenius than by Minor Size; yet the eminent size of the archetype has not seldom been reached by single Individuals in Seedorf, Robenhausen and other Swiss lake localities.[7]

I do not know the white cattle of the King of Sardinia otherwise than by the drawings given by different illustrated papers f. e. in the Field (the Country

Gentlemans News Paper) Jan. 10. 1863 and elsewhere. It is a tamed type not rarely kept in Umbria, but, as Prof. D. Filippi at Turin writes me, especially bred in Val di Chiana near Arezzo (the home of the extinct Bos trochoceros.)[8] Having seen no skull of that type it is difficult to judge about its affinities, but I am very much inclined to consider the so said Bos trochoceros as a mere race of Bos primigenius, to which belongs also the long horned cattle of Italy and Hungary; it seems very probable therefore, that the breed of Val di Chiana has the same origin.[9]

I should be very glad to know, what you think about the conclusions drawn from the facts observed in the milk-dentition of Horses and other animals, and exposed or rather hidden in my paper on fossil horses.[10] I admit willingly, that the tableau or pedigree (page 86) founded on the peculiarities and affinities of the dental system is a premature essay, but I feel convinced, that the considerations exposed in 78–90 can grow fertile, and that the identity of the milk-dentition of certain species with the dentition of the full age of other species historically precursors of the former, contains hints not to be overseen. I think that it is more than an accident, that the dentition of Anoplotherium and Dichobune is preserved to-day in the milk teeth of Moschus and Tragulus, that Merychippus has in the youth the dentition of the miocene Anchitherium, in the full age that of Equus., that the diluvial Equus fossilis inherits in its young age the dentition of the miocene Hippotherium, and the recent Equus Caballus that of the Equus fossilis etc. etc. (pg. 38. 57. 74. etc. 101 etc)[11]

Whishing Sincerely, that you may soon fully recover from your illness[12] and being, with much pleasure at your disposition as often as I can hope to serve you, | I remain, with much respect, | Dear Sir yours sincerely | L Rütimeyer

Basel 3 January 1865.

DAR 176: 227

CD ANNOTATIONS
1.1 Adding ... same origin. 3.9] *crossed blue crayon*
4.1 I should] *after opening square bracket blue crayon*
Top of letter: 'Milk Teeth of Recent & Extinct Forms—' *blue crayon*

1 CD's letter to Rütimeyer has not been found.

2 The reference is to 'Three forms of *Lythrum salicaria*'; Rütimeyer's name appears on CD's presentation list for this paper (see *Correspondence* vol. 12, Appendix III).

3 Rütimeyer refers to his projected monograph on the natural history of cattle, *Versuch einer natürlichen Geschichte des Rindes*, eventually published in two parts (Rütimeyer 1867a and 1867b). Copies of these papers are in the Darwin Pamphlet Collection–CUL. Rütimeyer published a preliminary summary of his research in Rütimeyer 1865, and announced the forthcoming publication of his 1867 monograph and its relationship to his previous work in Rütimeyer 1866, p. 11 n. CD's annotated copies of Rütimeyer 1865 and 1866 are in the Darwin Pamphlet Collection–CUL. In January 1865, CD was revising the chapters on domestic animals for *Variation* (see 'Journal' (Appendix II)); cattle are discussed in *Variation* 1: 79–93. Rütimeyer 1861, 1865, and 1866 are cited in *Variation*, 1: 79–81 (see nn. 6 and 9, below). CD received Rütimeyer 1867a and 1867b too late for inclusion in *Variation* (letter to Ludwig Rütimeyer, 4 May [1867] (*Calendar* no. 5527)).

[4] CD had arranged for a skull of one of the Chillingham cattle, an ancient breed, to be sent to Rütimeyer after reading his treatise on the animal remains found during the excavation of the Swiss lake-dwellings (Rütimeyer 1861; see *Correspondence* vol. 9, letter to Ludwig Rütimeyer, 5 December [1861], *Correspondence* vol. 10, letters to Ludwig Rütimeyer, 15 [and 16] January [1862] and 11 February [1862], and Rütimeyer 1867b, p. 132). CD had long been interested in the origin and habits of the Chillingham cattle (see *Notebooks*, Notebook D, p. 48, and *Correspondence* vol. 9). See also *Variation* 1: 81 and 83–4.

[5] Rütimeyer was professor of zoology and comparative anatomy at the University of Basel, Switzerland.

[6] The Chillingham cattle were kept on the Chillingham Park estate of Charles Augustus Bennet, sixth earl of Tankerville (see Ritvo 1992). They closely resembled the fossil form, *Bos primigenius*, a species domesticated in Switzerland during the Neolithic period. In *Variation* 1: 81, CD cited Rütimeyer for the information that the Chillingham cattle were less altered from the 'true primigenius type' than any other known breed and described the Chillingham cattle as 'semi-wild' descendants of *B. primigenius*, though 'much degenerated in size'; see also *Variation* 2: 119. Rütimeyer discussed *B. primigenius* and the descent of Chillingham cattle in Rütimeyer 1865, p. 54, and Rütimeyer 1867b, pp. 130–3, 146–9. A lithograph of the skull of a Chillingham bull, obtained from Lord Tankerville, appears in Rütimeyer 1867b, plate III.

[7] For a discussion of the relationship between living and fossil races of cattle in England and Switzerland see Rütimeyer 1867b, pp. 155–7.

[8] Rütimeyer refers to the illustration accompanying an article by Francis Trevelyan Buckland, 'Italian cattle at the Zoological Gardens', in the *Field*, 10 January 1863, pp. 28–9. The article described a white bull presented to the Zoological Society of London by Victor Emmanuel II, the king of Sardinia. Rütimeyer acknowledged the information of Filippo De Filippi, professor of zoology and director of the Museum of Zoology in Turin (*DBI*), and also a photograph of one of the cattle that he had received from CD in Rütimeyer 1867b, p. 143 n. In *Variation* 1: 83–6, CD discussed white cattle breeds in Britain, the Americas, and Africa.

[9] Rütimeyer eventually reached a different conclusion about the relationship between the fossil forms, 'Bos trochoceros' and *Bos primigenius*; in his 1866 essay on European cattle, he claimed that 'Bos trochoceros' was not a distinct race, but rather the female of an early domesticated form of *Bos primigenius* (see Rütimeyer 1866, p. 26). CD cited Rütimeyer 1866 in his description of the fossil forms of cattle in *Variation* 1: 81.

[10] The reference is to Rütimeyer's 1863 monograph on the comparative odontography of the horse. Rütimeyer argued that similarities between the milk-teeth and adult teeth of both existing and fossil forms of horse were indicative of lines of descent. In *Variation* 1: 51, CD cited Rütimeyer 1863 as suggesting that horses may have descended from more than a single species. CD's annotated copy is in the Darwin Library–CUL (see *Marginalia* 1: 718).

[11] The first three pages indicated by Rütimeyer are marked in CD's copy of Rütimeyer 1863 (see *Marginalia* 1: 718). Rütimeyer included a summary of his findings on the odontography of horses in Rütimeyer 1865, pp. 17–18; the section is scored in CD's copy of the paper (Darwin Pamphlet Collection–CUL). See also CD's annotation at the top of the letter. The discussion of comparative odontography is greatly expanded in Rütimeyer 1867a.

[12] See letter from T. H. Huxley, 1 January 1865, n. 9.

To T. H. Huxley 4 January [1865][1]

Down. | Bromley. | Kent. S.E.
Jan. 4th

My dear Huxley

Very many thanks for your Photograph, which is excellent, but it makes you look too black & solemn as if facing the bench of Bishops.—[2]

We were all charmed with M^rs^. Huxley "too sincere" note.[3] Oh that I should live to be called "Owen-like"! I was indeed innocent of concealing the context, for I did not read one line beyond the charming lines which I quoted, & they were enough for me![4]

How hard you are worked & I do wish that you had more leisure or at least not so many lectures. It is an absolute marvel to me how much you do.— I knew there was very little chance of your having time to write a popular treatise on Zoology;[5] but you are about the one man that could do it. At the time I felt it would be almost a sin for you to do it, as it would of course destroy some original work. On the other hand, I sometimes think that general & popular Treatises are almost as important for the progress of science as original work.— As for writing being a great labour to you, I can hardly swallow that. Your words on paper seem always to come out spontaneously. I have heard it hinted that you wrote the slashing leading article in the last Reader.[6] It is a capital article whether or no you wrote it. That is splendid about the pump & shoes—[7]

I am no great thing in health, but manage most days to do a little work.—

Our kindest remembrances to Mrs Huxley | Ever yours | C. Darwin

Imperial College of Science, Technology and Medicine Archives (Huxley 5: 211)

[1] The year is established by the relationship between this letter and the letter from T. H. Huxley, 1 January 1865.

[2] See letter from T. H. Huxley, 1 January 1865. The photograph has not been found.

[3] See letter from H. A. Huxley, 1 January 1865.

[4] Henrietta Anne Huxley had accused CD of behaving like Richard Owen in quoting lines from a Tennyson poem out of context. See letter from H. A. Huxley, 1 January 1865 and nn. 2 and 3.

[5] See letter from T. H. Huxley, 1 January 1865 and n. 3.

[6] Joseph Dalton Hooker had correctly suggested that Huxley wrote the *Reader* article (see letter from J. D. Hooker, 1 January 1865 and n. 6, and [T. H. Huxley] 1864b).

[7] In his *Reader* article, Huxley criticised what he considered to be the prevailing bias among statesmen and church leaders against the introduction of science into élite English schools, such as Eton. The reform of these schools was currently being considered by Parliament, following a report that had been prepared by the government-appointed Clarendon Commission, which contained recommendations for supplementing the traditional curriculum of classical languages and mathematics with other subjects, including the natural sciences (see Shrosbree 1988). Huxley cited a passage from one of the opponents of scientific education, who had drawn an analogy between teaching the physical laws of an air pump, and teaching the art of shoemaking ([T. H. Huxley] 1864b, p. 821):

> How will an Eton boy be the better for knowing how to make a pump? Doubtless it is a good thing to know how to make a pump; but it is also a good thing to know how to make shoes; and yet you do not propose to introduce shoemaking as a branch of liberal education.

From Hugh Falconer to E. A. Darwin 5 January [1865][1]

5 Jan^y^

Dear M^r^. Darwin

Your brother has been bored enough about the matter already—[2] It is of no importance—and he does not appear to be over strong at present.

I have no objection other than this.

It seemed to me sufficient that you saw it.[3]

I wanted to be set right about the Sentence—which standing alone—read so sharp[4]

Yours very Sinly | H Falconer

DAR 164: 24

[1] The year is established by the relationship between this letter and the letter from Hugh Falconer to E. A. Darwin, 3 January 1865.

[2] See letter from Hugh Falconer to E. A. Darwin, 3 January 1865 and nn. 1–3.

[3] On the back of this letter, Erasmus wrote: 'I thought you would like to see Falconers Eloge so I have sent it'. Erasmus evidently sent CD this letter, together with the letter from Hugh Falconer to William Sharpey, 25 October 1864 (*Correspondence* vol. 12).

[4] See letter from Hugh Falconer to E. A. Darwin, 3 January 1865 and n. 2. Falconer refers to a sentence from the anniversary address of Edward Sabine, which was delivered at the Royal Society of London on 30 November 1864. Sabine's address contained the controversial remark that *Origin* had not been among the grounds for awarding CD the Copley Medal. For a discussion of the controversy surrounding Sabine's address, see *Correspondence* vol. 12, Appendix IV.

To Hugh Falconer 6 January [1865][1]

Down
Jan. 6th.

My dear Falconer

I return your letter to Sharpey received this morning from my Brother.[2] Your letter is by far the grandest eulogium which I have ever received, or shall ever receive; and if one half—or one quarter be true and not exaggerated by your great kindness, I may well rest content that I have not laboured in vain. But I hope yet to do a little more work and justify your good opinion.

My dear Falconer | Yours most gratefully | Ch. Darwin

Copy
DAR 144: 38

[1] The year is established by the relationship between this letter and the letter from Hugh Falconer to E. A. Darwin, 5 January [1865].

[2] Letter from Hugh Falconer to William Sharpey, 25 October 1864 (*Correspondence* vol. 12). See letter from Hugh Falconer to E. A. Darwin, 3 January 1865 and nn. 1–3.

From William Darwin Fox 6 January [1865][1]

Delamere Ry | Northwich
Jan[y] 6

My dear Darwin

I must write you a few lines to thank you much for your sending me your paper upon the 3 forms of Lythrum Salicaria.[2] I have read it with very great pleasure, and shall look at my Garden Variety of it with much more interest than I ever did before, next year.

I forget whether I have told you that my wife brought me a very fine little Boy three weeks since to day.[3]

Most of our friends *condole* with us on the affliction but we are well contented to bear it. We have not one too many in our 16.[4] 14 are now at home—but we shall soon lose my eldest Boy who has taken his degree at Oxford, & will be ordained in March.[5]

I have now 5 Grandchildren also.

But I will not bother you with more reading. May every blessing be yours and Mrs Darwins in this new year | I am Your affec^{e} Cousin | W. D Fox

DAR 164: 183

[1] The year is established by the references to CD's paper on *Lythrum salicaria*, and to the birth of a son to Fox (see nn. 2 and 3 below).

[2] 'Three forms of *Lythrum salicaria*'. Fox's name appears on CD's presentation list for the paper (see *Correspondence* vol. 12, Appendix III).

[3] Gilbert Basil Fox, born 16 December 1864, was the twelfth child of William Darwin Fox and his second wife, Ellen Sophia. Fox and his first wife, Harriet, had five children, one of whom died in 1853 (*Darwin pedigree*).

[4] On learning that Fox and his wife were expecting a new baby, CD wrote to Fox: 'I never heard anything so awful as your sixteenth child!' (see *Correspondence* vol. 12, letter to W. D. Fox, 30 November [1864]).

[5] Samuel William Darwin Fox took his BA at Wadham College, Oxford, in 1864. His first clerical position was as curate of St Paul's, Manningham, Yorkshire, in 1865 (*Alum. Oxon.*, *Crockford's clerical directory*).

From Thomas Rivers 6 January [1865][1]

Nurseries, Sawbridgeworth, Herts. | *Great Eastern Railway.*[2]

Jan^{ry} 6/64

My dear Sir/

I tardily but warmly beg to thank you for the pamphlet you have recently sent me "On the sexual relations of Lythrum"[3]

Your minute powers of observation astonish me as does your perseverance While reading your brochure I could not help contrasting what you do with my own impatience in my small crossing efforts[4] it must have been born with you for no training would bring on such perseverance & patience

My Elms raised from three varieties of weeping elms have made nearly a foot in growth— at present they are all perfectly erect.[5]

I sincerely hope that your health is improving with age. for many long tedious years I was thin pale & delicate but on passing 55 I seemed to take a new hold of life & am now at 67 robust & vigorous. From the age of 30 to 55 I, although upwards of 6 feet in height, weighed only 10 stone I now weigh 14 & I assure you I enjoy & am very grateful for my uninterrupted good health

I am My dear Sir | Y^{rs}. very truly | Th^{os}. Rivers

M Carrière[6] has written to ask if the Nectarine has really produced a peach from seed. I could say yes with great truth—as I have done[7]

Postmark: JA 6 65
DAR 176: 163

1 Rivers misdated the letter. The date is established by the postmark.

2 In this and subsequent letters, the location of Rivers's nursery is followed on the letterhead by: '*Harlow Station is the most convenient for passengers*'.

3 'Three forms of *Lythrum salicaria*'. Rivers's name appears on CD's presentation list for the paper (see *Correspondence* vol. 12, Appendix III).

4 Rivers was noted for developing new varieties of fruit by crossing (see *Journal of Horticulture* n.s. 33 (1877): 342–4). CD had first written to Rivers in December 1862, requesting information on whether slight variations appeared suddenly from buds, much like 'sports' (see *Correspondence* vol. 10, letter to Thomas Rivers, 23 December [1862]). At that time, CD was beginning his draft chapter on bud-variation for *Variation* (*Correspondence* vol. 10, Appendix II). CD discussed bud-variation in peach trees in *Variation* 1: 340–1, 374–5.

5 CD had asked Rivers for information on the weeping elm (see *Correspondence* vol. 11, letter to Thomas Rivers, [14 February 1863]). Rivers's observations are cited in *Variation* 2: 19; CD noted the weeping habits of elms and other trees as instances of 'capricious' inheritance.

6 Elie Abel Carrière.

7 Rivers had raised a variety of peaches from the stones of nectarines (see *Gardeners' Chronicle*, 24 September 1859, p. 774, and 20 December 1862, pp. 1195–6). His results are cited in *Variation* 1: 340. CD had corresponded extensively with Rivers in 1863 on the relationship between peach and almond trees (see *Correspondence* vol. 11).

To Ray Society[1] [before 7 January 1865][2]

1141 In reference to Minutes 1118 & 1132, the consideration of the Translation of Gaertner's work on Bastardezeugung was then resumed, & a letter from Mr Darwin was read stating that the work was in no degree antiquated & that no one would be able to add to it notes of much value—[3] The production of an English Translation of this work being strongly recommended both by Mr Darwin & Dr Hooker | Resolved to accept the Translation of Gaertner's work for an early year.[4]

The Natural History Museum, London. General Library Manuscripts: L MSS RAY A: vol. 2, p. 106r: Minute 1141, 13th January 1865

1 The letter has not been found. The text given here has been taken from the minutes of the special council meeting of the Ray Society, 13 January 1865.

2 The date is established by the relationship between this letter and the letter to J. D. Hooker, 7 January [1865].

3 CD had previously written to the Ray Society, recommending a translation of Karl Friedrich von Gärtner's *Versuche und Beobachtungen über die Bastarderzeugung im Pflanzenreich* (Gärtner 1849; see *Correspondence* vol. 12, enclosure to letter to J. D. Hooker, 23 September [1864]). This letter was read at the special council meeting of the Ray Society on 4 November 1864 (see *Correspondence* vol. 12, letter to Ray Society, [before 4 November 1864]). On receiving a communication from the society's secretary,

Henry Tibbats Stainton, Joseph Dalton Hooker had asked CD to write a second letter of recommendation, evidently addressing concerns about the possible need for explanatory notes to accompany the translation (see letter from J. D. Hooker, 1 January 1865 and n. 1).

4 The translation was not undertaken (see Curle 1954, pp. 25–6).

To J. D. Hooker 7 January [1865][1]

Down
Jan 7th.

My dear Hooker.

I wrote to Stainton to say that Gärtner is by no means antiquated & no notes are required to the Translation.—[2] I return Thwaites' letter; you had better pass sub silentio about the natives being like in colour to the soil; for if true, I sh^{d}. think it was the dust of each district adhering to their perspiring bodies.—[3]

Thanks for telling me of the Cucurb; with adhering tendrils.—[4] Today I shall get my M.S. on Climbing plants from the Copyist after much delay from his having caught the scarlet Fever, & in a week or two I will run over it & send it to Linn. Soc.—[5] (But I think I will first consult Bentham,[6] whether Linn. Soc. will like expence of so large a paper with 12 woodcuts; for if too expensive for Linn. Soc. I w^{d} send it to Royal Soc.; though I sh^{d}. prefer the former.—)[7] See P.S.

I am sorry that you have not begun your Plant Book,[8] but I cannot say that I am surprised with the heaps of things which you have to do. Is New Zealand Flora finished?[9] If there is any Introduction *please be sure* & tell me.— If you discuss miscellaneous points, I think it w^{d}. be worth while to count how many plants there are with irregular corollas in comparison with, for instance, England.—[10]

I do not quite agree with what you say about the Reader;[11] it seems to me to give a splendid & interesting resume of all that is doing in all branches of science. It does seem a pity that Huxley sh^{d}. edit the scientific part. I read over again the slashing leading article & thought it excellent; but none of us could see its "withering & desolating" effects.[12] How capital that was about no more use in an Eton Boy knowing how to make a pump than a pair of shoes:[13] what a glorious profundity of ignorance it shows.— By the way I felt convinced that a Leader in the Reader 2 or 3 weeks ago, on Spiritualism was by Tyndall, who was called by the rapper the "Poet of Science".— It was a capital article; was it by Tyndall?[14]

You ask how I am; I have now had five pretty good days, but before that I spent fully a third of my time in bed, but had no actual vomiting. D^{r}. Jenner is exhausted as to doing me any good.[15] All Doctors seem to think that I am case of suppressed gout:[16] do you know of any good men hereafter to consult? I did think of trying Bence Jones;[17] but I know it is folly & nonsense to try anyone.— We are all well & I enjoy much having all the Boys at home:[18] they make the house jolly. I am glad that you are all right at your home. Our boy Horace has made a sudden start in power of walking & that, I think, is very good sign of real improvement in health.—[19]

Farewell my dear old friend | Yours ever | C. Darwin

P.S. Will you be so kind as to read the passage in red brackets & following to Bentham;[20] as it will save me writing, & whenever you write again within a fortnight, tell me in half-a dozen words his answer.—

I am *quite ashamed* to say that if my Copyist has not overcharged me, the M.S. will take about 102 pages of the Journal, though the pages of the Journal have so many lines & are in so small a type,— But I declare I do not think I have spun out my matter. But I repeat that I am ashamed & disgusted at the length of my paper.— The Royal is so rich, that it could afford to print it.—

DAR 115: 257a–c

[1] The year is established by the relationship between this letter and the letter from J. D. Hooker, 1 January 1865.

[2] CD's letter to Henry Tibbats Stainton has not been found; see, however, the letter to Ray Society, [before 7 January 1865] and n. 3. CD refers to his suggested translation of Karl Friedrich von Gärtner's *Versuche und Beobachtungen über die Bastarderzeugung im Pflanzenreich* (Gärtner 1849).

[3] CD refers to the letter from George Henry Kendrick Thwaites, enclosed with the letter from J. D. Hooker, 1 January 1865. Thwaites had speculated on the origins of the skin colour of native peoples in Ceylon, which seemed to him to resemble the colour of the soil.

[4] See letter from J. D. Hooker, 1 January 1865 and nn. 2 and 3.

[5] CD refers to 'Climbing plants', an extract of which was read at the Linnean Society on 2 February 1865. The copyist of CD's manuscript was the Down schoolmaster, Ebenezer Norman, who had worked in this capacity for many years (see *LL* 1: 153). An entry in CD's Account book–cash account (Down House MS) dated 15 March 1865 records a payment of £5 11*s.* to 'Norman for copying Science'. For CD's copy of the manuscript, in his own hand, see DAR 17 and 18.

[6] George Bentham, president of the Linnean Society.

[7] 'Climbing plants' was issued in June 1865 in a double number (nos. 33 and 34) of the *Journal of the Linnean Society (Botany)*, pp. 1–118. CD's paper appeared in two other forms, first as an offprint for the author, while commercial offprints were available from August 1865 (*Publishers' Circular*, 1 August 1865, p. 391; see also letter to J. D. Hooker, [29 July 1865]). See Freeman 1977, pp. 116–18. CD evidently considered sending the paper to the Royal Society for publication in its *Philosophical Transactions*.

[8] CD refers to Hooker's proposed book on the geographical distribution of plants. See letter from J. D. Hooker, 1 January 1865 and n. 13.

[9] CD refers to the second volume of Hooker's *Handbook of the New Zealand flora* (Hooker 1864–7). The work, which had been commissioned by the New Zealand government (see *Correspondence* vol. 11, letter from J. D. Hooker, 6 January 1863), was published in two parts. The first volume was issued in 1864, the second in 1867. See also *Correspondence* vol. 12, letter to J. D. Hooker, 8 October [1864] and n. 7, and R. Desmond 1999, p. 216.

[10] For the introduction to the second volume of Hooker 1864–7, Hooker used the same essay that he had written for Bentham and Ferdinand von Mueller's *Flora Australiensis* (see Bentham and Mueller 1863–78, 1: 1–xl). It included only a general description of the corolla parts of a flower (see Hooker 1864–7, 1: xiii–v). CD may have been anticipating a more theoretical introduction, like the one that Hooker had written to accompany his earlier work on New Zealand plants (Hooker 1853). For CD's interest in irregular flowers, see, for example, *Correspondence* vol. 11, letter to M. T. Masters, 6 April [1863].

[11] See letter from J. D. Hooker, 1 January 1865 and n. 14.

[12] CD refers to Thomas Henry Huxley and [T. H. Huxley] 1864b. See letter from J. D. Hooker, 1 January 1865 and n. 6. CD is paraphrasing Hooker's remarks.

[13] See letter to T. H. Huxley, 4 January [1865] and n. 7.

[14] CD refers to 'Science and the spirits', the lead article in the 10 December 1864 issue of the *Reader*, pp. 725–6 ([Tyndall] 1864). See the letter from J. D. Hooker, [8–18 January 1865], for Hooker's confirmation that John Tyndall was the author. The article contained a critical account of a séance, and included descriptions of 'table-rapping' and other phenomena commonly educed by spirit mediums. At the conclusion of the séance, the medium asked for the name by which Tyndall was known in the spiritual world. The alphabet was read, and knocking was heard from beneath the table after letters that eventually spelled: 'poet of science'. The efforts of Tyndall and other scientific practitioners to denounce spiritualism are discussed in Oppenheim 1985, pp. 327–30. For historical studies of Victorian spiritualism, see also Barrow 1986 and A. Owen 1989. CD had expressed scepticism about spiritualist phenomena in his letter to Susan Darwin, [26 April 1838] (*Correspondence* vol. 2).

[15] The physician William Jenner had been treating CD since March 1864 (see *Correspondence* vol. 12; see also letter from T. H. Huxley, 1 January 1865, n. 9). The last known letter from Jenner is [after 24 November] 1864.

[16] Henry Holland first diagnosed CD's condition as 'suppressed gout' in 1849 (see *Correspondence* vol. 4, letter to W. D. Fox, 6 February [1849]). According to Holland, many chronic nervous and dyspeptic symptoms were caused by the same accumulation of toxic substances that provoked outward attacks of gout. In medical literature of the period, gout was sometimes linked with stomach disorders such as dyspepsia and flatulence (see, for example, Holland 1855, p. 233, and Garrod 1863, pp. 263–4). The diagnosis of suppressed gout may also have been made by William Brinton and Jenner. CD had consulted Brinton in 1863 (see *Correspondence* vol. 11). In his own notes on his symptoms, CD wrote: 'Doctors puzzled, say suppressed gout Family gouty' (see Appendix IV). For a discussion of the diagnosis, see Colp 1977, pp. 109–10. On theories of 'suppressed gout' during the period, see R. Porter and Rousseau 1998. For a discussion of recent theories of CD's illness, see Colp 1998.

[17] Henry Bence Jones was a prominent London physician who had published works on gout and was recognised as an authority on stomach and renal diseases (*DNB*). CD began to consult Jones in July 1865 (CD's Account book–cash account (Down House MS), entry for 22 July 1865).

[18] CD's sons Francis and Leonard were attending Clapham School in 1865 (CD's Account book–classed account (Down House MS)); his son George was at Cambridge University (*Alum. Cantab.*). Emma Darwin's diary (DAR 242) records that Francis and Leonard returned to school on 2 February 1865, and suggests that George may also have been at Down for part of December 1864 and January 1865.

[19] CD's youngest son, Horace, aged 13, had been recovering from an intermittent illness since 1862 (see *Correspondence* vols. 10–12).

[20] CD bracketed the lines in the second paragraph of this letter from 'But I think' to 'the former.—' in red crayon.

From J. D. Hooker [8–18 January 1865][1]

Royal Gardens Kew

Dear Darwin

Bentham craves your paper however long—for the Linnean, & so do I.[2]

The worst of the Royal Socys. Proceedings is, that to *Naturalists* they are nought, they do not take them in, & if you print a paper in them, you are pestered by correspondents about it—[3] I wish the Royal & Linnean would join in publishing as Physical, & Biological sections of one body

I think the 2 last No^{s}. of Reader vast improvements The spiritualism Leader was Tyndalls. I had heard it rehearsed often, any time this *5 years* past, about which time ago it occurred.[4]

What the devil is this "suppressed Gout"? upon which Doctors fasten every ill they cannot name If it is *suppressed* how do they know it is gout.? if it is apparent, why the Devil do they call it suppressed? I hate the use of cant terms to cloak ignorance.[5]

Ever Yrs affec | J D Hooker.

DAR 102: 4–5

[1] The date range is established by the relationship between this letter and the letters to J. D. Hooker, 7 January [1865] and 19 January [1865].

[2] Hooker refers to George Bentham, 'Climbing plants', and the *Journal of the Linnean Society* (*Botany*). See letter to J. D. Hooker, 7 January [1865] and nn. 6 and 7.

[3] CD had considered sending his manuscript to the Royal Society (see letter to J. D. Hooker, 7 January [1865] and n. 7). The *Proceedings of the Royal Society of London* were published monthly and contained shorter papers and abstracts of papers read at the weekly meetings of the Society. Longer papers were published annually in the Royal Society's *Philosophical Transactions* (see *Record of the Royal Society of London*, pp. 178–9, 312–13). As a fellow of the Royal Society, CD received both publications. His copies of the *Proceedings* and *Transactions* are in the Darwin Library–CUL.

[4] Hooker refers to John Tyndall and a description by Tyndall in the *Reader* of a séance he had attended (see letter to J. D. Hooker, 7 January [1865] and n. 14).

[5] See letter to J. D. Hooker, 7 January [1865] and n. 16. Hooker had also expressed scepticism about the diagnosis of 'suppressed gout' in his letter of 2 July 1862 (*Correspondence* vol. 10).

To Thomas Campbell Eyton 9 January [1865?][1]

Down. | *Bromley.* | *Kent. S.E.*
Jan. 9th

Dear Eyton

I am uncommonly much obliged to you for your facts about breeding:[2] if any other striking case should occur to you, perhaps you will have kindness to inform me.

I am equally or even more obliged for promised measurements of web of feet of Otter Hounds: if you possibly can will you give me comparative with Fox Hounds or Harriers; for I have here no opportunity of seeing Dogs.[3]

With hearty thanks | Ever yours sincerely | Charles Darwin

American Philosophical Society (285)

[1] The year is conjectured from the relationship between this letter, the letter to F. T. Buckland, 11 December [1864], and the letter to T. C. Eyton, 29 December [1864?] (*Correspondence* vol. 12).

[2] Eyton's letter has not been found. The facts to which CD refers are probably cases of sterility between normally fertile mammals or birds; CD cited instances of sexual incompatibility communicated to him by Eyton in *Variation* 2: 162. CD requested such information in his letter to Eyton of 29 December [1864?] (*Correspondence* vol. 12).

[3] In his letter of 29 December [1864?] (*Correspondence* vol. 12), CD had asked Eyton to examine the feet of otter-hounds. Eyton may be the 'friend' cited in *Variation* 1: 39–40 as the source of information that English otter-hounds, as dogs with aquatic habits, have more skin between their toes than do other hounds; however, see also *Correspondence* vol. 12, letter to F. T. Buckland, 11 December [1864].

To Ray Society[1] [14–18 January 1865][2]

1146. Read a letter from Mr Darwin expressing his regret that the state of his health would not permit of his writing an Introductory Chapter to the Translation of Gaertner's work.[3]

The Natural History Museum, London. L MSS RAY A: vol. 2, p. 107r: Minute 1146, 3d February 1865

1 The letter has not been found. The text given here is taken from the minutes of the special council meeting of the Ray Society, 3 February 1865.

2 The date range is established by the minutes of the special council of the Ray Society, 13 January 1865 (see letter to Ray Society, [before 7 January 1865]), and the relationship between this letter and the letter to J. D. Hooker, 19 January [1865].

3 CD had recommended that the Society undertake a translation of Karl Friedrich von Gärtner's *Versuche und Beobachtungen über die Bastarderzeugung im Pflanzenreich* (Gärtner 1849). See letter from J. D. Hooker, 1 January 1865 and n. 1, and letter to Ray Society, [before 7 January 1865]; see also letter to J. D. Hooker, 19 January [1865].

From T. H. Huxley 15 January 1865

26 Abbey Place
Jany 15. 1865

My dear Darwin

Many thanks for Deslongchamps paper which I do not possess.[1]

I received another important publication yesterday morning in the shape of a small but hearty son,[2] who came to light a little before six— The wife[3] is getting on capitally and we are both greatly rejoiced at having another boy as your godson ran great risks of being spoiled by a harem of sisters[4]

I hope this will be the last for I really cannot afford any more—boys or girls[5]

The leader in the 'Reader' *is* mine & I am glad you liked it[6]—the more so as it has got me into trouble with some of my friends— However the revolution that is going on is not to be made with rose water

I wish if anything occurs to you that would improve the scientific part of the Reader you would let me know as I am, in great measure, responsible for it—[7]

I am sorry not to have a better account of your health—[8]

With kind remembrances to Mrs Darwin[9] & the rest of your circle

Ever Yours faithy | T H Huxley

DAR 166: 305

1 Huxley probably refers to Jacques-Armand Eudes-Deslongchamps, a French palaeontologist who specialised in marine invertebrates and reptiles (Sarjeant 1980–96). The paper that Huxley received may have been Eudes-Deslongchamps 1842, an essay on monstrous appendages in pigs that CD cited in *Variation* 1: 75–6. CD was revising the chapters on domestic animals for *Variation* (see 'Journal' (Appendix II)). No previous reference to the paper has been found.

2 Henry Huxley.

3 Henrietta Anne Huxley.

[4] Leonard, Huxley's second son and CD's godson, was born in 1860. The daughters were Jessie (born 1858), Marian (born 1859), Rachel (born 1862), and Nettie (born 1863). Huxley's first son, Noel, had died in 1860 (Clark 1968, pp. 376–7).
[5] Another daughter, Ethel, was born in 1866.
[6] [T. H. Huxley] 1864b. See letter from J. D. Hooker, 1 January 1865 and n. 6, and letter to T. H. Huxley, 4 January [1865] and n. 7.
[7] See letter from J. D. Hooker, 1 January 1865 and n. 14, and letter to J. D. Hooker, 7 January [1865].
[8] See letter to T. H. Huxley, 4 January [1865], and letter to J. D. Hooker, 7 January [1865].
[9] Emma Darwin.

From Charles Lyell 16 January 1865

Magdeburg:[1]
January 16, 1865.

My dear Darwin,—

I was so busy with the last chapters of my new edition of the 'Elements' before I left town a month ago,[2] that I did not reply to your kind letter[3] about my after-dinner speech on your Copley medal at the Royal Society anniversary.[4] I have some notes of it, and hope one day to run over it with you, especially as it was somewhat of a confession of faith as to the 'Origin.' I said I had been forced to give up my old faith without thoroughly seeing my way to a new one.[5] But I think you would have been satisfied with the length I went. The Duke of Argyll expresses in his address to the Edinburgh Royal Society very much what I have done ('Antiquity of Man,' p. 469), that variation or natural selection cannot be confounded with the creational law without such a deification of them as exaggerates their influence.[6] He seems to me to have put the difficulty pretty clearly, but on the other hand he has not brought out as fully as I should have liked him to have done, the great body of evidence so admirably brought to bear in your work, in proof of the bond of mutual descent, and the manner in which species and genera branched from common ancestors. He did not entertain this idea till he had read your book, and he is now evidently impressed with it, as I am; and he would, I think, go the whole length, were it not for the necessity of admitting, in order to be consistent, that man and the quadrumana came from a common stock. He does, indeed, in defiance of consistency, admit for the humming-birds what he will not admit for the *primates*, and Guizot's theology is introduced to support him;[7] but the address is a great step towards your views—far greater, I believe, than it seems when read merely with reference to criticisms and objections. The reasoning about materialism appears to me admirably put,[8] and his definition of the various senses in which we use the term 'law';[9] though, having only read the speech once, I am not yet able to judge critically on all these points. He assumes far too confidently that the colours of the humming-birds are for mere ornament and beauty.[10] I can conceive a meaning in your sense for the advantage of the creature, or of its friends and enemies, in every coloured ray of light reflected from the plumes.[11] We must indeed know far more than we do before we can dogmatise on the irrelevancy of particular colours to the well-being of a species. He ought also to define beauty, and tell us whether it is in reference to man or bird. I have no objection to the idea of beauty or variety for its own sake, but to assume it so positively is unphilosophical.

We have been about three weeks at Berlin, and I had some good geological talk with Ferdinand Roemer, Beyrich, Von Kœnen, Gustav Rose, Ewald, Dr. Roth, and Dove the meteorologist, besides Ehrenberg, Magnus, Lepsius, and Du Bois-Reymond,[12] and an animated conversation on Darwinism with the Princess Royal, who is a worthy daughter of her father, in the reading of good books and thinking of what she reads.[13] She was very much *au fait* at the 'Origin' and Huxley's book, the 'Antiquity,' &c. &c.,[14] and with the Pfahlbauten Museums which she lately saw in Switzerland.[15] She said after twice reading you she could not see her way as to the origin of four things; namely the world, species, man, or the black and white races. Did one of the latter come from the other, or both from some common stock? And she asked me what I was doing, and I explained that in recasting the 'Principles' I had to give up the independent creation of each species.[16] She said she fully understood my difficulty, for after your book 'the old opinions had received a shake from which they never would recover.' I shall be very glad to hear what you think of the Duke of Argyll's comments on the 'Origin'.[17] I think that your book is a vast step towards showing the methods which have been followed in creation, which is as much as science can ever reach, and the Duke, I think, has not fully appreciated the advance which has been made, even in his own mind.

I had hoped that a copy of the 'Elements'[18] would have been sent to you while I was still at Berlin. You will find much that is new, and nothing, I think, clashing with the 'Origin'. Please read my description of the Atlantis theory.[19] I fear I shall return and find the book still unborn, which is too bad of the printer. Please let me know how your health has been during the last four weeks.

Ever most truly yours | Charles Lyell.

P.S. In an article in the Berlin 'Punch' on the Pope's encyclical, in which all the innovations which trouble his Holiness are enumerated, 'Die Darwinische Lehre die uns alle Affen macht' was not forgotten.[20]

Dover: January 19.

K. M. Lyell ed. 1881, 2: 384–6

1 Magdeburg is a city on the Elbe River in central Germany (*Columbia gazetteer of the world*); the letter was evidently posted on Lyell's return to London from Berlin, where he had spent the Christmas holidays (K. M. Lyell ed. 1881, 2: 381).

2 The sixth edition of the *Elements of geology* (C. Lyell 1865) was published in January 1865 (*Publishers' Circular*, 1 February 1865, p. 60). An annotated copy is in the Darwin Library–Down (see *Marginalia* 1: 524–5).

3 The letter to Lyell has not been found.

4 An anniversary dinner was held following the 30 November meeting of the Royal Society at which CD was presented with the Copley Medal. In his letter of [before 30 November 1864] (*Correspondence* vol. 12), Erasmus Alvey Darwin had informed CD that Lyell was to attend the dinner and give his views on *Origin*. Lyell's after-dinner speech is published in Bartholomew 1975–6.

[5] In his speech, Lyell stated (Bartholomew 1975–6, p. 216):

> No intellectual condition can be more unsatisfactory than to have our old faith weakened, undermined or thoroughly unsettled without being able thoroughly to embrace the new one which is offered in its place. I cannot say that such is my present state of mind in regard to the Origin of Species, altho' I am not yet gone quite so far as Mr Darwin and some of his followers.

CD had been disappointed with the view of CD's transmutation theory and theory of human descent that Lyell published in *Antiquity of man* (C. Lyell 1863a). See *Correspondence* vol. 11, letter to J. D. Hooker, 24[–5] February [1863], letter to Charles Lyell, 6 March [1863], and letter from Charles Lyell, 15 March 1863.

[6] Lyell refers to the address delivered to the Royal Society of Edinburgh on 5 December 1864 by George Douglas Campbell, eighth duke of Argyll (G. D. Campbell 1864). Campbell distinguished between laws such as natural selection, which determine the success of existing species or modifications, and creational laws, which produce modifications or new forms of life. According to Campbell, CD had frequently forgotten this distinction in *Origin*, and had written 'of natural selection "producing" this and that modification of structure' (G. D. Campbell 1864, pp. 275–6). Campbell had sent Lyell a copy of his address; in his letter to Campbell of 25 January 1865 (I. E. Campbell ed. 1906, 2: 484), Lyell remarked, 'Your objection that Darwin has in some parts of his book made natural selection do more in the way of originating or creating than is admissable, or even consistent, with his own explanation of natural selection, was felt strongly by me. . . . Darwin is inclined to believe that he has made a greater step in the direction of discovering an originating cause or law than he has really made'. Lyell had made a similar point in *Antiquity of man* (C. Lyell 1863a, p. 469): 'If we confound "Variation" or "Natural Selection" with such creational laws, we deify secondary causes or immeasurably exaggerate their influence'.

[7] Campbell claimed that despite the close physical affinity between humans and the lower animals, human descent could not be explained on the same principles that might plausibly govern the descent of birds. 'We are conscious,' he wrote, 'of an amount and of a kind of difference between ourselves and the lower animals, which is, in sober truth, immeasurable, in spite of the close affinities of bodily structure' (G. D. Campbell 1864, p. 288). In support of his position, Campbell cited the French statesman and historian François Guizot, who maintained that humans could not have survived unless they had appeared with all of their faculties and powers fully developed (see G. D. Campbell 1864, p. 289, and Guizot 1864, p. 49). For Campbell's earlier reservations about *Origin*, see his presidential address to the Royal Society of Edinburgh on 3 December 1860 (G. D. Campbell 1860, pp. 371–6).

[8] In the conclusion of his address (G. D. Campbell 1864, pp. 291–2), Campbell stated that recent developments in science tended to show that 'all the realities of nature [were] in the region of the Invisible'; for example, current work in physiology suggested that 'material organs [were] merely the special forms built up and fashioned by the vital forces . . . for the discharge of special functions'. In his letter to Campbell of 25 January 1865 (I. E. Campbell ed. 1906, 2: 484), Lyell praised the address for having done 'a real service to the scientific and theological public by showing . . . that there is no tendency to materialism in the reasoning or speculations of modern naturalists and physicists, but quite the contrary'. For a discussion of materialism in the context of debates between scientific practitioners and theologians in Britain in the 1860s, see Barton 1987, pp. 130–3.

[9] Campbell distinguished between a notion of law that referred to an observed order of facts, and other notions of law that involved the operation of forces and the fulfilment of purposes (G. D. Campbell 1864, p. 270).

[10] According to Campbell, CD's theory of transmutation by natural selection was only able to account for changes that were of direct use to organisms in the struggle for existence, and was therefore incomplete (G. D. Campbell 1864, pp. 276–81). He argued (*ibid.*, p. 277) that humming-birds had a great many 'structures designed for mere ornament, and entirely separate from any other known or conceivable use'.

[11] In *Origin*, pp. 88–90, CD briefly discussed his theory of sexual selection, according to which ornamental characteristics developed as a result of competition between the male members of a species

for the possession of females. However, CD argued that for birds the process was 'more peaceful'. The colourful plumage of birds developed as a result of females choosing the most visually attractive partner. The subject of beauty in birds is also discussed in *Correspondence* vol. 9, letter to W. B. Tegetmeier, 14 April [1861], and letter from B. P. Brent, 29 May 1861.

12 Lyell refers to Ferdinand Römer, Ernst Beyrich, Adolf von Koenen, Gustav Rose, Julius Wilhelm Ewald, Justus Ludwig Adolph Roth, Heinrich Wilhelm Dove, Christian Gottfried Ehrenberg, Heinrich Gustav Magnus, Karl Richard Lepsius, and Emil Heinrich Du Bois-Reymond.

13 Victoria Adelaide Mary Louise, the eldest child of Queen Victoria and Prince Albert, had married Frederick William, who became crown prince of Prussia in 1861. She was noted for her enjoyment of intellectual debate and her patronage of the arts and sciences (Pakula 1996).

14 Lyell refers to *Origin*, to Thomas Henry Huxley's *Evidence as to man's place in nature* (T. H. Huxley 1863a), and to his own *Antiquity of man* (C. Lyell 1863a). In a letter of 9 January 1865 to another of her daughters, Alice, Queen Victoria expressed concern that the princess royal's reading of CD's works was undermining her religious beliefs (see Pakula 1996, pp. 215–16).

15 'Pfahlbauten', or ancient lake-dwellings built on wooden piles, had recently been excavated in Switzerland (see C. Lyell 1863a, pp. 17–29, Keller 1866, and first enclosure to letter from Charles Lyell to J. D. Hooker, [31 May 1865] and n. 12).

16 Lyell was currently preparing the tenth edition of *Principles of geology* (C. Lyell 1867–8; see K. M. Lyell ed. 1881, 2: 390). An annotated copy is in the Darwin Library–CUL (see *Marginalia* 1: 543–4). Lyell concluded his discussion of the theory of transmutation with the remark: 'Mr. Darwin, without absolutely proving this, has made it appear in the highest degree probable' (C. Lyell 1867–8, 2: 492). In previous editions, Lyell had argued against the transmutation of species (see, for example, C. Lyell 1853, p. 585).

17 See letter to Charles Lyell, 22 January [1865].

18 See n. 2 above.

19 In order to explain similarities between the Miocene plants of Europe and the modern plants of the Atlantic states of North America, Franz Unger and Oswald Heer had proposed the existence of a continent in the present basin of the Atlantic, over which plants had migrated eastward from America to Europe (Unger 1860, Heer 1855–9 and 1860). The Atlantis theory was criticised on botanical grounds by Asa Gray and later by Daniel Oliver, who argued that plants had migrated westward, across America and Asia (A. Gray 1858–9, Oliver 1862). Lyell discussed various criticisms of Heer's Atlantis theory in *Elements of geology* (C. Lyell 1865, pp. 265–73); he also mentioned CD's suggestion, presented in *Origin*, pp. 369–72, that plant migrations during a warmer, pre-glacial period (following the Miocene), accounted for the fairly uniform distribution of plants across polar regions and throughout subarctic and temperate regions as well (C. Lyell 1865, p. 270). Lyell noted that the Atlantis theory required 'a prodigious amount of subsidence in a comparatively brief period', and concluded: 'the theory which derives the American types from the east instead of the west seems by far the most natural' (C. Lyell 1865, pp. 270, 271). However, Lyell was not convinced that a single genus originated in one location, from which it dispersed, as did a species; he thought that this assumption, inherent in CD's transmutation theory, might weaken the arguments for either a westward or an eastward Miocene migration. Using the example of species of the walnut genus that had flourished in the European Eocene before the supposed migration of other walnut species from America, he cautioned the reader to bear in mind the assumption of a single origin of a genus. He added, however, that this was 'not the place to enter into a question so difficult and unsettled as that of the origin of species' (C. Lyell 1865, p. 272). For CD's discussion of 'single centres of creation', see *Origin*, pp. 351–6.

20 Pope Pius IX's encyclical was published 8 December 1864, in conjunction with the 'syllabus of [modern] errors' (*Catholic encyclopedia*, *Enchiridion symbolorum*). A commentary appeared in the Berlin satirical journal *Kladderadatsch*, 15 January 1865, p. 10. It referred to 'die Derwin'sche Lehre, dass schon unsere Grossältern Affen waren' ('the Darwinian theory, that even our grandparents were apes'). Lyell, probably quoting from memory, wrote: 'the Darwinian theory that makes apes of us all'.

To Henry Denny 17 January [1865][1]

Down. | *Bromley.* | *Kent. S.E.*
Jan 17th

Dear Sir

I am much obliged for the M.S. which I had quite forgotten & for your note.[2] I am glad to hear that you have resumed your long discontinued labour on the Anoplura,[3] for I always thought that they would lead to valuable & curious results.

Will you excuse me asking you to inform me whether the Chiloe pediculi[4] form a distinct species or well-marked variety? Is it the case that the Lice differ on different races of man; & can you believe, from any other evidence which you may possess, the statement by Mr Marshall about the Polynesian lice not living on a distinct race of man?[5] I shd. be grateful for any information on this head, *especially* if you would permit me to quote you as my authority.[6] But perhaps you will not have spare time to write to me on this head.—

Pray believe me | Dear Sir | Yours very faithfully | Ch. Darwin

Alfred Denny Museum, University of Sheffield

1 The year is established by the relationship of this letter to the letter from Henry Denny, 23 January 1865.

2 Denny's letter has not been found. The manuscript referred to was a portion of CD's *Beagle* zoology notes (DAR 30 and 31; see also R. D. Keynes ed. 2000). In 1844, Denny had initiated a correspondence with CD on the occurrence of lice on animals and on humans while pursuing research on Anoplura (lice; see *Correspondence* vol. 3, letter to Henry Denny, 20 January [1844], and Denny 1844). CD had provided Denny with *Beagle* lice specimens and, apparently, a page of notes on *Pediculus*, a genus of Anoplura, from his *Beagle* zoology notes (see *Correspondence* vol. 3, *Correspondence* vol. 4, letter to Henry Denny, 21 July [1847], and this volume, Supplement, letter to Henry Denny, [27 July – 10 August 1844]). The page of notes, clearly folded for enclosure in an envelope, is in DAR 31: 315 (see also R. D. Keynes ed. 2000, p. 283, and n. 4, below).

3 Following the publication of his account of the British Anoplura (Denny 1842), Denny received a grant from the British Association for the Advancement of Science to report on exotic Anoplura (*Report of the 14th meeting of the British Association for the Advancement of Science, held at York in 1844*, p. 392). The report was never published.

4 CD's 'Catalogue for specimens in spirits of wine' (Down House MS, Notebook 63.2, entry 1185) lists a *Pediculus* collected from the island of Chiloé, Chile; however, this specimen has not been found (see R. D. Keynes ed. 2000, pp. 283, 358, 410). The Denny collection at the University Museum, Oxford, contains four unidentified lice that may also have been collected by CD in Chiloé (see K. G. V. Smith 1987, pp. 29, 88).

5 Marshall (or Martial) was a surgeon on a British whaling vessel whom CD met in 1834 while on Chiloé during the *Beagle* voyage (see *Correspondence* vol. 3, letter to Henry Denny, 3 June [1844]). He has not been further identified. Marshall reported that lice found on the Sandwich islanders were unable to survive on the bodies of English sailors. His information was recorded in CD's *Beagle* zoology notes (DAR 31: 315 and R. D. Keynes ed. 2000, p. 283), and later published in *Descent* 1: 219–20 in a discussion of the unity of the human species. In his 'Catalogue of animals in spirits and wine' (Down House MS, Notebook 63.1, entry 646, R. D. Keynes ed. 2000, p. 340), CD noted:

> it would be interesting to compare these parasites with those inhabiting an European individual to observe whether they have been altered by transportation & domestication: It would be curious to make analogous observation with respect to various tribes of men.

[6] See letter from Henry Denny, 23 January 1865 and n. 2.

From Asa Gray 17 January 1865

Cambridge, Mass.
Jany, 17th. 1865.

Dear Darwin

Yours of 26th Dec. just received—long *en route*— must have crossed one from me,[1]—yet I am not sure.

Only the *separate* copy of your *Lythrum* paper came by the post,[2] (& that I have not yet read), so I suppose I have lost Scott's & Cruger's papers.[3] I am sorry that the Cuckoos are not more satisfactory.[4] I wonder that my letter to Dr. Brewer has brought no response[5]

You are mistaken in thinking the Fish-men here (in U.S.) are all *Agassizian*.[6] I understand there is a perfect *hatred* between all of them, (Gill, Girard of Washington, & Ayres of California) and Agassiz.[7] But I know nothing of the calibre of these people.

The new Herbm. building is finished & in occupation (costing Mr. Thayer $12250—in depreciated money, to be sure), and *perfectly satisfactory*.[8] But the supporting fund small at best—lacks $1000 or so of being filled.— will come in time, and I hope more.— for I want a curator.[9]

People have much & many things to give for now. At present we are feeding Savannah—while the rebels are starving our men (prisoners) in the interior of the country.[10]

Do you not begin to believe that we shall put down the rebellion, restore the Union, and do away with Slavery?[11]

Heartily do I wish you a prosperous year, and continually improving health—& power to work—and less discomfort— Also—tho' a small matter—I give you joy over the *Copley Medal*, which R.S. honors itself in giving to you.[12]

Ever | A. Gray

DAR 165: 146

CD ANNOTATION

7.1 Heartily . . . year,] '(Herbarium)' *added pencil*

[1] The letter to Gray has not been found. Gray last wrote to CD on 5 December 1864 (see *Correspondence* vol. 12).

[2] 'Three forms of *Lythrum salicaria*' was published in the *Journal of the Linnean Society* (*Botany*) 8 (1865): 169–96. Author's offprints of the paper were ready in December; Gray's name appears on CD's presentation list for the paper (see *Correspondence* vol. 12, Appendix III).

[3] Gray may refer to Scott 1864a and Scott 1864c; he also refers to Crüger 1864. These papers were communicated to the Linnean Society by CD, and published with CD's 'Three forms of *Lythrum salicaria*' in the 1865 volume of the society's journal. CD had already sent Gray a copy of Scott 1864b (see *Correspondence* vol. 12, letter to Asa Gray, 13 September [1864]). CD had expressed his interest in Crüger 1864 in his letter to Asa Gray, 25 February [1864] (*Correspondence* vol. 12).

[4] CD had requested information about cuckoos in an enclosure to his letter to Asa Gray of 29 October [1864] (*Correspondence* vol. 12). The enclosure has not been found. CD's query was evidently prompted by an article in the 8 October 1864 issue of the *Spectator*, which discussed the instinct of cuckoos (*Cuculus canorus*) to lay their eggs in other birds' nests as evidence of God's design in nature. For CD's views on the parasitic behaviour of cuckoos, see *Origin*, pp. 216–18, *Origin* 4th ed., pp. 260–62, and *Natural selection*, pp. 506–8. See also *Correspondence* vol. 12, letter from Asa Gray, 5 December 1864, which includes a note from Henry Bryant on the habits of American cuckoos.

[5] Gray had also written to Thomas Mayo Brewer for information about cuckoos (see *Correspondence* vol. 12, letter from Asa Gray, 5 December 1864).

[6] In his letter of 7 November 1864 (*Correspondence* vol. 12), Benjamin Dann Walsh had remarked on Louis Agassiz's popularity and influence in the United States. See also letter to B. D. Walsh, 4 December [1864]. Agassiz's extensive work in ichthyology is discussed in Lurie 1960. Agassiz was one of the leading opponents of Darwin's transmutation theory in America (see Lurie 1960, Winsor 1979, Morris 1997, and *Correspondence* vols. 8 and 9). Beginning in 1863, he began to use the resources of the museum of comparative zoology at Harvard to collect fish specimens from throughout the world, and in 1865 he led an expedition to the Amazon River with the object of gathering evidence against CD's transmutation theory (see Lurie 1960, pp. 336–7, and Winsor 1991, pp. 66–76).

[7] Gray refers to Theodore Nicholas Gill, Charles Frédéric Girard, and William Orville Ayres. Agassiz's controversial professional style and deteriorating relations with his students in the 1860s are discussed in Winsor 1991, pp. 27–42, 44–65.

[8] Construction of the new herbarium at Harvard University was largely funded by the Boston financier Nathaniel Thayer (see Dupree 1959, pp. 327–8). The building of the herbarium is discussed in the letters from Asa Gray, 16 February 1864 and 11 July 1864 (*Correspondence* vol. 12).

[9] Gray eventually appointed Horace Mann as curator in 1866 (see Dupree 1959, p. 337).

[10] Savannah, Georgia, had fallen to the Union army on 21 December 1864 (Denney 1992, pp. 506–7). Prison conditions are discussed in McPherson 1988, pp. 800–2; the suffering of Union prisoners is attributed largely to the shortage of resources and deteriorating economy in the South.

[11] CD and Gray corresponded at length on the American Civil War (see *Correspondence* vols. 9–12). CD had long been opposed to slavery. See *Journal of researches* 2d ed., pp. 499–500, Colp 1978, and Browne 1995, pp. 196–9, 213–14, 244–6. For CD's discussions with Gray on slavery, see *Correspondence* vols. 9–11 and this volume, letter to Asa Gray, 19 April [1865].

[12] CD was presented with the Copley Medal of the Royal Society on 30 November 1864 (see *Proceedings of the Royal Society of London* 13 (1863–4): 505, and *Correspondence* vol. 12, Appendix IV).

To Richard Kippist 18 January [1865][1]

[Down]

[sending 'by the carrier tonight a long paper for the Society'[2] and asking him to acknowledge its receipt]

LS incomplete[3]
Christie's, London (22 June 1988)

[1] The date 18 January is given in the sale catalogue. The year is established by the reference to the 'long paper' sent to the Linnean Society (see n. 2, below), and by the relationship between this letter and the letter to J. D. Hooker, 19 January [1865].

[2] The reference is to 'Climbing plants'. In the letter to J. D. Hooker, 7 January [1865], CD commented on the great length of the manuscript of 'Climbing plants', which he was about to send to the Linnean Society for publication. Richard Kippist was librarian of the Linnean Society (*DNB*).

[3] The original letter is complete and is described in the sale catalogue as being one page long.

To J. D. Hooker 19 January [1865][1]

Down
Jan. 19th

My dear Hooker

It is working hours, but I am trying to take a day's holiday, for I finished & despatched yesterday my Climbing paper.[2] For the last ten days I have done nothing but correct refractory sentences, & I loathe the whole subject like tartar-emetic.— By the way I am convinced that you want a holiday—& I think so because you took the Devil's name in vain so often in your last note.[3] Can you come here for Sunday; you know how I shd. like it, & you will be quiet & dull enough here to get plenty of rest.—[4]

I have been thinking with regret about what you said in one of your later notes, about having neglected to make notes on the gradation of characters in your genera.[5] But would it be too late? surely if you looked over names in series the facts would come back & you might surely write a fine paper "On the Gradation of important characters in the genera of Plants". As for unimportant characters, I have made their *perfect* gradation a very prominent point, with respect to the means of climbing, in my paper. I begin to think that one of the commonest means of transition is the same individual plant having the same part in different states: thus *Corydalis claviculata*, if you look to one leaf may be called a tendril-bearer,—if you look to another leaf it may be called a leaf-climber.[6] Now I am sure I remember some cases with plants in which important part, such as position of ovule, differs—differences in spire of leaves on lateral & terminal branches &c.—

There was not much in last Nat. Hist. R. which interested me except Colonial Floras & the Report on sexuality of Cryptogams. I suppose the former was by Oliver:[7] how extremely curious is the fact is of similarity of orders in Tropics.— I feel a conviction that it is somehow connected with Glacial destruction,[8] but I cannot "wriggle" comfortably at all on the subject.[9] I am nearly sure that Dana makes out that the greatest number of Crustacean forms inhabit warmer temperate regions.[10]

I have had an enormous letter from Leo Lesquereux (after doubts I did not think it worth sending you) on Coal Flora:[11] he wrote some excellent articles in Silliman against "Origin" views;[12] but he says now after repeated reading of the book he is a convert! But how funny men's minds are; he says he is chiefly converted because my books makes the Birth of Christ—Redemption by Grace &c plain to him!![13]

Ray Soc. will publish Gärtner; but consults me about a Translator & can give no advice, & asks me for Introductory Chapter, which I have felt compelled to decline & am sorry for.—[14]

Ever yours affect | C. Darwin

There is not one question in this letter! please to make a note of this little fact & carry it to my credit.

P.S. I by no means wished to send my paper on Climbers to Royal Soc. & suggested it solely to avoid expence for Linn. Soc.— I always preferred latter.[15]

DAR 115: 258a–c

[1] The year is established by the relationship between this letter and the letter from J. D. Hooker, 1 January 1865.

[2] 'Climbing plants'. See letter to Richard Kippist, 18 January [1865].

[3] See letter from J. D. Hooker, [8–18 January 1865].

[4] According to Emma Darwin's diary (DAR 242), Hooker did not visit Down House until 4 to 6 March 1865.

[5] See letter from J. D. Hooker, 1 January 1865.

[6] In 'Climbing plants', pp. 70–2, CD described the gradation from the first-formed leaves to reduced leaflets to tendrils on individual specimens of *Corydalis claviculata*, concluding: 'we here behold a plant in an actual state of transition from a leaf-climber to a tendril-bearer'. On pages 108 to 115, he discussed gradations in various twiners, leaf-climbers, and tendril-bearers, which suggested that the plants had developed from one form into another over long periods of time. CD's notes on *C. claviculata* are in DAR 157.2: 26–8. CD added a brief account of gradation in climbing plants to his discussion of 'modes of transition' in *Origin* 4th ed., p. 221. See also *Correspondence* vol. 12, letters to J. D. Hooker, [27 January 1864] and [8 February 1864], and letter to Daniel Oliver, 11 March [1864].

[7] CD refers to two review articles, 'New colonial floras' and 'Report on sexuality in the lower Cryptogamia', that appeared in the January 1865 issue of the *Natural History Review*, pp. 46–63 and 64–79. CD's annotated copies of these papers are in his collection of unbound journals, Darwin Library–CUL. Daniel Oliver, the botanical editor of the *Natural History Review* responsible for material on flowering plants, may have been the author of 'New colonial floras'. The author of 'Report on sexuality in the lower Cryptogamia' was probably Frederick Currey, the botanical editor responsible for material on non-flowering plants.

[8] In *Origin*, pp. 376–81, CD had proposed the widespread destruction of tropical plants, and the migration of temperate plants towards, and in some cases across, the equator during the glacial epoch. CD's hypothesis was an attempt to explain the range and affinities of species inhabiting the northern and southern temperate zones, and intertropical mountains; however, the author of 'New colonial floras' (*Natural History Review* n.s. 5 (1865): 59–60) noted that CD's hypothesis did not account for the similarity of vegetation in tropical regions of the globe.

[9] 'Wriggling' out of theoretical difficulties had long been a joke between CD and Hooker. For examples, see *Correspondence* vol. 10, letter from J. D. Hooker, 7 November 1862, n. 17, and *Correspondence* vol. 12, letter to J. D. Hooker, 5 April [1864] and n. 8.

[10] James Dwight Dana had corresponded with CD about the geographical distribution of Crustacea in 1856 and 1857 (see *Correspondence* vol. 6). In his work on Crustacea, Dana showed that there was a greater number of species in warmer temperate latitudes (Dana 1853a, p. 1498). An annotated copy of Dana 1853a is in the Darwin Library–CUL (see *Marginalia* 1: 178–9).

[11] Letter from Leo Lesquereux, 14 December 1864 (*Correspondence* vol. 12).

[12] Lesquereux published a series of papers between 1859 and 1863, entitled 'On some questions concerning the coal formations of North America' (Lesquereux 1859–63). In the third part of this series, published in the November 1860 issue of the *American Journal of Science and Arts*, pp. 367–84, Lesquereux claimed that the fossil flora of America's coal basins showed no trace of the intermediate forms that might support a theory of the transformation of species by successive variations. An annotated copy of this paper is in the Darwin Pamphlet Collection–CUL. The *American Journal of Science and Arts* was commonly known as 'Silliman's journal', after its founder, Benjamin Silliman.

[13] In his letter of 14 December 1864 (*Correspondence* vol. 12), Lesquereux apologised for having sought 'geological proofs in contradiction to [CD's] system'. He remarked that he had initially rejected *Origin* as 'hostile to every kind of religious faith', but that CD's book now satisfied him 'concerning some of the most obscure points of the biblical teachings; the miraculous birth of Christ, . . . the first species of a higher and Divine type; the doctrine of grace &c'.

[14] See letter to Ray Society, [14–18 January 1865] and n. 3. The Ray Society minutes for 3 February 1865 record a resolution by the Council to employ a translator for Gärtner 1849; however, the work was never undertaken (see Curle 1954, p. 26).

15 See letter to J. D. Hooker, 7 January [1865], and letter from J. D. Hooker, [8–18 January 1865]. Because of the cost of printing his lengthy paper paper 'Climbing plants', CD had considered sending it to the Royal Society rather than to the Linnean Society for publication.

From J. D. Hooker [20 January 1865][1]

Royal Gardens Kew

Dear Darwin

Not tomorrow, but tomorrow week if I can I shall be delighted to go to Down even though the Gooseberries will not be ripe![2] I have been worked to death by the Genera Plantarum.[3]

Ever yrs affec | J D Hooker

Friday.

DAR 102: 6

1 The date is established by the relationship between this letter and the letter to J. D. Hooker, 19 January [1865]; see also letter from Frances Harriet Hooker, [27 January 1865]. In 1865, 20 January was a Friday.

2 CD had invited Hooker to spend Sunday at Down House (see letter to J. D. Hooker, 19 January [1865] and n. 4). Visiting during the gooseberry season had evidently become a joke between Hooker and the Darwin family (see, for example, letter from J. D. Hooker, [after 17 June 1865] and n. 6).

3 Hooker was collaborating with George Bentham on a multi-volume work describing all the known vascular plant genera (Bentham and Hooker 1862–83). The first two parts had been published in 1862.

From John Scott 20 January 1865

Rungbee
Jany. 20th. 1865.

Sir,

I trust my letter of the 21st. December—the date of my arrival in Calcutta[1]—& also details of Experiments on Verbascum—registered on the 4th. Jany. have come duly to hand.[2] I will be anxious to hear what you think of the latter; when you have got time to look it over. I hope that I have not made any such mistakes in my calculations as that which you have drawn my attention to in Primula paper.[3] I do not know how it has escaped my notice— it was *very* **stupid** of me to have *overlooked* **it**. I will indeed be sorry to hear that you have been at much trouble with those which you have now so kindly taken in hand to correct for me. . . . . . I thank you most kindly for the high opinion you express with regard to my Primula paper. It of itself more than sufficiently repays me for all the time & labour spent in experimentation. I only hope that my Verbascums, may similarly interest you—for they also are the results of much time & labour.

I only received the copies of Primula paper, which you had so kindly forwarded. Those sent are quite sufficient to supply I think all whom I should like to present a copy to. Please accept my best thanks for the honour you have done me in sending copies of it to so many eminent authors.[4] I have also to thank you for copy of Edin. Courant, showing me that Prof. Balfour has not entirely forgot me.[5]

—*Truly* **indeed** *does he say* that **you** have advanced my *prospects in life*—for without your kind & helping hand my future looked dark & ominous in the extreme. I only trust that I may now soon be in a position where I may repay to you the pecuniary portion of my debt—& also by assiduous application try to prove myself not entirely an unworthy recipient of your many kindnesses.[6]

I think I mentioned in a P.S. to letter of the 21st. that on making a passing call at Dr. Anderson's on my way up to Calcutta, he immediately gave me a place in the Cinchona Department.[7] He also told me that in accepting it, I did not not need to consider it as permanently settled that I should remain in it, for the present

Incomplete
DAR 177: 114

1 The letter from John Scott has not been found. He had left his position at the Royal Botanic Garden, Edinburgh, in March 1864 and, on the advice of CD and Joseph Dalton Hooker, decided to seek employment in India (see *Correspondence* vol. 12). He departed for Calcutta on 28 August 1864 (see *Correspondence* vol. 12, letter from J. D. Hooker, 30 August 1864, and this volume, letter from J. D. Hooker, [26 January 1865]).

2 Scott sent CD a draft of a paper on *Verbascum* that he had prepared while sailing to India; CD passed the manuscript on to Hooker for comments, and it was eventually returned to Scott (see letter from J. D. Hooker, [10 March 1865], letter to J. D. Hooker, 16 [March 1865], and letter from John Scott, 21 July 1865). After a suggestion from CD in 1862, Scott began work on *Verbascum* to test the experiments on hybrid sterility performed by Karl Friedrich von Gärtner (see *Correspondence* vol. 10, letter to John Scott, 19 November [1862], and letter from John Scott, 17 December [1862], and *Correspondence* vol. 11, letter from John Scott, 21 September [1863]). In his letter of 16 May [1864] (*Correspondence* vol. 12), Scott remarked: 'the degree of sterility of unions between differently coloured varieties may be inversely proportionate to their colour affinities'. Scott published his results on *Verbascum* in Scott 1867. An annotated copy of Scott 1867 is in the Darwin Pamphlet Collection–CUL. CD cited some of Scott's findings in *Variation* 2: 100–7, *Origin* 5th and 6th eds. (see Peckham ed. 1959, p. 465), and *Cross and self fertilisation*, p. 330.

3 CD had commented on a draft of Scott 1864b in September 1863, praising it as an 'excellent memoir' and suggesting some alterations (see *Correspondence* vol. 11, letter to John Scott, 7 November [1863]). CD received a revised version of the paper from Scott in January 1864, together with a set of queries regarding further corrections, but asked Scott to let him send the paper to the Linnean Society for publication without further alterations (see *Correspondence* vol. 12, letter from John Scott, 7 January [1864], and letter from Emma Darwin to John Scott, 9 January 1864). Scott's paper contained calculations of the comparative fertility of different forms of Primulaceae species when self-pollinated or crossed with other forms (see Scott 1864b, pp. 103, 106). The letter in which CD noted the mistakes in Scott's calculations has not been found. An annotated copy of the paper is in the Darwin Pamphlet Collection–CUL. A second annotated copy is in the collection of unbound journals, Darwin Library–CUL. On Scott's *Verbascum* calculations, see the letter from J. D. Hooker, [10 March 1865] and n. 6.

4 CD's letter informing Scott that he had sent copies of his *Primula* paper, Scott 1864b, to others has not been found. CD had sent a copy of Scott 1864b, together with his own notes, summarising the main points of the paper, to Asa Gray (see *Correspondence* vol. 12, letter to Asa Gray, 13 September [1864]). Gray wrote a review of the paper in the January 1865 issue of the *American Journal of Science*

and Arts, pp. 101–4. CD also urged Hooker several times to read the paper (see *Correspondence* vol. 12, letter to J. D. Hooker, [20–]22 February [1864] and n. 6). He eventually sent Hooker a summary of the paper's main points, suggesting that he encourage Daniel Oliver to review it (see *Correspondence* vol. 12, letter to J. D. Hooker, 13 September [1864], and letter from J. D. Hooker, 16 September 1864). He sent the paper, with references to its important points, to Oliver after Oliver agreed to review it (letter to Daniel Oliver, 17 September [1864]). Oliver's brief notice of the paper appeared in the *Natural History Review* for October 1864, p. 640. CD also sent a copy to Benjamin Dann Walsh (letter from B. D. Walsh, 29 May 1865).

5 The item from the *Edinburgh Evening Courant* has not been identified. John Hutton Balfour was regius keeper of the Royal Botanic Garden, Edinburgh. Scott's relations with his superiors at the Garden were apparently strained after he failed to accept an appointment in Darjeeling offered to him by Balfour in May 1863 (see *Correspondence* vol. 11, letter from John Scott, [3 June 1863], and *Correspondence* vol. 12, letter from John Scott, 28 May 1864). CD and Hooker had suspected that Balfour was prejudiced against Scott because of Scott's support for CD's theory of transmutation (see *Correspondence* vol. 11, letter from J. D. Hooker, 10 June 1863, and *Correspondence* vol. 12, letter to J. D. Hooker, 5 April [1864], and letter from J. D. Hooker, 6 April 1864). Balfour had supplied Hooker with a character reference for Scott that praised his botanical abilities and diligence, while noting his sullen temper and neglect of more practical duties (see *Correspondence* vol. 12, enclosure to letter from J. D. Hooker, 6 April 1864). See also letter from John Scott, 28 May [1864] and the enclosed reference from Balfour.

6 On learning that Scott had resigned his position at the Royal Botanic Garden, Edinburgh, CD had offered him financial support (see *Correspondence* vol. 12, letter to John Scott, 9 April 1864). Scott initially declined (see *ibid.*, letter from John Scott, 5 May [1864]); however, when it became clear that his own resources were inadequate for a passage to India, Scott accepted CD's offer, promising to repay the sum once he had settled into a position abroad (see *ibid.*, letters from John Scott, 28 May [1864] and 8 June 1864). CD made three separate payments to Scott, totalling £115, in each case noting the entry in his account book as a 'gift' under the heading 'science' (CD's Account book–cash account (Down House MS)). CD also briefly considered employing Scott at Down (see *Correspondence* vol. 12, letter to J. D. Hooker, 7 April [1864]). He repeatedly praised Scott's abilities as an observer to Hooker (see, for example, *ibid.*, letter to J. D. Hooker, 22 [May 1864]), and Hooker, through his contacts in India, eventually helped in obtaining a position for Scott (see n. 7, below).

7 Hooker had written to Thomas Anderson, superintendent of the Botanic Garden at Calcutta, to enquire about employment for Scott (see *Correspondence* vol. 12, letters from J. D. Hooker, [4 June 1864] and [19 September 1864]). Scott was given a position at Rungbee, a newly established *Cinchona* plantation several miles south-east of Darjeeling (see letter from J. D. Hooker, [26 January 1865], *Cinchona cultivation in British Sikkim* (British parliamentary papers, Session 1866, 53: 645–51), and Markham 1880, pp. 389–90).

From Alfred Russel Wallace 20 January 1865

5 Westbourne Grove Terrace | W.
January 20th. 1865

My dear Darwin

I was pleased to hear a few weeks since that you were a little better in health & were again working, & I sincerely hope you are now getting over your severe illness.

For the last six months I have been doing absolutely nothing, & fear I shall not be inclined for work for some time to come. The reason is that I have suffered one of those severe disappointments few men have to endure. I was engaged to be married at 'Xmas, & had every reason to look forward to happiness, when at the last moment, when every thing was arranged, & even the invitations sent out by the lady's father, all was suddenly broken off! No cause has been given me except mysterious statements of the *impossibility* of our being happy, although her *affection* for *me remains unchanged.*[1]

Of course I can only impute it to some delusion on her part as to the state of her health. You may imagine how this has upset me when I tell you that I never in my life before had met with a woman I could love, & in this case I firmly believe I was most truly loved in return.

Scarcely any of my acquaintances know of this, but though we have met so little yet I look upon you as a friend, & as such hope you will pardon my boring you with my private affairs.

I send you two papers of mine of which I have lately got copies,[2] & with best wishes for your speedy restoration to perfect health | Believe me | Dear Darwin | Yours very sincerely | Alfred R. Wallace

Charles Darwin Esq—

DAR 106: B20–1

1 Wallace had been engaged to Marian Leslie, daughter of Lewis James Leslie (A. R. Wallace, notes for Wallace 1905 (Wallace family papers, private collection); Census returns 1861 (Public Record Office RG9/18: 2)). Wallace discusses the engagement in A. R. Wallace 1905, 1: 409–11. He married Annie Mitten, the eldest daughter of the bryologist William Mitten, in 1866 (A. R. Wallace 1905, 1: 411–12).

2 A. R. Wallace 1863b and 1864b. An inscribed and annotated copy of A. R. Wallace 1863b is in DAR 133: 10. An annotated copy of A. R. Wallace 1864b is in the collection of unbound journals, Darwin Library–CUL.

To Charles Lyell 22 January [1865][1]

Down. | Bromley. | Kent. S.E.
Jan 22

My dear Lyell

I thank you for your very interesting letter.[2] I have the true English instinctive reverence for rank & therefore liked to hear about the Princess Royal.[3] You ask what I think of the Duke's address[4] & I shall be glad to tell you.

It seems to me *extremely* clever like every thing that I have read of his; but I am not shaken; perhaps you will say that neither gods nor men could shake me. I demur to the Duke reiterating his objection that the brilliant plumage of the male humming bird could not have been acquired through selection, at the same time entirely ignoring my discussion (p. 93 3rd Edition) on beautiful plumage being acquired thro' *sexual* selection.[5] The Duke may think this insufficient, but that is another question. All analogy makes me quite disagree with the Duke that the differences in the beak, wing & tail are not of importance to the several species.[6] In the only two species which I have watched, the difference in flight & in the use of the tail was conspicuously great.[7]

The Duke who knows my orchis book so well might have learnt a lesson of caution from it, with respect to his doctrine of differences for mere variety or beauty.[8] It may be confidently said that no tribe of plants presents such grotesque & beautiful differences which no one until lately conjectured were of any use; but now in almost every case, I have been able to shew their important service.

It should be remembered that with humming birds or orchids a modification in one part will cause correlated changes in other parts.[9] I agree with what you say about beauty. I formerly thought a good deal on the subject & was led quite to repudiate the doctrine of beauty being created for beauty's sake.[10] I demur also to the Duke's expression of "new births":[11] that may be a very good theory but it is not mine,—unless indeed he calls a bird born with a beak $\frac{1}{100}$th of an inch longer than usual "a new birth"; but this is not the sense in which the term w^d^ usually be understood.

The more I work the more I feel convinced that it is by the accumulation of such extremely slight variations that new species arise. I do not plead guilty to the Duke's charge that I forget that natural selection means only the preservation of variations which independently arise.[12] I have expressed this in as strong language as I could use; but it w^d^ have been infinitely tedious had I on every occasion thus guarded myself. I will cry "peccavi"[13] when I hear of the Duke or you attacking Breeders for saying that man has made his improved Shorthorns or Pouter-pigeons or Bantams. And I c^d^ quote still stronger expressions used by agricuturalists. Man does make his artificial breeds, for his selective power is of such importance relatively to that of the slight spontaneous variations. But no one will attack Breeders for using such expressions, & the rising generation will not blame me.[14]

Many thanks for your offer of sending me the Elements;[15] I hope to read it all, but unfortunately reading makes my head whiz more than any thing else. I am able most days to work for 2 or 3 hours & this makes all the difference in my happiness.[16] I have resolved not to be tempted astray, & to publish nothing till my Vol. on Variation is completed.[17]

You gave me excellent advice about the foot-notes in my Dog chapter,[18] but their alteration gave me infinite trouble, & I often wished all the Dogs & I fear sometimes you yourself in the nether regions.

We (dictater & writer) send our best love to Lady Lyell[19] | yours affectionately | Charles Darwin

If ever you sh^d.^ speak with the Duke on the subject please say how much interested I was with his Address & tell him about Sexual Selection.

LS(A)
American Philosophical Society (304)

[1] The year is established by the relationship between this letter and the letter from Charles Lyell, 16 January 1865.

[2] Letter from Charles Lyell, 16 January 1865.

[3] Victoria Adelaide Mary Louise, Queen Victoria's daughter. Lyell had written about her interest in CD's theory in his letter of 16 January 1865.

[4] Lyell had discussed the opening address delivered to the Royal Society of Edinburgh on 5 December 1864 by George Douglas Campbell, eighth duke of Argyll (G. D. Campbell 1864); see letter from Charles Lyell, 16 January 1865.

[5] See letter from Charles Lyell, 16 January 1865 and nn. 10 and 11. CD's discussion of sexual selection in *Origin* 3d ed., pp. 92–5, was unchanged from that of the first edition. CD addressed Campbell's views on humming-birds in *Descent* 2: 151–3. CD's notes on G. D. Campbell 1864, dated December 1864, are in DAR 47: 20.

[6] Campbell argued that the variations in bills, wings, and tails of humming-birds were not such as to give competitive advantage in the struggle for existence (G. D. Campbell 1864, p. 281): 'It seems rather to have been a rule having for its object the mere multiplying of life, and the fitting of new forms for new spheres of enjoyment, according as these might arise out of corresponding changes in other departments of the organic world'.

[7] CD described the manner of flight of two species of humming-birds, *Trochilus forficatus* and *T. gigas*, which he saw in Chile, in *Zoology* 3: 110–12. See also *Journal of researches*, pp. 330–2, in which *Trochilus forficatus* is identified as *Mellisuga Kingii*, and R. D. Keynes ed. 2000, pp. 235–6, 245–6, 279.

[8] Against the view that natural structures were created for the sake of beauty or variety, CD argued in *Orchids*, pp. 346–60, that the unusual or beautiful forms of the orchid flowers facilitated pollination by insects. Campbell had reviewed *Orchids* in the October 1862 issue of the *Edinburgh Review* ([G. D. Campbell] 1862).

[9] CD discussed the principle of correlation of growth in *Origin*, pp. 11–12 and 143–50. According to this principle, when slight variations occur in one part of an organism, and are accumulated through natural selection, other parts of the organism are modified.

[10] In *Origin*, p. 199, CD briefly discussed the question of beauty in relation to selection, stating: '[some naturalists] believe that very many structures have been created for beauty in the eyes of man, or for mere variety. This doctrine, if true, would be absolutely fatal to my theory.' The discussion of beauty is expanded in *Origin* 4th ed., pp. 238–41. CD began to make notes on differing ideals of beauty in humans and on mate selection between 1838 and 1840 (see *Notebooks*, Notebook D, 99; Notebook M, 32; Notebook N, 26–9; and Old and useless notes, 8, 14, 22–4; see also Barrett 1980). See also *Correspondence* vol. 8, letter to G. H. K. Thwaites, 21 March [1860], and *Correspondence* vol. 12, letter to A. R. Wallace, 28 [May 1864] and n. 18.

[11] According to Campbell, CD's theory implied 'the possibility of new births being the means of introducing new species'. Campbell emphasised, however, that CD offered no explanation of such births (G. D. Campbell 1864, p. 286).

[12] See letter from Charles Lyell, 16 January 1865 and n. 6.

[13] *Peccavi*: 'I have sinned'.

[14] In *Origin*, p. 30, CD discussed the predominance of human selection in the production of domestic breeds, relative to the direct action of external conditions of life, habit, and simple variability: 'We cannot suppose that all the breeds were suddenly produced as perfect and as useful as we now see them. . . . The key is man's power of accumulative selection: nature gives successive variations; man adds them up in certain directions useful to him. In this sense he may be said to make for himself useful breeds.' He also gave numerous examples of animal breeders who habitually spoke of an animal's organisation as something they could 'model almost as they please' (*ibid.*, p. 31).

[15] C. Lyell 1865. See letter from Charles Lyell, 16 January 1865 and n. 2.

[16] After a period of improvement in the spring and summer of 1864, CD had a return to ill health toward the end of 1864 (see *Correspondence* vol. 12). See letter to J. D. Hooker, 7 January [1865].

[17] CD had begun *Variation* in 1860; after several interruptions, he most recently resumed work in September 1864 (see *Correspondence* vol. 12). *Variation* was published in 1868.

[18] CD sent Lyell the draft of his chapter on dogs for *Variation* in August or September 1860 (see *Correspondence* vol. 8, letter to Charles Lyell, 11 August [1860], and letter from Charles Lyell, 18 September 1860). Lyell had had difficulties accepting CD's view, first expressed in *Origin*, p. 17, that the various breeds of dogs had descended from several distinct species rather than from a single progenitor (see

Correspondence vol. 7). Lyell returned CD's manuscript at the end of September, and remarked in his letter of 25 September 1860 (*Correspondence* vol. 8): 'The case you make out seems very strong'. In his letter of 26 [September 1860] (*Correspondence* vol. 8), CD wrote that he was 'grieved' to hear from Lyell that material from the footnotes should be worked into the text, but would 'obey'. The manuscript has not been found. See *Variation* 1: 15–45.

19 Mary Elizabeth Lyell. The letter is in Emma Darwin's hand.

To Edward Walford 22 [January–April 1865?][1]

Down Bromley Kent
22[d]

My dear Sir

I should of course be proud to be one of your Series;[2] but I cannot spare a day to go to London; & I am not likely to be there soon: I may, however in course of summer, & will call on Mr. Edwards, if it be then not too late; & he can inform me.—[3]

My dear Sir | Yours very faithfully | Ch. Darwin

American Philosophical Society (237)

1 The date range is conjectured on the assumption that CD visited the photographer, Ernest Edwards, during his stay to London in November 1865, with a view to being included in Walford's photographic series (see nn. 2 and 3, below), and on the basis of CD's illness, which began at the end of April (see 'Journal', Appendix II).

2 Edward Walford had assumed editorial responsibilities for the series *Portraits of men of eminence in literature, science, and art, with biographical memoirs*, begun by Lovell Augustus Reeve (see Walford 1868 and Browne 1998, pp. 269–71). A portrait of CD taken by Edwards appeared in volume 5 of the series, published in 1866 (Reeve and Walford eds. 1863–7, 5: facing p. 49).

3 CD probably sat for Edwards during his visit to London between 8 and 20 November 1865. See letter from E. A. Darwin to Emma Darwin, 25 [November 1865] and nn. 2 and 3.

From Henry Denny 23 January 1865

Phil[l] Hall Leeds
Jan 23[d]/65

Dr Sir

With regard to Mr Marshalls statement respecting the Polynesian Lice,[1] it is a very *extraordinary* statement, & I cannot see any reason why the *Lice* of *one Human being* should not live upon another! It is a fact that the parasites of *one Genus* of animals do not live upon animals of another Genus. *Swine* upon the *Horse* or *Ass* or *Ass* upon the *Ox*—or *Horse*, Deer The same with Birds. Species of the same Genus as the various *Ducks* have frequently the same Louse again some of the smaller *Passerine* Birds. But the *Goose* & *Swan* have not the same Louse or that of the *Duck* or *Merganser*. As to the *identity* of the same *pediculus* on different Races of Men—this is a question of some perplexity. And the first question to decide is are the *different races* of men **varieties only** of *one Genus* or *Species*! I presume the former, & if so, the *differences* which exist & are constant in the characters of the Lice of these varieties, are not *greater* than we find in the races themselves, or in the *varieties* of the *Genus Canis*, or *Columba*, or Gallus. Yet the *same species* of Lice infest the different *varieties* of Fowl, Dogs & Pigeons.[2] Mr Andrew Murray has paid more attention to this subject, than anyone else in Britain whose paper I send you, from which you will perceive, there are well *marked* characters which distinguish the *anterior legs* & *antennæ* of the Lice of different Races as also the *colour*—, but as he truly observes, these are not greater, than those of the Races themselves ∴ if we cannot believe that they are any thing more than *varieties*, the same will apply to the Parasites, which also remain fixed![3] I think it very probable Mr Marshalls statement as to the *darker colour* & *difference* in *appearance*: to a *general observer*, may be *correct* enough![4] The questionable point is their *not living* upon the English Sailors!. (Dr Milligan,[5] who resided many years in Tasmania informed me that the aborigines of Tasmania stated to him, that until introduced by *Europeans* they had *never* seen the Louse on the Human body! & he had no reason to *doubt* their veracity! The natives wear no *clothes whatever* & the women shaved their heads &c— & the men although they wore long hair, so besmeared it with *fat* & *red ochre*, as to render it altogether unfit for the *abode* of even a *Louse*.

Mr Murray describes two varieties of *Australian* Lice, which differ in *colour* &c from the *European*.[6] are we to suppose the *Australian* Lice are an Introduced from some other race? or aboriginal. if the latter, It is difficult to believe the *Tasmanians* have been excepted from Parasites. I should rather have suspected the *smearing* of the Hair was to destroy or extirpate them! and if the women *shaved* the *hair* this very practice was adopted by the Egyptian Priests to keep themselves free from vermin!— I believe the Human races in every part of the world have Lice, which may offer slight modifications of character! But I would not commit myself to decide, whether we shall call these *specific* characters or not. The transition between species in some instances is so gradual that it becomes difficult to say where the *variety* or *Species*, begin or terminate!— I fear you will not obtain much information from this rambling note, but any grain is quite at your service.—

Believe me | Dr Sir | Yours very truly | Henry Denny

Charles Darwin Esq | &c—

P.S. I shall be glad of Mr Murrays Paper back again when you have done with it.

DAR 80: B150–1

CD ANNOTATIONS

0.3 Dr Sir … perplexity 1.9] *crossed red and blue crayon*

1.13 Yet the … & Pigeons. 1.14] *scored red crayon; square brackets around text, red crayon*

1.19 they are … with it. 5.1] *crossed red crayon*

[1] See letter to Henry Denny, 17 January [1865] and n. 5. CD had sought Denny's opinion on a statement made by Marshall, a surgeon on a British whaling vessel whom CD had met during the *Beagle* voyage. Marshall claimed to have observed that lice that normally lived on the bodies of Sandwich Islanders were unable to survive on the bodies of the English.

[2] In *Descent* 1: 219, CD mentioned on Denny's authority that the same species of lice (*Pediculus*) were found on different varieties of dogs, fowls, and pigeons. See CD's second annotation to this letter.

[3] See Murray 1860, pp. 571–3. Andrew Murray's paper addressed the argument that each human race possessing a distinct parasite must be a distinct species. CD cited Murray's paper in *Descent* 1: 219–20, remarking that in insects even slight structural differences, when constant, were generally regarded as of specific value, and adding (p. 220): 'the fact of the races of man being infested by parasites, which appear to be specifically distinct, might fairly be urged as an argument that the races themselves ought to be classed as distinct species'. CD went on to discuss other evidence bearing on the unity of the human species, noting that the problem was 'a hopeless endeavour' until some definition of the term 'species' was generally accepted, and suggesting ultimately that the category of 'sub-species' was 'most appropriate' (*Descent* 1: 228, 235). CD also supported the view that human races had descended from a common stock (*ibid.*, 2: 388).

[4] See R. D. Keynes ed. 2000, p. 283, and letter to Henry Denny, 17 January [1865] and n. 5. Marshall's description of the lice appears in *Descent* 1: 219–20.

[5] Joseph Milligan.

[6] Murray 1860, p. 576.

From J. D. Hooker [26 January 1865][1]

Kew
Thursday

Dr Darwin

John Scott has arrived at Calcutta after 116 days passage![2]—with the plants he took out in capital order—[3] Anderson has given him an appt. of £240 per annum at Dajeeling! on the spot.[4]

I am just out of bed having had a threat of Influenza, which has been expelled by Dovers powder.[5]

I shall take the 4pm. train on Saturday from Charing X. it arrives at Bromley 4.49.[6]

Ever yrs affec | J D Hooker

I shall dine before I leave.

DAR 102: 7

[1] The date is established by the relationship between this letter, the letter from J. D. Hooker, [20 January 1865], and the letter from F. H. Hooker, [27 January 1865]. In 1865, the Thursday following 20 January was 26 January.

[2] See letter from John Scott, 20 January 1865 and n. 1.

[3] Hooker had asked Scott to look after a Wardian case of plants on his voyage to Calcutta (see *Correspondence* vol. 12, letter from J. D. Hooker, [4 June 1864] and n. 19, and letter from John Scott, 8 June 1864).

[4] The reference is to Thomas Anderson. Hooker had written to Anderson about employment for Scott in India, and had been informed that Anderson needed a gardener for a proposed botanic garden at Darjeeling (see *Correspondence* vol. 12, letters from J. D. Hooker, [4 June 1864] and [19 September 1864]). See also letter from John Scott, 20 January 1865 and n. 7.

[5] Dover's Powder was a compound made from the root of *Cephaëlis ipecacuanha*, opium, and sulphate of potash. *C. ipecacuanha* (now *Psychotria ipecacuanha*) is a South American plant of the family Rubiaceae, which possesses emetic, diaphoretic, and purgative properties. The powder was often taken in a glass

of white wine posset, which was to be drunk while the patient was under warm covers and sweating (see Royle and Headland 1865, pp. 456–60).

6 Owing to continued sickness, Hooker was not able to visit Down. See letter from F. H. Hooker, [27 January 1865].

To John Edward Gray 27 January [1865][1]

Down. | Bromley. | Kent. S.E.
Jan 27—

My dear Gray

Many thanks for your very kind congratulations[2]

I am sorry to hear that you have been ill.

I thought when I saw you that you looked far from well.[3]

Believe me | yours very sincerely | Ch. Darwin

LS
American Philosophical Society (305)

1 The year is established by the reference to Gray's congratulations, which were probably for the Copley Medal. CD had been awarded the medal at the 3 November 1864 meeting of the council of the Royal Society (see *Correspondence* vol. 12).

2 Gray's letter has not been found.

3 CD may have seen Gray while he was visiting his cousin Sarah Elizabeth Wedgwood in London in August 1864 (see 'Journal' (Appendix II)). Joseph Dalton Hooker remarked in his letter of [23 November 1864] (*Correspondence* vol. 12) that Gray was suffering from disease of the prostate.

From Frances Harriet Hooker [27 January 1865][1]

My dear M^r. Darwin—

I am sorry to say that Joseph must give up his visit to you tomorrow— instead of being better, as he wrote you word yesterday,[2] he is worse, & seems to have a regular attack of influenza, so that he certainly cannot leave home tomorrow—

He bids me say how sorry he is—especially as he had hoped to have seen your boys at home for the holidays—[3]

With kind regards to M^rs. Darwin,[4] believe me | yours aff^tly. F H Hooker

Kew— W—
Friday

We have just heard that D^r. Falconer is very ill—with rheumatic fever & bronchitis—[5] They are very anxious about him—

DAR 104: 231–2

1 The date is established by the relationship between this letter and the letter from J. D. Hooker, [26 January 1865], and by the reference to Hugh Falconer's illness (see n. 5, below).

2 See letter from J. D. Hooker, [26 January 1865].

3 See letter to J. D. Hooker, 7 January [1865], n. 18.

4 Emma Darwin.

[5] Following an attack of 'acute rheumatism, with disease of the heart and lungs', Falconer died on 31 January 1865 (*DNB*).

From Henry Walter Bates 28 January 1865

Royal Geographical Society | 15, Whitehall Place, S.W.
Jan 28 1865

My dear Mr Darwin

The receipt of a letter from you was an unlooked-for pleasure.[1] I have had news of your health from time to time, having seized all opportunities of asking from persons likely to know, & had heard lately of your slight improvement. Let us hope it may continue.

You are very kind to enquire after my personal affairs. I have no doubt Dr Hooker[2] has kept you well acquainted with what is done & said in Natural History circles & the perusal of the Journals &c keeps you well-informed about the rest. Perhaps there may be a few Entomological items bearing upon Darwinian views, which have not yet fallen in your way.

I was much gratified on receiving the Berlin "Bericht for Entomology 1862" (you know the Natural History reports appear in Wiegmann's Archives) to find at the very commencement a flattering notice of our paper on the mimetic Butterflies.[3] These reports originally were written by Erichson & are now written by Gerstaecker.[4] They are usually very skilfully & not very mercifully done. Gerstaecker has seized all the essential points of my paper & repeats them with an evident bias in their favour.[5] Being the highest Entomological tribunal I think you will like to have the testimony of this "Bericht" to the absence of, at any rate, any important errors in my facts & arguments.

You will be glad to hear that I like my present position very much.[6] I should have preferred a Natural History appointment but I had no chance of one & the birth of one sweet little child with expectation if another forced upon me cogent arguments for accepting the first thing that offered.[7] I hope besides to do a little in improving this great Society & assisting Naturalists in travelling.[8] The Mantidæ monograph progresses, but about this I will speak in a future letter.[9]

My dear Mr Darwin | Yours sincerely | H W Bates

DAR 160: 79

[1] CD's letter has not been found.

[2] Joseph Dalton Hooker.

[3] A favourable report of Bates 1861 appeared in the article by Carl Eduard Adolph Gerstaecker, 'Bericht über die wissenschaftlichen Leistungen im Gebiete der Entomologie während des Jahres 1862', *Archiv für Naturgeschichte* 29 (1863): 315–18 (Gerstaecker 1863a). The journal was founded by Arend Friedrich August Wiegmann in 1835. CD had encouraged Bates to publish his researches on butterflies (see *Correspondence* vol. 9, letters to H. W. Bates, 4 April [1861] and 25 September [1861]), and gave a favourable review of Bates 1861 ('Review of Bates on mimetic butterflies'; see *Collected papers* 2: 87–92).

[4] The *Archiv für Naturgeschichte* contained essay reviews of works published in various fields of natural history during the previous year. Gerstaecker had been reviewing entomological literature for the journal since 1853. The entomological reviews had been written by Wilhelm Ferdinand Erichson from 1837 to 1848, and by Hermann Rudolph Schaum from 1849 to 1852.

[5] Gerstaecker praised Bates's essay on the entomology of the Amazon as a careful, sound, and skilled attempt to answer the question of the constancy of species in favour of the Darwinian theory (Gerstaecker 1863a, p. 316). In his article, Bates had argued that the similarity in outward appearance of certain rare forms of butterfly to other more numerous forms assisted the former in avoiding predators, and that such cases of mimicry in Amazonian butterflies offered proof of the theory of transmutation (Bates 1861, pp. 511–12). CD cited Bates's work in *Origin* 4th ed., pp. 502–6, as illustrative of the principle of natural selection.

[6] Bates had accepted an appointment as assistant secretary of the Royal Geographical Society in 1864 (see *Correspondence* vol. 12, letter from J. D. Hooker, [after 28 April 1864]). He had applied unsuccessfully for a position at the British Museum in 1862 (Woodcock 1969, pp. 250, 256–7).

[7] Bates had married Sarah Ann Mason on 19 January 1863 (see *Correspondence* vol. 11, letter from H. W. Bates, 24 January 1863, letter to H. W. Bates, 26 January [1863], and letter to J. D. Hooker, 30 January [1863]). She had given birth to a daughter during the previous year. Although no father was listed on the birth certificate, Bates accepted the child as his own when the couple married. They eventually had three sons and a second daughter (see Woodcock 1969, pp. 253–4, and Bates 1892, p. xxxvii).

[8] On Bates's work at the Royal Geographical Society, see Mill 1930.

[9] Bates had been privately commissioned to write a monograph on the Mantidae (praying mantids) in 1863; however, the work was never published (see *Correspondence* vol. 11, letter from H. W. Bates, 29 September 1863, and Curle 1954, p. 26).

To Henry Denny 28 January [1865][1]

Down. | Bromley. | Kent. S.E.
Jan. 28th

Dear Sir

I thank you very sincerely for your interesting note & answers to my queries.[2] I am also much obliged for Mr Murray's paper;[3] (returned by this Post) but the interest of this paper to my mind is much enhanced or rather almost wholly depends on a sentence in your note, viz. "Yet the same species of Lice infest the different varieties of Fowl, pigeon & dog." Now as I understand this, you have never observed distinct varieties of the same species of Pediculus on different domestic varieties.— If, as is probable, I allude to this subject, I shd. like to quote on your authority this statement.[4] But then (though I am ashamed to give so much trouble) I shd be glad to know whether you have often looked to the vars. of our domestic mammals & birds. It is not probable, but if you have ever received pediculi from the domesticated animals inhabiting distant lands, or recently brought from such lands & have found them the same, this wd. add immensely to the value of the statement, I presume you have not attended to the domestic vars. of other animals, besides Fowls, Pigeons & Dogs.[5] I fear that you will think me very troublesome; pray excuse me & believe me—

Yours very faithfully | Ch. Darwin

Alfred Denny Museum, University of Sheffield

[1] The year is established by the relationship between this letter and the letter from Henry Denny, 23 January 1865.
[2] See letter from Henry Denny, 23 January 1865.
[3] Murray 1860. See letter from Henry Denny, 23 January 1865 and n. 3.
[4] Denny's statement is cited in *Descent* 1: 219.
[5] See letter to Henry Denny, 23 March [1865].

To A. R. Wallace 29 January [1865][1]

Down. | Bromley. | Kent. S.E.
Jan 29

My dear Wallace

I must ease my mind by saying how much I admire the two papers you have sent me.[2] That on parrots contained most new matter to me, & interested me *extremely*;[3] That in the Geograph. Journal strikes me as an epitome of the whole theory of Geograph. distribution[4] the comparison of Borneo & New Guinea,—the relation of the volcanic outbursts & the required subsidence,—& the comparison of the supposed conversion of the Atlantic into a great archipelago seemed to me the 3 best hits.[5] They are both indeed excellent papers.—

Believe me yours very sincerely | Charles Darwin

Do try what hard work will do to banish painful thoughts.[6]

P.S. During one of the later French Voyages, a *wild* pig was killed & brought from the Aru Islands to Paris.[7] Am I not right in inferring that this must have been introduced & run wild. If you have a clear opinion on this head, may I quote you?[8]

LS(A)
British Library (Add. MS 46434, f. 49)

[1] The year is established by the relationship between this letter and the letter from A. R. Wallace, 20 January 1865.
[2] Wallace 1863b and Wallace 1864b. See letter from A. R. Wallace, 20 January 1865 and n. 2.
[3] Wallace 1864b discussed the distribution of parrots in the Malayan archipelago. Wallace presented his findings as evidence of a division of the region into two faunal territories (see Wallace 1864b, p. 277, and n. 4, below).
[4] Wallace 1863b argued that the islands of the Malay Archipelago were divided into two faunal regions, one linked to the continent of Asia, the other to Australia. The map accompanying the paper showed a line, later known as 'Wallace's line', separating Borneo, Java, and Sumatra from other islands to the south and east (see Camerini 1993). Following the observations of George Windsor Earl, Wallace noted that the boundary separating the two regions was a stretch of deep ocean, and that the islands on either side of the boundary were connected by shallow seas to the Asian or Australian mainland. Wallace added extensive zoological evidence of the distinctness of the two regions (Wallace 1863b, pp. 228–33). CD scored many passages in this section in his copy of Wallace's paper (DAR 133: 10). CD had discussed Earl's observations in his letter to Wallace of 9 August 1859 (*Correspondence* vol. 7). In *Origin*, p. 395, CD noted the relationship between the depth of the sea and faunal similarities on adjacent islands; in *Origin* 4th ed., p. 470, he added that the observations made by Earl of deep ocean separating two widely distinct mammalian faunas in the Malay region had been 'fully confirmed by Mr. Wallace's admirable researches'. CD had started making notes on the distribution of mammals in the Malay region in the late 1830s (*Notebooks*, Notebook B, 80, 82, and 241; DAR

205.3: 211). Observations on the depth of water between adjacent islands in the region appeared in the 1844 sketch of his species theory (see *Foundations*, p. 154). On the importance of the boundary between Asian and Australian faunal regions as a focal point for CD's efforts to account for the distribution of organisms according to his theory of descent, see Camerini 1993.

5 Using a comparison of Borneo and New Guinea, which were similar in soil, climate, and elevation, yet zoologically divergent, Wallace emphasised that the line separating the two faunal regions was not based on differences of physical geography or climate (Wallace 1863b, pp. 231–2). He accounted for the separation of Java, Sumatra, and Borneo from the Asian continent by the elevatory action of volcanoes, accompanied by subsidence of the adjacent land (*ibid.*, pp. 227–9). To illustrate further the contrast between these two regions, Wallace invited readers to imagine the continents of Africa and South America similarly connected by an archipelago of islands filling the channel of the Atlantic (*ibid.*, pp. 232–3).

6 See letter from A. R. Wallace, 20 January 1865 and n. 1.

7 CD refers to the Aru pig collected on the 1826 to 1829 voyage of the French ships *L'Astrolabe* and *La Zélée*, and described by Henri de Blainville (see Blainville 1841–55, 2: 130–1, 224). The specimen was later discussed by Hermann Engelhard von Nathusius, who suggested that the pig was a feral *Sus indicus*, a pig otherwise known only in a domestic state (Nathusius 1864, pp. 168–71). An annotated copy of Nathusius 1864 is in the Darwin Library–CUL (see *Marginalia* 1: 630–5, where Aru is given as 'Arn'). CD discussed the Aru pig in *Variation* 1: 67. CD may have been interested in the Aru pig as a possible example of reversion. Feral pigs that had come to resemble wild boars were the principal evidence for the view that all domesticated animals, when they run wild, revert to the character of their parent stock (see *Variation* 1: 77–8, 2: 33). In January 1865, CD was revising the chapters on domestic animals for *Variation* (see 'Journal' (Appendix II)); chapter 3 (*Variation* 1: 65–79) discusses pigs.

8 See letter from A. R. Wallace, 31 January [1865] and n. 5. The Aru Islands, located to the west of New Guinea, were included in Wallace's study of the physical geography of the Malay Archipelago. Wallace's paper contained accounts of the distribution of animals, including wild pigs, on various islands in the region (Wallace 1863b, p. 231).

To Abraham Dee Bartlett 30 January [1865][1]

Down. | Bromley. | Kent. S.E.
Jan. 30th

My dear Sir

You have two rabbits of mine from Porto Santo.[2] Will you be so good as to have one of them killed, taking great care that the skull & vertebræ are not broken, & sent as soon as you can, addressed,

C. Darwin Esq
Care of Down Postman
Bromley
Kent

per Rail

I shall be very much obliged if you will inform me whether you have got young from these rabbits with the females of any other breed?[3]

I want to beg one other favour; I want to examine under the microscope the tipped feathers of *Gallus Sonneratii*[4] Could you send me one or two?

Believe me my dear Sir | yours very faithfully | Ch. Darwin

LS(A)

Yale University Library, Manuscripts and Archives, Joseph Bradley Murray Collection (Box 1, Folder 4)

[1] The year is established by CD's request for the return of the Porto Santo rabbit, which he received in February 1865 (see *Variation* 1: 114).

[2] The Porto Santo rabbit was a feral form of a domesticated breed that had been introduced to Porto Santo, an island off the coast of Portugal, in the fifteenth century. It was distinct in size and colouring from the wild rabbits native to England. For CD's interest in the rabbits, see *Variation* 1: 112–15, *Notebooks*, Notebook QE, p. 18, *Correspondence* vol. 5, letter from R. T. Lowe, 19 September 1854, and *Correspondence* vol. 6, letter from T. V. Wollaston, [February 1856]. CD had received two live rabbits from Porto Santo in 1861, and placed them at the Zoological Gardens, Regents Park, London (see *Correspondence* vol. 9, letter to A. D. Bartlett, 26 May [1861]). The rabbits had since undergone modifications that made them resemble the English variety in colouring (see *Variation* 1: 114). CD was revising the chapters on domestic animals for *Variation* (see 'Journal' (Appendix II)). The material on rabbits appears in chapter 4 (*Variation* 1: 103–30). The Porto Santo rabbits are discussed in *Variation* 1: 112–15 and 2: 279.

[3] CD had asked Bartlett to cross the Porto Santo rabbits with other breeds (*Correspondence* vol. 9, letter to A. D. Bartlett, 26 May [1861]). In *Variation* 1: 115, CD cited Bartlett's observation that the Porto Santo rabbits would not mate with any other breed of rabbit at the Zoological Gardens. CD noted that, had the history of these rabbits not been known, most naturalists would have considered them a distinct species (see also *Correspondence* vol. 9, letter to P. L. Sclater, 2 June [1861]).

[4] CD discussed *Gallus sonneratii*, a species of fowl, in *Variation* 1: 233–4, noting that he had examined the feathers of some hybrids raised at the Zoological Gardens.

From A. R. Wallace 31 January [1865][1]

5, Westbourne Grove Terrace | W.
Jan. 31st.

Dear Darwin

Many thanks for your kind letter.[2] I send you now a few more papers. One on "*Man*" is not much in your line.[3] The other 3 are Bird lists, but in the introductory remarks are a few facts of Distribution that may be of use to you, & as you have them already in the Zool. Proc. you can cut these up if you want "extracts."[4]

I hope you do not very much want the Aru pig to be a domestic animal run wild,—because I have no doubt myself it was the species peculiar to the New Guinea fauna (*Sus papuensis.* Less.) a very distinct form.[5] I have no doubt it is this species though I did not get it myself there, because I was told that on a small island near called there "*Pulo babi*" (pig Is.) was a race of pigs (different from & larger than those of the large islands) which had originated from the wreck of a large ship near a century ago— The productions of the Aru Islands closely resemble those of New Guinea,—more than half the species of birds being identical, as well as about half of the few known mammals.

I am beginning to work at some semi-mechanical work, drawing up catalogues of parts of my collection for publication.[6]

I enclose my "carte".[7] Have you a photograph of yourself of any kind you can send me? When you come to Town next may I beg the honour of a sitting for my brother-in-law Mr. Sims, 76, Westbourne Grove.[8]

Yours very sincerely | Alfred R. Wallace

C. Darwin Esq.

P.S. Your paper on *Lythrum salicaria*, is most beautiful.[9] What a wonderful plant it is! I long to hear your paper on Thursday on *Tendrils* & hear what you have got out of them.[10] My old friend Spruce a good botanist & close observer could probably supply you with some facts on that or other botanical subjects if you would write to him.[11] He is now at Kew but almost as ill as yourself.

ARW.

DAR 106: B22–3

[1] The year is established by the relationship between this letter and the letter to A. R. Wallace, 29 January [1865].

[2] Letter to A. R. Wallace, 29 January [1865].

[3] Wallace refers to his ethnographic essay 'On the varieties of man in the Malay Archipelago' (Wallace 1864c). Wallace had sent CD a copy of his other recent paper on man (Wallace 1864a) in May 1864 (see *Correspondence* vol. 12, letters from A. R. Wallace, 10 May 1864 and 29 May [1864], and letter to A. R. Wallace, 28 [May 1864]).

[4] Wallace 1862, 1863a, and 1863c. Annotated copies of these articles are in the collection of unbound journals, Darwin Library–CUL. CD sometimes cut sections from articles and letters that were relevant to his research and placed these with other material on the same subject (see, for example, *Correspondence* vol. 12, letter from Hermann Crüger, 21 January 1864, n. 6, and letter from Asa Gray, 11 July 1864, n. 14).

[5] See letter to A. R. Wallace, 29 January [1865] and nn. 7 and 8. Wallace refers to the New Guinea *Sus papuensis*. It is described in Blainville 1841–55, 2: 130–1, as a distinct animal introduced to New Guinea, and resembling the pigs of Malabar. CD was apparently not convinced by Wallace's account of the Aru pig. In *Variation* 1: 67, CD relied on Hermann Engelhard von Nathusius's opinion that the Aru pig was probably *Sus indicus*; the pigs had been introduced and then become feral. See also letter to A. R. Wallace, 1 February [1865].

[6] Wallace 1865. A copy of the paper is in the collection of unbound journals, Darwin Library–CUL.

[7] The photograph has not been found.

[8] The photographer Thomas Sims made several portraits of Wallace (see Wallace 1905, 1: 324, 385).

[9] 'Three forms of *Lythrum salicaria*'. Wallace's name appears on CD's presentation list for the paper (see *Correspondence* vol. 12, Appendix III).

[10] Wallace refers to 'Climbing plants'. An abstract of the paper was read at the Linnean Society on 2 February 1865.

[11] Richard Spruce had travelled with Wallace in the Amazon basin from 1848 to 1850. Spruce continued to collect plants in South America until his return to England at the end of May 1864 (Wallace 1905, 1: 276–9; Spruce 1908, 1: xxxiv–xxxv, xlvi; *DNB*). CD had written to Spruce for botanical information in 1863, although this letter has not been found (see *Correspondence* vol. 12, letter from A. R. Wallace, 2 January 1864; see also letter from Richard Spruce to J. D. Hooker, 29 July 1864).

To A. R. Wallace 1 February [1865][1]

Down. | Bromley. | Kent. S.E.
Feb 1

My dear Wallace

I am much obliged for your photograph, for I have lately set up a Scientific Album;[2] & for the papers which I will read before long.[3] I enclose my own Photo, taken by my son & I have no other.[4] I fear it will be a long time before I shall be able to sit to a Photographer otherwise I should be happy to sit to Mr Simms[5]

Thanks for information about the Aru pig, which will make me very cautious.[6] It is a perplexing case, for Nathusius says the skull of the Aru resembles that of the Chinese breed & he thinks that Sus Papuensis has been founded on a young skull; D^e^Blainville stating that an old skull from New Guinea resembles that of the wild pigs of Malabar, & these belong to the S. scrofa type, which is different from the Chinese domestic breed. The latter has not been found in a wild condition[7]

Believe me dear Wallace | yours very sincerely | Ch. Darwin

LS
British Library (Add. MS 46434, f. 53)

[1] The year is established by the relationship between this letter and the letter from A. R. Wallace, 31 January [1865].
[2] Wallace enclosed his photograph with his letter of 31 January [1865]. CD's 'Scientific Album' has not been found.
[3] Wallace 1862, 1863a, 1863c, and 1864c. See letter from A. R. Wallace, 31 January [1865].
[4] The photograph, the first to show CD with a beard, was taken by William Erasmus Darwin in 1864 (see *Correspondence* vol. 12, letter from W. E. Darwin, [19 May 1864]). It is reproduced as the frontispiece to *Correspondence* vol. 12.
[5] In his letter of 31 January [1865], Wallace asked CD to sit for his brother-in-law, the photographer Thomas Sims.
[6] See letter from A. R. Wallace, 31 January [1865] and n. 5.
[7] CD refers to the descriptions of pigs by Hermann Engelhard von Nathusius and Henri de Blainville (Nathusius 1864, pp. 168–71, and Blainville 1841–55, 2: 130–1, 224). Nathusius maintained that the Aru pig was *Sus indicus* (known to Europeans as the Chinese breed), and that it had become feral after being introduced onto one or more of the Aru islands. The 'Sanglier d'Arrou' (or Aru pig) had been identified by Blainville as '*Sus scrofa* (ferus)'. Blainville also described 'Sangier des Papous' (the New Guinea pig, or *Sus papuensis*), as an introduced animal that resembled the pigs of Malabar. Blainville's description was based on an adult specimen; however, Nathusius noted that *Sus papuensis* was first established on the basis of a young specimen. Nathusius reported Blainville's findings, but added that the evidence establishing the New Guinea pig as a distinct species was open to question. CD did not mention *Sus papuensis* in *Variation*. He followed Nathusius in dividing all known pig breeds into two groups, the *Sus scrofa* group and the *Sus indicus* group (*Variation* 1: 65–79). CD believed the former had descended from the common wild boar; he noted that the wild species from which the latter had descended was unknown (*Variation* 1: 65).

To J. D. Hooker 2 February [1865][1]

Down
Feby 2^d^

My dear Hooker

I heard this morning of Falconer's death.[2] Poor fellow I am much grieved; It will be a great loss to science. What a lot of knowledge of all kinds has perished with him. He was always a most kind friend to me. So the world goes.—

But I write to ask you to write me ever so short a note to tell me when you are well.—[3] I got pretty well on Monday & had bad day again yesterday.— Whenever you can come here for a Sunday do pray come. I must take my chance of being well or ill; for I can rarely tell even a day before hand; but last week I had an extra bad time.—[4] I hope you have got over that horrid influenza, which makes a man feel so wretchedly ill.

Farewell | Yours affectionately | C. Darwin

DAR 115: 259

[1] The year is established by the reference to Hugh Falconer's death (see n. 2, below).
[2] After a short illness, Falconer died of heart-failure on 31 January 1865 (see letter from F. H. Hooker, [27 January 1865], and *DNB*).
[3] Hooker was recovering from an attack of influenza; before becoming ill, he had planned to visit Down House (see letter from F. H. Hooker, [27 January 1865]).
[4] Emma Darwin reported CD's sickness in her diary on 24, 25, and 27 January (DAR 242).

From J. D. Hooker 3 February 1865

Kew
Feby 3d/65.

My dear Darwin

I hope you are better— I am pretty well, but somehow not over strong & bothered with eczema in the lobes of the ear, for which I am put upon Mercury & Iodide of Iron by Startin.[1]

Poor old Falconer![2] how my mind runs back to those happiest of all my days, that I used to spend at Down 20 years ago.—when I left your house with my heart in my mouth like a school-boy.[3] We had heard he was ill on Wednesday or Thursday & sent daily to enquire, but the report was so good on Saturday that we sent no more, & on Monday night he died.[4] We had dined together at the Athenæum[5] just 10 days before. He took cold on the day of that awful fog—Rheumatic fever & Bronchitis. From the first his heart did not act, & his attendant, Dr Murchison,[6] took a gloomy view of his case— still on Friday he rallied & Dr M. thought the worst was over, or at least said so: on Saturday the heart again gave way & no stimulant sufficed to get it to act— on Sunday there was no more hope. He suffered terribly— he was fixed with pain, could not move a muscle, & the sweat rolled down his face & ears with agony. Poor dear old Falconer he had led the worst life for his temperament that was possible— At Post. Mort. his heart was found choked with fatty deposits. I go to the grave tomorrow with Thomson Bentham & many friends.[7]

What a mountainous mass of admirable & accurate information dies with our dear old friend.— I shall miss him greatly not only personally, but as a scientific man of unflinching & uncompromising integrity—& of great weight in Murchisonian & other counsels where ballast is sadly needed.[8] The inconceivability of our being born for nothing better than such a paltry existence as ours' is, gives me some hope of meeting in a better world. What does it all mean.— When we think what millions upon millions of lives & intellects it has taken to work up to a knowledge of gravity & Natural selection, we really do seem a contemptible creation intellectually & when we feel the death of friends more keenly the older we grow, we do strike me as being corporeally most miserable, for we have no pleasures to compensate fully for our griefs & pains: these alone are unalloyed.

To return to the present your paper went off extremely well last night.[9] Currey read it well,—right well.[10] Masters pointed out that all **terete** petioles had closed bundles.[11] & Bentham said that there were interesting *Legumes*, uniting *Nissolia* with others & *Aphaca*, which would afford you capital means of testing your views.[12] These last were most seductive & interesting.

Ever Yrs affec | J D Hooker.

DAR 102: 8–9

[1] Hooker was recovering from influenza (see letter from F. H. Hooker, [27 January 1865]). Three years earlier, CD had consulted James Startin, a leading skin-specialist in London, about his eczema (see *Correspondence* vol. 11, letter to J. D. Hooker, 26 [March 1863] and n. 2, and Plarr 1930; see also Colp 1977, pp. 75, 216–17 n. 25). For contemporary discussions on the use of mercury ointment for eczema, and on the use of iodide of iron, see Royle and Headland 1865, pp. 144–5, Copland 1866, p. 272, and Ringer 1869, p. 165.

[2] Hugh Falconer died on 31 January 1865 (*DNB*).

[3] On Hooker's and Falconer's visits to Down House in the 1840s, see *Correspondence* vol. 3, letters to J. D. Hooker, [21 November 1845] and [10 February 1846], n. 8, and Appendix III. See also L. Huxley ed. 1918, 2: 458–9, and Bowlby 1990, pp. 265–7, 275.

[4] Falconer started to feel feverish on 19 January; the attack 'developed into acute rheumatism, complicated with disease of the heart and lungs, which proved fatal' on the morning of Tuesday 31 January (see Charles Murchison's biographical sketch in Falconer 1868, pp. xlviii–xlix). Frances Harriet Hooker, in her letter of [27 January 1865], told CD that she and Hooker had just heard of Falconer's illness.

[5] The Athenaeum Club, 107 Pall Mall, London (*Post Office London directory* 1865).

[6] Charles Murchison was Falconer's friend and physician (*DNB*).

[7] Hooker refers to Thomas Thomson and George Bentham. Falconer was buried at Kensal Green, London, on 4 February 1865 (*DNB*).

[8] The reference is to Roderick Impey Murchison. For Hooker's opinion of what he described as 'Murchisonian science', see *Correspondence* vol. 11, letter from J. D. Hooker, 15 September 1863 and n. 16.

[9] An abstract of 'Climbing plants' was read at the meeting of the Linnean Society on 2 February 1865.

[10] Frederick Currey was the Linnean Society's secretary for botany (Gage and Stearn 1988, p. 59).

[11] Maxwell Tylden Masters was responding to the portion of the abstract of 'Climbing plants' on leaf-climbers, and particularly to the enlargement of the petiole (see 'Climbing plants', pp. 42–3, 47, 113–14). Masters referred to the internal structure of petioles that had become cylindrical (terete) and enlarged after clasping a twig or other object; the presence of 'closed bundles', defined as strands of 'specialized tissue, ... destitute of cambium, the procambium having become permanent tissue' (Jackson 1928, p. 57), indicated that the petiole no longer increased in diameter. In his letter of 7 February 1865, Masters wrote to CD detailing some of the comments he made at the meeting of the Linnean Society. See also n. 12, below.

[12] George Bentham, the president of the Linnean Society, was referring to CD's hypothesis that the legume *Lathyrus nissolia*, which has grass-like leaves and is the only *Lathyrus* without tendrils, had developed from an ancestor resembling *L. aphaca* in which the tendrils were not highly developed, and was a reversion to a primordial ancestor without tendrils ('Climbing plants', pp. 114–15). See also *Correspondence* vol. 12, letter to J. D. Hooker, [27 January 1864] and n. 23.

To [Robertson Munro?][1] 3 February [1865 or 1866?][2]

Feb 3rd

Dear Sir

I am much obliged for yr letter[3] & am particularly glad to hear that you are going to experimentise on Passifloras. & perhaps publish the results. If published I shd very much wish for a copy.—[4] I have b[een] sincerely grieved to read your letter on the inaccuracy of Mr John Scott; & I fear that you suspect him of recording experiments never made.[5] This would be a detestable crime I cannot think so ill of any human being as to believe in this without the fullest proof— May you not have been misled by the crosses having been made early or late in the day & on successive days.— I know I have sometimes made a few crosses day after day & when I added them up I have been surprised to find how high the total number amounted to. I can feel sure, however, from the tenor of your letter, that you will have no wish to be hard on a fellow labourer & will make your trial in a perfectly fair spirit for as you fairly say, Truth must come before every consideration.— It never occurred to me in the least to doubt Mr Scotts experiments, for I had collected information from several & quite independent sources showing that Passiflora is sometimes strangely infertile with its own pollen whilst fertile with that of other species.—[6]

With very best thanks for your letter | I am Dear Sir | Yours truly | C. Darwin

ADraftS
DAR 96: 16

[1] The recipient of the letter of which this is a draft may have been the Edinburgh nurseryman and florist, Robertson Munro (see nn. 4 and 5, below).

[2] The date range is conjectured on the basis of the allusion to John Scott's paper on sterility in *Passiflora* (Scott 1864c; see also n. 6, below), and CD's statement in *Variation* 2: 138 that he had been in communication with Munro in 1866 (see n. 4, below).

[3] The letter to which this is a draft reply has not been found.

[4] Munro published an account of sterility in *Passiflora* in 1868. There is a lightly annotated presentation copy of Munro 1868 in the Darwin Pamphlet Collection–CUL (G471). CD discussed Munro's work on *Passiflora* in *Variation*, stating that in 1866 he received some 'interesting details' about *Passiflora* from Munro (see *Variation* 2: 138). Munro conducted part of his research on *Passiflora* at the Royal Botanic Garden, Edinburgh, 'during several years' (Munro 1868, p. 400); and, by 1864, having pollinated and fertilised the self-sterile *Passiflora alata* in the garden with pollen from a plant at Keith Hall, Aberdeenshire, he was in a position to raise seedling plants from the union (Munro 1868, p. 300).

[5] Munro may have been familiar with John Scott and his work because Munro conducted some of his research at the Royal Botanic Garden, Edinburgh, where Scott was employed and carried out his experiments. In 1862, CD had encouraged Scott to test for self-sterility in *Passiflora*; CD was interested in evidence that some species were more easily fertilised by pollen of other species than by their own (see *Origin*, pp. 250–1, and *Correspondence* vol. 10, letter to John Scott, 11 December [1862], and Appendix VI). On Scott's experiments and his discussions with CD about *Passiflora* in 1863 and 1864, see *Correspondence* vols. 11 and 12. Scott's paper on sterility in *Passiflora* (Scott 1864c) appeared in the issue of the *Journal of the Proceedings of the Linnean Society* for 12 December 1864 (*General index to the Journal of the Linnean Society*, p. vi).

[6] CD discussed the self-sterility of *Passiflora* in *Variation* 2: 137–8, and referred to the work of Scott and Munro. CD also cited Mowbray 1822, Lecoq 1845, and Gärtner 1849, along with three other unnamed sources. See also *Correspondence* vol. 11, enclosure to letter to John Scott, 6 March 1863.

To Max Ernst Wichura 3 February [1865][1]

Down, Bromley, Kent

Dear & Respected Sir

From continued ill-health & other reasons I have only recently finished reading your work on Hybrid Willows, & I did not wish to thank you for your very kind present until I had read it.[2] Your work has greatly interested me in many ways. The extreme frequency of Hybrid Willows is quite a new fact to me & your explanation of their numbers on certain spots is remarkably ingenious.[3] I was also very glad to read your account of the development of pollen-grains in hybrids.[4] I see you doubt Gärtner's statements about reversion or Rückschäge;[5] on the other hand I observe that Naudin wishes to make it a general law that all Hybrids quickly revert to either parent form;[6] I suspect this great difference in your & Naudin's conclusions is that he has worked on cultivated varieties & you on wild forms.[7] You will find in the Chapter on Hybridism in the "Origin of Species," the view that Hybrids are sterile from not being perfectly accommodated to their conditions of life, briefly given;[8] but I have lately made observations which cause me to doubt in some degree the truth of this view.[9] I have been much interested & gratified by the remarks which you make towards the close of your volume on my views with respect to Species.[10]

With sincere respect I beg leave to remain, Dear Sir, Yours faithfully & obliged | Charles Darwin

LS
Autographia (dealers), San Rafael, California

1 The date 3 February is given in the sale catalogue. The year is established by the reference to Wichura 1865 (see n. 2, below), and by CD's 1865 discussions of Wichura 1865 with other correspondents (see, for example, letter to B. D. Walsh, 27 March [1865]).

2 Wichura 1865. There is a heavily annotated presentation copy of Wichura's *Die Bastardbefruchtung im Pflanzenreich erläutert an den Bastarden der Weiden* in the Darwin Library–CUL; it includes an abstract made by CD (see *Marginalia* 1: 871–3). See also letter to B. D. Walsh, 27 March [1865].

3 Though Wichura concluded that hybrid willows were generally rare in nature, he found instances where they outnumbered their parent species (Wichura 1865, pp. 64–5). CD wrote in the margin of this discussion: 'more Hybrids than pure parents!!' (see *Marginalia* 1: 873). Wichura also noted the rapid germination of hybrid seeds in bare spots (Wichura 1865, p. 65); CD wrote by this passage, 'Explains how these numbers come from Willows vegetating only in bare places'. See also CD's inserted abstract of the book; most of the abstract was written by Emma Darwin (*Marginalia* 1: 871).

4 For Wichura's description of the imperfections and irregularities in the development of hybrid pollen-grains, see Wichura 1865, pp. 32–40, 79–80, 89. In CD's abstract (see n. 3, above), Emma wrote for pages 38 and 39, 'pollen gets worse & worse in offspring of Hybrids inter se & in . . . the more complex hybrid'; CD made a similar annotation in Wichura 1865, p. 38 (see *Marginalia* 1: 871–3). Wichura's discussion of irregular pollen-grains included those in plants that were highly variable but not hybridised; see Wichura 1865, p. 89, where CD wrote in the margin: 'Plants which vary have often irregular pollen— I think some connexion between *sterility* & variability' (see also CD's abstract, and *Marginalia* 1: 872–3). In CD's discussion of the causes of variability in *Variation* 2: 268, he referred to Wichura's observations of irregular pollen-grains.

5 Wichura suggested that the reversions Karl Friedrich von Gärtner observed in his crossing experiments were due to his not taking enough care to prevent pollination of a hybrid offspring by a parent plant (see Gärtner 1849, pp. 437–41, Wichura 1865, pp. 27–8, and *Marginalia* 1: 871–3). There are

heavily annotated copies of Gärtner 1844 and 1849 in the Darwin Library–CUL (see *Marginalia* 1: 248–98).

6 CD probably refers to Charles Victor Naudin's most recent conclusions about reversions published in Naudin 1863; however, he had also read much of Naudin's relevant earlier work. See *Correspondence* vol. 12, letters to J. D. Hooker, 22 [May 1864] and n. 6, and 13 September [1864], and letter from C. V. Naudin, 6 December 1864 and n. 5. CD also cut out and annotated a note written by Asa Gray in the *American Journal of Science and Arts* (vol. 39, January 1865, pp. 107–8) about Naudin's recent observations of *Datura* hybrids returning to their parent forms (see DAR 205.7: 5). For discussions of Naudin's work and CD's assessment of it, see Geison 1969, pp. 404–1, Rheinberger 1983, Olby 1985, pp. 47–53, and J. Harvey 1997. There is an annotated copy of Naudin 1863 in the Darwin Pamphlet Collection–CUL.

7 See Wichura 1865, p. 28, and *Marginalia* 1: 872. The abstract CD inserted in Wichura 1865 indicates that he remembered Gärtner's stating that reversions occurred primarily in cultivated plants (see *Marginalia* 1: 871–2; see also *Origin*, p. 269). CD discussed this conclusion of Gärtner's, and the light it threw on the difference between Naudin's and Wichura's results, in his chapter on hybridism in *Origin* 4th ed., pp. 332–3, and in *Variation* 2: 49–50.

8 CD refers to chapter 8 of *Origin*, 'Hybridism' (pp. 245–78); for his conclusions about hybrid sterility, see pp. 254–67. Regarding the sterility of hybrids, CD wrote: 'it is scarcely possible that two organisations should be compounded into one, without some disturbance occurring in the development, or periodical action, or mutual relation of the different parts and organs one to another, or to the conditions of life' (*Origin*, p. 266).

9 CD refers to his experimental work on the dimorphic *Primula* and *Linum*, and the trimorphic *Lythrum*; he started this work in 1861, and was continuing to make crosses with dimorphic forms (see *Correspondence* vols. 9–12 and this volume). His results had led him to consider whether sterility might have been selectively acquired; when discussing his crosses of *Lythrum*, he wrote: 'I am now strongly inclined to believe that sterility is at first a selected quality to keep incipient species distinct' (see *Correspondence* vol. 10, letter to J. D. Hooker, 12 [December 1862]). CD's notes on hybridism in DAR 205.7: 155–65 state that he had nonetheless rejected natural selection as a cause of sterility by 1863 (see *Correspondence* vol. 10, Appendix VI). However, CD's comment to Wichura, and additions made to *Origin* 4th ed., pp. 310–26, 336–8, and *Variation* 2: 185–9, indicate that he was still deliberating about the causes of hybrid sterility, particularly in plants. Regarding Wichura's discussion of hybrid sterility, CD wrote in *Origin* 4th ed., p. 318:

> The above view of the sterility of hybrids being caused by two different constitutions having been confounded into one has lately been strongly maintained by Max Wichura; but it must be owned that the sterility, so like in every respect to that of hybrids, which affects the illegitimate offspring of dimorphic and trimorphic plants of the same species . . . makes this view rather doubtful.

CD was comparing the sterility of hybrids, or the offspring of hybrids, with the sterility of same-form unions among dimorphic and trimorphic plants; he evidently still thought the latter might be caused by natural selection (see *Origin* 4th ed., pp. 323–6). See also CD's annotations to Wichura 1865, p. 83, and CD's inserted abstract (see *Marginalia* 1: 871–2).

10 In the concluding chapter of his book Wichura briefly mentions CD's 'scharfsinnig' (sagacious) theory of natural selection (Wichura 1865, p. 82).

From Maxwell Tylden Masters 7 February 1865

Rye Lane | Peckham.
Feb. 7./65.

My dear Sir/

I had the pleasure of hearing a portion of the summary of your paper on twining plants at the last meeting of the Linn. Society and I hope you will not think me intrusive by writing to you on the matter—[1] As unfortunately I only heard part of the abstract read I do not know the extent of ground you travel over and hence perhaps the contents of this note may seem irrelevant, if so excuse me and throw aside this letter—

As an accidental occurrence, growth in a spiral direction takes place in two ways either as a twisting of the stem, branch or leaf as the case may be, upon itself in which instances the fibres of the organ are only secondarily and indirectly affected or the fibres of the stem etc themselves may be chiefly and primarily implicated— This condition is very often associated with the first but not necessarily so.— Moquin speaks of the first cases as "Enroulements" of the latter not very happily as Torsions.[2] The latter cases occur very generally in Plants with opposite or verticillate leaves e.g. Galium (for a specimen of wh. I am indebted to yourself)—Dipsacus—Valerian Equisetum, Labiates &c. &c.[3] Comparatively rarely does this spiralism affect plants with alternate leaves— As a consequence of this twisting of the stem it often happens that the leaves & branches instead of retaining their usual opposite, verticillate or decussate condition become placed one over the other along one edge—sometimes though more rarely in two rows— I enclose a drawing from a Dipsacus stem which I showed at the Linnean Society when I first began to study Botany and which puzzled me so much at the time that the description I gave of it though substantially correct I believe is by no means lucid—[4]

Some of these cases of spiral growth—if not all are dependent upon some obstacle to upward growth—resistance to motion—[5] Again they are very frequently associated with fasciation—

—Another point I noticed in the resumé of your paper was the assumption by a petiole of the characteristics of an axis so far as its wood circle was concerned—under certain conditions—[6] from this you drew the inference (quite in accordance with what I have always thought and indeed expressed in print) that there is a closer relationship between leaf & stem axis and appendage than is usually surmised.[7] In truth the arrangement of the woody bundles in a leaf stalk depends on or at least is associated with differences in form in the stalk itself thus when the petiole is cylindrical the woody bundles are arranged in a ring

instead of in a semi circle as they usually are in the channelled petioles

In all 'peltate leaved plants the petiole is cylindrical and the woody bundles arranged in a ring—[8]

In one of the recent volumes of the Brit Assoc. Reports there is an abstract of a paper on this subject by myself wherein I endeavour to show that the grooved petiole has reference to the protection of the axillary bud from undue pressure etc etc[9]

I have some thoughts of putting together my notes on spiralism as a teratological phenomenon so as to form a paper for the Linn. Soc. but I shall wait till I see the results of your labors in extenso—[10] meanwhile if by supplying references or in any other way I can be of the least assistance I shall be glad to do so if only to compensate for boring you with what may be irrelevant matters

faithfully y^rs.^ | Maxwell. T. Masters

DAR 171: 71

CD ANNOTATIONS

3.1 Some . . . motion— 3.2] *scored blue crayon*

3.1 Some . . . fasciation— 3.3] *crossed blue crayon*

4.6 In truth . . . petioles 4.12] *scored blue crayon*

4.9 cylindrical] *underl blue crayon*

Verso of last page: 'very interesting letter | Spiralism p 4— | Petioles— ⟨showed⟩ bundles are cylindrical— but, I think you will find that I have expressed myself with caution'[11] *blue crayon*; 'Put foot-note that I hear from Dr. Masters that is cylindrical [*underl blue crayon*] petiole—because petioles of Solanum apparently cylindrical & lose channelled structures, after clasping—[12] | Give the reference to B. Assoc Year & page.'[13] *ink*

[1] An abstract of 'Climbing plants' was read at the meeting of the Linnean Society on 2 February 1865. Joseph Dalton Hooker mentioned a comment Masters made at the meeting in the letter from J. D. Hooker, 3 February 1865.

[2] Moquin-Tandon 1841, pp. 178–82. CD's heavily annotated copy of Moquin-Tandon 1841 (*Eléments de tératologie végétale, ou histoire abrégée des anomalies de l'organisation dans les végétaux*) is in the Darwin Library–CUL (see *Marginalia* 1: 598–601); the only annotation on pages 178–82 is a scoring of the last sentence on page 182. CD's reading notebooks record that he read Moquin-Tandon 1841 in 1846 (see *Correspondence* vol. 4, letters to J. D. Hooker, 10 April [1846] and [May 1846], and Appendix IV). For CD's discussion of types of spiral growth in twining plants, see 'Climbing plants', pp. 5–7, 96–8.

[3] In 1864, CD sent Masters several specimens of abnormal plants, including one showing an example of torsion in *Galium*; in his reply, Masters mentioned a similar torsion in *Dipsacus* (see *Correspondence* vol. 12, letter from M. T. Masters, 19 September 1864). Examples of spiral torsion in *Galium* and other plants are discussed in Masters 1869, pp. 319–26, and an illustration of the *Galium* sent by CD is given on page 323. CD mentioned the hook-climber *Galium aparine* in 'Climbing plants', p. 105; see also DAR 157.1: 59. Masters cites CD's findings on spiral torsion in 'Climbing plants' in Masters 1869, p. 320 n. There is an annotated copy of Masters 1869 in the Darwin Library–CUL (see *Marginalia* 1: 571).

[4] Masters may refer to a paper he read at the Linnean Society on 6 March 1855 (Masters 1855). See *Correspondence* vol. 12, letter from M. T. Masters, 19 September 1864. The enclosure has not been found; Masters 1855 was published without an illustration. However, Masters discussed spiral torsion in *Dipsacus* in Masters 1869, pp. 320–1, with an illustration on page 321.

[5] CD included this statement in 'Climbing plants', p. 6 n., and prefaced it by observing that it was well known that stems occasionally became spirally twisted in a monstrous manner. The note was inserted into a discussion of the probability that stems twisted themselves to gain rigidity for additional strength and for easier upward growth around rough surfaces. CD added that Master's statement agreed with and might explain the normal axial twisting of twining plants, but that it did 'not preclude the twisting being of service to the plant by giving greater rigidity to the stem'. CD may have sent this note and others to the Linnean Society for inclusion in the manuscript of 'Climbing plants' that he had already sent on 18 January 1865 (see letter to J. D. Hooker, 19 January [1865]).

[6] In 'Climbing plants', pp. 42–3, CD noted the changes in a *Solanum jasminoides* petiole after it had clasped a stick. When examining cellular changes in a transverse slice of the petiole, he found that

a 'semilunar band' of woody vessels had grown and been converted into a complete ring of hard tissue in the clasping petiole; he included a diagram of this change in 'Climbing plants', p. 43 (see also CD's notes on these observations in DAR 157.1: 101–3). For a comment by Masters on petiole structure, made at the meeting during which the abstract of 'Climbing plants' was read, see the letter from J. D. Hooker, 3 February 1865 and n. 11.

7 For CD's notice of the relationship between leaves and the stem axis as indicated by the *Solanum* petioles, see 'Climbing plants', pp. 43, 47, and 113–14. CD had read the paper in which Masters came to a similar conclusion regarding leaves and stems (see Masters 1862a, pp. 211–14, and *Correspondence* vol. 10, letter to M. T. Masters, 26 February [1862]). CD had considered the relationship between plant parts in 1864, also in relation to his work with climbing plants (see *Correspondence* vol. 12, letter to Daniel Oliver, 11 March [1864], and letter from Daniel Oliver, 12 March 1864 and n. 9).

8 CD referred to Master's points on the general structure and shape of petioles, and related these to the changes in the *Solanum* petiole, in 'Climbing plants', pp. 43–4 n. For CD's addition of this note to the manuscript of 'Climbing plants', see n. 5, above.

9 Masters 1863b.

10 Only an abstract of 'Climbing plants' was read at a meeting of the Linnean Society on 2 February 1865; the paper was printed in full in the double issue of the *Journal of the Linnean Society of London* (*Botany*) for 12 June 1865 (*General index to the Journal of the Linnean Society*, p. vi). Masters included a section on 'Spiral torsion', in which he cited 'Climbing plants', in *Vegetable teratology* (Masters 1869, pp. 319–26).

11 Evidently these are notes for CD's reply; however, the letter to Masters has not been found.

12 See nn. 6–8, above. The leaf-climber *Solanum jasminoides* is described in 'Climbing plants', pp. 41–3.

13 CD did not cite Masters 1863b in 'Climbing plants'.

To Richard Kippist 8 February 1865

To the Librarian Linnean Soc. Burlington House

Please give Bearer all Parts of Transactions due to me.—[1]

Ch. Darwin

Down Feb. 8th 1865—

Linnean Society of London

1 As a fellow of the Linnean Society CD was entitled to receive, free of charge, all volumes, or parts of volumes, of the society's *Transactions* and *Journal* (*List of the Linnean Society of London*, 1865). CD's copies of the *Transactions of the Linnean Society of London*, for the 1860s and some other years, are in the Darwin Library–CUL. The bearer may have been George Snow, who ran a carrier service between London and Down every Thursday (*Post Office directory of the six home counties* 1866). In 1865, 8 February fell on a Wednesday.

To A. D. Bartlett 9 February [1865][1]

Down. | *Bromley.* | *Kent. S.E.*
Feb 9

Dear Sir

I wrote to you above a week ago to ask you to send me immediately the body of one of my Porto Santo Rabbits, together with some information.[2]

I have not received the rabbit nor any acknowledgement of my letter from you. I now request you will be so good as to write to me by return of post

Dear Sir | yours faithfully | Ch. Darwin

LS(A)
Cleveland Health Sciences Library (Robert M. Stecher collection: FF8)

[1] The year is established by CD's request for the body of the Porto Santo rabbit; see letter to A. D. Bartlett, 14 February [1865].
[2] See letter to A. D. Bartlett, 30 January [1865]. CD received the rabbit later in February (*Variation* 1: 114).

To J. D. Hooker 9 February [1865][1]

Down
Feb. 9th

My dear Hooker

I had not heard of poor Falconer's sufferings before receiving your note.[2] The thought has quite haunted me since. Poor fellow it is horrid to think of him,— I quite agree how humiliating the slow progress of man is; but everyone has his own pet horror, & this slow progress, or even personal annihilation sinks in my mind into insignificance compared with the idea, or rather I presume certainty, of the sun some day cooling & we all freezing. To think of the progress of millions of years, with every continent swarming with good & enlightened men all ending in this; & with probably no fresh start until this our own planetary system has been again converted into red-hot gas.— Sic transit gloria mundi,[3] with a vengeance.

I have been having 5 or 6 wretched days, miserable from morning to night & unable to do anything, but am much better today. How I wish I could beg borrow or steal your eczema, intensified a dozen fold; for this alone would do me good.[4] I hope Startin has done you good.[5] I wish I knew whether it was any earthly use consulting any doctor, for I can get nothing more out of Dr. Jenner.[6] Remember when *you can with ease*, you are pledged to come here. You cannot think how it pleases me to hear that you remember the gone-bye days at Down with pleasure;[7] they were indeed pleasant to me. What a loss the loss of nearly all Society has been to me & to my family; & now for the few years which I may live it will be worse.—[8]

I am much pleased to hear that my paper on Climbers went off well.[9] Masters has written to me & what he tells me accords with & supports what I have written.[10] I am rather surprised that the hypothetical case of L. nissolia was noticed:[11] the case interested me, because a few years ago I remember looking at this plant & concluding that it would be utterly impossible to even conjecture what could be the meaning of its strange leaves: it would be interesting to me to trace the graduated forms mentioned by Bentham,[12] but I must resist the temptation. T[he] diversified powers of movements in the Climbers,[13] & the gradations in structure struck me as

Hugh Falconer
Courtesy of the Moray Council Museums Service

Emma Darwin
Carte de visite by Elliot and Fry, London
© English Heritage Photo Library

Francis Darwin
Carte de visite by Bassano, London
© English Heritage Photo Library

the most interesting points in the subject.[14] Well, I ought to love the subject, for it helped me over many a weary hour, when I could do nothing else.[15]

Elizabeth Wedgwood has been here for a day,[16] & has told me a little news of you & Mrs Hooker,[17] to whom pray give kindest remembrances, & of a most pleasant luncheon at her house—

My dear old friend | Yours affect^{ly} | C. Darwin

DAR 115: 260

[1] The year is established by the relationship between this letter and the letter from J. D. Hooker, 3 February 1865.

[2] CD refers to the death of Hugh Falconer on 31 January 1865 and the letter from Hooker of 3 February 1865.

[3] *Sic transit gloria mundi*: 'So passes away the glory of the world' (H. P. Jones ed. 1900).

[4] On 7 February 1864, Emma Darwin's diary (DAR 242) recorded: 'C. very languid for 3 days'. CD believed that his attacks of eczema alleviated his other symptoms, and made him feel more alert (*Correspondence* vol. 12, letter from Emma Darwin to J. D. Hooker, 17 March [1864]).

[5] Hooker had consulted the skin specialist James Startin about his eczema, as had CD in 1862 (see letter from J. D. Hooker, 3 February 1865 and n. 1).

[6] CD also expressed dissatisfaction with the treatment of his illness by William Jenner, and by other doctors, in his letter to Hooker of 7 January [1865].

[7] See letter from J. D. Hooker, 3 February 1865.

[8] The excitement of social intercourse often made CD feel violently ill afterwards, so he avoided society as much as possible, except that of relations (see, for example, *Correspondence* vol. 2, letter to W. D. Fox, [25 March 1843], *Correspondence* vol. 5, letter to W. D. Fox, 7 March [1852], and *Correspondence* vol. 11, letter to J. D. Hooker, 3 January [1863]); however, in 1863 and 1864 CD was even more ill than usual, and at one stage could 'only stand very short visits even of the boys' (see *Correspondence* vol. 11, letter from Emma Darwin to J. D. Hooker, 26 December [1863]; see also *Correspondence* vol. 12, letter to F. T. Buckland, 15 December [1864]). For discussions of CD's illness, see Bowlby 1990, pp. 370–8, F. Smith 1990 and 1992, Browne 1998, and Colp 1998. See also Appendix IV.

[9] An abstract of 'Climbing plants' was read at the meeting of the Linnean Society on 2 February 1865. See letter from J. D. Hooker, 3 February 1865.

[10] In his letter of 3 February 1865, Hooker had mentioned a comment of Maxwell Tylden Masters on 'Climbing plants'. See letter from M. T. Masters, 7 February 1865.

[11] George Bentham had commented on CD's discussion of *Lathyrus nissolia*, which did not bear tendrils, and its relationship to tendril-bearing species of *Lathyrus* (see 'Climbing plants', pp. 114–15, and letter from J. D. Hooker, 3 February 1865 and n. 12).

[12] CD had raised seedling *Lathyrus nissolia* in 1857 (see *Correspondence* vol. 6, letter to J. D. Hooker, 14 July [1857]), and he examined flower specimens of the species in June 1863 for his research on heterostyly (DAR 111: A52); however, no other notes on *L. nissolia* have been found. Bentham had mentioned that he knew of other *Lathyrus* species that would be helpful to CD in testing his views, presented in 'Climbing plants', on the derivation of tendrils in the Leguminosae (see letter from J. D. Hooker, 3 February 1865 and n. 12).

[13] See 'Climbing plants', pp. 115–18.

[14] See letter to J. D. Hooker, 19 January [1865] and n. 6.

[15] For CD's work on climbing plants during periods of poor health, see, for example, letters to J. D. Hooker, 14 July [1863] and [27 January 1864] (*Correspondence* vols. 11 and 12).

[16] Sarah Elizabeth Wedgwood, Emma Darwin's sister (*Darwin pedigree*). Emma recorded in her diary (DAR 242) that Elizabeth arrived on 8 February and departed on 9 February 1865.

[17] Frances Harriet Hooker.

From August Schleicher 9 February 1865

Professor Dr. A. Schleicher | Jena

Excuse, Sir, my very bad English, but I prefer to show my rudeness in writing Your language insteed of causing to You any trouble with reading German.

With this letter I send to You a little pamphlet; perhaps You will think it not unworthy to be epitomised or translated by a Reviewer.[1]

This summer a Congress of botanists, horticulturists and seedsmen and a great exhibition shall take place at Erfurt.[2] In the program just published I read, that the first object of transactions is 'the Theory of Darwin'.[3]

Pray, accept my photography;[4] dare I hope to be favoured with Yours? With the best wishes for Your health and with sincere respect

Aug. Schleicher.
Jena, 9/2 65.

DAR 177: 53

[1] There is a lightly annotated copy of Schleicher 1864 in the Darwin Pamphlet Collection–CUL; Schleicher's article was a review of *Origin* that appeared in an agricultural journal. In 1863, Schleicher had sent CD another article of his that was later translated into English (Schleicher 1863 and 1869; see *Correspondence* vol. 11, letter to J. D. Hooker, [22–3 November 1863]). There are lightly annotated copies of Schleicher 1863 and 1869 in the Darwin Pamphlet Collection–CUL; Schleicher 1863 is inscribed by the author, and Schleicher 1869 is inscribed by the translator. In several letters to CD in 1864, Ernst Haeckel had mentioned Schleicher's interest in CD's transmutation theory (see *Correspondence* vol. 12, letters from Ernst Haeckel, 9 [July 1864], 10 August 1864, and 26 October 1864). For a discussion of Schleicher, Haeckel, and CD, see R. J. Richards 2001.

[2] A general agricultural congress was held in Erfurt, Germany, between 9 and 17 September 1865 (*Erfurt-Congress*).

[3] See *Erfurt-Congress*, p. 15. For several of the questions raised about the importance of CD's theory for horticulture, see *Nachtrag*, p. 4.

[4] The photograph Schleicher sent has not been found.

To A. D. Bartlett 14 February [1865]

Down. | Bromley. | Kent. S.E.
Feb 14

My dear Sir

I am very much obliged for your note,[1] answering so fully all my questions, & for the feathers of the Gallus.[2] The rabbit has arrived safely, but most unfortunately for me the entrails have been taken out.[3] If you catch the other, will you be so good as to send it unmutilated, to me "care of the Down Postman
Bromley Kent."

with my thanks believe me | yours faithfully, | Ch. Darwin

Endorsement: 'Darwin | February 14/65'
LS
American Philosophical Society (Getz 4771)

[1] The letter from Bartlett has not been found.
[2] See letter to A. D. Bartlett, 30 January [1865] and nn. 3 and 4.
[3] CD wanted to compare the weight of the Porto Santo rabbits with that of other rabbits (see *Variation* 1: 113, 127). No reply to this letter has been found; however, the weight of the rabbit given by CD in *Variation* 1: 113 indicates Bartlett did send the other rabbit with entrails included. CD also compared the colour of the rabbits sent to him in 1865 with that of the Porto Santo rabbits they had descended from (see letter to A. D. Bartlett, 30 January [1865] and n. 2). CD had earlier made osteological observations of Porto Santo rabbit specimens that had been preserved in wine and in brine (see, for example, *Correspondence* vol. 6, letter to W. D. Fox, 8 [June 1856]; see also *Variation* 1: 116, 120, 122, 126–9).

To J. D. Hooker 15 [February 1865][1]

Down
15th

My dear Hooker

I write again chiefly to tell Oliver not to trouble himself to inform me on Hildebrand's paper in Bot. Zeitung on Pulmonaria; as author has sent me a copy.—[2] Secondly to tell you *not* to come here, if you had thought of it, on Feb. 25th, as we shall have visitors & this would spoil for me your visit.—[3] Any other Saturday, when you can spare time wd suit us.—

Can you give me any notion what to subscribe for poor dear Falconer's bust: would 5 guineas be too much or not enough?—[4]

I meant to have shown you a paper by Heer,—, an address to some Helvetic Soc:—in which he discusses alpine & arctic Floras—[5] he does not allude to your paper, but considers Scandinavia as parent source, *from being oldest mountain range*—[6]

Wallace has published some splendid papers in Proc. Zoolog. Soc. & Geograph. Journal on distribution in Malay Arch.—[7]

I am reading & skimming through Lyell's new Edit. of Elements:[8] it is an astonishing monument of labour, knowledge & clear thought. What a wonderful man he is.— I wish some one wd concentrate similar knowledge & labour on a work on the whole science of Botany, excluding of course systematic Botany.[9] I do believe a work of great interest cd. be made. There is a redundance of elementary Treatises: but a summary of knowledge seems to me much wanted.—

Yours affect | C. Darwin

DAR 115: 261

[1] The date is established by the reference to the memorial fund for Hugh Falconer (see n. 4, below).
[2] CD refers to 'Dimorphismus von Pulmonaria officinalis' by the German botanist Friedrich Hildebrand (Hildebrand 1865, pp. 13–15). In his letter to J. D. Hooker, 10 December [1864] (*Correspondence* vol. 12), CD asked Hooker to ask Daniel Oliver to look out for discussions of CD's work on floral dimorphism in *Botanische Zeitung*; he especially wanted to see the paper by Hildebrand on dimorphic *Pulmonaria* (see also *Correspondence* vol. 12, letter from Friedrich Hildebrand, 21 June 1864, and letter to Friedrich Hildebrand, 25 June [1864]). Hildebrand 1865 appeared in two parts, the first part in the issue of *Botanische Zeitung* for 6 January 1865, pp. 1–6, and the second part in the issue for 13 January 1865, pp. 13–15. CD's annotated copy of Hildebrand 1865 is in the Darwin Pamphlet Collection–CUL

(pamphlets Q262 and G1236). CD cited the second part of the paper in 'Illegitimate offspring of dimorphic and trimorphic plants', p. 430, and in *Forms of flowers*, p. 101.

3 In his letter of 9 February [1865], CD had encouraged Hooker to visit Down House. On 25 February 1865, Emma Darwin recorded in her diary (DAR 242) that 'Ed. 3 Jossi & Horace' came; that is, Edmund Langton and Katherine Elizabeth Sophy, Margaret Susan, and Lucy Caroline Wedgwood visited, and Horace Darwin, who had been in London since 20 February, returned home. The 'Josselinas' was the family nickname for Caroline and Josiah Wedgwood III's daughters, with whom Edmund Langton was friendly (see *Correspondence* vol. 11, letter from Charles and Emma Darwin to W. E. Darwin, [4 May 1863]).

4 CD may have read the notice in the *Reader*, 11 February 1865, p. 167, of the subscriptions being raised for a memorial of Hugh Falconer in the form of a marble bust. Falconer died on 31 January 1864 (*DNB*).

5 CD refers to Oswald Heer's address to the Société Helvétique des Sciences Naturelles (Heer 1864). There is an annotated copy of Heer 1864 in the Darwin Pamphlet Collection–CUL.

6 In his essay 'Outlines of the distribution of Arctic plants' (Hooker 1860), Hooker supported CD's theory of plant migrations southward from the Arctic during a former cold period; CD used this theory to explain present distribution patterns of Arctic and alpine plants on isolated mountain tops and higher elevations in temperate and tropical regions (see *Origin*, pp. 365–82). Hooker's observations led him to conclude that the Arctic flora as a whole was largely Scandinavian, and that this Scandinavian flora was of great antiquity and was more widely distributed over the globe than the other Arctic floras (Hooker 1860, pp. 251, 253, 258). See *Correspondence* vol. 10, letters to J. D. Hooker, 25 February [1862] and 7 March [1862], and letter from J. D. Hooker, 27 February 1862; see also Browne 1983, pp. 117–29, 133–5. For Heer's statement that Scandinavia was a source for alpine plants of the northern hemisphere, see Heer 1864, pp. 24–6. On the back of Heer 1864, CD wrote: 'Very good pamphlet with objections [to CD's theory] very well put.— & forcible'.

7 CD refers to papers by Alfred Russel Wallace in the *Proceedings of the Zoological Society of London* and in the *Journal of the Royal Geographical Society* on geographical distribution in the Malay Archipelago, where Wallace had made observations between 1854 and 1862 (Wallace 1862, 1863a, 1863b, 1863c, and 1864b). Wallace had sent these papers to CD in January (see letters from A. R. Wallace, 20 January 1865 and 31 January [1865]). See also C. H. Smith ed., 1991.

8 CD refers to the sixth edition of Charles Lyell's *Elements of geology* (C. Lyell 1865), published in January 1865 (*Publishers' Circular*, 1 February 1865, p. 60). Lyell had offered to send CD a copy of the book in his letter of 16 January 1865. There is an annotated copy of C. Lyell 1865 in the Darwin Library–Down (*Marginalia* 1: 524–5).

9 Hooker and George Bentham were already collaborating on a multi-volume work on systematic botany (Bentham and Hooker 1862–83).

From J. D. Hooker [17 February 1865][1]

Kew
Friday

Dear Darwin

I had thought of going to you on the 25th. & was only waiting to see my way a little.[2] I find however that the examinations of Army Asst. Surgeons will prevent me,[3] & so I shall try & get down on the following Saty. if you are well enough to see me.[4]

I am puzzled about the Falconer affair, not what to subscribe, for I have no hesitation in thinking that £1 " 1. or £2 " 2 would be ample, adding if necessary afterwards to £5.5.0 the object being to secure a bust only—for which £100 or

£150 should be enough. But I like to see these subscriptions confined to definite objects & bodies—as far as possible,[5] & especially in Falconers case, because I suspect that Botanists will not subscribe with enthusiasm.—for this simple reason, that though he enjoyed for nearly 25 years the most magnificent pay as a Botanist, he did nothing Botanical for it—[6] his collections he let go to ruin, his mss he shut up & would let no one see.—[7] As Superintendent of the Gardens his name was a bye-word in Calcutta & a scandal elsewhere. Poor dear old F. these are very hard things for a friend to say—but so it is— it is impossible to exaggerate the mischief he did. He very nearly brought about the abandonment of the Calcutta Gardens, & to all my appeals at the India House for something to be done for Botany, *Falconer* was thrown in my teeth.[8] So my idea is that the Geolog. Soc. should form the subscription & confine it to their own Members,[9] & his friends afterwards subscribe for a copy of the bust if so disposed. I will subscribe to both.

I have not seen Heers paper nor Wallaces,[10] for indeed I have been & am hard worked this winter. I look to, & only look to! reading Lyell with great interest.[11] There is a good deal of snarling at him since Falconers death.[12] I do wish that he had only shown himself at the Funeral, but he is far from strong.

How I should like to write the book you want on Botany—[13] it has been as you know a dream of mine for a very long period. No one could do it so well as yourself!

Ever Yrs affec | J D Hooker

P.S. I have just heard from Anderson that he has given Scott the Curatorship of the Gardens with £240 a year & house[14]

DAR 102: 10–11

[1] The date is established by the relationship between this letter and the letter to J. D. Hooker, 15 [February 1865]; the Friday following CD's letter to Hooker of 15 [February 1865] was 17 February.

[2] In his letter of 15 [February 1865], CD had warned Hooker to visit on any Saturday other than 25 February, when the Darwins expected other visitors.

[3] Hooker was one of the examiners in science for medical officers in the armed services (L. Huxley ed. 1918, 1: 387). Examinations for army medical officers were held at Chelsea Hospital in February and March 1865 (*Statistical, Sanitary, and Medical Reports (Army Medical Department)* 7 (1865): 605–6).

[4] According to Emma Darwin's diary (DAR 242), Hooker visited Down House between 4 and 6 March 1865.

[5] CD had asked Hooker what would be an appropriate contribution to the fund being raised for a marble bust of Hugh Falconer (see letter to J. D. Hooker, 15 [February 1865]). For two additional memorials made to Falconer, see the letter from George Busk, 20 February 1865, n.4.

[6] Although Falconer had been superintendent of the botanic garden in Saharanpur, north-west India, from 1832 to 1842, and superintendent of the botanic garden in Calcutta and professor of botany at the Calcutta Medical College from 1848 to 1855 (*DNB*), he became known primarily for his geological and palaeontological research; however, see also nn. 7 and 8, below. A list of Falconer's botanical memoirs and reports is provided in Falconer 1868, 1: lv–lvi; see also R. Desmond 1994.

[7] When Falconer returned to Britain in 1843 (see n. 6, above), he brought back seventy chests of dried plants, which suffered damage from damp on the voyage and in the cellars of East India House, the headquarters of the East India Company, where they were deposited (see Falconer 1868, 1: xxxv). In preparing *Flora Indica*, Hooker and Thomas Thomson, referring to Falconer as 'one of the most

estimable, able, and accomplished of Indian botanists', also recorded that his 'magnificent' India collection, with 'voluminous notes' was the only herbarium of importance to which they could not procure access (Hooker and Thomson 1855, pp. 67–8). However, it is not clear whether the poor condition and inaccessibility of the collection was the fault of Falconer or the East India Company; in 1858, the year after the East India Company was dissolved, Hooker secured several collections for the Royal Botanic Gardens, Kew, from the East India Company's cellars. These collections had all suffered from long neglect, and included Falconer's (see R. Desmond 1995, pp. 200–1, McCracken 1997, p. 154, and R. Desmond 1999, p. 206).

8 For Hooker's mixed reports from India of Falconer's behaviour and work as superintendent of the Calcutta botanic garden, see *Correspondence* vol. 4, letters from J. D. Hooker, 13 October 1848, 3 February 1849, and 6 and 7 April 1850; see also Hooker 1854, 2: 244–7. For positive accounts of Falconer's work in India, see Falconer 1868, 1: xxxix–xl, R. Desmond 1992, pp. 98–100, 223, and McCracken 1997, pp. 8, 20. Hooker felt that the East India Company, which was effectively Falconer's employer in Calcutta, had treated Falconer far better than he and his colleague Thomson had been treated as collectors and botanists, and that the company had provided insufficient support and recognition of his and Thomson's labours (L. Huxley ed. 1918, 1: 355–8, 2: 145–6, and R. Desmond 1999, pp. 201, 217–19, 250–1). In 1869 and later years, Hooker still despaired over the support for and progress of the study of botany in India (see L. Huxley, ed. 1918, 2: 17, R. Desmond 1992, pp. 102, 323, McCracken 1997, p. 23, and R. Desmond 1999, p. 270). Hooker eventually published the seven-volume *Flora of British India* (Hooker 1872–97)

9 Falconer had been a fellow of the Geological Society of London since 1842 (*List of the Geological Society of London*).

10 In his letter to Hooker of 15 [February 1865], CD said that he wanted to show Hooker a section of Heer 1864 in which Oswald Heer discussed the distribution of Arctic and alpine plants; in the same letter CD recommended Alfred Russel Wallace's papers on geographical distribution in the Malay Archipelago (Wallace 1862, 1863a, 1863b, 1863c, and 1864b).

11 The reference is to the sixth edition of Charles Lyell's *Elements of geology* (C. Lyell 1865). For CD's opinion of this work, see the letter to J. D. Hooker, 15 [February 1865].

12 Hooker may refer to the quarrel between Lyell and Falconer over *Antiquity of man* (C. Lyell 1863a): Falconer had publicly denounced Lyell in the *Athenæum* for not having sufficiently acknowledged in *Antiquity of man* his own research and that of others (see *Correspondence* vol. 11, Bynum 1984, Grayson 1985, and L. G. Wilson 1996).

13 See letter to J. D. Hooker, 15 [February 1865].

14 Thomas Anderson, who was the superintendent of the botanic garden at Calcutta, had first given John Scott a temporary position at one of the *Cinchona* plantations near Darjeeling (see letter from John Scott, 20 January 1865 and n. 7, and letter from J. D. Hooker, [26 January 1865]). Anderson then offered Scott the curatorship at the botanic garden at Calcutta (see letter from John Scott, 21 July 1865).

From George Busk 20 February 1865

15 Harley St
Feb 20th 1865

My dear Darwin

Your cheque reached me in due course & I have to thank you for it.[1] It is proposed to have a meeting of poor Falconers friends & admirers next Saturday at Mr Fergusson's[2] to constitute a Committee to carry out the Memorial to be determined upon.— As it is desirable to have as many good names as possible in the list of supporters, I was just about to write to you to say that if you had no objection your name would be included in it. If you have will you kindly let me know by Saturday Morning.

The Memorial contemplated & I think from what has been said, very likely to be carried out—is

1st a Bust to be placed as afterwards determined in London—[3]

2. To found a Fellowship if possible, or if not a Scholarship or in default of that a prize—in Natural Science in the University of Edinburgh—[4]

Believe me | Yours truly | Geo Busk

DAR 160: 380

[1] CD sent the cheque as a contribution towards a memorial for Hugh Falconer (see letter to J. D. Hooker, 15 [February 1865], and letter from J. D. Hooker, [17 February 1865]). In his Account book–banking account (Down House MS), CD recorded sending Busk a cheque for £3 3*s.* 'for Falconer' on 18 February 1865; on 1 March 1865 a second payment of £7 7*s.* for 'Falconer Testimonial' is recorded. Busk was one of four people who had undertaken to receive subscriptions for a Falconer memorial (*Athenæum*, 11 February 1865, p. 201). See also nn. 3 and 4, below.

[2] James Fergusson.

[3] Timothy Butler was commissioned to sculpt a marble bust of Falconer; it was afterwards exhibited in the rooms of the Royal Society of London (*DNB*, s.v. Falconer, Hugh).

[4] Nearly £2000 was collected for Falconer's memorial fund; the money was used for a marble bust (see n. 3, above), and to establish the Falconer Memorial Fellowship at the University of Edinburgh, a fellowship for graduates in natural science or medicine, to encourage the study of palaeontology and geology (*DNB*). Falconer had been a medical student at the University of Edinburgh from 1826 to 1829 (*DNB*). By a separate subscription, another marble bust of Falconer was placed in Calcutta, at the Museum of the Asiatic Society of Bengal (see Charles Murchison's memoir in Falconer 1868, 1: l). CD's second contribution (see n. 1, above) may have been to the Falconer Memorial Fellowship or to the bust in India.

To Charles Lyell 21 February [1865][1]

Down. | Bromley. | Kent. S.E.
Feb 21

My dear Lyell

I have taken a long time to thank you very much for your present of the Elements.[2]

I am going thro' it all, reading what is new & what I have forgotten, & this is a good deal.

I am simply astonished at the amount of labour, knowledge & clear thought condensed in this work. The whole strikes me as something quite grand. I have been particularly interested by your account of Heer's work & your discussion on the Atlantic Continent:[3] I am particularly delighted at the view which you take on this subject; for I have long thought that Forbes did an ill-service in so freely making continents.[4]

I have also been very glad to read your argument on the denudation of the Weald[5] & your excellent resumé on the Purbeck beds,[6] & this is the point at which I have at present arrived in yr book.

I cannot say that I am quite convinced that there is no connection, beyond that pointed out by you, between glacial action & the formation of lake-basins:[7] but you will not much value my opinion on this head as I have already changed my mind some half-dozen times.[8]

I want to make a suggestion to you; I found the weight of your vol. intolerable, especially when lying down;[9] so with great boldness cut it into 2 pieces & took it out of its cover; now c^d not Murray,[10] without any other change, add to his advertisement a line saying "if bound in 2 Vols one shilling or 1/6 extra".

You thus might originate a change which w^d be a blessing to all weak handed readers.

Believe me my dear Lyell | yours most sincerely | Charles Darwin

Originate a second *real blessing* & have the edge of the sheets cut, like a bound book[11]

My love to Lady Lyell | E.D.[12]

LS(A)
American Philosophical Society (306)

[1] The year is established by the reference to C. Lyell 1865 (see n. 2, below).

[2] The sixth edition of Lyell's *Elements of geology* (C. Lyell 1865) was published in January 1865 (*Publishers' Circular*, 1 February 1865, p. 60). Lyell had offered to send it to CD in his letter of 16 January 1865. CD's annotated copy is in the Darwin Library–Down (see *Marginalia* 1: 524–5).

[3] See letter from Charles Lyell, 16 January 1865 and n. 19. Lyell preceded his discussion of Oswald Heer's Atlantis theory with a description of Heer's research on the Miocene plants and insects of Switzerland (see C. Lyell 1865, pp. 246–73).

[4] Edward Forbes had argued that former land-bridges connecting the Iberian peninsula, the Azores, and Ireland, explained the present distributions of plants and animals (see E. Forbes 1845 and 1846, and Browne 1983, pp. 115–17). Lyell mentioned Forbes's theory in C. Lyell 1865, p. 267. For CD's earlier discussions of Forbes's land-bridges, see, for example, *Correspondence* vol. 3, letter to J. D. Hooker, [13 March 1846], and *Correspondence* vol. 6, letters to Charles Lyell, 16 [June 1856] and 25 June [1856]. For CD's earlier discussions of Heer's hypothesised continent once connecting Europe and the Americas, see *Correspondence* vol. 8, letter to Charles Lyell, 20 November [1860], *Correspondence* vol. 9, letter to J. D. Hooker, 28 [December 1861], and *Correspondence* vol. 10, letter to J. D. Hooker, 4 November [1862]. See also *Origin*, pp. 357–8.

[5] Lyell discussed the denudation of the Weald, a district between the North and South Downs in Hampshire, Surrey, Sussex, and Kent, in C. Lyell 1865, pp. 351–74. In *Principles of geology* (C. Lyell 1830–3), Lyell argued that marine erosion was the main agent in the formation of the Weald, and he defended this position in C. Lyell 1865, dismissing Joseph Beete Jukes's and Andrew Crombie Ramsay's recent reinterpretations of the denudation of the Weald as being brought about by fluvial agents (see Jukes 1862a and Ramsay 1864; see also *Correspondence* vol. 12, letter from A. C. Ramsay, 10 July 1864, and Davies 1969, pp. 347–8). In C. Lyell 1865, p. 365, Lyell referred to 'recent investigation' having revealed the long series of ages over which the Weald had been eroded. In *Origin*, pp. 285–7, CD had calculated that 300 million years were required for marine denudation of the Weald. Owing to criticism, CD halved this period of time in *Origin* 2d ed., p. 287, and eliminated the entire discussion in *Origin* 3d ed. (see *Correspondence* vol. 8, letter to J. D. Hooker, 3 January [1860] and n. 16, and letter to Asa Gray, 3 April [1860], and *Correspondence* vol. 10, letter from J. B. Jukes, 25 May 1862 and n. 4).

[6] The Purbeck beds are discussed in *Elements of geology* (C. Lyell 1865, pp. 375–91). CD was particularly interested by the section on the imperfection of the fossil record with respect to the Mammalia,

pp. 377–84 (see *Marginalia* 1: 524–5). See *Origin*, pp. 279–311, for CD's discussion of the imperfection of the geological record.

[7] The formation of rock-basins had been a controversial topic since Ramsay had introduced his theory of their glacial origin in 1862 (Ramsay 1862). Although Lyell accepted that lakes were common in glaciated areas, he explained their development by the gradual movements of upheaval and subsidence, in combination with fluvial erosion (see C. Lyell 1865, pp. 168–74). See Davies 1969, pp. 304–7.

[8] CD had initially supported Ramsay's theory of the glacial origin of rock-basins (see *Correspondence* vol. 10, letter to Charles Lyell, 14 October [1862]), but had also been impressed by other opinions, including that of Roderick Impey Murchison (see Murchison 1864a and *Correspondence* vol. 12, letter to J. D. Hooker, [23 August 1864]). In the letter to J. D. Hooker, 22 October [1864], CD wrote: 'I have now come round again, to Ramsay's view for third or fourth time; but Lyell says when I read his discussion in the Elements I shall recant for fifth time'.

[9] There are in total 809 pages in C. Lyell 1865.

[10] John Murray published C. Lyell 1865.

[11] Sheets containing the printed pages would be folded and bound with the folds left uncut. CD preferred that the publisher cut the sheets, rather than the buyer (see also *Correspondence* vol. 11, letter to T. H. Huxley, 10 [January 1863]).

[12] Most of the letter is in the hand of Emma Darwin; she refers to Mary Elizabeth Lyell.

From J. D. Hooker [27 February 1865?][1]

Kew
Monday.

Dear Darwin

I shall leave town by the 4: PM—or 4:10 train on Saturday,[2] which arrives at Bromley at 4:49, if not convenient to send the tax cart,[3] I shall enjoy the walk & feel all the freer—[4] Do not worry about whether you can see me or no, I shall amuse myself—

Ever yrs affec | J D Hooker

DAR 102: 12

[1] The date is conjectured on the basis of the location of the letter in the archive: the letter was deposited with other 1865 letters in DAR 102 (Hooker's letters to CD, 1865–8), which is arranged chronologically. The arrangement is presumably CD's and has been preserved (see the preface to *Handlist of Darwin papers at the University Library, Cambridge*, p. 4 (Cambridge: University Library, 1960)). According to Emma Darwin's diary (DAR 242), Hooker visited Down House from 4 to 6 March 1865. In 1865, the Monday preceding 4 March fell on 27 February.

[2] In his letter to Hooker of 9 February [1865], CD encouraged Hooker to visit Down House when he could do so '*with ease*'. Hooker eventually arranged to visit from 4 to 6 March 1865 (see letter from J. D. Hooker, [17 February 1865] and n. 4).

[3] A two-wheeled, open cart drawn by one horse, so-called because it was subject to a reduced tax (*OED*).

[4] Down is six miles south-east of Bromley.

To William Bernhard Tegetmeier 27 February [1865][1]

Down. | Bromley. | Kent. S.E.
Feb 27.

My dear Sir

In a week's time I shall be ready to correct my M.S. on Fowls. Will you therefore have the kindness to return soon this M.S., which you have kept so long for me, with any corrections or errors pointed out.[2] You will remember that I have asked your permission to state that the M.S. has had the benefit of your supervision, & I hope you will grant me this favour.[3] In the course of a few weeks I shall put a number of specimens into the engraver's hands & as soon as they are finished I will return your skulls, which owing to my ill-health, I have retained for so unreasonable a period.[4]

Will you be so good as to register the M.S., when returned, & as there is no reason to put you to the most trifling expense I beg permission to enclose stamps.

Can you grant me one small favour, that is to enclose 2 or 3 spangled feathers of Hamburgs or any other breed, or more particularly any dark-tipped feathers of the Game breed as I want to compare them with some curious tipped feathers which have been sent me.[5]

Believe me my dear Sir | yours sincerely | Charles Darwin

I hope that you are well & that all things go on pretty well with you.— My health is better, but I much fear I shall ever remain that most wretched creature, a confirmed invalid.—[6]

LS(A)
The New York Botanical Garden Archives (Charles Finney Cox collection) (Tegetmeier 59)

[1] The year is established by the reference to correcting the chapter on fowls for *Variation* (see n. 2, below).

[2] Fowls are the subject of chapter 7 of *Variation* (*Variation* 1: 225–75). Having written the first draft of his manuscript on fowls in May 1861, CD sent it to Tegetmeier for review (*Correspondence* vol. 9, letter to W. B. Tegetmeier, 14 June [1861]). In November 1864 CD had started revising the earlier chapters of *Variation*; he continued working on the chapters on domestic animals until 22 April 1865 (see *Correspondence* vol. 12, Appendix II, and this volume, 'Journal' (Appendix II)).

[3] Tegetmeier's assistance with the manuscript, and his help in obtaining information and specimens, are acknowledged in *Variation* 1: 225 n.

[4] Tegetmeier had sent skulls of wild and domesticated fowls for CD to examine in April 1861 (see *Correspondence* vol. 9, letters to W. B. Tegetmeier, 25 February [1861] and 14 April [1861]). CD may have returned some of the specimens in 1864 (see *Correspondence* vol. 12, postscript to letter from W. B. Tegetmeier, [before 21 December 1864]), but he did not return the last of them until February 1867, retaining some so that they could be drawn as illustrations for *Variation* (see *Correspondence* vol. 14, letter to W. B. Tegetmeier, 9 July [1866], and letters to W. B. Tegetmeier, 6 January [1867] and 5 March [1867] (*Calendar* nos. 5347 and 5431)). In *Variation* 1: 260, CD wrote that nearly half of the fifty-three skulls he examined were lent to him owing to 'the kindness of Mr. Tegetmeier'. For the published engravings of some of the fowl skulls, see *Variation* 1: 261–5.

[5] CD had asked Abraham Dee Bartlett to send from the Zoological Gardens the 'tipped' feathers of *Gallus sonneratii* (grey junglefowl or Sonnerat's junglefowl), a wild species found in India. CD later thanked him for the feathers of the 'Gallus' (see letters to A. D. Bartlett, 30 January [1865] and 14 February [1865]). In *Variation* 1: 233–4, CD described the neck and loin feathers, or hackles, of *G. sonneratii*, and noted that he had examined the feathers of a cross between *G. sonneratii* and a 'red game-hen'. CD was interested in determining which wild species domestic breeds had descended from; as part of this effort, he described variations in *Gallus* feathers in *Variation* 1: 239–45 (see also *Variation* 1: 249–50). At one time, CD had several fowl breeds at Down House with which he performed crossing experiments; these included a spangled Hamburgh and a white game breed (see *Correspondence* vol. 7, letters to W. B. Tegetmeier, 21 [February 1859] and 9 April [1859]; see also J. Secord 1985, p. 526).

6 For CD's health in 1864 and early 1865, see the letter to Charles Lyell, 22 January [1865] and n. 16, and the letter to J. D. Hooker, 9 February [1865].

From W. B. Tegetmeier [28 February – 5 March 1865][1]

Muswell Hill. N

My dear Sir.

I was much grieved at receiving a letter from you not in your own hand writing—although the postscript was more assuring than the letter itself.—[2] I will carefully go over your MSS. and make any observations that occur to me either in the margin in pencil or on slips of paper—[3]

Thanks for your kind wishes I am getting on fairly, being very busy with new and tolerably profitable work,[4] but I have had the misfortune to cut completely across and through the cornea of my right eye—and I have as yet only partially recovered the sight, the eye being very intolerant of light—

I send you spangled feathers of crested birds—and pencilled feathers of the genuine dutch breed

Our North country genuine spangled fowls, have larger and better spangles, of which I will send you samples soon—

There are no spangled game now bred—nor am I aware of any well established of that particular variety existing—[5]

I am now editing the Poultry and some other departments of the Field, and should you feel disposed to use the paper in any way as a means of eliciting information—We (for in this case I may safely use the editorial pronoun as I speak for every one connected with the paper) should feel much pleased if we could in any manner assist you.[6]

I have seen no account of your paper on Tendrils[7] I should much like to read it if it is published[8]

If you could give me all next week for the reading of the MS. I should be glad as my accident caused my work to accumulate to a rather overwhelming extent

Trusting to hear that your health is improving | Believe me Very sincerely Yours etc | W B Tegetmeier

DAR 178: 64

CD ANNOTATIONS

2.2 I have had . . . eye— 2.3] *cross in margin, red crayon*
4.1 Our . . . soon— 4.2] *scored red crayon*
6.1 I am . . . Field,] *cross in margin, red crayon*
7.1 I have . . . published 7.2] *cross in margin, red crayon*

1 The date is established by the relationship between this letter and the letters to W. B. Tegetmeier, 27 February [1865] and 6 [March 1865].

2 Only the signature and postscript of the letter to W. B. Tegetmeier of 27 February [1865] were in CD's hand.

[3] The reference is to CD's manuscript on fowls for *Variation* (see letter to W. B. Tegetmeier, 27 February [1865] and n. 2).

[4] In 1864, Tegetmeier became the editor of the pigeon and poultry department of the *Field* (see *Correspondence* vol. 12, letter from B. P. Brent, 2 September 1864).

[5] See letter to W. B. Tegetmeier, 27 February [1865] and n. 5. CD described the game breed, the spangled Hamburgh breed, and the pencilled Hamburgh breed of fowls in *Variation* 1: 226–8. For a discussion of pencilling and spangling, see *ibid.*, pp. 243–4, and for a discussion of the feather colours of the game breed, see *ibid.*, pp. 251–2. CD noted the tipped feathers of a game breed variety in *ibid.*, p. 254.

[6] See n. 4, above. The letter was written on *Field* stationery with the heading crossed out. The *Newspaper press directory* for 1865 describes the *Field* as 'a gentleman's paper, devoted to sports, pastimes, natural history, and all country pursuits'. Published weekly, the *Field* had its offices at 346 Strand, London (*Newspaper press directory* 1865). For CD's long-standing relationship with plant and animal breeders, including poultry breeders, see J. Secord 1985.

[7] An abstract of 'Climbing plants' was read at the Linnean Society on 2 February 1865.

[8] 'Climbing plants' was published on 12 June 1865 as a double issue of the *Journal of the Linnean Society (Botany)* (*General index to the Journal of the Linnean Society*, p. vi). An author's offprint inscribed by CD was sent to Tegetmeier (Freeman 1977, pp. 116–17).

From William Erasmus Darwin [late February–May 1865][1]

Yellow Short styled P.	Yellow Long styled P.	Red Long Styled Primrose

Magnified 350—

Mem[2]

DAR 108: 89a

CD ANNOTATIONS

Left column: Yellow Short styled P.] 'Common Primrose | 1865' *added below, pencil*

Middle column: Yellow Long styled P.] 'Long styled' *underl pencil*; 'Common Primrose' *added below, pencil*

Right column: Red Long Styled Primrose] 'Red' *double underl pencil*; 'Plants which I have crossed this year'[3] *added below, pencil*

Bottom of page] 'for Oxlips'[4] *added red crayon*

Verso: 'Size of Pollen in long-styled Red Primrose' *pencil*; 'Use'[5] *added & circled pencil*

[1] The date is conjectured from the flowering season of the common primrose (*Primula vulgaris*) and from CD's annotations. See also nn. 3 and 5, below.

[2] This sketch of the pollen-grains of *Primula vulgaris* under magnification forms part of a series of diagrams of the pollen of Primulaceae made by William in the spring of 1864 and 1865 (see DAR 108: 83–89a and *Correspondence* vol. 12, letters and memoranda from W. E. Darwin, [15 March 1864], 22 March [1864], [30 April 1864], and [after 19 May 1864]). CD believed that differences in pollen size and in the size of some flower structures were an indication of heterostyly (see *Forms of flowers*, pp. 3, 244). The sketch is reproduced here at 33 per cent of its original size.

[3] In 1865 and 1866, CD made a series of experimental crosses with red (purple) long-styled primroses (*Primula vulgaris* var. *rubra*) raised from seed sent by John Scott in 1863 (see *Correspondence* vol. 11, letter from John Scott, [26 July – 2 August 1863], *Correspondence* vol. 12, letter from John Scott, 16 May [1864], and CD's notes in DAR 108 and DAR 110). See also this volume, letter to B. D. Walsh, 19 December [1865], and letter to J. D. Hooker, 22 December [1865]. CD published his results in 'Illegitimate offspring of dimorphic and trimorphic plants', pp. 420–3, *Variation* 2: 109 n., and *Forms of flowers*, pp. 224–8.

[4] CD may have intended to use the data on primrose pollen in his ongoing investigation into the differences between *Primula vulgaris* and *P. veris*, and their relationship to the common oxlip (now *P. veris* × *vulgaris*) and the Bardfield oxlip (*P. elatior*). He published his results in 'Specific difference in *Primula*', and *Forms of flowers*, pp. 55–75.

[5] In 'Illegitimate offspring of dimorphic and trimorphic plants', p. 421, CD described the pollen of the red (purple) long-styled primrose as 'of the small size proper to the long-styled form' but 'mingled with many minute and shrivelled grains'. See also note in DAR 108: 93, in which CD wrote that William made this observation after the pollen from the red long-styled primrose was sent to him.

From Benjamin Dann Walsh 1 March 1865

Rock Island, Illinois U.S.
March 1. 1865

My dear Mr. Darwin,

Your letter of Dec. 4 enclosing your photograph came duly to hand & by the same mail your second letter enclosing Westwood's photograph.[1] I am under great obligations to you for both. Westwood I never saw, but I have a very distinct recollection of your countenance when you were at Christ's, & if you were to shave clean & put on a wig, I should say you are but very little changed since then.[2] Immediately on the receipt of your letters, I wrote to several of my Eastern Correspondents on Westwood's affair of the Portraits,[3] & as soon as I received answers, I took occasion to write to him & communicate the results. I suppose it will probably be some little time yet before I receive a reply. There is a good portrait of Say[4] in existence, of which he can have a copy taken, if he desires it. Of Harris there is nothing but a photograph extant, which I rather wonder at, considering that the New England naturalists set great store by him & are apt to get wrathy if one picks any holes in his jacket.[5]

I received your very interesting Paper on Trimorphism in Lythrum salicaria &c,[6] & read it several times with renewed attention. I was much struck with one fact, which I should not have anticipated, viz. that in the same genus some species are trimorphic, some dimorphic & some monomorphic.[7] Hence we may infer analogically that in Cynips some species may be dimorphic, & some monomorphic,[8] though in *Apis, Vespa & Bombus* all the known species are dimorphic. I think there is some great mystery concealed in this matter, which time & patient observation & experiment will be certain to elucidate. I wish, if you have one, you would send me a copy of your Paper on *Primula*.[9] I saw long ago an abstract with very full extracts of your *Linum* paper in Silliman's Journal.[10] I quite agree with you in your note (p. 101)[11] on the irrelevancy of von Mohl's arguments.[12] How **can** men be so muddle-headed? Has Leersia oryzoides ever been observed in a state of nature?[13] If not, I do not think that we can argue conclusively from the habits of a tame organism to the habits of a wild organism. At this rate, because birds of prey will not procreate in confinement, we might infer that they never procreate in a state of nature, & consequently that eagles, like donkeys, must be immortal.

I sent you a month or two ago a Paper of mine on "Phytophagic Species", & by this mail I send you another on "Willow-galls".[14] You will begin to think that I spawn a pamphlet bi-monthly. But with me the winter is the time for writing, & in the summer I am out in the woods reading the great Book of Nature. There is one matter in this last pamphlet which I should like your opinion on, i.e. Wagner's procreative larvæ (pp. 571–4 & 641–4).[15] The more I turn the thing over in my mind, the more I don't believe a word of it. And yet I find that Sieboldt has to a certain extent endorsed the theory, by translating from the Danish a Paper on the subject.[16] All these men seem to me to have confounded two very different things, 1st the case of alternate generation where A produces B, & B produces A & so on ad infinitum & 2nd the (supposed) case of Cecidomyia, where, *in the same species*, A (the larva) sometimes produces A (the larva) & then dies, & sometimes produces nothing at all but becomes gradually developed into the imago B, which reproduces A & so on. Is there anything analogous to this in the known metagenetic transformations?[17] It strikes me like the theory of the schoolboy that sometimes the earth travelled round the sun & sometimes the sun travelled round the earth. But out here in the backwoods we know but very little on these great modern discoveries. I wish you would enlighten me.

I am delighted to find that you approve of the way in which I handled Agassiz.[18] I am told there is a notice of that Chapter in the London "Reader" of Dec 31, but have not seen the article.[19] Thank you for getting my Cynipidous theory noticed in the Nat. Hist. Review.[20] There has also appeared a short notice of that Paper in the Stettin "Zeitung", or whatever the German title of their Natural History Journal may be;[21] so it will get tolerably well ventilated, at all events, which is all I wish or expect. *Magna est veritas et prævalebit*,[22] as you by this time see with your great Theory. I told young Agassiz, who argued against your theory because so many naturalists disbelieved it,[23] that the wonder was, not that so many disbelieved but that in six years from the date of its promulgation so many believed; & I asked him how many naturalists believed in Cuvier's great theory six years after that was promulgated?[24]

Since I last wrote, I have read carefully through Agassiz's "Classification", which one of the New England Naturalists told me contained a most unanswerable refutation of Darwinism,[25] though he allowed that the argument in the "Methods of Study" was a complete failure.[26] The book bears neither title-page nor date, & so far from finding any refutation of your theory in it, I actually feel uncertain whether it was written & printed *before* or *after* your book was published. The line of argument is precisely the same as that in the "Methods of Study", & one book is nothing but an abridged re-hash of the other. I was astonished to find that he believes that the same identical species can be & has been created twice over in two separate localities & in two separate geological epochs. Does any other naturalist believe this absurdity? I should have thought that in that case the theory of Chances might have taught him, that we should be as likely to find recent species in the Devonian as in the Pliocene strata, & that we might expect to meet with as many European species in Australia as in North America. He gives me the impression all the time of a dishonest lawyer pettifogging a hard case. Sometimes he won't have it that there are *any* identical species in successive geological epochs— this was what he asserted roundly in a Lecture which he delivered last year in Rock Island,[27] & what he asserts by implication in the "Methods of Study" & sometimes he says that *there are* identical species in two distinct geological epochs, but that there was a separate creation for each batch. This reminds one of the Western lawyer, whose client was sued for a kettle which he had borrowed & returned with a large crack in it, & who put in three pleas:—1st. that his client had never borrowed the kettle, 2nd. that it was already cracked when he borrowed it, & 3rd. that it was perfectly sound when he returned it.

I was also much amused to find how he & I, from exactly the same premises, arrive at very opposite conclusions. Because animals have every mental faculty that Man has, only developed in a less degree, *I* draw the conclusion that neither men nor animals have any souls, & *he* draws the conclusion that both men & animals have got souls, which can & will exist in a future elysium independently of their bodies.[28] In that case, if we calculate up all the animals that have ever existed since the old Paleozoic times, the Agassizian elysium will have to be a pretty large one to hold all their souls. His idea of the soul of a Naturalist studying the souls of his favorite groups of animals—I suppose Agassiz will devote himself to the souls of Turtles & Fish—reminded me of a French parody of Virgil, which represented the soul of a Coachman in the Elysian shades busily cleaning the soul of a Coach with the soul of a brush.[29]

I have no possible chance out in this uncivilized region to get a sight of Bates's Paper on mimetic Lepidoptera;[30] when you see him, I wish you would tell him from me that I should be much obliged by anything from his pen. I have seen a review of his Book on the Amazonian insects in Silliman's Journal a year or two ago,[31] & was much interested in it.

I do not think that species of *Bombus* or of any other genus would copulate differently from their congeners; this would be a violation of what I have called the "Unity of Habits" in the same genus.[32] But I have recently verified in a great many species of *Bombus* the well-known fact, that the ♂ reproductive organs are invariable & differ remarkably in each, just as they do in the Dragon-flies, some being simple curved hooks, & in others the hooks being armed with one or more internal teeth, the thing aimed at being evidently to keep fast hold of the anus of the ♀ by the forceps. My idea is, that a slight variation in the shape of this forceps might be better suited to encite the passions of the ♀ & so be an advantage to the ♂. No ♀ Bombus can ever be **ravished**, nor indeed can any other ♀ insect that I know of. Before copulation can take place, the ♀ must direct her anus upwards, just as a ♀ cat does when she is rutting. But enough of this.

I have recently been working on a large Collection of Insects from the Rocky Mountains & have partially verified a fact which has been asserted by our best N.A. Coleopterist, D^r. J. L. LeConte, viz. that a very large number of genera of Coleoptera, which in the Atlantic States are uniformly winged, contain many wingless species on the Pacific seaboard, or rather in the whole region of country west of the Great American Desert which borders the eastern base of the Rocky Mountains.[33] Taken in connection with Woollaston's facts on the prevalence of apterous types in Coleoptera on the Madeira Islands,[34] & the further fact, which I have quoted on LeConte's authority in my "N. E. Naturalists" Paper, that the Coleopterous Faunas in California &c are exceedingly limited in that geographical extent,[35] does not this seem to indicate that at no very remote period, geologically speaking, this whole region of country was an Archipelago, like the Pacific Islands for example? I am going to get LeConte to enquire of Schaum & LaCordaire[36] & other European Coleopterists, whether Woollaston's facts, which I learnt from your book, are met with in other groups of Islands whose coleopterous fauna has been studied.[37] It is most remarkable how apterous genera, such as *Eleodes* the American representative of *Blaps*, which are scarcely at all represented on this side of the Great Desert, swarm both in individuals & in species when you get on the other side of it. Every body has noticed this. Even families which almost everywhere else are uniformly winged, contain here certain anomalous genera which are not only wingless but have the elytra soldered together. In the long-horned beetles, there is Say's genus *Moneilema*, which looks for all the world like an *Eleodes*, & I believe that LeConte's genus *Amphizoa* is nothing but the American representative of your Hydradephagous genus *Pelobius* with its elytra connate & its wings gone. Lacordaire refers it to *Dytiscidæ*; Schaum to *Carabidæ*; & LeConte makes it the type of a new family.[38] Its habits are at present unknown; but I feel confident that it is aquatic from the details of its structure.

Ever yours very truly | Benj. D. Walsh

P.S. I was glad to see the other day that you have the R.S. medal.[39]

Field Museum of Natural History, Chicago

CD ANNOTATIONS

2.1 I received ... immortal. 2.16] 'Unity of Coloration' *added*[40]

2.8 I wish . . . *Primula.* 2.9] *cross in margin*
2.10 I quite . . . arguments. 2.11] *double scored*
9.3 viz. . . . Mountains. 9.7] *double scored*

[1] See *Correspondence* vol. 12, letters to B. D. Walsh, 4 December [1864] and [4 December 1864?]. Walsh had requested photographs of CD and John Obadiah Westwood (see *Correspondence* vol. 12, letter from B. D. Walsh, 7 November 1864). The photograph of CD, taken by William Erasmus Darwin, is reproduced as the frontispiece to *Correspondence* vol. 12; see also letter from W. E. Darwin, [19 May 1864].

[2] Walsh was a student at Trinity College, Cambridge, when CD was at Christ's College from 1828 to 1831 (*Alum. Cantab.*; see also *Correspondence* vol. 12, letter from B. D. Walsh, 29 April – 19 May 1864 and nn. 1 and 3).

[3] See *Correspondence* vol. 12, letter from J. O. Westwood, 2 December 1864. CD enclosed Westwood's letter, which requested portraits of Walsh, Thomas Say, Thaddeus William Harris, and 'any other brother Entomologist of the New World', in his letter to B. D. Walsh, [4 December 1864?] (*Correspondence* vol. 12).

[4] Thomas Say.

[5] Thaddeus William Harris was the author of an influential report on the insects of Massachusetts (Harris 1841), and had been a leading figure in economic entomology in the United States (*DAB*).

[6] 'Three forms of *Lythrum salicaria*'. Walsh's name appears on CD's presentation list for this paper (see *Correspondence* vol. 12, Appendix III).

[7] See 'Three forms of *Lythrum salicaria*', pp. 189–91 (*Collected papers* 2: 123–5). CD noted that *L. graefferi* was trimorphic, like *L. salicaria*, and that *L. thymifolia* was dimorphic, while *L. hyssopifolia* had only one form.

[8] Walsh published a paper on dimorphism in *Cynips*, a genus of gall-wasps, in March 1864 (Walsh 1864a). See also n. 20, below.

[9] 'Dimorphic condition in *Primula*'.

[10] Extracts from CD's paper 'Two forms in species of *Linum*', with a commentary by Asa Gray, were published in the September 1863 issue of the *American Journal of Science and Arts*, pp. 279–84; see *Correspondence* vol. 11, letter from Asa Gray, 7 July 1863. The journal was known as 'Silliman's Journal' after its founder and first editor, Benjamin Silliman.

[11] Walsh refers to the footnote on pp. 191–2 of 'Three forms of *Lythrum salicaria*' (*Collected papers* 2: 125, 130–1), which begins with a reference to *Origin* 3d ed., p. 101, where CD expressed his belief that in all plant and animal reproduction distinct individuals at least occasionally cross together. See also n. 12, below.

[12] Hugo von Mohl had argued (Mohl 1863, pp. 309, 321) that the existence of small, closed, self-fertile flowers in *Viola*, *Oxalis*, and other genera was evidence against CD's view that no species was perpetually self-fertilised (see *Origin*, pp. 96–101, and *Orchids*, p. 359). CD argued that since these genera also produced ordinary flowers accessible to cross-pollination by insects and capable of producing seeds, Mohl's examples did not invalidate the theory that no species is perpetually self-fertilised (see 'Three forms of *Lythrum salicaria*', pp. 191–2 n. (*Collected papers* 2: 125, 130–1), and *Forms of flowers*, pp. 314, 320–1, 330). There is a discussion of Mohl 1863 in CD's manuscript version of 'Three forms of *Lythrum salicaria*' (DAR 27. 2: A44a–b); see also *Correspondence* vol. 12, letter to Asa Gray, 28 May [1864]. There is an annotated copy of Mohl 1863 in the Darwin Pamphlet Collection–CUL. For CD's observations on the small, closed flowers (later called cleistogamic) of *Viola* and *Oxalis*, see *Correspondence* vols. 10–12.

[13] In 'Three forms of *Lythrum salicaria*', p. 192 n. (*Collected papers* 2: 131), citing Duval-Jouve 1863 on *Leersia oryzoides*, CD stated that the completely enclosed flowers of this grass 'apparently offer the best case of perpetual self-fertilization'. CD later described his observations on the cleistogamic flowers of this species, from plants collected near Reigate, Surrey, and from specimens grown in his greenhouse, in *Variation* 2: 91 and *Forms of flowers*, pp. 333–5. See also *Correspondence* vol. 12, letter to J. D. Hooker,

13 April [1864], and letters from William Bennett, 29 April 1864 and 25 May 1864 and n. 1. CD's notes on this species, dated between July 1864 and July 1866, are in DAR 111: A37–40 and A58.

14 The first part of Walsh's article 'On phytophagic varieties and phytophagic species' (Walsh 1864–5) and his article 'On the insects, coleopterous, hymenopterous, and dipterous, inhabiting the galls of certain species of willow' (Walsh 1864b) were published in the *Proceedings of the Entomological Society of Philadelphia* in November and December 1864, respectively. There are heavily annotated copies of Walsh 1864–5 and 1864b, inscribed by Walsh, in the Darwin Pamphlet Collection–CUL.

15 Walsh 1864b, pp. 571–4 and 641–4, discussed observations made by the Russian entomologist Nikolai Petrovich Wagner of gall midges that propagated in the larval state. Wagner had described generation in the larva of the gall midge, *Miastor metraloas* (family Cecidomyiidae), in a monograph published in Russian (Wagner 1862); a preliminary description, which failed to identify the species, was written in 1861 and published in German in 1863 (Wagner 1863). Walsh was at first sceptical of Wagner's claims to have observed what later became known as paedogenesis; however, see Walsh 1866, p. 288. In CD's copy of Walsh 1864b (Darwin Pamphlet Collection–CUL), the pages discussing Wagner's work are lightly annotated.

16 Walsh refers to the German invertebrate zoologist Karl Theodor Ernst von Siebold's partial translation and discussion (Siebold trans. 1864) of Frederik Vilhelm August Meinert's article on reproduction in insect larvae (Meinert 1864). Meinert confirmed Wagner's observations of reproduction in the larva of *Miastor metraloas*. Siebold had also written an editorial note to Wagner 1863, pp. 513–14, stating that Wagner's observations had been confirmed by others.

17 Walsh refers firstly to alternation of generations and secondly to Wagner's observations of reproduction in *Cecidomyia*, which included larvae being reproduced asexually within the body of a parent larva (see n. 15, above). Wagner's observations of paedogenesis were subsequently confirmed by others (see *EB* 13: 429, and letter to B. D. Walsh, 19 December [1865]). In the nineteenth century the term metagenesis, or alternation of generations, was sometimes also applied to 'certain metamorphoses not now recognized as instances of alternation of generations' (*OED*).

18 See *Correspondence* vol. 12, letter to B. D. Walsh, 4 December [1864]. In Walsh 1864c Walsh had criticised Louis Agassiz's statement that the insects of the temperate zone of North America differed 'specifically throughout' from those of Europe, a view based on his belief in the theory of many contemporaneous local creations (see J. L. R. Agassiz 1850, pp. 187 n., 248). Walsh argued that scientific opinion was against Agassiz in that the majority accepted that identical forms existed in North America and Europe; he also argued that the similarities between the insects of the two continents were best explained in terms of genetic links (Walsh 1864c, pp. 218–23), and criticised Agassiz's attempts to refute CD's theory of the origin of species in his book, *Methods of study in natural history* (J. L. R. Agassiz 1863). There are annotated copies of J. L. R. Agassiz 1850 and J. L. R. Agassiz 1863 in the Darwin Library–CUL and the Darwin Library–Down, respectively (see *Marginalia* 1: 11–13), and an annotated copy of Walsh 1864c in the Darwin Pamphlet Collection–CUL. See also letter from Asa Gray, 17 January 1865 and n. 6.

19 The article, 'Mr. Walsh on Darwinism', appeared in the *Reader*, 31 December 1864, p. 837. The article favourably reviewed Walsh 1864c, quoting from it at length, including Walsh's view that Agassiz had 'totally misapprehended and misstated the Darwinian Theory, and appears never even to have given himself the trouble to read Darwin's book through' (Walsh 1864c, p. 223). See also n. 18, above.

20 CD had given Walsh's paper on dimorphism in *Cynips* (Walsh 1864a) to John Lubbock, who reviewed it in the January 1865 issue of the *Natural History Review*, pp. 138–41 (see *Correspondence* vol. 12, letter from B. D. Walsh, 29 April [– 19 May] 1864, letter to B. D. Walsh, 21 October [1864], and letter from John Lubbock, 3 November 1864). There is an annotated copy of Walsh 1864a, inscribed by Walsh, in the Darwin Pamphlet Collection–CUL.

21 The reference is to 'Ueber den wahrscheinlicher Dimorphismus der Cynipiden-Weibchen', which appeared in the October–December 1864 issue of *Entomologische Zeitung*, the journal of the entomological society of Stettin (Osten-Sacken 1864).

22 *Magna est veritas et praevalebit*: 'truth is great and will prevail' (*Chambers*).

23 Alexander Agassiz had entered into correspondence with Walsh on Darwinism following the publication of Walsh 1864c (see *Correspondence* vol. 12, letter from B. D. Walsh, 7 November 1864); for a discussion of Agassiz's views on CD's theory, see G. R. Agassiz ed. 1913, pp. 47–52, and Winsor 1991, pp. 147–54.

24 Walsh refers to Georges Cuvier's formulation of functional anatomy, which held that every organism represents an integrated system in which all parts mutually correspond and are adjusted to produce a definite function in relation to the organism's environment (Cuvier 1805), a formulation which became popular after its successful application to the reconstruction of extinct vertebrates from fragments of bones and teeth (Cuvier 1812). For a discussion of Cuvier's zoology and its reception, see Coleman 1964 and Appel 1987.

25 Louis Agassiz's 'Essay on classification' formed part 1 of the first volume of *Contributions to the natural history of the United States* (J. L. R. Agassiz 1857–62); there is an annotated copy in the Darwin Library–CUL (see *Marginalia* 1: 9–11). The essay was reprinted as a separate book in 1858 and 1859, before the appearance of *Origin*, and responded to debates generated by Baden Powell's *Essays on the spirit of the inductive philosophy* (Powell 1855). See Winsor 1991, pp. 6–9, 23–7. For CD's assessment of Agassiz's 'Essay on classification', see *Correspondence* vol. 7, letter to Asa Gray, 4 April [1858] and n. 6, and *Correspondence* vol. 9, letter to Cuthbert Collingwood, 14 March [1861] and n. 7.

26 The reference is to J. L. R. Agassiz 1863. See also nn. 18 and 19, above, and *Correspondence* vol. 12, letter from B. D. Walsh, 4 December [1864] and nn. 4 and 5.

27 Rock Island, Illinois. Louis Agassiz made a lecture tour through the American Midwest in 1864 (Lurie 1960, p. 341). See also *Correspondence* vol. 12, letter from B. D. Walsh, 7 November 1864.

28 In his 'Essay on classification' (J. L. R. Agassiz 1857–62, pp. 64–6), Agassiz discussed the possession of souls by animals and concluded: 'may we not look to a spiritual concert of the combined worlds [of animals and humans] and all their inhabitants in presence of their Creator as the highest conception of paradise?'

29 A reference to Charles Perrault's parody of Virgil's *Aeneid* (*Larousse du XXe siècle*).

30 Bates 1861. CD recommended this paper in his letter to Walsh of 4 December [1864] (*Correspondence* vol. 12). CD wrote an unsigned review of Bates's paper in the April 1863 issue of the *Natural History Review*, 'Review of Bates on mimetic butterflies'. See also letter from H. W. Bates, 28 January 1865 and nn. 3–5.

31 Bates 1861 was reviewed by Gray in the September 1863 issue of the *American Journal of Science and Arts* (A. Gray 1863).

32 See CD's query regarding *Bombus* copulation in *Correspondence* vol. 12, letter to B. D. Walsh, 4 December [1864]. Walsh argued that unity of habits was the 'very essence of the thing that we call a Genus, because Habits are correlated with Structure, and Structure makes the Genus' (Walsh 1864b, p. 570; see also *ibid.*, pp. 641–3).

33 A published statement of John Lawrence Le Conte's assertion has not been found; however, see Henshaw 1878 for a list of Le Conte's entomological publications.

34 Walsh refers to Thomas Vernon Wollaston. Wollaston had stated that 200 out of 550 species of beetle inhabiting the island of Madeira were wingless or possessed a wingless habit (see T. V. Wollaston 1856, pp. 82–7). See also *Origin*, pp. 135–6, and n. 37, below.

35 On pages 220–1 of his paper 'On certain entomological speculations of the New England school of naturalists' (Walsh 1864c), Walsh cited Le Conte's conclusions on the geographical distribution of Coleoptera within the United States. Parts of California, Oregon, and Washington were included in Le Conte's western or pacific district, an area characterised as having a small number of geographically localised species of Coleoptera (Le Conte 1859, pp. iii–v). CD scored this passage in his copy of Walsh 1864c in the Darwin Pamphlet Collection–CUL.

36 Hermann Rudolph Schaum and Jean Théodore Lacordaire. Lacordaire was in the process of publishing a multi-volume monograph on Coleoptera (Lacordaire 1854–75).

37 Walsh refers to CD's discussion of T. V. Wollaston 1856 in *Origin*, pp. 135–6. See also n. 34, above. CD had used Wollaston's facts on Coleoptera on the island of Madeira in *Origin* as an illustration

of natural selection. Arguing that individual insects with a strong tendency to use their wings would have a greater chance of being blown out to sea, whereas those with wings reduced in size or less conducive to flight would have a greater chance of survival, CD concluded that a combination of natural selection and disuse would produce wingless species over generations. See also *Natural selection*, pp. 291–3. CD probably discussed his theory with Wollaston in a missing letter that Wollaston received on 27 February 1855; Wollaston, however, interpreted the data in terms of adaptive use and disuse within the limits of what he thought of as an unchanging species type (see *Correspondence* vol. 5, letter from T. V. Wollaston, 2 March [1855] and nn. 11 and 12, and Wollaston 1856, pp. 81–7, 186–94). CD's annotated copy of T. V. Wollaston 1856 is in the Darwin Library–CUL (see *Marginalia* 1: 876–9).

38 Say described the genus *Moneilema* in Say 1823–4, pp. 403–4. Le Conte described the genus *Amphizoa* in Le Conte 1853, pp. 227–8, and used it to represent the distinctive family Amphizoidae (see Le Conte 1861–2, pp. 36–8); Lacordaire placed *Amphizoa* in the family Dytiscidae, noting that the genus had a singular mixture of the characteristics of Dytiscidae and Carabidae (Lacordaire 1854–75: 1, pp. 409–410); Schaum placed it in the family Carabidae (Schaum 1860, p. 53). 'Hydradephagous' refers to aquatic, predatory beetles; *Pelobius* was a European genus of beetles in the family Dytiscidae, with a head resembling the characteristics of the Carabidae, and was described in Schönherr 1806–17, 2: 27–8.

39 The Royal Society's Copley Medal was presented to CD on 30 November 1864 (see *Correspondence* vol. 12 and Appendix IV). The award was announced in the January 1865 issue of the *American Journal of Science and Arts*, p. 114.

40 See letter to B. D. Walsh, 27 March [1865] and n. 5.

From Frederick Ransome 6 March 1865

Rushmere Lodge, | Norwood Lane, S.
March 6th. 1865

My dear Sir—

I note that the time has expired when the note of hand for £100 was to have been redeemed[1]—and after the very great kindness and consideration you have shewn, I do not like again to trespass upon your patience—but owing to the late pressure in the money market our director felt it would not be desireable to make a call upon our Share holders[2]—and in Consequence they have not paid me the Balance due upon the purchase of Patent Rights &c amounting to several thousand pounds.—[3] I am however happy to inform you that matters are progressing most satisfactorily and a further call will shortly be made a portion of which will be paid to me on a/c and I shall then have much pleasure in redeeming the note of hand in your possession.[4]

If however you consider that I have already absorbed all your patience in this matter—I will at once make arrangements to provide the means, but if you will Consent to allow the matter to stand over for a few months it will be a great Convenience to me—more especially as I have recently removed my residence to the above address and have had additional calls upon my finances—

I must however request that you will allow me to pay Interest at the rate of 5% upon the loan—as I am to receive Interest from the Company at this rate—and it is only reasonable that I should repay it upon the portion of my borrowed Capital—

Awaiting the favour of your early reply[5] | I remain My dear Sir | Very faithfully yours | Fred[k] Ransome

Cha[s]. Darwin Esq
Down nr Bromley

DAR 99: 19–20

[1] CD held Ransome's bond for £100, which was due to be repaid on 4 March 1865. See *Correspondence* vol. 12, letters from Frederick Ransome, 4 March 1864 and 7 March 1864. The loan was made in connection with the Patent Siliceous Stone Company, of which Ransome was a founder (see nn. 2 and 3, below).

[2] Presumably the reference is to the shareholders of the newly formed Patent Concrete Stone Company (see n. 4, below); the director of the company has not been identified. CD had purchased shares in the original company, the Patent Siliceous Stone Company, in 1852, and provided several loans (CD's Investment book (Down House MS), pp. 59–60). See also *Correspondence* vol. 11, CD memorandum, 14 February 1863 and nn. 1 and 2, letter to D. T. Ansted, 15 April 1863, and letter from D. T. Ansted, 17 April 1863, and this volume, Supplement, CD Memorandum, July 1857 and n. 1.

[3] In 1848, Ransome invented the artificial sandstone that the Patent Siliceous Stone Company was founded to manufacture (*Modern English biography*; *Minutes of Proceedings of the Institution of Civil Engineers* 115 (1894): 402–4). He was also patentee and general manager of the Patent Concrete Stone Company (*Post Office London directory* 1866).

[4] Ransome was attempting to launch the Patent Concrete Stone Company at new premises (see letter from Frederick Ransome, 9 March 1865 and n. 3, and *Correspondence* vol. 14, letter to Frederick Ransome, [before 6 July 1866], and letter from Frederick Ransome, 6 July 1866). The company did not go into full-scale production until 1867 (*Engineering*, 28 June 1867, pp. 671–2).

[5] CD's reply has not been found; however, an entry in CD's Investment book (Down House MS), p. 60, dated 2 March 1866, records the receipt of £5 from Ransome, representing one year's interest on the loan. See also letter from Frederick Ransome, 9 March 1865.

To W. B. Tegetmeier 6 [March 1865]

Down
6th

My dear Sir

I forgot to ask you to return, if you can lay your hands on it, with other M.S.,[1] a page about crossing Pigeons.—[2]

Yours very sincerely | Ch. Darwin

Endorsement: 'March 6/65'
The New York Botanical Garden Archives (Charles Finney Cox collection) (Tegetmeier 29)

[1] In his letter to Tegetmeier of 27 February [1865], CD asked for the return of a manuscript of a draft chapter on domestic fowls written for inclusion in *Variation* (see *Variation* 1: 225–75). CD had sent the manuscript for Tegetmeier's comments in June 1861 (see *Correspondence* vol. 9, letters to W. B. Tegetmeier, 22 March [1861] and 14 June [1861]).

[2] CD probably refers to the manuscript sent with his letter to Tegetmeier of 27 [December 1862] (*Correspondence* vol. 10). The manuscript has not been found, but it was evidently a list of the crosses between different pigeon varieties carried out by CD and others; CD had also asked Tegetmeier to return it in 1863 (see *Correspondence* vol. 11, letter to W. B. Tegetmeier, 19 February [1863]). The list

is probably that referred to in *Variation* 1: 192 n. For a discussion of CD's pigeon experiments see *Correspondence* vol. 10, letter to W. B. Tegetmeier, 27 [December 1862] and n. 4.

From Frederick Ransome 9 March 1865

Rushmere Lodge, | *Norwood Lane, S.*
March 9/65

My dear Sir—

Pray accept my thanks for so kindly meeting my convenience respecting the payment of the £100 and also for the Cordial expression of congratulation at the encouraging prospects of our Company—[1] You may rely upon it the payment to you shall not be delayed longer than I can help—[2]

I think it would be desireable (if you would not object) to send your bankers a line to say you have heard from me and the matter is arranged between us.—at same time informing them that I have lately changed my residence—

If you should at any time be passing either here or our office (which is

2 Queen St Place
Southwark Bridge[3]

it would give me very much pleasure to shew you some of our productions.

Believe me My dear Sir. | ever faithfully yours | Fred^k. Ransome

Chas. Darwin Esq

DAR 99: 22–3

[1] The letter to Ransome has not been found. The reference is to the Patent Concrete Stone Company (see n. 3, below). See also letter from Frederick Ransome, 6 March 1865.

[2] Ransome had to ask for a further postponement in the repayment of the loan in 1866 (see *Correspondence* vol. 14, letter to Frederick Ransome, [before 6 July 1866], and letter from Frederick Ransome, 6 July 1866). The last entry relating to Ransome in CD's Investment book (Down House MS) is dated 4 March 1866.

[3] These were evidently the new premises of the Patent Concrete Stone Company; the company is not listed at this address until 1866 (*Post Office London directory* 1865–7).

From Rudolph Heine 10 March 1865

Sir!

I have read the paper from You: "on the origin of species, etc." with great pleasure and greater information.[1] But a law, very important by inheritance, i have not found asserted, the law, that to young animals are inherited in the utmost strength those organs, which will be utmost used and exercised by the father and the mother before the conception of the young animal and by the mother when pregnant with it.[2]

By this cause a young hunting dog, for example, is inherited with a finer nose, when the father before his uniting with the mother or when the pregnant mother were frequently used by hunting, and a colt is inherited with stronger muscles, when the father and the mother before their uniting or when the pregnant mother were frequently used by riding or carrying.

This law proves the frequently inheritance even of conditions no endurings for life-time, and it seems to be a cause, quickly to spread conditions, founded upon surprising development; and by this means efficiently to favour the origin of new varieties, species, genera and families.

I hope, You will be pleased to excuse me, for not better knowing the english language.

I am respectfully | Dr. Heine, | physician.

Bitterfeld (near Leipsic)
the 10 of Mars 1865.

To Mr. Charles Darwin. | author of the book: on the origin of species, etc.

DAR 166: 134

1 *Origin.*
2 CD discussed some effects of increased use or disuse in modifying the shape and structure of animal organs in *Origin*, pp. 134–9, and *Variation* 2: 295–303.

From J. D. Hooker [10 March 1865][1]

Kew
Friday.

Dear Darwin

I forgot to tell you that Currey asked me to ask you,[2] what you wished about the execution of the woodcuts,—[3] he says there was something unsatisfactory about the last. Will you write straight to him about them

F. Currey Esq
3 New Square Lincolns Inn
W. C.

he wants to put them in hand at once.

Thomson has gone over Scott's paper.[4] & the enclosed is his conclusion;[5] he is a scientific arithmetician & would I am sure solve the difficulty if soluble; the Denominators are obviously arbitrary & useless.[6]

I have read through Scott's paper, & am amazed at his industry & ability, but in its present state the paper is not fit for publication, & I think the experiments should have been repeated—[7] his discord with Gærtner either shakes the whole value of his system of experimentation, or shows that similar experiments give different results, or that one of the observers is wrong.[8]

You are the only judge of the value of such a paper:— it is awfully tedious to read. Much of it is simply a repetition of the Tables, & that part might be shortened—perhaps.

Ever yrs affec | Jos D Hooker
I return Scotts' paper.
Currey wants to know, who you would wish to execute the wood engravings—
See that proofs are sent to you of them

DAR 102: 13–14

[1] The date is established by the relationship between this letter and Hooker's visit to Down House between 4 and 6 March 1865 (see n. 2, below), and by the relationship between this letter and the letter to J. D. Hooker, 16 [March 1865]; the intervening Friday was 10 March.

[2] Frederick Currey was botanical secretary of the Linnean Society (Gage and Stearn 1988, p. 59). Hooker presumably intended to convey Currey's message when he visited Down House between 4 and 6 March 1865 (Emma Darwin's diary, DAR 242).

[3] The woodcuts were for 'Climbing plants', an abstract of which was read at the Linnean Society on 2 February 1865; Currey was evidently preparing the paper for publication in the society's journal. The paper included thirteen woodcuts. In 'Climbing plants', p. 49 n., CD wrote that figures 5 to 13 were drawn from living plants by his son, George Howard Darwin; in *Climbing plants* 2d ed., p. v, CD wrote that all the illustrations were drawn by George.

[4] The references are to Thomas Thomson and to a draft of Scott 1867. John Scott sent the manuscript and a separate (missing) letter on the same subject to CD from Calcutta on 4 January 1865 (see letters from John Scott, 20 January 1865 and 21 July 1865); CD may have given Hooker the manuscript during his visit to Down House between 4 and 6 March 1865. Thomson was based at the Royal Botanic Gardens, Kew, while collaborating with Hooker on material for a new *Flora Indica* (see L. Huxley ed. 1918, 2: 12). There is an annotated copy of Scott 1867 in the Darwin Pamphlet Collection–CUL.

[5] The enclosure has not been found; CD may have forwarded Thomson's criticisms to Scott (see letter to J. D. Hooker, 16 [March 1865]).

[6] In his published paper, in order to show the 'comparative fertility of the different unions' (Scott 1867, tables 2–7, pp. 153–62), Scott counted the number of capsules and seeds resulting from self-pollination in a species or variety of *Verbascum*, and the number resulting from cross-pollination with other species or varieties. He then calculated the number of seeds that would have been produced from each cross if twenty capsules had been formed and divided the number of seeds produced by cross-pollination between the various taxa by the number of seeds produced by a self-pollinated plant, and multiplied that figure by 1000. This yielded a figure for comparative fertility ranging from 1000 for a self-pollinated *Verbascum* species to 242 for the least fertile cross between two different species of *Verbascum*. In his draft manuscript, Scott may have made an error in the method of this last step in the calculation. In the letter from John Scott, 21 July 1865, Scott apologised for the '**singular blunderings**' in his calculations.

[7] In his letter of 21 July 1865, Scott revealed to CD that his manuscript had been written on the ship to India, from notes on experiments that he had made on *Verbascum* at the Royal Botanic Garden, Edinburgh (see n. 8, below). There is no indication in the published paper (Scott 1867) that further experiments were made in India. Scott sailed for India on 28 August 1864, arriving 21 December 1864 (see *Correspondence* vol. 12, letter to J. H. Balfour, 15 September [1864], and this volume, letter from John Scott, 20 January 1865).

[8] The reference is to Karl Friedrich von Gärtner's experiments with *Verbascum* in which crosses between differently coloured varieties of the same or of different species were found to be less fertile than parallel crosses between similarly coloured varieties (Gärtner 1844 and 1849). CD discussed these experiments in *Origin*, pp. 270–1, and encouraged Scott to attempt these and similar experiments in the letters to John Scott, 19 November [1862] and 11 December [1862] (*Correspondence* vol. 10). CD also sent Scott a digest of Gärtner's experiments (see Scott 1867, p. 164). On the importance of Gärtner's work for CD's theory of natural selection, see *Correspondence* vol. 10, Appendix VI. Scott's

experiments on *Verbascum* were made in 1863 and 1864 (see, for example, *Correspondence* vol. 11, letter from John Scott, 21 September [1863], and *Correspondence* vol. 12, letter from John Scott, 16 May [1864]). Scott's published conclusion from these experiments was that there was no clear relationship between affinities of colour and degrees of fertility in crossing varieties of distinct species of *Verbascum*, but that with varieties of the same species, similarly coloured varieties were more fertile when crossed than differently coloured varieties (see Scott 1867, pp. 165–6, 171–2). CD discussed Scott's results in *Variation* 2: 106–7.

From W. B. Tegetmeier 13 March 1865

Muswell Hill | N
Monday March 13/65

My dear Sir

I have the page respecting the pigeons,[1] and will forward it with the others.—[2] I have concluded my experiments with the mongrels and having carried them on to three generations and not finding any trace of sterility I do not think it worth while to pursue the experiments further[3]

I have got out a few interesting results as to colour and the non reversion to either of the original varieties.[4] I think as the Royal Society made a grant of £10 for the expenses I ought to send in a report or short paper.—[5]

I enclose feathers from the saddle or tail coverts of Spangled Hamburgh hen[6]

The cutting from the Field may perhaps interest you somewhat, being as the old writers used to say "written out of experience" and being very decidedly opposed to the instinctive theory respecting the homing of pigeons.—[7]

I am going over your MSS. but my eye still refuses a long course of exercise on white paper[8]

There is now being published in the Field a series of papers on the different varieties of pigeons, illustrated as the enclosed, each variety being done by a well known amateur of the breed.[9]

Trusting to hear of your continued improvement | Believe me | Very Truly Yours | W B Tegetmeier

C Darwin Esq

DAR 178: 63

[1] See letter to W. B. Tegetmeier, 6 [March 1865] and n. 2.

[2] See letter to W. B. Tegetmeier, 6 [March 1865] and n. 1.

[3] Tegetmeier had undertaken experiments to find out whether particular domestic varieties of fowls and pigeons produced sterile hybrids when crossed *inter se* (see *Correspondence* vol. 10, letter to W. B. Tegetmeier, 27 [December 1862], *Correspondence* vol. 11, letter from W. B. Tegetmeier, 18 February 1863, and letter to W. B. Tegetmeier, 19 February [1863]). Tegetmeier gave details of these experiments in his letter to CD of 29 June – 7 July 1863 (*Correspondence* vol. 11). See also *Correspondence* vol. 12, letter from W. B. Tegetmeier, [before 21 December 1864]. CD reported Tegetmeier's findings with respect to the fertility *inter se* of different domestic varieties of fowl in *Variation* 1: 242–3, and of domestic varieties of pigeon in *Variation* 1: 192.

[4] Presumably a reference to the results of crossing a Spanish cock and a silk hen; the cross yielded offspring resembling in colour the presumed ancestral breed *Gallus bankiva*, rather than either of the parent varieties. CD had requested that Tegetmeier perform the cross to repeat his own experiments on these varieties in 1859 and 1860 (see *Correspondence* vol. 10, letter to W. B. Tegetmeier, 27 [December 1862] and n. 7, *Correspondence* vol. 11, letter to W. B. Tegetmeier, 19 February [1863], and *Variation* 1: 242–3, 2: 40).

[5] On 1 December 1862, the council of the Royal Society of London resolved to grant £10 to Tegetmeier for 'experiments on the cross-breeding of pigeons' (Royal Society, council minutes, 1 December 1862). There is no record of a report on these experiments in the Royal Society archives.

[6] See letter to W. B. Tegetmeier, 27 February [1865], and letter from W. B. Tegetmeier, [28 February – 5 March 1865]. CD described the feathers of the spangled Hamburgh breed of domestic fowl in *Variation* 1: 227.

[7] The enclosure has not been found; however, Tegetmeier evidently enclosed a cutting of his article on the smerle or Antwerp carrier pigeon, published in the *Field*, 25 February 1865, p. 139 (see letter to W. B. Tegetmeier, 14 March [1865]). In the article, Tegetmeier attributed the homing ability of pigeons, 'usually attributed to some mysterious power or instinct', to their ability to navigate from familiar landmarks. The article formed part of a series on pigeon breeds (see n. 9, below).

[8] Tegetmeier had injured his right eye. See letter from W. B. Tegetmeier, [28 February – 5 March 1865].

[9] Tegetmeier refers to a series of articles published in the *Field*, 19 November 1864, p. 366, 3 December 1864, p. 395, 31 December 1864, p. 459, 18 February 1865, p. 115, 25 February 1865, p. 139, and 4 March 1865, p. 155; all the issues, except for that of 25 February 1865, are in DAR 138.3, and all but that of 19 November 1864 are lightly annotated. The last articles in the series were published in the *Field*, 1 April 1865, p. 228, and 15 April 1865, p. 258; CD's copy of the 1 April issue is in DAR 138.3. The series provided descriptions of the standard characteristics of specific breeds of pigeons and fowls, illustrated by wood-engravings; Tegetmeier contributed the articles on the English pouter pigeon and the smerle pigeon in the issues for 31 December 1864 and 25 February 1865, respectively. The enclosed article on the smerle pigeon (see n. 7, above), was illustrated with a wood-engraving by Luke Wells.

To W. B. Tegetmeier 14 March [1865][1]

Down. | Bromley. | Kent. S.E.
Mar 14.

My dear Sir

I am much obliged for your note & the feathers.[2] You must not on any account fatigue your eyes with the M.S.[3] I quite agree that you ought to send a short report to the Royal Soc. I am not surprized at the results.[4]

I do not like to trouble you but I should like to hear the date of the 1st No. of the Field containing the Articles on the Pigeon that I might order them.[5]

I have been much interested by y^{r} article on the Carrier; & the wood cut seems to me quite excellent.[6]

In my book I shall have two Chapters on Pigeons like that on Fowls, & I sh^{d} like to illustrate the former with about 6 woodcuts as large as a page;[7] I had intended applying to Harrison Weir but if Mr Wells knows other fancy pigeons **well**, I think I c^{d} not do better than to employ him.[8]

I was thinking that it w^{d} be desirable to have cuts of the head alone (of almost full size) of the Spanish, Polish, & Hamburg breeds of Fowls.—[9]

In the course of a month or two when your eye sight will be better I hope you will allow me to consult you on this subject.

I thought of H. Weir because I thought I c^d trust him as a known fancier to make me drawings of *highly developed* birds *not caricatured*.[10]

Believe me my dear Sir | yours very sincerely | Charles Darwin

P.S. Would it be worth your while (I am sure that you know that I intend nothing disrespectful to you) to aid me for a fee of two guineas in making terms with Mr Weir or Mr Wells for the desired woodcuts of 6 or 7 pigeons & of the Heads of 3 or 4 Cocks **& seeing to drawings** before they are cut **when on wood**, that they are satisfactory *& seeing proofs of the cuts*.[11] So that I might have your guarantee that they were faithfully executed. That would not give you a great deal of trouble, though it would be considerable.— I have so little strength that I am very anxious to economise work. I sh^d like *good* cuts of Pouter Carrier, Barb— Short-faced Almond Tumbler & according to cost also of Fantail, Jacobin & Turbit.[12]

LS(A)
The New York Botanical Garden Archives (Charles Finney Cox collection) (Tegetmeier 30)

[1] The year is established by the relationship between this letter and the letter from W. B. Tegetmeier, 13 March 1865.

[2] See letter from W. B. Tegetmeier, 13 March 1865.

[3] In his letter of [28 February – 5 March 1865], Tegetmeier mentioned that he had injured his right eye.

[4] See letter from W. B. Tegetmeier, 13 March 1865 and n. 5.

[5] See letter from W. B. Tegetmeier, 13 March 1865, and n. 9. Tegetmeier sent CD a packet containing five issues of the *Field* with his letter of 27 March 1865 (see letter from W. B. Tegetmeier, 27 March 1865 and n. 1); these items are in DAR 138.3 and most are lightly annotated.

[6] The reference is to Tegetmeier's article on the smerle or Antwerp carrier pigeon published in the *Field*, 25 February 1865, p. 139 (see letter from W. B. Tegetmeier, 13 March 1865 and n. 7). The article was illustrated by a wood-engraving by Luke Wells.

[7] CD refers to *Variation*, which was published in two volumes in 1868. Chapters 5 and 6 discussed variation in pigeons (*Variation* 1: 131–224); chapter 5, 'Domestic pigeons', included seven illustrations by Wells (see n. 6, above). For CD's work on pigeons, see J. A. Secord 1981.

[8] Harrison William Weir was a noted pigeon fancier, poultry breeder, and animal painter and engraver (*DNB*). In addition to providing a drawing of *Columba livia* from a dead bird, Wells was commissioned to prepare drawings from living representatives of six domestic breeds of pigeon (*Variation* 1: 135 n. 6). See n. 10, below. See also letter from W. B. Tegetmeier, 27 March 1865, letters to W. B. Tegetmeier 28 March [1865], 6 April [1865], and [7 April 1865], and letter to John Murray, 31 March [1865].

[9] Chapter 7 of *Variation* included illustrations of the heads of Spanish, Hamburgh, and Polish fowl (*Variation* 1: 226 and 228–9). The illustrations were drawn by Wells (see letter to W. B. Tegetmeier, [7 April 1865]).

[10] See n. 8, above. CD had met Weir in the course of procuring pigeon specimens and acknowledged his indebtedness to him in *Variation* (see *Correspondence* vol. 6, letter to W. B. Tegetmeier, 15 March [1856], and *Variation* 1: 132 n. 2). Weir had designed the colour plates for Wingfield and Johnson's *Poultry book* (edited by Tegetmeier; Wingfield and Johnson 1856–7); his illustrations were drawn from life and printed in colour under his supervision (*ibid.*, title page). An annotated copy of Wingfield and Johnson 1856–7 is in the Darwin Library–CUL (see *Marginalia* 1: 798–800). Weir had also designed the illustrations for Wingfield and Johnson 1853.

[11] According to CD's Classed account book (Down House MS), CD made a payment of £3 3*s*. to Tegetmeier for 'wood cuts' on 24 January 1866. See also letter from W. B. Tegetmeier, 27 March 1865 and nn. 3 and 4.

[12] The pigeon breeds illustrated in *Variation* were English pouter, English carrier, English barb, English fantail, African owl, and short-faced English tumbler (*Variation* 1: 137, 140, 145, 147, 149, 152). CD noted that these were drawn with 'great care' by Wells from living birds selected by Tegetmeier (*Variation* 1: 135 n. 6).

To Rudolph Heine 15 March [1865][1]

[Down]

I am pleased to hear that you have been interested in my work on the *Origin of Species*. Your law of inheritance I dare say is true & may have some bearing on the modification of Species; but I fear it would be difficult to prove it with sufficient clearness to allow of its use [. . .][2]

LS incomplete[3]
J. A. Stargardt

[1] According to the catalogue entry from which this text is taken, the letter is dated 15 March. The year is established by the relationship between this letter and the letter from Rudolph Heine, 10 March 1865.

[2] See letter from Rudolph Heine, 10 March 1865.

[3] The original letter is complete and is described in the sale catalogue as being two pages long.

To J. D. Hooker 16 [March 1865]

Down
16th

My dear Hooker,

Thanks for your note, & very many thanks about Scott's paper.[1] Give Thomson my best thanks; you & he have relieved me from much doubt, & I have returned Scott his paper with proper advice & explanations.—[2] I received a few days ago a letter from him, with a flourishing account of his prospects & full of gratitude. He sends messages to you & Thomson, which I enclose instead of copying.—[3]

On getting your note I wrote to Mr. Currey about woodcuts & told him I would pay for them.—[4]

I did so thoroughily enjoy your visit my dear old friend, & it did me no harm.— It was high luck being so well, for I have had no such brilliant days before or since.—[5]

My life is as uniform as that of a mill-horse; my sole work being the grinding & polishing of badly written sentences.—[6] You did not take Heer's pamphlet,[7] & I do not suppose you care to have it sent.—

Farewell | Yours affectly. | C. Darwin

Endorsement: 'March/65'
DAR 115: 264

[1] See letter from J. D. Hooker, [10 March 1865]. CD refers to John Scott and to the draft of Scott 1867.
[2] CD refers to Thomas Thomson. See letter from J. D. Hooker, [10 March 1865], and letter from John Scott, 21 July 1865. CD's letter to Scott has not been found.
[3] The letter from John Scott has not been found.
[4] See letter from J. D. Hooker, [10 March 1865] and nn. 2 and 3. CD's letter to Frederick Currey has not been found.
[5] Hooker had visited Down House between 4 and 6 March 1865 (Emma Darwin's diary (DAR 242)). For CD's health in early 1865, see letter to W. B. Tegetmeier, 27 February [1865] and n. 6.
[6] CD was working on his manuscript of *Variation* (see CD's 'Journal' (Appendix II)).
[7] Heer 1864. See letter to J. D. Hooker, 15 [February 1865].

From Francis Trevelyan Buckland 18 March 1865

37 Albany St | Regents Park
March 18 | 1865

My Dear Sir

Would you allow me to be bold enough to introduce to you my friend Mr Cholmondely Pennell, of the Admiralty with whose name you are probably acquainted.[1] He is anxious to have a few minutes conversation with you on a litterary matter.[2]

Yours most truly | Frank Buckland

C Darwin Esq FRS | &c

DAR 160: 359

[1] Henry Cholmondeley Pennell was a second-class clerk in the Naval department of the Admiralty, Whitehall (*British imperial calendar* 1865). Buckland may have assumed that CD had read contributions Pennell made to the *Athenæum* and the *Field*, both periodicals with which CD was familiar (see *Men and women of the time*, s.v. H. C. Pennell); Pennell had also recently published *The angler-naturalist: a popular history of British fresh-water fish, with a plain explanation of the rudiments of ichthyology* (Pennell 1863).
[2] Pennell had published poetry in addition to works on angling (*Men and women of the time*); in 1865 he was preparing a book on the pike (Pennell 1865).

From George Stewardson Brady 19 March 1865

22 Fawcett St, | Sunderland;
March 19. 1865.

Dear Sir,

You will, I feel sure, forgive my encroaching upon your leisure by asking for a little information on a subject mentioned in your book on the "Origin of Species".—

In a paragraph about the fertilization of clover by the humble-bee you mention that no other insect is so constructed as to be able to get at the nectar, and that consequently, but for its agency, the clover would probably become extinct.[1] A good while ago I quoted this passage in a lecture delivered to the Natural History

Society at Newcastle;[2] and afterwards a friend of mine (Mr Howse)[3] laughed the whole thing to scorn, saying that he had often seen the tubes of the corolla of the Clover *pierced* halfway down by other insects for the sake of getting at the nectar. I should much like to know whether you can give any explanation of this, or whether these punctures, supposing them to exist can be made by any other creatures or for other purposes.—[4] After all it is a matter of no consequence to *your theory*, nor do I see that it even much affects your assertion as to the fertilization of the clover, for I fancy that by the operation of piercing the corolla, any pollen adhering to the proboscis of an insect must be rubbed off before getting even inside the tube, so that fertilization would still be left to be effected by the longer proboscis of the bee.

It is due to you to mention why I should like an answer from you on this question. I have engaged to read a paper to our Geological Society at Sunderland on the "Origin of Species", and I know that my friend, who is a virulent *Anti-Darwinian*, intends to be there; and as I again quote this passage he will certainly attack it.

So that I shall esteem it a great favour if you will kindly give me something to say about it.

Believe me, | very truly yours | George S. Brady.

P.S. | I forward, for your acceptance a copy of a short paper of mine on Marine Entomostraca.[5]

DAR 160: 276

1 CD made these observations in *Origin*, pp. 73–4 and 94–5. Information received from Charles Hardy in 1860 had already led CD to doubt the truth of these statements and prompted him to make further observations on bees (see *Correspondence* vol. 8, letter from Charles Hardy, 23 July 1860, and letter to Charles Hardy, 27 July [1860], and *Correspondence* vol. 10, letters to John Lubbock, 2 September [1862] and [3 September 1862], letter to W. E. Darwin, [2–3 September 1862] and n. 5, and letter to Asa Gray, [3–]4 September [1862]). CD modified his statements in the third and fourth editions of *Origin* (see Peckham ed. 1959, pp. 183–4), but retained his overall conclusion that the humble-bee (*Bombus*) was the main pollinator of red clover and that most species of hive-bee were unable to reach the nectar of this plant in normal circumstances.

2 Brady refers to the Natural History Society of Northumberland, Durham, and Newcastle-upon-Tyne.

3 Richard Howse.

4 CD's reply has not been found. In *Origin* 4th ed. (1866), p. 107, CD noted that he had observed hive-bees sucking at the holes made by humble-bees at the base of the corollas of red clover. CD's notes on this subject are in DAR 49: 149.

5 Brady 1865. This article has not been found in the Darwin Pamphlet Collection–CUL.

From H. W. Bates 22 March 1865

Royal Geographical Society | 15, Whitehall Place, S.W.
March 22 1865

My dear Mr Darwin

It gave me great pleasure this morning to see a letter in your handwriting showing that you had recovered your previous tolerably good state of health.[1] The news will be welcome to all our friends when I tell them.

Pray do not send your copy of the Heliconidæ paper as I have still a few left & will send one to Mr Walsh,[2] to whom I am indebted for copies of his papers lately arrived. I have read every line of his with great pleasure,—just a little deadened perhaps by the diffuseness of his style & want of closeness in his otherwise just reasoning.

I will make use of this opportunity to tell you of a small discovery of mine lately made in working out the species of a very long genus of Longicorn Coleoptera. It is with regard to the abrupt & profound modifications in sexual parts between *very* closely-allied species. You will perhaps recollect my telling you some time ago of a series of dissections made by a friend of mine of the male organs in Chrysomelidæ he having found great differences from species to species & even separated what were previously considered to be varieties, on the ground of modifications in these organs.[3] In my Longicorns the parts in question are the *accessory* organs, that is the terminal abdominal segment from which the organ protrudes when necessary. The genus is Colobothea—one of those genera of which so many exist in all orders, in which the species seem to have been endlessly multiplied by nature. Nearly all the species can be distinguished by the form of the acessory organs in the sexes & I have found most profound modifications in what would be otherwise considered as *local* varieties. I have no doubt whatever that the species on dissemination over a wide area & breaking up into local varieties undergo modifications of these organs very readily & that this fact has operated greatly in the multiplication of species in nature for it is difficult to conceive the variously formed males of these closely-allied races to be equally adapted to their own females & to the females of their sister races. There is a physical obstacle here in the way of amalgamation with the parent or sister forms of segregated local varieties or races. I shall publish my notes on the subject in the Annals & Mag. Nat. Hist. in the course of my papers on Amazons Longicornes.[4]

Yours sincerely | H W Bates

I am expecting "cartes" by every post from my photographer & will send a pair on their arrival[5]

DAR 160: 80

[1] CD's letter has not been found; however, it was evidently written after CD received the letter from B. D. Walsh, 1 March 1865 (see n. 2, below).

[2] Benjamin Dann Walsh had written to CD that he was unlikely to find a copy of Bates's paper on mimetic Lepidoptera (Bates 1861, the 'Heliconidæ paper') where he was living; he requested CD to ask Bates for 'anything from his pen' (see letter from B. D. Walsh, 1 March 1865).

[3] The reference is to dissections made by Joseph Sugar Baly (see *Correspondence* vol. 9, letter from H. W. Bates, 18 March 1861 and n. 7, and *Correspondence* vol. 10, letter to H. W. Bates, 9 May [1862], and letters from H. W. Bates, 19 May 1862 and 14 June 1862). Chrysomelidae are a family of leaf-beetles. CD also discussed the morphology of the sexual organs of insects with Walsh (see *Correspondence* vol. 12, letters to B. D. Walsh, 21 October [1864] and 4 December [1864], and letter from B. D. Walsh, 7 November 1864, and this volume, letter from B. D. Walsh, 1 March 1865).

[4] Bates had begun publishing his study of Amazonian longicorn or long-horned beetles in the May 1861 issue of the *Annals and Magazine of Natural History* (Bates 1861–6); he described species of the genus

Colobothea in the issues for March, May, and August 1865 (*Annals and Magazine of Natural History* 15: 214–25, 382–94; 16: 101–8). For Bates's discussion of the role of markedly different male sexual organs in the development of closely allied species, see *ibid.* 15: 388–9; citing Baly's work on this topic, as well as his own, Bates added:

> The inevitable law of Natural Selection which governs the general process of the adjustment of the local races to new conditions will explain the changes of conditions of life in time; and the laws of variation, diversified in details as are the species themselves, will explain the rest.

[5] A photograph of Bates is reproduced facing p. 121.

From John Lubbock 22 and 26 March 1865

Chiselhirst
22 Mar./65

My dear Mr. Darwin

The rest of the books which you were kind enough to lend me are now on their way to Down.[1] I have kept them an unconscionable time, & it was very ungrateful of me not to have written to thank you. I was however really about to do so.

I shall be at High Elms[2] on Sunday & will try to see you. I should much like to do so, it seems an age since we have met. I am very glad to hear you are better.

My father is I fear far from well.[3]

The book is at length very nearly done.[4] The M.M.S. is all gone to the Printer. How about yours?[5]

Believe me, dear Mr Darwin, | Yours affec | John Lubbock

C Darwin Esq | FRS etc—

High Elms
26 Mar. 65

My dear Mr. Darwin

The preceding note speaks for itself. I stupidly omitted to post it & only found it yesterday. You must have thought me most uncourteous. I came up to Down this morning & hoped to have seen you, but was very sorry to hear from Parslow that you were unwell again[6]

Yours affec | John Lubbock

C Darwin Esq

DAR 170: 50

[1] Lubbock may refer to the first and third volumes of the *Narrative of the surveying voyages of His Majesty's ships Adventure and Beagle* (*Narrative*); volume 1 covered the first expedition, from 1826 to 1830, and volume 3 was CD's *Journal and remarks*, which was also published separately as *Journal of researches*. Lubbock had borrowed these in September 1864 (see *Correspondence* vol. 12, letter from John Lubbock, 2 September 1864 and n. 1).

[2] High Elms, the home of John William Lubbock, John Lubbock's father, was about a mile and a half from Down House.

[3] John William Lubbock died on 20 June 1865, having suffered from 'gout and general debility' since 1860 (*DNB*).

Darwin family at Down House. Photograph, mid 1860s. DAR 219.12: 9.
By permission of the Syndics of Cambridge University Library
Left to right: Leonard, Henrietta, Horace, Emma, Elizabeth, Francis and a school-friend

Charles Victor Naudin
Cabinet card by Bisson Frères, 1863
Courtesy of the Archives of the Gray Herbarium,
Harvard University, Cambridge, Massachusetts, USA

Benjamin Dann Walsh
From the *Transactions of the Illinois State Historical Society for the Year 1920*. By permission of
the Syndics of Cambridge University Library

4 Lubbock 1865, which had been in preparation since the end of 1863, was published in May (*Publishers' Circular* 28: 264, 284). See also *Correspondence* vol. 12, letter to John Lubbock, [1 January 1864] and n. 5, and letter from John Lubbock, 10 January 1864 and n. 4. There is an annotated copy of Lubbock 1865 in the Darwin Library–CUL (see *Marginalia* 1: 512–13).

5 The reference is to *Variation*; CD had been revising the early chapters since mid-November 1864 (see *Correspondence* vol. 12, Appendix II, and this volume, Appendix II). See also letter to Charles Lyell, 22 January [1865] and n. 17.

6 Joseph Parslow was the Darwins' butler. For CD's health in early 1865, see the letter to J. D. Hooker, 16 [March 1865] and n. 5.

To Henry Denny 23 March [1865][1]

Down Bromley | Kent
March 23^d

Dear Sir

I had not heard before in answer to my letter of Jan. 28th & now thank you sincerely for taking the trouble to write to me.[2]

I regret that you have had no opportunity to observe the lice of our various domestic animals from distant lands.— I am much interested by what you tell me about the aperea, for I had come to the conclusion from other reasons that the aperea is **not** the progenitor of our guinea-pigs.—[3]

I am not sure what you mean by the "stock-dove": properly that means the *C. œnas*; our domestic pigeon being *C. livia*, which has an enormous range.—[4] *C. œnas* has a wide range, but how wide I do not know; but as your specimen had not its habitat marked this is unimportant.—

If you thought it worth while to address a letter to the Council of the Zoolog. Soc. stating in some detail what you required, it is probable that proper orders would be issued; but I fear you w^d. have to pay frequent visits to the Gardens to get the orders attended to, but I daresay you would receive some specimens.— Mr Sclater w^d. feel interest in regard to the Birds—[5]

With my best thanks | I remain Dear Sir | Your's faithfully | Ch. Darwin

American Philosophical Society (120)

1 The year is established by the relationship between this letter and the letter to Henry Denny, 28 January [1865].

2 See letter to Henry Denny, 28 January [1865]. The letter from Denny has not been found.

3 In July 1844 CD had sent Denny specimens of lice taken from a *Cavia aperea* or wild guinea-pig. In his *Beagle* 'Catalogue for animals in spirits of wine' (Down House MS), CD had noted that it would be interesting to compare these specimens with the lice of a European domestic guinea-pig to observe whether lice had been altered by transportation or domestication (Notebook 63.1, entry 646; see also R. D. Keynes ed. 2000, p. 340). Denny found the specimens to be a different genus and species from the lice on the domestic variety of guinea-pig, which led him to doubt that the wild guinea-pig could be the progenitor of the domestic guinea-pig (see *Correspondence* vol. 3, letter to Henry Denny, 12 August [1844], and letter from Henry Denny, 30 October 1844). See also letter to Henry Denny, 17 January [1865] and nn. 2 and 5, and letter from Henry Denny, 23 January 1865. CD's conclusion that *C. aperea* was not the parent form of the domestic guinea-pig is discussed briefly in *Variation* 2: 152.

[4] CD discussed the possible relationship and crosses between *Columba oenas* and *C. livia*, or what he also called the rock pigeon, in *Variation* 1: 183, 193 n. See also J. A. Secord 1981, pp. 179–82.
[5] Philip Lutley Sclater was secretary of the Zoological Society of London from 1860 to 1903. He was an acknowledged expert on birds (*DSB*).

From Edward Perceval Wright 24 March 1865

10 Clare Street. | Dublin.
24th March 1865.

Dear Mr. Darwin

Many thanks for your subscription (10/6) to the Cybele Hibernica, which I received this morning— I do not think it will be published until September next—as so many new Collections are turning up however we have already more than 130 Subscribers—so that there will be no lack of funds.[1]

In looking over some Nat Hist notes of a friend of mine who spent many years in India—I found the following, which I thought would interest you— The Common Buffalo about Furreedpore—(Buf. Arnee)[2] is in the habit—during the inundation of the Ganges—of wading into the waters for the purpose of cropping the grass. This it does often in two & three feet of water, but Mr Dunlop[3] was surprised one day to find them swimming in from ten to 12 feet of water & every now & then *diving* down & coming up again with a mouthful of herbage— he then watched them over & over again & found dozens of them engaged in a similar pursuit—& on timing them he found they were always from 30 to 35 seconds under the water. I am not aware that this fact has ever been recorded—but I know the accuracy of my friends observations too well to have any doubt as to its truth— He is the same person who gave me the fresh-water Teredo that I described in the last part of the Linnean Transactions—[4]

I have come across an interesting work on the genera of Alcyonarian Actinozoa—in a form (from the W of Ireland) of Alcyonium which like Haimeia is solitary—& might be easily mistaken for an Anemone— In all its habits it is an Zoantharian polyp in its structure it is alcyonaria save the anomaly of its being not gregarious—[5]

Again thanking you for taking an interest in our Cy-Hibernica
Believe me to remain | very truly yours | E. Perceval Wright

Chas Darwin Esq

DAR 181: 174

CD ANNOTATIONS
1.1 Many . . . funds. 1.4] *crossed pencil*
2.13 Linnean Transactions—] *before closing square bracket pencil*
3.1 I have . . . Cy-Hybernica 4.1] *crossed pencil*
Top of letter: 'My Bear case[6] | Pigeons floating & drinking[7]' *pencil*; 'Gradation & Abnormal Habits.[8]' *red crayon*

[1] The reference is to D. Moore and More 1866; there is an annotated copy of this work in the Darwin Library–Down (see *Marginalia* 1: 598). David Moore, director of the Botanic Garden, Glasnevin, Dublin, and Alexander Goodman More gave a special acknowledgment to Wright, the director of the Natural History Museum, Dublin, for his advice and assistance in planning and carrying out their flora of Ireland. At the time of publication, the number of subscribers to the volume was 157; additionally, the British Association for the Advancement of Science granted the authors £25 towards the cost of publication (D. Moore and More, p. vi and list of subscribers at end of volume). CD's Account book–cash account (Down House MS) records a payment of £10 6*s*. for 'Cybele Hibernica' on 22 March 1865.

[2] Faridpur, situated on the floodplain of the Ganges, was in the former British province of Bengal, and is now a district and a town in Bangladesh. For the identification of the buffalo referred to, see the letter from E. P. Wright, 31 March 1865.

[3] A. A. Dunlop (see E. P. Wright 1864, p. 451); this individual has not been further identified.

[4] Wright had recently published a description of a new freshwater species of ship-worm (Teredo), '*Nausitora Dunlopei*' (E. P. Wright 1864).

[5] Wright refers to a specimen that he thought was a soft coral (an Alcyonaria or Octocorallia); however, he believed the specimen could be easily mistaken for an anenome (a Zooantharia or Hexacorallia). Wright later identified it as a new species of soft coral (Alcyonaria), '*Hartea elegans*' (E. P. Wright 1865).

[6] This and the following annotation refer to cases of diversified habits analogous to the buffalo's diversified feeding habits. The 'Bear case' refers to the description in *Origin*, p. 184, of the North American black bear swimming with its mouth wide open, 'like a whale', to catch insects. This observation, and CD's speculation that natural selection could eventually produce bears with more aquatic habits, was misunderstood, and drew so much ridicule that CD modified it in later English editions; in *Origin* 2d ed., he inserted 'almost' before 'like a whale' and omitted a following sentence on natural selection (see, for example, *Correspondence* vol. 7, letters to Charles Lyell, 25 [November 1859] and [10 December 1859], *Correspondence* vol. 8, letter to Andrew Murray, 28 April [1860], and Peckham ed. 1959, p. 333). CD continued to defend his 'Bear case', and later regretted making the modification to later editions of *Origin* (see *Correspondence* vol. 8, letter to W. H. Harvey, [20–4 September 1860], and letter to R. G. Whiteman, 5 May 1881 (*Calendar* no. 13146)). See also enclosure to letter from Samuel Butler, 1 October 1865 and n. 11.

[7] A reference to flocks of pigeons having been observed drinking from the Nile while they floated downstream, when they were unable to alight on the shore (see letter to Charles Lyell, 25 March [1865], and *Variation* 1: 181).

[8] 'Gradation & Abnormal Habits' may be a reference to a portfolio of notes into which CD placed Wright's letter; no such portfolio has been found. CD may have intended to use information from Wright's letter in later editions of *Origin*; some additions to his section on gradation and abnormal habits in *Origin*, pp. 179–86 ('On the origin and transitions of organic beings with peculiar habits and structure'), were made in the fourth, fifth, and sixth editions (see Peckham ed. 1959, pp. 328–37). CD included information on gradation and changes in habit in the new chapter 7 in *Origin* 6th ed., pp. 168–204 (see also letter to George Busk, 2 September [1871], and letter to J. D. Hooker, 16 September [1871] (*Calendar* nos. 7921 and 7949)). CD did not use the information provided here by Wright in any of his subsequent publications.

To Charles Lyell 25 March [1865][1]

Down. | Bromley. | Kent. S.E.
March 25th

My dear Lyell

I have been glad to see the enclosed correspondence & thank you for sending it & for your note.[2] I presume I may quote Miss Buckley about the roosting in trees (which is only new point) as "from information received through Sir C. Lyell". If you think I ought to name Miss. B; please tell me, otherwise I will quote as above.—[3]

In Upper Egypt where natives live in conical mud hovels, the pigeons regularly settle in flocks on low trees, but not on the Palms.[4] The Duke making such a point on this rests on the Lamarckian belief that everything in structure & habits *must* change:[5] I have put the case that such a change, if not selected or induced by compulsion, would be a downright difficulty on my notions.[6]

Here is a more curious case from compulsion the domestic pigeons settle on the Nile, & float down the stream, whilst they drink, in districts where the banks are absolutely perpendicular; so that they look like a flock of Gulls.—[7]

I have read most of H. Spencer's Biology & agree with you.[8] Some of his remarks are very clever & suggestive, but somehow I seldom feel any wiser after reading him, but often feel mistified. His style is detestable in my opinion; & no wonder as he dictates & never alters. Hooker agrees that his last no^r is best he ever wrote.—[9]

I finished your Elements with uncommon interest; but have nothing to remark:[10] I was, however, particularly struck by your summing up on the Laurentian stages.[11]

My health keeps much the same; but I have of late had fewer black days & generally do my two hours work & am making considerable progress in getting ready for press my "Domesticated animals & Cultivated Plants"[12]

My dear Lyell | Yours ever very sincerely | Ch. Darwin

American Philosophical Society (307)

[1] The year is established by the reference to the sixth edition of Lyell's *Elements of Geology* (C. Lyell 1865), which was published in January 1865 (*Publishers' Circular* 28: 60). See n. 10, below.

[2] The note from Lyell and the enclosures have not been found; however, the letter indicates that the enclosures included a message from Arabella Burton Buckley (see n. 3, below), and a letter from George Douglas Campbell, eighth duke of Argyll (see n. 5, below). The first extant letter in the correspondence between CD and Buckley is the letter to A. B. Buckley, 18 December [1871] (*Calendar* no. 6508).

[3] CD refers to observations of the roosting habits of 'half-bred' carrier pigeons; CD used this information, communicated 'through Sir C. Lyell from Miss Buckley', in *Variation* 1: 181 n. 2.

[4] CD quoted this observation, provided by Robert Scot Skirving, in a discussion of the habits of pigeons (see *Correspondence* vol. 8, 2d letter from R. S. Skirving, [1860?], and *Variation* 1: 181). See also n. 7, below.

[5] The reference is to remarks made on the perching habits of pigeons by Campbell in a letter to Lyell dated 22 March 1865; a portion of the letter was published in I. E. Campbell ed. 1906, 2: 486. CD also refers to Jean Baptiste de Lamarck's hypothesis of change in habits leading to progressive change in the structure of species (Lamarck 1809 and 1815–22). CD's annotated copies of the first volume of Lamarck 1809, and of Lamarck 1815–22, and later editions, are in the Darwin Library–CUL (see *Marginalia* 1: 476–80).

[6] See *Variation* 1: 180–1 and 222–3. CD discussed the Lamarckian theory of progressive development in *Origin* 3d ed., pp. xiii and 134–5, pointing out that his own theory of evolution by natural selection did not include a necessary law of change or progression (see Peckham ed. 1959, pp. 60, 222–3). CD made a further public statement emphasising this difference between Lamarckian and Darwinian evolution in his letter to the *Athenæum*, 18 April [1863] (*Correspondence* vol. 11). CD had corresponded

with Lyell on Lamarckian evolution on several occasions, and in particular following the publication of *Antiquity of man* (C. Lyell 1863a), which CD felt had insufficiently distinguished the two theories (see *Correspondence* vol. 11, letter to Charles Lyell, 12–13 March [1863], and letter from Charles Lyell, 15 March 1863).

7 This instance of behavioural change in pigeons observed by Skirving is cited in *Variation* 1: 181. CD was evidently collecting information on 'Gradation & Abnormal Habits'; see annotations to the letter from E. P. Wright, 24 March 1865 and nn. 6–8.

8 The reference is to Herbert Spencer's *Principles of biology* (Spencer 1864–7), which was issued in instalments as a continuation of *First principles* (Spencer 1860–2). See also *Correspondence* vol. 12, letter from A. R. Wallace, 2 January 1864, and letters to J. D. Hooker, [10 and 12 January 1864] and 3 November [1864]. CD's annotated copy of Spencer 1864–7 is in the Darwin Library–CUL (see *Marginalia* 1: 769–73).

9 CD refers to number 13 of Spencer's *Principles of biology*, which was published in January 1865 (Spencer 1864–7, 2: 1–80). This number contained the first part of Spencer's discussion of morphological development in plants and animals. Joseph Dalton Hooker may have discussed Spencer's work with CD when he visited Down House between 4 and 6 March 1865. For Hooker's opinion of earlier instalments, see *Correspondence* vol. 12, letters from J. D. Hooker, 24 January 1864 and 26[–8] October 1864.

10 C. Lyell 1865. There is an annotated copy of this work in the Darwin Library–Down (see *Marginalia* 1: 524–5). For CD's earlier comments on C. Lyell 1865, see the letter to J. D. Hooker, 15 [February 1865], and the letter to Charles Lyell, 21 February [1865].

11 C. Lyell 1865, pp. 578–80. In his discussion of the Laurentian series of rocks, north of the St Lawrence river, Quebec, Canada, Lyell summarised the researches of the stratigrapher William Edmond Logan. These researches were thought by some to provide evidence of life in the lower Laurentian strata dating from long before the Cambrian period. This would have confirmed CD's suggestion in *Origin*, p. 307, that the presence of phosphatic nodules and bituminous matter in some of the lowest azoic rocks indicated the existence of life before the Cambrian. For CD's interest in this discussion, see *Marginalia* 1: 525. See also *Correspondence* vol. 8, letter to Asa Gray, 25 April [1860] and n. 5, and *Correspondence* vol. 12, letter to A. C. Ramsay, 12 July [1864] and n. 8. CD modified the fourth edition of *Origin* (pp. 371–2) to incorporate Logan's observations of the *Eozoon* fossils in the Laurentian series (see Peckham ed. 1959, pp. 514–15). The findings of Logan and others were, however, controversial and by 1874 CD inferred that the Laurentian did not contain organic remains (see letter to J. D. Hooker, 25 March [1874], *Calendar* no. 9372). CD did not change later editions of *Origin*. See also O'Brien 1970, and Burkhardt 1974, pp. 43–5.

12 The reference is to *Variation*; CD had been revising the early chapters since mid-November 1864, and was currently working on the chapters on domestic animals (see CD's 'Journal' (*Correspondence* vol. 12, Appendix II, and this volume, Appendix II)).

From W. B. Tegetmeier 27 March 1865

The Field. | *346, Strand,* | *London, W.C.*
March 27 *1865*.

My dear Sir

I sent you a packet of **Fields** containing all the Pigeon articles published as I thought it would trouble you less than sending the dates of the papers[1] I shall publish another engraving and article this week and will forward you a copy[2]

Mr L Wells will be most happy to do any thing for you, he is most careful and conscientious in his work and economical in his charges[3]

I need not say that it will give me very great pleasure to assist you in any way, either by furnishing Mr Wells with typical specimens or by supervising his work[4]

For careful accurate detail I think he is the best man by far you could employ. If you compare the pouter and almond tumbler in the **Fields** sent with the Smerle, Barb and Carrier I think you will not hesitate to employ the draughtsman of the last three in preference.[5]

I am an intimate personal friend of M^r. H. Weir and admire his artistic skill but for your purpose as for mine in the Field, I am sure that accuracy is better than high art[6]

Mr Wells charges, for the drawing on the wood, about 30/ per subject, of the size used in the Field The Engraving cost rather more than the drawing.— Would any of the Field blocks suit you I could arrange for you to have stereos or electros at a moderate cost[7]

I am afraid you will think I have been very neglectful in not replying sooner to your last,[8] but I have been very busy during the last two weeks and did not see Mr Wells till late last week

Trusting to hear your health is improving | Believe me | Yours Very Truly | W B Tegetmeier

C Darwin Esq

P.S. | I send you a cutting from the Field with a marked paragraph, that may interest you. I am almost ashamed to send you such an article,[9] but you know that the best artists occasionally paint what are termed, coarsely enough "pot boilers" and therefore I hope to be forgiven for following their example

DAR 178: 65

1 See letter to W. B. Tegetmeier, 14 March [1865]. Tegetmeier sent the issues of the *Field* for 19 November 1864, 3 December 1864, 31 December 1864, 18 February 1865, and 4 March 1865; these issues are in DAR 138.3 and most are lightly annotated. Tegetmeier had sent a cutting of his article on carrier pigeons from the 25 February 1865 issue of the *Field* with his letter of 13 March 1865. The series of articles provided descriptions of the standard characteristics of specific breeds of pigeons and fowls, illustrated by wood-engravings (see letter from W. B. Tegetmeier, 13 March 1865 and n. 9).

2 CD acknowledged receipt of the issue of the *Field* for 1 April 1865 in his letter to Tegetmeier of [7 April 1865]; the issue is in DAR 138.3: 6 and is lightly annotated. Tegetmeier was editor of the poultry section of the *Field* (see letter from W. B. Tegetmeier, [28 February – 5 March 1865]).

3 The reference is to Luke Wells and his probable interest in providing drawings for *Variation*. See letter to W. B. Tegetmeier, 14 March [1865] and nn. 7–9, and 12.

4 See the postscript to the letter to W. B. Tegetmeier, 14 March [1865] and nn. 9, 11, and 12. Tegetmeier is acknowledged for providing Wells with living birds for the illustrations of pigeon breeds in *Variation* 1: 135 n. 6.

5 Tegetmeier refers to a series of illustrations of pigeon breeds published in the *Field* between November 1864 and March 1865 (see n. 1, above). The draughtsman of the illustrations of the almond tumbler and the pouter (*Field*, 19 November 1864, p. 366, and 31 December 1864, p. 459) has not been identified. Wells drew the illustrations of the barb, smerle, and carrier pigeons (*Field*, 18 February 1865, p. 115, 25 February 1865, p. 139, and 4 March 1865, p. 155).

6 Tegetmeier refers to Harrison William Weir. See also letter to W. B. Tegetmeier, 14 March [1865] and nn. 8 and 10. In his acknowledgment of Wells's drawings, CD wrote: 'It may be confidently asserted

that the characters of the six breeds which have been figured are not in the least exaggerated' (*Variation* 1: 135 n. 6).

7 See letter to W. B. Tegetmeier, 14 March [1865] and n. 11. Tegetmeier refers to the engraved wood blocks from which the illustrations were printed for the series on pigeon breeds in the *Field* (see n. 5, above). Stereotypes and electrotypes were forms of metal printing plate prepared from moulds of the wood blocks; stereotypes were the cheaper and more commonly used of the two, particularly for journal illustration, while electrotypes were usually reserved for higher-class work (Twyman 1998, pp. 54–5). For a description of the processes involved in stereotyping and electrotyping, see Glaister 1996, pp. 156–7, 460–1.

8 Letter to W. B. Tegetmeier, 14 March [1865].

9 The article has not been identified.

To B. D. Walsh 27 March [1865][1]

Down. | Bromley. | Kent. S.E.
Mar 27.

My dear Sir

I have been much interested by your letter.[2] I received your former paper on Phytophagic unity, most of which was new to me.[3] I have since received your paper on willow-galls: this has been very opportune as I wanted to learn a little about galls.[4] There was much in this paper which has interested me extremely, on gradations &c and on your "unity of correlation".[5] This latter subject is nearly new to me, though I collected many years ago some such cases with birds;[6] but what struck me most was when a bird-genus inhabits two continents the two sections sometimes display a somewhat different type of colouring. I sh^{d} like to hear whether this does not occur with widely ranging insect-genera?

You may like to hear that Wichura has lately published a book which has quite convinced me that in Europe there is a multitude of spontaneous hybrid willows. W^{d} it not be very interesting to know how the gall-makers behaved with respect to these hybrids?[7] Do you think it likely that the ancestor of Cecidomyia acquired its poison like Gnats. (which suck men) for no especial purpose; at least not for gall-making.?[8] Such notions make me wish that some one w^{d} try the experiments suggested in my former letter.[9] Is it not probable that Guest-flies were aboriginally gall-makers, & bear the same relation to them which Apathus probably does to Bombus?[10] With respect to Dimorphism you may like to hear that Dr Hooker tells me that a dioicous parasitic plant allied to Rafflesia has its 2 sexes parasitic on 2 distinct species of the same genus of plants;[11] so look out for some such case in the 2 forms of Cynips.[12] I have posted to you copies of my papers on Dimorphism.[13] Leersia does behave in a state of nature in the provoking manner described by me.[14] With respect to Wagner's curious discovery my opinion is worth nothing:[15] no doubt it is a great anomaly, but it does not appear to me nearly so incredible as to you: remember how allied forms in the Hydrozoa differ in their so-called alternate generations: I follow those naturalists who look at all such cases as forms of Gemmation;[16] & a multitude of organisms have this power or traces of this power, at all ages from the germ to maturity. With respect to Agassiz's views there were many, & there are still not a few, who believe that the same species is created on many spots.[17] I wrote to Bates & he will send you his mimetic paper, & I dare say others: he is a first rate man.[18]

Your case of the wingless insects near the Rocky mountains is extremely curious:[19] I am sure I have heard of some such case in the old world, I think on the Caucasus. W^d not my argument about wingless insular insects perhaps apply to truly Alpine insects; for w^d it not be destruction to them to be blown from their proper home?[20] I sh^d like to write on many points at greater length to you, but I have no strength to spare—

With every good wish believe me yours very sincerely | Charles Darwin

If you publish on wingless insects, kindly inform me or send, if you can, a copy.—

LS(A)
Field Musuem of Natural History, Chicago (Walsh 3)

1 The year is established by the relationship between this letter and the letter from B. D. Walsh, 1 March 1865.

2 Letter from B. D. Walsh, 1 March 1865.

3 The reference is to the first part of Walsh 1864–5, in which Walsh discussed the influence of insects' food-plants on insect variation and speciation, arguing that feeding on distinct food-plants could, over time, produce distinct species of insects, or what he called phytophagic species. CD's indexed and heavily annotated copy of the first part of Walsh 1864–5, inscribed by Walsh, is in the Darwin Pamphlet Collection–CUL. Walsh coined the term 'phytophagic unity' in Walsh 1864b, p. 635, to convey his observation that when the larvae of one species of gall-making insect feed on a given genus of plants, many more species of the same genus will also be found on that genus of plants. CD discussed Walsh's findings in both papers in *Origin* 4th ed., pp. 55–6 and 187. See also letter from B. D. Walsh, 1 March 1865 and n. 14.

4 Walsh 1864b; there is a heavily annotated copy of this paper in the Darwin Pamphlet Collection–CUL. CD cited Walsh 1864b in his discussion of galls in *Variation* 2: 282–5. CD used galls, swollen tissues that are produced on plants by insects, as an example of how external agencies may affect structure.

5 A handwritten index appended to CD's copy of the first part of Walsh 1864–5, in addition to CD's annotations on page 428, indicate that CD was particularly interested in Walsh's remarks on gradation in structure and colour from phytophagic varieties to phytophagic species (Walsh 1864–5, pp. 427–9; see *Origin* 4th ed., pp. 55–6). See also CD's handwritten index appended to his copy of Walsh 1864b, and Walsh 1864b, pp. 567, 642–3. CD also refers to Walsh's law of 'Unity of Coloration' (Walsh 1864b, pp. 635–6), according to which identical shades of colour and patterns of design in a group of species indicated a genetic connection. Emma Darwin, who wrote most of this letter at CD's dictation, evidently wrote 'correlation' in error, a mistake that was corrected to 'coloration' by Walsh.

6 CD had been considering variations of colour in birds for some time (see, for example, *Notebooks* and *Ornithological notes*); however, the particular cases he is referring to have not been identified. See also *Origin*, pp. 25, 132, 133, and *Variation* 1: 195–7. CD discussed gradations in colour, pattern, and structure in bird plumage as an indication of descent from pre-existing species in *Descent* 2: 135–53; CD's notes for this discussion, mostly dated between 1866 and 1868, part of a portfolio on sexual selection, are in DAR 84.1 and 84.2.

7 CD refers to Max Ernst Wichura and to Wichura 1865; see letter to M. E. Wichura, 3 February [1865]. There is an annotated copy of Wichura 1865 in the Darwin Library–CUL (see *Marginalia* 1: 871–3). CD's curiosity regarding gall-making insects and hybrid willows was also stimulated by Walsh's discussion of which gall-making insect species inhabited particular species of willows (see Walsh 1864b, pp. 544–6, 638–41). In Walsh 1866, which was the second part of Walsh 1864b, Walsh

investigated CD's query and made further observations on various gall-making insects inhabiting different species of willow. CD cited Walsh 1864b and 1866 in his discussion of galls (see *Variation* 2: 282–3). An annotated copy of Walsh 1866 is in the Darwin Pamphlet Collection–CUL.

8 *Cecidomyia* is a genus of gall-forming midges. Walsh discussed them in Walsh 1864b (see also letter from B. D. Walsh, 1 March 1865).

9 CD refers to his suggestion that Walsh experiment with inserting insect and other poisons into plant tissues (see *Correspondence* vol. 12, letter to B. D. Walsh, 21 October [1864] and n. 6). See *Variation* 2: 283–4 for CD's conclusions regarding the poisons and plant tissues.

10 Walsh discussed the guest-flies, or inquilines, whose larvae inhabited the galls of other insects in Walsh 1864b, pp. 619–34; he noted that guest gall-gnats usually lived in the galls of species to which they were closely allied (*ibid.*, p. 634). Walsh also discussed several different species of guest-flies in Walsh 1866, arguing that the inquilinous saw-flies, for example, were 'primordially identical' with the gall-making saw-fly (Walsh 1866, p. 279; see also pp. 237–8). *Apathus* (now known as *Psithyrus*) is a genus of humble-bee that closely resembles the humble-bee genus *Bombus* and lays its eggs in the latter's nests. Walsh had discussed *Apathus* and *Bombus* in Walsh 1864c, pp. 246–7; there is an annotated copy in the Darwin Pamphlet Collection–CUL. CD described the parasitic habits of *Apathus* in *Natural Selection*, p. 509.

11 Joseph Dalton Hooker may have given CD this information during his visit to Down House between 4 and 6 March 1865. See also letter to J. D. Hooker, 4 May [1865].

12 Walsh had published a paper on dimorphism in *Cynips*, a genus of gall-wasps, in March 1864 (Walsh 1864a). See also letter from B. D. Walsh, 1 March 1865 and n. 20.

13 CD refers to 'Dimorphic condition in *Primula*' and 'Two forms in species of *Linum*'; Walsh had requested that CD send the former after receiving a presentation copy of 'Three forms of *Lythrum salicaria*' (see letter from B. D. Walsh, 1 March 1865).

14 CD refers to his discussion of *Leersia oryzoides* in 'Three forms of *Lythrum salicaria*' (see letter from B. D. Walsh, 1 March 1865 and n. 13).

15 CD refers to Nikolai Petrovich Wagner's observations of paedogenesis in the larvae of *Miastor metraloas* (Wagner 1862 and 1863). See letter from B. D. Walsh, 1 March 1865 and nn. 15 and 17.

16 CD added this statement to *Origin* 4th ed., p. 518, noting Wagner's discovery (see n. 15, above). Gemmation, or budding, is a form of asexual reproduction whereby a new individual is formed by the protrusion and complete or partial separation of a part of the parent (*OED*); the nineteenth-century meaning of alternation of generations may have included gemmation (see letter from B. D. Walsh, 1 March 1865 and n. 17).

17 CD alludes to Walsh's discussion of Louis Agassiz's views expressed in J. L. R. Agassiz 1857–62 and 1863 (see letter from B. D. Walsh, 1 March 1865 and nn. 25–7).

18 CD refers to Henry Walter Bates and to Bates 1861.

19 See letter from B. D. Walsh, 1 March 1865 and nn. 33–5.

20 CD refers to his discussion of the development of wingless insects on the island of Madeira in *Origin*, pp. 135–6. See letter from B. D. Walsh, 1 March 1865 and nn. 34 and 37.

From Charles Cardale Babington 28 March 1865

Cambridge
28. March. 1865

Dear Darwin

I think that you will be glad to learn that the University has at last provided a proper room for a small Zoological Museum[1] and that the Philosophical Society contemplates offering its museum to the University to form part of that to be formed there.[2] As the Philos! Museum contains a large collection of Fish given by you I think it well to inform you of this.[3] The fish will be better taken care of and more useful there than where they now are.

Yours truly | Charles C. Babington—

I have got a good place for the Univ^{y} Herb. and am beginning to get it put to rights. It seems to be very rich and valuable.[4]

DAR 160: 9

1 The reference is to the new Cambridge University Museum of Zoology, situated on the New Museums Site, behind Pembroke Street, Cambridge, on the site of the old Botanic Garden; building work began in June 1863, and the museum was completed in February 1865 (Willis and Clark 1886, 3: 180).

2 In 1865 the Cambridge Philosophical Society donated its natural history collections to the university, and transferred its library from All Saints' Passage to new accommodation adjacent to the new Department of Zoology and Comparative Anatomy (A. R. Hall 1969, pp. 26–8).

3 The reference is to CD's collection of fish from the *Beagle* voyage. CD's classified annotated catalogue of 'Fish in Spirits of Wine', prepared in 1836, is in DAR 29.1: B1–20 (see also *Correspondence* vol. 1, Appendix II, and R. D. Keynes ed. 2000). A description of the collection, prepared by Leonard Jenyns, was published in *Zoology* between 1840 and 1842. See also D. M. Porter 1985, pp. 1005–6. The fish specimens are in the Cambridge University Museum of Zoology.

4 Babington had become professor of botany at Cambridge University in 1861, following the death of John Stevens Henslow. The arrangement of the herbarium collections had been postponed owing to a lack of sufficient space; much of the collection was stored in boxes and was inaccessible (Willis and Clark 1886, 3: 161, Babington ed. 1897, p. 354). According to his journal, Babington began moving the botanical collections from the old room by the Anatomical Museum to new rooms on the New Museums Site on 19 January 1865 (Babington ed. 1897, p. 203). The University Herbarium housed CD's *Beagle* plant collections (D. M. Porter 1985, pp. 1014–18, and D. M. Porter 1986).

To W. B. Tegetmeier 28 March [1865][1]

Down. | Bromley. | Kent. S.E.
March 28th

My dear Sir

I am heartily obliged for your kind proffer of assistance about the woodcuts, & for your note (*never* hurry yourself to answer me) & for your present of the Fields.[2] And lastly for your article on crossing;[3] which seems to me excellent: & if I discuss such subjects will certainly be very useful for me to quote.—

I quite agree with what you say about the woodcuts—that of the Pouter struck me as very poor; I did not so much dislike the Tumbler but my impression seems a poor one & the block apparently injured in the beak— the others are excellent, your Smerle, I think the best,—the Barb rather hard about the beak.—[4] I will certainly get Mr Wells to do what few drawings I may require.[5] I was thinking once of having a long folding Plate with cuts of 6 or 7 pigeons in a row; but I must first consult Mr Murray what he thinks.—[6] If I settle on folding plate; (otherwise they are a little too big) several of the Block would do excellently from the Field; but would the Editor approve?[7] & can you find out what I sh^{d} have to pay, as I must consider Mr Murray's & indeed my own interest?— If some time soonish you could let me hear what cost w^{d} be, I could decide.

Yours sincerely obliged | Ch. Darwin

Please do not forget to register M.S. of Fowls whenever you have completed the reading.—[8]

P.S. I have been measuring the cuts in the Field & they are too large to be introduced on a folded Plate, & a little too large for my pages. So take no trouble about the Field cuts.— After communicating with Mr Murray, I will write through you to Mr Wells about making fresh drawings.—

The New York Botanical Garden Archives (Charles Finney Cox collection) (Tegetmeier 31)

[1] The year is established by the relationship between this letter and the letter from W. B. Tegetmeier, 27 March 1865.

[2] See letter from W. B. Tegetmeier, 27 March 1865 and nn. 1, 4, and 7

[3] The article has not been identified.

[4] See letter from W. B. Tegetmeier, 27 March 1865 and n. 5.

[5] CD refers to Luke Wells (see letter from W. B. Tegetmeier, 27 March 1865).

[6] John Murray was CD's publisher. See letter to John Murray, 31 March [1865].

[7] Tegetmeier had offered to lend CD the engraved wood blocks from which the pigeon illustrations recently published in the *Field* had been taken (see letter from W. B. Tegetmeier, 27 March 1865 and n. 7). The editor of the *Field* was John Henry Walsh (Rose 1953).

[8] CD refers to his draft chapter on domestic fowls, written for inclusion in *Variation*, which Tegetmeier was reading and correcting (see letter to W. B. Tegetmeier, 27 February [1865]).

From H. W. Bates 29 March 1865

Royal Geographical Society | 15, Whitehall Place, S.W.
March 29 1865

My Dear Mr Darwin

Enclosed is the promised portrait.[1] Will you do me the favour to send me yours for my album;[2] the larger one you gave me in 1862 now hangs up in my study.[3]

I sent a copy of "Mimetic Butt^s^ to Mr Walsh.[4] We are overwhelmed here with North Pole business.[5]

Yours sincerely | H W Bates

DAR 160: 81

[1] See letter from H. W. Bates, 22 March 1865. A photograph of Bates is reproduced facing page 121.

[2] No reply to this request has been found; however, CD may have sent the photograph taken by William Erasmus Darwin in June 1864, which is reproduced as the frontispiece to *Correspondence* vol. 12 (see also *ibid.*, letter from J. D. Hooker, [11 June 1864]).

[3] CD had probably sent Bates a copy of the photograph taken by William Erasmus Darwin in 1861, which is reproduced as the frontispiece to *Correspondence* vol. 9 (see also *ibid.*, letter to Asa Gray, 11 April [1861]).

[4] The references are to Bates 1861 and Benjamin Dann Walsh. See letter from H. W. Bates, 22 March 1865 and n. 2.

[5] Bates refers to the efforts made by the council of the Royal Geographical Society to obtain government sponsorship for an expedition to the North Pole (see *Journal of the Royal Geographical Society* 33 (1865): clxxix–clxxxii); Bates was assistant secretary of the society.

From W. B. Tegetmeier [29 March – 7 April 1865][1]

⟨*one page missing*⟩

But I daresay you have noticed this yourself for nothing seems to escape you—

Pray do not hesitate to command me in any and every way that I can be of the slightest aid to you

My eye is getting on very well On saturday I wrote fifty of these small slips of MSS. and have not felt any inconvenience[2]

I mention this that you may have no scruples respecting me

Believe Me | My dear Sir Yours Very Truly | W B Tegetmeier

C Darwin Esq

[Enclosure]

*Note to. fol 1[3]

I think it is not worth while to mention the works of mine quoted. My edition of the Poultry Book was never completed owing to the failure of the publisher[4] and the Profitable Poultry has been long out of print.[5] I would rather you would say from "Information obtained from me" if you think fit.[6]

Incomplete
DAR 178: 62, 66

[1] The date range is conjectured from the relationship between this letter and the letters to W. B. Tegetmeier, 28 March [1865] and [7 April 1865]. See n. 2, below.

[2] Tegetmeier refers to his notes on CD's manuscript chapter on fowls, which he had been reading and correcting (see letters to W. B. Tegetmeier, 27 February [1865] and 28 March [1865], and letter from W. B. Tegetmeier, 13 March 1865). CD congratulated Tegetmeier on the improvement in his eye in his letter of [7 April 1865]. The enclosure to this letter may be one of the slips referred to.

[3] Tegetmeier refers to a note on the first folio of CD's manuscript chapter on fowls. The note contained CD's acknowledgment of Tegetmeier's assistance (see *Variation* 1: 225, n. 1).

[4] Tegetmeier refers to Wingfield and Johnson 1856–7, which he edited; there is an annotated copy in the Darwin Library–CUL (see *Marginalia* 1: 798–800). Eleven parts were published before the London publisher, William S. Orr & Co., failed.

[5] Tegetmeier 1854. An annotated copy is in the Darwin Library–CUL (see *Marginalia* 1: 803).

[6] CD's acknowledgment on the first page of the published version of the chapter on fowls was evidently based on the advice contained in this letter; Tegetmeier was acknowledged as the chief source of CD's information and thanked for reviewing the chapter (see *Variation* 1: 225, n. 1).

To John Murray 31 March [1865][1]

Down. | Bromley. | Kent. S.E.
March 31st

My dear Sir

After much delay from illness, I have at last made considerable progress in a Book with some such Title as enclosed, & which I hope will go to press this early autumn.[2] Hence it is time to think of the woodcuts. I do not want you to pledge yourself *even indirectly* to publish, until you see M.S.; but will you aid me about the cuts with advice, I being responsible for all outlay.—

1st. I want 29 cuts of bones (I am sorry so many are unattractive), chiefly skulls & 3 & 4 little bones on same cut.[3] Now I must have artist here to explain what is to be done; & I shd. much prefer Mr. G. B. Sowerby, as I am accustomed to him.[4] The specimens used have to be drawn right on the wood. But what can be done about cost. How can any agreement be made before drawings are made?—

2d. I want 7 cuts of size of Page of Origin, *well executed* of chief races of Pigeons (& *perhaps* 3 Heads of Cocks).[5] Now Mr. Wells has lately drawn for the Field *some* of the pigeons **excellently**:[6] he charges 30s for each drawing (as I hear from *sub-editor* of Field) & the cutting costs a little more for each.[7] But my cuts would be a *little* smaller. Should I ask him to send an estimate for drawings & cutting to you?? or trust to my friend the Sub-Editor.

3d. I have 5 drawings to be exactly copied,[8] so that this will be simple job—

4th I shall want 4 or 5 or 6 cuts of plants, & seeds,[9] which Mr Fitch of Kew[10] could alone do well: & I shd. think he might be trusted not to overcharge.—

Altogether I shd. wish for very nearly 50 cuts; but if you think this too much for Book (I **guess** about size of Origin) please advise me; for I could reduce the number to about 35.—[11]

Lastly kindly inform me whether you think size of Origin good size.—

Please remember that I do not at all wish you to pledge yourself to publish.—

My dear Sir | Yours very sincerely | Ch. Darwin

How about Orchis Book?[12] I suppose Origin has ceased selling. I shd be very sorry to have now labour of another Edition, but I suppose that will never be wanted.[13] I heard yesterday a new French Edit. is wanted[14] & this will give me some trouble.—

Of present book I have 7 chapters ready for press & all others very forward, except the last & concluding one.[15] Sir C. Lyell read the 1st. Chapt. in M.S. on Dogs, 3 or 4 years ago, & thought favourably, & as it appeared to me certainly too favourably of it.—[16]

P.S. In the "Dovecote & Aviary by Rev. E. S. Dixon" published by you, there are 2 cuts of wild & Dovecote pigeon, which are well done, & though not necessary, would be very useful for my book, would you have kindness to let me use them, if in your possession?—[17]

John Murray Archive (Darwin 131–5)

[1] The year is established by the relationship between this letter and the letter from John Murray, 1 April 1865.

[2] The enclosure has not been found; Murray advertised CD's forthcoming book in the *Reader*, 15 April 1865, p. 427. The notice is under the heading 'Mr. Murray's quarterly list of works in the press', and the title given is, 'Domesticated Animals and Cultivated Plants, or the Principles of Variation, Inheritance, Re-union, Crossing, Interbreeding, and Selection under Domestication'. See also letter

to John Murray, 2 June [1865] and n. 3. *Variation* was not published until 1868. See also letter to Charles Lyell, 22 January [1865] and nn. 16 and 17.
3 *Variation* was published with twenty-five figures of bones, including skulls, of rabbits, pigeons, fowls, and ducks.
4 George Brettingham Sowerby Jr drew illustrations for *South America* (1846), *Fossil Cirripedia* (1854), *Living Cirripedia* (1851, 1854), and *Orchids* (1862). In preparing many of these illustrations, Sowerby had attended CD at Down House (see, for example, *Correspondence* vol. 5, letter to W. D. Fox, 24 [October 1852] and n. 1, and letter to John Lubbock, 10 [September 1853] and n. 3, and *Correspondence* vol. 9, letter to John Murray, 21 September [1861] and n. 4).
5 See letter to W. B. Tegetmeier, 14 March [1865].
6 CD refers to the drawings by Luke Wells. See letters to W. B. Tegetmeier, 14 March [1865] and n. 6, and 28 March [1865], and letter from W. B. Tegetmeier, 27 March 1865 and n. 5.
7 See letter from W. B. Tegetmeier, 27 March 1865. William Bernhard Tegetmeier edited the poultry section and some other sections of the *Field* (see letter from W. B. Tegetmeier, [28 February – 5 March 1865]).
8 CD presumably refers to four illustrations of pigs (*Variation* 1: 69, 72, 75), and a fifth illustration of a rabbit (*Variation* 1: 108).
9 *Variation* has three botanical figures, illustrating pea pods, peach and almond stones, and plum stones (*Variation* 1: 328, 337, 345).
10 Walter Hood Fitch had become a freelance botanical artist in 1860; he still often worked for William Jackson Hooker and Joseph Dalton Hooker illustrating *Curtis's botanical magazine* (R. Desmond 1995). Fitch had prepared the illustrations for CD's paper 'Dimorphic condition in *Primula*' (see this volume, Supplement, letter to Daniel Oliver, 23 March [1861]).
11 *Variation* was published with forty-three illustrations.
12 Of the original printing of 1500 copies of *Orchids*, 580 remained unsold (see letter from John Murray, 1 April 1865 and n. 7). See also *Correspondence* vol. 10, letter to John Murray, 9 April [1862], and Freeman 1977, p. 112.
13 CD refers to the third edition of *Origin*, which was published in 1861 with a print-run of 2000. A fourth edition of *Origin*, with a print-run of 1500 copies, was published in 1866 (see Freeman 1977, pp. 78–9).
14 A French edition of *Origin*, translated by Clémence Auguste Royer, was published in 1862 (Royer trans. 1862; see *Correspondence* vol. 10). A second French edition, with notes added by CD, was published in 1866 (see Freeman 1977, p. 102). See memorandum from C. A. Royer, [April–June 1865].
15 When *Variation* was published in 1868, it contained twenty-eight chapters.
16 See *Correspondence* vol. 8, letter from Charles Lyell, 25 September 1860, and this volume, letter to Charles Lyell, 22 January [1865] and n. 18. CD refers to his draft of chapter 1 of *Variation*, 'Domestic dogs and cats' (*Variation* 1: 15–48).
17 The reference is to the illustrations of *Columba livia* and *Columba affinis* in Dixon 1851, pp. 150 and 163. An annotated copy of Dixon 1851 is in the Darwin Library–CUL (see *Marginalia* 1: 199–201). Edmund Saul Dixon's illustrations were not used in *Variation*.

From E. P. Wright 31 March 1865

⟨*10*⟩ *Clare Street.* | *Dublin.*
31st March 1865.

Dear Mr. Darwin.

I took the first opportunity of seeing my friend to find out if I understood him rightly when he told me it was the Arnee that he had found possessed of powers of diving—[1] He tells me that there can be no doubt of its being the animal called Arnee in India, & on shewing me the pairs of Horns—of a wild male & female shot by himself, I have no doubt it is the Bubalus Arna of Horsfield Catalogue of Mammalia in the Museum of the H. E. I. Co.[2] page 179—and the Bos Arnee of Shaw— between the wild Arnee & the domesticated form—there is hardly any perceptible difference—[3] The Wild Bull constantly visits the tame cows, inde⟨ed⟩ often makes himself very troublesome by so doing & it was on the entreaties of some of my friends tenants, to shoot a wild Bull who was in the habit of breaking down fences, that the one of which my friend possesses the Horns was shot—

Another little bit of information is—the *tongue* of these creatures is a favorite morsel with the Europeans—but during the inundations they are not eaten—& on M^r^. D's tasting one, he found the reason to be, the extremely strong prawn or shrimp flavor of the meat. This is owing to the quantity of Gammari eaten with the grass that the animals dive for, & of which numbers are to be found in the paunch, with the unruminated grass. I hope you will never think I can fancy that finding out any fact, for you, could be a trouble, & that you will think the explanation satisfactory—

The Horns in my museum of B. arnee were named by Hodgson[4] & it was with these I compared those of Dunlop.

Very truly Yours | E. Perceval Wright

DAR 181: 175

CD ANNOTATIONS

1.1 I took . . . shot— 1.11] *crossed pencil*

2.1 Another] *after opening square bracket pencil*

2.6 I hope . . . Dunlop. 3.2] *crossed pencil*

Top of letter: 'It is Bos Arnee which dives & swallows many freshwater Shrimps or crustaceans See End—' *pencil*; 'Odd Habits leading to *Transition*'[5] *red crayon*

[1] The friend was A. A. Dunlop. See letter from E. P. Wright, 24 March 1865 and n. 3. CD's letter to Wright asking for further information on Dunlop's observations has not been found.

[2] The reference is to Thomas Horsfield's *Catalogue of the mammalia in the Museum of the Hon. East-India Company* (Horsfield 1851). This zoological museum comprised specimens contributed by public servants attached as naturalists to missions and deputations on behalf of the Indian government, or presented by members of the civil and military services in India (*ibid.*, p. iii). The collections had passed under the control of the secretary of state for India in 1858, and were housed in the India Museum, Fife House, off Whitehall, London (see Moir 1988, pp. 34, 95).

[3] Horsfield 1851, p. 179, listed *Bubalus arna* as described in Hodgson 1847, pp. 709–11; Brian Houghton Hodgson had not yet decided if the domesticated 'Buffaloe' was a separate species from *Bubalus arna.* Horsfield included '*Bos Arnee*', described by George Shaw (Shaw 1800–26, 2 (pt. 2): 400–1), as a synonym for *Bubalus arna.* Now named *Bubalus bubalis*, the Asian water buffalo is usually thought to include both wild and domesticated animals; however, some still treat the domesticated *Bubalus bubalis* as a separate species from what they call the wild *Bubalus arnee* (see Nowak 1999, pp. 1147–51).

[4] Wright had been the director of the Zoological Museum of the University of Dublin since 1857 (*DNB*). Hodgson presented a specimen skull and horns to the Museum of the East India Company (Horsfield 1851, p. 179; see also n. 3, above).

[5] CD was interested in cases of diversified habits that might lead, by the gradual process of natural selection, to changes in structure (see letter from E. P. Wright, 24 March 1865 and nn. 6–8).

From Clémence Auguste Royer[1] [April–June 1865][2]

L'Election naturelle[3] est contenue pratiquement dans tout l'ensemble de ces lois,[4] et surtout dans tout ce qui concerne le régime des castes où l'on doit reconnaître des races probablement différentes et très inégales dans leur développement humain, du moins à l'origine.

Ce même principe de choix ressort des prescriptions et des conseils donnés par les legislateurs aux hommes des castes nobles dans le choix de leurs femmes.[5] Il leur est enjoint d'éviter de s'allier avec des femmes qui portent les signes néfastes tels que d'avoir les yeux chassieux ou autres difformités ou infirmités[6] Ce sont peut-etre les préceptes de ces lois qui ont transformé les instincts, les habitudes et les caractères physiques de la race indo-germanique.

Si tel est l'esprit général de ces lois, esprit qu'on retrouve dans tout leur ensemble, les paragraphes où il est plus explicitement exprimé sont les suivants.

liv. III. par. 6 et suiv. 10–14–19. [7]
liv. IX. par. 72–73 et 178[8]
liv X. par. 5 jusqu'à 49 – 66 à 73.[9]

A Mem
DAR 80: B44

CD ANNOTATION
Top of memorandum: 'Castes of India are *[apposite]* to breeding like domestic animals' *pencil*

1 For a translation of this memorandum, see Appendix I. There is an incomplete translation of this memorandum by Emma Darwin in DAR 80: B42–3.

2 The date range is established from the relationship between this memorandum and the letters in which CD and Emma Darwin mention that CD is working on the translation of the second French edition of *Origin* (see n. 4, below).

3 The phrase 'élection naturelle' appears to have been used as a translation of 'natural selection' first by Edouard Claparède (see *Correspondence* vol. 10, letter from Edouard Claparède, 6 September 1862). However, 'sélection' was already in use as a technical term in French zoology (*Grand Larousse de la langue française*). In the first French edition of *Origin*, Royer translated 'natural selection' as 'élection naturelle' (Royer trans. 1862). In the second French edition, 'sélection' was substituted for 'élection' at CD's request, although Royer expressed her disapproval of this change in a footnote (Royer trans. 1866, p. 95 n.; see also J. Harvey 1997, pp. 76–7). Royer's footnote referred to a paragraph that CD had inserted to defend the term natural selection against criticisms on various grounds, including its implication of conscious choice (*Origin* 3d ed., pp. 84–5). Royer's preference for the word 'élection', which may be translated as 'choice', reflected her belief in the control of nature by an intelligent power (C. Blanckaert in Tort 1996, p. 3746).

4 The memorandum is evidently a note for the second French edition of *Origin* (Royer trans. 1866). In the first edition of her translation (Royer trans. 1862, p. 57), Royer inserted a note comparing animal breeding practices to the strict marriage laws governing Hindu castes. This memorandum appears to be an expansion of that earlier note, and refers to the Hindu 'Laws of Menu' or 'Manu' described in the Sanskrit text Manava-Dharma-Sastra, which was translated into French in 1833 (Loiseleur Deslongchamps trans. 1833). CD probably made corrections to the manuscript of Royer trans. 1866 between April and June 1865 (see letter to John Murray, 31 March [1865], letter to J. D. Hooker, 10 [April 1865], and letters from Emma Darwin to H. E. Darwin, [6 June 1865] and [21 June

1865] in DAR 219.9: 27, 32). See also 'Journal' (Appendix II). CD evidently rejected the additions proposed in this memorandum, as they do not appear in the published version of Royer trans. 1866; there is a copy of this work in the Darwin Library–Down.

5 Royer also discussed marriage selection and its effect on human evolution in the first French edition of *Origin* (see Royer trans. 1862, Préface). For a discussion of Royer's translations of *Origin* and CD's reaction to her editorial insertions, see J. Harvey 1997, pp. 62–9, 76–9. See also *Correspondence* vol. 10, letter to Asa Gray, 10–20 June [1862], letter to Armand de Quatrefages, 11 July [1862], letter from Edouard Claparède, 6 September 1862, and letter to J. D. Hooker, 11 September [1862]. In a letter from Emma Darwin to H. E. Darwin, [6 June 1865], in DAR 219.9: 27, Emma reported that CD was at work 'on the verdammte Mlle Royer whose blunders are endless'.

6 The reference is to the Hindu 'Laws of Menu' (see n. 4, above). The text is also discussed in Lucas 1847–50; CD read Lucas 1847–50 in 1856 (*Correspondence* vol. 4, Appendix IV), and there is a heavily annotated copy of this work in the Darwin Library–CUL (see *Marginalia* 1: 513–23). CD refers to Lucas 1847–50 in an annotation to Emma's translation of this memorandum (see n. 1, above). 'Prosper Lucas about Beautiful men marrying beautiful in caste—'. The translation is also annotated 'Man' in red crayon, indicating that CD kept this material for possible use in *Descent*; however, he did not refer to it in any of his published works.

7 This section of book 3 of the 'Laws of Menu' discusses the rules of marriage, including caste restrictions for four named Hindu castes, and the signs of family or individual deformity or disease that one should avoid in a marriage (see W. Jones trans. 1869, pp. 40–1). CD wrote 'I have not looked' against this reference on Emma's translation of the memorandum (see DAR 80: B42–3).

8 In book 9 of the 'Laws of Menu', paragraphs 72 and 73 discuss conditions under which a marriage to a woman with an undetected physical defect, or a marriage contracted through deception, can be annulled; paragraph 178 designates a son 'begotten through lust' by a high-caste man on a low-caste woman as a 'living corpse' (see W. Jones trans. 1869, pp. 200–1, 211–12). On Emma's translation of the memorandum CD crossed through this and the subsequent citation (see n. 9, below) and added the word 'Nothing' (see DAR 80: B43).

9 These passages define the status of children born into various Hindu castes. The last passage cited (paragraphs 66 to 73) outlines the effect of the parents' good or bad conduct on the possibility of offspring from a mixed caste rising to a superior caste (see W. Jones trans. 1869, pp. 228–35). For CD's reaction to these passages, see n. 8, above.

From Lucy Caroline Wedgwood [April–May 1865?][1]

Leith Hill Place, | Dorking.
Tuesday

Dear Uncle Charles

I looked at 3 or 4 bunches of oxlips this morning.

We commonly find them either in shady hedge-banks, or else in the woods, where primroses grow;— I think *never* in the open fields with cowslips.[2]

In one case there were 8 or 10 plants growing in a group (some very small), but no primrose or cowslips anywhere near (tho in a primrosy wood) The others were mostly single plants or 2 together—none with primroses or cowslips very near.—[3]

Yr. affte. niece L C Wedgwood

DAR 108: 171–2

1 The date range is conjectured from the relationship between this letter and the letter from L. C. or M. S. Wedgwood to [Emma Darwin?], [May 1865], and from the flowering time of the common oxlip. See also n. 2, below.

[2] For several years CD had been asking his nieces in Surrey for help with botanical observations. He probably began to receive observations from Lucy Wedgwood and her two sisters, Katherine and Margaret, in 1862 (see *Correspondence* vol. 10, letter to K. E. S., L. C., and M. S. Wedgwood, 4 [August 1862]; see also *Correspondence* vol. 12, letter from L. C. Wedgwood, [6 June 1864]). He rarely found oxlips near Down House (see *Forms of flowers*, p. 71). He published information on where primroses, cowslips, and common oxlips grew in 'Specific difference in *Primula*', pp. 438 and 448, and *Forms of flowers*, pp. 57, 70, and 71, noting that in parts of Surrey, specimens might be found 'on the borders of almost every field and small wood' (*Forms of flowers*, p. 71).

[3] CD had long been interested in whether oxlips were the hybrid offspring of primroses (*Primula vulgaris*) and cowslips (*Primula veris*); see *Natural selection*, pp. 128–33, and *Origin*, pp. 49–50. He had been making observations and carrying out experiments since 1862 to test his initial view that the three forms were varieties descended from a common parent (see *Correspondence* vol. 10, letter to John Scott, 3 December [1862] and n. 13). Though CD seemed confident by 1864 of his experimental results indicating that the common oxlip was a hybrid resulting from the crossing of primroses and cowslips, he continued to test this conclusion over the next three years (see, for example, *Correspondence* vol. 12, letter to J. D. Hooker, 10 June [1864] and n. 17, and letter to Charles Lyell, 9 June [1867] (*Calendar* no. 5566)). CD's notes on these experiments, dated between 1862 and 1867, are in DAR 108 and DAR 157a. He published his conclusions in 'Specific difference in *Primula*', pp. 443–7, and *Forms of flowers*, pp. 55–75.

From John Murray 1 April 1865

50, Albemarle St. | W.
April 1st. /65

My Dear Sir

I answer without delay your obliging note of yesterday—[1]

I have no doubt your new work on Domestic Animals & Plants will be one of great value & interest & I shall be very proud to be the publisher—of it as of your "Origin"—[2] I do not think you could select a better size than it—& it is of some consequence that your new work shd range with the former—[3]

I am quite well satisfied that Mr Sowerby Mr Wells & Mr Fitch shd each—in his own department make the Drawings on wood of wch you stand in need—[4] I shd be sorry to stint you in number & beg you to put in hand what you think needful, only instructing the several artists to send me beforehand an estimate of their charge I wd urge that all the subjects be so reduced as to come well within the dimensions of the page of your Origin of Species—leaving a fair margin[5]

I fear I have parted with the cuts of Dixons book[6] I was not pleased with him & gladly washed my hands of it—but the Birds can be re-engraved.

I am going over to France for 3 weeks—& in my absence wd ask you to correspond with my Cousin & partner Mr Robert Cooke who will give you the latest intelligence of the state of the Editions of Origin & Orchids—[7]

Meanwhile if you are anxious to put the cuts in hand—I see no objection— When I return I will look after them & I will also ask you to let me see any portion of your MS wch may be ready.

I was sorry to hear from Sir Ch Lyell a poor account of your health— but hope you are better again

I am | My Dear Sir | Your faithful & obliged servt | John Murray

P S. As I have been prepared from the first for this work & look upon it as the Complement of the "Origin of S."[8] I propose—unless you object to announce it in my Quarterly List of New Works—wch will appear next week[9] This will give timely notice to those who possess the Origin—but I sh^{d} not think of publishing before November. I shall therefore put the Title which you have sent *at once* in our List, unless Mr Cooke hears to the contrary.

Chas Darwin Esq.

DAR 171: 332

CD ANNOTATION
9.5 I shall . . . contrary. 9.6] *double scored red crayon*

1 Letter to John Murray, 31 March [1865].
2 The firm of John Murray published *Variation* in 1868; they had first published *Origin* in 1859.
3 Though both works were published in octavo, the height of *Origin* was 198 mm, while *Variation* was 220 mm (see Freeman 1977, pp. 84, 124). See also letter to John Murray, 31 March [1865].
4 See letter to John Murray, 31 March [1865]. The references are to George Brettingham Sowerby Jr, Luke Wells, and Walter Hood Fitch.
5 See n. 3, above.
6 Dixon 1851. See letter to John Murray, 31 March [1865] and n. 17.
7 No correspondence with Robert Francis Cooke on the sales of *Origin* and *Orchids* has been found; however, the following is written at the bottom of the letter: '*on hand* | 150 Origin of Species | 580 Orchids'. It is probable that these figures were added by Cooke, and are the answers to CD's enquiry (see letter to John Murray, 31 March [1865]). The 150 copies of *Origin* remained from a printing of 2000 copies of the third edition, published in April 1861, and the 580 copies of *Orchids*, published on 15 May 1862, remained from a printing of 1500 (see Freeman 1977, pp. 78, 112, and *Correspondence* vol. 10, letter to John Murray, 9 April [1862]).
8 CD had announced in the introduction to *Origin*, p. 2, that he hoped to follow that work by 'publishing in detail all the facts, with references' on which his conclusions had been founded. CD had consulted Murray about his plan to publish a more extensive work on species on several occasions (see *Correspondence* vol. 7, letter to T. H. Huxley, 16 December [1859], and letter to John Murray, 22 December [1859], and *Correspondence* vol. 9, letters to John Murray, 3 March [1861] and 30 April [1861]). *Variation* included an expanded study of the subject of the first chapter of *Origin*.
9 See letter to John Murray, 31 March [1865] and n. 2.

To John Murray 4 April [1865][1]

Down, Bromley, Kent. S.E.
Ap. 4.

My dear Sir

I am much obliged for your kind note.[2] I will act on your advice and instructions about the wood-cuts. I am quite willing that you should insert the advertisement as you propose, and I presume that you approve of the title. But I must beg you to bear in mind that my health is extremely precarious and that it depends upon this and upon nothing else whether my book will be ready for the press in the autumn.[3] On your return I will send you a bundle of MS to look at, and you can be still free to publish or not as you like; but I shall be *much* pleased if we can agree to publish together.

Believe me my dear Sir | Yours very sincerely | Charles Darwin

Copy
DAR 143: 434

[1] The year is established by the relationship between this letter and the letter from John Murray, 1 April 1865.
[2] See letter from John Murray, 1 April 1865.
[3] CD recorded that he became ill on 22 April 1865 and did not resume work on *Variation* until 25 December (see 'Journal' (Appendix II)). The work was not published until 30 January 1868 (see Freeman 1977, p. 122). In his journal entry for 1867, CD wrote: 'owing to interruptions from my illness & illness of children; from various editions of Origin & papers especially Orchis book & Tendrils, I have spent 4 years & 2 months over it.—' (DAR 158).

To J. D. Hooker 6 April [1865]

My dear Hooker.

Have you the Botanische Zeitung for 1860 p. 204 (Dr. Alefeldt on Pisum.)[1] & can you lend it me? It is not in Linnean or Royal Socy.!! If you can lend it, please have it addressed to

C. Darwin
care of Down Postman
per Rail Bromley
Kent

I am tired today, & no heart to write, & indeed not a word to say. I work a little every day with groans & sighs & am as dull as a pig.— It is hopeless & useless but, *when you meet* Busk, ask him whether any man is better than Jenner for giving life to a worn-out poor devil.—[2]

Yours affect | C. Darwin

Down. 6th April.

Endorsement: '/65.'
DAR 115: 262

[1] The reference is to Alefeld 1860, which provided information on a study of around fifty varieties of pea (*Pisum*), and concluded, first, that all the varieties belonged to the same species, and second, that their ancestral form was the field pea, *P. elatius*. CD cited this paper and the first part of its findings in *Variation* 1: 326, as part of a discussion of cultivated varieties of culinary plants and their wild prototypes.
[2] In 1863 George Busk had given advice to CD on his stomach disorder, and recommended consulting William Brinton; both Brinton and William Jenner were experienced in the treatment of stomach complaints (see *Correspondence* vol. 11, letter from George Busk, [*c.* 27 August 1863]). CD consulted Brinton in November and December 1863, and began consulting Jenner in March 1864 (see *Correspondence* vol. 11, letter to J. D. Hooker, 5 [December 1863], and *Correspondence* vol. 12, letter from

William Jenner, 24 November 1864); however, by January 1865 CD had become discouraged with Jenner's treatment, and had started to enquire about other physicians (see letters to J. D. Hooker, 7 January [1865] and 9 February [1865]).

To W. B. Tegetmeier 6 April [1865][1]

Down. | Bromley. | Kent. S.E.
Ap 6th.

My dear Sir

I have heard from Mr Murray[2] & I shd be much obliged if you wd forward this note to Mr Wells.[3] I shall be much obliged to him if he will draw & have engraved on wood, with the same care as displayed in your drawing of the Smerle, figures of a *Pouter*, *Carrier*, *Short-faced Tumber*, *Barb Fantail*, *Turbit* & *Jacobin*.[4] These are to be first-rate birds, but not exaggerated. The enclosed paper shews the exact size of the type in the page (without the heading) of the book in which they are to appear.[5]

The figures may be upright, which is best, or transversely; but Mr Murray remarks that wood cuts look better with a little margin.[6] I observe that the figure of the Barb in the Field[7] is too large for the head reaches beyond the enclosed piece of paper.

The character of the beak ought in all cases to be particularly attended to.

I suppose Mr Wells can get an estimate or contract for the cutting on the wood, for Mr Murray wishes Mr Wells first to send an estimate to him of the *whole cost* of the 7 figures, & this ought to be the first thing done.

I can settle hereafter about drawings of fowls' head & some other things.[8] There is no great hurry, I suppose, (for I do not know how long wood cutters take) but I shall not go to press till August.[9]

My dear Sir | Yours very sincerely | Charles Darwin

P.S. | Please address

Messrs Murray & Cook[10]
50 Albemarle St. | W.

LS(A)
The New York Botanical Garden Archives (Charles Finney Cox collection) (Tegetmeier 32)

1 The year is established by the relationship between this letter and the letter from John Murray, 1 April 1865.

2 Letter from John Murray, 1 April 1865.

3 The note has not been found. Luke Wells was the artist recommended by Tegetmeier to prepare the pigeon illustrations for *Variation* (see letter to W. B. Tegetmeier, 14 March [1865], and letter from W. B. Tegetmeier, 27 March 1865).

4 CD refers to Wells's drawing of the smerle pigeon in the *Field* (see letter to W. B. Tegetmeier, 14 March [1865] and n. 6). In addition to providing a drawing of *Columba livia* from a dead bird, Wells was commissioned to prepare drawings from living representatives of six domestic breeds of pigeon (*Variation* 1: 135 n. 6). The breeds illustrated were the English pouter, English carrier, English barb, English fantail, African owl, and short-faced English tumbler (*Variation* 1: 137, 140, 145, 147, 149, 152).

[5] The enclosure has not been found.
[6] See letter from John Murray, 1 April 1865.
[7] CD refers to Wells's illustration of the barb pigeon in the *Field*, 18 February 1865. Tegetmeier sent CD a packet containing this and four other issues of the *Field* with his letter of 27 March 1865 (see letter from W. B. Tegetmeier, 27 March 1865 and n. 1); these items are in DAR 138.3 and are lightly annotated.
[8] See letters to W. B. Tegetmeier, 14 March [1865] and [7 April 1865].
[9] See letter to John Murray, 4 April [1865] and n. 3.
[10] John Murray had asked CD to correspond with his cousin and partner Robert Francis Cooke while Murray was in France for three weeks (see letter from John Murray, 1 April 1865).

From J. D. Hooker [7–8 April 1865][1]

Kew
Friday.

Dear Darwin

I have a German paper for you from Grisebach on the Geog. Distrib of W. Indian plants,[2]* very long, also a small bottle from Thwaites with some things I forget what.[3] I shall send them with *Strelitzia*.[4]

I cannot recollect any news for you, worth sending.— We have been much occupied of late with Garden Reform, we have been robbed much by our own people, & I am putting the whole establishment on a different footing, discharged two foremen—dismissed half a dozen gardeners & labourers & clapped one fellow in jail for 6 months. All this is not very agreeable work, but we have really a first rate Curator now, & I am anxious to put every-thing straight for him to go on without troubling me.[5] I am very proud of having picked him up. We have purchased £70 worth of orchids to make up our shameful losses, besides a great many other plants;[6] & I am actively corresponding abroad for new plants &c.

I have been living a long while on the gratification of my last visit to Down.[7]

Yesterday we had one of our "small club" dinners, in Albemarle Street, Lubbock, Huxley, Spottiswood, Spenser, Busk & self,[8] with Colenso[9] & H B Wilson[10] as guests, & a very pleasant evening it was: though I must confess I cannot go along with Colenso— his incessant prating about his own "affair" is quite wearisome: he really is in some respects a very weak man. On first coming in, he asked the name of our club— I said it has none— he replied, "I would call it the Zulu club". *& so on.*—[11] then he told us that they have witheld his salary &c. I thought Wilson a very superior man.[12]

I am concerned about Lubbock, his wife seems really to be very ill, & in a bad way[13]—& he is going into Parliament[14]—for which I am no less sorry— I grudge so good a man from Science—& have a presentiment that it will inaugurate a very trying life for him. I believe I am no end of way happier in avoiding every avenue to ambitious ends in my small walk of life—& so long as one's mind & time is fully occupied, there is nothing to regret in a life of mere drudgery. We (L & I) talk of going to Maidstone from Thursday to Monday but what with M^rs L^s health & my father being in bed with Influenza,[15] it may not come off at all!

I have actually stuck for want of something to say to you!—

Saturday Mg.

Yours of 6th. just arrived I send the Bot: Zeit: on Monday with the other things.[16]

Thwaites & Christy & G. Gray are chosen for R. S.[17] I do not know of anyone else we know. or care about.

I have written to Busk being utterly ashamed of always *forgetting* to ask him.[18]

Ever Yrs affec | J D Hooker

*Thomson[19] is digging out its essence so do not trouble to read it yet.

Endorsement: 'Apr 65'.
DAR 102: 15–16

CD ANNOTATIONS
5.1 I am . . . Parliament— 5.2] *double scored pencil*
10.1 I have . . . him.] *scored pencil*

[1] The date is established by the endorsement and by the reference to the arrival of CD's letter to Hooker of 6 April [1865]. The Friday after 6 April 1864 was 7 April.

[2] Hooker refers to August Heinrich Rudolf Grisebach and to Grisebach 1864; there is a copy of this work, with most of its pages uncut, in the Darwin Pamphlet Collection–CUL. Grisebach had completed a flora of the British West Indies in 1864 (Grisebach 1859–64); this was one of a series of colonial floras initiated by William Jackson Hooker, director of the Royal Botanic Gardens, Kew, with the sponsorship of the Colonial Office (see R. Desmond 1995, pp. 216–17, 370). CD was interested in new colonial floras for their potential to provide information on the history of plant distribution (see letter to J. D. Hooker, 19 January [1865] and nn. 7 and 8).

[3] The items sent by George Henry Kendrick Thwaites have not been identified; however, Thwaites was aware of CD's interest in dimorphic flowers and had earlier sent him specimens in spirits from Ceylon (now Sri Lanka; see *Correspondence* vol. 11, letter from G. H. K. Thwaites, 17 February 1863).

[4] CD had received information on the pollination of *Strelitzia reginae* from Roland Trimen in 1864 (see *Correspondence* vol. 12, letters to Roland Trimen, 13 May 1864 and 25 November 1864); during Hooker's last visit to Down (see n. 7, below), CD had probably asked for specimens to provide further evidence that this genus was pollinated by birds (see *Cross and self fertilisation*, p. 371 n., and letter to J. D. Hooker, 13 April [1865]).

[5] Hooker also referred to problems with employees, and to his new curator, John Smith (1821–88), in his letters of [2 April 1864] and 8 April 1864 (*Correspondence* vol. 12). Smith became curator of the Royal Botanic Gardens, Kew, in 1864 (see R. Desmond 1995, p. 430). The former curator, John Smith (1798–1888), had resented being overruled by Hooker and his father, William Jackson Hooker, in gardening matters; Hooker was assistant director of the Royal Botanic Gardens, Kew (see R. Desmond and Hepper 1993, pp. 18–19, and R. Desmond 1995, pp. 221, 425, 430). For more on Hooker's opinion of the new curator, see also *Correspondence* vol. 12, letters from J. D. Hooker, [4 June 1864] and 26[–8] October 1864.

[6] Hooker discussed orchid poisoning and other horticultural problems that had developed at the Royal Botanic Gardens, Kew, in his letters of [2 April 1864], [4 June 1864], and 15 June 1864 (*Correspondence* vol. 12).

[7] Hooker spent from 4 to 6 March at Down House (Emma Darwin's diary, DAR 242). See also letter to J. D. Hooker, 16 [March 1865].

[8] The reference is to the dining club, later called the X Club, established on 3 November 1864 at St George's Hotel, Albemarle Street, London, by Hooker and a group of close friends. The initial membership consisted of Hooker, John Tyndall, Thomas Henry Huxley, George Busk, Edward Frankland, Herbert Spencer, John Lubbock, and Thomas Archer Hirst; William Spottiswoode joined at

the December meeting. The members were united by friendship and a 'devotion to science, pure and free, untrammelled by religious dogmas' (quoted in Barton 1998, p. 411; see *ibid.* for an account of the club's history and significance).

9 John William Colenso.

10 Henry Bristow Wilson.

11 Between 1862 and 1865 Colenso, the bishop of Natal, had published five parts of a seven-part work of biblical criticism (Colenso 1862–79). In response to Colenso's questioning of church doctrine, Robert Gray, the bishop of Cape Town, tried him, found him guilty of erroneous teaching, and deposed him in 1863. Using the secular courts, Colenso defended his position within the Anglican Church and his right to publish his opinions. On 20 March 1865 Colenso won his appeal against his deposition and he later returned to his diocese; however, Gray refused to reinstate him (see Guy 1983, pp. 95, 122–144). Hooker and CD had supported Colenso's right to freedom of opinion and in 1864 had contributed to his defence funds. See *Correspondence* vol. 12, letter from E. A. Darwin, 1 February [1864], and letters from J. D. Hooker, 5 February 1864 and 16 February 1864 and nn. 17 and 18. For Colenso's connections with Hooker and other members of the X Club, see *Correspondence* vol. 12, letter from J. D. Hooker, 16 February 1864 and n. 19; see also Barton 1998, pp. 435–7.

12 Wilson was a liberal Anglican clergyman who had been accused of heresy as a result of the opinions he expressed in his contribution to the controversial volume *Essays and reviews*; he was eventually cleared of the charge. On the history of this controversy and the involvement of X Club members, see *Correspondence* vol. 9, Appendix VI, Brock and MacLeod 1976, Ellis 1980, and Barton 1998, pp. 434–5. CD had contributed to a fund organised for the defence of the essayists in 1861 (see *Correspondence* vol. 9, letter to John Lubbock, 1 [and 2] August [1861] and n. 5).

13 Ellen Frances Lubbock was pregnant; she had her sixth child, Rolfe Arthur Lubbock, in September 1865 (*Burke's peerage*).

14 On 29 March 1865 Lubbock was requested by the Liberal Registration Committee to stand as a parliamentary candidate for West Kent at the General Election to be held in July. Lubbock was confirmed as a Liberal candidate for this constituency at a meeting in Maidstone, Kent, on 20 April 1865 (John Lubbock's diary (British Library, Add. Ms. 62679)).

15 On William Jackson Hooker's illness, see the letter from J. D. Hooker, 12 April [1865].

16 See letter to J. D. Hooker, 6 April [1865] and n. 1, and nn. 2–4, above.

17 Hooker refers to Thwaites, Henry Christy, and George Robert Gray; they had been nominated for election as fellows of the Royal Society at a council meeting on 6 April 1865. The membership of Thwaites and Gray was formally approved at the annual general meeting held on 1 June 1865; Christy died before his election could be confirmed (Royal Society, council minutes, 6 April 1865; *Proceedings of the Royal Society of London* 14 (1865): 482; *DNB*).

18 See letter to J. D. Hooker, 6 April [1865] and n. 2.

19 Thomas Thomson.

To W. B. Tegetmeier [7 April 1865]

Down. | Bromley. | Kent. S.E.

My dear Sir

I have received the M.S.[1] all safe & have just begun to go over it, & I can clearly see that I am deeply indebted to you for many valuable suggestions & corrections. I have received another "Field" & very many thanks for it.—[2] It is good news that your injured Eye is recovering.—[3]

I enclose a separate note (which please look at) for Mr Wells;[4] will you have the kindness to make him *clearly understand* that *you* are to see the drawings on the wood, for your approval & likewise when engraved.—

Could Mr Wells draw for me the heads alone, (of size to fill up breadth of my page) of *Game*, *Spanish*, *Hamburgh* & *Polish*?[5] Could you aid him to see good birds? But as these figures are not indispensable, I sh^d^. like hear what the cost would be.— M^r^. Wells could write direct to me.—

I am much obliged to you for wishing to see my M.S. on Pigeons; but you do not know what a job it will be, for I have treated this subject at greater length than any other.[6] But you can read as much as your patience admits, & I shall be grateful for any corrections.[7] I fear that I shall want that M.S. back by about May 1st— Please pack it up in same manner as sent & let it be **registered**.—

I believe that some bones of the skull are wrongly named, as likewise was case with some in Fowls' skulls.—[8]

The Pigeon M.S is sent off this day.—

With cordial thanks | My dear Sir | Yours very sincerely | Ch. Darwin

(I enclose stamps for return of Pigeon M.S. *registered*.)

P.S. | If not inconvenient I sh^d^. like a blue Pouter drawn with the black wing-bars & sub-terminal tail bar well & carefully shown.—[9]

Endorsement: 'Posted April 7 1865'
The New York Botanical Garden Archives (Charles Finney Cox collection) (Tegetmeier 33)

[1] CD refers to the manuscript of his chapter on fowls for *Variation*, which Tegetmeier had been reading and correcting (see letters to W. B. Tegetmeier, 27 February [1865] and 28 March [1865], and letter from W. B. Tegetmeier, 13 March 1865). It is probable that Tegetmeier sent the manuscript with his letter of [29 March – 7 April 1865].

[2] The reference is to the issue of the *Field* for 1 April 1865 (see letter from W. B. Tegetmeier, 27 March 1865 and nn. 1 and 2), containing an article, illustrated by Luke Wells, on the trumpeter breed of pigeon.

[3] See letter from W. B. Tegetmeier, [29 March – 7 April 1865].

[4] The note has not been found. See also letter to W. B. Tegetmeier, 6 April [1865] and nn. 3 and 4.

[5] Chapter 7 of *Variation* included illustrations by Wells of the heads of Spanish, Hamburgh, and Polish fowl (*Variation* 1: 226, 228, 229).

[6] Tegetmeier may have expressed this wish in the missing portion of his letter to CD of [29 March – 7 April 1865]; CD had mentioned that he intended to have two chapters on pigeons in *Variation* in his letter to Tegetmeier of 14 March [1865]. CD had begun writing the first chapter on pigeons in June 1858, and completed the second chapter in June 1860 (see *Correspondence* vol. 7, Appendix II, and *Correspondence* vol. 8, Appendix II). CD may have returned to the pigeon chapters when he began revising the manuscript of *Variation* in November 1864 (see *Correspondence* vol. 12, Appendix II). These chapters became chapters 5 and 6 of *Variation* (see *Variation* 1: 131–224).

[7] Tegetmeier's assistance with CD's work on pigeons is acknowledged in *Variation* 1: 132, n. 2.

[8] Chapter 5 of *Variation* was illustrated with five engravings of pigeon skulls; CD provided a summary of the anatomical characteristics of the skulls in the accompanying text (see *Variation* 1: 163–5). CD's work for chapter 7, on fowls, included an examination of fifty-three skulls of various breeds; nearly half of the specimens had been provided by Tegetmeier (see *Variation* 1: 260–6). Tegetmeier, who had long been interested in variation in poultry skulls (see E. W. Richardson 1916, pp. 101–2), may have informed CD that some of the bones had been incorrectly named.

[9] The illustration of the English pouter by Wells in *Variation* 1: 137 shows the wing and tail bars.

From William Duppa Crotch 10 April 1865

Uphill House | Weston Super Mare
April 10/65

Dear Sir

Since I was in Shetland—3 or 4 years ago[1]—I have twice visited the Canary Islands in search of development & an Atlantis![2] & have brought back more than 40.000 specimens—mostly entomological—& some confirmed opinions—

Those on the mutation of species needed no confirmation—but with regard to the great continent, in part submerged—which connected America with Africa in the miocene & subsequent eras, I feel more confident than ever that the Sahara is the ocean bed which separated this land from Africa, & that migration worked Eastward of Greenland—across the telegraphic plateau,[3] then a portion of the dry land of Atlantis—rather than across Asia via Behring's Straits (which must have been a perfect Noah's Ark) as Asa Gray supposes—[4] The utter non-existence of infra-Saharan species in the Canary Islands, & continual occurrence of Transatlantic types is very surprising— I do not know how intimately you investigated these Islands & shall feel extremely obliged to you for any expression of opinion with regard to the Atlantis Hypothesis—[5]

Further than this I think Adhemar's theory of the Antarctic glaciers[6] not only possible but as probably explaining the submergence of the Western portion of Atlantis— I brought home with me about 50 sketches & hope to publish an account of doings, beholdings & theorisings shortly, which I should like to dedicate to you,[7] if you will permit me—since, in common with many others, I owe to you entirely the scientific adjustment of the crude Lamarckian doctrines[8] which I held, as the nearest approach to the truth.

Believe me | Yours sincerely | W. D. Crotch M.A. Oxon.

DAR 161: 274

[1] Crotch had written to CD in 1861 about physiological changes in Shetland horses and seagulls (see *Correspondence* vol. 9, letter from W. D. Crotch, 25 January 1861).

[2] Atlantis was the popular name given to Edward Forbes's hypothetical sunken Miocene continent in the Atlantic Ocean linking the Iberian peninsula with the Azores and with Ireland, providing a land-bridge over which species migration took place (E. Forbes 1845 and 1846). For extensive discussions of Forbes's Atlantis hypothesis, see *Correspondence* vols. 3, 5, and 6. Oswald Heer subsequently put forward an Atlantis hypothesis to provide a land-bridge between Europe, Africa, and North America (Heer 1855 and 1860). See also letter from Charles Lyell, 16 January 1865 and n. 19, and letter to Charles Lyell, 21 February [1865] and nn. 3–5.

[3] The transatlantic cable, making telegraphic communication between Britain and North America possible, was laid on the sea-bed plateau between Ireland and Newfoundland, Canada, in 1856 (see W. H. Russell 1865, especially p. 7).

[4] To account for the similarity between the plant species of the north-eastern United States and eastern Asia, Gray had posited a 'mingling' of species over the Behring Straits during a warmer temperate period both before and after the glacial age. See A. Gray 1858–9 and Dupree 1959, pp. 250–1.

[5] CD had long been critical of attempts to explain the geographical distribution of animals and plants by postulating former land-bridges (see letter to Charles Lyell, 21 February [1865] and n. 3). During

the 1850s, CD had carefully compared lists of species on the Canary Islands, Madeira, and the Azores, with the plants and animals of nearby landmasses (see *Correspondence* vols. 5 and 6). He summarised some of his findings, which were in opposition to the Atlantis theory, in *Origin*, pp. 357–8. See also *Natural selection*, p. 538.

6 The reference is to Alphonse-Joseph Adhémar and Adhémar 1842.

7 Crotch publicly announced his intention of writing a history of the Canary Islands and Madeira with a view to substantiating the Atlantis hypothesis in the *Entomologist* (*Entomologist* 2 (1864): 111); however, the work was never published.

8 For CD's views on the evolutionary theory of Jean Baptiste de Lamarck, see *Origin* 3d ed., p. xiii.

To J. D. Hooker 10 [April 1865][1]

Down
March 10th

My dear Hooker.

You speak of having nothing to say, but your note contained a *heap* of things which I liked to hear.[2] I daresay I shall hear tomorrow of the book &c &c & Strelitzia (for which hearty thanks) being at Bromley.[3] I suppose you would like all future numbers of "Can you forgive her"?[4] I entirely agree with what you say about J. Lubbock & parliament;[5] I am quite vexed to think of it; but it is the commonest of all kinds of ambition.— We heard yesterday a better account of Mrs. L.—[6] I cannot conceive how you have managed to detect so much roguery at Kew; you must have had odious work.—

When you write again, if you can remember, tell me who are authors of Review, on Linn: Transact,—, Planchon—& Sub-Species.—[7] I have not yet read all, but I am sure I shall be curious to know.— I dare not guess, for you say that my guesses are always wrong.— Nevertheless I am sure that you wrote a Leader in G. Chronicle on Selection &c as to be discussed in Germany.—[8]

This reminds me that I am, as it were, reading the "Origin" for the first time, for I am correcting for a 2nd. French Edition;[9] & upon my life, my dear fellow, it is a very good book, but oh my gracious it is tough reading & I wish it were done.—

I am in truth sorry that you had trouble & gave Busk, trouble by writing to him;[10] for there was no hurry; & I feel sure I must bear my load of daily discomfort, & be thankful that I can occupy myself for a few hours daily.

Farewell my good old friend | Yours affect | C. Darwin

Endorsement: '/65'
DAR 115: 263

1 The date is established by the relationship between this letter and the letter from J. D. Hooker, 12 April [1865], and by the endorsement. CD evidently wrote 'March' in error.

2 See letter from J. D. Hooker, [7–8 April 1865].

3 Hooker had promised to send Grisebach 1864, a specimen of *Strelitzia*, and other items. See letter from J. D. Hooker, [7–8 April 1865].

4 The reference is to the novel *Can you forgive her?* by Anthony Trollope (Trollope 1864–5), which was published in monthly parts from January 1864 to August 1865 (see N. J. Hall 1991, p. 560). CD may

have introduced Hooker to the novel when Hooker visited Down House from 4 to 6 March 1865 (Emma Darwin's diary, DAR 242).

5 The reference is to John Lubbock (see letter from J. D. Hooker, [7–8 April 1865]).

6 Hooker had mentioned the apparent illness of Ellen Frances Lubbock; she was in fact pregnant (see letter from J. D. Hooker, [7–8 April 1865] and n. 13).

7 CD refers to three unsigned review articles in the April 1865 issue of the *Natural History Review.* Joseph Reay Greene reviewed the 1863–4 *Transactions of the Linnean Society of London* ([Greene] 1865), George Bentham reviewed Planchon 1864a and 1864b ([Bentham] 1865), and Thomas Thomson's article 'Species and subspecies' was a review of Jordan 1864 ([T. Thomson] 1865). See letter from J. D. Hooker, 12 April [1865].

8 CD refers to an article published in the *Gardeners' Chronicle* on 25 March 1865, pp. 267–8, reporting that the 'scientific men of Germany' proposed discussing 'Darwin's theory', as it related to 'the production of new races of plants by selection', at the meeting of the second German horticultural congress, to be held at Erfurt in September 1865; the article went on to discuss definitions of species, races, and varieties.

9 The second French edition of *Origin* was translated by Clémence Auguste Royer; CD supplied notes for this edition (see Freeman 1977, p. 102, and Royer trans. 1866). See also memorandum from C. A. Royer, [April–June 1865].

10 CD had been seeking medical advice from George Busk. See letter to J. D. Hooker, 6 April [1865], and letter from J. D. Hooker, [7–8 April 1865].

From John Scott 10 April 1865

Rungbee
10th. April 1865.

Sir,

I duly received your letter along with the copies of my papers[1]—for all the trouble about which I can only express my great indebtedness.[2] I am quite ashamed at the time I have allowed to elapse in acknowledging the receipt of the above along with your own paper on the Lythrums[3]—which I have read with great interest— How completely the sexual phenomena of the Lythrums casts into the shade those of the Primula, &c, which we have of late regarded as so remarkable![4]

I have also to acknowledge receipt of Dr. Cruger's paper on Orchids. I see Dr. Cruger attributes the sterility of the male Catasetum to a deficiency in the conducting tissue.[5] My experiments results on the Gongoras & Acroperas illustrates this view, inasmuch as I have found a gradual decrease in the amount of fertility according to the amount I cut from the column in inserting the pollen-masses.[6] This is further illustrated by the fact that in those cases in which the capsule did not set when I had cut largely from the column—I invariably found that a sufficiency of pollen-tubes were developed, showing us that the power of stimulating the full developement of the ovules does not, at least in every case reside innately, & independently in the pollen.

I should have liked if Dr. Cruger had entered more fully into the fertilisation of Gongora He has said nothing as to the contraction of the stigmatic cavity! I have not as yet had an opportunity of examining **many** of the **species**—but all that I have seen, exhibit the same contracted characteristics—and of course it would have been utterly impossible for an insect—I am inclined to think—to insert the pollinia— Fertilisation in these could only have been effected in the manner described by Dr. Cruger by a large developement of the viscous matter around the mouth of the stigma as I formerly supposed.[7]

I should like much to have time & opportunities to re-commence experiments on the above. At present I have neither being kept very busy indeed with the Cinchonas[8] Whether or not I remain here much longer I really cannot say. I rather think that I will go down to the Calcutta Botanic Gardens, as Curator, Dr. Anderson having offered it me, & appears desirous that I should take it for a short time.[9] I expected that I would have known definitely ere this whether I should go or not, Dr. Anderson having promised to pay us a visit here in the latter end of March.

I trust my other letters and paper on Verbascums have come duly to hand,[10] as I have not had the pleasure of hearing from you since the date of their arrival. I have to express my best thanks for your remarks upon composition.[11] I can assure you, I well know the need of improving myself—but alas! for my powers. I am most completely awanting in the powers of expression, so that my highest source of happiness, affords me the greatest pain.

I was sorry to hear of the old bye law which prevents my election as an Associate of the Linnean Society.[12] I should have indeed liked much to have had that honour conferred upon me. But now of course it can't be helped. I am much pleased to find that Prof. A. Gray has noticed my paper—[13] I expected that he would have something to say on the **N.A.** *Hottonia inflata*.[14]

I am sorry to see by your last letter that your health was still weak. I sincerely trust that you are now as spring advances beginning to enjoy a greater share of this first of all comforts. I will be glad if I have a letter from you soon, & find that I can possibly be of any service to you here, should it be that I have to go to Calcutta. In case this should be, when you are pleased to write me again, will you kindly address to the care of Dr. Anderson as formerly.

And now with many thanks—which I as yet can alone offer—for all your kindness to me.[15]

I remain | Sir | Your obliged & obed Serv | J. Scott.

DAR 177: 115

CD ANNOTATION[16]

End of letter: 'Situation good | I am glad to hear that you had work so soon | Climbing Paper[17] | Hookers | Leersia[18] | Lagestroemia'[19] *pencil*

[1] The letter to Scott has not been found, but was written before CD wrote a letter to Scott on 11 March 1865 that is also missing (see letter from John Scott, 21 July 1865). Scott probably refers to copies of his papers in the *Journal of the Linnean Society (Botany)* (Scott 1864a and 1864c).

[2] Scott had sought CD's advice before preparing the manuscript of Scott 1864c (see *Correspondence* vol. 12, letter from John Scott, 19 March 1864 and n. 21). He also sent the completed manuscripts of Scott 1864a and 1864c to CD for comments (see *Correspondence* vol. 12, letters from John Scott, 28 March 1864, 5 May [1864], and first letter of 10 June [1864]). CD communicated both papers to the Linnean Society in June 1864.

[3] 'Three forms of *Lythrum salicaria*'.

[4] Scott is comparing CD's work on the trimorphic *Lythrum* with CD's work on the structure and functions of the reproductive organs in the dimorphic *Primula*, 'Dimorphic condition in *Primula*', which had stimulated Scott's study of this genus (Scott 1864b). For Scott and CD's discussions of the Primulaceae, see *Correspondence* vols. 10–12.

[5] CD first told Scott of Hermann Crüger's work and of the manuscript of Crüger 1864 in February or March 1864 (see *Correspondence* vol. 12, letter from John Scott, 19 March 1864 and n. 20). For Crüger's discussion of the sterility of the male *Catasetum*, see Crüger 1864, pp. 127–8. Conducting tissue is the portion of the style through which the pollen tubes pass from the stigma to the ovules.

[6] For Scott's experimental manipulations of *Gongora* and *Acropera*, see *Correspondence* vol. 10, letter from John Scott, 11 November 1862, *Correspondence* vol. 11, letters from John Scott, [after 12] April [1863] and 21 May [1863], and *Correspondence* vol. 12, letter from John Scott, 19 March 1864 and n. 16. CD cited Scott's observations on the pollination of these orchid genera in 'Fertilization of orchids', p. 153 (*Collected papers* 2: 150), and *Orchids* 2d ed., pp. 168–70.

[7] Crüger discussed the possible insect pollination of *Gongora* in Crüger 1864, pp. 130–1. CD and Scott had wondered how the stigmatic chamber of *Gongora* and *Acropera* flowers, which was very narrow, could allow for insect pollination; Scott had succeeded in pollinating *Gongora atropurpurea* only by manually applying pollinia and viscous material to the stigmatic chamber (see, for example, *Correspondence* vol. 12, letter from John Scott, 19 March 1864 and nn. 12–16, and *Orchids* 2d ed., pp. 167–71).

[8] Scott had taken a position at Rungbee, a *Cinchona* plantation near Darjeeling, India, in December 1864 (see letter from John Scott, 20 January 1865 and n. 7).

[9] Thomas Anderson was superintendent of the Calcutta Botanic Garden. See letter from J. D. Hooker, [17 February 1865] and n. 14.

[10] See letter from John Scott, 20 January 1865. The letters from John Scott of 21 December 1864 and 4 January 1865 have not been found. CD received Scott's manuscript on *Verbascum*, which was sent on 4 January 1865 (see letter from John Scott, 20 January 1865, letter from J. D. Hooker, [10 March 1865], and letter to J. D. Hooker, 16 [March 1865]). CD's reply to Scott, written on 11 March 1865, has not been found (see letter from John Scott, 21 July 1865).

[11] CD had previously advised Scott on his writing style, and may also have commented on the style of Scott 1864a and Scott 1864c in the letter to which this is a reply, which is missing (see n. 1, above, *Correspondence* vol. 10, letter to John Scott, 11 December [1862], and *Correspondence* vol. 11, letters to John Scott, 31 May [1863] and 1 and 3 August [1863]).

[12] In February 1864 George Bentham had suggested that Scott be put forward for election as an associate of the Linnean Society; however, a bylaw of the society limited the election of associates to those resident in the 'British Dominions' (1861 Charter and Byelaws of the Linnean Society of London, Library of the Linnean Society). The proposal was made following the reading and warm reception of Scott's paper on the Primulaceae (Scott 1864b) at the Linnean Society on 4 February 1864 (see *Correspondence* vol. 12, letters from J. D. Hooker, 5 February 1864, [before 9 February 1864], and 26[–8] October 1864, and letter to John Scott, 9 February [1864] and n. 9).

[13] Scott refers to Asa Gray's review of Scott 1864b entitled 'Dioico-dimorphism in the primrose family' (A. Gray 1865a); the review appeared in the January issue of the *American Journal of Science and Arts*. CD had encouraged Gray to write the review, and sent him a copy of Scott 1864b together with an abstract of its contents (see *Correspondence* vol. 12, letter to Asa Gray, 13 September [1864], and letter from Asa Gray, 3 October 1864).

[14] Scott began his paper on *Primula* with a discussion of the dimorphic species of the allied genus *Hottonia*, including observations on the North American species *H. inflata* (see Scott 1864b, pp. 78–9). Gray did not comment on these observations in his review, and mentioned the genus *Hottonia* only briefly (A. Gray 1865a, p. 101).

[15] For CD's support of Scott, financial and otherwise, see *Correspondence* vol. 12, letter to John Scott, 10 June 1864, n. 1, and this volume, letter from John Scott, 20 January 1865 and nn. 3, 4, and 6.

[16] CD's annotations appear to be notes for a letter to Scott that has not been found.

[17] CD probably intended to send Scott a copy of 'Climbing plants', an abstract of which was read before the Linnean Society on 2 February 1865; offprints of the paper were available from 12 June 1865 (see Freeman 1977, p. 117).

[18] In a missing letter to Scott (see n. 16, above), CD probably discussed his observations on the cleistogamic flowers of *Leersia oryzoides* (see letter from John Scott, 21 July 1865). In *Forms of flowers*, p. 335, CD noted that he had sent seeds from the cleistogamic flowers to Scott, who 'cultivated the plants in various ways, but they never produced perfect flowers'.

[19] CD had acquired a *Lagerstroemia*, a heterostyled member of the Lythraceae, from Hooker in 1863, and evidently wanted to learn more of Indian species (see letter from John Scott, 21 July 1865 and nn. 15 and 16).

From W. B. Tegetmeier 10 April 1865

Muswell Hill | London | N
Monday Ap 10 *1865*.

My dear Sir

The MSS arrived duly.—[1] Your estimate of the extent to which my patience would be taxed in reading it is very erroneous. I devoured it at one sitting.—but I will not return it for a few days with your permission as I want to look over several parts again—.

Mr Wells will undertake the drawings at 30^s^/ each on the wood and will take the greatest pains to do them with extreme accuracy.—

The reason that the Smerle and Trumpeter are more correct than the others, is that I told him to draw the birds as they actually existed, whereas the Barb and Carrier were drawn when my eye was injured and I did not see the sketches and the Owners insisted on Mr Wells seeing with their eyes. consequently the engravings are not correct.— fancy standards instead of actual transcripts being given—[2]

Mr Wells is not a business man and would rather have nothing to do with the wood engravers account. but I will see the Engravers and get an estimate sent in to Mr Murray[3]

Wells drew for our old friend Mr Yarrell so he may be trusted.[4]

I enclose a reply from Mr Zurhorst (the most successful and largest breeder of White Cochins) respecting the colour of the Chicken in the down.[5] I felt quite certain that Mr Brent was wrong.[6] I do not know whether he ever had any white cochins. but I should regard Mr Zurhorsts letter as conclusive, he is a very intelligent gentleman and an ardent admirer of this breed—

Like every one else he seems proud to be able to assist you even in the slightest and asks me if I am enquiring for "the great writer on species. if so I prostrate myself as a persian to the rising sun" and he further says the replies are from his own experience.

I have made several notes on the Pigeon MSS but will not send them until I return the MSS

Believe me very truly Yours | W B Tegetmeier

Pray do not trouble to write or dictate a reply to this it requires none and I shall be writing when I return the MSS

DAR 178: 67

[1] Tegetmeier refers to CD's manuscript of the chapters on variation in domestic pigeon breeds for *Variation* (see letter to W. B. Tegetmeier, [7 April 1865]).

[2] The reference is to drawings of pigeons by Luke Wells in the *Field* (see letter from W. B. Tegetmeier, 27 March 1865 and n. 5, and letters to W. B. Tegetmeier, 28 March [1865], and [7 April 1865] and n. 2).

[3] In his letter to W. B. Tegetmeier of 6 April [1865], CD asked that Wells send an estimate of the cost of the engravings to John Murray. See letter from W. B. Tegetmeier to John Murray and R. F. Cooke, 17 April 1865.

[4] William Yarrell had helped CD buy equipment for the *Beagle* voyage, and later assisted him with his pigeon work, advising him about breeds and providing introductions to dealers and breeders (for Yarrell's correspondence with CD, see *Correspondence* vols. 1–5). Yarrell introduced CD to Tegetmeier in 1855 (see *Correspondence* vol. 5, letter to W. B. Tegetmeier, 31 August [1855], n. 2). Yarrell's books on British birds and fishes were illustrated with hundreds of wood engravings made under his supervision (Yarrell 1836 and 1837–43).

[5] The enclosure has not been found, but Frederick William Zurhorst was cited on the colour of the down feathers of Cochin chicks in *Variation* 1: 249:

> In white fowls of various breeds the chickens are uniformly yellowish white. . . . This is also generally the case with the chickens of white Cochins, but I hear from Mr. Zurhost [*sic*] that they are sometimes of a buff or oak colour, and that all those of this latter colour, which were watched, turned out males.

See also n. 6, below.

[6] Bernard Peirce Brent is acknowledged in *Variation* 1: 249 for sending CD information on the colour of down feathers in the chicks of a number of breeds of domestic fowl; CD's correspondence with Brent on this subject has not been found. CD began studying variation in young fowl in 1855 (see *Correspondence* vol. 5, letter to W. D. Fox, 14 October [1855]) in an attempt to shed light on the ancestral similarities and dissimilarities of well-established breeds. CD's first letter to Tegetmeier on down feathers is dated 12 [May 1857] (*Correspondence* vol. 6); Tegetmeier probably made additional enquiries in the missing portion of his letter of [29 March – 7 April 1865] after reading CD's manuscript on 'Fowls' (see letter to W. B. Tegetmeier, [7 April 1865] and n. 1).

From J. D. Hooker 12 April [1865][1]

12th. April.

Dr Darwin

The Strelitzia & books never went till today.[2]

My Father is very unwell with Bronchitis & Influenza & we very uneasy about him, he being $79\frac{3}{4}$[3] Last week, after a hard mornings work in the Gardens, he went to the City in the afternoon, to see a nephew[4] ill of Typhus. it was the first hot day & knocked him up.

Many thanks for your note.[5] Bentham wrote on Planchon Thomson on sub-species & Greene of Cork on Linn. Trans.[6] I shall be curious for your criticisms— I did the Darwinian thing in G. C.—(right for once— exceptions makes rules)[7] I can think of nothing else nasty to say & so good bye

Henry Bence Jones
Photograph by Ernest Edwards, 1868
Courtesy of the Wellcome Library, London

John Chapman
From Haight 1940. By permission of
the Syndics of Cambridge University Library

Henry Walter Bates
Carte de visite by R. J. Whitlock, Birmingham
Courtesy of the Royal Geographical Society, London

John Lindley
From the *Journal of Horticulture,* 16 October 1902
By permission of the Syndics of Cambridge University Library

Ever Yrs affec | J D Hooker

I am egregiously delighted with your calm judgement on the "Origin".[8] Do you know I have reread some of my papers with the same result **& never was wrong once in my opinion**

DAR 102: 17

[1] The year is established by the reference to William Jackson Hooker's age; he was born on 6 July 1785 (*DNB*).

[2] See letter from J. D. Hooker, [7–8 April 1865], and letter to J. D. Hooker, 10 [April 1865].

[3] The reference is to William Jackson Hooker. See letter from J. D. Hooker, [7–8 April 1865].

[4] This nephew has not been identified.

[5] Letter to J. D. Hooker, 10 [April 1865].

[6] Hooker refers to the authors of three unsigned reviews in the April issue of the *Natural History Review*: George Bentham, Thomas Thomson, and Joseph Reay Greene. See letter to J. D. Hooker, 10 [April 1865] and n. 7.

[7] Hooker refers to his unsigned leader in the *Gardeners' Chronicle*, 25 March 1865, pp. 267–8; see letter to J. D. Hooker, 10 [April 1865] and n. 8.

[8] See letter to J. D. Hooker, 10 [April 1865].

To J. D. Hooker 13 April [1865]

Down
April 13th

My dear Hooker.

Here is a pretty thing, the Strelitzia has arrived safely this morning (specific name not given) but **no** Books & Bottle from Thwaites.[1] I hope they are not lost, but if so of course I will make good the Bot. Zeitung.—[2]

I shall be glad to hear news before very long of your Father: Bronchitis at near 80 is very serious.—[3]

Yours affect | C. Darwin

Your note received this morning made me laugh— I hope my judgment about the Origin is as good & unfailing as yours about your own papers![4]

It is worth while to pass your little finger, like a Birds beak, parallel to stamens of Strelitzia, as if to enter within the little petal at the base for nectar,, & see how neatly the two other oddly shaped blue petals open & expose pollen.—[5]

Endorsement: '/65.'
DAR 115: 266

[1] CD refers to a *Strelitzia* specimen, and to a bottle and some other items that Hooker had undertaken to forward from George Henry Kendrick Thwaites; Hooker had also mentioned that he would send Grisebach 1864 and an issue of *Botanische Zeitung* (see letters from J. D. Hooker, [7–8 April 1865] and 12 April [1865], and n. 2, below).

[2] CD had asked to borrow an issue of *Botanische Zeitung* for 1860 containing an article on *Pisum* (Alefeld 1860). See letter to J. D. Hooker, 6 April [1865].

[3] Hooker mentioned the illness of William Jackson Hooker in his letter to CD of 12 April [1865].

[4] See letter from J. D. Hooker, 12 April [1865].

[5] For CD's interest in the pollination of *Strelitzia*, see the letter from J. D. Hooker, [7–8 April 1865], n. 4.

To J. D. Hooker 17 April [1865]

Down
Apr 17th

My dear Hooker,

We have been quite gay: J. Lubbock came here on Friday to lunch & was very pleasant:[1] you will be glad to hear that he did not speak seriously of M^{rs}. L. & I suspect it is only case of family-wayishness.[2] The more I think of it, the sorrier I am about parliament, though yet I do not at all like the thought of his being beaten.[3] He says that he hears that his Book on Man will tell heavily against him.[4] He will be Sir John, before very long, for poor old Sir J. has heart-mischief & is dropsical.[5]

On Sunday we had a call from M. Laugel, geologist & litterateur, a very agreeable, clever, & charming man: just returned from N. America & very enthusiastic for the Federals & very sanguine for their future in every respect.[6]

I will keep Bot. Zeitung for about one week—[7] I have read N. Hist R. & guessed right to myself that Bentham wrote on Planchon;[8] I liked the article; but as I had just read the essay there was not much new to me.—[9] I have been very much struck by Thomson's article: it seems to me quite remarkable for its judgment, force & clearness. It has interested me greatly.[10] I had sometimes *loosely* speculated on what nomenclature would come to & concluded that it would be trinomial. What a name a plant will formally bear with the authors name after genus (as some recommend) & after species & subspecies! It really seems one of the greatest questions which can now be discussed for systematic Nat. Hist.[11] How impartially Thomson adjusts the claims of "hair-splitters" & "lumpers"[12] I sincerely hope he will pretty often write reviews or essays. It is an old subject of grief to me, formerly in geology, & of late in Zoolog. & Botany, that the very best men, (excepting those who have to write Principles & Elements &c)[13] read so little & give up nearly their whole time to original work; I have often thought that science would progress more if there was more reading. How few read any long & laborious papers. The only use of publishing such seems to be as a proof that author has given time & labour to his work.

Well farewell my dear old fellow— let me hear before very long how Sir William is—[14]

Yours affect | C. Darwin

Endorsement: '/65.'
DAR 115: 265

1 CD was writing on Easter Monday. Emma Darwin's diary (DAR 242) indicates that George Howard Darwin was home from Cambridge, and that Elizabeth, Francis, and Leonard Darwin were home from 13 to 24 April; William Erasmus Darwin may have been present as well, after arriving on 13 April. Horace and Henrietta Emma Darwin were living at Down. There is no mention in the diary of John Lubbock's visit.

2 Hooker had been concerned about the health of Ellen Frances Lubbock (see letter from J. D. Hooker, [7–8 April 1865] and n. 13).

[3] Lubbock was standing for election to Parliament as a Liberal Party candidate for West Kent; this constituency had returned two Conservative members of parliament at the last election in 1859 (*Dod's parliamentary companion* 1865). See also letter from J. D. Hooker, [7–8 April 1865] and n. 14.

[4] The reference is to *Pre-historic times* (Lubbock 1865), in which Lubbock discussed the archaeological evidence for the antiquity of humankind. Although in recent years the view that humans had a chronologically remote and 'savage' origin had become more widely accepted (see Grayson 1983, pp. 217–20), Lubbock's supporters believed that the book would harm his chances of election and urged him to delay its publication until after the election in July (see Hutchinson 1914, 1: 74).

[5] John William Lubbock died on 20 June 1865 (see letter from John Lubbock, 22 and 26 March 1865 and n. 3).

[6] The reference is to Auguste Laugel. In 1860, Laugel had sent CD a copy of his review of *Origin* in *Revue des Deux Mondes* (Laugel 1860). See *Correspondence* vol. 8, letter to Asa Gray, 25 April [1860] and letter to ?, 25 [April 1860?]. There is an annotated copy of Laugel 1860 in the Darwin Pamphlet Collection–CUL. Laugel described his 1864 trip to the northern states of the USA in Laugel 1865; he discussed his support for the Union during the American Civil War in Laugel 1863, Laugel 1864, and Laugel 1866.

[7] See letter to J. D. Hooker, 13 April [1865], n. 2.

[8] CD refers to George Bentham's review of Planchon 1864a and 1864b in the April 1865 issue of the *Natural History Review* ([Bentham] 1865); see letter from J. D. Hooker, 12 April [1865] and n. 6.

[9] CD evidently refers to Planchon 1864b; there is an annotated copy of this essay in the Darwin Pamphlet Collection–CUL.

[10] The reference is to [T. Thomson] 1865. Thomas Thomson's article was a critical review of Alexis Jordan's attempt to establish by experiment the permanence or variability of plant forms, with a view to providing a new system of classification (Jordan 1864). Near the start of the article, Thomson wrote: 'Darwin's marvellous observations have of late set all the world thinking about variation and variability, and have already stimulated to an enormous extent the observing faculties of naturalists' ([T. Thomson] 1865, p. 227). Thomson used his review of Jordan's work as a point of departure for discussing classification systems in taxonomic botany and zoology in general.

[11] Thomson discussed trinomial systems of botanical nomenclature in [T. Thomson] 1865, pp. 238–41. In 1842, CD had served on a committee of the British Association for the Advancement of Science to 'draw up a series of rules with a view to establishing the nomenclature of zoology on a uniform and permanent basis' (see *Correspondence* vol. 2 and McOuat 1996). CD developed his understanding of scientific nomenclature further in the course of his study of barnacles, which involved him in practical problems of classification (see *Correspondence* vol. 4, Introduction and Appendix II); CD's ideas developed further in discussions with Hooker on classification in botany (see, for example, *Correspondence* vol. 7, letter from J. D. Hooker, [25] February 1858). In 1842, the British Association rules of nomenclature had codified a binomial system in which each species was given two names, one to designate the genus and one the species, followed by the author's name in parentheses whenever the species name was joined to a new generic name (see Strickland *et al.* 1863). However, by the end of the century the consensus amongst taxonomists in Europe and North America that had allowed the adoption of these rules had broken down, and a trinomial system of naming a species by genus, species, and sub-species began to be adopted by some zoologists and botanists, to take account of geographic variation (see Mayr 1942, pp. 108–11, Stresemann 1975, pp. 264–7, Stace 1980, pp. 206–10, and Stevens 1994).

[12] [T. Thomson] 1865, pp. 226 and 235–6, discussed the different conclusions of 'local botanists', who made detailed observations of a limited number of plants inhabiting a small area, and of 'general botanists', who made extensive surveys of a wide range and large number of plants. The former, characterised here by CD as 'hair-splitters', tended to create more species according to the small differences they observed; the latter, the 'lumpers', tried to 'grasp the whole range of plant forms, and . . . came to overlook or under-estimate the importance of minute characters', thereby creating a 'bias in favour of combination, rather than separation'. Jordan, whose work Thomson was reviewing, was

well known for naming a large number of species (see *Correspondence* vol. 5, letters from H. C. Watson, 13 August 1855 and 2 October 1855, and *Correspondence* vol. 8, letter from H. C. Watson, 10 May 1860). For additional discussions on these two tendencies in systematics and their bearing on CD's theory, see, for example, *Correspondence* vol. 6, letter to J. D. Hooker, 1 August [1857], and letter from J. D. Hooker, [6 December 1857], and *Correspondence* vol. 7, letter to J. D. Hooker, 11 March [1858].

[13] CD refers to compilers of text-books or of works aiming to provide an overview of a subject. CD was presumably thinking of Charles Lyell, who was working on the tenth edition of *Principles of geology* (C. Lyell 1867–8), and whose *Elements of geology* had reached a sixth edition in January 1865 (C. Lyell 1865). CD had expressed his admiration for Lyell's work in this genre on several occasions (see, for example, *Correspondence* vol. 11, letter to J. D. Hooker, 24[–5] February [1863], and this volume, letter to J. D. Hooker, 15 [February 1865]).

[14] Hooker's father, William Jackson Hooker, was seriously ill (see letter from J. D. Hooker, 12 April [1865]).

From W. B. Tegetmeier to John Murray and Robert Francis Cooke 17 April 1865

Muswell Hill, N.
April 17 *186*5

Gentlemen[1]

My friend Mr Darwin has requested me to overlook the drawings of some pigeons and fowls for his forthcoming work.[2] I have seen Mr Wells the Artist who has made the drawings for **The Field**—[3] his terms would be 30^s^/. per subject for drawing on the wood.—full page size. as he declined to undertake the engraving, I have seen Messrs Butterworth & Heath who have engraved our blocks and obtained an estimate from them which I enclose[4]

I should prefer the drawings to be made by Mr Wells as he is exceedingly careful and truthful in his work. I am not at all particular as to the engraver, but as Messrs Butterworth & Heath engrave Mr Wells drawings very well I obtained the estimate from them, I should be glad of your authorization to enable Mr Wells to proceed with the work

I am Gentlemen | Yours obd^tly^ | W B Tegetmeier

Messrs Murray & Co

I have left a specimen of the blocks as drawn and engraved for "**The** *Field*"[5]

[Enclosure]

356 Strand W.C.
April 10, 1865

Dear Sir

In reference to the engravings for D^r^. Darwin's book about which you spoke to us today, we think the cost would be £2·10/– each— We understand them to be single birds, drawn in the style of those we are engraving for 'the Field'.[6]

We remain dear Sir | Yours faithfully | Butterworth & Heath

W. B. Tegetmeier, Esq.

John Murray Archive (Darwin 136–8)

[1] Murray had asked CD to send any correspondence to his cousin and partner, Robert Francis Cooke, during the first three weeks of April (see letter to W. B. Tegetmeier, 6 April [1865] and n. 10).
[2] See letters to W. B. Tegetmeier, 14 March [1865], 28 March [1865], and [7 April 1865]. Tegetmeier refers to *Variation*.
[3] The reference is to Luke Wells and to a series of illustrations of pigeons in the *Field* (see letters from W. B. Tegetmeier, 27 March 1865 and n. 5, and 10 April 1865, and letter to W. B. Tegetmeier, [7 April 1865], n. 2). The *Field* was an illustrated weekly paper that carried articles on country pursuits and natural history; Tegetmeier was editor of the poultry section (see letter from W. B. Tegetmeier, [after 27 February 1865]).
[4] Butterworth and Heath were a firm of wood-engravers with premises at 356 Strand, London (*Post Office London directory* 1865).
[5] Tegetmeier refers to the engraved wood blocks from which the illustrations for the series on pigeon breeds in the *Field* were printed (see letter from W. B. Tegetmeier, 27 March 1865 and nn. 5 and 7).
[6] Six of the engravings of pigeon breeds in *Variation* are signed by Butterworth and Heath (see *Variation* 1: 135, 137, 140, 145, 147, 152).

From R. F. Cooke to W. B. Tegetmeier 18 April 1865

50A, Albemarle Street | London, W.
April 18, 1865.

Sir

In reply to your letter of yesterday,[1] I beg to authorize you to place M^r. Darwin's Drawing in the hands of Mr. Wells the artist,[2] who done so nicely the illustrations in the Field and altho' M^r. Murray[3] has had no engravings done by Messrs. Butterworth & Heath, yet as they have been in the habit of engraving M^r. Wells' drawings, he will be very happy that he should undertake these for M^r. Darwin's book, on the terms mentioned, which no doubt is a fair average price.[4]

I am Sir | faithfully yours | Robt. Cooke

W. B. Tegetmeier, Esq.

Copy
DAR 148: 1

[1] See letter from W. B. Tegetmeier to John Murray and R. F. Cooke, 17 April 1865.
[2] Luke Wells.
[3] John Murray.
[4] See letter from W. B. Tegetmeier to John Murray and R. F. Cooke, 17 April 1865.

To Asa Gray 19 April [1865][1]

Down Bromley Kent
April 19th

My dear Gray

I have not written for a long time (& a good job too perhaps you will think), not since receiving your letter written on Jany 19th;[2] but I have often thought of you & often wished to write, but either had other things to do, or felt too tired. I have nothing particular to say now, but the grand news of Richmond has stirred me up to write.[3] I congratulate you, & I can do this honestly, as my reason has always urged & ordered me to be a hearty good wisher for the north, though I could not do so enthusiastically, as I felt we were so hated by you.—[4]

Well I suppose we shall all be proved utterly wrong who thought that you could not entirely subdue the South. One thing I have always thought that the destruction of Slavery would be well worth a dozen years war.[5] Two days ago a very charming man, enthusiastic for the north, called here, Mr. Laugel, & he does not believe that you will attack us & Canada.[6] I fear it will take many years before your country will shake down to its old routine.—

I received a little time ago a paper with good account of your Herbarium & Library,[7] & a long time previously your excellent review of Scotts Primulaceæ, & I forwarded it to him in India, as it would much please him.—[8] I was very glad to see in it a new case of Dimorphism (I forget just now the name of plant);[9] I shall be grateful to hear of any other cases, as I still feel interest on subject. I shd. be very glad to get some seed of your dimorphic Plantagos;[10] for I cannot banish suspicion that they must belong to the very different class like that of the common Thyme. How could the wind, which is agent of fert. with Plantago, fertilise "reciprocally dimorphic" flowers like Primula.[11] Theory says this cannot be, & in such cases of one's own theories I follow Agassiz & declare "that Nature never lies".[12] I shd even be very glad to examine the 2 dried forms of Plantago. Indeed any dried dimorphic plants wd. be gratefully received. You made capital remarks, with respect to Mohls little imperfect flowers, on flowers which rarely open—[13]

Did my Lythrum paper interest you?[14] I crawl on at rate of 2 hours per diem with Variation under Domestication;[15] & I have begun correcting proofs of my paper on "Climbing Plants". I suppose I shall be able to send you a copy in 4 or 5 weeks. I think it contains a good deal new & some curious points, but it is so fearfully long, that no one will ever read it. If, however, you do not *skim* through it, you will be an unnatural parent, for it is your child[16]

Believe me, my dear Gray, | Yours affectionately | Charles Darwin

Gray Herbarium of Harvard University (77)

[1] The year is established by the relationship between this letter and the letter from Asa Gray, 15 and 17 May 1865.

[2] Letter from Asa Gray, 17 January 1865; CD wrote '19th' in error.

[3] Richmond, Virginia, had fallen to Federal troops on 3 April 1865; General Robert E. Lee of the Army of Northern Virginia surrendered to General Ulysses S. Grant at Appomattox Courthouse, Virginia, on 9 April 1865. The other Confederate armies soon followed suit (see McPherson 1988, pp. 846–9, and Denney 1992, pp. 554–7).

[4] Tension between Britain and the northern states of the USA during the American Civil War, and Gray's staunch support of the North, had affected the correspondence between Gray and CD (see, for example, *Correspondence* vol. 10, letter to Asa Gray, 15 March [1862], and letter from Asa Gray, 31 March [1862], and *Correspondence* vol. 11, letters to Asa Gray, 2 January [1863] and 4 August [1863], and letter from Asa Gray, 27 January 1863). See also Colp 1978.

[5] For CD's correspondence with Gray about the American Civil War and his opposition to slavery, see the letter from Asa Gray, 17 January 1865 and n. 11.

[6] CD mentioned Auguste Laugel in his letter to J. D. Hooker, 17 April [1865]. On British fears that a victorious Union army would turn to new theatres of war, including Canada, in an effort to expand its territory, see Crook 1975, pp. 126–7, 168–171, and 177.

[7] The article has not been found. Gray's herbarium and library had been moved to the new Harvard University Herbarium (see Dupree 1959, pp. 327–9, and letter from Asa Gray, 17 January 1865). The building of the herbarium is discussed in the letters from Asa Gray, 16 February 1864 and 11 July 1864 (*Correspondence* vol. 12).

[8] CD refers to A. Gray 1865a, Scott 1864b, and John Scott. See letter from John Scott, 10 April 1865 and n. 13.

[9] At the end of his notice of Scott's paper, Gray wrote: 'We may here append the remark that the Thymelæaceous genus *Leucosmia* is dimorphous, and some species of *Drymispermum* exhibit one if not both of the two forms' (A. Gray 1865a, p. 104).

[10] CD probably refers to four North American species of heterostyled *Plantago* discussed in A. Gray 1856, p. 269. There is an annotated copy of A. Gray 1856 in the Darwin Library–CUL (see *Marginalia* 1: 348–51). CD discussed the pollination mechanisms of *Plantago* with Gray on several occasions (see *Correspondence* vol. 10, letter to Asa Gray, 26[–7] November [1862], and letter from Asa Gray, 29 December 1862, and *Correspondence* vol. 11, letter to Asa Gray, 11 May [1863] and n. 14). CD evidently did not obtain seeds of dimorphic *Plantago* from Gray (see letter from Asa Gray, 15 and 17 May 1865); his observations on dimorphism in this genus were made primarily with the locally available *P. lanceolata* (see *Forms of flowers*, pp. 306–7). CD's notes on *P. lanceolata*, dated from 28 April to 14 June 1863, are in DAR 109: A27–9 and DAR 49: 91; see also DAR 109: A50 and A64, and DAR 111: A51.

[11] Species of thyme were taken by CD to be representative of a class of dimorphism that he later called 'gyno-dioecism', in which plants bear either hermaphrodite or female flowers (see, for example, *Correspondence* vol. 10, letter to Daniel Oliver, 12 [April 1862], *Correspondence* vol. 11, letter to W. E. Darwin, [5 May 1863] and n. 6, and *Correspondence* vol. 12, letter to W. E. Darwin, 14 May [1864], and letter to Asa Gray, 28 May [1864]). See also CD's experimental notes on thyme dated from 1861 to 1864 in DAR 157a: 72 and DAR 109: A21–2, 28, and 45–6 v. CD discussed *Thymus* and *Plantago* as examples of gyno-dioecious plants in *Forms of flowers*, pp. 298–307. CD presumably refers to the unlikelihood that pollen on the short stamens of one form of dimorphic *Primula* species could be carried by the wind to the short style of the other form; he had demonstrated that the dimorphic condition was an adaptation to ensure cross-pollination by insects (see 'Dimorphic condition in *Primula*'). CD noted that wind played no part in the pollination of dimorphic *Linum* species in 'Two forms in species of *Linum*', pp. 77–8 (*Collected papers* 2: 100–1), and *Forms of flowers*, pp. 93–4.

[12] CD refers to Louis Agassiz. In his letter to Asa Gray, 1 January [1857] (*Correspondence* vol. 6), CD wrote: 'Lyell told me, that Agassiz having a theory about when Saurians were first created, on hearing some careful observations opposed to this, said he did not believe it, "for Nature never lied"'. See also *ibid.*, letter to J. D. Hooker, [21 March 1857] and n. 4.

[13] CD refers to Gray's review of a translation of Mohl 1863 in the January 1865 issue of the *American Journal of Science and Arts* (A. Gray 1865b). Hugo von Mohl had argued (Mohl 1863, pp. 309, 321) that the existence of small, imperfectly formed, closed, self-fertile flowers in *Viola*, *Oxalis*, and other genera was an argument against CD's view that nature was arranged to prevent perpetual self-fertilisation and to promote intercrossing (see *Origin*, pp. 96–101, and *Orchids*, p. 359). In his review, Gray agreed with Mohl that these examples stood as exceptions to CD's argument, but he did not consider Mohl to have overthrown the general principle. On CD's answer to Mohl's criticisms, see the letter from B. D. Walsh, 1 March 1865 and n. 12.

[14] Gray acknowledged receiving a copy of 'Three forms of *Lythrum salicaria*' in his letter of 17 January 1865. For CD's presentation list for this paper, see *Correspondence* vol. 12, Appendix III.

[15] CD refers to *Variation* (see letter to John Murray, 31 March [1865] and n. 2).

[16] CD's paper 'On the movements and habits of climbing plants' was published on 12 June 1865 in a double issue of the *Journal of the Linnean Society* (*Botany*) ('Climbing plants'). CD had an author's offprint separately printed by Taylor and Francis, and it was also published commercially in August 1865 by Longman, Green, Longman, Roberts & Green and by Williams and Norgate (see *Publishers' Circular*, 1 August 1865, p. 391, and Freeman 1977, pp. 116–18). CD apparently began sending out copies of the paper to correspondents in early June (see letter from Charles Kingsley, 14 June 1865). 'Climbing plants' begins with an acknowledgment of Gray's study of the movements of the tendrils of cucurbitaceous plants (A. Gray 1858) for stimulating CD's interest in the subject. CD had been regularly corresponding with Gray on this subject since 1863 (see, for example, *Correspondence* vol. 11, letters to Asa Gray, 26 June [1863] and 4 August [1863], and letter from Asa Gray, 1 September 1863, and *Correspondence* vol. 12, letter to Asa Gray, 28 May [1864]). See also *Autobiography*, p. 129.

From J. D. Hooker [19 April 1865][1]

Kew
Wednesday Nt.

Dear Darwin

I have nearly finished "Can you forgive, her", & have made up my mind that I cannot at all do so.—& don't care whether she minds it or no.[2]

What a jolly letter I got from you today![3] Lyell[4] came out most kindly to ask after my Father,[5] I read it to him, & he cocked his chin in the air & said, "Well—it is worth living on to be able to write such a letter as that"— I am so pleased at your opinion of Thomson's article.[6] I do wish he would do himself justice, for he is a *very* able man. I thought the article extraordinarily good, & am so little accustomed to see him exert his powers, that I did not think it was his—though I knew he had been writing such an article!

My Father is still very poorly—not worse certainly, but still in bed, & coughing much—

I echo all you say about non-reading men.—[7] & always did & do lift up my voice against geologists in particular— have I not roared it out at Down? It is the great fault of the best school of English scientific men—Tyndall, Ramsay[8] & many others.

I gnash my teeth when I think of Lubbock going into Parliament.[9] Lyell is dead against it & gave me the experience of the friends who advised him against it when it was proposed that there should be a university Member & that L. should stand for it, Grote especially.[10] The **awful** waste of time, of energy, of brain, of life, & all that makes life worth having.— always except a man goes in for Politics, Finance, or Self-aggrandizement— for such the uphill drag through mire of all kinds, Dinners, Committees, Deputations, Lady P$^{s.}$ receptions,[11] Levees—&c &c.— all this & more may be worth a man's undergoing who has a clear calling that way & a prospect of some 25 years political superiority or supremacy at the end of it.

I know Laugel[12] very well, he lives at Richmond, & sometimes comes with his wife to my informal dinners of 6.— you call him charming— so he is;—but Lord bless you you should see his wife—the most lovely & loveable little woman, in the way of Lady Lyell[13] 20 years ago; so fresh & nice & good, & crispy.

I hoped to have got away at Easter but of course could not & do not expect to get away now.

Ever Yrs affect | J D Hooker

All Burchells collections are coming here soon! most valuable & excellent.[14] Brazil & S. Africa.

DAR 102: 18–19

CD ANNOTATION

Top of letter: 'Peach'[15] *blue crayon, circled blue crayon*; 'Stone' *pencil*

[1] The date is established by the relationship between this letter and the letter to J. D. Hooker, 17 April [1865]; the Wednesday after 17 April 1865 was 19 April.

[2] In Anthony Trollope's novel *Can you forgive her?*, Alice Vavasor breaks off her engagement to the paragon John Grey, and becomes attached to her ruthless cousin, an aspiring politician (see Trollope 1864–5). See letter to J. D. Hooker, 10 [April 1865].

[3] Letter to J. D. Hooker, 17 April [1865].

[4] Charles Lyell.

[5] William Jackson Hooker. See letter from J. D. Hooker, 12 April [1865].

[6] Hooker refers to [T. Thomson] 1865 (see letter to J. D. Hooker, 17 April [1865]).

[7] See letter to J. D. Hooker, 17 April [1865].

[8] The reference is to John Tyndall and Andrew Crombie Ramsay.

[9] The reference is to John Lubbock; see letter from J. D. Hooker, [7–8 April 1865] and n. 14, and letter to J. D. Hooker, 17 April [1865].

[10] In 1861 Lyell declined the offer of the candidature for the University of London's representative in Parliament (see K. M. Lyell ed. 1881, 2: 343). George Grote, the historian, had been an active reform member of parliament for the City of London between 1832 and 1841, during which time he had published little (*DNB*). He retired to devote himself to writing his eight-volume *History of Greece* (Grote 1846–56).

[11] Hooker refers to Emily Mary Temple, wife of Henry John Temple, third Viscount Palmerston, who was then prime minister; Lady Palmerston's Saturday evening parties were well known in political circles (see *Modern English biography* and Pemberton 1954, pp. 82–3, 118).

[12] Hooker refers to Auguste Laugel (see letter to J. D. Hooker, 17 April [1865]).

[13] The reference is to Mary Elizabeth Lyell; Mme Laugel has not been further identified.

[14] William John Burchell had died in 1863, leaving his South African and South American plant collections and botanical manuscripts to the Royal Botanic Gardens, Kew (*DNB*, R. Desmond 1994). CD had been familiar with Burchell's work (see *Correspondence* vol. 1, letter to J. S. Henslow, [5 September 1831], letter to Catherine Darwin, 22 May – 14 July 1833, and Appendix IV).

[15] See letter to J. D. Hooker, 4 May [1865] and n. 8.

From Herbert Spencer 22 April 1865

88, Kensington Garden's Sq^{re}.
22nd. Ap^{l}. 1865

My dear Sir

We are getting our staff of the *Reader* into better working order; and are proposing forthwith to use all the means available for making a more decided impression, and establishing our position.[1] Profs Huxley and Tyndall, M^{r}. J. S. Mill[2] and myself, have severally agreed to write a few leading articles by way of giving the intended tone and direction.[3]

Among other means of making the public aware of the character of the *Reader*, we propose to obtain, so far as possible, occasional brief letters from the leading men of science, announcing such interesting novelties as admit of being understood by the general public, and are of fit nature to be quoted from our columns. I have a letter from Sir John Herschel consenting to aid us in this way. Sir Charles Lyell, too, has promised the like aid.

Can you in like manner give us, occasionally, the valuable help of your name?

I am aware that your health is so precarious that you are obliged to be very economical of your energies—a fact which I greatly regret. And I am fully impressed with the fact, that it would be wrong to ask you to undertake any other labours than those important ones which occupy you. The favor we ask, however, does not involve such a waste of labour in any appreciable degree. A letter of *a dozen lines* would suffice the purpose of giving us the weight of your name; and making it apparent that you joined in the effort to establish a scientific journal, and an organ of progressive opinion.[4] We are all just now making considerable sacrifices to do this; and feel the necessity of concentrating all possible means of insuring that success which we have not yet quite achieved.

very sincerely yours | Herbert Spencer

LS
DAR 177: 225

[1] The *Reader: a Review of Literature, Science, and Art* was a weekly paper, founded in 1863, that provided a forum for reviews of literary and scientific works. Its audience was 'upper-middle-class, highly educated, politically liberal, and philosophically radical' (Sullivan ed. 1984, p. 346). Spencer was instrumental in re-launching the paper with new proprietors and increased scientific content in November 1864, after both he and CD had become shareholders (see *Correspondence* vol. 12, letter to John Lubbock, 19 November [1864] and n. 4, and Spencer 1904, 1: 118–19). The paper had been faltering since Thomas Henry Huxley's article on science and church policy in the issue of 31 December 1864 offended many of the Christian-Socialist supporters of the paper and led some of the proprietors and editors to withdraw (see letter from J. D. Hooker, 1 January 1865 and n. 14, and letter from T. H. Huxley, 15 January 1865, A. Desmond 1994–7, 1: 331–2, 343, and Barton 1998, p. 440). For CD's opinion of the *Reader*, see the letter to J. D. Hooker, 7 January [1865] and n. 11.

[2] Spencer refers to Thomas Henry Huxley, John Tyndall, and John Stuart Mill.

[3] On the intended character of the *Reader* and its connection with Huxley, Spencer, Tyndall, and other members of the X Club, see Barton 1998, pp. 433 and 439.

[4] No correspondence from CD appears in the 1865 issues of the *Reader*. For CD's support for the *Reader*, see n. 1, above.

From Eliza Meteyard 25 April 1865

Wildwood | North End | Hampstead. N.W.
April 25. 1865

Dear Sir

Permit me to offer you, in the hope of your acceptance, the 1st Vol of the 'Life of Wedgwood'.[1] I regret I have to send the work in this partial form, but the envy and machinations of two unscrupulous men rendered the step imperative.[2] The other volume will appear as soon as possible.

I lay the book before you with a trembling hand, and with sincere and great humility of spirit. I know what the subject requires—and I would that I had the highest human ability to do justice to my conception of it. Yet where I fail—my publishers splendid justice to the work—must be some compensation[3]—and the veneration I have for the names of Darwin and Wedgwood, and the liberal point of view from which I judge them and their opinions—must make up—as I hope it will in your kindly eyes—for deficiencies of various kinds. A bigot—whether social, political, or religious, *could not* estimate the character of two such men. Men far greater than their generation.

I am using your valuable letters as I go on.[4] I shall not have space I fear to do full justice to the more scientific aspect of Wedgwood's labours and bent of mind in *this* work[5]—but—as I am going to carry on the details of these and other subjects, through a work I shall call 'Thomas Wedgwood and his Contemporaries'[6] I shall be able to find a still more fitting place for many interesting though severer truths. As soon as I have finished the Life of Josiah Wedgwood, I will make what further extracts I need from your valuable letters—and carefully return them. Meanwhile they are in excellent keeping.

I trust your health has improved, for missing that—we miss almost the best thing we hold in life. Do not please trouble yourself to write— a line from Mrs Darwin just to say the book has reached you safely—is all that is required.

Dear Sir | With deep respect & grateful obligation | your's faithfully | Eliza Meteyard

C. Darwin Esq.

DAR 171: 160

[1] The reference is to the first volume of Meteyard 1865–6, a biography of Josiah Wedgwood I. For CD's unfavourable assessment of the book, see B. Wedgwood and Wedgwood 1980, pp. 299–300.

[2] The individuals referred to have not been identified.

[3] Meteyard refers to the publishers Hurst & Blackett of 13 Great Marlborough Street, London (*Post Office London directory* 1865); the first volume of Meteyard 1865–6 included over a hundred illustrations, including many of Wedgwood ware.

[4] CD had provided Meteyard with letters between his paternal grandfather, Erasmus Darwin, and his maternal grandfather, Josiah Wedgwood I (see *Correspondence* vol. 11, letter to J. D. Hooker, [22–3 November 1863], and this volume, supplement, letter to Hurst & Blackett, 15 November [1863]); a number of extracts from the letters between Wedgwood and Darwin appear in Meteyard 1865–6. There is a collection of over a hundred letters from Erasmus Darwin to Wedgwood written between 1766 and 1794 in DAR 227.1. See also *Correspondence* vol. 9, Appendix V.

[5] Meteyard dealt briefly with Wedgwood's scientific interests in chapters 9 to 11 (Meteyard 1865–6, 2: 415–572). See also Schofield 1963, B. Wedgwood and Wedgwood 1980, and King-Hele 1999 for a discussion of Wedgwood's links with provincial scientific circles.

[6] Thomas Wedgwood was a son of Josiah Wedgwood. The title of the book was changed to *A group of Englishmen*; the book made only three references to the correspondence supplied by CD (Meteyard 1871, pp. 8, 254–5).

From George Busk 28 April 1865

15 Harley St
April 28 1865

My dear Darwin

Just as I was starting on a short trip to Belgium from which I have but just returned, Hooker wrote to me to ask whether I could suggest the name of any one who might be able to lay down any treatment for your complaint—[1] I am very sorry to have delayed so long answering his query & still more so to learn by a note from him yesterday that you are at present suffering a great deal.—

I dont know whether Dr Garrod (84 Harley St) has ever been consulted in your case,[2] but if not I think he has had as great experience in gouty affections, if yours should be anything of that kind, as anyone—[3] He is a sensible man & I dont know any one more to be recommended for a trial, which I most fervently hope may be of some benefit against your cruel foe—

Believe me | Yours very truly | Geo. Busk

DAR 160: 381

[1] See letter from J. D. Hooker, [7–8 April 1865] and n. 18; see also letter to J. D. Hooker, 10 [April 1865].
[2] Busk refers to Alfred Baring Garrod. There is no record of CD having consulted him.
[3] In his letter to J. D. Hooker, 7 January [1865], CD mentioned that 'All Doctors' thought he was suffering from 'suppressed gout'. CD added that William Jenner, who had been treating him since March 1864, no longer thought he could help. See also letter to J. D. Hooker, 6 April [1865] and n. 3, and Appendix IV.

From L. C. or Margaret Susan Wedgwood to [Emma Darwin?][1] [May 1865]

Oxlips.

Long styled	short styled	total
13	16	29

These are the minimum numbers, for 4 of the short styled plants & 1 long styled consisted each of from 2 to 8 small plants, each cluster here counted as one plant so close together that the leaves interlaced—[2]

We have the numbers in each cluster shd Uncle Ch. want them.

Endorsement: 'May 1865'
AL incomplete
DAR 108: 74

CD ANNOTATIONS
Below table: 'Shows from Heteromorphic union'[3] *red crayon*

Verso: 'Leith Hill Place[4] | May 1865 | Showing that both long- & short-styled' *ink*

[1] The letter is in the hand of either Lucy Caroline or Margaret Susan Wedgwood, CD's nieces, who, along with their sister, Katherine Elizabeth Sophy Wedgwood, had previously sent botanical observations (see letter from L. C. Wedgwood, [April–May 1865?] and n. 2). It is conjectured that this was part of a letter to Emma Darwin. CD published this information in 'Specific difference in *Primula*', p. 448, and *Forms of flowers*, pp. 70–1, stating that it came from a friend in Surrey (see n. 4, below).

[2] The observations are of the numbers of the two different forms of the dimorphic oxlip and are discussed in 'Specific difference in *Primula*', p. 448, and *Forms of flowers*, pp. 70–1. CD used these data as evidence that the crosses producing oxlips in nature would probably be between different forms of the two parent species (see n. 3, below). For CD's interest in the relationship between oxlips, primroses, and cowslips, see the letter from L. C. Wedgwood, [April–May 1865?] and n. 3.

[3] By a heteromorphic union, CD meant a cross in which long-styled or short-styled flowers of separate plants had the pollen of the other form placed on their stigmas; he referred to crosses in which pollen of the same flower form was placed on the stigma as homomorphic ('Dimorphic condition in *Primula*', p. 87; see also *Collected papers* 2: 120). CD later substituted the terms legitimate and illegitimate for heteromorphic and homomorphic respectively. The figures collected by his nieces indicated to him that the plants had been derived from heteromorphic (legitimate) crosses, because if they had been derived from homomorphic (illegitimate) crosses, 'either the long- or short-styled form would have greatly preponderated in number' ('Specific difference in *Primula*', p. 448; see also *Forms of flowers*, pp. 70–1).

[4] Leith Hill Place, Surrey, was the home of Josiah Wedgwood III, CD's brother-in-law, and father of Lucy Caroline and Margaret Susan Wedgwood (Freeman 1978).

To J. D. Hooker [1 May 1865][1]

Down
Monday

My dear Hooker

I am a little better, though sickness continues. I can read a little & am very curious to read Caspary's speech or paper at the recent Amsterdam Hort. Congress,[2] on the union of buds in Cytisus,[3] if published in German or French.—[4]

Have you a copy? How was Gard:– Chronicle able to report?[5] Can you lend it me? & send by Post?

yours affect | C. Darwin

Endorsement: 'May 2/65'
DAR 115: 267

[1] The date is established by the endorsement; the Monday preceding 2 May 1865 was 1 May.

[2] The reference is to Robert Caspary's paper on *Cytisus adami* (Caspary 1865), which was read at the Congrès International de Botanique et d'Horticulture held at Amsterdam from 7 to 12 April 1865 under the auspices of the Royal Horticultural Society of the Netherlands. In its report on the Congress, the *Gardeners' Chronicle*, 29 April 1865, p. 386, noted that 'Professor Caspary spoke at some length on . . . hybridisation produced by grafting, adducing in illustration . . . the Cytisus Adami.'

[3] CD had long been interested in the hybrid laburnum, *Cytisus adami* (now known as +*Laburnocytisus adamii*), a graft hybrid of the common yellow laburnum *C. laburnum* (now *Laburnum anagyroides*) and *C. purpureus*, a dwarf purple broom (Bean 1970–88, 2: 510–11). See, for example, *Correspondence* vols.

4–6, *Correspondence* vol. 10, letter to Daniel Oliver, 24 July [1862] and n. 15, and letter to Thomas Rivers, 28 December [1862], and *Correspondence* vol. 11, letter to Thomas Rivers, 7 January [1863] and n. 12. Branches of +*L. adamii* have red, sterile flowers that regularly revert to both parent species in the colour of their flowers. CD collected articles on seed production printed in the *Gardeners' Chronicle* over many years (CD's annotated copies of the *Gardeners' Chronicle* are in the Cory Library, Cambridge Botanic Garden). *Cytisus adami* is discussed in *Variation* 1: 387–91, 406–7, and 2: 37, 364–5. In appearing to display hybrid fusion and reversion, *C. adami* was an important case contributing to CD's view that there was no essential difference between asexual and sexual reproduction; see Olby 1985, pp. 76–8.

4 There is a lightly annotated copy of Caspary 1865 in the Darwin Pamphlet Collection–CUL. The pamphlet is a reprint from the *Bulletin du Congrès International de Botanique et d'Horticulture réuni à Amsterdam* (1866): 65–80; bound with it is an annotated handwritten English translation of the pamphlet. CD asked Caspary for a copy in February 1866; see *Correspondence* vol. 14, letter to Robert Caspary, 21 February [1866], and letter from Robert Caspary, 25 February 1866. Caspary 1865 is cited in *Variation* 1: 388–9, but CD noted that he owed most of his information to Caspary's letters to him.

5 A short report of the proceedings of the Congrès International de Botanique et d'Horticulture in Amsterdam appeared in the *Gardeners' Chronicle*, 29 April 1865, pp. 385–6 (see letter from M. T. Masters, 12 July 1865); the report was written by the principal editor of the *Gardeners' Chronicle*, Maxwell Tylden Masters. CD annotated this report in his copy of the *Gardeners' Chronicle* (see n. 3, above).

From T. H. Huxley 1 May 1865

Museum of Practical Geology
May 1 1865

My dear Darwin

I send you, by this post, a booklet none of which is much worth your reading, while of ninetenths of it you may say as the man did who had been trying to read Johnsons Dictionary "that the words were fine but he couldn't make very much of the story"[1]

But perhaps the young lady who has been kind enough to act as taster of my books heretofore will read the explanatory notice & give me her ideas thereupon[2] (always recollecting that almost the whole of it was written in the pre-darwinian epoch—)[3]

I do not hear very good accounts of you—to my sorrow—though rumours have reached me that the *opus magnum* is completely developed though not yet born—[4]

I am grinding at the mill & getting a little tired My belongings flourishing as I hope yours are

Ever | Yours faithfully | T H Huxley

DAR 166: 306

CD ANNOTATION

Verso of letter: 'Kew | Hooker | Hackel on Medusa'[5] *pencil*; 'Masters about Caspary'[6] *ink*; 'Cohn's address?'[7] *blue crayon*

1 The 'booklet' Huxley sent was *A catalogue of the collection of fossils in the Museum of Practical Geology, with an explanatory introduction* (Huxley and Etheridge 1865). Huxley refers to Samuel Johnson's *Dictionary of the English language* (Johnson 1755).

[2] CD's daughter Henrietta Emma Darwin read and commented on several of Huxley's publications (see, for example, *Correspondence* vol. 11, letter to T. H. Huxley, [before 25 February 1863], and letter from T. H. Huxley, 25 February 1863).

[3] The preface to T. H. Huxley and Etheridge 1865 was written before 1859, when *Origin* was published. CD had read the proof-sheets of the preface in 1857 (see *Correspondence* vol. 6, letter to T. H. Huxley, 16 December [1857], and letter to J. D. Hooker, 25 December [1857]). In the preface, Huxley argued that the fossil record did not provide evidence for the 'progressive development of life' (M. Foster and Lankester eds. 1898–1903, 3: 175); the preface is reprinted in Huxley's *Scientific memoirs* (*ibid.*, pp. 125–79).

[4] Huxley refers to the manuscript of *Variation*. See also letter to John Murray, 31 March [1865]. CD wrote in his journal that he became ill on 22 April 1865 and did not resume work on the manuscript until 25 December (see Appendix II). *Variation* was not completed until 1867; it was published on 30 January 1868 (see letter to John Murray, 4 April [1865] and n. 3, and Freeman 1977, p. 122).

[5] The reference is to Ernst Haeckel's paper on propagation in Medusae, a family that included the jellyfish (Haeckel 1865b). See letter to J. D. Hooker, 4 May [1865] and n. 10.

[6] The reference is to Maxwell Tylden Masters and to Robert Caspary (see letters to J. D. Hooker, [1 May 1865] and 4 May [1865]).

[7] Probably a reference to Ferdinand Julius Cohn, a privat-dozent at the University of Breslau. 'Cohn of Breslau' was mentioned in the *Gardeners' Chronicle*, 29 April 1865, p. 386, as a participant in the Congrès International de Botanique et d'Horticulture at Amsterdam; he gave a paper on marine algae (Cohn 1865). CD had read Cohn's work on the contraction of tissue in plants (Cohn 1860) in relation to his own work on insectivorous plants (see *Correspondence* vol. 10), and he also referred to it in 'Climbing plants', p. 112. Cohn is not mentioned in CD's letter to J. D. Hooker, 4 May [1865]; see, however, the letter to George Henslow, [2–5 November 1865], and the letter from George Henslow, 6 November 1865 and nn. 6–8. CD and Cohn apparently did not begin to correspond until the 1870s (see *Calendar*).

From J. D. Hooker 2 May 1865

Kew
May 2/65.

My dear Darwin

We feel very much the shock that poor Fitzroys death must be to you.[1] My old friend Davis writes to me that he had been very excited for some time, & fancied that many had ill-treated him.[2] also that Sir Rodk had decidedly snubbed him at the Geographical.[3] He had taken a house at Norwood, for a few weeks— his daughters went to knock at his door on Sunday mg. & receiving no answer the door was forced & he found dead with his throat cut.[4] Poor old Fitzroy— I am very sorry—for though I did not know him much I always regarded him in joint association with you, & I did admire his $\frac{1}{2}$-Scientific pluck, as a Meteorologist,[5] & his wonderfull kindness & goodness.

I hope to heavens that they will not appoint Glaisher to the post, or Maury, or any of those cattle, who seem to live on self glorification.[6]

My Father is considerably better, but not fit for work & will I hope soon go to the West.[7]

As soon as I get any notice of Caspary on *Cytisus* I will send it—[8] I have seen none yet.

Lyell & Lady[9] were out last night both very well & bright.—

I do not at all like Ramsays answer to Lyell— the note on p 1. is carping & uncalled for,[10] & the whole thing far from clear— I am still of opinion that they are fighting for a shadow, in the present state of the question, & that a *Rock basin* is a Geological desideratum apart from Volcanic action.— The idea of a *series* of lakes, like the alpine ones, being formed by faults is in my notion preposterous—[11]

Masters I suppose reported for G. C. from Amsterdam the Hort Soc. sent him.[12]

Ever yr affec | J D Hooker

DAR 102: 20–1

CD ANNOTATION

End of letter: 'Hackels case'[13] *pencil*

[1] Robert FitzRoy committed suicide on 30 April 1865 (*DNB*). FitzRoy was commander of HMS *Beagle* for the 1831 to 1836 surveying voyage, on which CD was his guest and companion (see *Correspondence* vol. 1).

[2] John Edward Davis had been a second master on HMS *Terror*, the companion ship of HMS *Erebus*, on which Hooker served as assistant ship's surgeon and naturalist during the Antarctic voyage, 1839–43 (*Modern English biography*, *DNB*). Davis may have come into contact with FitzRoy because they were both naval men and worked in the centre of government administration in Whitehall, London: FitzRoy had been head of the Meteorological Department at the Board of Trade and Davis was a naval assistant in the Hydrographic Department at the Admiralty (*Navy list* 1865). For an account of FitzRoy's final breakdown and suicide, see Mellersh 1968, pp. 270–84.

[3] Roderick Impey Murchison was president of the Royal Geographical Society. FitzRoy had been a fellow of the society since 1830 ('List of council, officers, and fellows', *Journal of the Royal Geographical Society* 34 (1864): xlvii), and had been awarded the society's Gold Medal in 1837 (*DNB*). On FitzRoy's connection with Murchison during the 1850s, see Stafford 1989, pp. 85–6. For Murchison's obituary of FitzRoy, in which he expressed his high estimation of FitzRoy's career, see *Proceedings of the Royal Geographical Society* 9 (1865): 215–18.

[4] The FitzRoys had taken Lyndhurst House, Upper Norwood, Surrey (Mellersh 1968, pp. 281–4; *Modern English biography*); their regular residence was 38 Onslow Square, Brompton, London (*Post Office London directory* 1865). FitzRoy had three daughters, Emily-Unah, Fanny, and Katherine, from his first marriage, to Mary Henrietta O'Brien; Emily-Unah died in 1856. He had a fourth daughter, Laura Elizabeth FitzRoy, from his second marriage, to Maria Isabella Smyth. See *Burke's peerage* 1868, s.v. Grafton, duke of, and *Correspondence* vol. 6, letter to Syms Covington, 22 February 1857.

[5] FitzRoy had instituted the first national system of weather forecasting with the intention of providing storm warnings for shipping; he described his methods of forecasting in *The weather book* (FitzRoy 1863). The forecasts were controversial, and exposed FitzRoy to considerable adverse comment within the scientific community (Anderson 1999, pp. 179, 182). There was, nevertheless, a popular demand for the forecasts (Burton 1986, pp. 170–3). For an account of FitzRoy's work at the Meteorological Department, and the debate over whether the forecasts represented reliable scientific knowledge, see also Mellersh 1968, pp. 262–7, 276–80.

[6] James Glaisher was head of the magnetic and meteorological department at the Royal Observatory, Greenwich, and secretary of the Royal Meteorological Society (*DNB*). Glaisher was frequently in the public eye for his scientific balloon ascents; see Tucker 1996. Matthew Fontaine Maury was the former superintendent of the United States Naval Observatory and Hydrographical Office; he had been in Britain since 1862 as an agent of the Confederate States of America (*DAB*; see also Burton 1986, pp. 149–51, 155). Robert Henry Scott eventually took over as head of the reconstituted Meteorological Office in February 1867 (see Burton 1986, pp. 169, 173–4, and Anderson 1999, pp. 206–7).

[7] William Jackson Hooker was suffering from bronchitis and influenza (see letters from J. D. Hooker, 12 April 1865 and [19 April 1865]).

[8] The reference is to Robert Caspary's paper on *Cytisus adami*, now known as +*Laburnocytisus adamii* (Caspary 1865). See letter to J. D. Hooker, [1 May 1865].

[9] Charles and Mary Elizabeth Lyell.

[10] Hooker refers to Ramsay 1865, Andrew Crombie Ramsay's response to Lyell's discussion of the formation of lake-filled rock basins in the sixth edition of *Elements of geology* (C. Lyell 1865). Ramsay had proposed a theory of the glacial origin of rock basins in a paper read before the Geological Society of London on 5 March 1862 (Ramsay 1862); he elaborated the theory in Ramsay 1864. Lyell, however, argued that ice was incapable of excavating large rock basins and maintained that their formation was accounted for by unequal movements of upheaval and subsidence (C. Lyell 1863a, pp. 311–19, and C. Lyell 1865, pp. 168–74). For an account of the controversy, see Davies 1969, pp. 303–9. On the first page of Ramsay 1865, p. 3 n., Ramsay criticised Lyell for appearing to claim to have been the first to notice the correlation between glaciated areas and rock basins in regions that Ramsay had already noted in Ramsay 1862. There is a lightly annotated copy of Ramsay 1865 in the Darwin Pamphlet Collection–CUL. CD had been supportive of Ramsay's theory of the origin of lake basins, but had also been impressed by some critiques of the theory; see, for example, *Correspondence* vol. 10, letter to A. C. Ramsay, 5 September [1862], *Correspondence* vol. 12, letter to A. C. Ramsay, 12 July [1864], letter from J. B. Jukes, 10 August 1864 and n. 2, and letter to J. D. Hooker, 22 October [1864], and this volume, letter to Charles Lyell, 21 February [1865].

[11] The theory that rock basins such as those in the Alps were formed by faults or crustal fractures was put forward by Roderick Impey Murchison (Murchison 1864a and 1864b). For Murchison's contribution to the debate on rock basins, see Davies 1969, pp. 306–7. CD had discussed Murchison's theory with Hooker in 1864; see *Correspondence* vol. 12, letter to J. D. Hooker, [23 August 1864], and letters from J. D. Hooker, 5 September 1864 and 16 September 1864. There is an annotated copy of Murchison 1864b in the Darwin Pamphlet Collection–CUL. In 1864, Hugh Falconer had propounded a tectonic theory of the origin of rock basins, of which Hooker and CD were critical (see *Correspondence* vol. 12, letter from J. D. Hooker, 9 [March] 1864, and letter from J. B. Jukes, 10 August 1864 and n. 2). See also *ibid.*, letter from J. D. Hooker, 2 December 1864.

[12] Hooker refers to Maxwell Tylden Masters and to his report on the Congrès International de Botanique et d'Horticulture held at Amsterdam in April 1865; the report was published in the *Gardeners' Chronicle*, 29 April 1865, pp. 385–6. CD was interested in a paper on the hybrid *Cytisus adami* (+*Laburnocytisus adamii*) mentioned in the report (see n. 8, above, and letter to J. D. Hooker, [1 May 1865]). Masters also reported on the Congress in the *Proceedings of the Royal Horticultural Society* 5 (1865): 92–7.

[13] CD's annotation refers to Ernst Haeckel. See letter to J. D. Hooker, 4 May [1865] and n. 10.

To J. D. Hooker 4 May [1865]

Down
May 4th

My dear Hooker

Sincere thanks for two most interesting notes.[1] I was astounded at news about FitzRoy; but I ought not to have been, for I remember once thinking it likely; poor fellow his mind was quite out of balance once during our voyage.[2] I never knew in my life so mixed a character. Always much to love & I once loved him sincerely; but so bad a temper & so given to take offence, that I gradually quite lost my love & wished only to keep out of contact with him. Twice he quarrelled bitterly with me, without any just provocation on my part.[3] But certainly there was much noble

& exalted in his character. Poor fellow his career is sadly closed. You know he was nephew to L^{d}. Castlereagh, & very like him in manners & appearance.—[4]

You will be glad to hear, I know my dear old friend, that my ten or 12 days of sickness has suddenly ceased, & has left me not much the worse: I feared it was the beginning of another six or nine months miserable attack & feared it much:[5] Jenner has been here,[6] & is evidently perplexed at my case; he struck me as a more able & sensible man, than he did before, for then I could not talk with him: I shall consult no one else.—[7] Please tell me how I am to designate the edge of enclosed peach-stone (to be thrown away) which is marked with ink??[8]

I do not suppose you will care to hear, apropos to your Rafflesia but in Siebold's "Parthenogenesis" of Bee p. 107, there are cases of the 2 sexes of the same Gall-insect being produced from different plants.—[9]

I have just received a paper from Häckel, which gives an astonishing case of propagation in a Medusa;[10] it is exactly as if tadpoles were of two sexes & regularly laid eggs, but also produced by *budding* frogs, these likewise being of 2 sexes & laying eggs!

Very many thanks to you for writing to me about FitzRoy.— Poor fellow how kind he was to me at first during the voyage.—

I think I will write in 2 or 3 weeks to Masters about Caspary & Cytisus.—[11] I am very glad to hear pretty good accounts of Sir William.[12]

Yours ever affectionately | C. Darwin

How charming Mad. Laugel must be![13]

Do *not* answer till convenient about the edge of the peach-stone.

Endorsement: '/65.'
DAR 115: 268a–b

1 See letters from J. D. Hooker, [19 April 1865] and 2 May 1865.

2 Robert FitzRoy committed suicide on 30 April 1865 (see letter from J. D. Hooker, 2 May 1865). On FitzRoy's breakdown in 1834, during the voyage of the *Beagle*, see *Correspondence* vol. 1, letter to Catherine Darwin, 8 November 1834. See also *Correspondence* vol. 8, letter to J. S. Henslow, 16 July [1860], Mellersh 1968, pp. 130–6, Bowlby 1990, pp. 152–8, and n. 3, below.

3 In his *Autobiography* CD recalled the incidents referred to; the first occurred early in the voyage, at Bahia, Brazil, and concerned slavery (pp. 73–4), and the other occurred during FitzRoy's breakdown at Concepción, Chile, in 1834 (p. 75). On CD's relationship with FitzRoy, see Barlow 1932. See also *Autobiography*, p. 76, *LL* 1: 332, and Mellersh 1968, pp. 62–136.

4 Robert Stewart, Viscount Castlereagh, was British foreign secretary from 1812 to 1822; he committed suicide on 12 August 1822 (*DNB*). FitzRoy's mother was Stewart's sister (Mellersh 1968, p. 19).

5 According to Emma Darwin's diary (DAR 242), CD vomited or suffered nausea every day from 21 April to 1 May 1865; he also recorded in his journal that he became ill on 22 April 1865 (see Appendix II). See also letter from George Busk, 28 April 1865, and letter to J. D. Hooker, [1 May 1865]. CD had been seriously ill for parts of 1863 and 1864 (see *Correspondence* vols. 11 and 12).

6 According to CD's Account book–cash account (Down House MS), CD paid his physician William Jenner £10 10*s*. on 3 May 1865, presumably the date of his visit to Down House.

7 CD had become dissatisfied with Jenner's treatment of him and had asked Hooker to recommend him another doctor (see letter to J. D. Hooker, 6 April [1865] and n. 2). Jenner had been treating CD since March 1864; on the course of treatment that he prescribed, see *Correspondence* vol. 12. George Busk had recommended another physician in his letter of 28 April 1865.

[8] At the end of the letter, Hooker wrote: 'ventral or placental suture of endocarp'. Hooker may have sent this information to CD in a reply; however, no letter referring to it has been found. Peach and almond stones are illustrated in *Variation* 1: 337; the information may have been used in giving instructions to the artist who produced the engravings, probably Walter Hood Fitch, a freelance botanical artist often used by the staff of the Royal Botanic Gardens, Kew (see letter to John Murray, 31 March [1865], and letter from John Murray, 1 April 1865).

[9] By 'your Rafflesia', CD may be referring to a comment of Hooker's that he passed on to Benjamin Dann Walsh, in his letter to Walsh of 27 March [1865]: 'With respect to Dimorphism you may like to hear that D[r] Hooker tells me that a dioicous parasitic plant allied to Rafflesia has its 2 sexes parasitic on 2 distinct species of the same genus of plants'. Hooker may have told CD this during his visit to Down from 4 to 6 March 1865; no letter containing this information has been found. There is an annotated copy of Siebold 1857 (*On true parthenogenesis in moths and bees*) in the Darwin Library–CUL (see *Marginalia* 1: 756–7). CD marked a passage on page 107 where Karl Theodor Ernst von Siebold refers to the work of a colleague: 'From a particular gall, he always reared nothing but female individuals of the Hymenopterous insect *Stomoctea*, but was much astonished when he obtained nothing but males of this insect from the pupa of a *Tenthredo*.'

[10] CD refers to Haeckel 1865b; there is an annotated copy of this work in the Darwin Pamphlet Collection–CUL. An English translation of the article appeared in the June 1865 issue of *Annals and Magazine of Natural History*, pp. 437–44; CD's annotated copy is in the Darwin Library–CUL. In *Variation* 2: 384 n. 41, CD wrote: 'Ernst Häckel has recently . . . observed the surprising case of a medusa, with its reproductive organs active, which produces by budding a widely different form of medusa; and this latter also has the power of sexual reproduction.' Haeckel had mentioned his ongoing work on the Coelenterata, a class of invertebrates that included sea anemones and jellyfish, in 1864 (see *Correspondence* vol. 12, letters from Ernst Haeckel, 9 [July 1864] and n. 14, and 26 October 1864).

[11] CD refers to Maxwell Tylden Masters and to Caspary 1865 (see letter to J. D. Hooker, [1 May 1865], and letter from J. D. Hooker, 2 May 1865).

[12] See letter from J. D. Hooker, 2 May 1865. Hooker's father, William Jackson Hooker, was suffering from bronchitis and influenza (see letters from J. D. Hooker, 12 April 1865 and [19 April 1865]).

[13] See letter from J. D. Hooker, [19 April 1865]. Madame Laugel was the wife of Auguste Laugel.

To Friedrich Rolle 6 May [1865][1]

Down. | Bromley. | Kent. S.E.
May 6. 1863

Dear Sir

Illness has prevented me from sooner thanking you for your very kind present of your "Der Mensch" &c for which I am very much obliged.[2] As I am far from strong I have as yet only cut the pages, but I can see that you have produced a full & valuable history of the subject & the search after such ample materials must have cost you much labour.

With my best thanks I beg leave to remain | Dear Sir yours faithfully | Charles Darwin

Copy
Forschungsinstitut und Natur-Museum Senckenberg, Frankfurt

[1] The year is established by CD's reference to having received the first instalment of Rolle 1866, which was issued in 1865 (see n. 2, below). See also *Correspondence* vol. 14, letter from Friedrich Rolle,

28 January 1866, in which Rolle states that he sent the first part of Rolle 1866 to CD the previous summer. The copyist has written '1863' in error.

2 Rolle's *Der Mensch* (Rolle 1866), a study of the development of human society in the light of *Origin* and recent discoveries in geology, was published in five parts; the first part was published on 1 May 1865 (*Börsenblatt für den Deutschen Buchhandel und die mit ihm verwandten Geschäftszweige* 32 (1865): 954), the second and third parts were issued on 17 November 1865 (*ibid.* 32 (1865): 2634), and the fourth and fifth parts were issued on 15 January 1866 (*ibid.* 33 (1866): 103). CD's annotated copy of Rolle 1866 is in the Darwin Pamphlet Collection–CUL. CD cited Rolle 1866 in *Descent* 1: 4 n. as one of several works that had concluded that the human race was 'the co-descendant with other species of some ancient, lower, and extinct form' (*ibid.*, p. 3). CD had begun a correspondence with Rolle, a German geologist and palaeontologist, in 1862, after Rolle sent him a copy of his popular exposition of CD's theory (Rolle 1863; see *Correspondence* vol. 10). For Rolle's work in popularising CD's views, see *Correspondence* vol. 11, letter from Friedrich Rolle, 26 January 1863, and letter to Friedrich Rolle, 30 January [1863].

To James Philip Mansel Weale 6 May [1865]

Down. | *Bromley.* | *Kent. S.E.*
May 6

Dear Sir

From your obliging letter of Feb. 19. it appears that you have many interesting subjects of investigation.[1]

I am sorry to say that the state of my health prevents my writing at any great length.

Your case of the mule is very curious but I have heard of an analogous instance.[2]

I do not believe there is any book which w^d aid you in naming the Cape orchids except extremely expensive illustrated works. Nor can I aid you with any suggestion about Zoological works. Few foreign countries are so well off botanically as the Cape will be with Harvey's work.[3]

Although I feel great interest in the fertilisation of orchids yet I w^d suggest to you not to spend too much time on this subject as I suspect botanists will think that enough has been written on it, unless indeed you find something quite new.[4]

You are probably aware that Robt Brown published in Linn: Transact: about 25 years ago a grand paper on the structure & fertilisation of Asclepias; yet I dare say there is a good deal to be made out in this genus, on its manner of fertilisation.[5]

I have seen a Hymenopter from America with its tarsi covered with the pollen masses of Asclepias;[6] but how they get placed on the stigma I do not know.

Your researches in the Caverns of S. Africa, if thoroughly carried out, w^d probably be very interesting.[7] I fear that you will find it impossible to name the plants & animals which you may observe until you can bring or send specimens to Europe.

With sincere thanks for your obliging letter & your tracings[8]

I am dear Sir | yours very faithfully | Charles Darwin

Postmark: MY 6 65
LS
American Philosophical Society (308)

1 A note on the cover of this letter reads, 'First letter from C.D^{n} 1865 in answer to one from P.E. 1864'. The letter was addressed to Port Elizabeth in South Africa, and forwarded to Adelaide, South Africa. No earlier letter from Weale has been found; however, a letter from Weale describing his natural history interests was published in the January 1865 issue of the *Natural History Review* (Weale 1865); Weale noted that he wrote to the editors of the *Natural History Review* 'with the hopes of drawing the attention of such men as Messrs. Darwin and Wallace to some interesting caves' (*ibid.* p. 145). CD communicated Weale's papers on the pollination of orchids and asclepiads (milkweeds) to the Linnean Society (Weale 1867 and 1870a–d), and cited him several times in the second edition of *Orchids*. Weale also provided CD with material for *Descent*, the fifth edition of *Origin*, and *Expression*. There is an annotated copy of Weale 1867 in the Darwin Pamphlet Collection–CUL.

2 The reference has not been identified. It may have concerned the inheritance of shoulder- and leg-stripes in horses and mules, as Weale reported seeing dun-coloured foals with leg-stripes in his letter in the January 1865 issue of the *Natural History Review* (Weale 1865, p. 146). The subject is discussed by CD in *Origin*, pp. 163–7, and in *Variation* 1: 56–61 and 2: 41–3 and 351, as an example of reversion to an ancestral type. CD had asked several of his correspondents to report any observations that they had made on this subject and added further examples in later editions of *Origin* (see *Correspondence* vols. 6–8 and 11, and Peckham ed. 1959, pp. 313–17). Weale is not cited in any of these discussions.

3 CD refers to William Henry Harvey and probably to W. H. Harvey and Sonder 1859–65. Harvey had corresponded with CD and was the author of a number of works on the taxonomy of South African plants; see, for example, W. H. Harvey 1838 and W. H. Harvey 1859–63.

4 Weale went on to produce five papers on orchid and asclepiad pollination, which CD communicated to the Linnean Society (see n. 1, above).

5 The reference is to Robert Brown and to Robert Brown 1831. CD was himself interested in pollination in Asclepiadaceae (see *Correspondence* vol. 6, letter to Asa Gray, 5 September [1857], *Correspondence* vol. 10, letter to H. G. Bronn, 30 June [1862], and *Natural selection*, p. 53), and he may have intended to study the family more closely; several Asclepiadaceae species were included in CD's list of hothouse plants (see *Correspondence* vol. 11, Appendix VI).

6 CD presumably refers to specimens of Hymenoptera sent by John Obadiah Westwood in 1861. See *Correspondence* vol. 9, annotations to letter from J. O. Westwood, 26 September 1861. CD made the same observation in the letter to Roland Trimen, 27 August [1863] (*Correspondence* vol. 11).

7 In his letter published in the January 1865 issue of the *Natural History Review* (Weale 1865), Weale referred to his wish to investigate some recently discovered caves in the Transvaal Republic, on the Vaal River basin, and in George District (see n. 1, above). George District was a small municipality surrounding the town of George, Cape Colony (*New school room map of South Africa*).

8 Weale's tracings have not been found.

From Bartholomew James Sulivan 8 May [1865][1]

Board of Trade. S.W.
May 8th.

My dear Darwin

I was going to write to you today thinking you would like to know that I saw the remains of poor FitzRoy laid in their last resting place on Saturday.[2] I was in Cornwall when I heard of it, having read a letter from him *the day before.* I had resigned my appointment a week before that—as three Doctors—consulted separately—told me I must give up, & ought to have done so sooner.[3]

FitzRoy wrote to me saying he thought I was right, and that he was anxious about what he ought to do. We have been urging him here for some time to take a long rest—& I have told him over & over again that he ought to do so, or give up entirely: but he wanted the strength of mind to decide, and I fear the idea of losing the Income for his family had much to do with it—[4] In his note he said he had been very ill but his wife[5] & a skilful Doctor, had—under god—saved his life, & he also remarked on never really feeling the value of blessings & particularly of health—until losing them. Though he trusted in his case it was only temporary—This I think shows he had no previous idea of taking away his own life, and that it could only have been done through the sudden impulse of insanity.

I came back Thursday night hoping to be in time for the funeral. I found a note from Mellersh[6] at Brighton asking me to tell him when it was to be—but I found no one knew any thing about it at the Board or Adm[ty] and I should not have known had not M[rs]. FitzRoys brothers[7] come to his office to lock up his papers &c & they told me it was early on Saturday at the Norwood Church, close to his house,[8] & they had told no one but relatives as they thought under the circumstances it ought to be very private—but they were sure it would be gratifying to M[rs]. FitzRoy if we were present. A French officer Capt Pigeard—of the Embassy—came to me afterwards to know when it was as he wished to attend on behalf of the French Navy, as they all thought so highly of him & many had recieved much kindness from him.[9] Mellersh came up in time & he & I—Capt Pegeard, & Babbington—FitzRoy's assistant,[10]—were the only strangers. His brother,[11] two of M[rs] FitzRoy's brothers[12]—Lord Campden[13] and Genl Wood & his brother[14] I think made up the old party. It was a very quiet and plain funeral, just what I think all funerals should be. Poor M[rs]. FitzRoy would go, & the two daughters were with her.[15] we all waited outside and walked after her carriage—& the same back the brothers only going into the house. It was a trying scene at the grave. Poor M[rs]. F and the girls looked dreadfully ill, & M[rs]. F gave way very much. The coffin was plain black wood with "Robert FitzRoy, born — — died — —" on a brass plate. You may suppose what a trial it was for me, and the thought of old times and scenes that would be mixed up with it all. The relatives were much pleased at the French Captain—showing such a desire to be present. I have no doubt many would have been there if they had known it, but no answer could be given to numerous letters to the office asking when it was to be, even from distant ports.

& Now about my own case. I remained abroad four months,[16] was not much better for two months while moving about in France & Italy—but after six weeks quiet rest at Vevey I got so well that when we went up to a mountain Pension for three weeks I was able to walk up mountains for hours without any return of pain or weakness in leg, or headache, & I returned as well as I ever felt but as I got to work again the symptoms soon returned, and by January I was very unwell. at one time I was a fortnight at home with severe shooting pains in back of head every few seconds, and afterwards for six weeks my leg gave way so in knee & below it that I could not walk from the station to office—& latterly—for the first time,—I feel my work a burthen to me. About a month since, the D[r] at Brighton, who sent me abroad, told me I must have perfect rest for my head: to make sure I consulted two others without their knowing what others said, and they all agreed, & said I ought to have given up before, & that if I went beyond the turning point I should go rapidly down hill, but that avoiding brain work & excitement I might live to a good age, as there was no disease of the heart, but only weakness of action so that the small vessels of head were congested— After the weather got warmer I got better, but with these opinions I felt I ought at once to give up— I go on till end of this month, and then go on half pay as a private gent. again. We mean to sell the house as I cannot afford to live near London, and also my girls are better by the sea; so we go to Bournemouth where we shall settle.[17]

I should like to run down to you for one night before I leave, to see you before we go. I am sorry to hear such a poor account of you. Why do you not give up all work and try a Summer in the Swiss Mountain air? My eldest boy[18] passed last July and is now Sub Lieut & Assist Surveyor of Firefly in Mediterranean. The second is now passing at Portsmouth.[19] The youngest[20] asked me to let him try the Cambridge local examination last December he was 15 & of course in the Junior batch.[21] out of 675 he was bracketed 9th. in the Mathematical Honors only 12 getting them; and of the 8 above him, 7 were so much older that they had gone up for the previous examn a year before, which also gave them a great advantage in experience. I hope therefore he will be able to win an open Scholarship at Cambridge if he lives to try. I always wished one boy to do that.

With very kind regards to Mrs. Darwin & your Daughters | Believe me dear Darwin | Yours most sincerely | B. J. Sulivan

DAR 177: 284

1 The year is established by the reference to the death of Robert FitzRoy; he committed suicide on 30 April 1865 (*DNB*).

2 FitzRoy was buried in the cemetery at St Luke's Church, Norwood, on 6 May 1865 (*DNB*). See also letter from J. D. Hooker, 2 May 1865, and letter to J. D. Hooker, 4 May [1865].

3 Sulivan had been suffering from health problems (see *Correspondence* vol. 12, letters from B. J. Sulivan, 18 March [1864] and 23 September [1864], and Sulivan ed. 1896, p. 378). He had been working as chief naval officer of the Marine Department at the Board of Trade (*DNB*).

4 FitzRoy had been head of the Meteorological Department at the Board of Trade (*DNB*). On FitzRoy's financial difficulties, see Anderson 1999, pp. 201–2.

5 Maria Isabella FitzRoy.

6 Arthur Mellersh and Sulivan had both served under FitzRoy on the *Beagle* during the 1831–6 voyage (see *Correspondence* vol. 1), Mellersh as a midshipman and Sulivan as second lieutenant (Freeman 1978).

7 The reference is to John George Smyth and Henry Smyth (*Burke's landed gentry* 1846, s.v. Smyth of Heath Hall).

8 See letter from J. D. Hooker, 2 May 1865 and n. 4.

9 Charles Pigeard was naval attaché at the French Embassy, London (*Post Office London directory* 1865). Elements of FitzRoy's weather-forecast service were based on work in France; FitzRoy had also been elected a corresponding member of the Académie des sciences in Paris in 1863 (see Burton 1986).

10 Thomas H. Babington was a senior clerk at the Board of Trade, where he worked as FitzRoy's assistant in the Meteorological Department (*British imperial calendar* 1865, p. 100; Mellersh 1968, pp. 288–9).

11 Sulivan refers to George FitzRoy of Grafton Regis, Northamptonshire (*Burke's peerage* 1970, s.v. Grafton, duke of).

12 See n. 7, above.

[13] Charles George Noel, Viscount Campden (*Burke's peerage* 1865).
[14] Sulivan refers to Lieutenant-general Thomas Wood. Wood had two brothers, David Edward Wood and Charles Alexander Wood; the Woods were cousins of FitzRoy (*County families* 1865; *Burke's peerage* 1878 s.v. Londonderry, Marquis of; *DNB*, s.v. FitzRoy, Robert, and Wood, David Edward).
[15] For FitzRoy's surviving daughters see letter from J. D. Hooker, 2 May 1865 and n. 4.
[16] Sulivan had taken six months' leave in 1864 because of poor health (see *Correspondence* vol. 12, letters from B. J. Sulivan, 18 March [1864] and 23 September [1864]). See also Sulivan ed. 1896, p. 378.
[17] The Sulivans retired to Bournemouth in Hampshire (*County families* 1872–90; see also letter from B. J. Sulivan, 31 May [1865]). Only two of Sulivan's daughters have been identified, Sophia Henrietta and Catherine Sabine Sulivan.
[18] James Young Falkland Sulivan.
[19] Thomas Edward Sulivan. Sulivan refers to the Royal Naval College in Portsmouth, Hampshire.
[20] Henry Norton Sulivan.
[21] The University of Cambridge established the Local Examination Syndicate in 1858. The aim was to use the authority of the University to promote a standardised measure of educational attainment. At centres throughout the country two examinations were held in December every year, one for boys under 16 and one for boys under 18. Papers relating to the syndicate are in the University Archives–CUL (Cambridge University Register (Local Examinations) 57.1: 1–46).

From Emma and Charles Darwin 13 May 1865

[Down]
May 13th. 1865

In case of my death, I desire that a yearly annuity of £50 be paid to J. Parslow for his faithful services & friendship, as long as he may live; or in case of his becoming unfit for service.—[1]

In case of my wifes sudden death, with no proper provision having been made for this annuity being paid, We desire our children to pay conjointly this annuity.—[2]

Ch. Darwin | Emma Darwin

DS
DAR 210.10: 26

[1] Joseph Parslow was the butler at Down House; he began in CD's service in Upper Gower Street, London, *circa* 1840 (see Browne 1995). Parslow retired in 1875 on a pension of £50 a year and the rent of his house, Home Cottage, Back Lane, Down (Freeman 1978).
[2] The document is folded in a small envelope bearing the words, 'Memorandum | Annuity for J. Parslow | May 1865.'

From Asa Gray 15 and 17 May 1865

Cambridge, Mass.
May 15, 1865—

My Dear Darwin

Your kind letter of the 19th ult. crossed a brief note from me.[1] I am too much *distracted* with work at this season to write letters on our affairs, and if I once begin, I should not know where to stop. You have always been sympathising and just, and I appreciate your hearty congratulations on the success of our just endeavors.[2] You have since had much more to rejoice over, as well as to sorrow with us. But the noble manner in which our country has borne itself should give you real satisfaction. We appreciate too the good feeling of England in its hearty grief at the murder of Lincoln.[3]

Don't talk about our "hating" you,—nor suppose that we want to rob you of *Canada*—for which nobody cares.[4]

We think we have been ill-used by you, when you thought us weak and broken.—& when we expected better things.[5] We have learned that we *must be strong* to live in peace & comfort with England,—otherwise we should have to eat much dirt. But now that we are on our feet again, all will go well, and *hatred* will disappear. Indeed, I see little of that. We do not even hate the Rebels, and may not even execute so much of justice as to convict of treason & hang their President, whom we have just caught,—but *I hope we shall*,—hang the leader & spare the subordinates.[6] We are now feeding the south, who starved our men taken prisoners.[7]

Slavery is thoroughly dead. We have a deal to do, but shall do it, I trust, and deserve your continued approbation. We have a load to carry—heavy, no doubt, but a young & re-invigorated country, with a future before it can do and bear, & prosper under what might stagger a full-grown, mature country of the Old world

I must look to the *Plantago* dimorphism: for, as you say, these plants, fertilised by wind, could gain nothing by being dimorphic. No dimorphic species grows very near here,—nor can I now get seeds of P. Virginica.[8] Perhaps a good look at even dried specimens, under your hints, may settle the matter.

I was exceedingly interested with the Lythrum paper (but had no time to write a notice of it.),[9] & I wait expectingly for your Climbing plants.[10] You are the very prince of investigators.

We hope presently to make Mrs. Wedgwood's acquaintance.[11]

In great haste, dear Darwin, | Your affectionate | A. Gray

My wife's cousin, *Brace* is in England, & hoping to see you.[12] I told him he must not take up your time, nor worry you with questions and talk. He is a good soul, but you will get nothing Scientific out of him. Social matters are in his way[13] | A.G.

P.S. 17th. May | I missed the post. I have since had the pleasure to meet Mrs. Wedgwood, and to talk with her of You and Yours. A. G.

Thanks for the Times.—apparently from you.[14]

DAR 165: 147

[1] See letter to Asa Gray, 19 April [1865]. Gray's letter has not been found.

[2] See letter to Asa Gray, 19 April [1865]. With the fall of Richmond, Virginia, the capital of the Confederacy, on 2 and 3 April, and the surrender of Robert E. Lee's Army of Northern Virginia on 9 April 1865, the American Civil War effectively came to an end (see McPherson 1988, pp. 844–52, and Denney 1992, pp. 553–4, 557).

[3] Abraham Lincoln was shot on 14 April 1865 and died of his wounds the following day (*DAB*, Denney 1992, p. 559). News of his death reached London on 26 April, and on 27 April *The Times* carried the official report of the murder and details of the shocked reaction from around the country; it also recorded that the House of Commons had signed an address of sympathy to the resident American minister (*The Times*, 27 April 1865, p. 7). An editorial said that the news would arouse sincere and profound sorrow throughout Europe (*ibid.*, p. 7).

[4] See letter to Asa Gray, 19 April [1865] and n. 6. The United States and Britain had been involved in several disputes concerning Canada's borders with the United States (see *Correspondence* vol. 11, letter to Asa Gray, 23 February [1863] and n. 19). On relations between Britain and the United States in respect of Canada after the end of the American Civil War, see Jenkins 1974–80, 2: 381–92; on the movement for the annexation of Canada by the United States, see Warner 1960, pp. 46–59.

[5] On the relations between Britain and the United States during the American Civil War, see Jenkins 1974–80. Gray may be referring in particular to relations between the two countries in 1861, after Britain's declaration of neutrality in May, which opened the way for the Confederacy to contract loans and purchase war materials (see *Correspondence* vol. 9, letter from Asa Gray, 31 December 1861, n. 12). Other incidents included the *Trent* affair in September 1861, and the *Alabama* affair in 1862 and 1863 (see *Correspondence* vol. 9, letter to Asa Gray, 11 December [1861], n. 3, and *Correspondence* vol. 11, letter from Asa Gray, 23 November 1863 and nn. 8 and 9). The Union Army suffered a major defeat at the first battle of Bull Run (Manassas) in July 1861 (see McPherson 1988, pp. 342–6).

[6] Jefferson Davis was president of the Confederate states during the American Civil War. Following the war, he was captured and held as a military prisoner for two years, then transferred to civilian custody after it was determined that he would be charged not in connection with the assassination of Lincoln but with treason. He was released on bail from civilian prison within a few days, on 13 May 1867, and on 5 December 1868 the indictment against him was dismissed. A universal amnesty was issued at Christmas 1868, but Davis declined ever to apply for a full pardon (see Davis 1991, pp. 651–63).

[7] See letter from Asa Gray, 17 January 1865 and n. 10.

[8] In his letter to Gray of 19 April [1865], CD had asked for dimorphic species of *Plantago*.

[9] 'Three forms of *Lythrum salicaria*' was read before the Linnean Society on 16 June 1864 and copies of the paper were issued to members of the society on 12 December 1864 (*General index to the Journal of the Linnean Society*). Gray's name appears on the presentation list for the paper (see *Correspondence* vol. 12, Appendix III). For CD's earlier discussions with Gray on *Lythrum*, see *Correspondence* vols. 10–12. Gray reviewed 'Three forms of *Lythrum salicaria*' in the May 1865 issue of the *American Journal of Science and Arts*, pp. 360–1, referring to CD as a 'prince of biological inquirers'.

[10] An abstract of 'Climbing plants' was read at a meeting of the Linnean Society on 2 February 1865, and CD had begun distributing offprints of the paper by June 1865 (see letter to Asa Gray, 19 April [1865] and n. 16). Gray reviewed 'Climbing plants' in two issues of the *American Journal of Science and Arts* (A. Gray 1865–6). In the first part of his review, p. 273, Gray remarked that CD's papers on dimorphic and trimorphic plants, and his work on *Orchids* and 'Climbing plants', showed 'a genius for biological investigation'.

[11] The reference is to Frances Wedgwood, wife of Francis Wedgwood, owner of the Wedgwood potteries; she made a number of visits to the United States following her first visit in 1859 (see B. Wedgwood and Wedgwood 1980, pp. 267–9, 291, 296).

[12] The reference is to Charles Loring Brace. In fact, Brace was a nephew of Gray's wife, Jane Loring Gray (Dupree 1959, p. 192). Brace did not visit CD (see letter from Asa Gray, 24 July 1865).

[13] Brace was a philanthropist who worked with the underprivileged children of New York City (Dupree 1959, p. 361). He sent CD a copy of his book, *The races of the old world. A manual of ethnology* in 1863 (Brace 1863; see *Correspondence* vol. 11, letter to C. L. Brace, 24 June [1863] and n. 2).

[14] This may have been the issue of 27 April 1865. See n. 3, above.

To John Chapman 16 May [1865][1]

Down. | Bromley. | Kent. S.E.
May 16

Dear Sir

Several months ago you were so good as to send me your pamphlet on the cure of Sea Sickness[2] & may have been surprized that I did not write to thank you, but by an odd chance I only observed a week ago that it came direct from you. I now want to ask you to pay me a professional visit here (any day but Friday next) to consider whether the ice treatment would be applicable in my case.[3] My sickness is not from mere irritability of stomach but is always caused by acid & morbid secretions.[4] I am anxious not to try any new treatment unless you have had experience in some similar cases leading you to think it adviseable.

This place is 6 hilly miles from Bromley where you will find flys waiting. The Victoria or Charing Cross stations are equally convenient in order to reach Bromley.

Will you have the kindness to let me know on what day you can come & about what hour?

I am dear Sir | yours very faithfully | Charles Darwin

P.S. If one of the bags for ice is so small that you can put it in your pocket I should be *much obliged* if you c^d bring one here, as I sh^d thus learn how to apply it, in case you think fit to make the trial.—

LS
University of Virginia Library, Albert and Shirley Small Special Collections Library, Darwin Evolution Collection (3314)

[1] The year is established by the record of Chapman's visit (see n. 3, below).
[2] CD refers to the first part of Chapman's *Functional diseases of the stomach* (Chapman 1864). CD's copy of the pamphlet has not been found in the Darwin Library–CUL or the Darwin Library–Down.
[3] According to Emma Darwin's diary (DAR 242), Chapman visited CD at Down House on 20 May 1865. Chapman's treatment consisted in the application of ice, placed in a specially designed ice-bag, on the spine. For a discussion of this treatment, see Colp 1977, pp. 82–4. CD's Account book–cash account (Down House MS) records a payment of £10 10*s*. to 'Dr Chapman' on 20 May 1865.
[4] CD had been seriously ill with a stomach disorder for parts of 1863 and 1864, his symptoms including persistent retching (see *Correspondence* vols. 11 and 12). In 1863 he consulted several specialists, including William Brinton, George Busk, and John Goodsir, and took hydropathic treatment at Malvern Wells, Worcestershire, under James Smith Ayerst, in an attempt to find a cure (see *Correspondence* vol. 11). Since March 1864 he had been a patient of William Jenner's (see letter to J. D. Hooker, 4 May [1865] and n. 7). During the first four months of 1865, CD often complained of indifferent or poor health (see, for example, letters to J. D. Hooker, 7 January [1865], 2 February [1865], and 16 [March 1865]). According to Emma Darwin's diary (DAR 242), CD vomited or suffered nausea every day from 21 April to 1 May 1865. CD described his symptoms, presumably for Chapman, in a manuscript dated 20 May 1865 (see Appendix IV).

From John Traherne Moggridge 17 May [1865][1]

14 Charles S^t | Paddington
May 17

Dear Sir

I send by rail a few plants of Lythrum Græfferi which I procured from *wet* places at Mentone—[2] The blossoming of this plant was so unusually late that I was unable to make any observations— I also enclose some bulbs of two species of Romulea.—[3]

Are you aware that *Pulmonarias* are dimorphic?—[4]
yours very sincerely | J. Traherne Moggridge
There are a few seeds & 3 plants of Primula marginata[5]

DAR 171: 202

[1] The year is established by the reference to *Romulea* (see n. 3, below).
[2] CD had first obtained specimens of the trimorphic *Lythrum graefferi* in 1862 in order to compare it with *L. salicaria* (see *Correspondence* vol. 10, letter to Daniel Oliver, [17 September 1862], and letter to J. D. Hooker, 6 October [1862]; the specimens are in DAR 142). He gave a brief description of the plant as having 'a close general correspondence in structure' with *L. salicaria* in 'Three forms of *Lythrum salicaria*', pp. 189–90 (*Collected papers* 2: 123; see also *Forms of flowers*, p. 165). His notes on *L. graefferi* are in DAR 27.2: 40–2. CD continued to perform crossing experiments with *L. salicaria* from 1865 to 1867 in order to establish the relationship between 'illegitimate unions' and hybrids (see letter to J. D. Hooker, 22 December [1865] and nn. 5 and 6). See also letter to M. E. Wichura, 3 February [1865] and n. 9.
[3] In July 1864, CD had asked Moggridge to send additional *Romulea* bulbs when he returned to Mentone the coming winter (see *Correspondence* vol. 12, letter from J. T. Moggridge, 15 July 1864 and n. 2, and letter to J. T. Moggridge, 1[7] July 1864).
[4] CD's son William Erasmus Darwin had discovered dimorphism in *Pulmonaria* in May 1863 (see *Correspondence* vol. 11, letters from W. E. Darwin, 4 May [1863] and 8 May [1863]). CD and William made extensive observations on *P. angustifolia* in 1864 (see *Correspondence* vol. 12). CD's notes on *Pulmonaria* are in DAR 110: A40–94 and B15–17. He discussed *P. angustifolia* in 'Illegitimate offspring of dimorphic and trimorphic plants', p. 431, and in *Forms of flowers*, pp. 104–10, 239, 252, and 287. CD and Moggridge first discussed dimorphism in July 1864 (see *Correspondence* vol. 12, letter from J. T. Moggridge, 15 July 1864, and letter to J. T. Moggridge, 1[7] July 1864).
[5] After an earlier examination of a short-styled specimen, CD had stated that there was 'hardly any doubt' that the species was dimorphic ('Dimorphic condition in *Primula*', p. 81; *Collected papers* 2: 48). Moggridge had enclosed specimens of the two forms of *Primula marginata* in his letter of 15 July [1864] (*Correspondence* vol. 12; see also *ibid.*, letter to J. T. Moggridge, 1[7] July [1864]). Moggridge described the species in *Contributions to the flora of Mentone* (Moggridge 1865–8, pl. 11).

From J. D. Hooker [26 May 1865][1]

Kew
Friday.

Dear old Darwin

It is a age since we corresponded.[2] I have been engrossed morning noon & night with business. My father has been away, he is much shaken, though quite as well as, at his age, can be expected.[3] My Herb^m^. Clerk died last week, a most serious loss to me, as he knew the ins & outs of the establishment from childhood, & was so trusty accurate & dependable: he was son of old Smith.[4]

Now we have to get rid of Curator of Pleasure Grounds,[5] & I want to take advantage of this to reorganize the whole establishment, which is worked to death,[6] & I dread a break down of our new Curator,[7] who, what with Garden duties & accounts, works 16 hours a day: as for myself who have never done less, this is all very well, but persons not accustomed to it cannot stand it— as matters stand neither he nor I could leave Kew a week.

Then all Burchell's enormous collections have come here,[8] & I have not only to train a new Clerk for myself, but a new Herbarium assistant— Oliver's defect is, that he is not a good utilizer of others labor, which is a sad draw-back both on his own acct:, & that of others, as he fails to train those under him & me.[9]

I am trying to look my future fairly in the face, but cannot see far ahead. My dear old Father piles duty on duty, & will neither give in nor give up. I do admire his gallantry, & I do not want to see him give up, but things *do* get into dreadful confusion, & I shall have a heavy day of reckoning

I send 2 letters of Willys which refresh me wonderfully— the complexity of blunders is charming.[10]

Lyell has sent me p. 112 of Principles to look over (Ch. VII. on changes of temperature)[11] much of the detailed argument seems to me a blunder altogether & that he is out of his depth.[12] I should like to talk it over with you Have you read Tylors book on Prehistoric man?[13] I am charmed with it—& rather disappointed with Lubbocks.[14] I wish he had not reclamated in re Lyell, whether right or wrong.[15]

My wife is poorly— I hoped she was to have given me a little daughter 7 months hence but she seems in an ambiguous way.[16] Pray say nothing of it in writing to me.

Ever yr affec | J D Hooker

DAR 102: 22–3

[1] The date is established by the relationship between this letter and the letter to J. D. Hooker, 1 June [1865]. The Friday before 1 June 1865 was 26 May.

[2] The last known correspondence between Hooker and CD was the letter from J. D. Hooker, 2 May 1865, and the letter to J. D. Hooker, 4 May [1865].

[3] William Jackson Hooker was nearly 80 years old; he had been suffering from bronchitis and influenza (see letter from J. D. Hooker, 2 May 1865 and n. 7).

[4] The reference is to Alexander Smith and to his father, John Smith (1798–1888). John Smith was curator of the Royal Botanic Gardens, Kew, until 1864 (see letter from J. D. Hooker, [7–8 April 1865], n. 5). Alexander Smith died on 15 May 1865 (R. Desmond 1995, p. 429).

[5] Alexander Williamson had been curator of the Pleasure Grounds since 1848; he retired in 1866 (see R. Desmond 1995, p. 432). On the history of the Pleasure Grounds, see R. Desmond and Hepper 1993, p. 13. The post of curator of the Pleasure Grounds was abolished by the Office of Works and Public Buildings (the government department responsible for the gardens) in 1865, and the duties of this post were transferred to the curator of the Botanic Gardens (R. Desmond 1995, p. 225).

[6] Hooker and his father, William Jackson Hooker, had been 'putting everything on a new footing of strict discipline' since the arrival of the new curator, John Smith (1821–88), in 1864 (Allan 1967, p. 212). See also letter from J. D. Hooker, [7–8 April 1865]. Hooker became director of the Royal Botanic Gardens, Kew, on 1 November 1865; on the changes in organisation that were a condition of Hooker's becoming director at Kew, see R. Desmond 1995, p. 225. On Hooker's work as director from November 1865, see also Turrill 1953, pp. 122–37.

[7] John Smith (1821–88) succeeded John Smith (1798–1888) as curator of the Royal Botanic Gardens, Kew, in 1864 (see R. Desmond 1995 and letter from J. D. Hooker, [7–8 April 1865], n. 5).

[8] Hooker refers to the plant collections of William John Burchell (see letter from J. D. Hooker, [19 April 1865] and n. 14).

[9] Daniel Oliver was keeper of the herbarium at the Royal Botanic Gardens, Kew (R. Desmond 1994).

[10] Hooker refers to his eldest son, William Henslow Hooker, aged 12 (Allan 1967, 'Hooker pedigree'); the enclosures have not been found. See letter to J. D. Hooker, 1 June [1865].

[11] Hooker refers to the ninth edition of Charles Lyell's *Principles of geology* (C. Lyell 1853); Lyell was preparing the tenth edition (C. Lyell 1867–8). Chapter 7 of the ninth edition became chapter 11 of the tenth edition (see C. Lyell 1867–8, 1: iv).

[12] Hooker refers to Lyell's argument in chapter 7 of the ninth edition of *Principles of geology* (C. Lyell 1853, pp. 102–12) that the relative positions of land and sea on the globe were a major influence on climate. He argued that the more land there was in the polar regions, and the more sea there was in the tropical regions, the more global temperatures would fall, and vice versa. For CD's earlier discussions of Lyell's theory, see, for example, *Correspondence* vol. 10, letter to A. C. Ramsay, 5 September [1862] and nn. 8 and 9, and *Correspondence* vol. 12, letter from J. D. Hooker, [2 April 1864] and n. 17, and letters to J. D. Hooker, 5 April [1864] and 23 September [1864]. For a discussion of Lyell's theory of climate, see Ospovat 1977.

[13] Hooker refers to Edward Burnett Tylor's *Researches into the early history of mankind* (Tylor 1865). There is a lightly annotated copy of Tylor 1865 in the Darwin Library–CUL (see *Marginalia* 1: 810–11).

[14] Hooker refers to John Lubbock's *Prehistoric times* (Lubbock 1865), a series of essays derived from five articles previously published in *Natural History Review* and a series of lectures delivered at the Royal Institution in 1863 (see *Correspondence* vol. 12, letter from John Lubbock, 10 January 1864 and n. 4). There is an annotated copy of Lubbock 1865 in the Darwin Library–CUL (see *Marginalia* 1: 512–13).

[15] In a note to the preface of Lubbock 1865, p. x, Lubbock charged Lyell with having made much use of his archaeological publications (particularly Lubbock 1861), in *Antiquity of man* (C. Lyell 1863c, pp. 11–16), 'extracting whole sentences verbatim, or nearly so', without proper acknowledgment. Lubbock went on to state that he had included the note lest readers think that it was he who had taken 'very unjustifiable liberties' with the work of Lyell. For two differing accounts of this dispute, see Bynum 1984, pp. 169–79, and L. G. Wilson 2002. For the full text of Lubbock's note, see the letter from Charles Lyell to J. D. Hooker, [31 May 1865], n. 3. See also Appendix V.

[16] Hooker's wife, Frances Harriet Hooker, did not give birth to another child until 1867 (Allan 1967, 'Hooker pedigree'). The Hookers' second daughter, Maria, had died in 1863 (see *Correspondence* vol. 11, letter from J. D. Hooker, [28 September 1863]).

To T. H. Huxley 27 May [1865][1]

Down. | *Bromley.* | *Kent. S.E.*
May 27th.—

My dear Huxley

I have taken a long time to thank you for the Catalogue, the introduction to which I read with much interest.[2] The cause has been that I have had a very bad month; but I know begin to hope that through the aid of an acquaintance of yours (Dr Chapman) my health will receive some considerable improvement from the application of ice to the spine.[3]

I write now to ask a favour of you, a very great favour from one so hard worked as you are. It is to read 30 pages of M.S, excellently copied out, & give me not lengthened criticism, but your opinion whether I may venture to publish it.[4] You may keep the M.S. for a month or two. I wd not ask this favour, but I *really* know no one else whose judgment on the subject wd be final with me.

The case stands thus; in my next book[5] I shall publish long chapters on bud—& seminal—variation, on inheritance, reversion, effects of use & disuse &c. I have also for many years speculated on the different forms of reproduction.[6] Hence it has come to be a passion with me to try to connect all such facts by some sort of hypothesis. The M.S which I wish to send you gives such a hypothesis; it is a very rash & crude hypothesis yet it has been a considerable relief to my mind, & I can hang on it a good many groups of facts.[7] I well know that a mere hypothesis, & this is nothing more, is of little value; but it is very useful to me as serving as a kind of summary for certain chapters. Now I earnestly wish for your verdict given briefly as "Burn it"—or, which is the most favorable verdict that I can hope for, "It does rudely connect together certain facts & I do not think it will immediately pass out of my mind" If you can say this much & you do not think it absolutely ridiculous I shall publish it in my concluding Chapter.[8] Now will you grant me this favour? You must refuse if you are too much over worked—

I must say for myself that I am a hero to expose my hypothesis to the fiery ordeal of your criticism

Believe me my dear Huxley | yours most sincerely | Ch. Darwin

LS(A)
Imperial College of Science, Technology and Medicine Archives (Huxley 5: 214)

[1] The year is established by the reference to T. H. Huxley and Etheridge 1865 (see n. 2, below).

[2] Huxley sent CD a copy of T. H. Huxley and Etheridge 1865, *A catalogue of the collection of fossils in the Museum of Practical Geology, with an explanatory introduction*, at the beginning of May (see letter from T. H. Huxley, 1 May 1865). CD had read the proof-sheets of the preface in 1857 (see *Correspondence* vol. 6, letter to T. H. Huxley, 16 December [1857], and letter to J. D. Hooker, 25 December [1857]).

[3] According to Emma Darwin's diary (DAR 242), John Chapman visited CD on 20 May 1865. See also letter to John Chapman, 16 May [1865], and Appendix IV. Chapman was proprietor and editor of the *Westminster Review*, to which Huxley had been a regular contributor (*Wellesley index*; see also A. Desmond 1994–7, 1: 185).

[4] The copy that CD sent to Huxley has not been found; however, the manuscript from which the copy was made is in DAR 51: C36–74. For a transcript of the original manuscript, which was titled 'Hypothesis of Pangenesis', see Olby 1963. This first statement of pangenesis differed in several respects from the version that was later published in chapter 27 of *Variation*; see Olby 1985, pp. 84–6.

[5] CD refers to *Variation*, which was published in 1868.

[6] CD's notebooks and manuscripts indicate that he developed his ideas about heredity and reproduction over a thirty-year period (see Kohn 1980, Hodge 1985, and Olby 1985). For an earlier indication that CD was attempting to develop a hypothesis connecting information on reproduction, see *Correspondence* vol. 6, letter to T. H. Huxley, [before 12 November 1857]. For a discussion of nineteenth-century debates on heredity and their bearing on the theory of transmutation by natural selection, see Gayon 1998.

[7] The hypothesis of pangenesis that CD outlined to Huxley suggested that each individual cell throws off minute particles (gemmules) that circulate in the bodily fluids and are capable of generating new cells, or remaining dormant until required. The gemmules are drawn together by mutual attraction and into the right places in the right order to, for instance, reconstruct (in organisms that have that capability) a tail or other limb where one has been torn off. Cells throw off gemmules at all stages of development and all gemmules circulate until required: thus embryo gemmules thrown off by embryo cells come together, when conditions are right, to form what CD called the germ (the female element) or the male element; the germ, if fertilised (or even if not, in the case of parthenogenesis), grows into a new embryo (see Olby 1963). Gemmules could also remain dormant for several generations before being sparked into activity. Thus CD thought pangenesis could explain both sexual and asexual reproduction, as well as reversion and the regrowth of body parts. For a discussion of the development of the hypothesis of pangenesis during the 1860s and the part played

by CD's contemporaries, including Herbert Spencer, Charles Victor Naudin, and Richard Owen, in shaping it, see Geison 1969.

[8] The manuscript discussed here was expanded and revised before being published in the penultimate chapter of *Variation*, 'Provisional hypothesis of pangenesis' (*Variation* 2: 357–404).

From T. H. Huxley 29 May 1865

Jermyn St.
May 29th 1865

My dear Darwin

I meant to have written to you yesterday to say how glad I shall be to read whatever you like to send me—[1]

I have to lecture at the Royal Institution this week[2] but, after Friday, my time will be more at my own disposal than usual; and, as always, I shall be most particularly glad to be of any use to you

Any glimmer of light on the question you speak of is of the utmost importance and I shall be immensely interested in learning your views— And of course I need not add I will do my best to upset them That is the nature of the beast—

I had a letter from one of the ablest younger zoologists of Germany, Haeckel,[3] the other day in which this passage occurs—

"The darwinian theory the establishment and development of which is the object all my scientific labours, has gained ground immensely in Germany (where it was at first so misunderstood) during the last two years—and I entertain no doubt that it will before long be everywhere victorious" And he adds, that I dealt far too mildly with Kölliker[4]

With kindest remembrances to Mrs Darwin & your family | Ever | Yours faithfully | T H Huxley

I am glad to hear the icebags are doing you good— I used to know Chapman very well in connexion with the Westminster[5]

DAR 166: 307

[1] CD had asked Huxley to read and comment on his pangenesis hypothesis (see letter to T. H. Huxley, 27 May [1865]).

[2] Huxley delivered the Friday evening lecture at the Royal Institution on 2 June 1865 (*Notices of the Proceedings at the Meetings of the Members of the Royal Institution of Great Britain* 4: 461–3); the lecture was on the methods and results of ethnology, and was later published in the *Fortnightly Review* (T. H. Huxley 1865).

[3] Huxley refers to Ernst Haeckel. Haeckel had written to CD in 1864 concerning the reception of Darwinian theory in Germany (see *Correspondence* vol. 12, letters from Ernst Haeckel, 9 [July 1864], 10 August [1864], and 26 October 1864). On the reception of CD's theory in Germany, see also *Correspondence* vol. 11, letter from Friedrich Rolle, 26 January 1863 and nn. 6–14, Corsi and Weindling 1985, Montgomery 1988, Junker 1989, Engels ed. 1995, and Nyhart 1995.

[4] In 1864 Rudolf Albert von Kölliker published a critical review of *Origin* (Kölliker 1864a and 1864b), which was translated into English in the *Reader* (Kölliker 1864c); see *Correspondence* vol. 12, letter from Ernst Haeckel, 10 August 1864 and n. 12. Huxley's answer to Kölliker's criticisms appeared in the issue of the *Natural History Review* for October 1864, and focused in particular on undermining Kölliker's

Robert FitzRoy
Oil painting presented to Greenwich Hospital (a naval Crown charity) by the surviving officers of HMS *Beagle* (from 1831 to 1836)

John Obadiah Westwood
From Reeve and Walford eds. 1863–7
By permission of the Syndics of Cambridge University Library

claim that CD was a teleologist ([T. H. Huxley] 1864c). Annotated copies of Kölliker 1864b and 1864c, and of CD's unbound copy of the *Natural History Review* for October 1864, are in the Darwin Library–CUL. For CD's enthusiastic response to Huxley's review of Kölliker, see *Correspondence* vol. 12, letter to T. H. Huxley, 3 October [1864] and n. 2, letter to Ernst Haeckel, 8 October [1864], and letter to J. D. Hooker, 8 October [1864]. CD had at first considered replying to Kölliker himself (see letters to J. D. Hooker, 28 August [1864] and [1 September 1864], and letter to T. H. Huxley, 3 October [1864] and n. 3). Haeckel later criticised Kölliker's review in Haeckel 1866, 1: 100–1 n.

5 Huxley refers to John Chapman and the *Westminster Review*. See letter to T. H. Huxley, 27 May [1865] and n. 3.

From William Allport Leighton 29 May 1865

Luciefelde | Shrewsbury
May 29. 1865.

My dear Sir,

Noticing the operations of a small Humble Bee, about the size of a horse bean, on the flowers of *Lupinus polyphyllus*, I was induced to examine more particularly the blossoms. I found that if the blossom be examined *before* the standard is *reflexed*, that the tube of stamens bears alternately five very large sagittate anthers, and five very small (comparatively) rounded or oblong anthers. (This is noticed by Vaucher)[1] The sagittate-anthered stamens are about $\frac{2}{3}$rds. the length of the pistil—*the pollen of their anther-cells is mature and is even now falling away*. On the contrary the small oblong-anthered stamens reach only in length to the base of the sagittate anthers & are about $\frac{1}{3}$rd. the length of the pistil. Moreover their pollen is not ready and their anther-cells are closed.

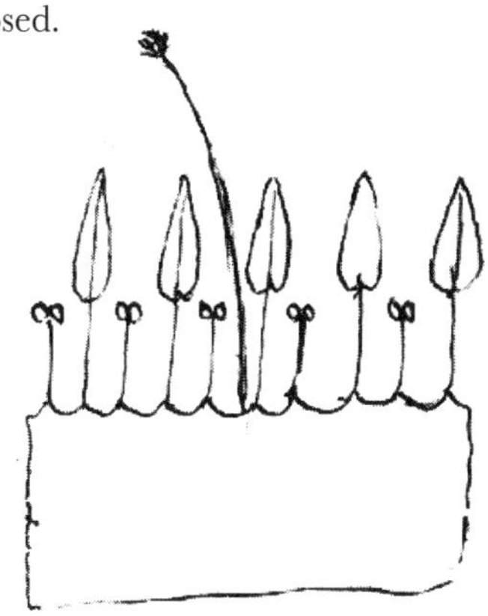

Examining the stigma I could not discover that as yet, any pollen was deposited on it. Nor did the microscope disclose any appreciable difference in the size or form of the pollen grains of the two sets of anthers.

Now if we examine the blossom **after** *the standard has become reflexed*, we find that the large sagittate-anthered stamens still maintain their relative length with the pistil—that their pollen has fallen & the anther cells & the entire stamens are withering away. But that the shorter stamens have become elongated to the length of the pistil,—four of them having their upper portions entirely covered with pollen—all trace of the anther cell having disappeared—and the 5th. or that opposite to the standard having withered away. The stigma also is now observed to have pollen grains on it.

When the blossoms were thus fully matured, I observed that the humble bee alighting on the alæ, bore them downwards by his weight, and in so doing at the same time caused a slight downward movement in the closed keel, whereby the pollen was extruded through the apex, and that the extruded pollen coming in consequence into contact with the body of the bee as he inserted his proboscis to the base of the stamens, became attached to it, & was carried away by him to other blossoms.

The following questions readily suggest themselves:— What becomes of the immense quantity of pollen of the large sagittate anthers? It can scarcely be useless—is it carried by insects or otherwise to fertilize other individual plants possessing pistils of similar relative length? Is the plant really fertilized by the shorter stamens when elongated to an equal relative length with the pistil, 〈an〉d is this fertilization effected by the agency of the bee?

The few plants in my garden presenting similar structure, I am unable to carry this investigation further. I therefore venture to send it to you, though in doing so I feel that I may be incurring the risk of "sending coals to Newcastle"[2]

Believe me, my dear Sir | Yours most truly | Willm. Allport Leighton

Charles Darwin Esq.

DAR 76: B95–6

CD ANNOTATION
Top of letter: '(About Lupines)' *pencil*

[1] The reference is to Jean-Pierre-Etienne Vaucher's *Histoire physiologique des plantes d'Europe* (Vaucher 1841), 2: 213, which CD read in 1862 (see *Correspondence* vol. 10). There is an annotated copy of Vaucher 1841 in the Darwin Library–CUL; CD noted the observation that lupins exhibit two sorts of anthers in the margin (see *Marginalia* 1: 813). There are notes on the different sets of anthers in *Lupinus nanus*, dated 5 August 1862, in DAR 76: B92.

[2] Leighton had been a boyhood friend of CD's and had attended botanical lectures and excursions with CD at Cambridge University; he and CD occasionally corresponded on botanical subjects (see, for example, *Correspondence* vol. 10, and Leighton's memoir of CD in DAR 112: B97–8). CD did not publish the observations in this letter and no reply to it has been found. However, CD made notes on Leighton's information, remarking both that self-fertilisation was inevitable, and that the pollen was in 'enormously greater quantity' than required for self-pollination (see DAR 76: B94). CD also preserved a newspaper cutting from the *Reader*, 10 February 1866, p. 154, that reported on Leighton's short article on pollination in *Lupinus polyphyllus* (Leighton 1866; the cutting is DAR 76: B93). CD had observed pollination in *Lupinus*, a genus in the family Leguminosae, in the garden at Maer Hall, Staffordshire, during the 1840s (see CD's note in DAR 77: 62c; see also DAR 49: 3–15). CD was interested in pollination in Leguminosae, since the occurrence of self-pollination in several species in this family appeared to present a problem for his general view that plants were generally adapted to facilitate crossing between individuals (see *Origin*, p. 97, *Natural selection*, pp. 68–71, and *Correspondence* vols. 4, 6, and 7). CD's later experiments with *L. luteus* and *L. pilosus* indicated that although these

species seeded freely when self-pollinated, crossing was nevertheless advantageous in that it conferred greater vigour on the seedlings of the second generation (see *Cross and self fertilisation*, pp. 147–50, 367, 396–7; CD's notes on these experiments are in DAR 78: 192–9).

From B. D. Walsh 29 May 1865

Rock Island. Illinois.
May 29, 1865

Chas. Darwin Esq
My dear Sir,
Many thanks for your letter of March 27.[1] I fear, in the present state of your health, I am troubling you too often with my scribblings; but the remedy is an easy one & in your own hands—throw them in the fire.

I was delighted with your *Linum* & *Primula* papers, & also with Mr. Scott's paper on *Primulaceæ*.[2] What a remarkable fact that is which he brings out, that the red variety of the common Primrose **absolutely refuses** to intercross with the normal form.[3] Such facts as these, it seems to me, knock the ground away completely from under the Creative Theory.

As to "Unity of Coloration" you have yourself given one very good example in the stripes which re-appear more or less in the several species of the genus *Equus*.[4] Another instance is found in the fawn of our common deer being *spotted* like many adult fallow deer, & Prof. Haldeman[5] writes me word that he has remarked on the rump in *cervus* & *antelope* "being frequently of a lighter tint than the general color". (*Freshwater Univalve Mollusca* No 7. Jan 1844 Planorbis p. 5)[6] There is another most remarkable one in the Owlet-moths (*Noctuadæ*), in almost every genus of which the "orbicular" & "reniform" spots are found, besides five transverse lines of color which Guenée calls respectively the "basal", the "transverse anterior", the "transverse posterior", the "subterminal" & the "terminal".[7] From the enormous number of species in Insects, an Entomologist is naturally led to notice such things more than other Naturalists. I am not sufficiently familiar with exotic species of Insects to say whether or not genera inhabiting two continents "sometimes display a somewhat different type of coloring".[8] But ever since I have been collecting in this country I have been struck almost every day by the same genera on both sides the Atlantic displaying the same coloration. For example, I have two N.A. species of *Panagæus* in my collection, (*P. fasciatus* Say, & *cruciger* Say) which imitate your *P. crux-major* both in design & color; yet with the exception of the *Lebia* & *Bembidium* groups, almost all other *Carabidæ* are of plain colors. If then there is no genetic connection between the N.A. & European species of *Panagæus*, & if (as I think we must concede) coloration is independent of structure, why should *Panagæus*, **on both sides of the Atlantic**, have four more or less confluent large red spots arranged in a quadrangle on its elytra?

My idea as to the mode in which *Cecidomyia* acquired its gall-producing poison,[9] is that originally they were all without it, as are to this day the Hessian fly (*C. destructor*)

& the Wheat-midge. (*Cec. tritice*). But that individuals by Variation acquired the power of secreting a minute portion of poison so as to irritate the plant slightly & cause a slight additional flow of sap, & a better nidus for the future larva,, whereby they gained an advantage over their fellows & so on according to your theory. Whether the Guest-flies are degraded Gall-flies, or in an incipient state of perfection, is perhaps a doubtful point, though I rather incline to the former hypothesis.[10]

As to Wagner's theory of Viviparous larvæ,[11] allowing it as you say to be a form of gemmation, ought we to expect to find a mode of reproduction characteristic of the Vegetable Kingdom, & hitherto only met with in Animals of very low type, in such highly organized & intellectual beings as Insects? Taking everything into consideration, I think Insects are superior to Fish, & could you yourself swallow down gemmiparous Fish without very good Ichthyological authority? I couldn't. Now if a Fish, which has no "penis intrans",[12] which inhabits the water still & which still breathes through gills, is not found to be gemmiparous, I am loath to believe that an insect can be so. Still, facts are stubborn things, & it may turn out that Wagner & Co are right, though I still believe he was deceived by parasitic larvæ.[13]

In a brief review of Wagner's book which I have lately seen reference is made to a genus *Daphnia* & Family *Daphnidæ*, in which *both males & females* are said to produce parthenogenic eggs in the summer, & another kind of eggs called "ephippia" in the winter.[14] To what class do these extraordinary animals belong? Can it be possible that true males generate eggs?

I bought a few weeks ago Lyell's Book on the Antiquity of Man,[15] & was highly delighted with it. I had no idea that Prof. Owen was capable of making such an ass of himself as he has done[16] ⟨*section excised*⟩

You have referred in the "Origin" to a (dipterous?) fly checking the propagation of cattle & horses in the southern parts of South America.[17] I find that in Gen. Páez's book on South America (p. 58) he says that in the Northern parts (Venezuela &c) a fly called 'gutano' oviposits in the umbilicus of newly born foals so constantly, that the herdsmen have to examine all the young foals & remove the larva, or otherwise the foal is sure to die.[18] May we not account for the extinction of the Horse in America by some such means as this?

Ever yours very truly, | Benj. D. Walsh

⟨*section excised*⟩ the so-called ⟨*remainder of line excised*⟩ I am as unwilling to believe in gemmiparous insects as in hermaphrodite insects. Both seem to me low forms of reproduction derived originally from the Vegetable Kingdom & dropped in the march of development at an early period.

Incomplete
DAR 47: 179, 179a; DAR 207: 18

CD ANNOTATIONS
4.1 My idea . . . larvæ. 5.11] *crossed pencil*

8.3 he says that ... as this? 8.7] *scored pencil*; 'Struggle for Existence' *added pencil*
On separate sheet: 'Unity of colour in Bembycidæ— | Panagæus in N. America coloured as in Europe— good | Law of Variation | My analogy of variation is nearly same as Walsh law of equable coloration, only I refer only to Variation'[19] *pencil*

[1] Letter to B. D. Walsh, 27 March [1865].
[2] CD had sent Walsh copies of 'Dimorphic condition in *Primula*', 'Two forms in species of *Linum*' (see letter from B. D. Walsh, 1 March 1865, and letter to B. D. Walsh, 27 March [1865]), and, evidently, an offprint of John Scott's study of reproduction in the Primulaceae (Scott 1864b).
[3] In Scott 1864b, pp. 97–102, Scott gave the results of experiments with *Primula vulgaris*, the common primrose, and a red and a white variety of primrose. His results suggested that the red primrose was infertile with pollen from the common and the white primrose, but fertile with its own pollen. If established, this would have been evidence in favour of CD's views on the origin of species. Thomas Henry Huxley, in several publications (most recently in T. H. Huxley 1863b), had argued that CD's views would not be beyond doubt until he could demonstrate 'the possibility of developing from a particular stock, by selective breeding, two forms, which should either be unable to cross one with another, or whose cross-bred offspring should be infertile with one another' (T. H. Huxley 1863b, p. 146). CD had also thought that Scott's results were remarkable (see *Correspondence* vol. 12, letter to Asa Gray, 13 September [1864]). However, he was unable to corroborate Scott's results regarding the sterility of the red variety when pollinated by the common primrose when he repeated the crossing experiments in 1865 and 1866 (see *Variation* 2: 109 n. and 'Illegitimate offspring of dimorphic and trimorphic plants', p. 420). CD's notes on these experiments are in DAR 108 and DAR 110. For further discussion of CD's and Scott's experimental work on intra-specific sterility, prompted by Huxley's challenge, see *Correspondence* vols. 10–12.
[4] For CD's interest in Walsh's law of unity of coloration, see his letter to Walsh of 27 March [1865] and nn. 5 and 6. In *Origin*, pp. 163–7, CD gave examples of the tendency of a number of species in the horse genus to develop stripes, especially in young animals and hybrids, and hypothesised that this was because they were descended from a striped ancestor. See also *Correspondence* vols. 7–9.
[5] Samuel Steman Haldeman.
[6] In his *Monograph of the freshwater mollusca of the United States*, Haldeman, in a discussion of *Planorbis*, a genus of pond-snails, commented: 'Certain uniform colors, lines, or spots, mark genera and families; in Cervus and Antilope the rump is frequently of a lighter tint than the general color' (Haldeman 184[2–]5, 'Genus Planorbis, *Müller*', p. 5).
[7] Walsh probably refers to Achille Guenée's three-volume monograph on the *Noctuadae* or '*Noctuélites*' (Guenée 1852). Guenée's terminology for describing the spots and lines of colour on the wings of this family of moths appears in Guenée 1852, 1: xlii–xliv and 3: fig. 1; however, Walsh appears to have translated Guenée's terms into a nomenclature of his own.
[8] See letter to B. D. Walsh, 27 March [1865].
[9] See letter to B. D. Walsh, 27 March [1865] and nn. 8 and 9.
[10] For CD's query as to whether guest-flies may have developed from gall-flies, and for Walsh's subsequent discussion in Walsh 1866, see the letter to B. D. Walsh, 27 March [1865] and n. 10.
[11] For CD's comment on Nikolai Petrovich Wagner's observation of paedogenesis (Wagner 1862 and 1863), see the letter to B. D. Walsh, 27 March [1865] and nn. 15 and 16.
[12] *Penis intrans*: internal penis (C. T. Lewis and Short 1969).
[13] For Walsh's eventual conclusion regarding paedogenesis, see the letter to B. D. Walsh, 19 December [1865], and Walsh 1866, p. 288.
[14] The reference is to the review of Wagner's article on viviparous gall-forming midge larvae (Wagner 1865) in the May 1865 issue of the *American Journal of Science and Arts*, pp. 362–3 (Minor 1865). Walsh was misled by a mistake in translation in the review. Wagner had drawn an analogy between the two phases of reproduction in *Aphis* and *Daphnia.* In both genera, males and females are produced asexually by parthenogenetic eggs in summer; they are also produced sexually in winter from eggs

carried by the female in a pouch, or ephippia (Wagner 1865, pp. 110–11). The review of Wagner 1865, however, mistranslated reproduction 'of both males and females' as reproduction 'in both males and females', thus implying that males produced eggs (Minor 1865, p. 362).

15 Walsh probably refers to the second American edition of Charles Lyell's *Antiquity of man* (C. Lyell 1863d).

16 In *Antiquity of man*, Lyell reviewed the long-running controversy between Richard Owen and, principally, Thomas Henry Huxley about the comparative anatomy of human and simian brains (C. Lyell 1863a, pp. 480–93). On the so-called hippocampus controversy, see A. Desmond 1982, Rupke 1994, and L. G. Wilson 1996. See also *Correspondence* vols. 9–11.

17 Walsh refers to *Origin*, p. 72.

18 The reference is to Ramón Páez's *Wild scenes in South America, or life in the llanos of Venezuela* (Paez 1863, pp. 58–9). Páez was a son of General José Antonio Páez (see letter from B. D. Walsh, 12 November 1865).

19 On Walsh's law of unity of coloration, see n. 4, above. For CD's discussion of analogous variation (the tendency of descendants of a common ancestor to vary in similar ways), see *Origin*, pp. 159–70, and *Variation* 2: 348–52. In *Variation* 2: 351–2, CD mentioned Walsh's 'Law of *Equable Variability*': 'if any given character is very variable in one species of a group, it will tend to be variable in allied species; and if any given character is perfectly constant in one species of a group, it will tend to be constant in allied species' (Walsh 1863, p. 213); adding that this law was similar to his statement in *Origin* that generic characters were less variable than specific characters (see *Origin*, p. 168). CD also comments on Walsh's law of equable variability in *Origin* 4th ed., p. 187.

From Edward Cresy 30 May 1865

Metropolitan Board of Works | Spring Gardens
30 May '65—

My dear Sir.

It is so long since I have had the pleasure of seeing or hearing of you that if you are well enough I propose to walk over & pay my respects next Monday, that being a leisure day here—.[1]

Should you not be able to see me I am sure M[rs] or Miss Darwin[2] will kindly drop me a line but I sincerely hope such may not be the case.

I was very much struck with the German naturalist⟨'s⟩ illustration of natural selection in the air breathing apparatus of crustacean⟨s⟩ in the Annals of Natural History I think that is a very strong case & the complexity of the argument to my mind gives great additional force—[3]

Yours very truly | E Cresy

C Darwin Esq.

DAR 161: 243

1 Cresy is listed in the *Post Office London suburban directory* 1865 as living in Alleyn Road, Lower Norwood, about eight miles from Down. There is no record of a visit by Cresy in 1865 in Emma Darwin's diary (DAR 242); his last recorded visit to Down was on 2 August 1863 (see *Correspondence* vol. 11, Appendix II). Cresy was principal clerk at the Metropolitan Board of Works (*Post Office London directory* 1865).

2 Emma and Henrietta Emma Darwin.

3 Cresy refers to an abstract (Anon. 1865b) of a French review of Müller 1864 (Anon. 1865a). The abstract focused partly on Fritz Müller's discussion of the genealogy of different species of air-breathing

crabs in his book, *Für Darwin* (Müller 1864). Müller's observations had demonstrated that the independent adaptations for breathing air in crab species from different families supported Darwin's theory of descent; for Müller's observations of these adaptations and CD's interest in them, see the letter to Fritz Müller, 10 August [1865] and n. 5.

To T. H. Huxley 30 May [1865][1]

Down. | Bromley. | Kent. S.E.
May 30

My dear Huxley

I thank you most sincerely for granting my request in so kind a manner.[2]

When ever you read the M.S. be so kind as to suspend your judgment until you have read the whole, & then turn the subject a little in your mind. I have thought of it much, more than appears in the M.S. & am becoming convinced that some such view will have to be adopted; but I see that it over throws in an uncomfortable manner one's common view on ordinary development.[3] The style of the M.S has to be improved.

You will have to take some of my facts & partial conclusions on trust, but the greater number of the facts will be quite as familiar to you as to me.

You will really do me a very great service & with *cordial* thanks believe me yours sincerely | Ch Darwin

LS
Imperial College of Science, Technology and Medicine Archives (Huxley 5: 217)

1 The year is established by the relationship between this letter and the letter from T. H. Huxley, 29 May 1865.

2 CD had asked Huxley whether he would read and comment on the manuscript of his pangenesis hypothesis (see letter to T. H. Huxley, 27 May [1865], and letter from T. H. Huxley, 29 May 1865).

3 CD may be alluding to the view expressed in his manuscript, that there is no essential difference between sexual and asexual reproduction or between reproduction and continual growth (see Olby 1963, pp. 253–6; see also letter to J. D. Hooker, 4 April [1867] (*Calendar* no. 5485)). This represented a radical departure from CD's earlier views on generation as expressed in his *Notebooks* (see Kohn 1980, pp. 83–7, and Hodge 1985, pp. 227–36). Olby has noted that CD's revised position on reproduction partly reflected contemporary work on parthenogenesis that contributed to the undermining of the distinction between buds and germ-cells during the 1860s (Olby 1985, pp. 73–5, 80). See also Churchill 1979, and Farley 1982, pp. 72–111. On precursors to CD's hypothesis of pangenesis, see Geison 1969, pp. 393–409.

From Charles Kingsley 30 May 1865

Eversley Rectory, | Winchfield.
May 30/65

My dear Sir

We desire much your photograph. I say honestly, there is no living man whom I have *not* seen, whom I wish to see as much as I do you: & failing that, to have some Eidolon[1] of you for myself & mine.

If you have any photograph of yourself—or if you can tell me where I can procure one, you will do me a real favour.

The more I think, & see, the more I find to make me thankful for your great book[2]

Yours faithfully | C Kingsley

DAR 169: 31

[1] *Eidolon*: image (*OED*).

[2] Kingsley refers to *Origin*; CD sent a presentation copy to him in 1859 (see *Correspondence* vol. 8, Appendix III). Kingsley was an acquaintance of Thomas Henry Huxley and John Lubbock and took an active interest in natural history (see F. Kingsley ed. 1877, Hutchinson 1914, 1: 23, and A. Desmond 1994–7, 1: 208–9). Kingsley's reaction to *Origin* was largely favourable (see, for example, *Correspondence* vol. 7, letter from Charles Kingsley, 18 November 1859, and *Correspondence* vol. 10, letter from Charles Kingsley, 31 January 1862); he believed, with Asa Gray, that natural selection and natural theology were not inconsistent with one another if one took the view that natural selection operated in accordance with divine purpose (see F. Kingsley ed. 1877, 2: 171–5, 346, and F. Kingsley ed. 1883, pp. 252–3). For a discussion of Kingsley's views, see Colloms 1975, pp. 236, 243–4, 252, and 265–6, and A. Desmond 1994–7, 1: 263, 289.

From Charles Lyell to J. D. Hooker [31 May 1865]

My dear Hooker

I send the correspondence[1] I am sorry to say that I think the effect of Lubbock having for more than a year had the opportunity of telling his own story to Huxley & Busk (& probably to many others) will make them always misunderstand the matter.[2] As to the general reader of Lubbocks book he has taken good care that he shall do the same—[3]

All one can do in such a case is to let those friends who know one be aware of what happened. I have not had much to do with Lubbock & the affair does not make me desire to be intimate.

sinly yrs | Cha Lyell

[Enclosure 1]

53 Harley Street
May 25. 1865

Copy

Dear Lubbock,

I have received a copy of your "Pre-historic Times" marked "from the author[4]

I like so much what I have seen of it & expect it to do so much good to the cause that it is a real disappointment to me to have to say that I think, after the full reply I gave both written & in conversation to a former letter of yours, that the wording of your note at page 10 of the preface about my borrowing of you without acknowledgement strikes me as unfriendly in its tone & as a decided overstatement of the facts.[5] It omits entirely the explanation which I gave you of what induced me to put in the note at page 11 which you point to as irreconcilable with the text especially at page 16.[6]

Your words are that I have "made much use of your earlier articles in the Nat. Hist. Rev. frequently indeed extracting whole sentences verbatim or nearly so." Now there are only three passages in which I have borrowed even any *expressions* from you.

The principal of these, that to which you particularly call my attention at page 16. (Lubbock p. 496) you will find fully given in a paper by Morlot printed at Berne in French in 1859 long before you first went to Denmark. (Morlot–Societé Vaudoise des Sciences Naturelles Tome VI. 1840 p. 276).[7] With Morlot's paper I was familiar when it first came out; indeed he sent me his MS of this & another paper that I might read them before they were printed. The same memoir was republished in English by the Smithsonian Institution 1861. (p 14)[8] before yours came out in the Nat. Hist. Rev, but I found with surprise as I told you before that though I have not borrowed a single fact or inference in this case from you there are some phrases which prove to me that I must have re-worded it after I had seen your paper. The next sentence (Lyell p 11. Lubbock p 491) relates to the duration of the bronze age as inferred from the number of bronze implements found & a gradual progress in the arts which they imply— I feel persuaded that I got the proofs of advance from the Danish archeologists but perhaps the corroboration founded on the *number* of the instruments I may have derived from your paper & it is the only single point which I believe you can make out in my second chapter which was not drawn from sources independent of you & antecedent to your first tour in Denmark.

In speaking of the skulls being like those of the Laplanders, their round shape & other characters, all given by Morlot I may have afterwards introduced from you the "prominent ridge" on the forehead.

In another case you may have deceived yourself by supposing that the coincidence as to the dimensions of the shell-mounds (Lyell p 12 Lubbock p. 493) implied that I borrowed from you, but you will find on comparison that I took the numbers from Morlot (Societé Vaudoise etc p 275–Smithsonian Inst p 13). which differed from yours; thus I give 1000 feet & you 900.

What I say of the bison & other Mammalia & of the Auk & Capercailzie was all from Danish sources & is given by Morlot, the words are not yours.

I mentioned to you that I conversed with Prof. Claparede[9] in 1859 & that I have a letter from him dated December 8. 1859 in which he gives me expressly for my use an abstract of what had been written by Steenstrup whom he had seen, & by Forchhammer & several other Danish antiquaries & naturalists whose papers he had read in the original language "Oversigt over det Konglike Danske Videnskabernes Selskabs Forhandlingen 1848–1851.[10]

I have already told you that my second chapter, the only one in which there is any coincidence between your Articles published in 1861–62[11] & mine, was set up in type before I read your Memoirs. It remained in type nine months while I was writing & printing the Glacial & Darwinian portions of my book & while I was travelling in Italy in 1862. During this long interval several new numbers of Keller's "Pfahlbauten"[12] appeared & I inserted from them & from your Danish paper a few additions & corrections without disturbing the paging. I could not make up my mind to break up the type of half my book & to delay its publication in order to introduce some new views & reasonings which these fresh contributions to science suggested. With difficulty I got room for the note in which I was determined to call attention to your "able papers" stating, what I meant as an apology for not doing justice to your reasoning that mine was written first. I had not room to explain more, but it would certainly have been better if I had done so at length in the preface As to my not citing you where I borrowed the expressions above alluded to, it was truly as you say "an inadvertence". I had forgotten it because I regarded those few words taken from you as absolutely unimportant

Let me now ask you whether there is a single fact or idea taken from your memoirs which was not antecedently published by the Danes & Swedes; if you can point out none surely it could not signify to the public or you which of the two authors borrowed from the other. But your note implies not only that I have frequently borrowed from you but that the unacknowledged obligations were of such value as to call for remonstrance & a disclaimer on your part of having been indebted to my publication.

I said in a former letter that it was not unnatural that you should allude to the anachronism of my note as compared to my text as it might possibly lead some few readers to suppose you had taken from me; but I regret & feel much surprised that you did not insert the short sentence which I gave you, saying 'that my early chapters were written long before I read your paper & that a few additions & corrections only were inserted from it, because I could not make up my mind to recast my work after so much of it was in type, which would have caused a serious delay of publication.

When you say that "you had reason to believe that I regretted the inadvertence of my statement" in the note at p 11. no reader will imagine that this was after I had given you frankly a full explanation both written & personal of what had occurred. They will simply suppose that you had learnt from a third person that I was sorry that the contradiction of dates in my note & text had been found out.[13]

I shall take a copy of this letter as I wish to show it to some of friends but I will wait a day or two to know whether you wish to have any explanation circulated with it, or intend to allow any more copies to be sent out by your publisher with the note at page 10 uncancelled or unmodified[14]

Believe me | dear Lubbock | Y^rs^. truly. | Cha Lyell

[Enclosure 2]

Chislehurst
29 May/65

My dear Sir Charles

I have read your letter with much surprise & regret.

You will find on examination that the three passages to which you refer are by no means the only ones in your book which agree exactly or almost exactly with mine. Compare for instance among others the passage about bronze castings in P. 10 that about the wild swan in P. 15, that about the Danish forests in P. 16. &c.[15]

The sentence about grain in Morlots paper to which you refer me bears no resemblance to mine.[16]

Of course we have neither of us any claim to originality in the matter. You might have obtained all the information in your Chapter on Danish Archæology, from Danish sources & from Morlots paper; but it is evident that you did not do so.

My object in the note to which you take exception, was however not to claim credit for myself, but simply to protect myself from the imputation of having quoted without acknowledgment from your work.

Some of my readers would assuredly have noticed the coincidences in my book & in the "Antiquity of Man",[17] & more particularly, in the face of the note to p. 11. of your work, they would almost inevitably have come to the conclusion that I had borrowed from you. That note gave me, & would I think give the Public an entirely different impression from that which, as it appears, you intended to convey.

As regards the letter to which you refer & which you think I ought to have quoted, I can only say that I would willingly have printed our correspondence, in my preface, if I had felt authorised to do so.[18]

I really cannot see that you have any reason to complain of me, & as I have the satisfaction of feeling that this is the opinion of several of our mutual friends I cannot but hope that on further consideration you will yourself arrive at the same conclusion.[19]

I am very glad that you like what you have seen of the book itself, & remain, dear Sir Charles, | Yours truly | John Lubbock

w Sir C Lyell B^{t}

[Enclosure 3]

53 Harley St
May 30/65.

Copy

Dear Lubbock

My proof-sheets on the Danish Kitchen-middens & the Swiss lake-dwellings are unfortunately destroyed, & I can only ascertain from the printer the exact date when they were put up in type.

They were first set up for the "Elements[20] in smaller type & afterwards, when my plan of publication was changed put into larger type, that of the Antiquity of Man, before I received your first paper. They were written before you went to Denmark, as they stand now, they make the same number of pages & almost the same number of lines as at first; a few lines shorter because I abridged one page in order to make room for a note acknowledging your papers.

It was impossible for me to read & re-read Morlot's papers, sent to me in English in his handwriting before you started for Denmark, to have a correspondence with him which I still retain & with Claparede as before stated, without hitting upon all those most striking points to which you allude; such for example as the wild swan proving that the kitchen-middens were going on during winter (Morlot p 16), that the succession of Danish forests was synchronous with the ages of stone, bronze & iron (Morlot pp 27 & 29),[21] that there was no grain or cereals etc

I told you before your book came out that I admitted that I must have seen your Danish paper & re-touched & re-worded several passages from your text but not so many as you think; I was surprised to find it so when you first called my attention to the fact. But, as I stated they were always in substance the same as now, which is quite intelligible as you say in your note just received "Of course we have neither of us any claim to originality in the matter. You might have obtained all the information in your chapter on Danish archæology from Danish sources & from Morlot's paper."

I fully admitted to you that you were quite right in pointing out that the co-incidences in your book & mine were not caused by your copying from me. It is the manner in which you worded your note at page X & your omission to allude to the verbal & written communications on the subject which I had with you of which I complain

You say that you did not feel authorized to print what I wrote & what I said when talking with you. How you could have felt any doubt is unintelligible to me; you might have asked me if you had any hesitation & you ought to have been glad to give my explanation which would have been satisfactory to those who know me better than you do.

Mutual friends, you say, think I have no reason to complain but I feel sure that they never saw my former letter to you before they read your note at page X.[22] Had they seen it they would have recommended you to insert my explanation of the true assertion in my note that my chapter was written before your first paper & they would also have suggested that you should state in the same note at page X that there was nothing on Danish archæology that we might not both have derived from common sources & that you were only anxious to show that you had not borrowed from me & not by implication to set up a claim of originality for discoveries taken without acknowledgement by me from you. Had they known all & allowed y^r. note to stand as it is they would have shown themselves no friends of mine still less of yours.

C. L.

Endorsement: May 31st / 65

Royal Botanic Gardens, Kew, Letters to J. D. Hooker, vol. 14, doc. 323; University of Edinburgh Library, Special Collections, Lyell 1, Gen. 113/3650–3, 3813–20, 3821–4

1 Lyell sent copies of the enclosed letters regarding his dispute with John Lubbock to CD and Thomas Henry Huxley as well as Hooker (letter from Charles Lyell to T. H. Huxley, 2 June 1865, Imperial

College of Science, Technology and Medicine Archives, *Huxley papers*, 6: 100; for CD's and Hooker's exchange of views on the letters, see the letter to J. D. Hooker, 1 June [1865], and the letters from J. D. Hooker, [26 May 1865], [2 June 1865], and [15 June 1865] and enclosure). Neither the letter to CD nor its enclosures have been found. Lyell later informed Huxley that he showed the correspondence to George Busk (letter from Charles Lyell to T. H. Huxley, 5 June 1865, Imperial College of Science, Technology and Medicine Archives, *Huxley papers*, 6: 104).

2 Lubbock consulted Busk and John Tyndall as well as Huxley and Hooker about his dispute with Lyell (letter from John Lubbock to Thomas Henry Huxley, 7 June 1865, Imperial College of Science, Technology and Medicine Archives, *Huxley papers*, 22: 64). Lubbock also consulted CD in person. In a letter to Henrietta Emma Darwin, [1 June 1865] (DAR 219.9: 28), Emma Darwin wrote:

> Papa has been greatly worried this mg by all the Lyell & Lubbock correspondence sent him by Sir C. & whereas after talking to John, he thought him not wrong, after seeing all the letters, he thinks he was quite wrong not to allude to Sir C's explanation of the matter. He has written to this effect to Sir C. & I suppose this will be shewn to John L.

3 Lyell alludes to the note Lubbock had written at the end of the introduction to *Prehistoric times* (Lubbock 1865, p. x):

> Note – In his celebrated work on the 'Antiquity of Man,' Sir Charles Lyell has made much use of my earlier articles in the 'Natural History Review,' frequently, indeed, extracting whole sentences verbatim, or nearly so. But as he has in three cases omitted to mention the source from which his quotations were derived, my readers might naturally think that I had taken very unjustifiable liberties with the work of the eminent geologist. A reference to the respective dates will, however, protect me from any such inference. The statement made by Sir Charles Lyell, in a note to page 11 of his work, that my article on the Danish Shell-mounds was published *after his sheets were written*, is an inadvertence, regretted, I have reason to believe, as much by its author as it is by me.

Lubbock refers to *Antiquity of man* (C. Lyell 1863c); the 'article on Danish shell-mounds' is Lubbock 1861. Lubbock published three further archaeological papers in the *Natural History Review* (Lubbock 1862a, 1862b, and 1863a). For Hooker's early consideration of this note, see the letter from J. D. Hooker, [26 May 1865] and n. 15.

4 The reference is to Lubbock 1865.

5 See n. 3, above. Lubbock had written to Lyell in March 1865 (letter from John Lubbock to Charles Lyell, 13 March 1865; University of Edinburgh, *Lyell 1* Gen. 113/3644–5); Lyell's reply has not been found.

6 The text of the note to which Lyell refers is as follows (C. Lyell 1863c, p. 11):

> Mr. John Lubbock published, after these sheets were written, an able paper on the Danish 'shell-mounds' in the October Number of the Natural History Review, 1861, p. 489, in which he has described the results of a recent visit to Denmark, made by him in company with Mr. Busk.

In his letter to Lyell of 13 March 1865 (see n. 5, above), Lubbock pointed out that some passages in Lyell's *Antiquity of man* (C. Lyell 1863c) were almost identical to sections of his own article on ancient Danish shell-mounds ('kjökkenmöddings', or kitchen-middens). He compared a passage from page 16 of Lyell 1863c with one from his own article (Lubbock 1861, p. 496). Lyell apparently replied, suggesting what Lubbock should say in the preface to Lubbock 1865. That letter has not been found, but Lyell's undated rough notes for it (University of Edinburgh, *Lyell 1* Gen. 113: 3646–8) contain the following comment:

> You can say that the early chapters were written many months & for some time in print before I availed myself of the information obtained from your paper to make a few additions & corrections but that I was unable to recast the first chapters which I must have done in order to do justice to all the new information which you had laid before the public.

In all following references within Lyell's letter where Lyell compares passages using only 'Lyell p. . . .' and 'Lubbock p. . . .' the references are to C. Lyell 1863c and Lubbock 1861.

[7] Lyell refers to Morlot 1859, in which Charles Adolphe Morlot had summarised information on kitchen-middens contained in Danish reports on the subject (see n. 10, below). Lyell or his copyist accidentally wrote '1840' for '1860' as the year of publication of the *Bulletin des Séances: Société Vaudoise des Sciences Naturelles.*

[8] The reference is to Morlot 1861b. The translation first appeared in the *Annual report of the Board of Regents of the Smithsonian Institution, showing the operations, expenditures, and condition of the institution for the year 1860* (Morlot 1861a). Morlot 1861b is an offprint of Morlot 1861a with different pagination.

[9] Edouard Claparède.

[10] Lyell refers to *Oversigt over det Kongelige danske Videnskabernes Selskabs Forhandlinger og dets Medlemmers Arbeider.* Notices of members' research were published in this journal but more extensive reports were published as occasional papers under the auspices of the Kongelige danske Videnskabernes Selskab (Royal Danish Academy of Sciences). Six reports by Johan Georg Forchhammer, Japetus Steenstrup, and Jens Jacob Worsaae were published as occasional papers with the title *Underslgelser i geologisk–antiquarisk Retning* between 1851 and 1855 (Forchhammer *et al.* 1851–5). In his book, Lyell does not specifically cite the Danish material, other than to say it was summarised in Morlot 1859 (C. Lyell 1863c, p. 8).

[11] Lyell refers to Lubbock 1861 and Lubbock 1862a; Lubbock 1862a was an article on ancient Swiss lake-dwellings.

[12] The reference is to Ferdinand Keller and to the first five reports in a series of six articles on ancient lake-dwellings in Switzerland (Keller 1856–66). Keller published the sixth article in 1866, and the whole series appeared in English translation that year (Keller 1866).

[13] Lyell claimed, in the note on page 11 of Lyell 1863c, that Lubbock had written his paper (Lubbock 1861) *after* Lyell had completed his chapter on Danish shell-mounds; see n. 6, above. Lyell's explanation of the discrepancy is given in the third enclosure (letter from Charles Lyell to John Lubbock, 30 May 1865).

[14] See n. 1, above. Lubbock and Lyell did not communicate directly after Lyell's letter of 30 May 1865, but on the advice of Hooker and Huxley, Lubbock had the note at p. x of Lubbock 1865 deleted from all subsequent print-runs of the first edition (letter from John Lubbock to J. D. Hooker, 23 June 1865, Royal Botanic Gardens, Kew, *Letters to J. D. Hooker* vol. 14, doc. 183–4).

[15] The pages refer to C. Lyell 1863c. For the passages in Lubbock's article that correspond to those in C. Lyell 1863c, pp. 10, 15, and 16, see Lubbock 1861, pp. 491–2, 500, and 502.

[16] See n. 7, above. The passages Lubbock refers to are Lubbock 1861, p. 496, Lyell 1863c, p. 16, and Morlot 1859, p. 276.

[17] Lubbock refers to Lubbock 1865 and C. Lyell 1863c.

[18] The letter from Lyell has not been found; Lyell's undated rough notes for the letter are in the University of Edinburgh (see n. 6, above).

[19] Lubbock apparently discussed with several friends, including Huxley, Busk, and CD, what course of action he should take and received support for his decision to append the note quoted in n. 3, above, to the preface of Lubbock 1865 (see, for example, letter from Thomas Henry Huxley to John Lubbock, 7 March 1865, British Library MSS ADD 49641, and the letter from John Lubbock to J. D. Hooker, 23 June 1865, Royal Botanic Gardens, Kew, *Letters to J. D. Hooker* vol. 14, doc. 183–4). In his letter to Hooker of 23 June 1865, Lubbock revealed that the note was written by an unnamed friend and added,

> Even Busk thought it was necessary for me to say something to prevent it being supposed that I was appropriating from Lyell; and I really do not see how it could have been put more mildly

Lubbock had CD's support, at least initially, as indicated by Emma Darwin in her letter to Henrietta Emma Darwin of [1 June 1865] (see n. 2, above).

[20] The sixth edition of *Elements of geology* (C. Lyell 1865) was published in January 1865 (*Publishers' Circular*, 17 January 1865, p. 3).

[21] Lyell refers to the pagination in Morlot 1861b (see n. 8, above).

[22] The reference is to Lubbock 1865, p. x.

From B. J. Sulivan 31 May [1865][1]

Board of Trade. S.W.
May 31st.

My dear Darwin

I am very sorry to hear such a poor account of your health—& I am half afraid of causing you any excitement by coming to see you.[2] do not hesitate to say if you would prefer my postponing it.[3] We leave so soon that I can only run down in the middle of the day for an hour or two. My wife[4] and I go tomorrow evening for a farewell visit of a night to a sister near Lewisham,[5] and I purpose if it is convenient to you going to Bromley by train that reaches there at 12 21.—and returning by the 3.54.—on Friday— would you send me a line to Mrs. Sweetlands 10 Granville Park, Blackheath SE—?[6] and I shall get it there on Friday morning. If you are not well enough do not hesitate to say so.

with our kind regards to Mrs. Darwin Believe me | very sincerely yours | B. J. Sulivan

My last note from Board of Trade as I finish this day.

DAR 177: 285

1 The year is established by the references to Sulivan's retirement and move to Bournemouth (see n. 5, below).
2 Sulivan had suggested coming to Down for a night in his letter of 8 May [1865]; no reply to his letter has been found.
3 There is no evidence that Sulivan visited CD during 1865.
4 Sophia Sulivan.
5 Sulivan retired from his post as the chief naval officer in the Marine Department at the Board of Trade and moved to Bournemouth, Hampshire (see letter from B. J. Sulivan, 8 May [1865] and n. 3).
6 Mrs Sweetland of 10 Granville Park, Blackheath (*Post Office directory of the six home counties* 1865), has not been further identified.

To J. D. Hooker 1 June [1865]

Down
June 1st

My dear Hooker

I am heartily sorry to hear that you are so overworked:[1] I suppose you cannot avoid it; but it really is enough to kill you.— Anyhow do not write to me; for I know better than you can tell me that you would very often write (& how often you have written) to amuse me if you were only quarter idle.—

I return Willie's letters;[2] there is something very charming in their simplicity & the Latin is splenditious;[3] but I am not a "bonus latinus".—[4]

I have had a shocking month with much sickness & have done nothing: I am now trying, at first with strong hope, now with weak hope, Dr. Chapman's ice-bags along the spine, which at least is comfortable.—[5]

I do not quite agree about Lubbock; or at least it would have been extraordinary generous to suffer the imputation of copying whole sentences from Lyell, without any acknowlegment.—[6] The passage in Lubbock, when I first read it made my blood grow cold, & I think, if I were in Lyell's place I would not publish again: his memory must have failed him for the time.—[7]

I have not seen Tylor;[8] but it is my great misery that I can hardly read a page without my head being affected. Some of Lubbock's book[9] has been read aloud to me & I like it very much: he writes admirably I think.

My dear old friend | Yours affect, | C. Darwin

Our children enjoyed their visit to Kew much.—[10]

P.S. Did you ever see a peloric *Antirrhinum majus*: if not, perhaps you w^d. like to see enclosed. I bought a plant & fertilised it with own pollen & *every seedling* is peloric as enclosed one,—all with 6 sepals, 6 divisions to corolla & 6 stamens.—[11] From what Lindley says on Order, it seems rather odd that they are *none* of them pentamerous.—[12]

I have received a correspondence from Lyell this morning with Lubbock.—[13] It is most unfortunate. Lubbock ought, I think, to have given in full Lyell's explanation.[14] No doubt Lyell took & forgot whole sentences from Lubbock— It is horrid— It will be our turn some day—perhaps we two shall tear each other's eyes out some day.[15] I begin to think that an author had better be kicked or spat upon rather than reclaim!

Endorsement: '/65.'
DAR 115: 269, 269b

1 See letter from J. D. Hooker, [26 May 1865].

2 Hooker had sent CD letters written by his son William Henslow Hooker; see letter from J. D. Hooker, [26 May 1865] and n. 10.

3 CD used the word 'splenditious' in two later letters to Hooker (see letters to J. D. Hooker, [22 January 1869] (*Calendar* no. 6568) and 7 August [1869] (*Calendar* no. 6855)). Henrietta Emma Darwin used a variant, 'splendicious', in a letter to William Erasmus Darwin dated [18 November 1862] (DAR 219.1: 67), which suggests that it may have been a family word .

4 Evidently CD used the phrase 'bonus latinus', to mean 'a good Latinist', but the expression has not been found elsewhere.

5 The reference is to John Chapman (see letter to John Chapman, 16 May [1865] and n. 3). CD began the ice treatment on 20 May 1865 (Emma Darwin's diary (DAR 242); see also Appendix IV). Emma Darwin, in a letter to Henrietta (letter from Emma Darwin to Henrietta Emma Darwin, [1 June 1865] (DAR 219.9: 28)), wrote:

> Papa had a good day yesterday in & out & to the sand walk. No effect from ice. Certainly no bad effect—

6 CD refers to John Lubbock, Charles Lyell, and C. Lyell 1863c. A dispute arose between Lyell and Lubbock when Lubbock published a note criticising Lyell's failure to acknowledge Lubbock's archaeological publications properly (see n. 7, below). For Hooker's view of the note, see the letter from J. D. Hooker, [26 May 1865] and n. 15. After writing his initial assessment of Lubbock's note, CD received a letter from Lyell enclosing some of the Lyell–Lubbock correspondence and modified his opinion of Lubbock's actions (see the postscript and n. 14, below).

7 CD refers to the note at the end of the preface to Lubbock 1865. For the text of the note, see the letter from Charles Lyell to J. D. Hooker, [31 May 1865], n. 3. Because Lubbock cancelled the note in late June 1865, it does not occur in all copies of the first edition of *Pre-historic times* (Lubbock 1865; see letter from Charles Lyell to J. D. Hooker, [31 May 1864] and n. 14). CD's annotated copy of Lubbock 1865 in the Darwin Library–CUL (see *Marginalia* 1: 512–13) includes the note.

[8] The reference is to Tylor 1865. See letter from J. D. Hooker, [26 May 1865] and n. 13.
[9] Lubbock 1865.
[10] Henrietta and George Howard Darwin went to London on 25 April 1865 (Emma Darwin's diary (DAR 242)).
[11] *Antirrhinum majus*, the common snapdragon, usually bears irregular flowers, with five sepals, stamens, and petals. CD described the peloric *Antirrhinum* in *Variation* 2: 59–60. CD's notes on the results of this artificial self-pollination, dated 20 May 1865, are in DAR 51: A23; he described the experiment in *Variation* 2: 70. CD's notes on his experiments in 1863, 1864, and 1865 on peloric flowers of *Antirrhinum majus* are in DAR 51: B18–21 and 23. See also *Correspondence* vol. 11, letter to John Scott, 20 [June 1863]. For CD's discussion of the results of these experiments and their bearing on prepotency, reversion, and his pangenesis theory, see *Variation* 2: 70–1 and 398–402.
[12] Lindley 1853, p. 681, indicates that members of the Scrophulariaceae, including *Antirrhinum*, have calyxes and corollas that are pentamerous or, by abortion (that is, malformation or incompletion of a part), tetramerous. A copy of Lindley 1853 is listed in CD's Library catalogue (DAR 240), but it has not been found in the Darwin Library–Down or the Darwin Library–CUL.
[13] Lyell's letter to CD has not been found, but see the letter from Charles Lyell to J. D. Hooker, [31 May 1865] and enclosures. References to specific letters in correspondence between CD and Hooker (see letter from J. D. Hooker, [2 June 1865], and letter to J. D. Hooker, [4 June 1865]) indicate that Lyell sent the same enclosures to CD, probably with a similar letter.
[14] See letter from Charles Lyell to J. D. Hooker, [31 May 1865] and n. 6.
[15] CD sometimes expressed concern that he might inadvertently have taken ideas from Hooker, especially in 1859, when *Origin* was published. See, for example, *Correspondence* vol. 7, letter to J. D. Hooker, 2 March [1859].

From T. H. Huxley 1 June 1865

Museum of Practical Geology
June 1 | 1865

My dear Darwin

Your M.S reached me safely last evening—[1]

I could not refrain from glancing over it on the spot and I perceive I shall have to put on my sharpest spectacles & best considering cap—

I shall not write till I have thought well on the whole subject[2]

Ever Yours | T H Huxley

DAR 166: 308

[1] Huxley refers to CD's manuscript of the section on pangenesis for *Variation* (see letter to T. H. Huxley, 27 May [1865] and n. 4).
[2] No letter from Huxley commenting on CD's manuscript has been found, but see the letter to T. H. Huxley, 12 July [1865], for CD's response to Huxley's critique.

From George Maw 1 June 1865

Benthall Hall. | n.r Broseley.
June 1st. 1865

Dear Sir

I have to day come across a very remarkable case of animal monstrosity with the particulars of which you will I think be interested

On the 3rd. or 4th. of May 1863 a sow belonging to Mr. Wood of the Star Hotel Shiffnal[1] brought forth a litter of eleven pigs all of which were well formed excepting the 2nd. born which is formed in several points like an Elephant. It has a distinct proboscis depending from the nose. Ears & mouth shaped exactly like an elephants & in one or two other respects deviating from the usual formation of a young pig. it is supposed to have been brought forth alive tho' never seen alive as the mother overlaid it— There was an interval of two or three hours between its birth & the birth of the succeeding nine healthy pigs all of which were delivered in the usual rapid succession.

The Mother has had previous & subsequent litters all of which have been healthy & well formed.

The history of the gestation of the litter to which this pig belonged is very remarkable & I have been very careful in sifting & ascertaining the probable correctness of the following facts.

In July 1863, the Sow was put to the Boar & one or two days afterwards (I cannot clearly ascertain the exact interval) Some Elephants belonging to Edmund's menagerie[2] were quartered at the Star Hotel one of these had a peculiar antipathy to pigs & on going up the hotel yard endeavoured to reach the sow with its trunk. The ostler tells me the pig appeared quite terror striken & to this is attributed the singular malformations in its young—[3] The young pig has been preserved in spirits & is now in possession of a chemist in the town— I believe it has not yet been seen by any one who would be likely to appreciate its scientific importance. The present owner wants £2 for it & if you think it is of sufficient importance to place in one of the London museums I shall be very happy to purchase it for this purpose & present it to the museum of the R C of Surgeons[4] or wherever you think it will be better located. It seems a pity it should be left in the obscurity of a small country town. I am not acquainted with Professor Owen[5] or would have written to him on the subject of placing the pig in the museum of the College of Surgeons

One point I am quite certain of; that it is not a made up specimen nor do I see any reason to doubt the fact of the Elephants visit soon after the sow received the boar, however we may view its possible connection with the monstrosity—

I have recently been investigating some deposits in N Wales which underlie the Boulder clay drift that I fully expect will prove to be of Tertiary age.—[6] They consist of white black or variegated sands alternating with Curious white clay & various colored Pipe clays & Earthy lignite— The general series is very much like many of the tertiary beds of Dorsetshire

The little remnants are preserved in pockets in the mountain limestone of Flintshire & Carnarvonshire & belong to a formation that must once have had a wide distribution in N Wales. I have observed it over a distance of 30 miles from East to West— I hope next week to make a careful examination of the district & shall try & find some organic remains.[7]

If you think the monstrous pig will be acceptable at one of the London museums will you kindly favor me with a line by an early post that I may secure it.

I am D[r] Sir | very truly your's | Geo Maw

C Darwin Esq.

DAR 171: 100

[1] John Wood was the proprietor of the Star Hotel, Market Place, Shiffnal, near Shrewsbury in Shropshire (*Post Office directory of Gloucestershire, with Bath, Bristol, Herefordshire, and Shropshire* 1863).

[2] James 'Barney' Edmonds, who married a niece of George Wombwell, took over the running of Wombwell's Windsor Castle Menagerie, one of three Wombwell menageries; the name was changed to Edmonds' Menagerie in the 1860s after the death of Wombwell's widow (see Turner 1995, pp. 42, 136–7).

[3] The ancient belief in the effect of maternal imagination on the development of the foetus was well established by the early eighteenth century, when it was used to account for a wide variety of malformations (Todd 1995, pp. 47–52). In the nineteenth century, the work of Isidore Geoffroy Saint-Hilaire began a new era in the study of monstrous development, bringing it more in line with the general study of embryogenesis (D. Wilson 1993, p. 173). For more on popular belief in the effect of maternal imagination, see Huet 1993, D. Wilson 1993, and Todd 1995; for the scientific debate on the significance of monsters in an evolutionary context, see E. Richards 1994.

[4] Maw refers to the Hunterian Museum of the Royal College of Surgeons.

[5] Richard Owen was the curator of the Hunterian Museum from 1836 to 1856. Although Owen no longer held the post, he would still have been a useful person to consult on the subject of specimens in his capacity as superintendent of the natural history collections at the British Museum (see Rupke 1994, pp. 30–1).

[6] The results of Maw's investigation were published in Maw 1865a. He concluded that, in the absence of fossil evidence, the deposits underlying the boulder-clay drift of North Wales were probably restricted to the Cretaceous or Tertiary periods (*ibid.*, p. 202). He encouraged other investigators to examine the deposits in order to find organic remains that would clarify their true geological age (*ibid.*, p. 204). Prior to 1863 the correspondence between CD and Maw mostly concerned botanical and zoological subjects (see *Correspondence* vols. 9 and 10). In 1863, Maw wrote to CD about his interest in the drift deposits of the Severn Gorge near his home (see *Correspondence* vol. 11, letter from George Maw, 25 April 1863). He was elected a member of the Geological Society of London, with CD as one of the nominating members, in February 1864 (see *Correspondence* vol. 11, letter to the Geological Society of London, [*c.* 28 December 1863] and n. 1).

[7] In a brief report summarising Maw 1865a given at the British Association for the Advancement of Science meeting in September 1865, Maw referred to these deposits as a 'widely distributed series of unfossiliferous deposits' and stated that they closely resembled 'some of the Tertiary deposits in the south of England' (Maw 1865b, p. 68).

From J. D. Hooker [2 June 1865][1]

Kew
Friday Nt.

Dear Darwin

I am *most* anxious to hear what you say to the Lyell & Lubbock correspondence.[2] My wife & I have gone over the letters with the books & Lubbocks original papers[3] & can find but little that fairly appears to be *plagiarism.*— So far as the Danish part is concerned, 2 sentences, of very few lines, not involving any new facts nor theory, are all we can find.[4] Do ask Hetty to enlighten me if there is any thing worse,[5] as I should like to join you in healing this ugly breach.[6] As matters appear to me, Lyell should have said frankly & at once, I had either forgotten that I copied your expressions, or I did so quite unwittingly—& I should have quoted you more pointedly in a foot note & in the preface.[7]

Lubbock on the other hand has taken up the matter in a very evil spirit. The first 9 lines of his note are more than uncourteous, they are rude & insulting— The last 5 lines are hardly intelligable—& do not appear to concern the public, at least as they stand without explanation.[8] They imply that Lyell told a fib to screen himself— Now really between two Scientific men, not to talk of between 28 & 70 this is too bad:[9] & except Lubbock can prove malice or a perverse intention on Lyells part to crush him (as Owen tried to crush Huxley)[10] it I think requires an apology.

This comes of *your* divine art of Compilation!.[11] Both, as it appears to me were making capital compilations, & from precisely the same sources & to illustrate the same subject And I am astonished at Lubbocks reclamation, even under his own exaggerated? idea of the mischief intentionally done him.

One statement of Lubbocks letter of 29th. May is explicit & may be cleared up— Lubbock says "You might have obtained all your information from Danish sources & from Morlots paper, but it is evident you did not do so".[12]

This is giving Lyell's statement, in his letter of May 25, the lie direct—& is inconsistent with Lubbocks opening sentence, "of course neither of us claim originality."[13]

No excuse can be made for Lubbocks not quoting Lyells correspondence,—Lyell may fairly attribute that to malevolence.[14]

This is altogether far worse than the Falconer affair,[15] & will take a deal of forgetting & forgiving. And now my dear D. shall I tell you what is at the bottom of it all?—perhaps you wont believe it— it is just this—that Lady Lyell will not call on Mrs Busk nor invite the Busks to her parties.[16] This the Lubbocks' & Huxley's resent.[17] You never agreed with me about the Lyells position respecting their Scientific reunions—but I always told you they were playing with the fire, & would assuredly burn their fingers.—[18] Here is Busk, an FRS, a Secy of Linn. Soc.—Hunterian Profr.—elected by Committee in Athenæum Examiner to the Army Medical Board, & God knows what all, besides being a universal favorite—is called on by Lyell to be pumped dry of his knowledge; living in the same street for years with the Lyells',[19] & never otherwise noticed by them.— His wife, a most thoroughly accomplished clever person, excellent wife & mother, really scientific, & the kindest & most hospitable charitable person alive, more of a Lady than others asked to Lady Ls. soirées— I do say that the Busks' must feel this to be social ostracism & nothing else. It is all very well to say that an Englishman house is his castle, & that it is no one elses affair who is invited to the house & so forth—but Lady Lyells Soirees are quasi public.— Every Englishman & foreigner of distinction—friend or stranger is invited, & the Owens & Carpenters & Busks **alone**, of people of their scientific standing, that live in London, are excluded[20]—for it is *exclusion* in such a case & remarked by every one to be so. The nobility are wiser in their generation & invite those they hate, because it is "de rigeur"!

Many thanks for the Peloria, every seedling being Peloric is a very important fact which I hope you will make something of.[21]

My wife has had a miscarriage but is recovering.[22] I am ordered to take her to the country as soon as she can be moved. I think of Yorkshire.

I have got through a deal of troublesome work here, & caught a most promising youth to be assist in the Herbarium, the other day.[23]

Ever yrs affec | J D Hooker

DAR 102: 24–7

[1] The date is established by the relationship between this letter and the letter to J. D. Hooker, 1 June [1865]. In 1865, the first Friday after 1 June was 2 June.

[2] Evidently both Hooker and CD received at about the same time letters from Charles Lyell with copies of his correspondence with John Lubbock (see letter from Charles Lyell to J. D. Hooker, [31 May 1865] and enclosures).

[3] See enclosures to letter from Charles Lyell to J. D. Hooker, [31 May 1865]. The books referred to are C. Lyell 1863c and Lubbock 1865; the 'original papers' are Lubbock 1861, 1862a, 1862b, and 1863a. Hooker's wife was Frances Harriet Hooker.

[4] See enclosures to letter from Charles Lyell to J. D. Hooker, [31 May 1865]; see also Bynum 1984, pp. 175–6, for a comparison of some of the parts of Lubbock 1861 and C. Lyell 1863c that could be taken as evidence of plagiarism by Lyell.

[5] Henrietta Emma Darwin was interested in the dispute between Lyell and Lubbock. While Henrietta was on holiday in Wales, Emma Darwin wrote to her about CD's worries over the dispute (see letter from Charles Lyell to J. D. Hooker, [31 May 1865] and n. 2). CD later sent her a letter on the subject from Thomas Henry Huxley to Hooker (see enclosure to letter from J. D. Hooker, [15 June 1865]; see also letter to J. D. Hooker, [17 June 1865] and n. 3). For an earlier criticism by Henrietta of C. Lyell 1863a, see *Correspondence* vol. 11, letter to Charles Lyell, 12–13 March [1863].

[6] Both Lubbock and Lyell appealed further to Hooker to help resolve their disagreement (letter from John Lubbock to J. D. Hooker, 23 June 1865, Royal Botanic Gardens, Kew, Letters to J. D. Hooker, vol. 14, doc. 183–4, and letter from Charles Lyell to J. D. Hooker, 5 June 1865, *ibid.*, doc. 325). Lubbock may have consulted CD again in person, as they were neighbours (see letter from Charles Lyell to J. D. Hooker, [31 May 1865], n. 2), but no correspondence from him on the subject has been found. No further correspondence between Lyell and CD on the subject has been found, although Lyell apparently did arrange for a proof copy of the revised preface to C. Lyell 1863c to be sent to CD (see letter from J. D. Hooker, 13 July 1865 and n. 9).

[7] There is no preface to the first edition of *Antiquity of man* (C. Lyell 1863a). The prefaces to the second edition (C. Lyell 1863b) and to the first printing of the third edition (C. Lyell 1863c), make no mention of Lubbock 1861. In a footnote in the second chapter (C. Lyell 1863a, p. 11), Lyell had written:

> Mr. John Lubbock published, after these sheets were written, an able paper on the Danish 'shell-mounds' in the October Number of the Natural History Review, 1861, p. 489, in which he has described the results of a recent trip to Denmark, made by him in company with Messrs. Busk, Prestwich, and Galton.

In the second edition and the earlier print runs of the third edition, the footnote is almost the same except for the omission of the names of Joseph Prestwich and Douglas Strutt Galton, who had not been with Lubbock and Busk in Denmark; Lyell corrected this error in the second and third editions after Lubbock had pointed it out to him (letter from Charles Lyell to John Lubbock, 20 February 1863, in BL MSS ADD 49640; the letter is reprinted in Hutchinson 1914, 1: 62). On the advice of Hooker and Huxley (see letter from J. D. Hooker, [15 June 1865] and n. 13), Lyell did change the footnote and his preface in print runs of the third edition made after August 1865, leading to variant versions of C. Lyell 1863c (see Appendix V for the text of the altered part of the preface to C. Lyell 1863c and the revised note; see also Bynum 1984 and Grayson 1985 for a comparison of the first three editions of *Antiquity of man*). The next, and last, edition of *Antiquity of man* did not appear until 1873, by which time Lyell had made substantial revisions.

8 Hooker refers to the note in Lubbock 1865, p. x (for the text of the note, see the letter from Charles Lyell to J. D. Hooker, [31 May 1865], n. 3; see also letter to J. D. Hooker, 1 June [1865] and n. 7).

9 In June 1865, John Lubbock was 31 and Charles Lyell was 67.

10 Hooker is probably referring to the 'Hippocampus controversy', the public argument over differences between human and ape brains that lasted from about 1860 to 1862 with Huxley and Richard Owen as the principal disputants. See *Correspondence* vols. 8, 9, 10, and 11; see also Rupke 1994, pp. 270–86, and L. G. Wilson 1996, for an account of the controversy.

11 When discussing *Antiquity of man* (C. Lyell 1863a) with Hooker, CD had praised the usefulness to science of compilers and compilations: 'You know I value & rank high Compilers being one myself!' (see *Correspondence* vol. 11, letter to J. D. Hooker, 24[–5] February [1863]).

12 See letter from Charles Lyell to J. D. Hooker, [31 May 1865] and enclosures. In the letter from John Lubbock to Charles Lyell, 29 May 1865, Lubbock referred to an article by Charles Adolphe Morlot (Morlot 1859), which summarised earlier Danish research (see Forchhammer *et al.* 1851–5), making it available to a wider audience. Lubbock's charge of plagiarism was based on a comparison of certain passages in Lubbock 1861 and C. Lyell 1863c (see n. 4, above), but Lyell later argued to Hooker: 'any one well read up in the Danish memoirs especially in Morlots abstract of them would never know without close comparison whether there was greater coincidence between my version and Lubbocks than was consistent with both of them coming from a common and older source—' (letter to J. D. Hooker, 4 June 1865, Royal Botanic Gardens, Kew, *Letters to J. D. Hooker*, vol. 14, doc. 324). See n. 13, below.

13 In his letter to Lubbock of 25 May 1865, Lyell claimed that there were only three passages where he 'borrowed even any *expressions* from [Lubbock]' (see letter from Charles Lyell to J. D. Hooker, [31 May 1865] and enclosures).

14 In his letter to Lubbock of 25 May 1865, Lyell asked why Lubbock did not include in Lubbock 1865 the explanation Lyell had given for inserting the note on page 11 of C. Lyell 1863c (see letter from Charles Lyell to J. D. Hooker, [31 May 1865] and n. 6).

15 Hugh Falconer had attacked Lyell for failing to acknowledge his and Prestwich's research sufficiently in *Antiquity of man* (C. Lyell 1863a). See *Correspondence* vol. 11 and this volume, letter from J. D. Hooker, [17 February 1865] and n. 12.

16 The reference is to George and Ellen Busk. A reason for the Busks' exclusion may have been that Ellen Busk was known as a freethinker and religious sceptic (see Jensen 1977, A. Desmond 1994–7, 1: 341, and L. G. Wilson 2002, pp. 78–80).

17 The reference is to John and Ellen Frances Lubbock and Thomas Henry and Henrietta Anne Huxley. Lubbock had known Busk since 1856 and considered him one of his closest friends and advisors (Hutchinson 1914, 1: 39–40). Huxley met Busk in 1850 and became closely attached to both him and Ellen Busk (Jensen 1977, p. 316).

18 No correspondence between CD and Hooker has been found in which the Lyells' position respecting their scientific reunions is discussed.

19 George and Ellen Busk lived at 15 Harley Street; Charles and Mary Elizabeth Lyell lived at 53 Harley Street (*Post Office London directory* 1865).

20 The reference is to George and Ellen Busk (see n. 16, above), Richard and Caroline Owen, and William Benjamin and Louisa Carpenter. The Owens may have been excluded because Lyell had publicly quarrelled with Richard Owen (see Bynum 1984, pp. 154–9). The reason for the Carpenters' exclusion is not known.

21 CD had sent Hooker a peloric specimen of *Antirrhinum majus* (see letter to J. D. Hooker, 1 June [1865] and n. 11). All the surviving seedlings produced by pollinating the peloric plant with its own pollen were peloric.

22 See letter from J. D. Hooker, [26 May 1865] and n. 16.

23 The reference is to William Botting Hemsley (R. Desmond 1994). See letter from J. D. Hooker, [26 May 1865] and n. 4.

To John Murray 2 June [1865]

Down. | Bromley. | Kent. S.E.
June 2^{d}

My dear Sir

I write a line to say that there is no chance of my being ready for publication this autumn.[1] I have entirely lost the last five or six weeks from illness: I am now better but know not in the least whether I shall keep well enough to work. This is a great evil. I am, of course, answerable for the few woodcuts in hand.—[2] I am never idle when I can do anything.—

My dear Sir | Yours very sincerely | Ch. Darwin

In your advertisement in "Reader" of my Book "*Reunion*" was printed for "*Reversion*"[3]

Endorsement: '1865—June 2^{d}'
John Murray Archive (Darwin 130)

[1] CD was preparing the manuscript of *Variation* for publication. He started working on it intermittently in 1860; from 16 November 1864 he had been revising the early chapters (see *Correspondence* vol. 12, Appendix II). In March 1865, he had written to Murray saying that he hoped the book would go to press in early autumn (see letter to John Murray, 31 March [1865]). CD agreed to Murray's proposal to announce the work as a forthcoming publication, but warned that whether or not the book would, in fact, be ready by autumn would depend on his health (see letter from John Murray, 1 April 1865, and letter to John Murray, 4 April [1865] and n. 3).

[2] See letter from John Murray, 1 April 1865, letter to John Murray, 4 April [1865], and letter from W. B. Tegetmeier to John Murray and R. F. Cooke, 17 April 1865.

[3] Murray had advertised *Variation* in the *Reader*, 15 April 1865, p. 427 (see letter to John Murray, 31 March [1865] and n. 2). He advertised the book again in the 1 August 1865 issue of *Publishers' Circular*, p. 386, under the title: 'Domesticated Animals and Cultivated Plants, or the Principles of Variation, Inheritance, Reversion, Crossing, Interbreeding, and Selection under Domestication'.

To W. B. Tegetmeier 2 June [1865][1]

Down. | Bromley. | Kent. S.E.
June 2^{d}

My dear Sir

I have become a little uneasy for fear that you might have despatched the Pigeon M.S & I never received it.—[2] I am in no hurry but sh^{d}. be glad of it before very long.—

I have lost **five** weeks by illness, & done nothing but I hope soon to resume some work.

Your hints in my Fowl M.S. were very useful.[3]

Have you published on their skulls, besides the paper in Zoolog. Proc.?[4]

I cannot avoid quoting *occasionally* Wingfield Book, when I have to allude to statements quoted from Mr Hewitt & Ballance.—[5]

If you have any communication with Mr Zurhorst, pray give him my cordial thanks for his kind information.—[6]

This w[d] be experiment worth trying, cross a young white Pouter which always breeds true, or red barb, *once* & *for first time* with a blue barred pigeon; then pair her, after she has reared her Mongrel offspring, with her own breed & observe whether her subsequent progeny is affected by the first impregnation.—[7]

My dear Sir | Yours very sincerely | Ch. Darwin

Have you seen D[r] Chapuis's French work on Belgian Carriers: he has sent it to me.—[8]

The New York Botanical Garden Archives (Charles Finney Cox collection) (Tegetmeier 56)

[1] The year is established by the relationship between this letter and the letter from W. B. Tegetmeier, 10 April 1865.

[2] CD sent his manuscript of the section on pigeons for *Variation* to Tegetmeier on 7 April 1865 (see letter to W. B. Tegetmeier, [7 April 1865], and letter from W. B. Tegetmeier, 10 April 1865).

[3] Tegetmeier had read and commented on CD's manuscript of the section on fowls for *Variation* (see *Variation* 1: 225–75; CD acknowledged Tegetmeier's assistance in a note on page 225). See letter to W. B. Tegetmeier, 27 [February 1865], and letter to W. B. Tegetmeier, [7 April 1865].

[4] A notice of Tegetmeier's presentation of living specimens and preparations of skulls of Polish fowl at the 25 November 1856 meeting of the Zoological Society of London appeared in the *Proceedings of the Zoological Society of London* (Tegetmeier 1856). No other paper on fowls by Tegetmeier has been found. CD discussed the perforations in the skulls of Polish fowls in *Variation* 1: 262–5 and 2: 332–3, and cited Tegetmeier 1856 in *Variation* 1: 262.

[5] Tegetmeier had begun editing and revising Wingfield and Johnson 1853, but the project was never finished because the publisher went out of business. An unbound annotated copy of the eleven instalments that were published (Wingfield and Johnson 1856–7) is in the Darwin Library–CUL (see *Marginalia* 1: 798–800). Tegetmeier had asked CD to cite him personally, instead of Wingfield and Johnson 1856–7, if CD thought fit (see letter from W. B. Tegetmeier, [29 March – 7 April 1865] and n. 6). Edward Hewitt and Charles Ballance, well-known poultry fanciers, are cited in Wingfield and Johnson 1856–7 and in *Variation*. Owing to the delay in publishing *Variation*, CD was able to use Tegetmeier 1867 in all his references to poultry, including information from Hewitt and Ballance. Tegetmeier 1867 was originally published in fifteen instalments in 1866 and 1867. CD's unbound annotated parts are in the Darwin Library–CUL (see *Marginalia* 1: 800–3). See also *Correspondence* vol. 14, letter from W. B. Tegetmeier, 22 January [1866].

[6] Frederick William Zurhorst had supplied Tegetmeier with information on the colour of chicken down, presumably for CD (see letter from W. B. Tegetmeier, 10 April 1865 and n. 5).

[7] There is no evidence that Tegetmeier ever performed such an experiment on CD's behalf. In *Variation* 1: 405, CD noted that Félicien Chapuis claimed that the influence of a first male is sometimes perceived in succeeding broods (Chapuis 1865, p. 59); CD warned that the statement required confirmation.

[8] The reference is to Chapuis 1865. CD's annotated copy of this work is in the Darwin Library–CUL (see *Marginalia* 1: 165).

To J. D. Hooker [4 June 1865][1]

Sunday

My dear Hooker

I fully agree to every word you say on the controversy, except that Lubbock's printed note does not seem to me insulting & is only too intelligible.[2] But I should state that I read the correspondence[3] only once & compared only two passages;[4] for such work now upsets me. I did not sufficiently notice the passage in Lubbocks letter which in fact gives Lyell the lie.[5] This I think he ought to apologize for. It has also occurred to me that Lyell's statement that his M.S. was set up in type is in fact no valid excuse for his not having distinctly quoted Lubbock, as it is manifest he altered whole sentences in type[6] Lyell's word of course requires no corroboration but I read Morlot's pamphlet immediately it came out,[7] & I found Lyell quite au fait with his facts, which must have been before Lubbock wrote. I doubt whether you or I or any one c[d] do any good in healing this breach. Time alone could do it if it ever can be done.

All that you tell me about the Busk affair is very curious & pitiable in such a pair as the Lyells.[8] I especially agree with what you say of the blame to Lubbock in not having quoted Lyell's letter to him on the subject.[9]

Have you read the latter half of Lubbocks book?[10] it seems to me admirable. Very many thanks for your letter which has greatly interested me.[11] If you can do any thing to heal the breach it will be a real good service. I can hardly offer to join you just now as I am very weak after 5 weeks sickness—[12]

yours affectionately | Ch. Darwin

LS
DAR 115: 270

[1] The date is established by the relationship between this letter and the letter from J. D. Hooker, [2 June 1865]. In 1865, the Sunday following 2 June was 4 June.

[2] CD refers to John Lubbock's accusing Charles Lyell of plagiarism, and to Lubbock's note in Lubbock 1865, p. x (see letter from Charles Lyell to J. D. Hooker, [31 May 1865], n. 3, for the text of the note in Lubbock 1865, and letter from J. D. Hooker, [2 June 1865], for Hooker's impression of the controversy).

[3] See letter from Charles Lyell to J. D. Hooker, [31 May 1865] and enclosures and n. 1.

[4] See letter from J. D. Hooker, [2 June 1865] and n. 4.

[5] In his letter to CD of [2 June 1865], Hooker had referred to a letter from Lubbock to Lyell dated 29 May 1865 (see enclosure to letter from Charles Lyell to J. D. Hooker, [31 May 1865]). Hooker argued that Lubbock, in that letter, gave 'Lyell's statement, in his letter of May 25, the lie direct'; that is, accused Lyell of lying when he wrote in his letter to Lubbock of 25 May 1865, 'there are only three passages in which I have borrowed even any *expressions* from you' (see letter from J. D. Hooker, [2 June 1865] and nn. 12 and 13).

[6] CD is referring to the note on page 11 of C. Lyell 1863c; see letter from J. D. Hooker, [2 June 1865] and n. 7. CD apparently noticed other places in C. Lyell 1863c where Lyell had made alterations (see Bynum 1984, p. 178, for a recent assessment of Lyell's conduct).

[7] The reference is to Charles Adolphe Morlot and Morlot 1859. CD's annotated copy of Morlot 1859 is in the Darwin Pamphlet Collection–CUL. See letter from J. D. Hooker, [2 June 1865], n. 12.

[8] CD refers to George and Ellen Busk and their exclusion from social gatherings hosted by Charles and Mary Elizabeth Lyell; Hooker claimed that the dispute that arose between Lubbock and Lyell was at least partly due to the resentment felt by Lubbock over this treatment of the Busks by the Lyells (see letter from J. D. Hooker, [2 June 1865] and n. 16).

[9] See letter from J. D. Hooker, [2 June 1865] and n. 13; see also letter from Charles Lyell to J. D. Hooker, [31 May 1865] and n. 2.

[10] CD refers to Lubbock 1865. See letter from J. D. Hooker, [26 May 1865] and n. 14.

[11] See letter from J. D. Hooker, [2 June 1865].

[12] CD recorded in his journal that he had been ill since 22 April 1865 (see letter to John Chapman, 16 May [1865] and n. 4, and Appendix II).

To [Richard Kippist][1] 4 June [1865][2]

Down. | *Bromley.* | *Kent. S.E.*
June 4

My dear Sir

I am much obliged for your very kind note.[3] I am glad to say I am much better, tho' still far from strong & living in my bedroom.[4] I was extremely stupid about the books. The Vol. for 1843 is right; & will you be so good as to have ready for the carrier on Thursday m$^{g.}$[5] the other vol for 1844 p. 295 a paper by Dutrochet.[6] The wrong vol shall then be returned[7]

My dear Sir | yours sincerely | Ch. Darwin

LS
American Philosophical Society (309)

1 The recipient is likely to have been Richard Kippist, the librarian of the Linnean Society. Similar requests for books addressed to Kippist in CD's hand are also in the collection of the American Philosophical Society (see, for example, *Correspondence* vol. 11, letter to Richard Kippist, 13 May [1863 or 1868]).

2 The year is established by CD's reference to Dutrochet 1843 and Dutrochet 1844 (see n. 6, below).

3 The letter from Kippist has not been found.

4 According to CD's 'Journal' (Appendix II), he had been ill since 22 April 1865.

5 A carrier service to and from London was operated every Thursday by George Snow, the village coal-dealer (*Post Office directory of the six home counties* 1862).

6 CD refers to Dutrochet 1843 and Dutrochet 1844, articles in *Comptes rendus hebdomadaires de séances de l'Académie des Sciences* on the twining characteristic of certain plants. CD may have been checking the proofs of 'Climbing plants', which appeared in the double issue of the *Journal of the Linnean Society* (*Botany*), 12 June 1865; he had started correcting the proof-sheets in April 1865 (see letter to Asa Gray, 19 April [1865]). A note in pencil on the verso of the first page of CD's manuscript copy of 'Climbing plants' (DAR 17 and 18) reads (DAR 17.1: B2 v.), 'The second note* has not been sent': this may refer to the references to Dutrochet 1843 and 1844 that were to appear in 'Climbing plants', p. 2 n. The reference to Dutrochet 1844 that appeared in the first note on page 2 of 'Climbing plants' is on a separate scrap of paper on p. 28 v. of the manuscript (facing p. 29). CD's copy of the proofs for 'Climbing plants' is in DAR 213.2.

7 Two volumes of the *Comptes rendus* were published in 1844, volumes 18 and 19. CD evidently ordered volume 18 in error.

To George Maw 4 June [1865]

Down. | *Bromley.* | *Kent. S.E.*
June 4

My dear Sir

The monstrosity of a proboscis-like prolongation of the snout occurs much more frequently as stated by Isidore Geoffroy than with any other animal; & therefore I presume is not rare.[1] Should you wish to present the specimen to the Coll. of Surgeons you had better write to enquire whether it w^{d} be acceptable, to—W. H. Flower Esq the Curator who occupies Professor Owen's former place.[2] I have attended a little to the belief of the effects of the imagination of the mother on the offspring, & must believe that the coincidence of the visit of the elephant & the birth of the monster was a simple accident.[3]

If you can find organic remains in the beds under the Drift in N. Wales, you will make a very interesting discovery.[4]

Pray excuse brevity as I am far from well[5] & believe me my dear Sir | yours sincerely. | Ch. Darwin

LS
Postmark: JU 6 65
Royal Horticultural Society, Lindley Library, Maw Correspondence, 9

[1] Maw had written about a case of animal monstrosity in his letter of 1 June 1865. CD refers to Isidore Geoffroy Saint-Hilaire and to Geoffroy Saint-Hilaire 1836, 3: 353. CD marked the sentence referring to pigs in his copy of the volume and also noted that a trunk was a common aberrant occurrence in pigs. CD's annotated copy is in the Darwin Library–CUL (see *Marginalia* 1: 312–14). CD mentioned the frequency of the occurrence of a monstrous proboscis in pigs in *Variation* 2: 57–8 and cited Geoffroy Saint-Hilaire 1836.

[2] William Henry Flower became curator of the Hunterian Museum of the Royal College of Surgeons in 1861. See also letter from George Maw, 1 June 1865, n. 5.

[3] See letter from George Maw, 1 June 1865 and n. 3.

[4] See letter from George Maw, 1 June 1865 and n. 6. In 1842, CD had visited many locations in north Wales and had written on the evidence of ancient glaciers there ('Ancient glaciers of Caernarvonshire'). In a chapter written for a scientific manual, CD had emphasised the importance of using fossil remains to determine the age of geological deposits (Herschel ed. 1851).

[5] Emma Darwin's diary (DAR 242) records CD's health on 4 June 1865 as 'languid and uncomf all day slight sick.' See also letter to [Richard Kippist], 4 June [1865] and n. 4.

To John Chapman 7 June 1865

Down. | Bromley. | Kent. S.E.
June 7. 1865

My dear Sir

On the whole I can give a pretty good report especially of the 2 last days; but I had acid vomiting in the evening of June 3rd & 4th.[1]

I think I have gained in general vigour; but I have had 3 or 4 fits of flatulence every day.

It is certain that the ice does not stop either flatulence or sickness at the time. I enclose a table of rate of pulse just before & after the ice.[2] You may see in this table how wrong I was in saying that my pulse is now generally about 60. I am certain that formerly this was the case; but am not so certain whether this has been the rate during the last 2 years since I have been weaker; yet generally I feel sure it has been 60 or under. If you have known cases in which the ice has accelerated the pulse, I shd feel hardly any doubt that this has been so with me.

I apply the ice 3 times daily but never for 1° 30′ each time for tho' it does not contract my chest or chill me, yet after about 1° 15′ I generally get weary of it. I *could* bear it for 1. 30 or I cd put it on 4 times a day if you desired. I always like it much for the 1st hour. With respect to my pulse I think it has become a little irregular with the accelerated action.

Pray do not trouble your self to write unless you have fresh orders to give.

Pray believe me my dear Sir | yours very faithfully | and obliged | Ch. Darwin

LS

University of Virginia Library, Albert and Shirley Small Special Collections Library, Darwin Evolution Collection (3314)

1 CD had been undergoing Chapman's ice treatment since 20 May 1865; Chapman had last visited CD on 28 May 1865 (Emma Darwin's diary (DAR 242)). See letter to John Chapman, 16 May [1865] and nn. 3 and 4, and letter to J. D. Hooker, 1 June [1865] and n. 5. The entries in Emma Darwin's diary (DAR 242) include the notes 'sickness at night' for 3 June 1865, 'slight sick' for 4 June, and 'good day' for both 5 and 6 June. CD had suffered from acid vomiting for some time (see *Correspondence* vol. 12, letter to J. D. Hooker, [20–]22 February [1864]).

2 The table of pulse rates has not been found, and Emma Darwin's diary (DAR 242) does not record CD's pulse rate during the ice treatment, except for 10 June 1865, which includes the note 'Very languid & low pulse'. The list of symptoms that CD prepared for Chapman (see Appendix IV) included acid vomiting and flatulence.

From Edward Cresy 9 June 1865

Metropolitan Board of Works | Spring Gardens

9 June '65—

My dear Sir.

Until yesterday I was unable to verify the reference for which Mrs Darwin asked me in her P.S.—[1] it is "The Darwinian hypothesis supported by observations on Crustacea by Fritz Muller of Desterro"—contained in the Annals & Magazine of Natural History No 89—May. '65— P 410. Article 44—[2]

The article purports to be an abstract of the Bibliotheque Universelle 1865 Bulletin Scientifique, p 154—[3] If I can be of service by so doing I will gladly ask Walter White to let me see the article itself as I have no doubt they have it at the Royal Society—[4] The argument struck me very much as a piece of circumstantial evidence does in a criminal trial where a hundred little immaterial facts are linked together to demonstrate irresistably the main conclusion—

I was very sorry to hear from Mrs Darwin you have again been so ill—[5] I wish your disciples could each club together a fragment of their health to repair yours—It is lamentable to think how much good work we are losing which we should otherwise possess—

I am sorry to say I missed hearing your paper on the climbing plants although looking out for it & never knew it was read till I saw the brief abstract in the papers.[6]

You would have been amused to have heard Huxley on Friday at the Royal Institution on Ethnology boldly proclaiming his faith in Darwinism & pitching Noahs ark in good set terms to the four winds—[7] He never lacks le courage de ses opinions—

With kindest remembrances to Mrs Darwin & your daughter[8]
I remain | Very truly yours | E Cresy

C Darwin Esq.

How is George getting on—[9] I hear of him occasionally from Carpmael—[10]

DAR 161: 244

[1] The letter to Cresy has not been found.

[2] Cresy refers to Anon. 1865b, which was an abstract of Anon. 1865a, a review of Müller 1864. See letter from Edward Cresy, 30 May 1865 and n. 3.

[3] The reference is to Anon. 1865a.

[4] Walter White was assistant secretary and librarian of the Royal Society of London (*DNB*).

[5] See letter to George Maw, 4 June [1865] and n. 5, and letter to John Chapman, 7 June 1865 and n. 1.

[6] Cresy refers to 'Climbing plants', which was read before the Linnean Society on 2 February 1865. The publication that contained a brief abstract of CD's paper has not been identified.

[7] Thomas Henry Huxley delivered the Friday evening lecture at the Royal Institution on 2 June 1865; he spoke on the methods and results of ethnology (*Proceedings of the Royal Institution of Great Britain* 4 (1865): 461–3). The text of the lecture was later published in the *Fortnightly Review* (T. H. Huxley 1865). In the lecture, Huxley argued that the application of CD's views on descent to ethnology could reconcile the views of those who believed in a common origin of man and those who favoured multiple ancestral forms. He also alluded to earlier work on the application of CD's theory to ethnology by Alfred Russel Wallace (see A. R. Wallace 1864a; for CD's exchange of views with Wallace on the topic, see *Correspondence* vol. 12, letter to A. R. Wallace, 28 [May 1864], and letters from A. R. Wallace, 10 May 1864 and 29 May [1864]). For Huxley's earlier involvement in ethnological disputes in 1864, see *Correspondence* vol. 12, letter from J. D. Hooker, 9 [March] 1864 and n. 23.

[8] Cresy refers to Emma Darwin and Henrietta Emma Darwin.

[9] George Howard Darwin was an undergraduate at Cambridge. He matriculated at Trinity College in October 1864 (*Alum. Cantab.*).

[10] William Carpmael was a civil engineer who worked with Cresy and whose son, Ernest, was an undergraduate at Cambridge at this time. Ernest matriculated at St John's College in October 1863 (*Alum. Cantab.*; see *Correspondence* vol. 11, letter from Edward Cresy, 27 April 1863 and n. 8). George stayed with Ernest when he competed for a scholarship at Trinity College in April 1864 (see letter from G. H. Darwin to H. E. Darwin, [4 April 1864] (DAR 251: 2236)).

From Charles Kingsley 10 June 1865

Eversley Rectory, | *Winchfield.*
June 10/65

My dear Sir

M^rs. Kingsley[1] & I send you many thanks for your photograph; & my one in return[2]

Yours ever faithfully | C Kingsley

DAR 169: 32

[1] Frances Eliza Kingsley.

[2] Kingsley had requested CD's photograph in a letter to CD of 30 May 1865. CD probably sent the photograph taken by his son William Erasmus Darwin in 1864 (see *Correspondence* vol. 12, letter from W. E. Darwin, [19 May 1864] and n. 8). This photograph is reproduced as the frontispiece of *Correspondence* vol. 12. The photograph of Kingsley has not been found in the Darwin Archive–CUL.

To John Lubbock 11 June [1865][1]

Down. | *Bromley.* | *Kent. S.E.*
June 11

My dear Lubbock

The latter half of your book has been read aloud to me, & the style is so clear & easy (we both think it perfection) that I am now beginning at the beginning.[2] I cannot resist telling you how excellently well in my opinion you have done the very interesting chapters on savage life.[3] Though you have necessarily only compiled the materials the general result is most original. But I ought to keep the term original for your last chapter which has struck me as an admirable & profound discussion.[4] It has quite delighted me for now the public will see what kind of man you are, which I am proud to think I discovered a dozen years ago.[5]

I do sincerely wish you all success in your election & in politics;[6] but after reading this last chapter you must let me say Oh dear Oh dear Oh dear[7]

yours affectionately | Ch. Darwin

P.S. You pay me a superb compliment but I fear you will be quizzed for it by some of yr friends as too exaggerated.[8]

LS
DAR 263: 7 (English Heritage: 88206456)

1 The year is established by the reference to Lubbock 1865 (see n. 2, below).

2 CD refers to Lubbock 1865, an annotated copy of which is in the Darwin Library–CUL (see *Marginalia* 1: 512–13).

3 Chapters 11 to 13 (pp. 335–472) in Lubbock 1865 are titled 'Modern savages'. Chapters 11 and 12 described the culture and technological achievements of many groups of people, ranging from the native inhabitants of Australia and New Zealand to the aboriginal peoples of North America. Lubbock gave brief accounts of the appearance and customs of a number of different peoples that he termed 'savages', the chief criterion for inclusion being the non-use of metals. He also limited his accounts to those groups that had been 'carefully observed by travellers' (*ibid.*, p. 337). In chapter 13 Lubbock compared the different groups of 'modern savages' and contrasted all 'savage' peoples with 'civilised' societies. CD frequently cited a later edition of Lubbock's book (Lubbock 1869), especially on savages, in *Descent.*

4 CD refers to chapter 14 (pp. 473–92), 'Concluding remarks', in Lubbock 1865. In this chapter, Lubbock argued in favour of CD's theory of transmutation in general but concluded that an ability to control the environment exempted humanity to a large degree from the effects of natural selection (*ibid.*, p. 481).

5 While Lubbock was a teenager, CD encouraged his father, John William Lubbock, to give him a microscope. CD lent the boy specimens of lower Crustacea from his own collection, and John Lubbock's earliest scientific work was based on these (see *Correspondence* vol. 5, letter to George Newport, 24 July [1851], n. 6, and letter to J. D. Dana, 27 September [1853] and n. 4).

6 Lubbock was standing for Parliament in the constituency of West Kent in the election of 22 July 1865 (see *The Times*, 18 July 1865, p. 6, and 24 July 1865, p. 6).

7 CD and Joseph Dalton Hooker deplored the loss to science that they predicted would result from Lubbock's entering politics. See letter from J. D. Hooker, [7–8 April 1865] and n. 14, and letters to J. D. Hooker, 10 [April 1865] and 17 April [1865].

8 In Lubbock 1865, p. 481, Lubbock wrote: 'Thus, then, the great principle of Natural Selection, which is to biology what the law of gravitation is for astronomy, not only throws an unexpected light on

the past, but illuminates the future with hope.' Lubbock also wrote: 'The thoughts of a Shakespeare or a Tennyson, the discoveries of a Newton or a Darwin, become at once the common property of mankind' (*ibid.*, p. 487).

From John Lubbock 12 June [1865][1]

Chislehurst
June 12

My dear M^r^. Darwin

Many thanks for your very kind letter.[2] I am delighted with your praise, & am especially glad that you like the last Chapter about which I was a little afraid.[3] The Morning Advertiser devoted a leading article the other morning to abusing me for writing the book & the Times for reviewing it favourably.[4]

You do not mention how you are. I hope the ice Continues to do you good.[5]

Ever dear Mr Darwin | Yours affec | John Lubbock

C Darwin Esq

DAR 170: 45

1 The year is established by the relationship between this letter and the letter to John Lubbock, 11 June [1865].

2 See letter to John Lubbock, 11 June [1865].

3 Lubbock refers to chapter 14 (pp. 473–92) of Lubbock 1865 (see letter to John Lubbock, 11 June [1865] and n. 4).

4 The review in *The Times* appeared on 6 June 1865, p. 9. The reference to the 'Morning Advertiser' has not been identified.

5 CD had been undertaking an ice treatment for his illness under the direction of John Chapman; see letter to John Chapman, 7 June 1865 and n. 1.

From Charles Kingsley 14 June 1865

Eversley Rectory, | Winchfield.
June 14/65

My dear Sir

I have been reading with delight & instruction your paper on Climbing plants.[1]

Your explanation of an old puzzle of mine—Lath. Nissolia—is a masterpiece.[2] Nothing can be more conclusive. That of the filament at the petiole-end of the Bean is equally satisfactory.[3]

Ah that I could begin to study Nature anew, now that you have made it to me a *live* thing; not a dead collection of names[4]

But my work lies elsewhere now.[5] Such work nevertheless helps mine at every turn. It is better that the division of labour sh^d^. be complete, & that each man should *do* only one thing, while he looks on, as he finds time, at what others are doing, & so gets laws from other sciences w^h^. he can apply—as I do—to my own.

Yours ever faithfully | C Kingsley.

DAR 169: 33

1 'Climbing plants' was published on 12 June 1865 in a double issue of the *Journal of the Linnean Society (Botany)*. No presentation list for this paper has been found.
2 CD's explanation of *Lathyrus nissolia* as a case of reversion (it has grass-like leaves instead of tendrils like other species of *Lathyrus*) is in 'Climbing plants', pp. 114–15; see also *Correspondence* vol. 12, letter to J. D. Hooker, [27 January 1864] and n. 23, and this volume, letter from J. D. Hooker, 3 February 1865 and n. 12.
3 CD's discussion of the suppression of tendrils in the common bean is in 'Climbing plants', p. 114.
4 Kingsley was a keen amateur natural historian who had lectured and written on the subject (see, for example, C. Kingsley 1855 and 1890). He had first written to CD praising *Origin* in 1859 (see letter from Charles Kingsley, 30 May 1865 and n. 2).
5 Kingsley was rector of Eversley, Hampshire, became chaplain to the queen in 1859, and was appointed professor of modern history at Cambridge University in 1860 (*DNB*).

From J. D. Hooker [15 June 1865][1]

Kew
Thursday

Dr Darwin

I am wonderfully taken with Tylor's book & anxious to know your opinion of it—which seems to me quite exceptionally good.[2]

I have an admirable note from Huxley about this Lyellian affair, of the demerits of which he seems have a perfect appreciation. I enclose it.[3]

I have just begun your climbers.[4]

My wife does not recover fast,[5] & still complains of a good deal of pain. We hope to get away to Teesdale on the 26th. the Benthams going with us.[6]—sleeping at York the first night.

Ever Yrs affec | J D Hooker

[Enclosure]

Jermyn S^t.
June 12th. 1865

My dear Hooker

I did not reply to your note last week[7] as I was in hopes after all that you might have come upon our ploy, which was very successful always excepting your own and Mrs Hooker's absence which was a great regret to all of us. I trust our not having you with us is no evidence that Mrs Hooker is any worse.

As you say, this Lyello-Lubbockian business is not a pleasing shindy. I have known all about it from the first from Lubbock & have given such advice as I thought would tend most towards a peaceful solution of the difficulty—[8] Latterly Lyell has been to me & I have found it very difficult to deal honestly with both sides without betraying the confidence of either or making matters worse

The candle is a very small one & by no means worth the game— and I should have absolutely dissuaded Lubbock from taking any notice of the small plunder that had taken place, if it had not been for that unlucky note in which Lyell (innocently I do believe but very stupidly) expressly affirmed he had got nothing from Lubbock:[9] & thereby (as it was obvious somebody had copied from somebody)—threw the onus upon Lubbocks shoulders— This rankled in Lubbocks mind & like all quiet and mild men who do get a grievance he became about twice as 'wud'[10] as Berserks like you & me. It was as much as I could do to get him to write to Lyell for an explanation before coming out with a preface to which what you have seen is milk & water.[11] I hoped that Lyell would see he was in a mess & would set the whole affair straight with half a dozen words of frank explanation as he might have done— Instead of that came a long windy affair looking at the whole business from an exclusively Lyellian point of view and really dictating to Lubbock what he should do to get Lyell out of the scrape— Of course the Lubbockian furnace got seven times hotter and it is a mercy nothing worse came of it than what you saw in Lubbocks preface[12]

I am very glad you backed up my advice to Lyell.[13] I must say he has behaved fairly enough since I put my finger into the pie and has done all I asked him to do— Lubbock I hope, will also comply with my (or rather I should say *our* Tyndall & Busk[14] being his advisers with me) advice to cancel his own note & then all trace of a shindy which one will be glad to forget will have been wiped out.[15]

I don't mind fighting to the death in a good big row but when A and B are supplying themselves from C's orchard I don't think it is very much worthwhile to dispute whether B filled his pockets directly from the trees or indirectly helped himself to the contents of A's basket— If B *has* so helped himself he certainly ought to say so like a man: but if I were A, I would not much care whether he did or not

Lyell has been horribly disgusted about it— but I am not sure the discipline may not have opened his eyes to new & useful aspects of nature

Give our kindest regards ⟨to⟩ Mrs Hooker and say how gri⟨eved⟩ we are to hear of her illness

Ever yours faithfully, | T H Huxley

DAR 102: 28; Imperial College of Science, Technology and Medicine Archives (Huxley 2: 131)

CD ANNOTATION[16]

End of letter: 'Kingsley— any one you know? | Wimmer Book? | How to | Histry of Indian Mutiny.—' *pencil*

1 The date is established by the date of the enclosure; in 1865, the first Thursday after 12 June was 15 June.

2 Hooker first mentioned Tylor 1865 in his letter of [26 May 1865], claiming that he was 'charmed' with it; see also letter to J. D. Hooker, 1 June [1865].

3 For Hooker and CD's discussion of the Charles Lyell–John Lubbock plagiarism affair, see the letters from J. D. Hooker, [26 May 1865] and [2 June 1865], and the letters to J. D. Hooker, 1 June [1865] and [4 June 1865]; see also Appendix V. Thomas Henry Huxley, whose letter to Hooker is enclosed, had been aware of the dispute from its earliest stages (see n. 8, below).

4 Hooker refers to 'Climbing plants' (see letter from Charles Kingsley, 14 June 1865, n. 1).

[5] Frances Harriet Hooker was recovering from a miscarriage (see letter from J. D. Hooker, [2 June 1865] and n. 22).

[6] Teesdale, which is in County Durham, just across the border from Yorkshire, was a familiar destination for the Hookers; see *Correspondence* vol. 12, letter from J. D. Hooker, [26 or 27 April 1864] and n. 18. Hooker refers to George and Sarah Bentham. Bentham was working with Hooker at Kew on the *Genera plantarum* (Bentham and Hooker 1862–83).

[7] Huxley refers to the letter from J. D. Hooker to T. H. Huxley, 6 June 1865 (Imperial College of Science, Technology and Medicine Archive, Huxley 3: 109). In this letter, Hooker informed Huxley that he had recommended to Lyell that he should follow Huxley's advice to cancel the note on page 11 of C. Lyell 1863c (for the text of the note, see the letter from Charles Lyell to J. D. Hooker, [31 May 1865], n. 6) and add an explanation in the preface stating exactly what information he obtained from Lubbock 1861 (see Appendix V for the text of the altered part of the preface to C. Lyell 1863c and the revised note).

[8] In the letter from T. H. Huxley to John Lubbock, 7 March 1865 (British Library MSS ADD. 49641), Huxley advised Lubbock to ask Lyell for an explanation of the note in C. Lyell 1863c (see n. 7, above) and to inform Lyell about the note he intended to publish. Huxley promised to consult with him further at 'the Club' (presumably a reference to the X Club; see letter from J. D. Hooker, [7–8 April 1865] and n. 8).

[9] The note Huxley mentions is in C. Lyell 1863c, p. 11 (see letter from Charles Lyell to J. D. Hooker [31 May 1865], n. 6).

[10] Wud (or wood): 'Extremely fierce or violent' (*OED*).

[11] Huxley had written to Lubbock on 7 March 1865 (British Library MSS ADD 49641; see n. 8, above). Following Huxley's advice, Lubbock wrote to Lyell on 13 March 1865 (University of Edinburgh Library, Special Collections, Lyell 1, Gen. 113/3644–5), asking for Lyell's authorisation to include a note explaining that Lyell 'through an inadvertence' had copied from Lubbock's article (Lubbock 1861).

[12] Lubbock later told Hooker that he had originally intended to print parallel passages from Lubbock 1861 and C. Lyell 1863c to prove his claim of plagiarism (letter from John Lubbock to J. D. Hooker, 23 June 1865, Royal Botanic Gardens, Kew, Letters to J. D. Hooker, vol. 14, doc. 183–4).

[13] In a letter from J. D. Hooker to T. H. Huxley of 6 June 1865 (Royal Botanic Gardens, Kew, Letters to J. D. Hooker, vol. 14, doc. 97–8), Hooker told Huxley that he had written to Lyell telling him to follow Huxley's advice, cancel his note (in C. Lyell 1863c), and say precisely what information he got from Lubbock, citing specific pages. See also letter from J. D. Hooker, [2 June 1865], n. 7.

[14] John Tyndall and George Busk, along with Lubbock, were members of the X Club (see n. 8, above). It is likely that Club meetings were a convenient venue where informal discussions on the dispute could be held. See Barton 1998.

[15] Lubbock did cancel his note, though he still had serious reservations. In a letter to Huxley of 7 June 1865 (Imperial College of Science, Technology and Medicine, London, Huxley Papers 22: 64), Lubbock wrote:

> I should be quite satisfied with anything you & Tyndall & Busk settle & don't think we are very likely to quarrel.
>
> I must say if Lyell states that he "wrote the text of Ch. 2 nearly verbatim as it now stands," he will be falling into his old error.
>
> If he does this I must still put in something to protect myself from the charge of copying.

Lubbock wrote to Hooker on 23 June 1865 (Royal Botanic Gardens, Kew, Letters to J. D. Hooker, vol. 14, doc. 183–4), still defending the note, but concluded, 'It is a hateful business, & sometimes I almost wish that I had run the risk of misinterpretation.'

[16] CD's annotations relate to subjects he discussed in his letter to Hooker of [17 June 1865]. 'Wimmer' is presumably a reference to Christian Friedrich Heinrich Wimmer; see letter to J. D. Hooker, [17 June 1865], n. 8; 'How to' is probably a reference to Prichard 1864; see letter to J. D. Hooker, [17 June 1865], n. 9.

To J. D. Hooker [17 June 1865][1]

Down Bromley | Kent.
Sat.

My dear Hooker

I am very glad to have seen Huxley's letter which is capital.[2] As Etty is much interested in the controversy & I thought it c^{d} not signify I have sent H's letter to her,[3] telling her to return it immediately to you. How witty it is!

I am pleased to hear that you are reading my climbing paper[4] for I thought you w^{d} not have time & it is awfully long. Of all men in the world Kingsley has written to me a note full of interest about it & especially about the Lathyrus.[5] I suppose you do not know any one who from having attended to such subjects w^{d} care for a separate copy.[6]

Remember you have Max Wichura's book on Hybrids.[7] Not that I want it back in any hurry. There was a capital resumé of it in the Reader some time ago.[8] We have read your Indian novel & I liked it very much. We want to read a Hist of the Indian mutiny. Can you recommend one?[9] I am extremely glad you are going an excursion with Mrs Hooker & we both very much hope it may do her good.[10] I have had a very bad 6 weeks with much vomiting & fear that the ice will not do much for me.[11]

Yours affectionately | Ch. Darwin

E' D's[12] love to Mrs Hooker & is very sorry to hear how unwell she continues

LS
DAR 115: 271

1 The date is established by the reference to the letter from Charles Kingsley, 14 June 1865 (see n. 5, below). The first Saturday following 14 June 1865 was 17 June.

2 See enclosure to letter from J. D. Hooker, [15 June 1865].

3 CD refers to Henrietta Emma Darwin. She was on holiday in Wales from 29 May 1865 to 22 June 1865 (Emma Darwin's diary (DAR 242)), and mentioned the letter from Thomas Henry Huxley in a letter to Emma Darwin that can be dated 19 June 1865 (DAR 245: 26). She wrote:

> Many thanks for D^{r}. H. it is very amusing. But can such a thing have had influence. I wish people weren't so foolish.

The controversy referred to is the dispute between Charles Lyell and John Lubbock. See Appendix V.

4 'Climbing plants'.

5 See letter from Charles Kingsley, 14 June 1865 and n. 2. Although Kingsley pursued an interest in natural history, he was perhaps better known for lending religious support to CD's views (see, for example, *Correspondence* vol. 7, letter from Charles Kingsley, 18 November 1859).

6 See annotations to letter from J. D. Hooker, [15 June 1865]. No presentation list for 'Climbing plants' has been found.

7 Early in the year, CD had received a presentation copy of Wichura 1865 from Max Ernst Wichura, and had evidently lent it to Hooker (see letter to M. E. Wichura, 3 February [1865], and letter to J. D. Hooker, 22 and 28 [October 1865]). CD's copy, which is annotated, is in the Darwin Library–CUL (see *Marginalia* 1: 871–3).

8 An abstract of Wichura 1865 was published under the title 'Vegetable hybrids' in the *Reader*, 3 June 1865, p. 631. In his annotations to the letter from J. D. Hooker, [15 June 1865] CD wrote 'Wimmer Book?'; the article in the *Reader* mentions Christian Friedrich Heinrich Wimmer. Wichura cited

Wimmer, not only for his published work (Wimmer 1853), but also for giving him access to the garden in which Wimmer grew many varieties of *Salix* (see also Wichura 1865, pp. 1–2).

9 See CD's annotations to the letter from J. D. Hooker, [15 June 1865], where he writes, 'How to | Histry of Indian Mutiny.—'. 'How to' is probably a reference to the novel *How to manage it* (Prichard 1864), set during the Indian Mutiny or Sepoy War of 1857–8. The novel received a favourable review in the *Athenæum*, 21 January 1865, p. 86.

10 Frances Harriet Hooker had suffered a miscarriage, and the Hookers were planning to travel to Teesdale, County Durham (see letter from J. D. Hooker, [15 June 1865] and nn. 5 and 6).

11 CD had been undergoing John Chapman's ice treatment since 20 May 1865 as a cure for his frequent bouts of sickness (Emma Darwin's diary (DAR 242)). See letter to John Chapman, 7 June 1865 and n. 1.

12 Emma Darwin.

To Charles Kingsley [17 June 1865][1]

Down Bromley Kent
Sat.

My dear Sir

I must thank you cordially for your note which has pleased me much.[2] I did not think that any one w^{d} have noticed the case of the Lathyrus,[3] which interested me because I remember looking at it many years ago in a Lord Dundreary state of mind.[4] It appears to me that we have looked at many things from the same point of view; at least I remember well when reading your capital paper on the great fir woods of Hampshire being surprised at your remarks on the presence of this or that weed shewing how exactly the same train of thoughts had often passed thro' both our minds.[5]

I ought to have thanked you for sending me your photograph which I am extremely glad to possess.[6]

Pray believe me my dear Sir | Yours very faithfully | Ch. Darwin

LS
Cleveland Health Sciences Library (Robert M. Stecher Collection: FF10)

1 The date is established by the relationship between this letter and the letter from Charles Kingsley, 14 June 1865; 17 June was the first Saturday after 14 June 1865.

2 See letter from Charles Kingsley, 14 June 1865.

3 See letter from Charles Kingsley, 14 June 1865 and n. 2.

4 Lord Dundreary was a celebrated comic part in Tom Taylor's play *Our American cousin* (Taylor 1869) The character was described as an English aristocrat with a 'well-bred air married to a vacant stare' (see Tolles 1940). See also *Correspondence* vol. 10, letter to W. E. Darwin, 26 April [1862]. Joseph Dalton Hooker wrote of Dundreary as 'a far more scientific character than I anticipated' (*Correspondence* vol. 11, letter from J. D. Hooker, [28 March 1863]).

5 The reference is to 'My winter garden' (Kingsley 1858), in which Kingsley praised the inexhaustible number of subjects of interest to a natural historian within a relatively small area. He considered, for example, 'What makes *Erica ciliaris* grow in one soil, and the bracken in another?' or, 'Why did that one patch of *Carex arenaria* settle in the only square yard for miles and miles which bore sufficient resemblance to its native sandhill by the sea-shore, to make it comfortable?' (*ibid.*, p. 411).

6 See letter from Charles Kingsley, 10 June 1865. The photograph of Kingsley has not been found in the Darwin Archive–CUL.

From J. D. Hooker [after 17 June 1865][1]

Dear Darwin

"Kaye's Indian Mutiny"[2] is the best I believe; but I have not read it.

My wife is I think really improving,[3] & has been downstairs. I am so glad you liked Huxleys letter[4] I thought it sensible shrewd witty & full of honest good feeling. I am not an admirer of Kingsley's but am glad he has the sense to understand you[5]

If Etty lets the Gooseberry season go bye without inviting me I will kill her.[6]

I have no news. Dont forget Tylor's book.[7]

Ever Yr aff | J. D. Hooker

DAR 102: 29

[1] The date is established by the relationship between this letter and the letter to J. D. Hooker, [17 June 1865].

[2] The reference is to the first volume of Kaye 1864–76. CD had asked Hooker to recommend a good history of the Indian Mutiny, after having read a novel set against the background of the uprising (see letter to J. D. Hooker, [17 June 1865] and n. 9).

[3] Frances Harriet Hooker had recently suffered a miscarriage; see letter from J. D. Hooker, [15 June 1865] and n. 5.

[4] Hooker enclosed a letter from Thomas Henry Huxley with his letter of [15 June 1865]. See also letter to J. D. Hooker, [17 June 1865].

[5] Hooker refers to Charles Kingsley (see letter to J. D. Hooker, [17 June 1865] and n. 5).

[6] Hooker refers to Henrietta Emma Darwin. Hooker's fondness for visiting during gooseberry season was something of a family joke. Francis Darwin, in a manuscript of reminiscences about his father, wrote (DAR 140.3; for typescript, see Robert Brown 1996, p. 8):

> Hooker too used to send him beautiful ones [bananas] from Kew— They were christened "Kew gooseberries" being I suppose I think considered a return for the gooseberry feast for which Hooker was supposed to come every year to Down

See also letter from J. D. Hooker, [20 January 1865], and letter to J. D. Hooker, 26 [May? 1867], *Calendar* no. 5552).

[7] The reference is to Tylor 1865. See letter from J. D. Hooker, [15 June 1865] and n. 2.

From Charles Victor Naudin[1] 18 June 1865

Muséum | *d'Histoire Naturelle,* | *Culture.* | Paris,
18 Juin 65.

Cher Monsieur Darwin,

M. Decaisne[2] et moi nous avons reçu les exemplaires de votre excellent travail sur les plantes grimpantes que vous avez bien voulu nous envoyer, et je viens, en son nom et au mien, vous en faire nos sincères remerciments.[3]

A mon tour, j'espère être bientôt en mesure de vous offrir un exemplaire de mon mémoire sur l'hybridation, qui est sous presse en ce moment.[4] Par la même occasion, je compte vous envoyer une petite note que j'ai faite, il y a une dizaine d'années, sur les vrilles (tendrils) des Cucurbitacées, que peut-être vous ne connaissez pas (Annales des sc. nat. tom. . . .),[5] et où je crois avoir démontré que ces organes sont les nervures d'une feuille appartenant à un rameau avorté, et non, comme l'a prétendu M. Payer, une dépendance de la feuille normalement développée.[6]

Je suis extrêmement occupé, en ce moment, non seulement de mes recherches cucurbitologiques et hybridologiques, mais encore de la rédaction d'un traité général d'horticulture appliqué à tous les climats de la France.[7] J'ajoute, avec regret, que, ce qui m'occupe beaucoup aussi, c'est de lutter contre une névralgie très douloureuse de la face, que ne me laisse de repos ni jour ni nuit.[8]

Ce serait un bonheur pour moi, Cher Monsieur, d'apprendre que votre santé est entièrement rétablie. Je le désire du fond du coeur, pour vous d'abord, puis pour la Science, qui a tout à gagner à ce que vous puissiez long-temps encore continuer à la servir.

J'ai reçu, avec le plus grand plaisir, le beau portrait que vous m'avez envoyé, et que je conserve précieusement comme un souvenir de vous, et aussi comme souvenir de M. votre fils, dont il prouve l'habilité en photographie.[9]

Veuillez donc, Cher Monsieur, agréer avec nos remerciments collectifs, l'assurance des sentiments distingués avec lesquels j'ai l'honneur d'être | votre dévoué serviteur et confrère, | Ch. Naudin | M^{bre}. de l'Institut. | au Muséum d'Hist. naturelle

DAR 172: 8

1 For a translation of this letter, see Appendix I.
2 Joseph Decaisne.
3 The reference is to 'Climbing plants'. No presentation list for 'Climbing plants' has been found.
4 The article referred to is Naudin 1865a. CD's annotated presentation copy is in the Darwin Library–CUL (see *Marginalia* 1: 638–9).
5 The reference is to Naudin 1855. No copy of this article has been found in the Darwin Pamphlet Collection, but the article had been mentioned to CD by Joseph Dalton Hooker (see *Correspondence* vol. 11, letter from J. D. Hooker, [31 July 1863] and n. 7). There was a long debate among a number of botanists, including Naudin and Jean-Baptiste Payer, in the Société Botanique from February to July 1857 on the origin of the tendril in the Cucurbitaceae (*Bulletin de la Société Botanique* 4 (1857): 109–111, 142–6, 322–4, 744–56, 787–8). CD mentions these discussions in a footnote to 'Climbing plants', p. 73. For some of CD's inquiries about tendrils, see *Correspondence* vol. 12, letter from Daniel Oliver, [before 31 March 1864] and nn. 3 and 5, and letter to Daniel Oliver, 11 March [1864] and nn. 3, 6, and 9.
6 In Payer 1845, p. 164, the tendril is described as a modified leaf. During the 1857 discussions (see n. 5, above), Payer added that the tendril in the Cucurbitaceae was analogous to the stipule of the leaf in the Rosaceae (Guillard 1857, pp. 145–6). Naudin had attempted to show that the tendril was a modification of the first leaf of an aborted shoot; that is, a 'mixed organ' consisting of a branch at its base and a leaf at the top (Naudin 1855, p. 9). In 'Climbing plants', CD mentioned Naudin's view of the tendril as 'part leaf and part branch', referring to the 1857 discussions but not citing Naudin specifically. CD then cited Payer on the authority of Miles Joseph Berkeley (who referred to Payer as 'Payen' in his article in *Gardeners' Chronicle* 1864, p. 721) as providing the 'most probable' explanation, that the tendril was 'a separate portion of the leaf itself' ('Climbing plants', p. 73).
7 The work referred to is Decaisne and Naudin 1862–71.
8 Naudin's nervous disorder had left him totally deaf and in constant pain. He had already been forced to give up teaching, and eventually resigned from the Muséum in 1868 after two operations on his facial nerves failed to resolve the problem completely (letter from C. V. Naudin, 29 March 1868, *Calendar* no. 6068; see also J. Harvey 1997).

9 At Naudin's request, CD had sent a photograph taken by his son William Erasmus Darwin in 1864 (see *Correspondence* vol. 12, letter to C. V. Naudin, 8 December [1864]; the photograph is reproduced as the frontispiece of *Correspondence* vol. 12). Naudin had sent a photograph of himself to CD in December 1864 (see *Correspondence* vol. 12, letter from C. V. Naudin, 6 December 1864 and n. 3). Although no photograph of Naudin has been found in the Darwin Archive–CUL, it is likely that the photograph reproduced in this volume is the same as the one CD received.

From W. B. Tegetmeier 20 June 1865

Muswell Hill | N
June 20/65

My dear Sir

Your kind letter pained me very much. You should not trouble to write or dictate letters to me when you are not as well as usual.

A single line from your amanuensis to acknowledge receipt of MSS on pigeons would have been enough.[1]

I have written to Mr Zurhorst and if he cannot try the experiment I can readily do so.—[2]

Please oblige me by not answering this.

And believe me | Yours very sincerely | W B Tegetmeier

C Darwin. Esq.

The Pigeon Voyageur Belge arrived safely.[3] Many thanks

DAR 178: 68

1 The most recent extant letter from CD to Tegetmeier is the letter to W. B. Tegetmeier, 2 June [1865]. It appears that he wrote a later letter in which he acknowledged receipt of the manuscript of the section on pigeons for *Variation*.

2 In his letter of 2 June [1865], CD had suggested a crossing experiment with pigeons. Tegetmeier refers to Frederick William Zurhorst, who had answered earlier queries from CD through Tegetmeier. See letter from W. B. Tegetmeier, 10 April 1865 and n. 5.

3 CD had mentioned *Le pigeon voyageur belge* (Chapuis 1865) to Tegetmeier (see letter to W. B. Tegetmeier, 2 June [1865] and n. 8), and probably sent it in response to a request from Tegetmeier; no letter from Tegetmeier making such a request has been found.

From Henry Holland 25 June [1865][1]

Brook Street
June 25

My dear Charles,

I have again to thank you for your kindness, in sending me your very valuable paper on the Climbing Plants.[2]

It is a subject new to me except in the most common facts; & I have read with great interest the researches, by which you raise it to the character of a branch of natural science.— We shall soon have to speak of Animal & Vegetable *Instincts* as having community in nature & origin.

I heard, & with regret, from Erasmus,[3] that you had been less well again, But I know so little of details, that I can do no more than express my earnest wish that you may be speedily relieved.

My best remembrances to Mrs Darwin.[4] Ever your's very truly | H Holland

I shall get some, (though less than usual), of sea-voyage this autumn.— first to the Orkneys with my second Son—[5] then to Portugal, to see the Exhibition at Oporto.[6] Crossing the B. of Biscay will give me sorrowful recollection of poor Falconer, with whom & Busk I crossed it last Autumn on our way to the Gibraltar Caves[7]

DAR 166: 246

[1] The year is established by the reference to 'Climbing plants', which was published in June 1865. No presentation list for the paper has been found.
[2] 'Climbing plants'.
[3] Erasmus Alvey Darwin.
[4] Emma Darwin.
[5] Francis James Holland.
[6] The 'Exhibition of Arts and Industry' opened in Oporto, Portugal, on 18 September 1865 (see *The Times*, 26 September 1865, p. 10).
[7] Holland, Hugh Falconer, and George Busk had visited Gibraltar in October 1864 to examine Genista Cave, Windmill Hill, and the fossil remains recently recovered from it. A report, in the form of a letter from Busk and Falconer to the Governor of Gibraltar, had recently been published in the *Quarterly Journal of the Geological Society* (Busk and Falconer 1865; see also Falconer 1868, 2: 554–63). See *Correspondence* vol. 12, letter from Henry Holland, 4 November [1864] and n. 6. Falconer died of heart-failure on 31 January 1865 (*DNB*).

From Thomas Rivers 6 July 1865

Nurseries, Sawbridgeworth, Herts, | Great Eastern Railway,
July 6/65

My dear Sir/

I have delayed thanking you for your kind present of "Movements & Habits of Climbing Plants.[1]

What patient research & watching! I am more than ever surprised at what you do & can do. I enclose a little sketch of a climbing French bean.[2] My son[3] had sown in his private garden a row of Haricot beans a variety received from France this is not a climbing bean but in a good season puts forth very slender climbing shoots. He was much amused at what he said was instinct one of the beans in the row put forth a climbing stem this was in front of a plum tree nailed to the wall but which had many young shoots protruding from it all of a sudden— he had been looking into your book— he observed the bean incline itself to the N. E. at a sharp angle, & lay hold of the shoot to which it is now attached— the fact is interesting & curious & so I have thought it worthy of communicating to you. I again thank you

I am My Dr Sir | Y^rs^. very truly | Th^os^. Rivers

I have the peach Almond in fruit this season[4]

DAR 176: 164

[1] The reference is to 'Climbing plants' (see letter from Charles Kingsley, 14 June 1865, n. 1). It is likely that CD sent Rivers an author's offprint (see Freeman 1977, pp. 116–17, for the publication history of the first edition of 'Climbing plants').

[2] The sketch has not been found.

[3] T. Francis Rivers.

[4] Rivers refers to a variety of fruit tree that he grew as part of his investigation into the relationship between the peach and almond. CD's interest in this relationship was first mentioned in a letter to Rivers in 1863 (*Correspondence* vol. 11, letter to Thomas Rivers, 11 January [1863]). Rivers sent CD a peach tree and an almond tree (see *Correspondence* vol. 11, letter from Thomas Rivers, 21 January 1863), and later provided him with fruits from which drawings of a series of peach and almond stones were made for *Variation* (see *Variation* 1: 337–8 for the illustrations and CD's acknowledgment of Rivers's help). In *Variation*, CD noted that Rivers cultivated a variety from France called the 'Peach-almond' (*Variation* 1: 338). CD concluded that the peach was probably a descendant of the almond (*Variation* 1: 339, 2: 218; see also letter from Thomas Rivers, 6 January [1865] and n. 7, for Rivers's view of the relationship between the peach and the nectarine).

To B. D. Walsh 9 July [1865]

Down Bromley | Kent
July 9

My dear Sir

I must send you a line to thank you for your interesting letter of May 29th.[1]

I have been ill during the last two months & have done no scientific work.

Many thanks for the case of Panagæus; this genus is almost sacred to me from old Cambridge recollections.[2]

Thanks also for the extract from Gen. Paez' work.[3] With respect to Daphnia it is a common genus of Entomostracan crustaceans[4] & the facts to which you allude are well known, excepting the case, if true, of the males producing one form of eggs, which must I think have been discovered since I attended to the Genus.[5]

I sent you a few weeks ago a paper by me on climbing plants;[6] but I doubt whether you will find it worth reading.

I am very much pleased that you like Bates' paper.[7] Pray read his Travels.[8]

Believe me | my dear Sir | yours very sincerely | Ch. Darwin

LS
Postmark: JY 10 65
Field Museum of Natural History, Chicago (Walsh 4)

[1] See letter from B. D. Walsh, 29 May 1865.

[2] In his letter of 29 May 1865, Walsh told CD about two North American *Panagaeus* species with the same coloration as the *Panagaeus crux-major* that CD knew in Britain. CD often referred nostalgically to this particular beetle, especially in connection with an incident from his student days at Cambridge. He had spotted a *P. crux-major* while holding two other beetles, put one of the others in his mouth so that he could catch the *P. crux-major*, and ended up losing all three when the beetle in his mouth squirted acid down his throat (*Correspondence* vol. 3, letter to Leonard Jenyns, 17 October [1846], and *Autobiography*, pp. 62–3).

[3] The reference is to Paez 1863. See letter from B. D. Walsh, 29 May 1865 and n. 18.
[4] Walsh had asked CD what class *Daphnia* belonged to (see letter from B. D. Walsh, 29 May 1865). Entomostraca was the name used to refer to all crustaceans other than Malacostraca (Leftwich 1973). In modern classification *Daphnia* is in the class Branchiopoda, order Diplostraca, sub-order Cladocera.
[5] See letter from B. D. Walsh, 29 May 1865 and n. 14. CD had last discussed the genus when John Lubbock sent him the manuscript of a paper on reproduction in *Daphnia* (Lubbock 1857) for comment (see *Correspondence* vol. 6, letter to John Lubbock, 27 October [1856]). CD did not publish on the genus.
[6] The reference is to 'Climbing plants'. No presentation list has been found.
[7] The reference is to Bates 1861 (see letter from B. D. Walsh, 1 March 1865 and n. 30). Walsh's comments on Bates 1861 must have been in one of the missing parts of his letter of 29 May 1865.
[8] The reference is to *The naturalist on the river Amazons* (Bates 1863), which CD had encouraged Henry Walter Bates to write (see *Correspondence* vols. 9 and 10, and Bates 1863, 1: iv). CD was very enthusiastic about this work (see *Correspondence* vol. 11, letter to H. W. Bates, 30 April [1863]). CD's annotated copy of Bates 1863 is in the Darwin Library–CUL (see *Marginalia* 1: 35–7).

From Charles and Emma Darwin to J. D. Hooker [10 July 1865][1]

Down
Monday

My dear Hooker

I do not know whether you are at Kew. When you can find time I sh^{d}. much like to hear how Teesdale suited you & M^{rs}. H.[2] I have had a very bad time, with incessant vomiting, but have been better for last 4 days— The ice to spine did nothing.[3] I fear I am much too weak to risk seeing you, even if gooseberries could tempt you.[4]

We are reading aloud on your recommendation Tylor & are **deeply** interested by it: what a clever man— do you know who & what he is?—[5]

Have you read last no^{r} of H. Spencers Biology:[6] I sh^{d} very much like to hear what you think of it, especially about the umbellifers.[7] When writing the Origin I consulted you & you did not think that the compactness of the head accounted for the difference of the outer flowers; but then I do not think we took into account the difference between a globular & flat corymb.[8] I am not satisfied by his views on irregular flowers; the peloric Gloxinia being upright seems a good argument; but then the peloric Snap dragon I find forms the same angle with the stem as does the very irregular common flower.[9]

The last number of the Nat. Hist review seems to me very good. That on H. Spencer I suppose is by Masters.[10] Let me have a short note from you whenever you can spare time.

All scientific work has been stopped with me for the last $2\frac{1}{2}$ months.[11]

yours affectionately | Ch Darwin

Remind Oliver to have the kindness to let me hear if there are any articles in German periodicals concerning my subjects.[12] what is Oliver doing?

Dear Dr Hooker[13]

I do hope Charles is making a little progress in spite of frequent returns of the sickness but there is a degree of vigour about him on the well days which makes me hope that his constitution is making a struggle. If he conquers this sickness I do hope you will be able to come & see him before long & I am sure there is nobody in the world he cares so much to see. He had one terribly bad week. We liked Dr Chapman so very much we were quite sorry the ice failed for his sake as well as ours.[14]

Will you give my love to Mrs Hooker. I trust she is stronger[15]

AL(S)
DAR 115: 272

[1] The date is established by the relationship between this letter and the letter from J. D. Hooker, 13 July 1865. The Monday before 13 July 1865 was 10 July.

[2] Joseph Dalton and Frances Harriet Hooker and George and Sarah Bentham travelled to York on 26 or 27 June 1865, and then to Middleton-in-Teesdale, County Durham (see letter from J. D. Hooker, [15 June 1865], and Jackson 1906, p. 202).

[3] CD had been following John Chapman's ice treatment since 20 May (see letter to J. D. Hooker, [17 June 1865] and n. 11). Emma Darwin recorded in her diary (DAR 242) that CD suffered several bouts of vomiting and flatulence throughout May and June, but indicated that his health was good on 6, 9, and 10 July 1865 and good for part of 7 and 8 July, though with some sickness.

[4] See letter from J. D. Hooker, [after 17 June 1865] and n. 6.

[5] Hooker first recommended Tylor 1865 to CD in May (see letter from J. D. Hooker, [26 May 1865] and n. 13) but ill health had prevented CD from reading it earlier. CD made several references to Tylor 1865 in *Descent* and cited the second edition, Tylor 1870, three times in *Expression*. There are annotated copies of Tylor 1865 and 1870 in the Darwin Library–CUL (see *Marginalia* 1: 810–11). For more on the career of Edward Burnett Tylor and the relation of his work in anthropology to Darwinism, see Leopold 1980 and Stocking 1987.

[6] The most recent instalment of Herbert Spencer's *Principles of biology*, part 14, was published in June 1865 (see Spencer 1864–7, 1: Preface). Spencer's list of subscribers included CD and Hooker (Spencer 1904, 2: 484). CD's instalments of Spencer 1864–7 are in the Darwin Library–CUL as a bound volume (see also *Marginalia* 1: 769–73).

[7] The inflorescences of the Umbelliferae (also referred to as the Apiaciae in Lindley 1853) are discussed in Spencer 1864–7, 2: 156–7.

[8] No correspondence between CD and Hooker on this subject has been found; however, in *Origin*, p. 145, CD wrote, 'in the case of the corolla of the Umbelliferae, it is by no means, as Dr. Hooker informs me, in species with the densest heads that the inner and outer flowers most frequently differ'. He prefaced this by stating (p. 144):

> I know of no case better adapted to show the importance of the laws of correlation in modifying important structures, independently of utility and, therefore, of natural selection, than that of the difference between the outer and inner flowers in some Compositous and Umbelliferous plants.

Spencer 1864–7, 2: 156–7, gives several examples of Umbelliferae species, some with flat inflorescences and differently shaped inner and outer flowers (when the flowers are crowded together), and others with globular inflorescences that do not have differently shaped flowers. Moreover, he argues that in species where the umbellules themselves are clustered, the tendency towards greater irregularity of outer flowers is increased as one moves to the periphery of the cluster. Spencer thus attempts to show that a causal relationship exists between the shape of the inflorescence and the shape of individual flowers within it and concludes (*ibid.*, p. 157):

> Considering how obviously these various forms are related to the various conditions, we should be scarcely able, even in the absence of all other facts, to resist the conclusion that the differences in the conditions are the causes of the differences in the forms.

The above passage was scored by CD in his copy of part 14. He also wrote on the back cover, 'Hooker disagrees about Umbellifers' (see *Marginalia* 1: 770). CD's statements in *Origin* were not substantially changed in later editions.

9 Spencer suggested that the shape of the flower of *Gloxinia erecta* was related to the angle it formed relative to the stem. The peloric regular flower, which was radially symmetrical, was associated with an upright attitude, while the normal irregular flower, which was bilaterally symmetrical, inclined at an acute angle from the stem (Spencer 1864–7, 2: 151–2). CD had kept Hooker informed about his recent crossing experiments on peloric and common *Antirrhinum majus* as well as sending him flower specimens (see letter to J. D. Hooker, 1 June [1865]).

10 CD refers to Maxwell Tylden Masters. The article on Spencer's *Principles of biology* in the issue of *Natural History Review* for July 1865, pp. 373–85, was unsigned. The review focused on aspects of plant teratology, a subject that Masters had studied extensively (see Masters 1869, and letter from M. T. Masters, 7 February 1865). CD's unbound copy of the July 1865 issue of *Natural History Review* is in the Darwin Library–CUL. The article is lightly annotated with an additional note written by CD on a separate sheet at the back of the issue.

11 CD recorded in his Journal that he had been ill since 22 April 1865 (see Appendix II; see also n. 3, above). By 'scientific work' CD usually meant writing (see letter to T. H. Huxley, 12 July [1865]); he was still making botanical experiments (see, for example, letter to J. D. Hooker, 1 June [1865], where he mentions his work on snapdragons).

12 Daniel Oliver was an editor of phanerogamic botany for the *Natural History Review*. He often referred German and French language publications on botanical topics to CD (see, for example, *Correspondence* vol. 12, letter from Daniel Oliver, 14 June 1864).

13 Emma Darwin added her own note to Hooker on the back page of CD's letter.

14 See n. 3, above. For CD's comments on the effects of the ice treatment, see the letter to John Chapman, 7 June 1865, and the letter to J. D. Hooker, [17 June 1865].

15 Frances Harriet Hooker had suffered a miscarriage (see letter from J. D. Hooker, [2 June 1865]).

To T. H. Huxley 12 July [1865][1]

Down Bromley | Kent
July 12

My dear Huxley

I thank you most sincerely for having so carefully considered my M.S.[2] It has been a real act of kindness. It w^{d} have annoyed me extremely to have republished Buffon's views, which I did not know of but I will get the book;[3] & if I have strength I will also read Bonnet.[4] I do not doubt your judgment is perfectly just & I will try to persuade myself not to publish.[5] The whole affair is certainly much too speculative; yet I think some such view will have to be adopted, when I call to mind such facts as the inherited effects of use & disuse &c.[6] But I will try to be cautious Any how I shall have plenty of time for consideration for my health has been so bad of late that I have written nothing during the last 2 months.

Again accept my sincere thanks & believe me my dear Huxley yours very truly | Ch. Darwin

you have been very good to take so much trouble.—

P.S. I read with much interest your article in the Fortnightly Rev. & quite agree on all the points on which I c^{d} judge.[7] As usual you do me much honour.[8]

P.S. 2^{d}.— Will you be so kind as to return the M.S of the so-called Pangenesis.[9]

LS(A)

Imperial College of Science, Technology and Medicine Archives (Huxley 5: 219)

1 The year is established by the reference to CD's manuscript of the section on pangenesis for *Variation*, which CD had submitted to Huxley for criticism (see letter from T. H. Huxley, 1 June 1865).

2 See letter to T. H. Huxley, 27 May [1865] and n. 4. Huxley's letter commenting on the manuscript on pangenesis has not been found.

3 CD refers to the views of Georges Louis Leclerc, comte de Buffon, as expressed in his *Histoire naturelle* (Buffon *et al.* 1749–82). CD discussed Buffon's 'organic molecules' in reference to his own theory of pangenesis in *Variation* 2: 375 n. 29. He wrote:

> If Buffon had assumed that his organic molecules had been formed by each separate unit throughout the body, his view and mine would have been closely similar.

For CD's theory of pangenesis as described in the manuscript he sent to Huxley, see the letter to T. H. Huxley, 27 May [1865], n. 7, and Olby 1963.

4 The reference is to Charles Bonnet. In *Variation* 2: 375 n. 29, CD contrasted his view of the nature of the germ or gemmule with that of Bonnet (Bonnet 1779–83, 5: 334). Unlike Bonnet, CD did not believe gemmules were pre-formed but held that they were continually produced throughout an organism's lifespan with some passed on from earlier generations. For more on the theories of Buffon and Bonnet, see Roe 1981, pp. 15–18, 42–3, and *passim*.

5 CD greatly expanded and made considerable changes to the manuscript on pangenesis before it was published in *Variation* 2: 357–404 (see Olby 1963, p. 232, for a comparison of the two versions).

6 For CD's discussion of the inherited effects of use and disuse in the manuscript on pangenesis, see Olby 1963, p. 237.

7 CD refers to T. H. Huxley 1865. See letter from Edward Cresy, 9 June 1865 and n. 7. A lightly annotated copy of T. H. Huxley 1865 is in the Darwin Pamphlet Collection–CUL.

8 Huxley described CD's theory of descent as 'the key to ethnology . . . reconciling and combining all that is good in the Monogenistic and Polygenistic schools' (T. H. Huxley 1865, p. 275). He noted that although CD had not applied his theory to ethnology, others such as Alfred Russel Wallace had, and he concluded that natural selection could account for racial variation in humans (*ibid.* p. 276). For Wallace's views on the subject, see *Correspondence* vol. 12, letter from A. R. Wallace, 10 May 1864, and for CD's response, see *ibid.*, letter to A. R. Wallace, 28 [May 1864].

9 See letter to T. H. Huxley, 27 May [1865] and n. 4.

From M. T. Masters 12 July 1865

Rye Lane Peckham. S.E.
July 12. 1865

My dear Sir/

I was present when D^r. Caspary's paper was read[1] and saw the drawings he exhibited—to wit Cytisus adami[2]—a similar case in a Robinia—and a "piebald" orange or *citrus*—the fruit of w^h. blended the characteristics of two species stripe wise![3]

I was so struck with the importance of the paper and with the singular results obtained that I requested him to send me the published account (an old friend of mine) I have no doubt he will do so I shall be writing to him in a few days as I have a letter of his unanswered and will remind him of the promise— I will take care to forward the paper to you when I get it.—[4]

Our good friend D^r. Hooker[5] may possibly have told you that in a few months I am to undertake the editorship of the Gardener's Chronicle in conjunction with

T. Moore and other practical men[6] I hope we shall be favored with communications and enquiries from you on any subject relating to practical Botany Horticulture or rural affairs generally—[7]

a wide range and so far as possible a high tone are what I shall aim at—

Hoping occasionally to hear from you | believe me—in haste but with great respect | faithfully y[rs] | Maxwell. T. Masters

DAR 171: 72

[1] Masters refers to a paper by Robert Caspary (Caspary 1865) that was delivered at the Congrès International de Botanique et d'Horticulture held in Amsterdam from 7 to 12 April 1865. CD read about the paper in *Gardeners' Chronicle* (29 April 1865, p. 386) and asked Joseph Dalton Hooker for information on obtaining it (see letter to J. D. Hooker, [1 May 1865] and nn. 3 and 4, and letter from J. D. Hooker, 2 May 1865 and n. 12; see also CD's annotations to the letter from T. H. Huxley, 1 May 1865). CD's lightly annotated copy of Caspary 1865, along with a handwritten translation, annotated by CD, is in the Darwin Pamphlet Collection–CUL.

[2] CD discussed Caspary's research on *Cytisus adami* (now +*Laburnocytisus adamii*) in *Variation* 1: 387–91, 406–7, and 2: 37, 364–5 (see letter to J. D. Hooker, [1 May 1865] and n. 3).

[3] There is no mention of *Robinia* in Caspary 1865. The orange tree to which Masters refers is discussed in Caspary 1865, pp. 2–5, and mentioned by CD in *Variation* 1: 391, where he notes that it produces 'compound fruit', with the bitter orange and citron being 'either blended together, both externally and internally, or segregated in various ways'.

[4] CD probably received a copy of the paper via Caspary in 1866 (see letter to J. D. Hooker, [1 May 1865] and n. 4).

[5] Joseph Dalton Hooker.

[6] John Lindley, the editor and one of the founders of the *Gardeners' Chronicle*, was listed on the title page of the magazine as editor until his death on 1 November 1865 (see *Gardeners' Chronicle*, 4 November 1865, p. 1033), but had most likely given up the actual editing of the magazine before then owing to ill health (see *Gardeners' Chronicle*, 18 November 1865, p. 1083). Masters and Thomas Moore, Lindley's assistant editor since 1841, probably shared editorial duties during Lindley's illness. Masters's official editorship commenced in November 1865, while Moore became co-editor from 1866 (see *Gardeners' Chronicle*, 8 June 1907, p. 368; see also letter from M. T. Masters, September 1865).

[7] CD often sent letters to the *Gardeners' Chronicle*, asking for information from readers, reporting on his own observations, or summarising facts sent to him by other botanists (see, for example, *Correspondence* vol. 9, letter to *Gardeners' Chronicle*, [before 15 June 1861]). CD's annotated copies of the *Gardeners' Chronicle* from 1841 to 1871 are in the Cory Library, Cambridge Botanic Garden.

From J. D. Hooker 13 July 1865

High Force Inn | Middleton Teesdale
July 13/65.

Dear old Darwin

Your letter gladdened me as your hand writing always does, though the account of your state is anything but gladsome[1] Thank M[rs] Darwin very much for her part.[2] This is just the place for my wife who gains ground rapidly, but loses it, or much of it, every now & then & is bloodless for a time. nevertheless, when weak,[3] Neuralgia does not attack her here, as at Kew, & she walks several miles, & eats like a Trojan.

This is a wonderfully pretty place, especially the banks of the river. the moors are loathsome & the hills low broad rockless treeless & contemptible.[4] I am studying the moraines all day long with as much enthusiasm as I am capable of after laying in bed till 9 eating heavy breakfasts & looking forward to dinner as the summum bonum of existence.[5]

I am reading Lubbock very carefully, it is excellent; most excellent, so well written sagacious & discriminating.[6] How well & wisely he treats C. Lewis' blunder & hastyness & the errors of others too.[7] The Savage chapters are tantalizing for their brevity & incompleteness.[8] Why did you not mention Lyells apologetic pages in your letter—to me—[9] I suppose because the subject was most distressing— reason enough, for so it is— I am not satisfied with his apology at all— it is not handsome at all, & from an old Prince of science to a young aspirant, is not liberal I think. I am not surprised at Lubbock having been *riled* up originally: but cannot alter my opinion of the note,[10] 1. that the accusation is too broad & loose (exaggerated I fear), & 2. the latter paragraph is a bitter sarcasm though wholly unintentionally so I quite believe— Do consider it. he first accuses Lyell of extracting *whole sentences* without acknowlegdment & then attributes the denial of the robbery to a *regretted inadvertence*! i.e. Lyell first robs, then prearranges a fib to hide it, *before he is accused* mind you—& lastly when accused regrets his wilful uncalled for fib, as an *inadvertence*! by Jove if this is not a climax of villainy what is? all an ignorant bye-stander could say is, that his conscience is of a piece with his conduct. I am taking an ignorant byestanders view of the case— of course I know well that Lyells conduct has in justice no such interpretation.— I hope Lubbock will take no more notice of the affair, it is not worth it, & never was. worth anything beyond a mere assurance that he was first, though apparently second in the field.[11]

I know a little of Tylors brother— he is an FGS. & FLS. that Edward Forbes' big heart embraced. The family are gas fitters &c &c, hardware tradesmen most wealthy & respectable, once quakers if not now quakers.[12] I never saw this author. I am so glad you like the book.[13] it is to me the most interesting since your orchid work[14] that I have read, & like it I can always dip back with gusto, & be refreshed.

I have read vol: 2 of Leckie (not vol 1 yet) & am extremely interested & instructed.[15] The next holiday I get I must read Buckle,[16] of which I have only read part of vol 2. I am also reading Tom Payne who I think has been pretty largely plundered by Grey & Colenso.[17]

We have the Benthams here & my cousin Inglis Palgrave.[18] I have not begun Giffords book[19] regularly, but dipped here & there. it seems awfully conceited, vague & meagre of accurate topographical or scientific information, & very diffuse, but wonderfully clever as a literary composition. & no doubt full of valuable historical matter— he is a queer fellow, eccentric like all his Grandfather's grandchildren; with brains enough for us all if he would put them to work systematically.[20]

I had the most enjoyable dinner I ever sat down to out your home or perhaps Lubbocks at Frank Palgraves the other day, the guests being Jowett Maurice & Gifford.— Browning & Stanley came afterwards.[21]

Can you afford to subscribe to the Palestine Exploration fund.[22] I am deeply interested in the proper expenditure of the money & with Spottiswood Grove & some other good men I hope to see it well worked.[23] I have given £3.3 (annually) for 3 years.

Herbert Spencer observations on umbellifer are utterly wrong—his reasoning partially right.[24] How very odd, that the Antirrhinum case occurred to me as to you; I think *that* part of his reasoning, all or in great part wrong.[25] he makes too little of Insect agency in shaping flowers, & of hereditary influence— too much of light air & unconquerable love of symmetry in the plant— still there is much in the line of argument. He sends me the proof sheets to comment on.[26] I have not seen them since— he did not know that you had taken up the modifying effects of insect agency in impregnation as a general law, that is.[27]

I have not seen Nat Hist. Review which will not pay, is all but defunct I hear! & cannot go on beyond the year.[28]

Oliver is working at African plants & his lectures[29]

Ever yrs affec | J D Hooker

My Charlie has taken hooping cough at school, he returns on 19th. which puts all our arrangements out as we must keep him apart, which is difficult during Willys holidays.[30]

DAR 102: 30–3

1 See letter from Charles and Emma Darwin to J. D. Hooker, [10 July 1865] and n. 14. The first page of the letter is written in pencil in CD's hand. The last letter CD had written to Hooker in his own hand was that of 1 June [1865].

2 See letter from Charles and Emma Darwin to J. D. Hooker, [10 July 1865] and n. 13.

3 Frances Harriet Hooker was recovering from a miscarriage (see letter from Charles and Emma Darwin to J. D. Hooker, [10 July 1865] and n. 15).

4 The Hookers had travelled to Middleton-in-Teesdale in County Durham on 26 or 27 June 1865 (see letter from J. D. Hooker, [15 June 1865]). The High Force Inn was located in Forest and Frith, one of the four townships of the town of Middleton-in-Teesdale. The town is on the river Tees (*Post Office directory of Northumberland* 1858).

5 Hooker's observations on moraines in the Tees valley were published in the *Reader*, 15 July 1865, p. 71 (Hooker 1865).

6 Previously, Hooker had told CD that he was disappointed with Lubbock 1865 (see letter from J. D. Hooker, [26 May 1865] and n. 14).

7 John Lubbock pointed out that in G. C. Lewis 1862, pp. 455 and 467, George Cornewall Lewis treated the travels of the Carthaginian Himilco and the Greek Pytheas as mythical; Lubbock discussed the evidence for their voyages having taken place as described (Lubbock 1865, pp. 36–43).

8 Chapters 11 to 13 in Lubbock 1865 (pp. 335–472) were titled 'Modern savages' (see letter to John Lubbock, 11 June [1865] and n. 3). The purpose of Lubbock's brief sketches of 'modern savages' was to 'throw light on the ancient remains found in Europe, and on the condition of the early races which inhabited our continent' (*ibid.*, p. 336).

9 In the letter to Hooker of [10 July 1865], CD made no mention of having received the proofs for a new section to be added to the preface of the third edition of *Antiquity of man* (C. Lyell 1863c). In response to Lubbock's allegation of plagiarism against him (see letter to J. D. Hooker, 1 June [1865], n. 6), Charles Lyell changed a footnote as well as adding the so-called 'apologetic pages' to the preface of copies of C. Lyell 1863c printed after June 1865 (letter from Charles Lyell to Thomas

Henry Huxley, 8 June 1865, Imperial College of Science, Technology and Medicine Archives, *Huxley* 6: 111; see also letter from J. D. Hooker, [2 June 1865], n. 7), resulting in variant versions of C. Lyell 1863c (see Bynum 1984, pp. 176–8). Lyell, who was in Kissingen, Bavaria, apparently arranged for copies of proofs of the revised preface to be sent to both CD and Hooker, among others (letter from Charles Lyell to Thomas Henry Huxley, 28 June 1865, Imperial College of Science, Technology and Medicine Archives, *Huxley* 6: 117). See Appendix V for the revised note and the 'apologetic pages' added to the preface of C. Lyell 1863c.

10 Hooker had once before mentioned to CD that he found the dispute between Lyell and Lubbock the more upsetting because of the difference in their ages (see letter from J. D. Hooker, [2 June 1865] and n. 9). For Hooker's initial impression of the note in the preface to Lubbock 1865 accusing Lyell of plagiarism, see the letter from J. D. Hooker, [2 June 1865]; for the text of the note, see the letter from Charles Lyell to J. D. Hooker, [31 May 1865], n. 3.

11 Hooker refers to the fact that, though Lubbock may have published on Danish shell-mounds before Lyell (see Lubbock 1861 and C. Lyell 1863a), Lubbock's own work was predated by that of Charles Adolphe Morlot (Morlot 1859). Morlot 1859 was a summary of Forchhammer *et al.* 1851–5. See enclosures to letter from Charles Lyell to J. D. Hooker, [31 May 1865], and Appendix V.

12 After reading part of Tylor 1865, CD had asked Hooker whether he knew anything about the author, Edward Burnett Tylor (see letter from Charles and Emma Darwin to J. D. Hooker, [10 July 1865] and n. 5). Tylor's brother, Alfred Tylor, was a geologist, a fellow of the Geological Society of London and the Linnean Society, and in charge of the family brassfoundry business. Evidently, he was one of many encouraged in the study of natural history by Edward Forbes (*Literary Gazette*, 25 November 1854, pp. 1016–18). Both brothers had been educated at the School of the Society of Friends, Grove House, Tottenham (*DNB*).

13 Tylor 1865.

14 The reference is to *Orchids*, published in 1862.

15 William Edward Hartpole Lecky's *History of the rise and influence of the spirit of rationalism in Europe* (Lecky 1865) had brought him to public attention in 1865 (*DNB*). The central theme of the work was the development of reason and decline of superstition in human society.

16 Hooker refers to Henry Thomas Buckle's *History of civilisation in England* (Buckle 1857–61). Lecky claimed to have been greatly influenced by his reading of Buckle (see Lecky 1865, 2: 188). CD had read and admired the first volume of Buckle 1857–61 in 1858 (see *Correspondence* vol. 4, Appendix IV, 128: 23; see also *Correspondence* vol. 7, letters to J. D. Hooker, 23 February [1858] and 31 March [1858]).

17 Hooker probably refers to Thomas Paine's *The age of reason* (Paine 1794 and 1795). Hooker's statement that Paine had been 'plundered by Grey & Colenso' may refer to Asa Gray's pamphlet on natural theology and natural selection (A. Gray 1861) and the biblical critique of John William Colenso, bishop of Natal (Colenso 1862–79). In his pamphlet, Gray echoes Paine's argument in the first part of *The age of reason* that the true revelation of God to man is embodied in the natural world (Paine 1794, pp. 22–31, A. Gray 1861, pp. 35–44; see also J. R. Moore 1979, pp. 339–40, on Gray's belief in divine immanence). Colenso follows Paine in his focus on the Pentateuch and belief in the importance of a commonsense, critical approach to the study of the Bible (see, for example Colenso 1862–79, 2: vi; see also Davidson and Scheick 1994 for a detailed analysis of Paine's biblical criticism).

18 The reference is to George and Sarah Bentham and to Robert Harry Inglis Palgrave, whose mother, Elizabeth Turner, was the sister of Hooker's mother, Maria (Allan 1967).

19 Hooker refers to William Gifford Palgrave's book, *Narrative of a year's journey through central and eastern Arabia* (Palgrave 1865).

20 William Gifford Palgrave, an elder brother of Robert Harry Inglis Palgrave, travelled across central Arabia in 1862 and 1863 (*DNB*). In 1864 Hooker mentioned to CD William Gifford Palgrave's plan to publish on his travels, but remarked that he had 'made no observations of the smallest scientific value' (see *Correspondence* vol. 12, letter from J. D. Hooker, 24 January 1864). Palgrave's paternal grandfather was Meyer Cohen (*DNB* s.v. Francis Palgrave).

21 Hooker refers to Francis Turner Palgrave, Benjamin Jowett, John Frederick Denison Maurice, William Gifford Palgrave, Robert Browning, and Arthur Penrhyn Stanley. He also refers to John Lubbock.

22 The expedition to study the natural history of Palestine was organised under the direction of Henry Baker Tristram and left England in November 1863. Tristram reported on its progress at the British Association for the Advancement of Science meeting at Bath in 1864 (see Tristram 1864a, 1864b, and 1864c). Notices about the progress of the expedition and some early results were published in *Natural History Review* 4 (1864): 152–3, 311–12, 638–9. The Palestine Exploration Fund, set up to finance the expedition, received support from the British Association for the Advancement of Science until 1875 (see Howarth 1931, p. 201). In 1860 Hooker had taken part in an expedition to Syria and Lebanon, during which time he visited Jerusalem (see *Correspondence* vol. 8, letter to J. D. Hooker, 2 September [1860] and n. 1; see also R. Desmond 1999, pp. 210–16).

23 Hooker refers to William Spottiswoode and William Robert Grove. In 1865, Grove was president elect of the British Association for the Advancement of Science, while Spottiswoode was president of its sectional committee for mathematics and physics (see *Report of the thirty-fifth meeting of the British Association for the Advancement of Science held at Birmingham in September 1865*, pp. xxx–xxxi).

24 Hooker refers to Herbert Spencer's view that the size and shape of flowers in umbellifers was related to the density and shape of the umbellules (see letter from Charles and Emma Darwin to J. D. Hooker, [10 July 1865] and n. 8).

25 Spencer had argued from observation of peloric *Gloxinia erecta* flowers that flower shape was causally related to the angle formed between stem and flower stalk. CD noted, however, that both peloric and common flowers of *Antirrhinum majus* were positioned at the same angle to the stem (see letter from Charles and Emma Darwin to J. D. Hooker, [10 July 1865] and n. 9).

26 Hooker had been receiving proofs from Spencer throughout the publication of the instalments of Spencer 1864–7; his assistance was acknowledged in the preface to the published volumes (see *Correspondence* vol. 12, letter from J. D. Hooker, 24 January 1864 and n. 5).

27 CD had emphasised the relationship between insects and flower structure in *Orchids*, giving many examples of flower forms adapted for cross-pollination by insects. An example of the sort of 'general law' Hooker had in mind may be CD's statement in 'Three forms of *Lythrum salicaria*', p. 175 (*Collected papers* 2: 111): 'I have found no exception to the rule, that when the stamens and pistil are bent, the bending is exactly to that side of the flower which secretes nectar . . . or . . . to that side where the structure of the flowers allows the easiest access to it.'

28 CD mentioned the review of Spencer's latest instalment of *Principles of biology* in the July 1865 issue of *Natural History Review* in the letter from Charles and Emma Darwin to J. D. Hooker, [10 July 1865], and also commented on the high quality of that issue. The *Natural History Review* ceased publication with volume 5 in 1865 (see A. Desmond 1994–7, 1: 342–3).

29 In his letter to Hooker of [10 July 1865], CD had asked about Daniel Oliver's activities. Oliver, keeper of the herbarium and library at the Royal Botanic Gardens, Kew, was professor of botany at University College London. Hooker may also refer to the series of lectures that Oliver gave to the garden staff at Kew (see Britten 1917, p. 91). At this time, Oliver was also working on his *Flora of tropical Africa* (Oliver 1868–77; see Jackson 1917).

30 Hooker refers to his sons, Charles Paget Hooker, aged 10, and William Henslow Hooker, aged 12.

From T. H. Huxley 16 July 1865

Jermyn St
July 16th | 1865

My dear Darwin

I have just counted the pages of your M.S. to see that they are all right and packed it up to send you by post, registered, so I hope it will reach you safely—[1] I should have sent it yesterday but people came in & bothered me about post time

I did not at all mean by what I said to stop you from publishing your views and I really should not like to take that responsibility[2]

Somebody rummaging among your papers half a century hence will find Pangenesis & say "See this wonderful anticipation of our modern Theories—and that stupid ass, Huxley, prevented his publishing them"

And then the Carlylians of that day will make me a text for holding forth upon the difference between mere vulpine sharpness & genius—[3]

I am not going to be made a horrid example of in that way— But all I say is publish your views—not so much in the shape of formed conclusions—: as of hypothetical developments of the only clue at present accessible—and don't give the Philistines more chances of blaspheming than you can help

I am very grieved to hear that you have been so ill again—

Ever | Yours faithfully | T. H. Huxley

DAR 166: 309

1 Huxley refers to the manuscript of CD's section on pangenesis for *Variation*, which CD had sent to him for criticism (see letter to T. H. Huxley, 12 July [1865]).

2 The letter containing Huxley's opinions on the manuscript has not been found; however, see the letter to T. H. Huxley, 12 July [1865], in which CD said he would try to persuade himself not to publish his views.

3 The reference is to Thomas Carlyle and his admirers. In his well-known book *On heroes, hero-worship, & the heroic in history*, Carlyle compared the virtuous man's ability to understand nature with that of the immoral man. The latter he likened to a fox and said: 'what such can know of Nature is mean, superficial, small' (Carlyle 1841, p. 173). CD had read many of Carlyle's books, including Carlyle 1841 (see *Correspondence* vol. 4, Appendix IV, DAR 119: 10b). Huxley's views on Carlyle are discussed in Paradis 1978, pp. 47–71.

To T. H. Huxley [17 July 1865][1]

Down
Monday

My dear Huxley

Forgive my writing in pencil, as I can do so lying down— I have read Buffon:—whole pages are laughably like mine.[2] It is surprising how candid it makes one to see one's views in another man's words— I am rather ashamed of the whole affair, but not converted to a no-belief— What a kindness you have done me with your "*vulpine sharpness*".—[3]

Nevertheless there is a fundamental distinction between Buffon's views & mine. He does not suppose that each cell or atom of tissue throws off a little bud; but he supposes that the sap or blood includes his "organic molecules", *which are ready formed*, fit to nourish each organ, & when this is fully formed, they collect to form buds & the sexual elements:—[4] It is all rubbish to speculate as I have done; yet, if I ever have strength to publish my next book, I fear I shall not resist "pangenenesis", but I assure you I will put it humbly enough—[5] The ordinary course of development of beings, such as the echinodermata, in which new organs are formed at quite remote spots from the analogous previous parts, seems to me extremely difficult to reconcile on any view, except the free diffusion in the parent of the germs or gemmules of each separate new organ; & so in cases of alternate generation.—[6] But I will not scribble any more. Hearty thanks to you, you best of critics & most learned man.

Yours most truly | C. Darwin

I was delighted to see the 3rd Editn. advertised of your "Man".[7]

Your last note made us all laugh.— The future rummager of my papers will I fear, make widely opposite remarks—[8]

Imperial College of Science, Technology and Medicine Archives (Huxley 5: 221)

[1] The date is established by the relationship between this letter and the letter from T. H. Huxley, 16 July 1865. The Monday after 16 July 1865 was 17 July.

[2] Huxley had suggested that Georges Louis Leclerc, comte de Buffon, had anticipated aspects of CD's hypothesis of pangenesis (see letter to T. H. Huxley, 12 July [1865] and n. 3). In *Variation* 2: 375, CD mentioned the second volume of Buffon 1749–82, citing the pages that refer to the theory of organic molecules.

[3] See letter from T. H. Huxley, 16 July 1865 and n. 3.

[4] For a brief description of CD's hypothesis of pangenesis, see the letter to T. H. Huxley, 27 May [1865] and n. 7. For CD's comparison of his own and Buffon's theory of organic molecules, see *Variation* 2: 375.

[5] *Variation* was published in 1868. Pangenesis was discussed in chapter 27 (*Variation* 2: 357–404). For a description of the differences between the final version of the section on pangenesis and the original manuscript version, see Olby 1963, p. 252.

[6] The development of Echinodermata is not discussed in the manuscript on pangenesis that CD sent to Huxley. Evidently, the phenomenon of new organs developing in different locations from analogous previous parts occurred to CD as an additional support to his theory after he received Huxley's remarks on his manuscript; he added the discussion of alternate generation, including his example of echinoderm growth, to the final version of the section on pangenesis (see *Variation* 2: 366–7).

[7] The reference is to *Evidence as to man's place in nature* (T. H. Huxley 1863a). A pirated American edition was published in 1863 (see *Correspondence* vol. 11, letter from T. H. Huxley, 2 July 1863 and n. 14) and a second English edition was published in 1864 (*English catalogue of books*, vol. 2), but no third edition appeared in 1865. However, a 'Select list of works' published by Williams and Norgate in May 1865 contained a notice of the 'Third Thousand' of *Evidence as to man's place in nature*. The list appeared at the back of Lubbock 1865, a book that CD had been reading around this time (see letter to John Lubbock, 11 June [1865]).

[8] See letter from T. H. Huxley, 16 July 1865.

From John Scott 21 July 1865

Royal Botanic Gardens | Calcutta
July 21st. 1865.

To, C. Darwin Esq.
Sir,

I would have written you long ago in answer to yours of March 11th.[1] but I have always deferred doing so in hopes of having something or other of scientific interest to communicate. Disappointed in this—as I have never got fairly settled to work since I arrived in India[2]—I will write a few notes by way of informing you, to whom I owe all—how it fares with me here.

I left Rungbee—as I think I mentioned in my last letter I had hopes of doing—in the latter end of May for Calcutta.[3] Dr. Anderson kindly giving me the choice of remaining in charge of the Cinchona, or going down as Curator of the Botanic Gardens, Calcutta.[4] I accepted the latter, glad to embrace the opportunity of being again in a situation where I might advantageously follow out those subjects—in my spare moments—which had previously engaged much of my time & attention:[5] though with a free conscience I think I may justly say that certain of the Edinburgh folk *falsely charged* me with allowing these subjects to engage my time to the sacrifice of my proper work. I think Mr. McNab, innerly at least—will be inclined now to give me this much credit—as I know that much more work has devolved upon him since I left the gardens, & further the person who succeeded me though not having all the duties to attend to that I had; has nevertheless failed to perform even these as he has been dismissed . . . I will not trouble you however with further details on these matters, and I have merely noticed them by way of vindicating myself to you, my best benefactor.[6]

I see plainly that my Verbascum paper is wholly unfit in its present state for publication[7] I was afraid of this when I sent it you, as I stated in a letter* which I had posted in time for the same mail as that by which I despatched the registered papers of Verbascum.[8] I would not have posted these latter so hurriedly, as I was fully aware that they would require much correction, having drawn them up from my notes in the course of my voyage, and while suffering almost unremittingly from sea-sickness. I really thought that I was never going to get free it—the slightest breeze was sufficient to upset me, & I did not get fairly free of this sensation until we had a second time crossed the line & were getting well up for the Sunderbunds.[9] As I said on this account I would not have posted the papers until I had time to examine them—though I can't understand how I have made such **singular blunderings** in my calculations. I am truly sorry and ashamed that I should have taken up your valuable time with such, as also that of Drs. Hooker & Thomson[10]—but as I had Dr. Andersons orders to get ready at once to go to Rungbee, I thought that it might be long ere I had time to look over it, & correct it, so I sent it thinking that should it be unfit for publication you might extract any matter from it you thought fit for your own use. As you had often said you were anxious to see results of further experimentation amongst varieties of Verbascums. . . . .[11] Just now I will not attempt to do anything with it: but whenever I have time to spare I will commence work a fresh on these subjects as well as on that of the other kinds of dimorphism manifested in the vegetable Kingdom & perhaps I will get material worthy of publication.[12]

In talking of this matter, might I take the liberty of asking, if you could spare me a few seeds of *Leersia oryzoides*.[13] I should like to make experiments on fertilisation in closed flowers.[14] We have here one *Leersia—aristata—* I have not seen it flower yet, so I am not aware whether it presents the above phenomena.

I was somewhat astonished to find that all the plants of *Lagerstroemia*, *elegans*; *Indica* & *var alba*, are what may be termed *long-styled*, this organ projecting beyond the longest stamen— the number of the latter organs are very variable especially in the first named species. . . . All the plants of both species in the gardens are sterile.[15] I have by way of experimenting tried a few flowers by artificial fertilisation on L. *Indica*, & var *alba* but with no success— As for *L. elegans*, I have never found good pollen on it— from deformity of anthers, conjoined with its sterility, I am inclined to regard it as a hybrid. When I have time however, I will have a little work on these. You will be interested to hear that the *L. Regineæ*, though I have seen only one form in the garden—a *long-styled* also—produces an abundance of fruit, the capsules well filled with good seed.[16]

In your last letter you say that though still weak your health was improving,[17] earnestly do I wish that you have since had no relapse, and are now enjoying an ordinary share at least, of that, perhaps one of the most desirable of Nature's boons.

I remain | Sir | Very respectfully yours | John Scott

*This letter seems never to have come to hand. I had it posted by one of the hotel servants— But this I understand now is an insecure mode. As these beings often take the stamps from the letter & destroy the latter.

DAR 109: B120a–b

CD ANNOTATION
3.9 As I . . . flowers. 4.3] *crossed pencil*

1 This letter has not been found (see n. 10, below).

2 Scott arrived in Calcutta on 21 December 1864 (see letter from John Scott, 20 January 1865).

3 Scott had been working on a *Cinchona* plantation near Darjeeling since his arrival in India (see letters from John Scott, 20 January 1865 and n. 7, and 10 April 1865).

4 Thomas Anderson was superintendent of the Calcutta Botanic Gardens. Joseph Dalton Hooker had written to Anderson to ask for assistance in finding employment for Scott (see letter from J. D. Hooker, [26 January 1865], n. 4).

5 Scott had undertaken a number of experiments at CD's suggestion when he was foreman of the propagating department at the Royal Botanic Garden, Edinburgh. See *Correspondence* vols. 10, 11, and 12.

6 James McNab was curator of the Royal Botanic Garden, Edinburgh, and had been Scott's immediate supervisor. McNab had found fault with Scott for spending too much time on his own experiments to the detriment of his regular duties (see *Correspondence* vol. 11, letters from John Scott, 22 May 1863 and [3 June 1863]). Scott left his post in Edinburgh in March 1864 because he felt he had been treated unfairly by his superiors (see *Correspondence* vol. 12, letter from John Scott, 28 March 1864). Hooker had written to John Hutton Balfour, the keeper of the Royal Botanic Garden, Edinburgh, asking for a reference for Scott, which he sent on to CD. Balfour praised Scott's industriousness and research abilities but repeated McNab's complaint about Scott's neglect of his garden work (see *Correspondence* vol. 12, letters from J. D. Hooker, [4 April 1864], and 6 April 1864 and enclosure).

7 Hooker and Thomas Thomson had gone over Scott's manuscript and found fault with some of his calculations. See letter from J. D. Hooker, [10 March 1865] and nn. 4, 6, and 8, and letter to J. D. Hooker 16 [March 1865].

8 Scott's letter has not been found (see letter from John Scott, 10 April 1865 and n. 10).

9 Scott refers to the Sundarbans, the seaward fringe of the Ganges delta in India (*EB*).

10 In his letter to Scott of 11 March 1865, CD had forwarded Hooker's and Thomas Thomson's criticisms of Scott's manuscript (see letter to J. D. Hooker, 16 [March 1865]). The letter to Scott has not been found.

11 CD first encouraged Scott to repeat Karl Friedrich von Gärtner's experiments on the relative fertility of varieties of *Verbascum* in 1862 (see letter from John Scott, 20 January 1865, n. 2).

12 The results of Scott's experiments on *Verbascum* were summarised by CD in *Variation* 2: 106–7. Scott published his results in Scott 1867. For a discussion of the significance of the experiments, see *Correspondence* vol. 12, letter to John Scott, 20 May [1864] and n. 5.

13 *Leersia oryzoides* was of interest to CD as an example of a self-fertilising unopened flower (later called cleistogamic). He read about the plant in a paper by Daniel Oliver ([Oliver] 1864) and asked Hooker about obtaining seeds; he later procured specimens with the help of William Bennett (see *Correspondence* vol. 12, letter to J. D. Hooker, 13 April [1864] and nn. 9 and 10, and letter from William Bennett, 29 April 1864 and nn. 2 and 4). CD's observations were published in 'Three forms of *Lythrum salicaria*', pp. 191–2 n. (*Collected papers* 2: 131). CD had probably written to Scott about his observations on *Leersia*, but the letter has not been found (see annotations to letter from John Scott, 10 April 1865 and n. 18). See letter from B. D. Walsh, 1 March 1865 and n. 13.

14 Scott received seeds of *Leersia oryzoides* from CD and reported the preliminary results of his experiments early in 1867 (see letter from John Scott, 22 January 1867, *Calendar* no. 5376). In *Forms of flowers*, p. 335, CD mentioned Scott's work on *Leersia*. Scott did not publish on *Leersia*.

15 CD had acquired *Lagerstroemia indica*, a member of the Lythraceae, from Hooker in 1863 because he thought it would be a good subject for his research on dimorphism (see *Correspondence* vol. 11, letter to J. D. Hooker, 15 and 22 May [1863] and n. 13, and letter from J. D. Hooker, [31 July 1863], and *Correspondence* vol. 12, letter to J. D. Hooker, [15 May 1864] and n. 4). CD's annotations to Scott's letter of 10 April 1865 indicate he had earlier asked about *Lagerstroemia*, and still wanted an answer. CD's notes on *L. indica* are in DAR 109: B116–17 and DAR 27.2: A17 v.

16 CD cited Scott's comments on the sterility of long-styled *Lagerstroemia indica* in *Forms of flowers*, pp. 167–8. He concluded that the evidence for heterostyly in *L. indica* was 'curiously conflicting', and also cited Scott's comments on *L. reginae* (*Forms of flowers*, p. 168).

17 See n. 1, above.

From Asa Gray 24 July 1865

Cambridge, Mass.
July 24, 1865.

My Dear Darwin

I had heard, thro' Hooker, that you had been poorly again, and I think that a letter, signifying my sorrow was written and has crossed yours just received.[1] I fancy you as now getting much better again. I am glad you *did* not see Brace (tho' sorry you *could* not): he is a great *talker*, or rather *questioner*, and would have exhausted you terribly.[2]

Here, at length are some seeds of *Specularia perfoliata*,[3] from Dr. Engelmann.[4] I have sown half, and send you the rest. I am reading in snatches, your admirable paper on Climbing Plants—as yet only 88 pages of it,—and am watching, with great interest, all the climbers I have at hand. What a nice piece of work you have made of it![5]

I see you explain & illustrate at length the double turn of a caught tendril. Is it not enough to say that, with both ends fixed, if it shortens say by the contraction of one side it must, by mechanical necessity turn its coil different ways, from a neutral point.[6]

I am vexed to have no *Adlumia* in the garden this year[7]—i.e. no seedlings came up last year—from the drought, I suppose. I may meet with it, in the country—the Western part of New York—where my wife and I are going—two days hence—to rusticate for 3 weeks—[8] She needs it.—tho' I hardly do. The first thing I do will be to write, for Sill. Journal an analysis of your great paper,[9] And when I return, if not too late, I mean to give 2 or 3 lectures from it to my University Class.

Ere this Mrs. Wedgwood should be back from Canada, but I have not learned that she is so.[10] She was to let me know, and we would have a day on the shore where Mr. Loring lives in summer—a pretty bit of country.[11] But it is now too late.

I wish she could have been here on Friday, when we welcomed back our Harvard men who had been in the war, over 500 of them—and remembered those who had died for their country. What a day we had!

Jefferson Davis richly deserves to be hung.[12] We are all willing to leave the case in the hands of the Government, who must take the responsibility. If I were responsible, I would have him tried for *treason*—the worst of crimes in a republic—convicted, sentenced to death,—and then I think I should commute the penalty, not out of any consideration for him, but from policy, and for his more complete humiliation. The only letters I have received expressing a desire to hang him, are from rebeldom itself—from Alabama. You see slavery is *dead*, **dead**,—an absolute unanimity as to this.[13] The Revolted States will behave as badly as they can, but they are so thoroughly whipped that can't stir, hand or foot,—and we are disbanding all our armies— a corporal's guard is enough to hold South Carolina—[14]

Seriously, there are difficult questions before us,—but only one result is possible—the South must be renovated, and Yankeefied.

Well— take good care of yourself, and let me know that you are again in comfortable condition

Ever Your affectionate friend | A. Gray

DAR 165: 148

[1] Neither letter has been found. Gray refers to Joseph Dalton Hooker.

[2] Gray had given his wife's nephew, Charles Loring Brace, an introduction to CD (see letter from Asa Gray, 15 and 17 May 1865 and nn. 12 and 13).

[3] Gray had sent CD seeds of *Specularia perfoliata*, which CD referred to as *Campanula perfoliata*, in 1863, but CD had inadvertently destroyed them and requested more seeds in 1864 (see *Correspondence* vol. 12, letter to Asa Gray, 28 May [1864] and n. 16; see also *Correspondence* vols. 10 and 11 for correspondence between Gray and CD on *Specularia*). CD discussed the cleistogamic flowers of *S. perfoliata* in *Forms of flowers*, p. 330.

[4] George Engelmann, a physician, plant collector, and botanist, had spent time working with Gray in 1856 at the Harvard Botanical Garden (*ANB*).

[5] CD had been inspired to study climbing plants after reading a short paper on coiling tendrils by Gray (A. Gray 1858; see letter to Asa Gray, 19 April [1865] and n. 16).

[6] Gray refers to 'Climbing plants', pp. 95–8.

[7] Gray had sent CD specimens and seed of *Adlumia cirrhosa* in 1857. At the time, CD was interested in the species because it formed fruits without insect pollination, a fact noted in *Cross and self fertilisation*, p. 366. Gray had disagreed with CD about whether *A. cirrhosa* was ever pollinated by insects. See *Correspondence* vol. 6, letter from Asa Gray, [August 1857], and letter to Asa Gray, 29 November [1857]. CD's observations on *A. cirrhosa* as a leaf-climber are in 'Climbing plants', pp. 44, 111.

[8] Jane Loring Gray and Asa Gray may have been going to Gray's family home in Sauquoit, New York (Dupree 1959).

[9] *The American Journal of Science and Arts* was commonly known as 'Silliman's journal', after its founder, Benjamin Silliman. The first part of Gray's review of 'Climbing plants' (A. Gray 1865–6) appeared in *American Journal of Science and Arts* 2d ser. 40 (1865): 273–82.

[10] The reference is to Frances Wedgwood (see letter from Asa Gray, 15 and 17 May 1865 and n. 11).

[11] Charles Greely Loring, Gray's father-in-law, had a summer home in Beverly, Massachusetts (Dupree 1959).

[12] Jefferson Davis was president of the Confederate states during the American Civil War. See letter from Asa Gray, 15 and 17 May 1865 and n. 6.

[13] Over the course of the American Civil War, Gray and CD frequently discussed their strong opposition to slavery (see, for example, *Correspondence* vol. 11, letter from Asa Gray, 1 September 1863, and this volume, letter to Asa Gray, 19 April [1865]). For CD's opposition to slavery, see also *Journal of researches*, pp. 27–8, Colp 1978, and Browne 1995, pp. 196–9, 213–14, 244–6.

[14] Gray refers to the fact that the Southern states had been crushed economically as well as militarily by the end of the war (McPherson 1988, pp. 818–19). South Carolina, generally viewed in the North as the origin of the rebellion, had been so devastated by General William Sherman's campaign in early 1865 that only a token force was thought to be required to prevent further revolt (*ibid.*, pp. 825–30).

To J. D. Hooker [29 July 1865][1]

Down
Saturday.

My dear Hooker

You were a good man to write me so pleasant a letter from Yorkshire.[2] I live such a life, I have never heard of Palestine Explo: Fund;[3] as you are interested in it, I am sure it must be good, & not merely to map the Temple,[4] so I will with pleasure subscribe 3 or 5 guineas (whichever **you** like) for the 3 years—[5] But you must tell me to whom to send money.—

Lyells corrected pages came when I was extra miserable;[6] I read them & threw them away & now to my surprise find that I have no clear recollection about them,—only a feeling that I was disgusted with everything in world—[7]

I was glad to read your article on Glaciers &c in Yorkshire—[8] you seem to have been struck with what *most* deeply impressed me at Glen Roy (wrong as I was on whole subject)[9] viz. the marvellous manner in which every detail of surface of land had been preserved for enormous period—[10] This makes me a little sceptical whether Ramsay, Jukes &c are not a little overdoing sub-äerial denudation.[11] In same Reader there was a striking article on English & foreign men of science; & I think unjust to England, except in pure Physiology; in Biology Owen & R. Brown ought to save us, & in Geology we are most rich.—[12]

It is curious how we are reading same books— we intend to read Leckie[13] & certainly to reread Buckle, which latter I admired greatly before.—[14] I am heartily glad you like Lubbock's book so much.—[15] It made me grieve his taking to Politicks, & though I grieve that he has lost his Election,[16] yet I suppose now that he is once bitten he will never give up Politicks, & Science is done for. Many men can make fair M.P.s, & how few can work in Science like him.—

I have been reading a pamphlet by Verlot on variation of flowers,[17] which seems to me very good, but I doubt whether it would be worth your reading. It was published originally in Journal d'Hort.[18] & so perhaps you have seen it. It is very good plan this republishing separately for sake of foreigners buying, & I wish I had tried to get permission of Linn. Soc for my Climbing paper, but it is now too late.—[19]

Do not forget that you have my paper on Hybridism by Max Wichura.[20]

I hope you are returned to your work refreshed like a giant by your huge breakfasts— How unlucky you are about contagious complaints with your children.—[21]

I keep very weak & had much sickness yesterday, but am stronger this morning—

Yours ever affect | C. Darwin

Can you remember how we ever first met; it was in Park St. but what brought us together??[22] I have been rereading a few *old* letters of yours, & my heart is very warm towards you.—

DAR 115: 273

[1] The date is established by the reference to the general election in West Kent, 22 July 1865 (see n. 16, below). The Saturday following 22 July 1865 was 29 July.

[2] See letter from J. D. Hooker, 13 July 1865. Although Hooker and CD both make reference to Yorkshire, Middleton-in-Teesdale, where the Hookers stayed, was in County Durham, just over the border from Yorkshire.

[3] Hooker had asked CD whether he could afford to subscribe to the Palestine Exploration Fund (see letter from J. D. Hooker, 13 July 1865 and n. 22).

[4] CD refers to the Temple of Jerusalem. An ordnance survey map of Palestine was completed under the auspices of the Palestine Exploration Fund in 1881, together with memoirs on topography, archaeology, plants, and animals (*EB*).

[5] Although there is no record of any payment for 1865 or 1866, there is a payment of £6 6*s*. to the Palestine fund in July 1867 (CD's Classed account books (Down House MS)).

[6] Hooker had asked CD why he had not commented on the revised preface and note that Charles Lyell had written for *Antiquity of man* (C. Lyell 1863c). Lyell apparently arranged for proofs to be sent to both CD and Hooker (see letter from J. D. Hooker, 13 July 1865, and n. 9). No letter to or from Lyell discussing these pages has been found. Lyell had sent CD copies of his correspondence with John Lubbock on the issue of plagiarism, but his letter to CD has not been found (see letter from Charles Lyell to J. D. Hooker, [31 May 1865] and enclosures; see also letter to J. D. Hooker, 1 June [1865] and n. 13).

[7] CD recorded in his journal that he had been ill since 22 April 1865 (see Appendix II). For CD's list of symptoms, see Appendix IV; see also Bowlby 1990, p. 379.

[8] Hooker 1865. Hooker had told CD that he was spending time studying the local moraines while on holiday (see letter from J. D. Hooker, 13 July 1865).

[9] In his 1839 paper, 'Parallel roads of Glen Roy', CD had argued that the 'roads' were ancient beaches formed as the land gradually rose from the sea. CD was eventually convinced by the observations of Thomas Francis Jamieson that the 'roads' were beaches of a glacial lake (see *Correspondence* vol. 9,

letter to T. F. Jamieson, 6 September [1861], and Appendix IX; see also *Correspondence* vol. 10, letter to A. C. Ramsay, 14 December [1862]).

10 In his letter in the *Reader* (Hooker 1865), Hooker described the moraines in the upper part of the Tees valley. He asserted that the contour of this part of the Tees valley was due to glacial action, but qualified this by adding: 'I do not thereby imply that the Valley, as such, is wholly due to subærial glacial action.' He concluded that the topographical features of the area were essentially unchanged since the glacial period.

11 Subaerial denudation is erosion resulting from fluvial and glacial action, in contrast to erosion from marine action (Challinor 1978). Andrew Crombie Ramsay and Joseph Beete Jukes had both rejected marine denudation as an explanation of topographical features, arguing instead in favour of glacial and fluvial action (see Jukes 1862a and 1862b and Ramsay 1862 and 1865). Another geologist who favoured the theory of subaerial denudation was Archibald Geikie (see Geikie 1863 and 1865). See Davies 1969, pp. 317–55, for a discussion of the role of these three geologists in the revival of fluvialism. CD had corresponded with Ramsay and Jukes about their theories (see *Correspondence* vol. 10, letters from J. B. Jukes, 25 May 1862 and 30 May 1862, letter from A. C. Ramsay, 26 August 1862, and letter to A. C. Ramsay, 5 September [1862]). CD's annotated copies of Jukes 1862b and Ramsay 1865 are in the Darwin Pamphlet Collection–CUL. CD had not been entirely convinced by the subaerial explanation, however, and continued to express doubts about it to Hooker (see *Correspondence* vol. 12, letter from J. D. Hooker, 26[–8] October 1864, and letter to J. D. Hooker, 3 November [1864]; see also this volume, letter to Charles Lyell, 21 February [1865] and n. 8).

12 CD refers to the article 'British and Foreign Science' in the *Reader*, 15 July 1865, pp. 61–2. The article claimed that England compared unfavourably with Germany and France in most fields of science. The author attributed this failing to the 'small social consideration paid to a scientific career', poor scientific education, and the absence of career structure in the sciences. The article noted the lack of laboratory facilities in physics and the biological sciences, and purposely omitted geology because it was 'not an abstract science', and because its 'popularity and success' in England were 'due to special causes not far to seek'. CD refers to the anatomist Richard Owen and the botanist Robert Brown.

13 Hooker told CD that he had read the second volume of Lecky 1865 (see letter from J. D. Hooker, 13 July 1865 and n. 15).

14 Buckle 1857–61. See letter from J. D. Hooker, 13 July 1865 and n. 16.

15 Lubbock 1865; see letter from J. D. Hooker, 13 July 1865. CD had written to John Lubbock praising the book (see letter to John Lubbock, 11 June [1865] and n. 2).

16 John Lubbock stood for West Kent in the general election of 1865. He was defeated on 22 July. The election results were reported in *The Times*, 24 July 1865, p. 6.

17 Verlot 1865. There is an annotated copy in the Darwin Library–CUL (see *Marginalia* 1: 817–19).

18 Bernard Verlot's paper, 'Sur la production et la fixation des variétés dans les plantes d'ornement', was first published in instalments in *Journal de la Société Impériale et Centrale d'Horticulture* (Verlot 1864). It was the prize-winning essay in a competition held by the Société Impériale et Centrale d'Horticulture. CD cited Verlot 1865 frequently in *Variation*, especially on colour variation in plants (see, for example, *Variation* 1: 386 and 2: 20, 70).

19 The London publishing firms of Longman, Green, Longman, Roberts & Green and Williams & Norgate offered commercial offprints of 'Climbing plants' from August 1865 (*Publishers' Circular*, 1 August 1865, p. 391; see also Freeman 1977).

20 Wichura 1865 (see letter to J. D. Hooker, [17 June 1865] and n. 7).

21 See letter from J. D. Hooker, 13 July 1865. In March 1864 Charles Paget Hooker had caught another contagious illness, ringworm, at school (see *Correspondence* vol. 12, letter from J. D. Hooker, 9 [March] 1864).

22 CD probably refers to a meeting in May or June 1839, when he and Hooker were introduced in the street by Robert McCormick, who had been with CD on the *Beagle* voyage between 1831 and 1832, and who was to become Hooker's superior on the *Erebus* voyage, 1839–43 (*LL* 2: 19, L. Huxley ed. 1918, 1: 41–4, Browne 1995, p. 419). However, there is evidence that suggests they first met briefly

at the Hunterian Museum on 22 January 1839, when Hooker, William Jackson Hooker, and Asa Gray were visiting London together (J. L. Gray ed. 1893, 1: 117; D. M. Porter 1993, p. 9). CD and Hooker began corresponding in November 1843 (see *Correspondence* vol. 2, letter to J. D. Hooker, [13 or 20 November 1843]).

To William Bowman 30 July [1865?][1]

Down. | Bromley. | Kent. S.E.
July 30

My dear Mr Bowman

I am very much obliged for your kind note.[2] I will attend to your information whenever I am able to publish my book;[3] but I have lately lost many months from continued illness.

I thank you most sincerely for your kindness to my son[4] & I had intended writing several months ago to thank for; but put it off from being so unwell until it was too late.

Pray believe me | my dear Mr Bowman | yours very sincerely | Ch. Darwin

LS
American Philosophical Society (301)

[1] The year is conjectured from the reference to CD's illness of 'many months', and the reference to CD's book, presumably *Variation*. See letter to John Murray, 2 June [1865] and n. 1.
[2] This letter has not been found.
[3] William Bowman, a leading ophthalmic surgeon, supplied CD with information about inherited imperfections of the human eye (see *Variation* 2: 8–10, 79).
[4] It is not known what Bowman's kindness was, or to which of CD's sons it was shown.

To Fritz Müller 10 August [1865]

Down. | Bromley. | Kent. S.E.
Aug 10

Dear Sir

I have been for a long time so ill that I have only just finished hearing read aloud your work on species.[1] And now you must permit me to thank you cordially for the great interest with which I have read it. You have done admirable service in the cause in which we both believe. Many of your arguments seem to me excellent, & many of your facts wonderful. Of the latter nothing has surprised me so much as the two forms of males.[2] I have lately investigated the cases of dimorphic plants, & I should much like to send you one or two of my papers if I knew how.[3] I did send lately by post a paper on climbing plants as an experiment to see whether it w^{d} reach you.[4]

One of the points which has struck me most in your paper is that on the differences in the air-breathing apparatus of the several forms. This subject appeared to me very important when I formerly considered the electric apparatus of fishes.[5]

Your observations on Classification & Embryology seem to me very good & original[6] They shew what a wonderful field there is for enquiry on the development of Crustacea; and nothing has convinced me so plainly what admirable results we shall arrive at in Natural History in the course of a few years.

What a marvellous range of structure the Crustacea present & how well adapted they are for your enquiry! Until reading your book I knew nothing of the Rhizocephala; pray look at my account & figures of Anelasma; for it seems to me that this latter Cirrepede is a beautiful connecting link with the Rhizocephala.[7]

If ever you have any opportunity, as you are so skilful a dissector, I much wish that you w^d^ look to the orifice at the base of the first pair of cirri in Cirripedes, & at the curious organ in it & discover what its nature is; I suppose I was quite in error, yet I cannot feel fully satisfied at Krohn's observations.[8] Also if you ever find any species of Scalpellum, pray look for complemental males; a German author has recently doubted my observations for no reason except that the facts appeared to him so strange.[9]

Permit me again to thank you cordially for the pleasure which I have derived from your work & to express my sincere admiration for your valuable researches. Believe me | Dear Sir with sincere respect | yours very faithfully | Ch. Darwin

P.S. I do not know whether you care at all about plants but if so I sh[d] much like to send you my little work on the Fertilization of Orchids & I think I have a German Copy.[10]

Could you spare me a Photograph of yourself, I sh[d] much like to possess one?—

LS(A)
Endorsed: 'Received Octbr. 8: 1865.'
British Library (Loan MS 10, no. 1)

[1] The reference is to Müller 1864, a study of the Crustacea that was supportive of CD's theory of transmutation. CD had received a presentation copy from Müller before 21 November 1864, but found the German difficult to understand and had it translated (see *Correspondence* vol. 12, letter to Ernst Haeckel, 21 November [1864] and n. 6). The translation was probably made by Camilla Ludwig, who had been governess at Down House between 1860 and 1863. CD's Classed account book (Down House MS) records, under the heading 'Science', a payment to Ludwig of £5 11*s*. 6*d*. on 10 June 1865 for translation. An annotated copy of Müller 1864 is in the Darwin Library–CUL (see *Marginalia* 1: 609). CD's annotated copy of the later translation by William Sweetland Dallas (Dallas trans. 1869) is in the Rare Books Collection–CUL (see *Marginalia* 1: 609).

[2] In Müller 1864, Müller discussed the existence of two forms of male in some species of Crustacea and argued that this dimorphism could be seen as a result of natural selection. Müller's example was a species of the genus *Tanais*. Müller found that one form of male developed enlarged olfactory filaments, making it better at finding females, while the other form had enlarged chelae (claws), giving it an advantage in clasping females (see Müller 1864, pp. 13–19 and Dallas trans. 1869, pp. 20–9). The absence of intermediate forms was explained by the lack of any advantage conferred by such hypothetical forms. CD added this information to *Origin* 4th ed., pp. 50 and 288.

[3] CD sent Müller 'Two forms in species of *Linum*', and 'Three forms of *Lythrum salicaria*'; the last-named concerns trimorphism (see letter from Fritz Müller, 5 November 1865 and n. 2).

[4] CD refers to 'Climbing plants', which was published on 12 June 1865 in the *Journal of the Linnean Society* (*Botany*). No presentation list for 'Climbing plants' has been found, but it is likely that CD sent

Müller an author's offprint. Müller resided at Destêrro on Santa Catarina Island off the coast of the state of Santa Catarina, Brazil. Destêrro is now Florianópolis.

5 Müller compared the structure of the air-breathing apparatus in certain non-aquatic species of two families of true crabs (Brachyura): shore crabs (Grapsoidae, now Grapsidae), and ghost and fiddler crabs (Ocypoda, now Ocypodidae). He observed that the structure and position of the apparatus differed in various ways in each family and argued that the differences could be explained only by assuming that each family adapted to an air-breathing existence independently (see Müller 1864, pp. 24–6, and Dallas trans. 1869, pp. 35–7). In *Origin*, pp. 193–4, CD wrote that the electric organs in fish of widely differing species appeared difficult to explain within the framework of his theory, but conjectured that they might be explained as analogous features, that is, structures with similar functions that have developed independently in response to similar habits. In the fourth edition of *Origin*, pp. 226–8, he added Müller's observations on these crab families.

6 For Müller's observations and remarks on crustacean embryology, see Müller 1864, pp. 31–65; his remarks on classification are on pp. 65–74 (see also Dallas trans. 1869, pp. 47–96, pp. 97–109). Müller noted that the use of embryological characteristics, although helpful in determining classificatory schemes, was by no means foolproof. He gave numerous examples of species or genera in which embryological similarity did not translate into adult similarity, or in which embryonic and adult stages were similar but the intermediate nauplius (larval) stages were disparate. He then devised a hypothetical classificatory scheme for the Crustacea to show the pitfalls of placing too great an emphasis on embryonic similarity (Müller 1864, pp. 73–4; Dallas trans. 1869, p. 109). In the fourth edition of *Origin*, p. 494, CD added a reference to Müller's classificatory scheme.

7 Rhizocephala is the name coined by Müller (see Müller 1862a and 1862b) for a highly unusual group of crustaceans that as adults are naked and limbless parasites attached to their crab hosts by means of root-like tubes, and that appear from the outside as sausage-like sacs. Müller had noted their similarity, in pre-adult stages, to cirripedes, placing them, not within what Darwin had described as the sub-class Cirripedia (barnacles), but alongside it (Müller 1864, p. 59; Dallas trans. 1869, pp. 88–9). CD had described *Anelasma* as a 'curious cirripede' that lived 'parasitic, with its peduncle imbedded in the skin of sharks' (*Living Cirripedia* (1851), p. 170). When he described the species *A. squalicola*, he could not find a cement gland or ducts, but noted that the peduncle had root-like branched filaments that, he conjectured, helped the animal to remain embedded in the shark's skin (*Living Cirripedia* (1851), p. 174). After reading Müller's account of Rhizocephala, CD may have considered the branching filaments in *Anelasma* to be an intermediate form between the cement ducts of other cirripedes and the root-like tubes of Rhizocephala.

8 In *Living Cirripedia* (1851), pp. 53–5, CD had described the empty orifice, which seemed to contain a nerve at its mouth, as possibly an auditory sac. His description was criticised by August David Krohn, who identified the structure as a female genital aperture (Krohn 1859, pp. 362–3). In a letter to the *Natural History Review*, [before 10 October 1862] (*Correspondence* vol. 10), CD referred to Krohn's paper, and asked whether any reader would 'endeavour to discover ova within the so-called auditory sack' ('"Auditory-sac" of cirripedes'; *Collected papers* 2: 87; see also *Correspondence* vol. 8, letter to Charles Lyell, 28 [September 1860] and nn. 4–6). For a discussion of the problem of the orifice see Newman 1993, pp. 368–71.

9 CD refers to his discovery in 1848 of the small 'complemental' males attached to the bodies of hermaphrodites in the barnacle genus *Scalpellum* (see *Living Cirripedia* (1851), pp. 281–93, and letter from Henry Holland, 2 January 1865, n. 5). CD did not name in his published writings the German author who doubted this discovery; however, in 1863 the German zoologist Carl Eduard Adolph Gerstaecker had questioned CD's account of the existence of complemental males in his *Handbuch der Zoologie* (see Gerstaecker 1863b, p. 410). CD referred to the criticism in a letter to *Nature*, 20 November [1873] ('Males and complemental males of certain cirripedes'), and in his *Autobiography*, pp. 117–18. For CD's discussion of complemental males in the genera *Scalpellum* and *Ibla* see *Living Cirripedia* (1851), pp. 281–93 (see also Newman 1993, pp. 377–82, for a discussion of CD's work on complemental males). See also *Correspondence* vol. 12, letter to J. D. Hooker, 26 November [1864] and n. 4.

[10] CD refers to *Orchids*, which was translated into German in 1862 by Heinrich Georg Bronn (Bronn trans. 1862).

From Fritz Müller[1] 12 August 1865

Desterro, Brazil,
12. August 1865.

Verehrter Herr!

Vor wenigen Tagen empfing ich Ihre Abhandlung über Kletterpflanzen,[2] und ich beeile mich, Ihnen für diese werthvolle Gabe meine Dankbarkeit auszudrücken. Ich habe sie mit dem grössten Interesse gelesen und bin sehr froh, dass meine Aufmerksamkeit auf diese merkwürdigen Pflanzen gelenkt worden ist, die in unserer Flora so ausserordentlich häufig sind. In ein paar Tagen sammelte ich, nachdem ich Ihre Arbeit gelesen hatte, in der Umgegend von Desterro die folgenden Gattungen von kletternden Pflanzen: *Spirally twining*: Dioscorea, Alströmeria (subgen. Bomarea,[3] Boussingaultia, Hebanthe,[4] (Goruplerena),[5] Aristolochia, Micania,[6] Manettia, Condylocarpon, Echites, Prestonia, Oxypetalum und einige andere Asclepiadeen, Tournefortia, Ipomoea (zahlreiche Arten), Calonyction, Cuscuta, Thunbergia (eingeführt), Mendozia, Cocculus (?), Davilla, Anchietia, Tetrapterys, Peixotoa, Brachypteris (?),[7] Thryallis, Dicella, Dalechampia, Mucuna, Caravillia;[8] *leaf-climbers*: Clematis, Fumaria (eingeführt) Tropaeolum; Ranken-träger (*Tendril-bearers*): Smilax, Mutisia, Strychnos, Haplolophium, Bignonia (zahlreiche Arten), Cissus, Passiflora (6 Arten), Momordica, Sechium (kultiviert) und verschiedene andere Cucurbitaceen, Serjania, Paullinia, verschiedene Papilionaceen; Hakenklimmer (*Hook-climbers*): Galium (Relbunium Endlicher)[9] Büttneria, Rubus; Wurzelklimmer (*Root-climbers*): Vanilla, verschiedene Aroideen, Piper. Ich kann noch hinzufügen die Gattungen Lygodium, Blumenbachia, Solandra, Combretum, Tragia, die ich früher in anderen Theilen unserer Provinz sah.[10] Da ich früher den windenden Pflanzen keine besondere Aufmerksamkeit widmete, so bin ich zur Zeit nicht im Stande, Ihnen Aufklärung über ihre Gewohnheiten zu geben.

Incomplete[11]
Möller 1915–21, 2: 72–3.

[1] For a translation of this letter, see Appendix I. According to Alfred Möller, all Fritz Müller's letters to CD were written in English (see Möller ed. 1915–21, 2: 72 n.); most of them have not been found. Many of the letters were later sent by Francis Darwin to Möller, who translated them into German for his *Fritz Müller: Werke, Briefe und Leben* (Möller ed. 1915–21). Möller also found final drafts of some Müller letters among the Fritz Müller papers and included these in their original English form (*ibid.*, 2: 72 n). Where the original English versions are missing, the published version, usually appearing in German translation, has been used. See also n. 11, below.

[2] 'Climbing plants'. See letter to Fritz Müller, 10 August [1865], n. 4.

[3] *Bomarea* is now considered to be a separate genus of the family Alstroemeriaceae.

[4] *Hebanthe* is a synonym of *Pfaffia*.

[5] This is probably a misspelling or incorrect transcription of *Gomphrena* (Amaranthaceae). Although *Gomphrena* has few climbers, there is one climbing species, *G. vaga*, which is found in the vicinity of Destêrro (Florianópolis).

[6] A misspelling of *Mikania* (see letter from Fritz Müller, [12 and 31 August, and 10 October 1865] and n. 18).
[7] A misspelling of *Brachypterys*.
[8] Probably a misspelling or incorrect transcription of *Canavalia*.
[9] Müller refers to *Galium*, subgenus *Relbunium*, as described in Endlicher 1836–42, p. 523. *Relbunium* is now considered to be a distinct genus.
[10] Müller resided in Destêrro on Santa Catarina Island off the coast of the state of Santa Catarina, Brazil. Destêrro is now Florianópolis.
[11] Part of the missing portion of this letter was submitted by CD, together with extracts from other letters from Müller to CD on climbing plants, to the Linnean Society. The resulting paper was read on 7 December 1865, and published in the 29 November 1866 issue of the *Journal of the Linnean Society (Botany)* (see letter from Fritz Müller, [12 and 31 August, and 10 October 1865], and Müller 1865b). When CD was preparing the letters for publication, he informed Müller that he had 'slightly modified the arrangement of some parts & altered only a few words' (letter to Fritz Müller, 17 October [1865]; see also letter to Fritz Müller, 9 December [1865]; for more on the modifications made to Müller's letters, see the letter to J. D. Hooker, 3 October [1865], and the letter to Daniel Oliver, 20 October [1865]).

From Fritz Müller [12 and 31 August, and 10 October 1865][1]

In your Paper on the "Movements and Habits of Climbing-Plants," you say that you have seen no tendrils formed by the modification of branches, and you even seem to entertain some doubt whether such tendrils exist.[2] In the genus *Strychnos*, the tendrils are called by Endlicher *ramuli cirriformes*, and I have now satisfied myself that they really are of this nature.[3] On the branches of upright shoots of a *Strychnos* which grows here, the tendrils are disposed in a very regular manner. On the branches, the leaves of the first, third, fifth, &c. pairs are horizontal, those of the second, fourth, and sixth pairs are vertical in relation to the main axis; and it is from the angles of every under leaf of these latter pairs that the tendrils spring. Now, on the points commonly occupied by tendrils, true branches are sometimes developed. The leaves from the angles of which the tendrils spring are often much reduced in size, while in other cases they are but little or not at all changed. Each tendril bears near its tip a pair of rudimentary leaves; and whilst very young the tendrils are straight, but soon become curved downwards and rolled into a helix, whether they have clasped a support or not. This *Strychnos* is a very inefficient climber; the short stiff tendrils but rarely catch anything.

A member of the Hippocrateaceæ, probably a *Tontelia*, is likewise a branch climber. One of its branches, three feet in length, had not as yet developed leaves, and resembled a gigantic tendril, with most of its lateral branchlets already grasping neighouring objects. From the angles of the tendril-branches, other branches arise, which as far as I have seen, are not sensitive, and never clasp anything. This latter arrangement must be serviceable to the plant; for such branches grow upright without being arrested in their course, whilst the plant is secured by the tendril-branches.

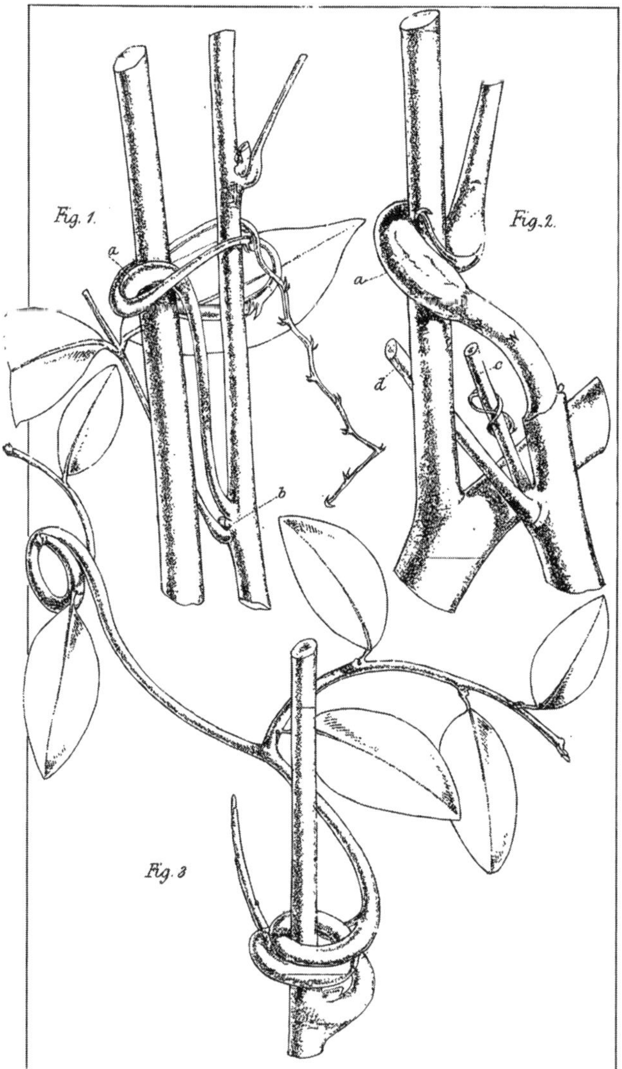

Engraved illustration to the letter from Fritz Müller,
[12 and 31 August, and 10 October 1865]
Reproduced from *Journal of the Linnean Society* (*Botany*) 9 (1866), tab. 9

Caulotretus, one of the Leguminosæ, offers another case of tendrils being formed from modified branches.[4] In the species which I observed, the branches bear tendrils only in the angle of their first leaf, and this leaf is always rudimentary. In young shoots it might, at first sight, be thought that tendrils spring from the axils of all their leaves. In this plant every tendril appears to consist of two parts, separated by a small swelling—the inferior being straight, the superior curved, with its end rolled into a helix. But what appears to be the inferior part of a tendril is in fact the first internode of a young branch, the swelling being its terminal bud, and the tendril really springs from this young branch, from the angle of its first squamiform leaf, but nevertheless accompanied by two stipules. The end of the tendril very soon rolls up into a helix; but it does not lose by this the faculty of catching a support; on the contrary I know of no other tendrils which become entangled with small objects so easily as these rolled-up, highly elastic tendrils of the *Caulotretus*.

By far more interesting than the tendrils of *Strychnos* and *Caulotretus* are those (Pl. IX. figs. 1 and 2) of a climbing Papilionaceous plant with a woody stem, which from its general aspect I suppose to belong to the *Dalbergieæ*, Benth.[5] They consist of thin, slender, flexible, leafless branches, with numerous (12–25) internodes, armed with sharp, hard, hook-like stipules. The young, soft, herbaceous shoots of this plant which rise from the ground are leafless. I saw one, seven feet high, which in its lower half was naked, while the upper half bore about a dozen tendrils stretched out in every direction. The oldest of these tendrils were from nine to twelve inches long, and armed with from twelve to sixteen pairs of sharp hooks: at the sides of the younger tendrils there were large, foliaceous, deciduous stipules, and at their bases very small bract-like leaves. The hooks of the tendrils are evidently stipules, which so often in this family assume the form of hooks or spines; in fact, while in the older tendrils they are strongly curved, and have a hard, sharp, darkly coloured apex, at the summits of the younger ones they are straight, soft, and green, resembling in this early state the much larger stipules at the bases of the tendrils. Afterwards, on the summit of the shoot, true leaves are developed at the bases of the tendrils instead of the small rudimentary ones; and finally, when the plant has reached the light, and spreads over the upper surface of a thicket or tree, the tendrils disappear. The inverse may be observed when the plant sets out on the conquest of a new dominion, a neighbouring tree for instance. Then a branch bearing only leaves begins to produce on its tip tendrils supported by leaves, and finally, growing rapidly to a long slender shoot, it produces only tendrils, the leaves being replaced by small squamæ. Thus in this plant, the branches assume four different shapes:—1st, tendrils, leafless, armed with hook-like stipules; 2nd, long, slender, leafless shoots, bearing tendrils and broad deciduous stipules; 3rd, branches with leaves, from the axils of which tendrils spring; and 4th, branches bearing only leaves without tendrils. Between the leaf and the tendril there is an accessory bud (fig. 1 *b*), which often developes into a branch; these branches issuing from the accessory buds seem never to produce tendrils. The tendrils, after having clasped a support, thicken partially where they are in contact with it (fig. 1*a*). Tendrils which have caught nothing behave in different ways. Some wither and fall after contracting irregularly. Others likewise become flexuous, or contract into a spire, or occasionally into a helix, but remain, thickening somewhat and becoming ligneous and rigid. Others produce branches from one or some of their internodes: this also occurs, and perhaps more frequently, with tendrils which have found a support; in this case the tendrils thicken much, and sometimes attain a diameter of more than one inch (fig. 2*a*, a thickened tendril clasping a branch of a *Psidium*; *b*, branch issuing from the tendril; *c*, tendril-bearing branch; *d*, branch from an accessory bud, without tendrils). Lastly, the tendrils even transform themselves into true branches: in this case they may remain nearly straight or become but little flexuous, and at their ends they produce leaves; the first of these leaves have sometimes hook-like persistent stipules, like those of the tendril, while the stipules of the following leaves are deciduous like those on other branches. These tendrils often become much elongated. I saw a shoot, almost all the tendrils of which were developed into serpentine branches; and under each of these branches there was a straight branch from an accessory bud. One of the tendrils was thirty inches long; it had twenty-five pairs of hooks, and at the tip three short internodes with leaves and destitute of hooks; from its seventeenth internode a branch arose. Excepting their hook-like stipules, by which they may be easily recognized, the branches formed by tendrils resemble in almost every respect the ordinary branches; but, as far as I have seen, they never produce tendrils, nor do the branches which spring from an internode of a tendril or (as I have already stated) from an accessory bud.

If we restrict the name of tendrils to filamentary organs used *exclusively* for climbing, those of the present plant would be excluded;[6] for after having done their work as tendrils, they may be transformed into, and do all the work of, branches.

While in this plant the highly modified tendrils may be changed again into true branches, in two other plants which I have seen, the branches themselves, without having suffered any modification, act as tendrils. One of these plants belongs to the *Dalbergiæ*. Many of its branches had clasped small branches of a tree. These tendril branches, as they may be called, had not continued to grow beyond the support; and where they touched it, most of them had thickened: some showed a tendency to spiral contraction, forming a semicircle between the support and the stem. The plant does not twine. I may add that another genus, belonging to the same section of the Leguminosæ, namely *Hecastaphyllum*,[7] is also a branch climber.

The second plant above referred to is a *Securidaca* (Polygalaceæ), and a most powerful climber (fig. 3). Its branches often curve in a very odd and complicated manner. Thus I saw a thin branch, which with its lateral twigs had become curved like ribs into semicircles (about four inches in diameter), imitating the bones of the thorax; from the twigs sprang secondary branchlets, which were very regularly curved, twisted together, and formed into a sort of network around the middle hollow space. When the branches wind round a support, they thicken and become more rigid, like true tendrils; but even these thickened parts may bear leaves or secondary branches. In the preceding plant the branches seem to be arrested in their longitudinal growth when they clasp a support; in the present plant they continue to grow, and the same branch may successively catch different objects. The branches which project freely from a thicket are rather thin and slender: with their twigs spreading all in the same horizontal plane and diminishing in length towards the extremity of the branch, and with their leaves arranged in two horizontal rows, they apparently form gigantic bipinnate leaves; and when covered with their bluish-purple flowers, this *Securidaca* is one of the most elegant and magnificent plants of our flora.

From the last two plants it is but one step to the primordial and simplest condition of branch climbers, exhibited by the numerous species which scramble up a thicket without twining and without the aid of rootlets, hooks, or tendrils.

Thus we can trace in the development of branch climbers the following stages:—[8]

1. Plants supporting themselves only by their branches stretched out at right angles—for example, *Chiococca*.

2. Plants clasping a support with their branches unmodified—*Securidaca* (*Hippocratia* according to Endlicher, Gen. Plant. No. 5700, "arbores v. frutices, ramis contortis scandentes").[9]

3. Plants climbing with the tendril-like ends of their branches. According to Endlicher (Gen. Pl. No. 5745), this is the case with *Helinus* ("ramulorum apicibus cirrhosis scandens").[10]

4. Plants with highly modified tendrils, which may, however, be transformed again into branches—for example, the above-mentioned Papilionaceous plant.

5. Plants with tendrils used exclusively for climbing—*Strychnos*, *Caulotretus*.

I will here add a few miscellaneous observations. You describe some species of *Bignonia* in which the tips of the tendrils become enlarged and adhesive after remaining for a short time in contact with some object;[11] but the trifid tendrils of *Haplolophium*, one of the Bignoniaceæ, terminate (without having come into contact with any object) in smooth shining disks, which, however, after adhesion, sometimes become considerably enlarged.[12] In *Cardiospermum* you state that the common peduncle which bears the subpeduncles with the flower-buds and the pair of short tendrils, although it spontaneously revolves, does not bend on contact or contract spirally;[13] hence it may be worth mentioning, as showing a difference in the action of the tendrils in related genera, that in *Serjania* the common peduncle contracts spirally when the single tendril which it bears has clasped, as frequently happens, the plant's own stem.[14]

With respect to spirally twining plants, you state that though the *Hibbertia dentata* sometimes revolves in one direction and sometimes in the other, yet it invariably twines from left to right.[15] But in another genus belonging to the same family, namely the *Davilla*, the stem twines indifferently from left to right or from right to left; and I once saw a shoot, ascending a tree about five inches in diameter, reverse its course in the same manner as so frequently occurs with *Loasa*.[16] Although individuals, as we have just seen, in some few cases twine in opposite directions, and you are able to give only two cases of species within the same natural order thus twining.[17] But a *Mikania* growing here twines from right to left, whilst the *Mikania scandens* described by you twines in an opposite direction;[18] and I believe that there are species of *Dioscorea* which twine in opposite directions. Lastly, with respect to the thickness of the support which can be ascended by spirally twining plants, I have lately seen a trunk about five feet in circumference which was thus ascended by a plant apparently belonging to the Menispermaceæ.[19]

Notes on some of the climbing-plants near Desterro, in South Brazil. By Herr Fritz Müller, in a letter to C. Darwin. [Read 7 December 1865.] *Journal of the Linnean Society (Botany)* 9 (1866): 344–9.

[1] The text is composed of parts of three letters edited by CD for publication in the *Journal of the Linnean Society* (*Botany*) (see letters to Fritz Müller, 17 October [1865] and 9 December [1865]). The texts of the original three letters cannot be distinguished in this version; the dates are established by the relationship between these letters and the letters to Fritz Müller, 17 October [1865] and 9 December [1865]. See letter from Fritz Müller, 12 August 1865, n. 11.

[2] In 'Climbing plants', pp. 110–11, CD had concluded that although all tendrils performed identical functions, they consisted of modified leaves, or modified flower-peduncles, or possibly modified branches and stipules; he also questioned whether the difference between foliar and axial organs could be 'of so fundamental a nature' as was generally supposed (*ibid.*, pp. 113–14; for CD's discussions on the derivation of tendrils, see *Correspondence* vol. 12, letters from Daniel Oliver, [28 January – 8 February 1864], 12 March 1864, and [before 31 March 1864], letter to Daniel Oliver, 11 March [1864], and letters to J. D. Hooker, [27 January 1864], [8 February 1864], and 3 November [1864]). Müller may be referring to CD's remark on page 112 of 'Climbing plants', that he knew 'nothing of any transitional states', if such occurred, between branches and tendrils; however, CD gave as an example *Lophospermum*, in which branches spontaneously revolve and are sensitive to contact, to suggest that such modifications were possible; see also *ibid.*, p. 10. In *Climbing plants* 2d ed., p. 196, he added that such transitional states had 'been fully described by Fritz Müller'. CD also revised his description of tendril formation. In 'Climbing plants', p. 48, he wrote, 'True tendrils are formed by the modification of leaves with their petioles, of flower-peduncles, perhaps also of branches and stipules'; in the second edition the sentence ends, '. . . flower-peduncles, branches, and perhaps stipules' (see *Climbing plants* 2d ed., p. 84 and n.).

[3] Müller refers to the description of *Strychnos* in Stephan Ladislaus Endlicher's *Genera plantarum* (Endlicher 1836–42, p. 575), 'ex axilla ramulem simplicem, cirrhiformem exserente' (putting forth a simple, tendril-like branch from the axil). CD mentioned *Strychnos* in *Climbing plants* 2d ed., pp. 84–5.

[4] CD mentioned *Caulotretus* in *Climbing plants* 2d ed., pp. 84–5 n.

[5] Müller included sketches of the plants he described with his letter of 31 August 1865 (see letter to J. D. Hooker, 3 October [1865] and n. 3). When CD informed Müller that he would submit Müller's observations on climbing plants to the Linnean Society he wrote, 'As the Soc. is rather poor I fear that it will give only a few wood cuts from your truly admirable sketches' (letter to Fritz Müller, 17 October [1865]). The printed version included one woodcut (Plate IX) of *Dalbergia* and *Securidaca*, which is reproduced above. The original sketches have not been found.

[6] In 'Climbing plants', p. 48, CD had given the following definition:

> By tendrils I mean filamentary organs, sensitive to contact and used exclusively for climbing. By this definition, spines or hooks and rootlets, all of which are used for climbing, are excluded.

[7] The correct spelling is *Hecastophyllum*; these plants are now included in the genus *Dalbergia*.

[8] CD reproduced this list of stages in the development of branch climbers, with slight modification, in *Climbing plants* 2d ed., pp. 84–5 n. Müller's list, tracing the gradation from branch to tendril, echoes CD's description of the possible stages of development from leaf to tendril and from flower-peduncle to tendril in 'Climbing plants', pp. 111–12.

[9] Müller refers to the description in Endlicher 1836–42, p. 1091, of *Hippocratea*, 'Arbores v. frutices, ramis contortis scandentes' (trees or shrubs climbing with twisted branches).

[10] Müller refers to the description in Endlicher 1836–42, p. 1102, of *Helinus*, 'ramulorum apicibus cirrhosis scandens' (climbing by the tendril-like (curled) tips of branchlets).

[11] In 'Climbing plants', pp. 57–9, 103–4, and 113, CD discussed the development of adhesive discs in *Bignonia capreolata*. For more on CD's work on this genus and on others with adhesive discs, see *Correspondence* vol. 12, letters to J. D. Hooker, 4 December [1864] and n. 13, and 10 December [1864] and n. 6. CD's notes on the development of adhesive discs in *Bignonia capreolata* are in DAR 157.1: 140.

[12] CD added Müller's observation on the discs in *Haplolophium* to *Climbing plants* 2d ed., p. 102 n., and amended his discussion of adherent discs on page 179 to take Müller's observation into account.

[13] CD's observations on *Cardiospermum halicacabum* are in 'Climbing plants', pp. 87–9, 92, 112, and 113. CD's notes on the species are in DAR 157.2: 68.

[14] CD added Müller's comment about spiral contraction in the peduncle of *Serjania* to *Climbing plants* 2d ed., p. 152 n.

[15] *Hibbertia dentata* is described in 'Climbing plants', pp. 14, 16, 20–1, and 116. CD's notes on the species are in DAR 157.1: 31, 49, and 57.

[16] Müller's observations on the twining of *Davilla* are given in *Climbing plants* 2d ed., p. 36 n.

[17] In 'Climbing plants', pp. 18 and 20, CD had described the reversal in the direction of twining of *Loasa aurantiaca* and *Scyphanthus elegans*. His notes on these species are in DAR 157.1: 49 and 54; other notes on *Loasa* are in DAR 157.1: 55–6.

[18] CD had noted in 'Climbing plants', p. 18, that *Mikania scandens* moved 'against the sun'; in *Climbing plants* 2d ed., pp. 33–4, he cited Müller's remark on the Brazilian species of *Mikania* twining in the opposite direction. CD's notes on *Mikania* are in DAR 157.1: 2 and 48.

[19] The final sentence was evidently a late addition by CD to the printed version of the paper, which was published in November 1866, from information contained in a later letter from Müller (see *Correspondence* vol. 14, letter from Fritz Müller, 13 February 1866). In 'Climbing plants', p. 21, CD had noted: 'Our English twiners, as far as I have seen, never twine round trees, excepting the Honeysuckle', and on page 22 he remarked: 'The tropical twiners, on the other hand, can ascend thick trees.' CD had previously discussed the topic with Joseph Dalton Hooker (see *Correspondence* vol. 12, letter from J. D. Hooker, 2 December 1864 and n. 10, and letter to J. D. Hooker, 4 December [1864]). In *Climbing plants* 2d ed., p. 37 n., he added Müller's observation on the thickness of the trunk around which a twining plant might climb.

To Asa Gray 15 August [1865][1]

Down. | Bromley. | Kent. S.E.
Augt 15

My dear Gray

I was much pleased to get your letter of July 24th—[2] Now that I can do nothing, I maunder over old subjects & your approbation of my Climbing paper gives me *very* great satisfaction.[3] I made my observations when I could do nothing else & much enjoyed it, but always doubted whether they were worth publishing.—

I demur to it not being necessary to explain in detail about the spires in caught tendrils running in opposite directions; for the fact for a long time confounded me & I have found it difficult enough to explain the cause to 2 or 3 persons.[4] One botanist has published that he could detect a difference of structure in the tendrils at the points of reversal of the spire![5] Very many thanks for Specularia seed.[6]

We continue to be deeply interested on American affairs; indeed I care for nothing else in the Times.[7] How egregiously wrong we English were in thinking that you could not hold the South after conquering it. How well I remember thinking that Slavery would flourish for centuries in your Southern States![8] My women read much aloud to me,[9] & I have lately heard three Books, worth your attention—*Lubbock* Prehistoric Man— *Tylor* early History of Civilization, which is admirable; & Lecky's Rationalism, which also strikes me as very well worth reading.—[10]

Mrs Wedgwood has enjoyed her American visit greatly, & has received the usual wonderful amount of American hospitality.—[11] This is a longer note than I have written for many weeks, so farewell.

I am trying a starving system of cure; eating very little of anything, & that almost exclusively bread & meat.[12]

Yours affectionately | Ch. Darwin

Gray Herbarium of Harvard University (87)

[1] The year is established by the relationship between this letter and the letter from Asa Gray, 24 July 1865.

[2] Letter from Asa Gray, 24 July 1865.

[3] CD refers to Gray's praise of 'Climbing plants' (see letter from Asa Gray, 24 July 1865 and n. 5).

[4] Gray had objected to the length of CD's explanation of the bi-directionality of spirals in caught tendrils, suggesting that a simple explanation based on mechanical necessity would have been sufficient (see letter from Asa Gray, 24 July 1865; see also 'Climbing plants', pp. 95–8).

[5] CD is probably referring to Isidore Léon. In an article on spiral motion in tendrils (Léon 1858), Léon wrote that he had observed under the microscope that, in a transverse section, cells of relatively greater diameter occurred on the outside edge of the tendril at those points where the spiral changed direction, whereas the larger cells would normally be towards the centre of the tendril (*ibid.*, p. 681). Daniel Oliver had sent CD an abstract of Léon 1858 in 1864 (see *Correspondence* vol. 12, letter from Daniel Oliver, [1 April 1864]). CD cited Léon 1858 in 'Climbing plants', p. 96, but Léon's observation on the structural changes in tendrils is not mentioned.

[6] See letter from Asa Gray, 24 July 1865 and n. 3.

[7] CD had earlier complained about the coverage of the American Civil War in *The Times*, which consistently adopted a hostile tone towards the Union cause (see, for example, *Correspondence* vol. 11, letter to Asa Gray, 23 February [1863] and nn. 22–3; for more on the bias of *The Times* in reporting the war, see Brogan ed. 1975 and Jenkins 1974–80, 2: 46–50). Emma Darwin cancelled their subscription to *The Times* at one point because of her indignation with the reporting of the war (see *Correspondence* vol. 12, letter to Asa Gray, 29 October [1864] and n. 13).

[8] For CD's earlier doubts about whether the North could prevail and whether slavery could be stopped, see, for example, *Correspondence* vol. 12, letter to Asa Gray, 13 September [1864] and n. 10. Reports in *The Times* had suggested that Southern independence was inevitable (Brogan ed. 1975, p. xvii;). For CD's views on slavery and Gray's comments on the defeat of the South, see the letter from Asa Gray, 24 July 1865 and nn. 13 and 14. See also Colp 1978.

[9] CD refers to Emma Darwin, Henrietta Emma Darwin, and possibly Elizabeth Darwin. In a letter to J. D. Hooker of 1 June [1865], CD had mentioned that he could hardly read a page without his head being affected. See also *Correspondence* vol. 12, letter to Asa Gray, 29 October [1864]).

[10] CD refers to Lubbock 1865, Tylor 1865, and Lecky 1865. CD and Joseph Dalton Hooker had already discussed these books (see letters to J. D. Hooker, [10 July 1865] and n. 5, and [29 July 1865] and nn. 13 and 15).

[11] The reference is to Frances Wedgwood (see letter from Asa Gray, 24 July 1865 and n. 10).

[12] CD had evidently abandoned John Chapman's ice treatment (see letter to J. D. Hooker, [17 June 1865], and letter from Charles and Emma Darwin to J. D. Hooker, [10 July 1865]). He consulted

Henry Bence Jones in July and August 1865, recording payments of £1 1s to Jones on 22 and 28 July 1865 and 13 and 30 August 1865 in his Account book–cash account (Down House MS). Jones recommended a strict diet (see letter to J. D. Hooker, 27 [or 28 September 1865]). Emma Darwin's diary (DAR 242) records 'began diet' on 24 July 1865; 'left off diet' on 29 July; 'began diet' on 17 August, and 'began regular diet' (crossed out) on 27 August.

To J. D. Hooker 16 August [1865]

Down. | Bromley. | Kent. S.E.
Aug 16th

My dear Hooker

I have this minute seen in the Times your Father's death. I did not even know that he was ill.[1] I hope earnestly that he did not suffer much at the close of his long life. He must have passed on the whole a happy & certainly a most active life. As I had not heard for some time from you, I have been continually thinking of you & had a presentiment of some ill news; but did not fear your Father.—

Do *not* write soon, but you know how anxious I shall be to hear of you, before very long.— I am bad, so no more.

My best of old Friends | Yours affectionately | Ch. Darwin

Endorsement: '/65.'
DAR 115: 274

[1] William Jackson Hooker, Hooker's father, died on 12 August 1865. The notice of his death appeared in *The Times*, 15 August 1865, p. 1. Hooker had last mentioned his father's health in a letter to CD of [26 May 1865], when he wrote of his father: 'he is much shaken, though quite as well as, at his age, can be expected'. According to the account in L. Huxley ed. 1918, 2: 68, W. J. Hooker had been active until his final illness, escorting a party of visitors through the Royal Botanic Gardens, Kew, only the day before he fell ill.

From F. H. Hooker [17 August 1865][1]

My dear Mr. Darwin

I write for Joseph, as I am sorry to tell you he is quite unable to do so himself— he imprudently slept on the floor beneath an open window one evening, & brought on a sharp attack of rheumatic fever— he has now been confined to his bed for a week, & we hope the worst is past— he has suffered terribly, but is easier now, though even yet scarcely able to turn in bed without assistance

Sir Wm. was taken ill early on the Wedy. morning, in the throat, a disease which is epidemic in Kew—[2] (Lady H.[3] Joe & I have all had it) he became rapidly worse— Lady H. was at Yarmouth,[4] but joined him on Thursday, & the accounts seemed better until Saty. afternoon, when he rapidly sank— Poor Joseph was so prostrated by rheumatism as to be utterly unable to be with him the last day, &

of course could not attend the funeral yesterday—[5] He desires his love to Etty, & thanks for her letter, which he received the first day of his illness—[6]

Believe me | Yours affect[ly]. | F H Hooker

DAR 104: 233–4

[1] The date is established by the reference to William Jackson Hooker's funeral (see n. 5, below).

[2] The reference is to W. J. Hooker, Joseph Dalton Hooker's father. For accounts of W. J. Hooker's last illness and death, see L. Huxley ed. 1918, 2: 68, and Allan 1967, pp. 209–10, and Yaldwyn and Hobbs eds. 1998, pp. 61–2.

[3] Maria Hooker, wife of W. J. Hooker.

[4] Maria Hooker's family home was in Yarmouth (Allan 1967, p. 32).

[5] W. J. Hooker's funeral was on 16 August 1865 (Jackson 1906, p. 203).

[6] The reference is to Henrietta Emma Darwin. The letter from her to J. D. Hooker has not been found.

From F. H. Hooker [21 or 28 August 1865][1]

My dear M[r]. Darwin

I think that Joseph progresses on the whole towards recovery[2]—but it is certainly very slow progress, & he is very susceptible to the weather— today being very mild and damp, with heavy rain in the morning, has not suited him, & he is not quite so well— he is, however, downstairs for the first time, & sitting in the study— I hope a few days more may see us at Yarmouth—

With Joseph's love, I am | Y[rs]. affty | F H Hooker

Monday

DAR 104: 241

[1] The date is established by the relationship between this letter and the letters from F. H. Hooker, [17 August 1865] and 6 September 1865. The Hookers went to Notting Hill on 2 September (letter from F. H. Hooker, 6 September 1865). There were two Mondays between 17 August and 2 September 1865: 21 and 28 August.

[2] The reference is to Joseph Dalton Hooker (see letter from F. H. Hooker, [17 August 1865]).

From E. A. Darwin 24 August [1865][1]

Hengwrt Dolgelly
24 Aug.

Dear Charles

The enclosed is an allotment of Monmouth shares.[2] Jos says they do not belong to Caroline & I therefore presume they belong to Emma,[3] but am not able to refer to anything here.

yours affec. E D

DAR 105: B36

[1] The year is established by the address. Hengwrt was a house in the small market and resort town of Dolgelly (now Dolgellau) in north-west Wales, rented for the summer of 1865 by the Hensleigh Wedgwoods (*Emma Darwin* (1904), 2: 207). Henrietta Emma Darwin and George Howard Darwin had spent some time there in June with some of their Wedgwood cousins (letter from H. E. Darwin to Emma Darwin, DAR 245: 26; this letter should be dated 19 June 1865).

[2] The reference is to shares in the Monmouth Canal and Railway Company.

[3] Erasmus Alvey Darwin refers to Josiah Wedgwood III, to his wife, Caroline Sarah, who was CD's sister, and to Emma Darwin. Both Emma and CD had owned Monmouth Canal and Railway shares. Emma's shares were in Erasmus's name, as he was a trustee, along with Josiah, of Emma's property (see *Correspondence* vol. 3, letter to Solicitor?, 1 October 1844). CD acquired further shares in 1847 (see *Correspondence* vol. 4, letter from Henry Allen Wedgwood, 21 February 1847 and nn. 7 and 9) and in 1852 Emma's shares were transferred to CD (see *Correspondence* vol. 5, letter to Josiah Wedgwood III, 20 November 1852 and n. 3). CD recorded the 1852 transfer of Emma's shares in his Investment book (Down House MS), p. 29. He recorded the payment of dividends on his shares in *ibid.*, pp. 5–8, 72–3. There is no record of any new shares in the Monmouth Canal and Railway company being acquired by Emma or CD in 1865.

From Ellen Frances Lubbock [27 August – 1 September 1865][1]

Dear M^r. Darwin

John is still away. I expect him home on Saturday or Sunday.[2] Just at present he is with Prof. Steenstrup in the country about Pressigny & Pont le Voye.[3]

He seems quite overwhelmed with the number of flints he has found—[4] he says they filled two carriages in one day. I cannot imagine what it will cost to bring home his luggage, nor where we shall put the contents when they arrive.

I believe Prof. Steenstrup will return with him, for the B. Ass. next week.[5]

I am very sorry indeed to hear so bad an account of your health for the last few months. I am sure one of the things John is most looking forward to in returning to High Elms is the chance of seeing you now & then when you are feeling pretty well.[6]

Please remember me most kindly to M^rs. Darwin & to Etty,[7] & believe me yours most sincerely | Ellen Lubbock

DAR 170: 10

[1] The date is established by the references to the British Association for the Advancement of Science meeting and the return of John Lubbock from France (see nn. 2 and 5, below). The letter was written in the week before the British Association meeting, which started 6 September 1865, that is, in the week beginning 27 August, and before Lubbock's return on 'Saturday or Sunday', that is, 2 or 3 September 1865.

[2] Ellen Lubbock refers to the return of her husband, John Lubbock, on 2 or 3 September 1865. See n. 3, below.

[3] Lubbock was with Japetus Steenstrup in France at le Grand Pressigny, about forty miles south of Tours, and Pontlevoy in the Loire Valley (*Times atlas*).

[4] Steenstrup and Lubbock were investigating prehistoric artefacts, including flints, the chipped or flaked stone implements sometimes found in geological formations with fossils of extinct mammals.

[5] The British Association for the Advancement of Science meeting in 1865 was held at Birmingham, between 6 and 13 September. Steenstrup and Lubbock presented a paper 'On the flints of Pressigny

le Grand', a brief synopsis of which appeared in the *Report of the thirty-fifth meeting of the British Association for the advancement of science; held at Birmingham in September 1865*, Transactions of the sections, p. 129. They presented a similar paper to the Ethnological Society in June 1866 (Steenstrup and Lubbock 1866).

6 High Elms, the Lubbock family home, was a short distance from Down House. John Lubbock made it his residence following the death of his father, John William Lubbock, on 20 June 1865 (*The Times*, 23 June 1865, p. 12).

7 Emma Darwin and Henrietta Emma Darwin.

From Fritz Müller[1] 31 August 1865

Desterro, Brazil,
31. August 1865.

Verehrter Herr!

Ich hoffe, Sie werden einen Brief richtig empfangen haben, in welchem ich Ihnen für die Uebersendung Ihrer anregenden Schrift über Kletterpflanzen dankte.[2] Ich habe seitdem einige Beobachtungen über diesen Gegenstand gemacht, welche wie ich denke, von einigem Interesse für Sie sein werden.

Zu der Liste von Kletterpflanzen, welche in den Hecken und Dickichts bei Desterro wachsen, kann ich die Gattungen Lygodium, Valeriana, Tragia, Caulotretus hinzufügen. Von drei anderen habe ich die Blüthen noch nicht gesehen und weiss nicht, zu welchen Gattungen sie gehören.

Incomplete[3]
Möller 1915–21, 2: 73–4.

1 For a translation of this letter, see Appendix I. See also letter from Fritz Müller, 12 August 1865, n. 1.

2 'Climbing plants'. See letter from Fritz Müller, 12 August 1865.

3 For the missing portion of the letter, see the letter from Fritz Müller, [12 and 31 August, and 10 October 1865]. See also letter from Fritz Müller, 12 August 1865, n. 11.

From M. T. Masters September 1865

41 Wellington Street | Strand. W.C.
September 1865

Private

My dear Sir/

You may have heard through our excellent friend D^r. Hooker[1] that I am about to assume the Editorship of the Gardeners' Chronicle with a staff of practical men as colleagues.[2]

I hope we may continue to be favored from time to time with notices enquiries and what not from your pen and I shall be greatly obliged for any hints or suggestions for the improvement of the paper and the extension of its utility that you may have to make.[3] D^r. Hooker has promised me his support and I am very anxious to secure an *influential* class of contributors and correspondents—

Pray excuse me for troubling you and believe me | faithfully y^rs | Maxwell. T. Masters

DAR 171: 73

[1] Joseph Dalton Hooker.
[2] In an earlier letter to CD, Masters had mentioned that he would assume the position of editor of the *Gardeners' Chronicle* in a few months' time (see letter from M. T. Masters, 12 July 1865 and n. 6).
[3] CD often sent letters to the *Gardeners' Chronicle* (see letter from M. T. Masters, 12 July 1865 and n. 7).

From E. A. Darwin 1 September [1865][1]

Hengwrt | Dolgelly
Sept 1

Dear Charles

I send you a Lanc & York but I am not absolutely certain whether it belongs to you—[2] Please to give me an answer of 'all right'.[3]

Yours affec. | E D

DAR 105: B35

[1] The year is established by the address (see letter from E. A. Darwin, 24 August [1865], n. 1).
[2] CD's Investment book (Down House MS) index lists 'Lancash & York' shares under trust property of Emma Darwin. In 1856, 720 shares of East Lancashire stock were acquired, and an interest payment of £22 3*s*. was recorded on 4 September 1865 (see *ibid.*, pp. 78–9). The East Lancashire and the Lancashire & Yorkshire were rival railway companies that merged on 13 August 1859 (see Bairstow 1993, p. 3).
[3] No further correspondence on Lancashire & Yorkshire shares has been found.

From F. H. Hooker 6 September [1865][1]

My dear M^r. Darwin

Joseph made no progress to recovery at Kew—[2] & so by D^r. Quain's[3] advice, we removed to Notting Hill on Sat^y., where we are staying with the Campbells—[4] Joseph bore the little journey very well, & was better on Sunday—

Monday was a bad day, with much fever, which pulled him down very much—but he is now better again, & says he is certainly gaining strength— But he is very much reduced, & very unequal to any, even the slightest exertion—

Buxton will probably be the next move, when he can bear the journey—[5]

He asks if you saw the article of M^r. Croll in the last Reader on the displacement of the Earth's axis(?) by the ice of the glacial epoch?—[6] & also do you know that the Reader has been sold to the Anthropologicals?[7]

Sir C Lyell[8] told him this,—if true, Joe thinks it a great disgrace to all parties concerned—

Sir Charles was looking very well, & Joseph was delighted to see him, though he unfortunately had a very severe fever attack at the time—

We wish much to know how you are— Will you kindly let the Lubbocks know about this account of Joseph?[9] as I am overwhelmed with letter writing—

Joe's love & mine to M^rs. Darwin, Etty & Lizzy[10] & yourself—

Believe me | Your's aff^tly. | F H Hooker

D^r. Campbell's
104. Lans-downe Road | Notting Hill. W.
Sep. 6.—

DAR 104: 239–40

1 The year is established by the reference to James Croll's article in the *Reader* (see n. 6, below), and by the relationship between this letter and the letter from F. H. Hooker, [17 August 1865].

2 Hooker had contracted rheumatic fever while attending his sick father, William Jackson Hooker. See letters from F. H. Hooker, [17 August 1865] and [21 or 28 August 1865], and Allan 1967, p. 209. The Hookers lived at 55 The Green in Kew, Surrey (Allan 1967, p. 199).

3 Richard Quain was a London physician with premises at 56 Harley Street (*Post Office London directory* 1865, *DNB*).

4 Archibald Campbell and his wife lived in Lansdowne Road, Notting Hill, London (*Post Office London directory* 1865). Hooker and Campbell had become friends while Hooker was travelling in India (Allan 1967, p. 166).

5 Buxton, Derbyshire, was a popular spa town and centre of hydropathy in the latter half of the nineteenth century (Leach 1987, pp. 76–7).

6 The reference is to Croll's article 'On the physical cause of the submergence of the land during the glacial epoch' in the *Reader*, 2 September 1865, pp. 270–1. Croll hypothesised that the weight of the ice-sheet that formed during a glacial period caused a shift in the earth's centre of gravity, resulting in a rise in sea-level in northern latitudes.

7 The *Reader* was purchased by Thomas Bendyshe, who was a member of the Anthropological Society of London, in August 1865 (Sullivan ed. 1984, p. 3). In December 1864, it had been purchased by a consortium of scientific men and established as the Reader Limited Company (see *Correspondence* vol. 12, letter to John Lubbock, 19 November [1864] and nn. 2 and 4, and this volume, letter from Herbert Spencer, 22 April 1865 and n. 1). The journal was sold to Bendyshe following the retirement of the editor, William Fraser Rae, due to illness (Byrne 1964, p. 65). On the controversies surrounding the Anthropological Society and the opposition of some of its leading members to CD's theory of transmutation, see *Correspondence* vol. 12, letter from J. D. Hooker, 9 [March] 1864 and n. 23, and letter from A. R. Wallace, 10 May 1864 and n. 7.

8 Charles Lyell.

9 John and Ellen Frances Lubbock were neighbours of CD.

10 Emma, Henrietta Emma, and Elizabeth Darwin.

To Edward Cresy 7 September [1865][1]

Down, Bromley, Kent
Sept. 7th.

Dear Cresy

As you have been so kind as to give me advice about the career of my son George, I have thought that you would permit him to call on you for a little additional advice.[2] We were in hopes to have been able to see you here for a Sunday, but I have been confined to my bed-room with almost constant sickness for the last four months, and have seen nobody and done nothing.[3] If I should get a little better I will still try to persuade you to come, though I might only be able to see you for a very short time, as it is a real pleasure to me to see you.

Dear Cresy, believe me | Yours very sincerely | Charles Darwin

Copy
DAR 143: 324

[1] The year is established by the relationship between this letter and the letter from Edward Cresy, 10 September 1865.

[2] George Howard Darwin matriculated at Trinity College, Cambridge, in October 1864 (*Alum. Cantab.*). Cresy may have given CD advice about George's pursuing a career as an engineer (Henrietta Emma Darwin, in a letter to George that can be dated February–March 1866 (DAR 245: 271) commented: 'Cresy said it wd be very good thing for engineer to have a little practice in public speaking'). Cresy was the principal clerk at the Metropolitan Board of Works, Spring Gardens, London (*Post Office London directory* 1865).

[3] CD recorded in his Journal that his sickness began around 22 April 1865 (see Appendix II).

From Edward Cresy 10 September 1865

Bude. Cornwall.
10 Septr '65—

My dear Sir.

Your note[1] reached me here yesterday where I have been spending a month most agreeably walking and sketching or rather smudging to be accurate—& I must say that though roughish the civilisation of Cornwall is much pleasanter than the want of it in New Brunswick[2] I am reminded of my last summer's trip by a blue nose brig with a regular blue nose skipper & blue nose crew[3] having just come to grief here he takes it easy saying it t'aint a place & didn't ought to be down in the chart.

I am truly sorry I was out or rather away when George called but hope he will look me up again before he returns to Cambridge.[4] I shall be at the office on Thursday & Friday & perhaps Saturday but shall ask him to drop me a line to say which day he will call as I have to look after some out of door matters.—[5] But it would be much pleasanter if his time allows if he will come & dine with me & sleep at my house any day after the 24th. as I shall then be back for good & till then shall be standing off & on—[6]

We are both greatly concerned to hear you have been so unwell.[7] Although you have many active disciples still your own illness is a national calamity especially when we are all so anxious to see your views extending— It is satisfactory though to see how they are extending. I brought Vogt's lectures on man just published by the Anthropological Society with me to read here & see that although strongly opposed to Lamarck he confesses to complete conversion to Darwinism—[8] If you have not read him you will be amused at his Huxleyan outspokenness,[9] still more at the excessive naïveté of the translator Dr James Hunt who struck out a few passages as likely to offend pious people but to mollify the savan's indignation at his want of courage he his printed the omitted matter in the Appendix![10] though why these should horrify people more than much which he has left in puzzles me to see.

I must tell you of a very beautiful because remarkably homely illustration of your solution of one of the difficulties of Nat selection viz the hexagonal combs. We had a plumb pudding for dinner & from the cook having exactly hit the right tenacity & thinness of crust it came up a complete sphere mapped out with hexagonal articulations, not here & there a pasty hexagon but the whole covered with exact regular hexagons— By guesswork each plumb should have punched just a swelling & then a round hole for itself like a round shot through a plank but I suppose the strain came to equally on all parts of the crust so the spherical plumbs laid the foundation, without any instinctive knowledge of a series of regular hexagonal combs—[11]

Ask George to write to Spring Gardens.[12] My wife desires kindest remembrances to yourself & M^rs^ & Miss Darwin.—[13] she has benefitted immensely here—

Very truly yours | E Cresy

C Darwin Esq

DAR 161: 245

[1] Letter to Edward Cresy, 7 September [1865].

[2] New Brunswick is one of the Maritime Provinces of eastern Canada. Cresy had visited the United States and Canada in the summer of 1864 (see *Correspondence* vol. 12, letter from Edward Cresy, 29 July 1864).

[3] 'Blue-nose': Nova Scotian (*OED*).

[4] CD had asked if his son George Howard Darwin could visit Cresy to discuss his future career (see letter to Edward Cresy, 7 September [1865] and n. 2).

[5] Cresy was the principal clerk at the Metropolitan Board of Works (*Post Office London directory* 1865).

[6] Cresy resided in Alleyn Road, Lower Norwood, Surrey (*Post Office London suburban directory* 1865).

[7] Cresy refers to himself and his wife Mary Louis Cresy. See letter to Edward Cresy, 7 September [1865] and n. 3.

[8] Carl Vogt's *Lectures on man* (Vogt 1864) was translated and edited by James Hunt and published for the Anthropological Society of London by Longman, Green, Longman, and Roberts. CD's annotated copy is in the Darwin Library–CUL (see *Marginalia* 1: 824). Vogt relates how, when he opposed earlier theories of transmutation, such as that of Jean Baptiste de Lamarck, he was 'much prejudiced by received opinions', and claimed that his gradual acceptance of CD's version of the theory was the result of 'continued self-instruction' (Vogt 1864, pp. 446–7). See also *Correspondence* vol. 12, letter from Ernst Haeckel, 9 [July 1864].

[9] Thomas Henry Huxley had expressed his enthusiastic support for CD in print on numerous occasions (see, for example, T. H. Huxley 1860b) and had publicly defended CD's theory against some of its critics (see A. Desmond 1994–7, 1: 260–5, 276–81). In a letter to CD, Cresy had recently referred to Huxley's 'boldly proclaiming his faith in Darwinism' in a lecture at the Royal Institution (see letter from Edward Cresy, 9 June 1865 and n. 7).

[10] A one-page appendix was placed at the end of Vogt 1864. It contained passages that Hunt, the president of the Anthropological Society, had excised from the main body of the text, such as the observations that Byzantine artists gave the Madonna ape-like arms, or that the ape brain most similar to that of humans belonged to the 'devil's ape' (*Ateles belzebuth*, a spider monkey; see Vogt 1864, p. 470).

[11] In *Origin*, pp. 224–35, CD had explained how the hexagonal cell produced by hive-bees could have developed by a simple principle of gradation. He considered how the cells of the Mexican *Melipona* bee, which had flat sides at points where the spherical shapes intersected, were intermediate between the simple rounded cells of the humble-bee and the perfect hexagonal cells of the hive-bee. He argued that the hive-bees' ability to make perfectly hexagonal cells was not so difficult to explain as it might appear, because the hexagonal walls were only formed when equidistant circular cells intersected at flat walls of wax added by the bees. Thus, the hexagonal shape, the most efficient for optimal storage, was a by-product of the spacing of circular cells. CD's notes on the structure of bee-cells are in DAR 48B.

[12] The address of the Metropolitan Board of Works (*Post Office London directory* 1865).

[13] Emma and Henrietta Emma Darwin.

From F. H. Hooker 13 September [1865][1]

My dear Mr. Darwin

I can give you a much better account of Joseph now—[2] he is decidedly gaining ground, & this morning he was even able to take a tiny turn in the garden— So that I trust he is really making progress at last—

We are talking of going to Buxton as soon as he can bear the journey—& if he has no relapse, I am beginning to hope we may manage it next week—

Willy is here today, en route for School this afternoon— & Charlie & Harriet[3] commenced school life at Brighton last week— Brian is under the care of his godmother, Miss Hawthorn—[4] so we are easy about all the children—

Joseph has this morning received the offer of the Directorship—[5]

We were sorry to hear the Lubbocks came in for a railway accident, but have heard no particulars—[6]

Believe me | Yours affectly. | F H Hooker

104, Lansdowne Road | Notting Hill. W.
Sept. 13—

DAR 104: 235–6

[1] The year is established by the reference to Joseph Dalton Hooker's appointment as director of the Royal Botanic Gardens, Kew (see n. 5, below).

[2] Hooker had been suffering from rheumatic fever (see letter from F. H. Hooker, 6 September [1865] and n. 2).

[3] William Henslow Hooker, Charles Paget Hooker, and Harriet Anne Hooker were the three eldest Hooker children.

[4] Brian Harvey Hodgson Hooker. Miss Hawthorn was a daughter of Robert Hawthorn, the vicar of Stapleford (see *Correspondence* vol. 8, letter from J. S. Henslow to J. D. Hooker, 10 May 1860, and *Correspondence* vol. 11, letter from J. D. Hooker, 1 October 1863).

[5] J. D. Hooker succeeded his father, William Jackson Hooker, as director of the Royal Botanic Gardens, Kew. He took up his appointment on 1 November 1865 (Allan 1967, p. 211; R. Desmond 1999, p. 221).

[6] John and Ellen Frances Lubbock were in a train derailment on 9 September 1865 on their way to Birmingham, for the meeting of the British Association for the Advancement of Science. The derailment was reported in *The Times*, 11 September 1865, p. 7. Lubbock was uninjured but Ellen suffered deep cuts to her arm and hand. See Hutchinson 1914, 1: 77–9, for a detailed account of the

accident in a letter from John Lubbock to his mother, dated by Lubbock 'Sunday, 10th July 1865' in error for Sunday 10 September.

From A. R. Wallace 18 September 1865

9, St. Mark's Crescent, | Regent's Park, N.W.
Septr. 18th. *1865*

Dear Darwin

I should have written before to thank you for the copy of your paper on climbing plants, which I read with great interest;[1] I can imagine how much pleasure the working out must have given you. I was afraid you were too ill to make it advisable that you should be bothered with letters.

I write now in hopes you are better, to communicate a curious case of *variation* becoming at once *hereditary*, which was brought forward at the Brit. Ass^{n}.[2]

I send a note of it on the other side, but if you would like more exact particulars, with names & dates and a drawing of the bird, I am sure Mr. O'Callaghan[3] would send them you.

I hope to hear that you are better & that your new book is really to come out next winter.[4]

Believe me | Yours very faithfully | Alfred R. Wallace

C. Darwin Esq.

Note

Last spring Mr. O'Callaghan was told by a country boy that he had seen a blackbird with a topknot; on which Mr. O'C. very judiciously told him to watch it & communicate further with him. After a time the boy told him he had found a blackbird's nest & had seen this crested bird near it & believed he belonged to it— He continued watching the nest till the young were hatched. After a time he told Mr O'C. that two of the young birds seemed as if they would have topknots. He was told to get one of them as soon as it was fledged. However he was too late and they left the nest, but luckily he found them near & knocked one down with a stone, which Mr. O'C. had stuffed & exhibited. It has a fine crest, something like that of a Polish fowl but *larger* in proportion to the bird, & very regular & well formed. The male must have been almost like the Umbrella bird in miniature, the crest is so large & expanded.

ARW.

DAR 106: B25–6

CD ANNOTATIONS[5]

Top of letter: 'Your Journal | Paper ['in Linn Soc.' *del*] Annals | I saw your paper in Linn Trans' *pencil*
End of letter: 'British Assoc Birmingham 1865' *pencil*

[1] The reference is to 'Climbing plants'. It is likely that CD sent Wallace an author's offprint (see Freeman 1977, pp. 116–17, for the publication history of the first edition of 'Climbing plants'); however, no presentation list for 'Climbing plants' has been found.

[2] Wallace refers to an exhibit, and a report entitled 'Remarks on a curious preserved specimen of the blackbird', a notice of which appeared in *Report of the thirty-fifth meeting of the British Association for the Advancement of Science; held at Birmingham in September 1865*, Transactions of the sections, p. 92.
[3] Patrick O'Callaghan was the author of the report to the British Association (see n. 2, above).
[4] *Variation* was not published until 1868. Wallace and CD occasionally corresponded on topics relating to CD's work on *Variation* (see, for example, letter from A. R. Wallace, 31 January 1865 and n. 5). On CD's delay in finishing *Variation*, see the letter to John Murray, 2 June [1865] and n. 1.
[5] CD's annotations are notes for his reply to this letter. See letter to A. R. Wallace, 22 September [1865].

To Fritz Müller 20 September [1865][1]

Down. | Bromley. | Kent. S.E.
Sep. 20.

My dear Sir

I am very much obliged for your interesting letter written in such wonderfully good English about climbing plants.[2] The case of Haplolophium is new to me & I am glad to have seen the tendril of Strychnos.[3] I do not suppose I shall attend any more to climbing plants, but I sh^{d} like to hear if you ever meet with 2 species of the same genus, twining in opposite directions[4] I should further like much to hear whether any twiners can ascend thick trunks.[5] How wonderfully rich you are in climbing plants! As I see you know much more about plants than I do, I sent off by the post yesterday a German copy of my Orchis book & two papers on Dimorphism.[6] The latter will I think interest you & perhaps one Chapter, for instance that on Catasetum in my Orchis book *w*^{d} be worth your reading to shew how perfect the contrivances are.[7] Your remarks on the spines *in* Acasta are quite new to me & seem very probable.[8] In my last letter I alluded to Anelasma; I am not sure, but I think I speculated on the relation of the branching filaments to cement tubes, but rejected the idea on account of the apparent continuity of the filaments with the outer membrane of the Capitulum.[9] Perhaps I may have made some mistake for your view now seems to me probable.[10] My specimen unfortunately had been removed out of the Sharks flesh.

The difficulty which you quote from A. Agassiz on the embryology of the Echinodermata is quite beyond me & I sh^{d} think w^{d} be just the subject for you.[11] Any how the difficulty is quite as great to L. Agassiz on his views of classification as to us on descent & modification, & that is some comfort.[12]

Does it not often strike you that Natural History is rendered extremely interesting by such views as we both hold. This frequently occurred to me when reading your work. I am sorry to say my health keeps so weak that I am not able to do any scientific work. If you write again I sh^{d} very much like to hear whether you intend to remain long at Desterro & how you like your new home I have always heard the Island is most beautiful[13] Have you ever read my Journal of Researches or Travels & if not w^{d} you like to have a copy.[14]

With sincere respect, | My dear Sir | yours very faithfully | Ch. Darwin

LS
British Library (Loan MS 10, no. 2)

[1] The year is established by the relationship between this letter and the letter from Fritz Müller, 12 August 1865.

[2] See letter from Fritz Müller, 12 August 1865 and n. 1, and letter from Fritz Müller, [12 and 31 August, and 10 October 1865]. CD had not yet received Müller's second letter on twining plants of 31 August 1865 (see letter to Fritz Müller, 17 October [1865]).

[3] The information on *Haplolophium* and *Strychnos* is contained in the section of the 12 August 1865 letter from Fritz Müller that was later published by CD in the *Journal of the Linnean Society* (*Botany*) 9 (1866): 344–9 (see Müller 1865b and letter from Fritz Müller, [12 and 31 August, and 10 October 1865] and n. 1; see also the letter from Fritz Müller, 12 August 1865 and n. 11). Müller included drawings of some of the climbing plants he described in his letters to CD (see letter to Fritz Müller, 17 October [1865]), and some of these were reproduced in Müller 1865b. He also sent CD specimens of *Haplolophium*, *Strychnos*, and *Caulotretus*; these are in DAR 142: 44–6.

[4] In 'Climbing plants', p. 20, CD had noted that there was no known instance of two species of the same genus twining in opposite directions. CD added the information he received from Müller on this topic to *Climbing plants* 2d ed., p. 36 n. (see letter from Fritz Müller, [12 and 31 August, and 10 October 1865] and nn. 16–18).

[5] CD's observations on twiners ascending the trunks of trees are in 'Climbing plants', pp. 21–2 (see letter from Fritz Müller, [12 and 31 August, and 10 October 1865] and n. 19).

[6] CD refers to Bronn trans. 1862, 'Two forms in species of *Linum*', and 'Three forms of *Lythrum salicaria*'; the last is on trimorphism.

[7] *Catasetum* is discussed in chapter 6 of *Orchids*, pp. 211–85 (Bronn trans. 1862, pp. 130–53). For more on pollination in *Catasetum*, see *Correspondence* vol. 12, letter from Hermann Crüger, 21 January 1864, and letter to Daniel Oliver, 17 February [1864].

[8] The letter from Müller of 12 August 1865 is incomplete. The section of the letter containing Müller's observation on spines in the barnacle *Acasta* has not been found. *Acasta*, classified by CD as a subgenus of *Balanus*, is a barnacle usually found attached to sponges (see *Living Cirripedia* (1854), pp. 176–7; for the modern classification, see Newman 1996, pp. 502–3). CD had described the cirri in many species of *Acasta* as remarkable because of the existence of spines on the fourth pair of cirri (*Living Cirripedia* (1854), p. 306). Müller's observation almost certainly concerned his discovery of a species of sponge-dwelling *Balanus*, *Balanus armatus*, that had spines on the third pair of cirri. In a letter to Max Johann Sigismund Schultze of 10 November 1864, Müller had discussed his observations on this species (the letter is published in Möller ed. 1915–21, 2: 58). Müller's description of *B. armatus* (in current taxonomy usually placed in synonymy with *B. trigonus*) was published in Müller 1867; an English version appeared the next year (Müller 1868a). In the article, Müller referred to the spines as an example of analogous variation similar to the air-breathing apparatus in some crabs that he had observed earlier (Müller 1867, pp. 344–5, and Müller 1868a, pp. 404–5; see also letter to Fritz Müller, 10 August [1865] and n. 5).

[9] In *Living Cirripedia* (1851), pp. 169–80, CD had noted that he found neither cement nor cement glands in *Anelasma squalicola*, but concluded that the organism probably became attached in the first instance in the normal way, that is by cement glands, but then was 'retained in its place, by being so deeply imbedded in the shark's body, and perhaps by the root-like branched filaments' (*ibid.*, p. 174). CD eventually concluded that the filaments were not modified cement ducts but originated in the outer membrane, which he described as continuous from the capitulum (the main body of the barnacle, which remained on the exterior of the host) to the peduncle (the stalk-like portion, which became embedded at the site of attachment); see *ibid.*, pp. 173, 380–1, and pl. IV, figs. 2 and 3. CD did not encounter Rhizocephala when he worked on barnacles, but, after reading Müller 1864, he had suggested to Müller that *Anelasma* might represent a connecting link between Rhizocephala and other cirripedes (see letter to Fritz Müller, 10 August [1865] and n. 7). CD evidently reconsidered the origin

of the branched filaments of *Anelasma*, seeing them as an intermediate stage between the cement ducts of other cirripedes and the root-like tubes in Rhizocephala.

10 Müller's observations on the relation between branched filaments in *Anelasma* and the root-like tubes of Rhizocephala are in the missing section of the letter from Fritz Müller, 12 August 1865 (see letter from Fritz Müller, 10 October 1865). Since Müller would not have received CD's letter of 10 August [1865] when he wrote his own letter to CD on 12 August 1865, it appears that the possibility of *Anelasma* as a connecting link occurred to each of them independently. When *Für Darwin* (Müller 1864) was translated into English, some additional material was added, including Müller's hypothesis on the relation of branched filaments to cement ducts and root-like tubes (see enclosure to letter from Fritz Müller, 22 April 1868, *Calendar* no. 6140, and Dallas trans. 1869, pp. 135–40).

11 The section of Müller's letter of 12 August 1865 in which he discussed Alexander Agassiz's observations on echinoderm embryology has not been found. Agassiz's letter to Müller, 9 March 1865, which contains these observations, is reproduced in Winsor 1976, pp. 189–94. In *Origin*, p. 449, CD had argued that two groups of animals that passed through similar embryonic stages did so because they shared a common ancestor, and concluded: 'community in embryonic structure reveals community of descent'. The problem Agassiz raised was the similarity in the larvae of sea urchins and brittle stars (Echinoids and Ophiurans). Agassiz pointed out that there was nothing in the adult forms to suggest any affinity and moreover that if they did share a common ancestor it was surprising that there were no connecting links in the fossil record, given that, with few exceptions, one would expect their skeletons to be well-preserved. Müller admitted in his reply (letter from Fritz Müller to Alexander Agassiz, 29 June 1865, in Möller ed. 1915–21, 2: 64–7) that the case was a real difficulty for Darwinian theory (see Winsor 1976, pp. 152–4 and *passim*, for a more detailed analysis of the case).

12 CD refers to Louis Agassiz and probably to his 'Essay on classification' (J. L. R. Agassiz 1857–62, vol. 1), in which Agassiz based his system of classification on embryological as well as adult morphological similarities. In *Origin*, p. 449, CD had referred to Agassiz's belief that it was a law of nature that 'ancient and extinct forms of life should resemble the embryos of their descendants, —our existing species'. The existence of similar larval forms in apparently disparate adult genera was therefore no more explicable within Agassiz's system based on 'prophetic' (embryonic) types (see Winsor 1976, pp. 146–8) than within a Darwinian one based on common descent. CD's annotated presentation copy of J. L. R. Agassiz 1857–62, vol. 1, is in the Darwin Library–CUL; see also *Marginalia* 1: 9–11.

13 Destêrro (now called Florianópolis) is on Santa Catarina Island, just off the coast of south-eastern Brazil (*Columbia gazetteer of the world*).

14 CD refers to the *Journal of researches*. While on the *Beagle* voyage, CD travelled down the eastern coast of South America, stopping in Rio de Janeiro and exploring parts of the interior of Brazil (see *Journal of researches*, pp. 21–43).

From F. H. Hooker 22 September [1865][1]

Buxton—
Sep[t]. 22—

My dear M[r]. Darwin

Many thanks for your letter,[2] which I waited to answer until we should be here— I can now tell you that we accomplished the journey very successfully in two days, passing the first night at Derby, & arriving here yesterday— We are at present located in a boarding house, but we do not like it, & are going out this morning to look for lodgings— The noise & publicity does not suit Joseph, & he was very tired last night—[3] However he is really wonderfully better, & beginning to walk about comfortably—but he is still very stiff in his joints, & can only move slowly— It is horribly cold here, but I suppose we shall get used to that in time—but it is a great contrast to Notting Hill—[4]

Dr. Tyndall, who was at Birmingham, told us that so far from Lady Lubbock having kept her room, she was sitting in the sections all Monday!— the accident having happened on Saturday—[5]

Joseph has read Phillips' address,[6] & thought it in matter extremely washy, but no doubt as delivered, was very well suited to the occasion— He was astonished at no allusion whatever being made to Sir J. Lubbock's book.[7]

With Joseph's love, | I am | Yours affectly. | F H Hooker

Have you read Geikie's book on Scotland?[8] Joseph was pleased with parts, but disappointed on the whole.

DAR 104: 237–8

1 The year is established by the relationship between this letter and the letter from F. H. Hooker, 13 September [1865].

2 The letter from CD has not been found.

3 The Hookers had travelled to Buxton, Derbyshire, so that Joseph Dalton Hooker might recuperate from an attack of rheumatic fever (see letter from F. H. Hooker, 6 September [1865] and nn. 2 and 5.

4 On the advice of his doctor, J. D. Hooker and his wife had left their home at Kew and stayed with friends in Notting Hill, London, until Hooker was well enough to travel to Buxton (see letters from F. H. Hooker, 6 September [1865] and 13 September [1865]).

5 John Tyndall was a vice-president of Section A (mathematics and physics) at the 1865 meeting of the British Association for the Advancement of Science, held at Birmingham, from 6 to 13 September 1865 (see *Report of the thirty-fifth meeting of the British Association for the Advancement of Science; held at Birmingham in September 1865*, p. xxxi). Ellen Frances Lubbock had been injured in a railway accident while travelling with her husband, John Lubbock, to Birmingham (see letter from F. H. Hooker, 13 September [1865] and n. 6). Apart from the accident, there may also have been some surprise that Ellen Lubbock attended meetings as she was pregnant. She gave birth to her sixth child, Rolfe Arthur Lubbock, on 19 September 1865 (*Burke's peerage*).

6 John Phillips delivered the presidential address to the British Association on 6 September 1865 (Phillips 1865; the address was also printed in the *Reader*, 9 September 1865, pp. 293–6).

7 In his presidential address, Phillips described the progress of science in various disciplines over the period since the British Association was founded. In discussing archaeology and palaeontology, Phillips referred to Charles Lyell's *Antiquity of man* (C. Lyell 1863a), but made no reference to John Lubbock's *Pre-historic times* (Lubbock 1865; see Phillips 1865, p. lxi).

8 The reference is to Archibald Geikie's *The scenery of Scotland viewed in connexion with its physical geology* (Geikie 1865).

To A. R. Wallace 22 September [1865][1]

Down. | Bromley. | Kent. S.E.
Sep 22

Dear Wallace

I am much obliged for your extract;[2] I never heard of such a case, though such a variation is perhaps the most likely of any to occur in a state of nature &

be inherited, inasmuch as all domesticated birds present races with a tuft or with reversed feathers on their heads. I have sometimes thought that the progenitor of the whole class must have been a crested animal.

Do you make any progress with your journal of travels?[3] I am the more anxious that you sh^{d} do so as I have lately read with much interest some papers by you on the Ouran Outang &c—in the Annals of which I have lately been reading the latter volumes.[4] I have always thought that Journals of this nature do considerable good by advancing the taste for Natural history; I know in my own case that nothing ever stimulated my zeal so much as reading Humboldt's Personal Narrative.[5]

I have not yet rec^{d} the last part of Linn. Tran. but your paper at present will be rather beyond my strength,[6] for tho' somewhat better I can as yet do hardly anything but lie on the sofa & be read aloud to. By the way have you read Tyler & Lecky.[7] Both these books have interested me much. I suppose you have read Lubbock.[8] In the last Chap. there is a note about you in which I most cordially concur.[9] I see you were at the Brit. Assoc. but I have heard nothing of it except what I have picked up in the Reader.[10] I have heard a rumour that the Reader is sold to the Anthrop. Soc.[11] If you do not begrudge the trouble of another note (for my sole channel of news thro' Hooker is closed by his illness)[12] I sh^{d} much like to hear whether the Reader is thus sold. I sh^{d} be very sorry for it as the paper w^{d} thus become sectional in its tendency. If you write tell me what you are doing yourself.

The only news which I have about the Origin is that Fritz Müller published a few months ago a remarkable book in its favour[13] & 2^{ndly} that a 2nd French edition is just coming out[14]

Believe me dear Wallace | yours very sincerely | Ch. Darwin

LS
British Library (Add. MS 46434 f. 56)

1 The year is established by the relationship between this letter and the letter from A. R. Wallace, 18 September 1865.

2 See letter from A. R. Wallace, 18 September 1865.

3 In his letter to CD of 2 January 1864, Wallace mentioned that he had started work on a 'small book' of his travels, and that he hoped to finish it by Christmas 1864 (see *Correspondence* vol. 12, letter from A. R. Wallace, 2 January 1864 and n. 10). Wallace later recalled that he spent much of 1867 and 1868 writing *The Malay Archipelago* (A. R. Wallace 1869); he spent the preceding three years in preparatory work related to his collections (A. R. Wallace 1905, 1: 405–6).

4 Three articles by Wallace on orang-utans appeared in the *Annals and Magazine of Natural History* in 1856 (A. R. Wallace 1856a, 1856b, and 1856c; see also letter to J. D. Hooker, 27 [or 28 September 1865], n. 12).

5 CD's seven volumes of Humboldt 1814–29, in various editions, all annotated, are in the Darwin Library–CUL (see *Marginalia* 1: 415–20). On 17 December 1840, CD recorded having read Humboldt 1814–29 (see *Correspondence* vol. 4, Appendix IV, 119: 10a). On the inspiration CD derived from reading Alexander von Humboldt, see *Correspondence* vols. 1–3, and *Autobiography*, pp. 67–8.

6 The reference is to 'On the phenomena of variation and geographical distribution as illustrated by the *Papilionidae* of the Malayan region' (A. R. Wallace 1864d). A summarised version of the paper appeared in the *Reader*, 16 April 1864, pp. 491–3 (see *Correspondence* vol. 12, letter to Asa Gray, 28 May

[1864], and letter to A. R. Wallace, 28 [May 1864] and n. 4, for CD's enthusiastic response to the summary of Wallace 1864d).

7 CD refers to Tylor 1865 and Lecky 1865, both of which had been recommended by Joseph Dalton Hooker (see letters from J. D. Hooker, [26 May 1865] and 13 July 1865). CD had also recommended these books to Asa Gray (see letter to Asa Gray 15 August [1865]).

8 Lubbock 1865.

9 CD probably refers to John Lubbock's remark about Wallace's independent discovery of natural selection. In a passage scored in CD's own copy in the Darwin Library–CUL (Lubbock 1865, p. 479; see *Marginalia* 1: 512–13), Lubbock writes:

> [Wallace] shows that the true solution of this difficulty lies in the theory of Natural Selection, which with characteristic unselfishness he ascribes unreservedly to Mr. Darwin, although, as is well known, he struck out the idea independently and published it, though not with the same elaboration, at the same time.

10 The *Reader* gave extensive coverage of the meeting of the British Association for the Advancement of Science held in Birmingham from 6 to 13 September 1865. The issue for 9 September 1865, pp. 291–6, covered the opening of the meeting and printed the presidential address of John Phillips (Phillips 1865), while that of 16 September 1865, pp. 321–5, reported on the sections. Numerous reports on papers given in the sections followed in subsequent issues. For a complete listing of articles related to the meeting see the index in the *Reader* 6 (1865): vii. Wallace was a member of the committee for Section D, zoology and botany (*Reader*, 16 September 1865, p. 325).

11 See letter from F. H. Hooker, 6 September [1865] and n. 7.

12 Joseph Dalton Hooker had suffered an attack of rheumatic fever shortly before his father's death on 12 August 1865 and had not personally corresponded with CD since then (see letter from F. H. Hooker, [17 August 1865]).

13 Müller 1864. CD began corresponding with Fritz Müller after reading Müller's book, a detailed study using evidence gathered from Crustacea to support CD's theory of transmutation (see letter to Fritz Müller, 10 August [1865]).

14 The second French edition of *Origin*, translated by Clémence Auguste Royer, with notes by CD, was published by V. Masson et Fils, Guillaumin, in 1866 (Freeman 1977). See also letter from C. A. Royer, [April–June 1865] and n. 4. CD's copy of Royer trans. 1866 is in the Darwin Library–Down House.

From Jeffries Wyman 22 September 1865

Cambridge [Massachusetts]
Sept 22 1865

My Dear Sir

Through the kindness of Dr Gray I have had an opportunity of reading your admirable paper on twining plants, which is so full of admirable observations on vegetable motions.[1] These are all of great interest on account of their physiological import, & I have always believed that they were the result of contractility of cells, analogous to those of the lowest animals.[2] Not however because this contractility has been actually demonstrated, but because it is the most reasonable to suppose that it exists. The movements of stems, but especially of petioles or tendrils, on the application of an irritation, are so analogous to those of some of the lower animals, that if one were not prepossessed with the idea that the motions of plants & animals could not be alike, we should at once admit their likeness. However I do not propose to discuss the question, but only to mention an experiment which gives *a* mechanical explanation of the double spiral of tendrils after the free end has become attached. The experiment shows, that if the cells on one side of a tendril were to shorten, & this in a straight line, the tendril (its ends being fixed to moveable supports) would at once assume the form of a double spiral.[3]

Take a piece of elastic cord (a) 3 inches long & a piece of twine (b) of the same size, 4 or 5 inches long, & tie their two ends together—

Stretch the elastic cord until the two are of the same length & secure them in this condition by fastening the ends, taking care that the two are exactly parallel & in contact. Wind both with fine thread, closely. In this way we have a compound cord contractile (through the elasticity of the india rubber) on one side & passive on the other. Hold the ends of this cord between the fingers, stretch it to its full length, & then allow the two ends to approach each other slowly. The cord at once begins to assume the form of a double spiral exactly reproducing the form of the tendril.

I do not find that the mathematicians can give an analysis of this motion, at any rate did not when asked what it *ought* to be a priori, & have not after they have seen it. Of course they will make it out, but the problem is not an easy one. I hope you will excuse me for troubling you with this, for although the experiment illustrates the spirals of the tendril & shows how they *might* be made it does not prove how the actually are made.

With great respect | Sincerely yours | J Wyman

Charles Darwin Esq

DAR 181: 190

1 CD had sent a copy of 'Climbing plants' to Asa Gray, who evidently lent it to Wyman (see letter to Asa Gray, 19 April [1865]).
2 For Wyman's views on contractility in plant cells, see Wyman 1854.
3 CD had attempted to explain the bi-directionality of the spirals of caught tendrils in 'Climbing plants', pp. 95–8. Gray suggested that CD might have explained the phenomenon simply as a mechanical necessity (see letter from Asa Gray, 24 July 1865 and n. 6); however, CD maintained that his explanation was necessary because the phenomenon was not easy to understand (see letter to Asa Gray 15 August [1865] and n. 4). CD did not alter this section in the second edition of *Climbing plants* (see pp. 163–9).

From J. D. Hooker [26 September 1865][1]

7 Terrace Road | Buxton
Tuesday

Dr Darwin

Out of the utter idleness of my mind I write to you you dear blessed ultima thule of my fatuous correspondence—to whom I can write in my folly, as well as in my sorrows & prosperity— Don't you see I am better.[2] I have quite got the use of my legs, walked full 5–6 miles yesterday & am getting up strength & flesh: if the Rheumatism would only leave my arm & finger joints I should be all right. Night sweats are still bad & most horrid, especially with linen sheets. We are most comfortably lodged, my wife is a splendid nurse;— we are blessed in the weather & *pertiklarly* in a goodish Cook!

We have read Uncle Silas isnt it creepy?, & crawly too, one should have a brandy bottle & sal volatile to get through it in safety alone.[3] how splendidly the interest is kept up. Then I took the Mill on the Floss, & am ravished with it; what a clever person the authoress is, I like it even better than Adam Bede.—[4] how evidently the authoress belongs to the class of life of her heroines, with whom first love is an animal passion with nothing to elevate it. How splendid are her analysis of the mixed motives of human action in the young but not in the old & yet how vividly she represents the acts & conversation of the old.

Then I took a dose of Jamieson's paper on the Glacial period of Scotland[5] & wrote him a long letter praising it. Still I am sure there was a time when the contour of submerged Scotland was ploughed by Icebergs moving in definite directions (S.W. to N.E. or rather vice versa). Given a submerged Gt Britain a hundred miles or so off Victoria Land & the Bergs would plough it in a direction SW to N.E.—Bergs, some of them 10 miles long & 700 ft below water! I can fancy no other explanation of the parallelism of the great Scotch valleys but this: & as there are *not* more things in Heaven & Earth than are dreamt of in &c[6] it follows as a matter of course

Then I read the Carrs of Carrlyon[7]—utter trash after G. Eliot— by the way I met Lewis at dinner & was not fascinated— Such a connection as his must tend to destroy a man's manners in public; the very sense of its exciting prejudice must insensibly react in a man's manners.[8]

Have you read Geikie, I liked it much but was not wholly satisfied somehow.[9] Have you read "Frost & Fire"[10] I have not, but want to; the author is a brand plucked from the burning! a son of a man my father knew well & I a little, Campbell of Islay an utter spendthrift who left his son a beggar, having run through a magnificent property that was sold by Tontine some 16 years ago—[11] The young man, bred to the utmost fashion & luxury, insisted on working for his bread instead of hanging onto his multitudinous relations. I have scarce seen him for years but was struck with his occupying his holiday tours with Geological observations—& I read his Nova Scotia book (I forget its name) with a good deal of interest.[12]

Col. Strachey is back, you did not know him I think. he called often on me in Town & is full of Science— he is in Phil. Club[13]

Lyell called on me every 2 days & hugely I enjoyed it— he is a wonderful man. I am so glad the Lubbock affair is settled, at any price—[14] it was a miserable affair whichever way regarded— What will Lubbock do now. I hear that Huxley was the great Scientific friend who backed his Parliamentary aspirations.[15] Tyndall[16] was dead against it. Thank God they did not ask me. I should have blurted out awfully against it. What madness of Lady L.[17] to have urged it; she would be the first to suffer.

How strange is the difference between the loss of an aged parent & child— my father has been my companion as well as parent for 25 years our intimacy has never been broken, our aims have been one as much as those of father & son ever could by possibility be but I have to *reflect* on his loss before I realize it & swell with grief.[18] How different in my childs case![19] I cannot see that it is *altogether* natural though it is so in the main. Is my grief for him more selfish than that for my child, I cannot feel it to be so— I do suppose we have a pure nature, independent of conditions (& of Darwinism applied!) but what it is we can only hope to know if we realize a future state. What did you think of Tylor? & Leckie[20] who is glorious, but I am astounded at his supposing that none but rationalistic views are held by most religious people including teachers of the X—[21] he never goes to X that is clear.

Ever Yrs aff | J D Hooker

What about the Reader?[22]

DAR 102: 34–6a

1 The date is established by the relationship between this letter, the letter from F. H. Hooker, 22 September [1865], and the letter to J. D. Hooker, 27 [or 28 September 1865]. The Tuesday between 22 September and 28 September 1865 was 26 September.

2 Hooker had been recovering from rheumatic fever. See letter from F. H. Hooker, 22 September [1865] and n. 3.

3 Hooker refers to Le Fanu 1864. The reviewer in the *Athenaeum*, 7 January 1865, pp. 16–17, concluded, 'The reader will be frightened at his own shadow as he goes to bed after finishing this book.'

4 The references are to Marian (or Mary Anne) Evans, who wrote under the name George Eliot, and to Eliot 1860 and Eliot 1859.

5 'On the history of the last geological changes in Scotland' by Thomas Francis Jamieson (Jamieson 1865). A brief notice of this paper also appeared in the *Philosophical Magazine* 29 (1865): 326. In 1861, CD had been convinced by Jamieson's glacial explanation for the parallel roads of Glen Roy (see *Correspondence* vol. 9, letter to T. F. Jamieson, 6 September [1861], and Appendix IX). He and Hooker nevertheless retained some doubts about whether glacial or fluvial action could account for all topographical features previously thought to have been the result of marine denudation (see letter to J. D. Hooker, [29 July 1865] and n. 11).

6 Hooker paraphrases Hamlet's statement, 'There are more things in Heaven and Earth, Horatio, than are dream't of in our Philosophy'; Shakespeare, *Hamlet* 1.3 (London: Isaac Iaggard and Ed. Blount, 1623).

7 Aïdé 1862.

8 George Henry Lewes was estranged from his wife, Agnes, with whom he had stopped living around 1851. In 1853, he and Marian Evans began a relationship that lasted until his death. For more on Lewes's unconventional lifestyle, see Ashton 1991.

9 Hooker refers to Geikie 1865. See also letter from F. H. Hooker, 22 September [1865] and n. 8.

10 J. F. Campbell 1865a.

11 Hooker refers to John Francis Campbell and his father, Walter Frederick Campbell. Walter Campbell was laird of the Hebridean island of Islay, but went bankrupt in 1848 following a period of depressed agricultural prices, a series of harvest failures, and the emergence of potato blight. The estate was not disposed of by tontine, but was sequestered and a trust set up on behalf of his creditors, for whose joint benefit it was eventually sold in 1853 (Storrie 1997, pp. 141–60).

12 J. F. Campbell 1865b.

13 Richard Strachey, a soldier and naturalist, had been in India; he was elected to the Philosophical Club of the Royal Society on 24 April 1865 (Bonney 1919, p. 59). The club was set up in 1847 to promote the scientific objects of the Royal Society, increase attendance at evening meetings, and encourage contribution of papers. Hooker was a founding member and CD was elected in 1854 (*ibid.*, pp. 13–14, 39).

14 Hooker refers to the dispute that arose between John Lubbock and Charles Lyell over alleged plagiarism. See letter from Charles Lyell to J. D. Hooker, [31 May 1865] and enclosures, and Appendix V.

15 Hooker refers to Thomas Henry Huxley. Lubbock unsuccessfully stood for Parliament in the general election of 1865 (see letter to J. D. Hooker, [29 July 1865] and n. 16).

16 John Tyndall.

17 Ellen Frances Lubbock.

18 Hooker's father, William Jackson Hooker, died on 12 August 1865 (*DNB*).

19 Hooker had been deeply grieved by the death of his daughter Maria Elizabeth at the age of 6 in 1863 (see *Correspondence* vol. 11, letter from J. D. Hooker, [28 September 1863] and n. 2). On Victorian attitudes to the death of children, see R. Keynes 2001, pp. 185–92.

20 Tylor 1865, Lecky 1865. See letter to A. R. Wallace, 22 September [1865] and n. 7.

21 Hooker uses 'X' to refer to the church (see *Correspondence* vol. 10, letter from J. D. Hooker, [10 March 1862] and n. 6). William Edward Hartpole Lecky had argued that the Protestant rejection of the idea of the miraculous paved the way for the rise of rationalism in Europe, citing the views of German theologians like David Friedrich Strauss as characteristic of the rationalist spirit of Protestantism (see Lecky 1865, 1: 182–5).

22 Hooker refers to the sale of the *Reader*. See letter from F. H. Hooker, 6 September [1865] and n. 7, and letter to A. R. Wallace, 22 September [1865].

From Julius von Haast 27 September 1865

Christchurch N. Z.
Septr 27. 1865

Dear Mr Darwin

From our mutual friend Dr Hooker I hear from time to time from you and of the complete recovery of your health,[1] whilst your excellent papers in the Linnean Journal show me, that you are again at work.[2] I need scarcely say, that it is with great delight and instruction that I follow your researches. In the mean time I have always given myself the pleasure to send you a copy of all my reports, so that you may be able, if your time should allow you, to see what I am doing.[3]

Unfortunately I am so much in the field, that I never have as yet found an opportunity to lay a more detailed account of my journeys before the public, but hope to do so soon.[4]

Dr Hooker & Prof. Ramsay[5] have been so kind as to propose me as a candidate to the Royal Society of London and I should be very thankful to you if you would lend me your powerful assistance towards attaining this goal of my greatest ambition.[6]

I know very well that there are many scientific men far more deserving than I am, who solicit such a distinction, but I may say with Göthe: "Nur die Lumpen sind bescheiden"[7] I know your kind co-operation will go a long way and I am certain, that if you will assist me, that my chances of being elected will become very great.—

I have also written to Sir Ch^s Lyell and Prof Tyndall,[8] requesting them to give me their help too.

Two months ago I returned from the West coast of this Province, where some very extensive goldfields, the existence of which I pointed out already three years ago, are now worked.[9] In going down the Coast towards Mt Cook, I found that the whole coast was formed by morainic accumulations for about *60 miles*, where in the often vertical cliffs, washed by the sea these beds can be easily studied.[10]

By this mail I send you a lecture written for our Mechanic Institute in which you will find some details about it,[11] but after my return from the West coast, for which region I shall start in a few days I shall try to write a more detailed account with Sections etc for the Geological Society of London.—[12]

It is unnecessary to tell you, how highly I shall be delighted to hear again from you. I asked in one of my last letters the favour to receive your photograph, which I shall highly prize; may I reiterate that request?[13]

With my best wishes for your welfare believe me, dear Mr Darwin | yours very faithfully | Julius Haast

Ch^s. Darwin Esqr | FRS. etc

DAR 166: 8

[1] Joseph Dalton Hooker had corresponded with Haast since 1861 (see *Correspondence* vol. 10, letter from J. D. Hooker, 20 September 1862 and n. 4).

[2] CD's most recent papers in the *Journal of the Proceedings of the Linnean Society* (*Botany*) were 'Climbing plants' (in the double issue dated 12 June 1865), 'Three forms of *Lythrum salicaria*' (12 December 1864), and 'Two forms in species of *Linum*' (13 May 1863).

[3] Haast sent CD two short papers, J. F. J. von Haast 1862a and 1862b, in November 1862 (see *Correspondence* vol. 11, letter to Julius von Haast, 22 January 1863 and n. 1), and an account of his geological expedition across the Southern Alps of New Zealand's Middle Island (now South Island) to the west coast in May 1863 (see *ibid.*, letter from Julius von Haast, 13 May 1863 and n. 2). He also sent CD his 1861 geological report made for the provincial government of Nelson province (J. F. J. von Haast 1861; see *Correspondence* vol. 11, letter to J. D. Hooker, 10 [November 1863]; CD's lightly annotated presentation copy is in the Darwin Pamphlet Collection–CUL), and a manuscript copy of his paper on the ground parrot, *Strigops habroptilus*; see *ibid.*, letter from Julius von Haast, 21 July [–7? August 1863] and n. 13, and letter from Emma Darwin to Julius von Haast, 12 December [1863] and n. 4. For a complete list of Haast's publications, see H. F. von Haast 1948, pp. 1088–100.

[4] In his capacity as provincial geologist, Haast published numerous papers on the geology of Canterbury province (see H. F. von Haast 1948, pp. 1088–100, and the *Royal Society catalogue of scientific papers*). He published a more general account (J. F. J. von Haast 1879) in 1879; a copy is in the Darwin Library–Down House.

[5] Andrew Crombie Ramsay.

[6] With the support of Hooker, Ramsay, CD, and others, Haast was proposed for fellowship of the Royal Society in 1866; he was elected in the following year (see H. F. von Haast 1948, p. 505; *Proceedings of the Royal Society of London* 16 (1867–8): 1). See also letter to Julius von Haast, 26 December [1865].

[7] Haast quotes from Johann Wolfgang von Goethe's poem 'Rechenschaft'. The full line reads, 'Nur die Lumpen sind bescheiden, Brave freuen sich der Tat.' ('Only knaves are furtive, the worthy rejoice in their deeds.')

[8] Charles Lyell and John Tyndall.

[9] Haast explored the west coast of Canterbury province from 29 March to 15 July 1865 (see J. F. J. von Haast 1879, pp. 65–107; the west coast of Canterbury province is now Westland province). He made a number of reports to the provincial government on various aspects of the journey (H. F. von Haast 1948, p. 1091). Haast had surveyed the Southern Alps, on the west coast of Canterbury province, from January to May 1862, during which time he also searched for gold-bearing deposits (see *Correspondence* vol. 11, letters from Julius von Haast, 5 March 1863 and 21 July [– 7? August] 1863, and H. F. von Haast 1948, pp. 199–219).

[10] For Haast's description of the moraines, see J. F. J. von Haast 1879, pp. 93–4. For CD's interest in Haast's earlier report on New Zealand glaciers, see *Correspondence* vol. 12, letters from J. D. Hooker, 15 June 1864 and n. 10, and 2 December 1864 and n. 28.

[11] Haast's 'Lecture on the west coast of Canterbury' was delivered to members of the Mechanics' Institute, Christchurch, on 25 September 1865 and printed in the *Lyttelton Times* on 28 and 29 September 1865 (H. F. van Haast 1948, p. 1092). No copy of the lecture has been found in the Darwin Pamphlet Collection–CUL.

[12] J. F. J. von Haast 1867.

[13] Haast had first requested a photograph of CD in his letter of 21 July [– 7? August] 1863 (*Correspondence* vol. 11).

To J. D. Hooker 27 [or 28 September 1865][1]

Down. Bromley Kent
Thursday 27th

My dear Hooker

I had intended writing this morning to thank Mrs. Hooker most sincerely for her last & several notes about you,[2] & now your own note in your hand has rejoiced me.[3] To walk between 5 & 6 miles is splendid, & with a little patience you must soon be well. I knew you had been very ill, but I hardly knew how ill, until yesterday, when Bentham (from the Cranworths) called here & I was able to see him for 10 minutes.[4] He told me, also, a little about the last days of your Father.[5] I wish I had known your Father better; my impression is confined to his remarkably cordial, courteous & frank bearing.[6] I fully concur & understand what you say about the difference of feeling in the loss of a father & child.[7] I do not think anyone could love a Father much more than I did mine & I do not believe three or four days ever pass without my still thinking of him, but his death at 84 caused me nothing of that insufferable grief, which the loss of poor dear Annie caused.[8] And this seems to me perfectly natural, for one knows for years previously that one's Father's death is drawing slowly nearer & nearer; whilst the death of one's child is a sudden & dreadful wrench.

What a wonderful deal you read: it is a horrid evil for me that I cannot read hardly anything, for it makes my head almost immediately begin to sing violently.[9] My good women-kind read to me a great deal, but I dare not ask for much science & am not sure that I could stand it.[10] I enjoyed Tylor *extremely* & the first part of Lecky; but I think the latter is often vague & gives a false appearance of throwing light on his subject, by such phrases, as "spirit of the age" "spread of civilization" &c.—[11] I confine my reading to $\frac{1}{4}$ or $\frac{1}{2}$ hour per day in skimming through the back volumes of the Annals & Mag. of N. H. & find much that interests me.—[12]

I miss my climbing plants very much, as I could observe them when very poorly. By the way Asa Gray approves of my paper on them.[13] Did I ever tell you that I have put myself under Bence Jones, & I am sure he has done me good by rigorous diet. I have been half starved to death & am 15 lb lighter, but I have gained in walking power & my vomiting is immensely reduced.[14] I have now hopes of again some day resuming scientific work, which is my sole enjoyment in life.—

I did not enjoy the "Mill on the Floss" so much as you, but from what you say we will read it again.[15] Do you know Silas Marner:[16] it is a charming little[17] It is a charming little story if you run short & like to have it we c^{d} send it by post.

We are now reading some American novels viz. Rutledge & Christine.[18] The latter we c^{d} lend you. We like them much but I think they are of too mild a nature for you, for I remember you were once awfully indignant at some novel I recommended. We have almost finished the first vol of Palgrave & I like it much; but did you ever see a book so badly arranged.[19] The frequently of the allusions to what will be told in the future are quite laughable. I know nothing about the Reader but hope soon to hear something from Wallace on the subject.[20] Did you read Miller's address to the chemical section at Birm.?[21] it seemed to me very good. I have not attempted Jamieson or Geikie or Frost & Fire;[22] perhaps the latter may not be too strong a dose for Emma,[23] who, however, is not very tolerant of science; Jamieson some day I shall certainly read to myself, & then there is Wallace's great paper,[24] which will be a tough job. By the way I was very much pleased by the footnote about Wallace in Lubbock's last chapter.[25] I had not heard that Huxley had backed up Lubbock about parliament.[26] I suspect Lady L.[27] is very ambitious, & to the world how grand parliament & politicks appear. Did you see a sneer some time ago in the Times, about how incomparably more interesting Politicks were compared with Science even to scientific men.[28] Remember what Trollope says in "Can you Forgive her" about getting into parliament, as the highest earthly ambition.[29] Jeffreys in one of his letters, I remember, says that making an effective speech in parliament is a far grander thing than writing the grandest History.[30] All this seems to me a poor short-sighted view. I cannot tell you how it has rejoiced me once again seeing your hand-writing, my best of old friends.

Yours affectionately | Ch. Darwin

Endorsement: 'Sept/65'
DAR 115: 275

[1] The date is established by the endorsement. In 1865, 27 September was a Wednesday, not a Thursday.
[2] See letters from F. H. Hooker, [17 August 1865], [21 or 28 August 1865], 6 September [1865], 13 September [1865], and 22 September [1865].
[3] Letter from J. D. Hooker, [26 September 1865].
[4] George Bentham, a colleague of Hooker's at Kew, had been visiting CD's neighbours, Robert Monsey and Laura Rolfe (Lord and Lady Cranworth; Jackson 1906, p. 204).
[5] Bentham visited the Hookers at Kew on 16 August 1865; Hooker's father, William Jackson Hooker had died on 12 August 1865 (Jackson 1906, p. 203).
[6] CD had maintained a cordial acquaintance with W. J. Hooker (see *Correspondence* vol. 4, letter to W. J.

Hooker, [*c.* February 1849], and *Correspondence* vol. 7, letter to W. J. Hooker [30 July 1858]).

[7] See letter from J. D. Hooker, [26 September 1865].

[8] CD's father, Robert Waring Darwin, died in 1848 (see *Correspondence* vol. 4, letter from Catherine Darwin, [13 November 1848]). Anne Elizabeth, CD's eldest daughter, died in 1851, aged 10 (see *Correspondence* vol. 5). For more on the effect of her death on CD, see Bowlby 1990, pp. 291–8, A. Desmond and Moore 1991, pp. 275–87, and R. Keynes 2001, pp. 180–98.

[9] For more on CD's symptoms brought on by reading, see Appendix IV.

[10] Emma, Henrietta Emma, and Elizabeth Darwin often read to CD (see letter to Asa Gray, 15 August [1865] and n. 9, and letter to A. R. Wallace, 22 September [1865]).

[11] CD refers to Tylor 1865 and Lecky 1865; Hooker had recommended them in his letters of [26 May 1865] and 13 July 1865.

[12] CD's bound volumes, many with annotations, of the *Annals and Magazine of Natural History*, ser. 1: vols. 1–20 (1838–47); ser. 2: vols. 1–14 (1848–54), ser. 3: vols. 15–16 (1865), are in the Darwin Library–CUL. In addition, there are unbound issues for the years 1855 and 1866 to 1873 in the collection of unbound pamphlets, Darwin Library–CUL, many of which are annotated.

[13] CD refers to 'Climbing plants'. For Gray's comments on the paper, see the letter from Asa Gray, 24 July 1865.

[14] CD first consulted Henry Bence Jones in July 1865 (see letter to Asa Gray, 15 August [1865], n. 12). From 22 August 1865, Emma Darwin's diary (DAR 242) records CD's weight at weekly intervals. His weight on 22 August was 11 st. 4 lb and on 26 September it was 10 st. 10 $\frac{1}{2}$ lb. He had numerous bouts of sickness in August but only two in September. See also Appendix IV.

[15] Eliot 1860. See letter from J. D. Hooker, [26 September 1865] and n. 4.

[16] Eliot 1861.

[17] The letter is in CD's hand until this point, and is the first known letter in his own hand since his letter to Hooker of 16 August, which he curtailed complaining of ill health. Emma Darwin continued the letter, repeating the words 'It is a charming little'; CD resumed writing and completed the letter from the words 'I have not attempted Jamieson or Geikie'.

[18] M. C. Harris 1860, Bullard 1856.

[19] Palgrave 1865. For Hooker's opinion of the book, see the letter from J. D. Hooker, 13 July 1865 and n. 19.

[20] On the purchase of the *Reader*, see the letter from F. H. Hooker, 6 September [1865] and n. 7. After hearing about the sale, CD had asked for more information from Alfred Russel Wallace (see letter to A. R. Wallace, 22 September [1865]).

[21] An abstract of William Hallowes Miller's address as president of the chemical section of the British Association meeting at Birmingham in 1865 was printed in the *Reader*, 23 September 1865, pp. 350–2. Miller had discussed the place of science in modern education and reviewed recent developments in chemistry (*Report of the thirty-fifth meeting of the British Association for the Advancement of Science; held at Birmingham in September 1865*, Transactions of the sections, pp. 22–7).

[22] Jamieson 1865, Geikie 1865, J. F. Campbell 1865a. See letter from J. D. Hooker, [26 September 1865] and n. 5.

[23] Emma Darwin.

[24] CD refers to 'On the phenomena of variation and geographical distribution as illustrated by the *Papilionidae* of the Malayan region' (A. R. Wallace 1864d). See letter to A. R. Wallace, 22 September [1865] and n. 6.

[25] CD refers to Lubbock 1865. For the text of the note, see the letter to A. R. Wallace, 22 September [1865], n. 9.

[26] CD refers to Thomas Henry Huxley and John Lubbock. See letter from J. D. Hooker, [26 September 1865] and n. 15.

[27] Ellen Frances Lubbock.

[28] CD may be referring to an article in *The Times*, 11 September 1865, p. 6, that reported on a meeting of the statistics and political economy section of the British Association for the Advancement of Science

meeting held at Birmingham (referred to in the article as the British Association for the Promotion of Science). The author of the article wrote:

> It is curious to see the avidity with which every opportunity is laid hold of to quit the passionless domain of pure science, and to plunge into the stormy hemisphere of politics.

[29] The passage to which CD refers is probably the following, in which Alice Vavasor writes to accept the marriage proposal of her cousin (Trollope 1864–5, 1: 256):

> Dear George, let me have the honour and glory of marrying a man who has gained a seat in the Parliament of Great Britain! Of all positions which a man may attain that, to me, is the grandest.

[30] CD refers to Francis Jeffrey. CD read the *Life of Lord Jeffrey* (Cockburn 1852) in 1853 (see CD's reading notebooks, *Correspondence* vol. 4, Appendix IV, 128: 5). The second volume of Cockburn 1852 contains a large selection of letters; however, the letter to which CD refers has not been identified.

To Samuel Butler 30 September [1865][1]

Down. | Bromley. | Kent. S.E.
Sep 30

My dear Sir

I am much obliged to you for so kindly sending me your "Evidence &c—"[2] We have read it with much interest. It seems to me written with much force, vigour, & clearness; & the main argument is to me quite new. I particularly agree with all you say in your preface.[3]

I do not know whether you intend to return to New Zealand & if you are inclined to write I should much like to know what your future plans are.[4]

My health has been so bad during the last five months that I have been confined to my bedroom. Had it been otherwise I would have asked you, if you c^d have spared the time to have paid us a visit; but this at present is impossible & I fear will be so for some time.

With my best thanks for your present I remain

my dear Sir | yours very faithfully | Charles Darwin

LS
British Library (Add. MS 34486, f. 58)

[1] The year is established by the relationship between this letter and the letter from Samuel Butler, 1 October 1865.

[2] CD refers to Butler's anonymously published pamphlet *The evidence for the resurrection of Jesus Christ as given by the four Evangelists, critically examined* ([Butler] 1865). Butler's main argument, based on conflicting accounts of the resurrection in the four Gospels, was that Christ did not die on the cross and rise from the dead, but rather survived the crucifixion. The publication has not been found in the Darwin Pamphlet Collection–CUL. Much of the essay was later incorporated into *The fair haven* ([Butler] 1873).

[3] In his preface, Butler accuses English commentators of 'intellectual cowardice' for refusing to discuss discrepancies in accounts of the resurrection in the four Gospels, in contrast to German critics. Butler had been unable to find a commercial publisher for the pamphlet and had had it privately printed.

[4] Butler had gone to New Zealand in 1859 and, after some success in sheep ranching, returned to England in late 1864 (*DNB*). CD had first learned about Butler through his anonymously published article 'Darwin on the origin of species: a dialogue', which appeared in the *Press* of Christchurch, New Zealand, on 20 December 1862 (see *Correspondence* vol. 11, letter to an editor, 24 March [1863]

and n. 3). Butler's grandfather, also Samuel Butler (1774–1839), had been headmaster at Shrewsbury School when CD was there between 1818 and 1825 (Freeman 1978).

From Robert Swinhoe [before 1 October 1865?][1]

parent stock prove unfruitful— It is rare to induce the Mallard to tread the female Muscovy, and I have never seen the result. But the Chinese assure me that when thus a hybrid is produced, it is in characters scarcely distinguishable from the Muscovy breed.[2]

I have a curious story of Dog-morality which may be of some use to you— *Sodomy* is a *common* crime among Chinese Dogs— I have heard cases of it at Foochow, and its occurrence is well known in Peking. Query whether the Dog apes his lord and master? For the prevalence of this crime in Peking above all places in China is too well known to need comment. The story that I am about to relate has been told to me by M^r. T. Watters a gentleman in this Consulate who has lately joined me from Peking,[3] and who was an eye-witness of the facts it contains. M^r. Watters was walking on the Wall of Peking when his attention was attracted to two Chinese Dogs below. One was a rough big brute, and the other a comely smaller animal— The big one commenced to caress the smaller one, but the latter at first objected. At last after many affectionate appeals the smaller dog yielded. Whilst in the act of sodomy several other dogs came up and barked angrily and bit at them. They (the other dogs) then retired for a time as if for consultation, and having evidently watched for the termination of the crime, returned just as the operation was finished. They then rushed upon the big dog. One seized him by the head, two others by the hind legs, and a fourth caught and entirely bit off his *penis*. The mutilated Dog ran away howling. Mr. Watters saw the Dog several days afterwards, so that the animal did not die from the effect of the laceration. It appears not to be an unusual punishment for Dogs to inflict on one another for this crime. Besides M^r. Watters others in Peking have seen the same inflicted, and one often sees both in Peking and Tientsin Dogs going about with their genital organs mutilated, caused, as the Chinese declare, in nearly every case as an award for this offence against the community of Dogdom.[4] As to immoral acts on the part of Dogs I suppose almost every observing man must have had cases fall within his experience. I myself when a lad at home have seen the House-dog committing the sin of Onan[5] with his mouth as he lay on the rug at the door—

Some time ago I reported to you that I had noticed incipient spurs on the wings of dried specimens of the *Merula albiceps* & other thrushes.[6] I now think that I must have been mistaken, as in a *fresh* specimen of an immature *M. albiceps* lately shot I can find no indication of such spur; and the young bird ought surely to show it if the old bird possessed it in any observable degree.

I find that there is a wild *Bos* in this island, which appears from Chinese accounts to be the stock of the small cattle that are reared in the farmsteads of Formosa. I do not observe any appreciable difference between the domestic Cattle here and those of south China, and I would therefore infer that the small yellow cow of South China has had its race derived from a stock that formerly existed (and perhaps still exists) in the mountains of the Southern Provinces, instead of being a small derivative of *Bos taurus*. I will not go so far as to say that the wild *Bos* of South China (if such there be) is identical with the wild *Bos* of Formosa, or even to say that differences may not be found on anatomical comparison between the respective domestic breeds. I have not even compared the two latter together, but only speak from what I recollect of the China breed. The Formosan Cow is small extremely graceful, with short usually upright horns, and straight back— The bull is not much larger, but is remarkable for a shoulder hump. The colour of the *Bos* is usually of a yellowish-brown of lighter or deeper hue, with white belly. A few individuals occur nearly black, but the hair is not much subject to albinistic or melanistic variation, as in thoroughly domesticated animals.[7] I intend collecting full particulars on the wild as well as domestic cattle of Formosa, and submitting a Paper on the subject to the Zoological Society.[8] I shall further attempt to get live specimens to England.

Believe me, | Dear Sir, | Your's truly | Robert Swinhoe.

C. Darwin, Esqr.

Incomplete
DAR 177: 328

[1] The date is conjectured on the assumption that the missing portion of the letter contained the information on ducks referred to in the letter to P. L. Sclater, 6 January [1866] (*Correspondence* vol. 14; see n. 2, below). Based on other Swinhoe correspondence in the 1860s, the approximate length of time for a letter to reach England from Taiwan was three months.

[2] Swinhoe refers to the mallard (*Anas platyrhynchos*; then *A. boschas*), and to the Muscovy duck, a species native to tropical America (*Cairina moschata*; then also *Anas moschata*). In his letter to P. L. Sclater of 6 January [1866] (*Correspondence* vol. 14), CD wrote that Swinhoe had informed him 'of a domestic race of duck in China as perhaps descended from Anas pœcilorhyncha', the spot-billed duck, a close relative of the mallard. Swinhoe had earlier informed CD of a supposed 'thorough race' of duck that had developed from hybrids of the Muscovy duck and Chinese domestic duck (see *Correspondence* vol. 10, letter from Robert Swinhoe, 12 November 1862). CD had long been interested in crosses with the Muscovy duck (see *Correspondence* vol. 2, letter to W. D. Fox, [25 January 1841], and *Correspondence*, vol. 11, letter to W. D. Fox, 9 March [1863] and n. 1). Several notes on the topic are in DAR 205.7: 192 and 201. CD did mention the Muscovy duck, or 'musk-duck', in *Variation* 1: 190 and 2: 40, but did not mention Chinese duck species or breeds. For CD's interest in determining if domestic breeds of ducks had derived from *Anas boschas*, see *Variation* 1: 276–87.

[3] Thomas Watters had been stationed in the Peking consulate as a student interpreter in 1863; in August 1865 he was an interpreter in Taku, south-east of Peking at the mouth of the Pei-ho river. He served as acting consul under Swinhoe in Taiwan from March to October 1866 (*Foreign office list*), and thus may have visited the consulate in Taiwan prior to this appointment.

[4] Swinhoe's information does not appear in CD's discussion of dogs in *Variation*, *Descent*, or *Expression*.

[5] Onan, the second son of Judah, whose 'sin' was coitus interruptus (see Gen. 38: 4–10); however the term onanism more commonly refers to masturbation (*OED*).

[6] In his letter to CD of 4 April 1864 (*Correspondence* vol. 12), Swinhoe had described a tubercle on the carpal edge of the wing of *Turdus albiceps* (now *T. poliocephalus niveiceps*) and noted its appearance on several other species of thrush. Swinhoe had believed the tubercle to be an abortive wing-spur (see

also Swinhoe 1864, pp. 363–4). *Merula* was a synonym of *Turdus* in the nineteenth century (see Peters *et al.* 1931–87, 10: 177).

[7] For CD's interest in varieties of domestic cattle and their descent from wild species, see the letter from Ludwig Rütimeyer, 3 January 1865, nn. 3, 4, and 6. CD did not include Chinese breeds in his discussion of cattle in *Variation* 1: 79–93.

[8] No paper by Swinhoe on the cattle of Formosa (now Taiwan) has been found.

From Samuel Butler 1 October 1865

15. Clifford's Inn E.C
Oct. 1. 1865.

Dear Sir,

I knew you were ill and I never meant to give you the fatigue of writing to me—[1] Please do not trouble yourself to do so again. As you kindly ask my plans I may say that though I very probably may return to New Zealand in three or four years I have no intention of doing so before that time.

My *study* is art,[2] and any thing else I may indulge in is only by play; it may cause you some little wonder that at my age I should have started as an art student, & I may perhaps be permitted to explain that this was always my wish for years, that I had begun six years ago, as soon as ever I found that I could not conscientiously take orders; my father[3] so strongly disapproved of the idea that I gave it up and went out to New Zealand staid there for five years—worked like a common Servant though on a run of my own and sold out little more than a year ago thinking that prices were going to fall—which they have since done. Being then rather at a loss what to do, and my capital being all locked up, I took the opportunity to return to my old plan and have been studying for the last 12 months unremittingly— I hope that in three or four years more I shall be able to go very well by myself, and then I may go back to N.Z or no as circumstances shall seem to render advisable.

I must apologise for so much detail but hardly knew how to explain myself without it.

I always delighted in your origin of species as soon as I saw it out in N.Z—not as knowing anything whatsoever of natural history, but it enters into so many deeply interesting questions, or rather it suggests so many that it thoroughly fascinated me. I therefore feel all the greater pleasure that my pamphlet should please you however full of errors it may be.[4]

The first dialogue on the origin which I wrote in the Press called forth a contemptuous rejoinder from (I believe) the Bishop of Wellington—[5] (please do not mention the name, though I think that at this distance of space & time I might mention it to yourself) I answered it with the enclosed which may amuse you.[6] I assumed another character because my dialogue was in my hearing very severely criticised by two or three whose opinion I thought worth having, and I deferred to their judgement in my next.[7] I do not think I should do so now. I fear you will be shocked at an appeal to the periodicals mentioned in my letter,[8] but they form a very staple article of bush diet, and we used to get a good deal of superficial knowledge out of them. I feared to go in too heavy on the side of the origin because

I thought that having said my say as well as I could I had better now take a less impassioned tone: but I was really exceedingly angry.

Please do not trouble yourself to answer this and believe me | Your's most sincerely | S. Butler

[Enclosure]
To the Editor of *The Press.*
Sir,

In two of your numbers you have already taken notice of Darwin's theory of the origin of species; I would venture to trespass upon your space in order to criticize briefly both your notices.

The first is evidently the composition of a warm adherent of the theory in question; the writer overlooks all the real difficulties in the way of accepting it, and, caught by the obvious truth of much that Darwin says, has rushed to the conclusion that all is equally true. He writes with the tone of a partisan, of one deficient in scientific caution, and from the frequent repetition of the same ideas manifest in his dialogue one would be led to suspect that he was but little versed in habits of literary composition and philosophical argument. Yet he may fairly claim the merit of having written in earnest. He has treated a serious subject seriously according to his lights; and though his lights are not brilliant ones, yet he has apparently done his best to show the theory on which he is writing in its most favourable aspect. He is rash, evidently well satisfied with himself, very possibly mistaken, and just one of those persons who (without intending it) are more apt to mislead than to lead the few people that put their trust in them. A few will always follow them, for a strong faith is always more or less impressive upon persons who are too weak to have any definite and original faith of their own. The second writer, however, assumes a very different tone. His arguments to all practical intents and purposes run as follows:

Old fallacies are constantly recurring. Therefore Darwin's theory is a fallacy.

They come again and again, like tunes in a barrel-organ. Therefore Darwin's theory is a fallacy.

Hallam made a mistake, and in his *History of the Middle Ages*, p. 398, he corrects himself.[9] Therefore Darwin's theory is wrong.

Dr. Darwin in the last century said the same thing as his son or grandson says now[10]—will the writer of the article refer to anything bearing on natural selection and the struggle for existence in Dr. Darwin's work?—and a foolish nobleman said something foolish about monkeys' tails. Therefore Darwin's theory is wrong.

Giordano Bruno was burnt in the year 1600 A.D.; he was a Pantheist; therefore Darwin's theory is wrong.

And finally, as a clinching argument, in one of the neighbouring settlements there is a barrel-organ which plays its psalm tunes in the middle of its jigs and waltzes. After this all lingering doubts concerning the falsehood of Darwin's theory must be at an end, and any person of ordinary common sense must admit that the theory of development by natural selection is unwarranted by experience and reason.

The articles conclude with an implied statement that Darwin supposes the Polar bear to swim about catching flies for so long a period that at last it gets the fins it wishes for.

Now, however sceptical I may yet feel about the truth of all Darwin's theory, I cannot sit quietly by and see him misrepresented in such a scandalously slovenly manner. What Darwin does say is that sometimes diversified and changed habits may be observed in individuals of the same species; that is, that there are eccentric animals just as there are eccentric men. He adduces a few instances and winds up by saying that "in North America the black bear was seen by Hearne swimming for hours with widely open mouth, thus catching—almost like a whale—insects in the water." This and nothing more. (See pp. 201 and 202.)[11]

Because Darwin says that a bear of rather eccentric habits happened to be seen by Hearne swimming for hours and catching insects almost like a whale, your writer (with a carelessness hardly to be reprehended in sufficiently strong terms) asserts by implication that Darwin supposes the whale to be developed from the bear by the latter having had a strong desire to possess fins. This is disgraceful.

I can hardly be mistaken in supposing that I have quoted the passage your writer alludes to. Should I be in error, I trust he will give the reference to the place in which Darwin is guilty of the nonsense that is fathered upon him in your article.

It must be remembered that there have been few great inventions in physics or discoveries in science which have not been foreshadowed to a certain extent by speculators who were indeed mistaken, but were yet more or less on the right scent. Day is heralded by dawn, Apollo by Aurora, and thus it often happens that a real discovery may wear to the careless observer much the same appearance as an exploded fallacy, whereas in fact it is widely different. As much caution is due in the rejection of a theory as in the acceptation of it. The first of your writers is too hasty in accepting, the second in refusing even a candid examination.

Now, when *The Saturday Review*, *The Cornhill Magazine*, *Once a Week*, and *Macmillan's Magazine*, not to mention other periodicals, have either actually and completely as in the case of the first two, provisionally as in the last mentioned, given their adherence to the theory in question,[12] it may be taken for granted that the arguments in its favour are sufficiently specious to have attracted the attention and approbation of a considerable number of well-educated men in England. Three months ago the theory of development by natural selection was openly supported by Professor Huxley before the British Association at Cambridge.[13] I am not adducing Professor Huxley's advocacy as a proof that Darwin is right (indeed, Owen opposed him tooth and nail),[14] but as a proof that there is sufficient to be said on Darwin's side to demand more respectful attention than your last writer has thought it worth while to give it. A theory which the British Association is discussing with great care in England is not to be set down by offhand nicknames in Canterbury.

To those, however, who do feel an interest in the question, I would venture to give a word or two of advice. I would strongly deprecate forming a hurried opinion for or against the theory. Naturalists in Europe are canvassing the matter with the utmost diligence, and a few years must show whether they will accept the theory or no. It is plausible; that can be decided by any one. Whether it is true or no can be decided only among naturalists themselves. We are outsiders, and most of us must be content to sit on the stairs till the great men come forth and give us the benefit of their opinion.

I am, Sir, | Your obedient servant, | A.M.

DAR 106: A1–2; Butler 1923, pp. 198–201

[1] See letter to Samuel Butler, 30 September [1865].

[2] Butler studied painting at the School of Art, Bloomsbury, London. From 1867 he continued his art education at Heatherley's School of Art, Newman Street, London (*DNB*, Butler 1923, p. xl).

[3] Thomas Butler. CD had known Thomas since they were both at Shrewsbury school. Thomas shared CD's early interest in beetles and, in the summer of 1828, they collected insects at Barmouth (see *Correspondence* vol. 1, letter to W. D. Fox, [30 June 1828], and Appendix I).

[4] Butler refers to [Butler] 1865. See letter to Samuel Butler, 30 September [1865] and n. 2.

[5] Butler refers to his anonymous article 'Darwin on the origin of species: a dialogue', which appeared in the *Press* of Christchurch, New Zealand, on 20 December 1862. CD had been sent a copy of the article anonymously and had sent it to an unidentified periodical for possible publication (see *Correspondence* vol. 11, letter to an editor, 24 March [1863?] and n. 3). CD's Scrapbook of reviews contains an annotated copy of the article (DAR 226.1: 131–2). The article is reprinted in Butler 1923, pp. 188–95. Charles John Abraham became the first bishop of Wellington, New Zealand, in 1858. The unsigned article, 'Barrel-organs', a response to Butler's article of 20 December 1862, was printed in the *Press*, 17 January 1863. It is not certain that Abraham was, in fact, the author (see Butler 1923, p. 188).

[6] The enclosure was returned to Butler (see letter to Samuel Butler, 6 October [1865]); the letter printed in the *Press*, 21 February 1863, is reproduced here.

[7] See enclosure. Butler, writing as 'A.M.', ostensibly criticises both himself as the author of the original article of 20 December 1862 and the writer of the critical response of 17 January 1863. A response to this letter, signed 'The Savoyard' appeared on 14 March 1863. Further letters appeared in the *Press* on 18 March, 11 April, and 22 June 1863, the first and last of which were signed 'A.M.', while the April letter was signed '"The Savoyard," or player on Barrel-organs' (all the letters are reprinted in Butler 1923, pp. 198–207).

[8] See the enclosure and n. 12, below, for the names of the periodicals.

[9] The reference is to Hallam 1818. In the article of 17 January 1863, the author, in fact, refers to a separately published supplement which appeared after the fourth edition of the book was published (Hallam 1848).

[10] Erasmus Darwin, CD's grandfather, discussed transmutation in *Zoonomia* (E. Darwin 1794–6; see King-Hele 1999, pp. 297–301, for a synopsis of the theory).

[11] Samuel Hearne was an explorer and Canadian administrator. The observation CD cited was recorded in Hearne 1795, p. 370. In the first edition of *Origin*, CD had written 'like a whale' but changed it to 'almost like a whale' in the second edition. He also deleted a qualifying sentence in which he wrote, 'I can see no difficulty in a race of bears being rendered, by natural selection, more and more aquatic in their structure and habits.' For more on the change to this passage, see the letter from E. P. Wright, 24 March 1865, n. 6. The passage appears on page 184 in the first two editions, but on page 202 in the third edition, which appeared in April 1861 (Freeman 1977).

[12] For publication details of the reviews of *Origin* in *Saturday Review*, *Cornhill Magazine*, and *Macmillan's Magazine*, see *Correspondence* vol. 8, Appendix VII. No review has been found in *Once a Week*. For more on the review in *Saturday Review*, see Bevington 1941, pp. 283–5.

[13] Thomas Henry Huxley was the president of section D, zoology and botany, at the meeting of the British Association for the Advancement of Science held in Cambridge from 1 to 8 October

1862. He delivered an inaugural address to the section entitled 'On the conditions and prospects of biological science' (*Parthenon* 1 (1862): 749). *The Times*, 3 October 1862, p. 7, reported that Huxley had 'emphatically affirmed that Mr. Darwin's work was as perfect in its logical method as it was accurate in its scientific facts' (see *Correspondence* vol. 10, letter from T. H. Huxley, 9 October 1862 and n. 1).

14 Richard Owen had presented a paper at the British Association meeting, 'On the characters of the Aye-aye, as a test of the Lamarckian and Darwinian hypothesis of the transmutation and origin of species', in which he argued against CD's theory (R. Owen 1862; see *Correspondence* vol. 10, letter from T. H. Huxley, 9 October 1862 and n. 6).

From T. H. Huxley 2 October 1865

26 Abbey Place | St Johns Wood
Oc^{t} 2nd 1865

My dear Darwin

'This comes hoping you are well' & for no other purpose than to say as much. I am just back from seven weeks idleness at Littlehampton with my wife & children the first time I have had a holiday of any extent with them for years.[1]

We are all flourishing—the babies particularly so—and I find myself rather loth to begin grinding at the mill again. There is a vein of laziness in me which crops out uncommonly strong in your godson[2]—who is about the idlest, jolliest young four year old I know—

You will have been as much grieved as I have been about dear old Hooker—[3] According to the last accounts however he is mending & I hope to see him in pristine vigour again before long

My wife is gone to bed or she would join me in kindest regards & remembrances to Mrs Darwin & your family—

Ever yours faithfully | T H Huxley

DAR 166: 310

1 Littlehampton, a coastal resort in Sussex, was a popular holiday destination for Huxley, his wife Henrietta Anne Huxley, and their six children, Jessie, Marian, Leonard, Rachel, Nettie, and Henry, who ranged in age from 7 years to 9 months (see L. Huxley 1900, 1: 302).

2 Huxley refers to Leonard Huxley. See A. Desmond 1994–7, 1: 290–1.

3 Joseph Dalton Hooker had suffered a severe attack of rheumatic fever in August and had recently been recuperating at Buxton, Derbyshire (see letter from F. H. Hooker, [17 August 1865], and letter from J. D. Hooker, [26 September 1865]).

From A. R. Wallace 2 October 1865

9, St Mark's Crescent, | Regent's Park,
Oct 2nd. *1865*

Dear Darwin

I was just leaving Town for a few days when I received your letter[1] or should have replied at once.

The "Reader" has no doubt changed hands, & I am inclined to think for the better.[2] It is purchased I believe by a gentleman who is a fellow of the Anthropolog. Society but I see no signs of its being made a special organ of that Soc.

The Editor (& (I believe Prop^r.) is a Mr. Bendyshe the most talented man in the Society,[3] & judging from his speaking, which I have often heard, I should say the articles on "*Simeon & Simony*", "*Metropolitan Sewage*" and "*France & Mexico*" are his, & these are in my opinion superior to anything that has been in the "Reader" for a long time;[4] they have the point & brilliancy which are wanted to make leading articles readable & popular. The articles on *Mill's Pol. Economy* & on *Mazzini*, are also first rate.[5] He has introduced also the plan of having two & now *three* important articles in each number,—one Political or Social, one literary & one Scientific. Under the old regime they never had an editor above mediocrity, except *Masson*;[6] there was a want of unity among the Prop^trs., as to the aims & objects of the Journal; & there was a want of capital to secure the services of good writers.[7] This seems to me to be now all changed for the better, & I only hope the rumour of that *bète noir* the Anthropological Soc., having any thing to do with it, may not cause our best men of science to withdraw their support & contributions.[8]

I have read *Tyler* & am reading *Lecky*.[9] I found the former somewhat disconnected, & unsatisfactory from the absence of any definite result or any decided opinion on most of the matters treated of.

Lecky I like much, though he is rather tedious & obscure at times. Most of what he says has been said so much more forcibly by *Buckle* whose work I have read for the second time with increased admiration although with a clear view of some of his errors.[10] Nevertheless his is I think unapproachably the grandest work of the present century, & the one most likely to liberalise opinion.

Lubbock's book is very good, but his concluding chapter very weak.[11] Why are men of science so dreadfully afraid to say what they think & believe?

In reply to your kind enquiries about myself, I can only say that I am ashamed of my own laziness. I have done nothing lately but write a paper on *Pigeons* for the "Ibis",[12] & am drawing up a Catalogue of my Collection of Birds.[13]

As to my "Travels" I cannot bring myself to undertake them yet, & perhaps never shall, unless I should be fortunate enough to get a wife who would incite me thereto & assist me therein,—which is not likely.[14]

I am glad to hear that the "Origin" is still working its *revolutionary way*, on the Continent.[15] Will Mueller's book on it, be translated?[16]

I am glad to hear you are a little better. My poor friend Spruce is still worse than you are, & I fear now will not recover. He wants to write a book if he gets well enough.[17]

With best wishes Believe me | Dear Darwin | Yours very faithfully | Alfred R. Wallace

C. Darwin Esq.

DAR 106: B27–30

[1] Letter to A. R. Wallace, 22 September [1865].

[2] CD had asked Wallace for confirmation of the rumour that he heard from Joseph Dalton Hooker that the *Reader* had been sold to members of the Anthropological Society of London (see letter to A. R. Wallace, 22 September [1865]; see also letter from F. H. Hooker, 6 September [1865] and n. 7).

[3] Thomas Bendyshe was both owner and editor of the *Reader* after August 1865 (Sullivan ed. 1984, p. 3).

[4] Wallace refers to three unsigned leading articles in the *Reader*. 'Simeon and simony' was on the selling of positions in the Church of England (*Reader*, 9 September 1865, pp. 279–80). 'Metropolitan sewage' discussed what to do with London's sewage output (*Reader*, 16 September 1865, pp. 307–8). 'France and Mexico' reported on Archduke Maximilian of Austria's becoming emperor of Mexico (*Reader*, 30 September 1865, pp. 363–4).

[5] The review article 'Political economy for the million' discussed the new 'people's edition' issued by the London publisher, Longman, Green, & Co., of works by John Stuart Mill (*Reader*, 23 September 1865, p. 346). Bendyshe had been expelled from the Conservative Club, London, for voting for Mill at the Westminster election of 12 July 1865 (*DNB*). 'Guiseppe Mazzini', the leading article of the same issue, discussed the career of the Italian revolutionary (*Reader*, 23 September 1865, pp. 335–6).

[6] David Masson was the editor-in-chief of the *Reader* from April until the summer of 1863 (Sullivan ed. 1984, p. 349). He was instrumental in increasing the scientific content of the magazine and giving greater coverage to scientific events (Byrne 1964, p. 63).

[7] From December 1864 to July 1865 there was no editor-in-chief of the *Reader* (Sullivan ed. 1984, p. 349). Editorial responsibilities, especially for the scientific content, were divided among a number of persons, including William Fraser Rae and Thomas Henry Huxley (Byrne 1964, p. 63). For more on the ownership and operation of the *Reader*, see the letter from J. D. Hooker, 1 January 1865, n. 14, the letter from Herbert Spencer, 22 April 1865, n.1, and the letter from F. H. Hooker, 6 September [1865], n. 7.

[8] The Anthropological Society of London had been founded by James Hunt in January 1863. Hunt had been a member of the Ethnological Society of London from 1856 and joint secretary from 1860 before resigning in 1863 to found his own competing organisation. While the Ethnological Society numbered many leading Darwinians among its members, the Anthropological Society was strongly opposed to Darwinian views. For the rivalry between the two organisations, see Stocking 1987, pp. 248–54; see also *Correspondence* vol. 12, letter from J. D. Hooker, 9 [March] 1864 and n. 23, and letter from A. R. Wallace, 10 May 1864 and n. 7. Bendyshe was an active member of the Anthropological Society, often translating works from foreign languages to be published by the society (*DNB*; Stocking 1971, p. 376). Wallace himself had been a member of the Anthropological Society when it was first founded and still attended meetings occasionally as well as publishing articles in the *Anthropological Review* (see, for example, A. R. Wallace 1864a; see also Stocking 1987, p. 248).

[9] CD recommended Tylor 1865 and Lecky 1865 to Wallace in his letter to Wallace of 22 September [1865]. For CD's opinion of the books, see the letter from Charles and Emma Darwin to J. D. Hooker, [10 July 1865], and the letter to Asa Gray, 15 August [1865].

[10] In Buckle 1857–61, Henry Thomas Buckle described the advance of civilisation in England. He linked scientific advancement and economic growth with political liberalism in a progressive view of history. CD was also a great admirer of Buckle's work (see letter from J. D. Hooker, 13 July 1865, n. 16).

[11] The reference is to Lubbock 1865; see letter to John Lubbock, 11 June [1865] and n. 4.

[12] Wallace's article, 'On the pigeons of the Malay Archipelago', appeared in *Ibis*, October 1865, pp. 365–400. CD's lightly annotated presentation copy is in DAR 133: 11.

[13] Wallace's catalogue 'On the raptorial birds of the Malayan Archipelago' appeared in *Ibis*, January 1868, pp. 1–27, and April 1868, pp. 215–16. He had already published a number of bird lists based on his collection but never completed a comprehensive catalogue (see C. H. Smith ed. 1991 for a complete bibliography of Wallace's writings). Wallace later wrote that after completing a series of papers on his collections of birds and insects, he had so many interesting areas of research to pursue that he no longer wished to spend time on the mere description of his collections (A. R. Wallace 1905, 1: 404).

[14] CD had recently asked Wallace how his book on his travels was progressing (see letter to A. R. Wallace, 22 September [1865] and n. 3). Earlier, Wallace had written to CD about the sudden

breaking off of his engagement (see letter from A. R. Wallace, 20 January 1865 and n. 1). See also A. R. Wallace 1905, 1: 386.

15 In his letter to Wallace of 22 September [1865], CD had mentioned Fritz Müller's book (Müller 1864) and also the forthcoming second edition of the French translation of *Origin* (Royer trans. 1866).

16 An English translation of Müller 1864 appeared in 1869 (Dallas trans. 1869).

17 Wallace had met Richard Spruce in July 1849 while they were both collecting in the Amazon basin. Spruce developed an intestinal illness while collecting *Cinchona* seeds and plants in Ecuador in 1860; the illness eventually left him unable even to sit up for more than short periods of time. He continued to travel in South America for the next four years, but with ever-worsening health, before returning to England. He received a Civil List pension in 1865. In 1869, his condition improved enough for him to resume his botanical work. See Spruce 1908, 1: xxxiv–xxxv; *Journal of Botany* 32 (1894): 51–2.

To J. D. Hooker 3 October [1865]

Down.
Oct 3^{d}

My dear Hooker

As you asked about the Reader I have thought that you w^{d} like to see the enclosed letter which please return some time[1] When you feel strong enough to read a moderately long letter accompanied with beautiful drawings by Fritz Müller in Brazil on certain climbing plants I shall beg you to read it, for I am in utter perplexity whether it is worth communicating to Linn. Soc.[2] The chief point of interest is, it describes a sub-division of which I saw no instances namely branches being converted into tendrils, but there is of course no sort of hurry. I feel that I am not botanist enough to judge whether it is worth communicating.

Before very long let us hear how you go on. yours very affectionately | C. Darwin

LS
Endorsement: '1865'
DAR 115: 276

1 CD enclosed the letter from A. R. Wallace, 2 October 1865, in which Wallace confirmed that the *Reader* had been sold to a member of the Anthropological Society of London (see letter from J. D. Hooker, 6 October 1865; see also letter from F. H. Hooker, 6 September [1865] and n. 7).

2 CD refers to the letter from Fritz Müller of 31 August 1865, which he had recently received (see letter to Fritz Müller, 17 October [1865] and n. 1). Eventually, CD sent part of that letter to the Linnean Society, along with extracts from two other letters from Müller (see letter from Fritz Müller, [12 and 31 August, and 10 October 1865] and n. 1). For more on the nature of the correspondence between CD and Müller, see the letter from Fritz Müller, 12 August 1865, nn. 1 and 11. The article included a plate by Müller illustrating three figures of branch-tendrils (see p. 217), but Müller's original drawings have not been found.

From Charles Shaw 3 October 1865

Admiral FitzRoy Testimonial Fund. | *55, Charing Cross, London, S.W.*[1]
Oct. 3^{d}. *1865*

Sir,

I have to acknowledge the receipt of your kind note requesting a statement of the objects of the "FitzRoy Testimonial", & the largest amount subscribed in one sum.— [2]

In consequence of Admiral FitzRoy haveing expended the whole of his patrimony, amounting to £6100, & which was never reimbursed to him;—[3] three children by his first wife—two of them daughters, the son a young Lieut. in the Navy, with nothing but his pay—are literally without means.—[4]

M^rs. FitzRoy (his second wife) & her young daughter[5] have a small provision under her marriage settlement.— It is therefore contemplated to raise by this fund a provision for the family of the late Admiral FitzRoy.—

The largest amount at present subscribed is £100 & the smallest ten shillings. I regret I have not a printed statement to send, but I hope the information I have given is what you require[6] I may add that the £3,000 granted by Government will do no more than pay the existing debts.

I have the honour to be | Sir, | Your obedient Servant | Charles Shaw | Hon Sec.

Charles Darwin Esq | &c &c &c

DAR 177: 147

1 The *Post Office London directory* 1865 lists the 'Royal Agricultural Benevolent Institution' and the 'Welsh Charity School, Charles Shaw, esq. sec' at 55 Charing Cross. Evidently, Shaw used his business address for the Admiral FitzRoy Testimonial Fund.

2 CD's letter has not been found. Robert FitzRoy committed suicide on 30 April 1865 (*DNB*; see also letter from J. D. Hooker, 2 May 1865, letter to J. D. Hooker, 4 May [1865], and letter from B. J. Sulivan, 8 May [1865]).

3 FitzRoy had used his own funds to hire vessels and equipment to support his official missions, expecting the outlay to be reimbursed by the government. In 1833, for example, he had bought a large schooner in the Falklands Islands (see *Correspondence* vol. 1, letter to Caroline Darwin, 30 March – 12 April 1833 and n. 8). For more on FitzRoy's expenditures and the Admiralty's refusal to reimburse him, see Mellersh 1968, pp. 104–6, 130–6.

4 FitzRoy had three daughters, Emily-Unah, Fanny, and Katherine, by his first wife, Mary Henrietta O'Brien. Emily-Unah died in 1856 (see letter from J. D. Hooker, 2 May 1865 and n. 4). His only son, Robert O'Brien FitzRoy, entered the navy in 1853, and became a captain in 1872 (*Modern English biography*).

5 FitzRoy married Maria Isabella Smyth, his second wife, in 1854. They had one daughter, Laura Elizabeth FitzRoy (*Burke's peerage* 1878, s.v. Grafton, duke of).

6 CD recorded a payment of £100 to 'Admiral Fitz Roy Testimonial' under the heading 'Charities' in his Classed account book (Down House MS), on 4 October 1865.

To T. H. Huxley 4 October [1865][1]

Down. | Bromley. | Kent. S.E.
Oct 4

My dear Huxley

I was rejoiced to see your hand writing with a good account of your family & of y^r own profound idleness, which must be a novelty to you.[2] I have done nothing since May 1st which is enough almost to break my heart; but I am slowly getting better, & this I owe entirely I believe to Dr Bence Jones who has half starved me to death & reduced my weight 15lb but done me wonderful good.[3]

Poor dear Hooker seems to have had a most serious illness, but he seems as vigorous as ever in mind, judging from a long & pleasant letter I recd from him a few days ago.[4]

I suppose you know about the sale of the "Reader"; but I have heard nothing beyond the bare fact of the sale to Mr Bendyshe.[5] I do not know whether we old shareholders are paid off.[6] I shall regret extremely if the Reader fails as a newspaper for general science.[7]

Pangenesis in my mind is rising a little more into favour after the shock it received from your criticism;[8] but you did me an immense service in making me estimate its value at a very low figure, as indeed must be the case with any mere hypothesis.

I hope you are grappling with your work like a refreshed giant & I dare say you did not quite resist looking at marine animals at little Hampton.[9]

My dear Huxley | yours most sincerely | Ch. Darwin

E.D's love to Mrs Huxley & the dear little people.[10]

LS

Imperial College of Science, Technology and Medicine Archives (Huxley 5: 223)

1 The year is established by the relationship between this letter and the letter from T. H. Huxley, 2 October 1865.

2 See letter from T. H. Huxley, 2 October 1865.

3 CD first consulted Henry Bence Jones in July 1865 (see letter to Asa Gray, 15 August [1865] and n. 12; see also letter to J. D. Hooker, 27 [or 28 September 1865] and n. 14).

4 CD refers to the attack of rheumatic fever suffered by Joseph Dalton Hooker in August 1865 and to the letter from J. D. Hooker, [26 September 1865]. See letter from F. H. Hooker, 6 September [1865] and nn. 2 and 5.

5 The reference is to Thomas Bendyshe (see letter from A. R. Wallace, 2 October 1865 and n. 3).

6 CD's Account book–banking account (Down House MS) records an investment of £80 in the Reader Limited Company on 24 November 1864, but there is no record that any money was repaid to him after the sale of the company (see also *Correspondence* vol. 12, letter to John Lubbock 19 November [1864]).

7 CD had been an enthusiastic subscriber to the journal, recommending it highly (see, for example, *Correspondence* vol. 12, letter to Roland Trimen, 13 May 1864). The last issue of the *Reader* appeared on 12 January 1867 (Sullivan ed. 1984, p. 349).

8 CD had sent Huxley his manuscript of his hypothesis of pangenesis, asking him whether it was suitable for inclusion in *Variation* (see letter to T. H. Huxley, 27 May [1865] and nn. 4–8). Huxley's reply has not been found; however, see the letters to T. H. Huxley, 12 July [1865] and [17 July 1865], and the letter from T. H. Huxley, 16 July 1865.

9 The Huxley family had spent their holiday in Littlehampton, West Sussex (see letter from T. H. Huxley, 2 October 1865 and n. 1).

10 The reference is to Emma Darwin, and to Henrietta Anne Huxley and the Huxley children (see letter from T. H. Huxley, 2 October 1865, n. 1). The letter is in Emma's hand.

To Samuel Butler 6 October [1865][1]

Down. | *Bromley.* | *Kent. S.E.*
Oct 6

My dear Sir

I thank you sincerely for your kind & frank letter which has interested me greatly.[2]

What a singular & varied career you have already run. Did you keep any journal or notes in N. Zealand for it strikes me that with your rare powers of writing you might make a very interesting work descriptive of a colonist's life in N. Zealand.[3]

I return your printed letter which you might like to keep.[4] It has amused me, especially the part in which you criticize yourself.[5] To appreciate the latter part fully I ought to have read the Bishop's letter which seems to have been very rich.[6] You tell me not to answer your note but I c[d] not resist the wish to thank you for your letter.

With every good wish | believe me my dear Sir | yours sincerely | Ch. Darwin

LS
British Library (Add. MS 34486, f. 56)

[1] The year is established by the relationship between this letter and the letter from Samuel Butler, 1 October 1865.
[2] See letter from Samuel Butler, 1 October 1865.
[3] In 1863, Butler's father, Thomas Butler, published *A first year in Canterbury settlement* (reprinted in Butler 1923), a book compiled from Samuel Butler's journal and letters to his father (see Butler 1923, p. xxi).
[4] Butler had sent CD a copy of his letter to the *Press*, Christchurch, New Zealand (see enclosure to letter from Samuel Butler, 1 October 1865).
[5] Butler had written a previous letter to the *Press*, which had received a critical reply; Butler then wrote again as though he were a third party, commenting on both letters (see enclosure to letter from Samuel Butler, 1 October 1865 and n. 7).
[6] CD refers to Charles John Abraham, the bishop of Wellington, and to the letter signed 'Barrel-organs', written as a response to Butler's original article of 20 December 1862 (see letter from Samuel Butler, 1 October 1865 and n. 5).

From J. D. Hooker 6 October 1865

7 Terrace Road | Buxton
Oct 6th /65.

Dear Darwin

I should have answered your last ere this, if only to say how glad I am that Jones has done some good,[1] Oh that it may last— First as to my ignoble self, I really improve fast & steadily & if I could but get rid of the slight stiffness & pains in all my joints would be well— they go slowly & will all be gone in time[2] I wrote to the Board asking an extension of leave till 20th & they volunteer till end of month—[3] this good feeling is unusual in Boards & gratifies me proportionally, it is a good augury (or whatever it is called)

Now for Novels— I read Silas Marner the other day & did not enjoy it—[4] after the quaking excitement of Uncle Silas & the love scenes of Mill on Floss, S.M, read flat & awfully Eliotian: too didactic & prosy without plot enough or incident enough— (how comparative all our feelings are!).[5] Have you read "Trevlyn Hold"[6] it is really very good: we both tried Scarsdale,[7] & found it execrable trash—& now for a confession I have read Clarissa Harlowe![8] I feel that this is self damnatory & can only plead my illness & the tedium of a Watering place. as however "frank confession is good for the soul," I will tell you The first 5 volumes are simply illegible, so dull so poor, so attenuated; that had I stopped there I should have considered the former popularity of the book as one of those things which "no fellow can be expected to understand" as Uncle Sam has it:[9] the 6th & 7th (horresco referens)[10] opened my eyes however; though to me they had no merit or interest whatever as a tale, I could quite understand the deep interest they must have had in an artificial & vicious age when alone such compositions could be put by mothers into the hands of virtuous daughters, with injunctions to *study them* & the *immense* good they may have done. In an age when men of fashion had no honor & when the prejudices of Education or absence of it & want of public journals kept women in the dark as to the means men employed, & when maudlin sensational writing *did* act on the brain in a way it does not now; it is obvious to me that Richardsons works must have frightened hosts of young women into caution at any rate, & stimulated a few to good works.[11] Be this as it may, there is no doubt I suppose that his works were perused by thousands as standard literature for young ladies in 1750–1770; & that the change of manners was so rapid, that in 1780 I find by the life of Reynolds (I am ashamed of owning that I have been reading a solid book) both Richardsons & Fieldings works were considered as too coarse for young ladies.[12]

I could not get beyond the first volume of Palgraves book, he is awaiting orders still at Cairo.[13] I must read Millers address,[14] I missed it. Trollope is the only Novelist I know who talks of Parliament as such a stunning walk & enviable life.[15] I can quite feel the abounding self-love that would follow a telling speech (& oh how nice self-love is) & that to rise to Gladstones, or Derbys or even Dizzys heights would be irresistable to most men;[16] but for a really able man, like Lubbock, to be 3^{d} rate in the house is to me an intolerable idea, & I do not see how he can be anything higher without he actually proposes to abandon business, science, & domestic happyness.[17] As to Jeffrey he speaks from Edinbro' & no doubt thought, in common with his townsmen that the Edinbro law court, (I forget its name) where he was at the top of the tree, was next thing to the H. of Commons.[18] There local allusions & local ideas & prejudices, expressed in strong broad Scotch, carried the day. Had he gone into Parliament he would have had to unlearn for 3 years; he never suspected this.[19] I quite agree that his view is poor & short-sighted

Many thanks for enclosed of Wallaces[20] I did not think either "Simeon & Simony" nor "France & Mexico" very good,[21] the first my wife condemned, the second I thought actually poor & pointless.— so much for opinions I thought the old Reader bad enough & this worse in as much as it has less real Science[22] As to calling Anthropologists a *bete noire* to Reader why so it is, only last number they had some 3 or 4 columns of Review of the Anthrops publications, & in a former N° condemned the Brit. Assoc for refusing an Anthrop. section.[23] Wallaces judgment of Tylor is unfair, the work is confessedly imperfect & fragmentary & must be so in present state of knowledge[24] I doubt if Buckle will liberalise opinion so much as Lecky.[25] It is all very easy for Wallace to wonder at Scientific men being afraid of saying what they think—[26] he has all "the freedom of motion in vacuo" in one sense, had he as many kind & good relations as I have, who would be grieved & pained to hear me say all I think, & had he children who would be placed in predicaments most detrimental to childrens minds by such avowals on my part, he would not wonder so much. Wallace is not a man of large sympathies, nor very charitable I think, & is certainly awfully cold & dry at times; yet he is essentially large minded, & *very* able I hope you saw Seemann's sneers at the "Origin" in his Report of the German Congress,[27] & trembled accordingly.

We leave this on Friday next for Lea Hurst, near Matlock, Mr Nightingale's where we stay quietly till Monday,[28] it is warmer than this: then we go to Liverpool to visit an Uncle & home by Chester, to Kew about the 20th. What a heap of Darwins & Wedgwoods are here![29]

I am gratified by your expressions about my father—[30] he was one of the most truly liberal & modest men I ever knew— he had not an atom of self in him, always thought nothing of himself & never took any self seeking steps to raise himself in the estimation of the Government or of scientific men. With $\frac{1}{10}$th of the exertion that Murchison displayed, he would have had honors & titles showered on him: & I *hate* the Rl. Socy for never recognizing the obligations science is under to him.[31] He never received any honor distinction or reward from the Crown or Govt. for all his public services, because he never would put himself into the way of them.[32] I thought the boast of the R. S. was that they sought out such as had similar claims upon science. I know I am not agreed with but I will not give in

Send Fritz Mueller paper to Kew & I will see to it, if I can.[33]

Ever Yr affec | J D Hooker

DAR 102: 37–42

[1] In his last letter, CD had informed Hooker of the improvement in his health since he consulted Henry Bence Jones (see letter to J. D. Hooker, 27 [or 28 September 1865] and n. 14).

[2] Hooker was recovering from a severe attack of rheumatic fever (see letter from F. H. Hooker, [17 August 1865], and letter from J. D. Hooker, [26 September 1865]).

[3] The Board of Works and Public Buildings had jurisdiction over the Royal Botanic Gardens, Kew. Hooker took up his appointment as director of Kew on 1 November 1865 (Allan 1967, p. 211; R. Desmond 1995, p. 223). He succeeded his father, William Jackson Hooker, who had died on 12 August 1865 (see letter to J. D. Hooker, 16 August [1865] and n. 1).

[4] Hooker refers to Eliot 1861. CD had recommended the novel to Hooker, offering to lend his copy (see letter to J. D. Hooker, 27 [or 28 September 1865]).

[5] Hooker refers to Le Fanu 1864 and Eliot 1860 (see letter from J. D. Hooker, [26 September 1865] and nn. 3 and 4). In contrast to Hooker, CD preferred *Silas Marner* (Eliot 1861) to *Mill on the Floss* (Eliot 1860); see letter to J. D. Hooker, 27 [or 28 September 1865].

[6] Wood 1864.
[7] Kay-Shuttleworth 1860.
[8] Hooker refers to *Clarissa* (S. Richardson 1747–8). CD had read Samuel Richardson's seven-volume work as an undergraduate at Cambridge and thought it 'the most glorious novel ever written'. See *Correspondence* vol. 1, letter to W. D. Fox, [3 January 1830].
[9] 'Uncle Sam' may be a reference to Samuel Richardson.
[10] *Horresco referens*: I shudder to relate.
[11] For more on the cautionary themes in Richardson's novels, see Doody 1974, pp. 128–50.
[12] Hooker refers to a passage in a biography of Joshua Reynolds (Leslie and Taylor 1865, 2: 203) in which the works of Richardson and Henry Fielding are mentioned.
[13] Hooker refers to Palgrave 1865, and to his cousin, William Gifford Palgrave. See letter to J. D. Hooker, 27 [or 28 September 1865] and n. 19. Palgrave was in the British diplomatic service and travelled extensively in the Middle East (*DNB*).
[14] William Hallowes Miller delivered the presidential address to the chemical section of the British Association for the Advancement of Science meeting at Birmingham in 1865 (see letter to J. D. Hooker, 27 [or 28 September 1865] and n. 21).
[15] CD had remarked on the view expressed in *Can you forgive her?* (Trollope 1864–5) that attaining a seat in parliament was the highest earthly ambition (see letter to J. D. Hooker, 27 [or 28 September 1865] and n. 29). Parliamentary life was a central theme of a number of Anthony Trollope's novels, many of which were written around the time of his own failed attempt to enter parliament (*DNB*).
[16] William Ewart Gladstone was chancellor of the Exchequer and, from October 1865, leader of the House of Commons; Edward George Geoffrey Smith Stanley, fourteenth earl of Derby, was leader of the Conservative party and had been prime minister from 1858 to 1859; Benjamin Disraeli was leader of the opposition, and had been chancellor of the Exchequer and leader of the House of Commons during Stanley's ministry (*DNB*).
[17] John Lubbock stood for West Kent in the general election of 1865. He was defeated on 22 July (*The Times*, 24 July 1865). Both Hooker and CD were dismayed at the thought of Lubbock entering politics, thinking that his work in science would suffer (see letter from J. D. Hooker, [7–8 April 1865] and n. 14, letter to John Lubbock, 11 June [1865], and letter to J. D. Hooker, [29 July 1865]).
[18] Hooker refers to Francis Jeffrey, who was judge of the Court of Sessions in Edinburgh from 1834 to 1850 (*DNB*). See letter to J. D. Hooker, 27 [or 28 September 1865] and n. 30.
[19] Jeffrey had, in fact, been a member of parliament and lord advocate from 1831 to 1834 (*DNB*; see letter to J. D. Hooker, 27 [or 28 September 1865]). Jeffrey's biographer records that he was not considered an effective speech-maker in parliament because of the weakness of his voice and tendency to speak either too quickly or too slowly (Cockburn 1852, 1: 314, 356–8).
[20] CD had enclosed the letter from A. R. Wallace of 2 October 1865 with his letter to J. D. Hooker, 3 October [1865].
[21] Hooker refers to two of the articles in the *Reader* mentioned by Wallace as probably the work of Thomas Bendyshe (see letter from A. R. Wallace, 2 October 1865 and n. 4).
[22] Wallace had informed CD that the new editorial policies of the *Reader* under Bendyshe would apparently include equal emphasis on political, literary, and scientific articles (see letter from A. R. Wallace, 2 October 1865). In fact, the scientific content declined noticeably during Bendyshe's editorship (Byrne 1964, p. 113).
[23] The first volume of the periodical publication *Memoirs read before the Anthropological Society of London*, which was published in 1865, was reviewed in the *Reader*, 30 September 1865, pp. 377–80. In a review of the events at the British Association for the Advancement of Science meeting, the rejection of the proposal of a new anthropological section was critically discussed in the *Reader*, 9 September 1865, pp. 292–3. The article, signed J.N.L., was probably by Joseph Norman Lockyer (North 1997, 5: 4066). Lockyer had reported on British Association meetings for the *Reader* since 1863 (Byrne 1964, p. 85).
[24] Wallace had given a negative opinion of Tylor 1865 in his letter to CD of 2 October 1865. For Hooker's earlier recommendation of Tylor 1865, see the letters from J. D. Hooker, [26 May 1865] and n. 13, and [15 June 1865].

25 For Wallace's views on Lecky 1865 and Buckle 1857–61, see the letter from A. R. Wallace, 2 October 1865.

26 Wallace's complaint had been particularly directed at the final chapter of Lubbock 1865 (see letter from A. R. Wallace, 2 October 1865).

27 A report of the meeting of German natural scientists and physicians at Hanover from 18 to 24 September 1865 appeared in the *Athenæum*, 30 September 1865, pp. 435–6. Hooker evidently surmised that the article, signed 'B.', was by Berthold Carl Seemann. Seemann's name appears on the list of members attending the meeting (*Amtlicher Bericht über die vierzigste Versammlung deutscher Naturforscher und Ärzte zu Hannover im September 1865*, p. 26), and he reported on the meeting in the *Journal of Botany*, November 1865, p. 359. The *Athenæum* article noted the popularity of an anti-Darwinian talk and remarked that other 'independent speakers' were 'beginning to perceive the fallacy of the Darwinian hypothesis' (*Athenæum*, 30 September 1865, p. 435).

28 Lea Hurst, about two miles from the spa town of Matlock, Derbyshire, was the seat of William Edward Nightingale (*DNB*).

29 Hooker's Liverpool uncle was his mother's brother Dawson William Turner, headmaster of the Royal Institution School in Liverpool (L. Huxley ed. 1918, 1: 19). CD's grandfather, Erasmus Darwin, had settled in Derby, near Buxton, in 1783 after his second marriage. Two of CD's half cousins, Edward Levett Darwin and Reginald Darwin, lived in Buxton (*Post Office directory of Cambridgeshire, Derbyshire, Nottinghamshire, Leicestershire, & Rutlandshire* 1864). Since Erasmus Darwin had recommended the waters at both Buxton and Matlock to Josiah Wedgwood I, the area had been a popular retreat for the Wedgwood family (King-Hele 1999, p. 110).

30 In his letter to J. D. Hooker of 27 [or 28 September 1865], CD wrote of his admiration for William Jackson Hooker, Hooker's father, who had died in August (see letter from F. H. Hooker, [17 August 1865]).

31 W. J. Hooker was never honoured by the Royal Society of London, unlike Roderick Impey Murchison, who received the Copley Medal in 1849. For Hooker's opinion of Murchison, see the letter from J. D. Hooker, 3 February 1865 and n. 8.

32 W. J. Hooker was honoured for his services to botany in 1836 when he was made Knight of the Order of Hanover (Allan 1967, pp. 94–5).

33 CD had asked whether Hooker would review a letter he had recently received from Fritz Müller, which contained observations on climbing plants (see letter to J. D. Hooker, 3 October [1865] and n. 2; see also letter from Fritz Müller, [12 and 31 August, and 10 October 1865]).

To Jeffries Wyman 8 October [1865][1]

Down Bromley | Kent.
Oct. 8

My dear Sir

I am very much obliged for your letter.[2] I have tried the little experiment with the string, & it answered well;[3] as did a very long narrow strip of elastic paper after having been scraped or compressed by being rubbed over a knife. I cannot see the difficulty of the mathematicians, the slip of paper when held at both ends, tends to form a regular bow, but as every part tends to contract into a small circle, if the contraction be not quite regular one part can pass by the side of the other, & the whole will break up into a set of circles or spires, which will be reversed at the 2 ends on the self-twisting principle which I have explained.

I am aware that you are much interested on the movement of plants for I have read your excellent paper on the bursting of certain gourds.[4] I suppose you know

Cohn's paper on the contraction of the stamens of certain Compositæ:[5] I have seen the phenomenon & was much interested by the paper.

I formerly made numerous observations shewing what an extraordinary small pressure is sufficient in certain cases to excite movement & as I believe contraction in the cells of some plants; I likewise found that certain re-agents such as Strychnine &c had a powerful influence on the movements.[6] But my health has been so weak for several years that I have not been able to publish these observations,[7] & I hardly know why I have mentioned them to you.

With many thanks for your note; & with sincere respect believe me | My dear Sir | Yours sincerely | Charles Darwin

LS(A)
The Harvard Medical Library in the Francis A. Countway Library of Medicine, Boston

[1] The year is established by the relationship between this letter and the letter from Jeffries Wyman, 22 September 1865.
[2] Letter from Jeffries Wyman, 22 September 1865.
[3] Wyman had described a method whereby the spiral coiling of a tendril could be demonstrated; CD had written a lengthy account of the phenomenon in 'Climbing plants', pp. 95–8 (see letter from Jeffries Wyman, 22 September 1865 and n. 3).
[4] CD refers to Wyman 1854, a short paper on contractility in the capsules of two plants (see *Correspondence* vol. 11, letter to Daniel Oliver, [before 27 November 1863] and n. 2).
[5] CD refers to the article by Ferdinand Julius Cohn, 'Ueber contractile Gewebe im Pflanzenreich' (Cohn 1860). CD's heavily annotated copy of the article is in the Darwin Pamphlet Collection–CUL. An English abstract of the article was published in the *Annals and Magazine of Natural History* 11 (1863): 188–202. CD referred to both the original article and the abstract in *Insectivorous plants*, pp. 256, 364.
[6] CD is probably referring to his experiments with the insectivorous plant *Drosera rotundifolia*, which he began in 1860 (see *Correspondence* vol. 8). In 1862 he performed experiments to test the effect of various chemicals, including strychnine, on leaf movement (see *Correspondence* vol. 10, letter to J. D. Hooker, 26 September [1862] and n. 1). CD's notes on the 1862 experiments are in DAR 54: 29–49.
[7] CD returned to his work on insectivorous plants in 1872 (*LL* 3: 322), and published his findings in 1875 as *Insectivorous plants.*

From Fritz Müller[1] 10 October 1865

Desterro, Brazil,
10. October 1865.

Verehrter Herr!

Gestern empfing ich Ihren freundlichen Brief vom 10. August[2] und fühle mich sehr geschmeichelt durch das Interesse, welches Sie an meinem kleinen Buch über Arten genommen haben;[3] ich fühle mich reich belohnt, wenn es Ihnen einiges Vergnügen gemacht hat. Ich danke Ihnen herzlich, dass Sie mir Ihre Photographie gesandt haben,[4] ich bin sehr glücklich, sie zu besitzen, und hoffe bald im Stande zu sein, Ihnen die meinige zu senden; zur Zeit habe ich keine. Ich nehme mit Dank Ihr Anerbieten an, dass Sie mir Ihr Werk über Orchideen senden wollen, ich würde es sehr gerne lesen; da Bücher immer etwas von ihrem Interesse durch Uebersetzung verlieren, so würde ich die englische Ausgabe der deutschen vorziehen.[5] Als Ihr

Brief ankam, schrieb ich grade die letzten Zeilen einer kleinen Abhandlung über das Holz von Kletterpflanzen. In der Voraussetzung, dass es Sie etwas interessieren wird, lege ich die Abhandlung diesem Briefe bei.[6] Wenn Sie davon Gebrauch gemacht haben, so schicken Sie sie bitte an Prof. Max Schultze von der Universität Bonn, oder wenn Sie einen Uebersetzer haben, könnte die Arbeit vielleicht in einer englischen Zeitschrift veröffentlich werden.[7] Prof. Max Schultze wird Ihnen seiner Zeit eine kleine Abhandlung über eine bermerkenswerthe Spongie schicken, welche in seinem Archiv für mikroskopische Anatomie veröffentlich wird.[8] Als ich im vergangenen Jahre anfing, die Spongien unserer Küste zu studiren, begegnete ich grossen Schwierigkeiten bei der Anwendung Ihrer Theorie auf jene Klasse. Es giebt, wie Sie wissen, kalkige spiculae bei manchen Spongien und kieselige bei anderen; ihre Formen sind bisweilen so ähnlich, dass man kaum vermuthen konnte, sie seien unabhangig von einander entstanden; aber auf der andern Seite schien es sehr unwahrscheinlich, dass sich kieselige spiculae schrittweise sollten in kalkige umgewandelt haben oder umgekehrt. Am wahrscheinlichsten schien mir die Annahme, dass beide sich herleiteten von einer Urform der spiculae, welche aus organischer Substanz gebildet waren und sich später bei einigen der Nachkommen mit kalkiger Substanz, bei anderen mit kieseliger inkrustirten. In dieser Ansicht wurde ich bestärkt durch die Entdeckung der in meiner Abhandlung beschriebenen Art, welche sehr schöne hornige spiculae von beträchtlicher Grösse hat. . . .[9] [missing text][10]

. . . Meine Kinder hatten eine einzelne Pflanze von Linum usitatissimum gezogen, und bei dieser bemerkte meine Tochter Rosa,[11] dass die Spitze des Stengels, welche vor dem Oeffnen der Blüthen etwas geneigt war, eine eigene revolutirende Bewegung hatte, welche der Sonne folgte, oder, da wir hier auf der südlichen Halbkugel leben, derjenigen der Uhrzeiger entgegengesetzt war.[12] Ich habe die Thatsache bestätigt gefunden und gesehen, dass ein Umlauf in ungefähr acht Stunden vollendet wurde; die Bewegung war ziemlich unregelmässig; bisweilen hielt sie eine Stunde lang still, dann wieder war sie so schnell, dass sie bei Innehaltung dieses Zeitmaasses den Kreislauf in 2–3 Stunden vollendet haben würde. Die folgenden Tage waren sehr windig, so dass ich meine Beobachtungen nicht fortsetzen konnte.

Was Anelasma betrifft, so werden Sie aus meinem ersten Briefe ersehen haben, dass auch ich diese Form betrachte als ein schönes verbindendes Glied zwischen den echten Cirripedien und den Rhizocephalen[13] oder den Cirripedia suctoria (?), wie sie von Lilljeborg genannt werden,[14] obwohl sie weder Cirri noch irgendwelche "Suctorial apparatus" besitzen.

Mit dem Wunsche, dass dieser Brief Sie in guter Gesundheit treffen möge, bin ich, werther Herr, in aufrichtiger Hochachtung | treulichst der Ihrige | Fritz Müller.

Incomplete[15]
Möller 1915–21, 2: 74–6.

[1] See letter from Fritz Müller, 12 August 1865, n. 1. For a translation of this letter, see Appendix I.
[2] See letter to Fritz Müller, 10 August [1865].
[3] Müller refers to Müller 1864 (see letter to Fritz Müller, 10 August [1865], n. 2).

[4] CD had probably sent the photograph taken by his son William Erasmus Darwin in 1864. The photograph is reproduced as the frontispiece to *Correspondence* vol. 12.

[5] CD had offered Müller a copy of the German translation of *Orchids* (Bronn trans. 1862) in his letter to Müller of 10 August [1865]. In his letter of 20 September [1865], CD wrote that he had sent it.

[6] The reference is to the manuscript of an article later published in *Botanische Zeitung* (Müller 1866). CD's annotated copy of the published article, with a separate sheet of notes in his hand, is in the Darwin Pamphlet Collection–CUL.

[7] Max Johann Sigismund Schultze was the editor of the *Archiv für mikroskopische Anatomie* and had been a friend of Müller's since 1846, when they were both students at the University of Greifswald (see Möller ed. 1915–21, 3: 25). CD later told Müller that he had immediately forwarded the manuscript of Müller's essay on the wood of twining plants to Schultze (see letter to Fritz Müller, 9 December [1865]). In a letter of 12 December 1865, Müller asked Schultze to try to get the essay published in a botanical journal (see Möller ed. 1915–21, 2: 74, 78).

[8] The reference is to Müller 1865c. An offprint of the article, with a note in CD's hand on the front cover, is in the Darwin Pamphlet Collection–CUL.

[9] Müller named the sponge *Darwinella aurea* (see F. Müller 1865c, p. 344). The genus *Darwinella* belongs to the order Dendroceratida, whose members are non-calcareous, and which is characterised by a skeleton composed entirely of spongin fibres (horny spicules) and lacking siliceous spicules (Bergquist 1978, p. 176).

[10] The letter is incomplete. CD extracted the sections of this letter on climbing plants and included them in an article that appeared in the *Journal of the Linnean Society* (*Botany*). This article is reproduced as the letter from Fritz Müller, [12 and 31 August, and 10 October 1865]; see also letter from Fritz Müller, 12 August 1865, n. 11.

[11] Rosa Müller.

[12] Müller later published the observation on *Linum usitatissimum* (common flax; Müller 1870, p. 137); CD cited the article in *Movement in plants*, p. 203. CD's annotated copy of Müller 1870 is in the Darwin Pamphlet Collection–CUL.

[13] Müller's observations on *Anelasma* were in a missing section of the letter from Fritz Müller, 12 August 1865; see, however, the letter to Fritz Müller, 20 September [1865] and nn. 9 and 10.

[14] The Swedish naturalist Wilhelm Lilljeborg coined the name 'Cirripedia Suctoria' to describe the genera *Liriope* and *Peltogaster*, from 'their sucking their nourishment from the animal to which they are attached' (see Lilljeborg 1859 and Lilljeborg 1860, p. 173). *Peltogaster* is now classified as a genus of Rhizocephala; *Liriope* is a genus of the order Isopoda, and is a hyperparasite (parasite upon a parasite) of some Rhizocephala.

[15] See n. 10, above.

To Frederick William Farrar 11 October [1865][1]

Down. | *Bromley.* | *Kent. S.E.*
Wed. Oct 11

Dear Sir

I am very much obliged to you for your kind present of your work on Language.[2] No subject in my opinion can be more interesting, & I hope soon to read your book, but I am not sure that I shall be able as my health is at present weak—

From some of your papers which I have read I feel certain that your opinions on all subjects would be most liberal & fair.[3]

With my best thanks | I remain yours faithfully | Ch. Darwin

LS
American Philosophical Society (Getz 4913)

1 The year is established by the relationship between this letter and the letter from F. W. Farrar, 6 November 1865.

2 The reference is to *Chapters on language* (Farrar 1865). CD's presentation copy is in the Darwin Library–Down (see *Marginalia* 1: 226). CD later cited Farrar's book in *Descent* 1: 56 n.

3 CD may be referring to papers by Farrar that appeared in the *Transactions of the Ethnological Society of London* and the *Anthropological Review* (Farrar 1864a, 1864b, and 1864c). CD referred to Farrar 1864b in *Descent* 1: 65 n.

To J. T. Moggridge 13 October [1865][1]

Down Bromley
Oct. 13.

My dear Sir

I am especially obliged to you for your beautiful plates and letter press;[2] for no single point in natural history interests and perplexes me so much as the self-fertilisation of the Bee orchis.[3] You have already thrown some light on the subject and your present observations promise to throw more.[4]

I formed two conjectures first, that some insect during certain seasons might cross the plants, but I have almost given up this; nevertheless pray have a look at the flowers next season. Secondly I conjectured that the Spider and Bee orchises might be a crossing and self-fertile form of the same species.[5] Accordingly I wrote some years ago to an acquaintance asking him to mark some Spider orchises and observe whether they retained the same character;[6] but he evidently thought the request as foolish as if I had asked him to mark one of his cows with a ribbon to see if it would turn next Spring into a horse. Now will you be so kind as to tie a string round the stem of half a dozen Spider orchises and when you leave Mentone[7] dig them up and I would try and cultivate them and see if they kept constant; but I should require to know in what sort of soil and situations they grow. It would be indispensable to mark the plant so that there could be no mistake about the individual. It is also just possible that the same plant would throw up at different seasons, different flower scapes, and the marked plants would serve as evidence.

With many thanks, my dear Sir, | Your's sincerely | Ch. Darwin

P.S. | I send by this post my paper on climbing plants parts of which you might like to read.[8]

Copy
DAR 146: 374

1 The year is established by the reference to 'Climbing plants' (see n. 8, below).

2 Moggridge sent CD three plates numbered XLIII to XLV, with descriptive text, from his book *Contributions to the flora of Mentone* (Moggridge 1865–8). The plates showed different forms of *Ophrys insectifera*, under which name Moggridge included *O. aranifera* (the spider ophrys), *O. arachnites* (the late spider ophrys), *O. scolopax*, and *O. apifera* (the bee ophrys). See also *Correspondence* vol. 12, letter to J. T. Moggridge, 19 June [1864] and n. 7. The plates and text, annotated by CD, are in the Darwin Pamphlet Collection–CUL. In 'Fertilization of orchids', p. 145 (*Collected papers* 2: 142), and in *Orchids* 2d ed., p. 58, CD referred to these plates and discussed Moggridge's view that the above-mentioned species were all varieties of *O. insectifera.*

[3] CD had remarked on the self-pollination of the bee orchis or ophrys (*Ophrys apifera*) in *Orchids*, pp. 63–72. He noted that, in addition to the adaptations that facilitated self-pollination, this orchid retained features associated with insect pollination, suggesting that 'perhaps at very long intervals of time', one individual of *O. apifera* was crossed by another (*ibid.*, p. 71; see also *Correspondence* vol. 9, letter to A. G. More, 17 July 1861 and n. 2). In *Orchids* 2d ed., p. 58, CD speculated on whether this orchid would ever either revert to a cross-pollinating form or become extinct. CD's notes on *Ophrys* are in DAR 70.

[4] In 'Observations on some orchids of the south of France', Moggridge had observed two forms of *Ophrys scolopax*, one at Mentone that was never self-pollinated, and another at Cannes, where all the flowers he observed were self-pollinated (Moggridge 1864). See *Correspondence* vol. 12, letter to J. T. Moggridge, 19 June [1864] and n. 6.

[5] CD evidently now doubted his earlier conclusion (see above, n. 3) about the possibility of cross-pollination in *Ophrys apifera*; he refers here also to the late spider ophrys (*O. arachnites*), calling it the spider orchid. In 1863 he had asked Philip Henry Gosse to observe whether insects ever removed pollinia from this species, referring to it as his 'greatest puzzle' (see *Correspondence* vol. 11, letter to P. H. Gosse, 2 June [1863] and nn. 12 and 13). He added that he had recently confirmed the view, earlier expressed in *Orchids*, p. 73, that *O. arachnites* was not a mere variety of *O. apifera* as it was not capable of self-pollination. For CD's speculations on the relation between these two species, see *Correspondence* vol. 10, letter to A. G. More, 7 June 1862 and n. 2. In *Orchids* 2d ed., p. 57, CD returned to the view that insects might 'at rare intervals transport the pollinia from plant to plant' in *O. apifera*; there are also relevant annotations on CD's copy of Moggridge 1869.

[6] The acquaintance may have been George Chichester Oxenden, who had provided CD with many orchid specimens, but no letter making such a request has been found (see however *Correspondence* vol. 10, letter from G. C. Oxenden, 8 July 1862 and n. 2).

[7] Owing to chronic ill health, Moggridge spent most winters at Mentone (now Menton), a town on the French Riviera near the Italian border (R. Desmond 1994).

[8] The reference is to 'Climbing plants', published in a double issue of the *Journal of the Linnean Society (Botany)* on 12 June 1865. No presentation list for this paper has been found.

From J. T. Moggridge 14 October [1865][1]

32 Eastbourne Terrace
Oct. 14

Dear Sir

Your most kind & gratifying letter arrived last night, accompanied by your pamphlet on climbing plants.[2]

Thank you very much for it, & also for two other similar tho smaller treatises which after much delay reached me from my old home at Swansea—[3]

I shall take the greatest pleasure in endeavouring to execute your commission; &, should I be able to manage it, I shall send an instalment coming into flower by rail or private hand as well.[4]

With many warm thanks | believe me | yours very sincerely | J Traherne Moggridge.

DAR 171.2: 203

[1] The year is established by the relationship between this letter and the letter to J. T. Moggridge, 13 October [1865].

[2] 'Climbing plants' (see letter to J. T. Moggridge, 13 October [1865] and n. 8).

[3] CD probably sent copies of his papers 'Three forms of *Lythrum salicaria*' and 'Two forms in species of *Linum*' to The Willows, Swansea, the address Moggridge had written from in the summer of 1864 (see *Correspondence* vol. 12, letter from J. T. Moggridge, 19 July [1864]). Moggridge's name appears on CD's presentation list for 'Three forms of *Lythrum salicaria*' (see *Correspondence* vol. 12, Appendix III).

[4] CD had asked Moggridge to send him live specimens of spider orchids (*Ophrys arachnites*; see letter to J. T. Moggridge, 13 October [1865]).

From Hermann Kindt 17 October 1865

Bampton, Penrith
17 October '65.

Sir,

You may, perhaps, remember your kindly complying with a wish of mine respecting an autograph of yours for a German friend;[1] and I should feel greatly indebted to you if you would grant me a similar request, i.e. this time for publicity. I have forwarded to you a journal, which I beg you will not take the trouble of returning, for which I should be happy to have a page in your handwriting for a facsimile reproduction.[2] In case you will kindly comply with my wish, will you please to extract a longer passage from your "Origin of Species", written on a piece of paper in the size of a page of our publication and signed with your esteemed name. As the journal also appears on the Continent, your numerous friends there would be much pleased to see your handwriting thus reproduced. You will also oblige me by looking over the enclosed biographical notice intended for a similar purpose and by altering dates and such passages as you cannot approve of.[3]

I shall receive your kind answer and favour with the warmest gratitude, and ever remain, with the expression of greatest admiration for your enlightening works, | Sir, | your's very faithfully | Hermann Kindt

Charles Darwin Esq. | M.A., F.R.S., etc

DAR 169: 15

[1] Kindt had requested CD's autograph, together with a few lines of his handwriting, for an unnamed German friend. CD sent his signature and copied the last sentence of *Origin* (see *Correspondence* vol. 12, letter from Hermann Kindt, 16 September 1864, and letter to Hermann Kindt, 17 September 1864).

[2] Kindt refers to the *Autographic Mirror* (*L'Autographe Cosmopolite*). The journal was a bilingual publication containing facsimile reproductions of letters and other documents by prominent people, together with biographical sketches of the authors. No copy of this journal has been found in the Darwin Pamphlet Collection–CUL, the Darwin Library–CUL, or the Darwin Library–Down.

[3] For the passage CD sent, see the letter from Hermann Kindt, 23 October 1865, n. 2; the biographical notice is in the *Autographic Mirror* 3 (1865): 82–3.

To Fritz Müller 17 October [1865]

Down. | Bromley. | Kent. S.E.
Oct 17.

My dear Sir

I received about a fortnight ago your second letter on climbing plants, dated Aug. 31st.[1] It has greatly interested me, & it corrects & fills up a great hiatus in my paper.[2] As I thought you could not object, I am having your letters copied & will send the paper to the Linnæan Soc.[3] I have slightly modified the arrangement of some parts & altered only a few words, as you write as good English as an Englishman. I do not quite understand your account of the arrangement of the leaves of Strychnos, & I think you use the word Bracteæ differently to what English Authors do; therefore I will get Dr Hooker to look over your paper.[4]

I cannot of course say whether the Linnæan Soc. will publish your paper, but I am sure it ought to do so. As the Soc. is rather poor I fear that it will give only a few wood cuts from your truly admirable sketches.[5] In my last letter I suggested your reading the Chap. on Catasetum in my Orchid book but I accidentally forgot to give the reason, viz that Dr Crüger of Trinidad fully confirms all that I say, & has sent me Humble bees with the pollinia adhering to their hairy bodies.[6] Hence the force of the ejection. He further states that, as I anticipated the bees come to gnaw the fleshy labellum.[7] I have now no doubt that all the extraordinary protuberances on the labella of exotic orchids serve to attract insects like the flesh round the stone of the cherry. But it is hardly worth while my troubling you with these details. If your paper is printed I will send you copies.

Believe me my dear Sir | yours very sincerely | Ch. Darwin

My health keeps so weak, that I can do nothing.

LS(A)
Endorsement: 'Received Novbr 23–65'
British Library (Loan 10: 3)

1 CD refers to the letter from Fritz Müller, 31 August 1865. Müller's first letter on climbing plants was the letter from Fritz Müller, 12 August 1865. See the letter from Fritz Müller, [12 and 31 August, and 10 October 1865], for the missing portion of the letter of 31 August 1865 containing Müller's botanical observations.

2 Müller's letters contained information on branch-climbers; CD had not found tendrils modified from branches in the course of his research for 'Climbing plants' (see letter to J. D. Hooker, 3 October [1865]).

3 Parts of Müller's first two letters, along with further observations from Müller's third letter of 10 October 1865, were submitted by CD to the *Journal of the Linnean Society (Botany)* and are reproduced in this volume as the letter from Fritz Müller, [12 and 31 August and 10 October 1865].

4 See letter to J. D. Hooker, 3 October [1865]. CD later sent the letter to Daniel Oliver, who made corrections (see letter to Daniel Oliver, 20 October [1865], and letter from Daniel Oliver, 23 October 1865). The term 'bracteæ' does not appear in the published version of the Müller letter; the description of *Strychnos* mentions 'rudimentary leaves' at the tip of each tendril (see Müller 1865b, p. 344, and letter from Fritz Müller, [12 and 31 August, and 10 October 1865]).

5 Only one plate accompanied the published paper (see Müller 1865b and letter from Fritz Müller, [12 and 31 August, and 10 October 1865] and n. 5).

6 See letter to Fritz Müller, 20 September [1865]. CD's discussion of the genus *Catasetum* is in 'Three sexual forms of *Catasetum tridentatum*', and in *Orchids*, pp. 211–48. In his paper 'A few notes on the fecundation of orchids and their morphology', Hermann Crüger had argued that the adaptations he found in orchids supported CD's theory of transmutation (Crüger 1864, p. 131). Crüger sent CD several specimens, including a humble bee, along with his paper, which CD submitted to the *Journal of the Linnean Society (Botany)*; see *Correspondence* vol. 12, letter from Hermann Crüger, 21 January 1864,

and letter to Daniel Oliver, 17 February [1864]. CD's two annotated copies of Crüger 1864 are in DAR 70: 144–5; both copies have portions cut out. CD also annotated the article in his unbound issue of the *Journal of the Linnean Society* (*Botany*) 8 (no. 31), which is in the Darwin Archive–CUL.

7 In *Orchids*, p. 220, CD had proposed that an insect gnawing through the *Catasetum* labellum would touch the antennae of the rostellum, causing the pollinium to eject and forcing the pollen masses onto the insect. Crüger sent CD specimens of bees belonging to three species of *Euglossa* that he saw gnawing the inside of the labellum of *Catasetum*, thus confirming CD's hypothesis (see 'Fertilization of orchids', p. 154 (*Collected papers* 2: 151); see also *Orchids* 2d ed., pp. 205–6). See *Correspondence* vol. 12, letter from Hermann Crüger, 21 January 1864 and nn. 1 and 5. In *Orchids* 2d ed., pp. 270–1, CD added similar observations by Fritz Müller on the labellum of *Oncidium*.

From Edward Cresy 18 October 1865

Metropolitan Board of Works | Spring Gardens
18 Oct '65.

My dear Sir,

Our chairman has been asking me what I thought would be a fair allowance to his son Harry at Cambridge to cover everything.[1] He has hitherto paid college expenses, bills &c himself, but Harry tells me his father from his antecedents has not the remotest idea of what is really necessary or unnecessary and the Chairman says he would much rather let Harry make his own arrangements if he knew what would be a proper & reasonable sum to allow him— Carpmael tells me he pays the bills himself & gives his son what he wants—[2] It has occurred to me that having had the experience of two at Cambridge you would be able to arrive at a very sound conclusion on this knotty point—[3] & from what I have the pleasure of knowing of both of them, that neither would be extravagant or in any way likely to give a wrong result—

I was very glad to see George & hope he will look me up at Christmass.[4]

I was very sorry to hear from him that your health was still so bad but hope you are improving again—[5] I must apologise for giving you so much trouble but I know Mrs Darwin or your daughter will kindly take pen in hand & write the figures which you will advise them—[6]

Yours very truly | E Cresy

Charles Darwin Esq.

DAR 161: 246

1 John Thwaites was the chairman of the Metropolitan Board of Works (*Post Office London directory* 1865). His son, Henry John Bardwell Thwaites, matriculated at Trinity College, Cambridge, in October 1865 (*Alum. Cantab.*).

2 William Carpmael was a civil engineer who worked with Cresy and whose son Ernest matriculated at St John's College, Cambridge, in October 1863 (see letter from Edward Cresy, 9 June 1865, n. 10).

3 William Erasmus Darwin, CD's eldest son, received his BA degree from Christ's College, Cambridge, in 1862; George Howard Darwin matriculated at Trinity College in October 1864 (*Alum. Cantab.*).

4 CD had asked Cresy whether he would advise George on his future career (see letter to Edward Cresy, 7 September [1865] and n. 2). George had tried to visit Cresy in September 1865, but found that he was away on holiday (see letter from Edward Cresy, 10 September 1865).
5 For more on CD's illness in 1865, see the letter to Asa Gray, 15 August [1865] and n. 12, and Appendix IV.
6 Cresy refers to Emma Darwin and Henrietta Emma Darwin.

To Edward Cresy 19 October [1865][1]

Down, Bromley, Kent
Oct: 19th.

Dear Cresy

My father allowed my brother and self at Cambridge £300 per annum.[2] I should have allowed my son the same, but having five of them I told them that was too much, so I have given them £260; but my eldest son got a scholarship worth about £20,[3] and I expect I shall have to allow George something extra; as it is, I pay for his private tutor, and when he stays beyond term time all additional expenses.[4]

George enjoyed his visit with you, and I thank you cordially for your valuable advice.[5]

My health steadily improves but only slowly.[6]

Dear Cresy, | Yours sincerely | Ch. Darwin

Copy
DAR 143: 325

1 The year is established by the relationship between this letter and the letter from Edward Cresy, 18 October 1865.
2 CD refers to his father, Robert Waring Darwin, and his brother, Erasmus Alvey Darwin. Robert Waring Darwin had also given CD £200 to settle his debts at the end of his final term (see *Correspondence* vol. 1, letter to W. D. Fox, [11 May 1831]).
3 William Erasmus Darwin is listed as a scholar at Christ's College, Cambridge, in 1859 (*Cambridge University calendar* 1860, p. 311).
4 CD's Classed account book (Down House MS) records, under the heading 'Boys & William', payments to George of £140 on 10 October 1864, £65 on 1 January, 1 April, and 2 October 1865, and £18 on 24 October 1865; payments to 'Routh private tutor' of £8 on 31 January 1865 and £16 on 3 July 1865 are also recorded. George's tutor was Edward John Routh, a highly successful mathematics coach at Cambridge (*Alum. Cantab.*, *DNB*).
5 See letter from Edward Cresy, 18 October 1865 and n. 4.
6 See letter from Edward Cresy, 18 October 1865 and n. 5.

To Asa Gray 19 October [1865][1]

Down. | Bromley. | Kent. S.E.
Oct. 19th

My dear Gray

I received yesterday your article on Climbers[2] & it has pleased me in an extraordinary & even silly manner. You pay me a superb compliment & as I have just said to my wife I think my friends must perceive that I like praise, they give me such hearty doses.[3]

I always admire your skill in reviews or abstracts, & you have done this article excellently & given the whole essence of my paper.— I daresay you are right about the climbing roses; I never thought about spring shoots behaving differently from others.[4] I have had a letter from a good zoologist in S. Brazil, F. Müller, who has been stirred up to observe climbers & gives me some curious cases of *Branch*-climbers, in which branches are converted into tendrils, & then continue to grow & throw out leaves & new branches & then lose their tendril character.[5]

I am certainly better & have not vomited for above 5 weeks. This is certainly due to not having eaten any thing but toast & meat for the last 2 months. But I cannot recover mental strength & I do no regular work.[6] I have had some flowers crossed for me this summer & have lately been counting seeds.[7]

You might like to hear that Mitchella behaves exactly like the Cowslip.[8] Did I ever tell you that a year or two ago I ascertained that Pulmonaria offers a curious case. The long-styled form being absolutely sterile with its own pollen, whilst the short-styled is almost perfectly with its own pollen.[9] I have also ascertained that plants raised from Dimorphic species fertilized by their own pollen, are themselves generally sterile, & are often dwarfs, so that they offer the closest analogy with Hybrids; the first cross & the product both being more or less sterile; this seems to me a very curious fact.[10]

I do not know when I shall be able to publish any of these results, for I have resolved, whenever able to do any thing, to publish my next book.[11] I have not heard very lately from Hooker, who I believe returns tomorrow to Kew. His illness has been very serious.[12] To me the loss of his correspondence has been very great.

My wife has read aloud to me Stephens's 2 books on Central America.[13] What a remarkably pleasant writer he is! & how singularly deficient in the spirit of a naturalist. Who is he & is he still alive[14] we liked his books extremely.

With cordial thanks for all your kindness & you are enough to stimulate a dead man to work.

Yours affect | C. Darwin

LS(A)
Gray Herbarium of Harvard University (93)

[1] The year is established by the reference to A. Gray 1865–6; see n. 2, below.

[2] CD refers to the first of a two-part review Gray had written of 'Climbing plants' in the *American Journal of Science and Arts* (A. Gray 1865–6). The first part of the review was published in September 1865 and the conclusion in January 1866.

[3] Gray described CD's botanical works as showing 'a genius for biological investigation, and a power of turning common materials and ordinary observations to high scientific account', which had not been surpassed 'since the days of Hunter and Charles Bell' (A. Gray 1865–6, pp. 273–4). Gray referred to the surgeons and anatomists John Hunter and Charles Bell.

[4] In his review, Gray noted that CD had difficulty understanding how the shoots of climbing roses could 'get under a trellis close to a wall' ('Climbing plants', p. 105). Gray commented that he had observed (A. Gray 1865–6, p. 282):

> the strong summer-shoots of Michigan Rose (*Rosa setigera* Mx., *R. rubifolia* R. Br.), trained on a latticed wall, are strongly disposed to push into dark crevices and away from the light; they would, many of them, pretty surely place themselves under the trellis, and the lateral shoots of the next spring would emerge as they seek the light.

CD added Gray's explanation in a note to *Climbing plants* 2d ed., p. 184.

[5] CD had recently received a second letter from Fritz Müller on Brazilian climbing plants, written after receiving a copy of CD's 'Climbing plants' in August 1865 (see letters from Fritz Müller, 12 August 1865 and n. 11, [12 and 31 August, and 10 October 1865], and 31 August 1865; see also letter to Fritz Müller, 17 October [1865]). Müller paid particular attention to branch climbers because CD had not observed any among the species he had studied for 'Climbing plants'. CD added many of Müller's observations to *Climbing plants* 2d ed.

[6] CD attributed his improved health to the strict diet recommended by Henry Bence Jones (see letter to T. H. Huxley, 4 October [1865] and n. 3). By 'regular work' CD typically meant writing; he was engaged in experimental work at the time (see n. 7, below; see also Bowlby 1990, pp. 378–80).

[7] Henry Lettington, the under-gardener CD once described as a 'skilful crosser', probably made the crosses (see *Correspondence* vol. 12, letter to J. D. Hooker, 19 [April 1864] and n. 4). In 1865, CD conducted crossing experiments with primroses (*Primula vulgaris*) and cowslips (*P. veris*) grown from seeds he had received from John Scott (see *Correspondence* vol. 11, letter from John Scott, [26 July – 2 August 1863], and this volume, memorandum from W. E. Darwin, [late February–May 1865], and letter from B. D. Walsh, 29 May 1865, n. 3). He counted seeds from crosses he made of long-styled purple or red primroses with common wild primroses, and of red cowslips with common cowslips. CD's notes on these crosses are in DAR 108: 89b–110. He continued crossing experiments on these plants in 1866 and 1867 (notes in DAR 108: 111–24). He reported his findings in 'Illegitimate offspring of dimorphic and trimorphic plants', pp. 416–30, and in *Forms of flowers*, pp. 224–38.

[8] CD may be referring to the relative rates of fertility of legitimate and illegitimate crosses in *Mitchella* and the cowslip. Gray had sent CD plants of *Mitchella repens* early in 1863 (see *Correspondence* vol. 11, letter to Asa Gray, 2 January [1863] and n. 5). CD's results of crosses with *M. repens* are reported in *Forms of flowers*, pp. 125–7.

[9] CD and his son William had made observations on *Pulmonaria* in 1863 and in April and May 1864 (see *Correspondence* vol. 11, letter from W. E. Darwin, 8 May [1863] and n. 4, and *Correspondence* vol. 12, letters from W. E. Darwin, 14 April 1864 and 18 April [1864], and letter to W. E. Darwin, 14 May [1864]). CD's notes on *Pulmonaria*, made between 1864 and 1866, are in DAR 110: A40–94 and B15–17; his findings are discussed in 'Illegitimate offspring of dimorphic and trimorphic plants' and in *Forms of flowers*, pp. 101–10, 239, 252, and 287.

[10] CD was intrigued by the similarity between hybrid sterility and the sterility of plants that were the result of illegitimate unions among dimorphic and trimorphic plants (see letter to Max Wichura, 3 February [1865] and n. 9).

[11] CD hoped to resume work on *Variation*, which was not published until 1868 (see letter to John Murray, 2 June [1865], n. 1).

[12] Joseph Dalton Hooker had suffered an attack of rheumatic fever in August, around the time of his father's death (see letter from F. H. Hooker, [17 August 1865]). He had been offered leave until the end of the month, but told CD he was returning to Kew about 20 October 1865 (see letter from J. D. Hooker, 6 October 1865); Daniel Oliver later told CD that he was expected to return to Kew on 26 October (see letter from Daniel Oliver, 23 October 1865 and n. 6).

[13] The reference is to John Lloyd Stephens and to his travel narratives, *Incidents of travel in Central America, Chiapas and Yucatan* (Stephens 1841), and *Incidents of travel in Yucatan* (Stephens 1843).

[14] Stephens died in 1852 (*DAB*).

To Daniel Oliver 20 October [1865][1]

Down. | Bromley. | Kent. S.E.
Oct 20

Dear Oliver

I have rec^d a letter from an excellent Zoologist, Fritz Müller in Brazil, on climbing plants, & it appears to me to be worth sending to the Linn. Soc.—[2] As he is not a professed botanist, I had asked Hooker to read it over,[3] but as on his return he will be so busy I sh^d be very much obliged if you will do me this kindness. As it will not take you more than ten minutes, I will assume that you will do so, & send a copy by this post. All that I want, is for you to see that there is no glaring error. It seems to me, in my ignorance, that he uses the word "Bracteæ" wrongly; if so will you correct it.[4]

Also I cannot understand his description of the leaves of Strychnos (a specimen is enclosed): if you do not understand I c^d alter the sentence in some such way as follows. "The first, 3rd & 5th pair of leaves are placed differently from the 2nd, 4th &c & the former alone bear on one side tendrils in their axes".[5] You need not waste your time in answering, for I know you are very busy, but return me the M.S. with corrections in pencil which can be rubbed out.

Dear Oliver | yours very sincerely | Ch. Darwin

P.S. If Hooker is in the Herbarium tell him I shall write in a few days & that I refrain from doing so now out of virtue, as he must be overwhelmed with business.[6]

LS(A)
DAR 261.10: 63 (English Heritage: 88206046)

[1] The year is established by the relationship between this letter and the letter from Daniel Oliver, 23 October 1865.

[2] CD probably refers to the letter from Fritz Müller of 31 August 1865. The botanical observations contained in the letter, along with extracts from two other letters from Müller, were sent by CD to the Linnean Society (see letter to J. D. Hooker, 3 October [1865] and n. 2).

[3] See letter to J. D. Hooker, 3 October [1865]. Owing to illness, Joseph Dalton Hooker had been away from the Royal Botanic Gardens, Kew, since 2 September 1865 (see letters from F. H. Hooker, [17 August 1865] and 6 September [1865]).

[4] See letter to Fritz Müller, 17 October [1865] and n. 4.

[5] In the published version of Müller's letter, the passage on *Strychnos* reads (letter from Fritz Müller, [12 and 31 August and 10 October 1865]; Müller 1865b, p. 344):

> On the branches, the leaves of the first, third, fifth, &c. pairs are horizontal, those of the second, fourth, and sixth pairs are vertical in relation to the main axis; and it is from the angles of every under leaf of these latter pairs that the tendrils spring.

[6] Oliver wrote in his letter to CD of 23 October 1865 that he had that morning heard that Hooker was returning on 26 October 1865.

To J. D. Hooker 22 and 28 [October 1865]

Down
Sunday 22^{d} & Saturday 28th.

(This letter is mere idle talk & you need not read till so inclined.) N.B.)

My dear Hooker

I hope you are established at Kew, & relieved of your stiff joints & not more overwhelmed with business than you might naturally expect—[1] Mind I write to amuse myself & expect no answer beyond a line in a week or two's time just to tell me how you are.— You wrote me a splendidly long & good letter just before you left Buxton.[2] You speak with regret that the Royal Soc. never publickly honoured your Father; but do you not think that is accounted for by the fewness of the Botanists ever on the Council?[3] I think it would aid in fairness of bestowal of medals; if the list of all that have received them were printed at end of List of Members, as is done by Geolog. Soc^{y}.—[4] I have been wading through the Annals & Mag. of N. Hist. for last 10 years, & have been interested by several papers; chiefly, however, translations,; but none have interested me more than Carter's on lower vegetables, infusoria & Protozoa.[5] Is he is as good a workman as he appears? for if so he would deserve a Royal Medal.—[6] I know it is not new; but how wonderful his account of the spermatozoa of some diœcious alga or conferva, swimming & finding the minute micropyle on a distinct plant & forcing its way in![7] Why, these zoospores must possess some sort of organ of sense to guide their locomotive powers to the small micropyle, & does not this necessarily imply something like a nervous system, in the same way as Complemental male cirripedes have organs of sense & locomotion & nothing else but a sack of spermatozoa?[8]

I fully agree with your remarks on Wallace's remarks & on the man himself:[9] I fear he will not do what he ought in science. As for the Anthropologists being a bête noir to scientific men, I am not suprised, for I have just skimmed through the last Anthro: Journal, & it shows, especially the long attack on Brit: Assoc: a curious spirit of insolence, conceit, dullness & vulgarity.[10]

I have read with uncommon interest Travers' short paper on the Chatham I.s. I remember your pitching into me with terrible ferocity because I said I thought the seed of Edwardsia might have been floated from Chili to N. Zealand; now what do you say my young man to the three young trees of the same size on one spot alone of the Island & with the cast up pod on the shore?[11] If it were not for those unlucky wingless birds, I c^{d} believe that the group had been colonized by accidental means; but as it is, it appears by far to me the best evidence of continental extension ever observed; the distance I see is 360 miles. I wish I knew whether the sea was deeper than between N. Z. & Australia[12] I fear you will not admit such a small accident as the wingless birds having been transported on ice-bergs. Do suggest, if you have a chance, to any one visiting the Islands again to look out for erratic boulders there.[13] How curious his statement is about the fruit trees & bees! I wish I knew whether the clover had spread before the bees were introduced.—[14]

A newspaper has been sent me from N. Zealand with a savage anonymous attack on Haast for geological plagiarism;[15] perhaps you have received or w^d. not care to see it.— I saw in Gard. Chron. the sentence about the Origin dying in Germany, but did not know it was by Seeman.—[16] I sh^d. not be surprised, if it were a bit of revenge; for he had the impudence to ask me for a Testimonial for some Professorship, which I felt compelled to refuse.—[17] Talking of the Origin, a Yankee has called my attention to a paper attached to Dr Well's famous Essay on Dew, which was read in 1813 to Royal Soc. but not printed, in which he applies most distinctly the principle of N. Selection to the races of man.—[18] So poor old Patrick Matthew, is not the first, & he cannot or ought not any longer put on his Title pages "Discoverer of the principle of Natural Selection"![19]

Do you know who wrote the article in July Quarterly on Bates, Wallace & you?[20]

You will like much better the 2^d vol. of Palgrave:[21] I hardly know why but I have liked the whole very much.— We are reading (but you will have no time now to read) Buckle, & like it *extremely* though we disagree with him every other page, & Emma incessantly gets into a rage with him.—[22] As you do not like Silas Marner, I will not like much the Mill on the Floss;[23] it is certainly most clever; but almost all the persons are odious, & there is no one so charming as Dolly.—[24]

You borrowed many months ago Max Wichura on Hybrid Willows: the book is of value to me, as being marked. so some time please see about it.—[25]

I heard only lately of the Subscription for FitzRoy & wrote to the Hon. Secretary to enquire purpose of subscription & was glad to hear it was money for his family.[26] The Sec. told me that the 3000, granted by Government w^d nearly all go to pay debts & his children were left penniless by his first marriage! Yet poor FitzRoy started with £20,000 as he told me. What a melancholy career he has run with all his splendid qualities.—

My health improves a little, but very slowly: I can, however, do no work & have great daily discomfort. God knows whether I shall ever do work again. On Nov. 7th we go for a week to 6 Queen Anne St to see D^r. B. Jones:[27] if you are well enough, which I know is very doubtful & are in London, you must call & let me hear when.— I heard two days ago from Oliver a not very good account of you & that you w^d. not return till Thursday.—[28]

My dear old friend | Yours affectionately | C. Darwin

Endorsement: 'October 1865'
DAR 115: 277

1 Hooker was expected to return to Kew on 26 October 1865, after more than a month's absence due to illness (see letter from Daniel Oliver, 23 October 1865 and n. 6).

2 See letter from J. D. Hooker, 6 October 1865.

3 The Royal Society of London had traditionally honoured achievement in the sciences with medals, conferred annually or biannually by vote of the council (see *Record of the Royal Society of London*, Appendix IV, and MacLeod 1971a). The distribution of medals among practitioners of different disciplines was a contentious issue, and reflected long-running disputes about the representation of 'natural' versus 'physical' science in the council (see, for example, *Correspondence* vol. 6, letter to J. D.

Hooker, 8 April [1856], and *Record of the Royal Society of London*, pp. 116–17). After Hooker was awarded a Royal Medal in 1859, he wrote to George Bentham about his indignation at the failure of the Royal Society to award his father, William Jackson Hooker, a Copley Medal. He was particularly opposed to the award being confined to 'great discoveries, or great generalizations', because much important taxonomic work was thereby excluded (letter to George Bentham, 20 November 1859; reprinted in L. Huxley ed. 1918, 1: 418). William Jackson Hooker never received a medal from the Royal Society; he was, however, awarded a knighthood in 1836 for his contributions to botany (see Allan 1967, pp. 94–5).

4 The annual report printed in the *Quarterly Journal of the Geological Society of London* included a list of all the winners of the society's Wollaston medal.

5 For CD's copies of the *Annals and Magazine of Natural History*, see letter to J. D. Hooker, 27 [or 28 September 1865], n. 12. Most of CD's copies of the journal for the years 1856 to 1864 have not been found, but volumes 15 and 16, for 1865, are in the Darwin Library–CUL. Between 1856 and 1865 Henry John Carter published about twenty articles in the *Annals and Magazine of Natural History*, all of which dealt with the structure and reproduction of microscopic organisms. CD's copy of Carter's article 'Conjugations of *Navicula serians*, *N. rhomboides*, and *Pinnularia gibba*' in vol. 15 is lightly annotated.

6 Two Royal Medals were awarded annually to representatives of both 'natural' and 'physical' sciences, based on the recommendation of the Council of the Royal Society, for 'contributions to the advancement of Natural Knowledge' (*Record of the Royal Society of London*, p. 349). CD had been awarded the medal in 1853 for his work on coral reefs and on Cirripedia (*Abstracts of Papers Communicated to the Royal Society of London* 6 (1850–4): 355–6). Carter received the medal in 1872 for his work on Infusoria, Rhizopoda, and Spongiadae (*Proceedings of the Royal Society of London* 22 (1871–2): 31).

7 In his article 'On fecundation in *Eudorina elegans* and *Cryptoglena*', Carter describes both micro-organisms as algae that, during the reproductive cycle, have free swimming 'spermatozoids' that seek out individual 'female cells'. He describes the 'point of ingress' (micropyle) of the cell and the 'amalgamation' of the spermatozoid and female cell (Carter 1858, p. 240).

8 CD had found parasitic males in the sacs of female or hermaphrodite individuals in the barnacle genera *Ibla* and *Scalpellum*. Many of these complemental males lacked stomach and mouth parts and several had no penis. See *Living Cirripedia* (1851), pp. 281–93.

9 See letter from J. D. Hooker, 6 October 1865. CD had enclosed the letter from A. R. Wallace, 2 October 1865, with his letter to J. D. Hooker, 3 October [1865].

10 On the disputes between members of the Anthropological Society of London and some of CD's scientific friends and supporters, see the letter from A. R. Wallace, 2 October 1865, and the letter from J. D. Hooker, 6 October 1865. An unsigned article in the October 1865 issue of the *Anthropological Review*, pp. 354–71, described the refusal of the British Association to accept a resolution put forward by the president of the Anthropological Society, James Hunt, proposing that a separate section be created for anthropology. The author was critical not only of the association's refusal to create the new section but also of the perceived preferential treatment given to papers from members of the rival Ethnological Society. For more on the rivalry between the two societies, see Stocking 1987, pp. 245–57. CD's copy of the October 1865 issue of the *Anthropological Review* is in the Darwin Library–CUL.

11 Henry Hammersley Travers had discovered three trees of a species known only in New Zealand and Chile, *Edwardsia microphylla*, in the Chatham Islands, which are in the South Pacific, 534 miles east of Christchurch, New Zealand (*Columbia gazetteer of the world*). Travers also found a seed of the same species on a sawn plank of a Totara tree (*Podocarpus totara*), together with logs of white and red pines, on the coast of the island, and suggested that the seed of *Edwardsia* had been carried by ocean currents from New Zealand (Travers 1864, p. 143). CD had long believed that certain problems of plant distribution could be explained by oceanic migration, suggesting icebergs, wind, driftwood, and birds as possible vectors of transmission (*Foundations*, pp. 57–255; for CD's discussion of plant distribution in *Origin*, see pp. 356–92). In his *Flora Novæ-Zelandiæ* (Hooker 1853–5, 1: xix), Hooker had also considered the difficulty of explaining the existence in two widely separated locations of rare local

species, but had dismissed the idea of transportation by aerial or oceanic currents, claiming seeds were too heavy to be airborne and could not stand exposure to salt water. CD and Hooker often discussed the topic in their correspondence, and at one point CD joked: 'I believe you are afraid to send me a ripe Edwardsia pod for fear I sh[d] float it from N. Zealand to Chile!!!' (*Correspondence* vol. 6, letter to J. D. Hooker, 3 June [1857]). CD's annotated copy of Travers 1864, in his unbound issue of the 3 October 1865 number of the *Journal of the Linnean Society* (*Botany*), including a separate sheet of notes referring to the article pasted on the back cover, is in the Darwin Archive–CUL.

12 In his paper (Travers 1864, p. 141), Travers noted that the kiwi (*Apteryx*) had once inhabited the Chatham Islands. CD could not explain the presence of the flightless bird by his migration theory (see n. 11, above), and was forced to admit that the bird's presence was strong evidence for the existence of a former land-bridge connecting the islands to New Zealand. Hooker favoured Edward Forbes's theory that oceanic islands were once connected to the major continents but CD had reservations about it (E. Forbes 1846; see, for example, *Correspondence* vol. 5, letter to J. D. Hooker, 27 May [1855] and n. 2, and *Correspondence* vol. 11, letter to J. D. Hooker, [28 August 1863] and n. 7). In *Origin*, pp. 357–8, CD remained highly sceptical of Forbes's theory in its broadest form, but admitted the possibility that there had been some land bridges. CD was interested in the depth of the ocean as an indicator of a boundary between different faunal regions (see letter to A. R. Wallace, 29 January [1865] and nn. 4 and 5).

13 In *Origin*, p. 363, CD discussed the possibility of icebergs as agents of seed transmission and noted that the presence of 'erratic boulders', that is, fragments of rock not native to the location where they were found, in the Azores suggested that icebergs had once drifted to the islands, possibly bringing non-native seeds. See also *Correspondence* vol. 7, letter to J. D. Hooker, 26 [April 1858] and n. 4.

14 Travers had noted that many introduced plants, including white clover, were spreading rapidly, and also that the importation of bees had resulted in good fruit production from introduced European trees and shrubs (Travers 1864, p. 144). In correspondence with Hooker, CD had suggested that the 'remarkable absence' of papilionaceous plants in New Zealand might be explained by the absence of hive bees (see *Correspondence* vol. 7, letter to J. D. Hooker, 12 January [1858]). In *Origin*, p. 97, CD had argued that the fertility of papilionaceous plants was greatly reduced if bees were excluded. See also *Correspondence* vol. 12, letter to J. D. Hooker, 4 December [1864] and n. 20. Some of CD's notes on the subject are in his Experimental book (DAR 157a: 33–8; see also *Natural selection*, pp. 68–71).

15 The article CD refers to has not been identified. Over the course of his career, Julius von Haast was involved in a number of well-publicised disputes, but most occurred after 1866 (see, for example, the discussion of his conflict with Frederick Wollaston Hutton over glacial theory in H. F. von Haast 1948, pp. 665–72).

16 CD refers to a brief article in the *Gardeners' Chronicle*, 7 October 1865, p. 938, which summarised a longer report in the *Athenæum*, 30 September 1865, pp. 435–6, and to Berthold Carl Seemann. The article mentioned the reading of two papers with opposing views of CD's theory at a meeting of German naturalists and claimed that German enthusiasm for Darwinism was beginning to cool. Hooker had mentioned the report in the *Athenæum* to CD, referring to 'Seeman's sneers' at CD's *Origin* (see letter from J. D. Hooker, 6 October 1865 and n. 27).

17 No correspondence between CD and Seemann regarding a testimonial for a professorship has been found. Seemann proposed CD for membership in the Academia Caesarea Leopoldino-Carolina Naturae Curiosorum (a German academy of naturalists) in 1857 (see *Correspondence* vol. 6, letter to J. D. Hooker, 11 September [1857]).

18 CD's informant was Charles Loring Brace, a nephew of Asa Gray's wife, Jane Loring Gray. In the historical sketch at the beginning of *Origin* 4th ed., p. xv, CD cited Brace for calling his attention to William Charles Wells's paper, 'An account of a white female, part of whose skin resembles that of a Negro'. The paper was published together with Wells's 'Essay on dew' and another essay in W. C. Wells 1818. CD's copy of the book is in the Darwin Library–Down. CD referred again to Wells's work in *Descent* 1: 243. For a discussion of Wells's paper, see K. D. Wells 1973.

[19] In a letter in the 7 April 1860 issue of *Gardeners' Chronicle*, pp. 312–13, Patrick Matthew claimed that he had anticipated CD's discovery of the principle of natural selection in his book, *Naval timber and arboriculture* (Matthew 1831). Matthew's political pamphlet, *Schleswig-Holstein, etc.* (Matthew 1864) had 'Solver of the species problem' on the title page (Dempster 1996, p. 34). Matthew told CD he would send him a copy of the book, but it has not been found in the Darwin Library–CUL or the Darwin Library–Down (see *Correspondence* vol. 12, letter from Patrick Matthew, 6 June 1864 and n. 3).

[20] CD refers to an unsigned article by William Houghton, 'Gleanings from the natural history of the tropics', which reviewed A. R. Wallace 1853, Hooker 1854, and Bates 1863, as well as other works ([Houghton] 1865; see *Wellesley index*).

[21] Hooker had written to CD that he could not get beyond the first volume of Palgrave 1865 (see letter from J. D. Hooker, 6 October 1865 and n. 13).

[22] CD refers to Buckle 1857–61 and to Emma Darwin (see letter from J. D. Hooker, 13 July 1865, n. 16).

[23] CD refers to Eliot 1860 and 1861; see letters from J. D. Hooker, [26 September 1865] and 6 October 1865, and letter to J. D. Hooker, 27 [or 28 September 1865].

[24] CD refers to Dolly Winthrop, a character in *Silas Marner* (Eliot 1861): 'a woman of scrupulous conscience, so eager for duties, that life seemed to offer them too scantily unless she rose at half-past four' (*ibid.*, p. 158).

[25] CD had received Wichura 1865 at the beginning of 1865 (see letter to M. E. Wichura, 3 February [1865]); he evidently lent it to Hooker shortly after. He had asked for its return in a letter to Hooker of [17 June 1865]. CD's annotated presentation copy is in the Darwin Library–CUL (see *Marginalia* 1: 871–3).

[26] Robert FitzRoy had died on 30 April 1865 (see letter from J. D. Hooker, 2 May 1865, and letter to J. D. Hooker, 4 May [1865]). CD's letter to Charles Shaw, the honorary secretary of the FitzRoy testimonial fund, inquiring about the subscription has not been found; however, see the letter from Charles Shaw, 3 October 1865 and n. 6.

[27] Six Queen Anne Street was the London residence of CD's brother, Erasmus Alvey Darwin; Henry Bence Jones had been treating CD since July (see letter to Asa Gray, 15 August [1865], n. 12, and letter to J. D. Hooker, 27 [or 28 September 1865]). Emma Darwin's diary (DAR 242), records a trip to London from 8 to 20 November 1865, but does not mention a visit to Bence Jones. CD's Account book–cash account (Down House MS) records two payments to Bence Jones for £12 12*s.*, one on 31 October 1865 and the other on 28 November 1865.

[28] Daniel Oliver informed CD that Hooker was expected to return to the Royal Botanic Gardens, Kew, on 26 October 1865; see letter from Daniel Oliver, 23 October 1865.

From Hermann Kindt 23 October 1865

Bampton | Penrith
23. October 1865.

Dear Sir,

Many warm thanks for your so kindly complying with my request as well as for the corrections of, and additions to, my biographical sketch, which latter I will correct accordingly.[1]

I shall have great pleasure in sending you the number of our journal containing your autograph as soon as the latter will have been reproduced.[2]

With renewed thanks for your courteous attention believe me, with greatest esteem and admiration, | your's faithfully, | Hermann Kindt

Charles Darwin Esq. M.A., F.R.S, etc

DAR 169: 16

[1] In his letter of 17 October 1865, Kindt had asked CD to send an autograph and to correct an enclosed biographical sketch.

[2] The autograph by CD appeared in the *Autographic Mirror* (*L'Autograph Cosmopolite*) 3 (1865), no. 262. CD's copy of the journal has not been found. For the autograph CD reproduced a paragraph from the last chapter of *Origin* 3d ed., p. 514:

> I have now recapitulated the chief facts and considerations, which have thoroughly convinced me that species have been modified, during a long course of descent, by the preservation or the natural selection of many successive slight favourable variations. I cannot believe that a false theory would explain, as it seems to me that the theory of natural selection does explain, the several large classes of facts above specified. It is no valid objection that science as yet throws no light on the far higher problem of the essence or origin of life. Who can explain what is the essence of attraction of gravity? No one now objects to following out the results consequent on this unknown element of attraction; notwithstanding that Leibnitz formerly accused Newton of introducing "occult qualities & miracles into philosophy."— Charles Darwin p 514 3d Edit of "Origin"

The biographical sketch is in *Autographic Mirror* (*L'Autograph Cosmopolite*) 3 (1865): 82–3.

From Daniel Oliver 23 October 1865

Royal Gardens Kew
Oct. 23. 65

My dear Sir/

I have looked over Müller's paper & return it by this post.[1] It will be an interesting thing for an evening Meeting.—[2] I have corrected his *bracteæ*,[3] &—assisted by Mr Bentham[4]—the haziness of the paragraph about upwards & downwards, &c.[5] Dr. Hooker returns on Thursday— I have a note this morning. He does not lose the stiffness in his joints yet.—[6]

The scrap of *Strychnos* cannot be named with any certainty.[7] Young shoots are often so different from those on the flowering branch

Ever very sincerely yrs | D Oliver

DAR 173: 30

[1] CD had sent Oliver a letter from Fritz Müller on Brazilian climbing plants that he hoped to submit for publication to the Linnean Society; see letter to Daniel Oliver, 20 October [1865] and n. 2.

[2] Müller's paper was read at the Linnean Society meeting of 7 December 1865. The printed version of the paper is reproduced in this volume as the letter from Fritz Müller, [12 and 31 August, and 10 October 1865].

[3] See letter to Daniel Oliver, 20 October [1865] and n. 4.

[4] George Bentham.

[5] No paragraph containing the terms 'upwards' and 'downwards' occurs in the printed version of Müller's paper (see above, n. 2). It is likely that the terms were used by Müller in his description of the positions of leaves and tendrils and their relationships to the stem axis. See letter to Fritz Müller, 17 October [1865], and letter to Daniel Oliver, 20 October [1865] and n. 5.

[6] Joseph Dalton Hooker had been recovering from an attack of rheumatic fever (see letter from F. H. Hooker, [17 August 1865], and letter from J. D. Hooker, [26 September 1865]). The Thursday following 23 October 1865 was 26 October.

[7] CD had enclosed a specimen of *Strychnos* in his letter to Oliver of 20 October [1865].

To Daniel Oliver 24 October [1865][1]

Down. | Bromley. | Kent. S.E.
Oct 24

Dear Oliver

I am extremely obliged to you for all the trouble you have taken in correcting so fully the paper which I sent you, & which I will now transmit to Linn. Soc.[2]

Dear Oliver | yours very sincerely | Ch. Darwin

LS
DAR 261.10: 60 (English Heritage: 88206043)

[1] The year is established by the relationship between this letter and the letter from Daniel Oliver, 23 October 1865.

[2] Fritz Müller's paper (Müller 1865b; reproduced in this volume as the letter from Fritz Müller, [12 and 31 August, and 10 October 1865]) was read at the Linnean Society meeting of 7 December 1865. Before the paper was published, CD added material from the letter from Fritz Müller, 10 October 1865 (see letter to Fritz Müller, 9 December [1865]).

From W. D. Fox [before 26 October 1865][1]

⟨*two pages missing*⟩ with you then.

I took my 2d Boy Charles to Oxford last week. I sho⟨uld⟩ much h⟨a⟩ve preferred Cambridge—but a⟨s⟩ he was fortunate enough to g⟨a⟩in a Studentship at Ch⟨rist⟩ ⟨Chur⟩ch[2] I had no ⟨ ⟩ ⟨*half a page missing*⟩ ⟨ ⟩ think of you, my p⟨ ⟩ ⟨ ⟩, with your sad health, while I ⟨ha⟩ve a perfectly enjoyabl⟨e⟩ life— tho my lungs are rather ⟨r⟩otten

I met a splen⟨di⟩d man for a couple of d⟨ay⟩s l⟨ast⟩ week

AL incomplete
DAR 164: 204 (fragile)

[1] The date is established by the relationship between this letter and the letter to W. D. Fox, 25–6 October [1865].

[2] Charles Woodd Fox matriculated at Christchurch College, Oxford, on 13 October 1865 (*Alum. Oxon.*).

To W. D. Fox 25–6 October [1865]

Down. | Bromley. | Kent. S.E.
Oct. 25

My dear Fox

It is a long time since we have had any communication & I should like much to hear how you are yourself as well as Mrs Fox[1] & all your family.

I know you will wish to hear about me. I have had a bad time for the last 6 months & have been able to do no scientific work. I have put myself under Dr Bence Jones's care & he has stopped my vomiting by a scanty diet of toast & meat; but I cannot recover my strength.[2] I know you are half a Doctor therefore I thought you wd like to hear these details.

My Sister Susan's health has lately been much failing & I fear from the last accounts that her state is becoming serious.[3] As for the rest of us we are pretty well, at least for Darwins, & that is not saying much. My children however are all well I am thankful to say

Oct 26. I had dictated thus far yesterday, & now by an odd chance your kind & welcome letter has arrived.[4] I heartily rejoice at the good account you give of yourself, patriarch as you are with your half dozen grandchildren I congratulate you most sincerely on the brilliant success of your son & my Godson.[5] My eldest son[6] is established as a banker at Southampton & is fairly well contented with his lot. My second son George has just passed his Little go at Trin. Coll.[7] He has a turn for mathematics, & means to try for a scholarship there this Easter.[8] As for myself I have not gone out of my grounds for the last 12 months; but on Nov. 7. we go for a week to Erasmus's in order that Dr Bence Jones may see me two or three times.[9] I hope this may coincide with part of your visit, as I sh^{d} like extremely to see you again for a few minutes,—more than that I fear I c^{d} not stand.

My wife desires to join in very kind remembrances to Mrs Fox | My dear old friend | Yours affectionately | C. Darwin

Postmark: OC 27 65
LS(A)
Christ's College Library, Cambridge (Fox 146)

1 Ellen Sophia Fox.
2 According to Emma Darwin's diary (DAR 242), CD began his diet under the supervision of Henry Bence Jones on 24 July 1865. See also letters to Asa Gray, 15 August [1865], n. 12, and 19 October [1865] and n. 6.
3 Susan Elizabeth Darwin lived at the Mount, the Darwin family residence in Shrewsbury.
4 See letter from W. D. Fox, [before 26 October 1865].
5 CD refers to Charles Woodd Fox. See letter from W. D. Fox, [before 26 October 1865] and n. 2.
6 William Erasmus Darwin.
7 'Little-go': 'the popular name ... for the first examination for the degree of B.A., officially called ... "The Previous Examination" at Cambridge' (*OED*). George Howard Darwin matriculated at Trinity College in October 1864.
8 George received a prize for achieving first class in the Easter term examination, 1866 (*Cambridge University calendar* 1867, p. 397).
9 See letter to J. D. Hooker, 22 and 28 [October 1865] and n. 27.

From Alfred Newton 27 October 1865

Magdalene College, Cambridge.
27 Oct. 1865—

My dear Sir,

I venture to hope you will kindly excuse my troubling you with a request—

There seems every probability of a Professorship of Zoology and Comparative Anatomy being established in this University—in which case I intend to offer myself as a Candidate for the post—[1] But since I took my degree I have been so much away—having for nine years held the Travelling Fellowship of this College[2]—that I am known personally to but very few of the resident Masters of Arts & others who form the Elective body—[3] Many of my friends therefore think that my chance of success will be materially strengthened by testimonials from a few of the most eminent naturalists— I therefore take the liberty of asking whether you will do me the great favour of giving me a recommendation—

You yourself can testify in the most remarkable degree to the enormous advantages which travelling in distant and varied foreign countries confers upon a student of Zoology—and nearly all my voyages and journeys have been made with special Zoological objects—[4] You also have honoured with your notice some few of the various contributions I have been able to make to scientific journals—[5]

I consequently hope you may be able to speak favourably of my qualifications for the important office to which I aspire—

I sincerely trust that whether from health or any other reason this letter may cause you no inconvenience to answer—and I remain | With very deep respect | Yours truly | Alfred Newton

Charles Darwin Esqre. F.R.S. | &c. &c. &c.

DAR 172: 43

[1] Until 1866, comparative anatomy had been taught at the University of Cambridge by the professor of anatomy. The new professorship was founded in the spring of 1866 and two candidates, one of whom was Newton, competed for the post (see A. F. R. Wollaston 1921, p. 133).

[2] Newton was elected in the spring of 1853 to the Drury Travelling Fellowship of Magdalene College, Cambridge, and held the post until March 1863 (A. F. R. Wollaston 1921, pp. 12, 75).

[3] Professorships were decided by a vote in the Senate of the university. Candidates canvassed the membership for votes, and it was customary to seek testimonials from prestigious non-members (A. F. R. Wollaston 1921, p. 133).

[4] Newton had made numerous ornithological expeditions to various locations including the West Indies, Lapland, Iceland, and Spitsbergen (A. F. R. Wollaston 1921).

[5] CD had been particularly interested in Newton's paper 'On an illustration of the manner in which birds may occasionally aid in the dispersion of seeds' (A. Newton 1863). Newton had sent CD the foot of a red-legged partridge (*Caccabis rufa*) with a ball of earth attached containing seeds (see *Correspondence* vol. 11, letters from Alfred Newton, 21 March 1863 and 31 October 1863). With his letter of 31 October 1863, Newton sent CD a copy of A. Newton 1863, part of which is in DAR 205.9: 366 and is annotated in CD's hand. CD planted the seeds and was able to grow eighty-two plants; he included the information in *Origin* 4th ed., p. 432 (see *Correspondence* vol. 12, letter to Alfred Newton, 29 March [1864] and nn. 3 and 4). CD later referred to a paper of Newton's on hybrid ducks (A. Newton 1860) in *Variation* 2: 157.

To Alfred Newton 29 October [1865]

Down. | Bromley. | Kent. S.E.
Oct. 29

My dear Sir

As I have always held, if you will permit me to say so, a very friendly feeling & respect towards you, it is with sincere regret, I may say with pain, that I feel I cannot give you a testimonial for the intended Professorship, of which I had not previously heard;[1] for in my opinion in the present state of Zoological Science it would be indispensable that a teacher should have especially studied comparative anatomy & Histology, & likewise have attended much to the invertebrate animals which exhibit such different types of structure.

If I am not mistaken, you have not published on these subjects, & have chiefly attended to Birds, including their habits, instincts, distrubution and other such philosophical points.[2] I do not suppose my testimonial would be of much value, & with not a few members of the University it w^{d} be absolutely injurious, nevertheless it would have been a great satisfaction to me to have given you a testimonial, had it been in my power.

Pray believe me | my dear Sir | yours very sincerely | Ch. Darwin

LS
Endorsement: 'Oct. $\frac{29}{30}$ /65'
CUL (Alfred Newton papers)

[1] See letter from Alfred Newton, 27 October 1865 and n. 1.
[2] A full bibliography of Newton's publications, all of which were on ornithological subjects, is contained in the appendix to A. F. R. Wollaston 1921.

From Alfred Newton 30 October 1865

Magd: Coll: Cambridge
30 Oct. 1865.

My dear Sir,

I beg leave to assure you that no apology is needed for declining my request that on the present occasion you should give me a testimonial—[1] I have to thank you much for the courteous terms in which your refusal is couched— In justice to myself, however, I hope you will allow me to state that I am well aware of the absolute necessity there is for a would-be Professor of Zoology to be acquainted with Comparative Anatomy. It is true that I have hitherto published little or nothing of value bearing upon this subject—but it is not the less true that for more than twenty years Comparative Anatomy has formed an important part of my studies—and that on this basis I have always endeavoured to ground any investigations I have made into Zoology.

I may be entirely wrong in my opinion, but I have long thought that the difficulties which meet Zoological students are the same both in kind and degree whatever be the branch of the science investigated. I have also thought that an attempt to master one branch of the science was a good sort of training for one who aspired to teach the remainder to others. This has been the reason why I have chiefly, though not exclusively, confined my attention to ornithology.

Of histology, I freely confess, I know nothing—perhaps in the present case it is of less importance, because I expect it to form the subject of some of the lectures of the Professor of Human Anatomy & *Physiology*

You are no doubt right in saying that a recommendation from you might possibly be injurious to me as regards a few members of the University. This idea had occurred to me before—but it also seemed to me that the cause of truth was better than success, and as I had never cared to conceal my opinions on former occasions,[2] I did not think it honest to do so now.

With deep respect | believe me to remain | Yours very truly | Alfred Newton

C. Darwin Esqre. F.R.S.

DAR 172: 45

[1] See letter from Alfred Newton, 27 October 1865, and letter to Alfred Newton, 29 October [1865]. For more on Newton's election to the professorship, see A. F. R. Wollaston 1921, pp. 133–5.

[2] Newton had been an early supporter of CD's views. Four days after the publication of the paper by CD and Alfred Russel Wallace on the tendency of species to form varieties (C. Darwin and Wallace 1858), he communicated his favourable impression in a letter of 24 August 1858 to a fellow ornithologist, Henry Baker Tristram (the letter is reproduced in part in A. F. R. Wollaston 1926, pp. 115–17). Newton commented favourably on *Origin* in A. Newton 1860, and supported CD's dispersion theory in A. Newton 1863. He suggested that natural selection was the cause of an apparent increase in the numbers of Pallas's sand-grouse in its native territory in A. Newton 1864 (see *Correspondence* vol. 12, letter from Alfred Newton, 2 April 1864 and n. 6).

From George Henslow 1 November 1865

10 South Crescent | Bedford Sq: W.C
Nov 1/65

Dear Sir,

My Brother in-law, Joseph Hooker,[1] has given me your address in order that I might be able to write to you on one or two little botanical matters about w^{h}. I should like to ask permission to consult you.

While at Margate in the Summer,[2] I was casually examining the fls of *Medicago sativa* which is largely cultivated there & discoverd a remarkable irritability in the stamens, viz: that when the alæ are slightly touched in a definite way the essential organs spring violently upwards, & by a simultaneous curving of the filaments, the anthers are dashed against the Vexillum, or any object (e.g. a pin) which may have caused the essential organs to rise.[3]

Have you ever observed it? ⟨N⟩either J. D. Hooker nr Prof Oliver[4] were aware of the fact, & the former suggested my making a few notes upon that & a few other plants for the Linnean[5] & I thought I should first like to hear if you had made any observ^{ns} in the same direction.

I, of course, imagined that this uprising had something to do with cross fertilization, (following y^{r} valuable generalization)[6] & examined as many other *Leguminosæ* as I could meet with, to see if there was anything analogous in others. But I find that in other cases (Onobrychis, Lupinus, trifolium, Lotus, &c &c) the *carina* falls by the downward pressure of any body introduced which causes the protrusion of the essential organs.[7]

On returning home, & taking up y^{r} *Origin of Species*, I opened at a sentence w^{h} I had quite forgotten.— Where you say that "so necessary are the ⟨visits⟩ of bees to Papilionaceous flowers that ... their fertility is greatly diminished if these visits be prevented".[8] Will you kindly tell me *where* you have published y^{r} remarks as I should much like to read them.[9]

another little observation is *Salvia* which, I think, seems constructed to secure cross fertilization. by depressing the abortive anther (w^{h} forms the short arm *to the levers*,) the fertile one is brought down, as I conceive on to the *back* of an insect, that would strike against (with its head) the abortive anth.

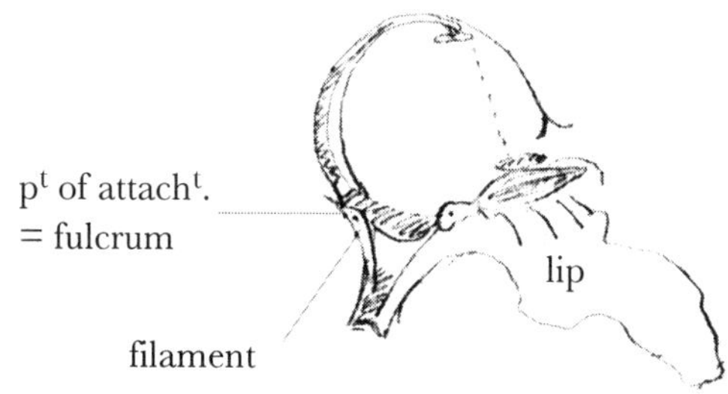

The stigma is so situated as to strike the back of the insect when inserted into another fl.—

All this is Hypothetical as I have *never seen* an insect do it.

⟨I shou⟩ld much like to hear if ⟨you⟩ have ever made any observati[ons] on this.

Crocus is another as affords (by Bees) means of cross fertilization

If you would kindly give me any inform^{n}. or refer me to any work bearing on these, I should be much obliged.

In mean time, | I beg to remain | Y^{rs} very faithfully | George Henslow

C Darwin Esq.

DAR 166: 150

1 Joseph Dalton Hooker had married George Henslow's elder sister, Frances Harriet Henslow, in 1851 (*DNB*). George Henslow's father, John Stevens Henslow, had been CD's mentor at Cambridge (see Browne 1995, pp. 117–18, 128–39, and Walters and Stow 2001, pp. 78–107).

2 Margate is a seaside resort on the Kent coast.

3 The alae are the lateral petals of a papilionaceous flower; the vexillum is the standard or large posterior petal (Jackson 1928).

4 Daniel Oliver.

5 Henslow's observations on *Medicago sativa* (alfalfa) were read at the 16 November 1865 meeting of the Linnean Society; his observations on *Indigofera* (family Leguminosae) were read at the meeting of 19 April 1866. The papers were published in the *Journal of the Linnean Society* (*Botany*) 9 (1867): 327–9 and 355–8 (G. Henslow 1865 and 1866).

6 Henslow refers to the 'general law of nature' that CD formulated in *Origin*, p. 97: 'no organic being self-fertilises itself for an eternity of generations; but ... a cross with another individual is

occasionally—perhaps at very long intervals—indispensable'. CD had first enunciated the principle in his letter to the *Gardeners' Chronicle*, [before 13 November 1858] (*Correspondence* vol. 7; see n. 9, below). That nature 'abhors perpetual self-fertilisation' is also one of the main conclusions of CD's work on orchids (see *Orchids*, p. 359). See also letter to J. T. Moggridge, 13 October [1865] and n. 3.

7 In his 1865 paper, Henslow noted that the stamens in *Genista* and *Cytisus* also sprang upward, as did those in *Medicago sativa*; however, he added that intercrossing between distinct flowers was more generally facilitated 'by the depression of the carina with or without that of the alæ' (G. Henslow 1865, p. 329).

8 The reference is to *Origin*, p. 97. The full sentence reads: 'So necessary are the visits of bees to papilionaceous flowers, that I have found, by experiments published elsewhere, that their fertility is greatly diminished if these visits be prevented.' CD began to consider the possibility of insect agency in the cross-pollination of papilionaceous flowers in Leguminosae in 1856 (see *Correspondence* vol. 6, letter to George Bentham, 26 November [1856] and n. 2). He experimented with leguminous plants between 1857 and 1859, recording the results in his Experimental notebook (DAR 157a: 33–8, 88–9), and in the manuscript of his 'big species book' (see *Natural selection*, pp. 68–71). For CD's interest in the subject, see *Correspondence* vols. 6 and 7.

9 CD reported his research on cross-pollination in kidney beans (*Phaseolus vulgaris*) and other leguminous plants in a letter to the *Gardeners' Chronicle*, [before 13 November 1858] (*Correspondence* vol. 7). The letter appeared in the 13 November 1858 issue of the journal under the title 'On the agency of bees in the fertilization of papilionaceous flowers, and on the crossing of kidney beans'; it was also published in the *Annals and Magazine of Natural History* 2 (1858): 459–65 (*Collected papers* 2: 19–25). CD's observations on *Phaseolus* are cited in G. Henslow 1865, p. 329.

To F. W. Farrar 2 November [1865][1]

Down. | *Bromley.* | *Kent. S.E.*
Nov. 2.

Dear Sir

As I have never studied the science of language it may perhaps be presumptuous, but I cannot resist the pleasure of telling you what interest & pleasure I have derived from hearing read aloud your volume.—[2]

I formerly read Max Müller & thought his theory (if it deserves to be called so) both obscure & weak;[3] & now after hearing what you say, I feel sure that this is the case & that your cause will ultimately triumph.[4]

My indirect interest in your book has been increased from Mr Hensleigh Wedgwood, whom you often quote, being my brother in law.[5]

No one could dissent from my views on the modification ⟨o⟩f species with more courtesy ⟨t⟩han you do.[6] But from the ⟨te⟩nor of your mind I feel an entire & comfortable conviction (& which cannot possibly be disturbed) that if your studies led you to attend much to general questions in Natural History, you w^d come to the same conclusions that I have done.

Have you ever read Huxley's little book of Six Lectures[7] I w^d gladly send you a copy if you think you would read it.

Considering what Geology teaches us, the argument fo⟨r⟩ the supposed immutability of specific Types seems to me much the same as if, in a nation whic⟨h⟩

had no old writings, some wise ol⟨d⟩ savage was to say that his language had never changed; but my metaphor is too long to fill up.

Pray believe me dear Sir yours very sincerely obliged | Ch. Darwin

LS

University of Virginia Library, Albert and Shirley Small Special Collections Library, Darwin Evolution Collection (3314)

[1] The year is established by the relationship between this letter and the letter from F. W. Farrar, 6 November 1865.

[2] Farrar had sent CD a copy of *Chapters on language* (Farrar 1865; see letter to F. W. Farrar, 11 October [1865] and n. 2).

[3] Friedrich Max Müller had published two volumes on the science of language, based on his lectures at the Royal Institution of Great Britain (Max Müller 1861 and 1864). Max Müller argued that all human languages descended from a common set of 'phonetic types' or 'roots', which originated in the faculty of giving 'articulate expression to the rational conceptions' of the mind (see Max Müller 1861, pp. 342, 369–72). CD read the first volume of Max Müller's lectures in 1862; he expressed dissatisfaction with Max Müller's account of the origin of language, and thought the work contained 'covert sneers' at himself (see *Correspondence* vol. 10, letter to J. D. Hooker, 4 November [1862], and letter to Asa Gray, 6 November [1862] and n. 5). On Max Müller's theory of language, see Schrempp 1983, Knoll 1986, and Radick 2000. His debate with Farrar is discussed in Alter 1999, pp. 79–96. His broader project of comparative philology and its bearing on debates over human origins and racial theory are discussed in Stocking 1987.

[4] Farrar argued that human language originated in the imitation of natural sounds, such as animal noises, and instinctual interjections, such as 'ah' (Farrar 1865, pp. 88–103, 109–18). Max Müller had dismissed such views, calling them the 'bow-wow' and 'pooh-pooh' theories of language (Max Müller 1861, pp. 344–56; but see also Max Müller 1864, pp. 88–92). In *Descent* 1: 56, CD cited Farrar 1865 and other works in opposition to Max Müller, concluding: 'I cannot doubt that language owes its origin to the imitation and modification, aided by signs and gestures, of various natural sounds, the voices of other animals, and man's own instinctive cries.'

[5] In his *Dictionary of English etymology*, Hensleigh Wedgwood defended the view that the original roots of language derived from the imitation of natural sounds (H. Wedgwood 1859–65, 1: iii–xviii). Farrar cited many examples from Wedgwood's *Dictionary* that supported the imitative theory of language against Max Müller's criticisms (see, for example, Farrar 1865, p. 132). For CD's favourable assessment of Wedgwood's *Dictionary*, see *Correspondence* vol. 8, letter to Charles Lyell, 10 January [1860], and letter to J. M. Rodwell, 5 November [1860]. See also *Correspondence* vol. 6, letter from Hensleigh Wedgwood, [before 29 September 1857], and *Correspondence* vol. 8, letter from Hensleigh Wedgwood, [January? 1860]. In a letter to H. E. Darwin, [17 March 1865] (DAR 219.9: 25), Emma Darwin wrote: 'This m$^{\underline{g}}$ [CD] enjoyed a talk with Uncle H[ensleigh] upon Max Müller for one thing who H. says is all wrong & has partly converted Papa to that opinion.'

[6] Farrar's book contained a brief allusion to CD's theory of transmutation (Farrar 1865, p. 51), described as the possibility that:

> myriads of centuries ago, there may have been a near genetic connection between the highest of the animals and the lowest of the human race . . . It is not yet proved that there was; we believe that there was *not*; but, nevertheless, the hypothesis is neither irreverent nor absurd.

[7] CD refers to the single-volume edition of Thomas Henry Huxley's 'On our knowledge of the causes of the phenomena of organic nature', a course of lectures to working men delivered at the Museum of Practical Geology in 1862 (T. H. Huxley 1863b; see L. Huxley ed. 1900, 1: 206). CD's annotated copy of the separately published lectures (T. H. Huxley 1862) is in the Darwin Library–CUL (see *Marginalia* 1: 425). CD had repeatedly praised the lectures (see, for example, *Correspondence* vol. 11, letter to T. H. Huxley, 10 [January 1863]).

To George Henslow [2–5 November 1865][1]

[Note on *Medicago lupulina*, communicated by Mr. Darwin:—]

"I covered a number of plants with a net (which I know does not injure their seeding), and left others close by uncovered, and these were visited incessantly by bees. I could not compare their relative fertility with accuracy, on account of the easy shedding of the seed; but I gathered 150 not quite ripe pods of both; those from under the net weighed 77 grains, and those visited by the bees weighed 101 grains. No doubt the difference in the weight of the seed would have been considerably greater, as the pod is formed independently of the number of seeds."[2]

Incomplete[3]
Journal of the Linnean Society (Botany) 9 (1867): 328 n.

1 The date range is established by the relationship between this letter and the letters from George Henslow, 1 November 1865 and 6 November 1865.

2 CD reported these observations of *Medicago lupulina* in response to a request from Henslow, who had noted an 'irritability in the stamens' of *M. sativa* that seemed to facilitate cross-pollination by insects (see letter from George Henslow, 1 November 1865 and nn. 5–9, and G. Henslow 1865). CD's observations were also published in *Cross and self fertilisation*, p. 368. CD's notes on *M. lupulina*, dated 1859, are in DAR 157a: 45.

3 CD's original letter to Henslow has not been found; it evidently contained recommendations of several published works, and possibly information on other Leguminosae (see letter from George Henslow, 6 November 1865 and nn. 3 and 6). Observations made by 'Mr. Darwin and others in *Trifolium*, *Lupinus*, *Phaseolus*, &c.', are cited in G. Henslow 1865, p. 329. CD's notes on cross-pollination by insects in *Trifolium*, *Phaseolus*, and other leguminous plants, made between 1857 and 1859, are in DAR 157a: 33–8 and 45–8.

From J. D. Hooker [3 November 1865][1]

Kew
Friday.

Dear old Darwin

I am back & well, all but stiff joints—better than before I was ill.[2]

I am up in heaps with work, & find I shall have a desperate fight to get scientific assistance, I will not give in however— I am prepared to improve the Gardens enormously & will do so, but if therein scientific character of the Establishment is to go down one iota, I shall intimate that I only hold the post with a view to retirement when able.[3]

My *elevation* brings me no increase of income but a higher scale of living;[4] as I now feel it my duty to give up Examinerships &c that yielded upwards of £300—[5] But I have no fear of not carrying my point, which is a properly educated assistant to be under Oliver.[6]

The Curator is in future to be my Asst. *in Garden duties*,[7] Oliver' with increased Salary, *in scientific matters*. an excellent arrangement, as there is no one able to be my assistant in both, nor are the functions compatible in any but one who like myself has grown with growth of the Establishment, & been educated to it. In the conversation I had with the Board they "let the cat out of the bag" in informing me, that, they abolished the Assistant Directorship because they knew of no one fitted for it,! not only an unintentional compliment to me, but an admission by implication that neither could they find another person fit to be Director!—[8] I took no notice, but have it in hand as "one for his nob." if needs be.[9]

You see "my Dander is up", as the Yankees say—but pray say nothing about this, fighting battles before byestanders is only a shade better than in the dark—& one gains nothing by appearing to be in opposition.

A thousand thanks for your long kind letter,[10] Why will you run your head against an "ingenious wriggler" I can answer you on all points anent the Chatham Isld.[11]

I have no idea who wrote trashy article on Bates &c.[12]

Carters observations are wonderful but want verification.[13] I think I remember his once describing Amoeba or Actinophrys as products of vegetation (origin in Chlorophyll)[14]

Wallace has turned table turner I am told[15] Travers Chatham Isld facts are too nice,—the three Edwardias all of one size along side their pod all to hand Oh Oh Oh The Forest trees & bees sounds very post-Darwinian! The Edwardsia may have floated from S. America, Good!—[16] did the Apteryx come from that Quarter too?—[17] you might at least have suggested N.Z. for S. America. Why will you break your head against a Wrigglers?

I hope to call before you leave[18] | J H

DAR 102: 43–6

[1] The date is established by the relationship between this letter and the letter to J. D. Hooker, 22 and 28 [October 1865], and by the date of Hooker's return to Kew (see n. 2, below). The Friday following 28 October 1865 was 3 November.

[2] Hooker had been suffering from rheumatic fever (see letter from J. D. Hooker, [26 September 1865] and n. 2). He returned to his duties at the Royal Botanic Gardens, Kew, on 26 October (see letter from Daniel Oliver, 23 October 1865).

[3] Hooker officially became director of the Royal Botanic Gardens, Kew, on 1 November 1865; he had served as assistant director under his father, William Jackson Hooker, since 1855 (R. Desmond 1995, pp. 223, 225). For Hooker's view of the difficulties of Kew administration, and his efforts to promote scientific botany and horticulture at the garden, see *Correspondence* vol. 12, letter from J. D. Hooker, 20 April 1864, and this volume, letters from J. D. Hooker, [7–8 April 1865] and nn. 5 and 6, and [26 May 1865].

[4] Hooker's annual salary as director was £800 in 1868 (Whitaker 1868, p. 102). The position included a residence at 49 Kew Green (R. Desmond 1995, p. 191). As assistant director his salary had been £400, and included a residence at 55 Kew Green (R. Desmond 1995, p. 200, and R. Desmond 1999, p. 206).

[5] Hooker was an examiner in botany for the medical service of the Indian Army, and for the Society of Apothecaries in London; he had resigned his position as examiner in botany at University of London the previous year (L. Huxley ed. 1918, 1: 385, 537).

[6] Daniel Oliver was librarian and keeper of the herbarium at Kew; in 1865 he was also made keeper of the museum (R. Desmond 1995, p. 225).

[7] A new curator, John Smith (1821–88), had been appointed in 1864 (see letter from J. D. Hooker, [7–8 April 1865] and n. 5, and R. Desmond 1995, p. 226).
[8] Hooker refers to the commissioners of the Board of Works and Public Buildings, which had had jurisdiction over Kew since 1851 (see R. Desmond 1995, pp. 180, 225).
[9] 'One for his nob' is an expression derived from cribbage, implying to hold in reserve (see *OED*).
[10] See letter to J. D. Hooker, 22 and 28 [October 1865].
[11] See n. 16, below. The ability to 'wriggle' out of difficulties had long been a joke between CD and Hooker (see letter to J. D. Hooker, 19 January [1865] and n. 9).
[12] [Houghton] 1865. See letter to J. D. Hooker, 22 and 28 [October 1865] and n. 20.
[13] CD had praised the articles by Henry John Carter in the *Annals and Magazine of Natural History* (see letter to J. D. Hooker, 22 and 28 [October 1865] and nn. 5–7).
[14] Hooker refers to Carter's article 'Transformation of the vegetable protoplasm into Actinophrys' (Carter 1857).
[15] Alfred Russel Wallace began attending séances in London in the summer of 1865 (see Raby 2001, pp. 184–7). His efforts to establish a scientific basis for spiritualism are discussed in A. R. Wallace 1905, 2: 275–350. See also letter to J. D. Hooker, 7 January [1865] and n. 14.
[16] Hooker refers to CD's discussion of Henry Hammersley Travers's paper on the Chatham Islands and its pertinence to the dispersal of *Edwardsia microphylla* and other species from South America across the Pacific Ocean (see letter to J. D. Hooker, 22 and 28 [October 1865] and nn. 11 and 12). CD had also mentioned the connection Travers made between the importation of bees and the increased production of fruit trees on the islands. CD and Hooker had long debated the means of species dispersal and geographic distribution.
[17] *Apteryx* is a genus of flightless birds, commonly known as kiwis, found in New Zealand. See letter to J. D. Hooker, 22 and 28 [October 1865] and n. 12.
[18] CD expected to be in London for a week from 7 November (see letter to J. D. Hooker, 22 and 28 [October 1865] and n. 27).

From Fritz Müller[1] 5 November 1865

Desterro, Brazil,
5. November 1865.

Verehrter Herr!

Ich habe Ihnen zu danken für Ihren freundlichen Brief vom 20. September und auch für Ihr Orchis-Buch und zwei Abhandlungen über Linum und Lythrum, welche ich mit der letzten französischen Post richtig empfing.[2] Ich habe diese verschiedenen Abhandlungen mit dem grössten Interesse gelesen. Wie merkwürdig verwickelt sind doch die Geschlechtsverhältnisse bei Lythrum salicaria! Die einzige Pflanze von dieser Familie, welche ich bis jetzt gefunden habe (eine Nesaea, Unter-Gattung Heimia), ist monomorph;[3] aber es mögen vielleicht dimorphe Arten unter den verschiedenen Cuphea unserer Flora sein.[4] Wir sind hier sehr reich an Orchideen, und ich hoffe, ich werde einige von den wunderbar vollkommenen Anpassungen selbst sehen können, welche Sie in Ihrem Buch beschreiben.[5]

Es giebt in den Wäldern des Festlandes viele Kletterpflanzen, welche sogar die riesigsten Stämme besteigen, aber ich kann mich nicht entsinnen, ob windende Formen dabei sind; im nächsten Monat beabsichtige ich eine Reise nach dem Itajahy-Fluss zu machen und bei dieser Gelegenheit werde ich meine Aufmerksamkeit auf diesen Punkt richten.[6] Bezüglich Dalbergia, von der Sie anführen, dass sie dicke Bäume besteigt, habe ich einigen Zweifel, ob sie ein ächter Winder ist oder nicht vielmehr ein Zweigklimmer wie manche andere Pflanzen desselben Stammes; d. h. ob sie mit Hülfe der schwingenden Bewegung ihrer Schösslinge klettert oder dadurch, dass sie sich nach der Seite krümmt, welche zufällig einen Stamm berührt.[7] Die letztere Bewegung in Verbindung mit dem Wachsthum des Schösslings nach oben würde bewirken, dass er spiralig aufsteigt, wie ein ächter Winder. Eine von unsern Zweig-klimmenden Dalbergieen, welche ich zuerst ohne Blätter fand und die nun junge Blüthenknospen hat, scheint eine Triptolemaea zu sein, und diese Gattung setzt Endlicher unmittelbar hinter Dalbergia.[8]

Sie fragen, ob nicht Naturgeschichte ausserordentlich anziehend gemacht wird durch solche Ansichten, wie wir beide haben.[9] Ganz sicherlich! Seitdem ich Ihr Buch über die Entstehung der Arten[10] gelesen habe und seitdem ich zu Ihren Ansichten bekehrt worden bin, sind mir viele Thatsachen, auf die ich früher mit Gleichgültigkeit sah, ausserordentlich bemerkenswerth geworden; andere, welche früher reine bedeutungslose Wunderlichkeiten erschienen, haben eine hohe Bedeutung erlangt, und so hat die ganze Natur ihr Angesicht verändert. Ich werde deswegen niemals im Stande sein, meine tiefe Dankbarkeit entsprechend auszudrücken, noch die grosse Verbindlichkeit, die mir Ihnen gegenüber obliegt.

Ich will eine einzige Thatsache anführen, welche mich sehr verwirrte, bis sie mir durch Ihr Buch verständlich wurde. In den Spalten granitischer Felsen an der Ostküste unserer Insel fand ich eine Art von Apium, welche in ihrem allgemeinen Aussehen und noch mehr in ihrem Geruch der wilden Sellerie ausserordentlich ähnelte, wie ich sie in früheren Jahren auf den sumpfigen Wiesen an der Ostseeküste gesammelt habe; dieses Apium war in Gesellschaft von einer Tetragonia, welche gleich schien mit Tetragonia expansa, die in Deutschland unter dem Namen neuseeländischer Spinat kultivirt wird, und an demselben Fleck war die sandige Küste bedeckt von einer Scaevola, welche in unserer Flora, soweit ich weiss, die einzige Gattung ist aus einer fast ganz auf Neuholland[11] beschränkten Familie. Jetzt, durch Ihr Buch, verstehe ich, warum Küstengattungen viel leichter einen so weiten Verbreitungsbezirk haben als andere Pflanzen.[12]

Ich beabsichtige bis 1870 oder 1871 in Desterro zu bleiben und dann nach dem Itajahy-Fluss zurückzukehren, wo ich von 1852–1856 lebte und wo ich eine kleine Kolonie besitze.[13] Die Landschaft auf unserer Insel ist sehr schön; sogar Reisende, welche die Inseln des Stillen Oceans, Java, u. s. w. besucht hatten, sagten mir, dass unsere Insel einer der schönsten Punkte wäre, welche sie je gesehen hätten. Unglücklicherweise hat die Vegetation jetzt viel von ihrer früheren Grossartigkeit verloren; die Urwälder sind beinah vollständig verschwunden, und viele unserer Hügel sind jetzt fast ausschliesslich bedeckt von niedrigen Büschen einer unansehnlichen Dodonaea.

Herr L. Agassiz erforscht augenblicklich den Amazonenstrom; wie ich aus einem in einer Rioer Zeitung veröffentlichten Briefe entnehme, so hofft er, dass die geographische Verbreitung der Fische in jedem Strom schlagende Gründe gegen die Umwandlungstheorie liefern wird.[14] Es scheint mir unmöglich, dass eine einigermassen vollständige Kenntniss von der geographischen Verbreitung der Fische in jenem mächtigen Strom und seinen zahlreichen Zuflüssen in ein paar Jahren erworben werden kann, und ich kann mir nicht denken, dass irgend ein entscheidender Einwand aus einer unvollkommenen Erkenntniss derselben abgeleitet werden kann.

Mit dem Wunsche, dass Ihre Gesundheit recht bald wieder gut werde, bin ich, werther Herr, in der aufrichtigsten Hochachtung | immer treulichst der Ihrige. | Fritz Müller.

Möller 1915–21, 2: 76–7.

[1] See letter from Fritz Müller, 12 August 1865, n. 1. For a translation of this letter, see Appendix I.

[2] In his letter of 20 September [1865], CD said that he had sent Müller the German edition of *Orchids* (Bronn trans. 1862), and two papers on dimorphism, 'Two forms in species of *Linum*' and 'Three forms of *Lythrum salicaria*'.

[3] When CD described *Nesaea verticillata* (family Lythraceae) as trimorphic in *Forms of flowers*, p. 167, he added a note that Müller had observed a homostyled species of the genus in Santa Catarina, Brazil.

[4] In *Forms of flowers*, p. 168, CD described three species of *Cuphea* (family Lythraceae) as possessing traits of heterostyly, suggesting that they may have reverted from a heterostyled condition. CD's notes on *Cuphea* are in DAR 109: B2. Müller found only homostyled species of *Cuphea* in Brazil (see Müller 1868b, p. 113).

[5] Müller published papers on orchids, and is cited extensively in 'Fertilization of orchids', and *Orchids* 2d ed.

[6] CD had asked Müller for information on twining plants ascending thick trunks in his letter of 20 September [1865]. In 'Climbing plants', p. 21, CD had noted that English plants, with the exception of honeysuckle (*Lonicera periclymenum*), never twined around trees, while tropical climbers were able to ascend thick trees. In *Climbing plants* 2d ed., p. 37 n., CD added Müller's observations of root-climbers and tendril-bearers ascending thick trees in southern Brazil. See also *Correspondence* vol. 12, letter to J. D. Hooker, 26 November [1864], and letters from J. D. Hooker, 29 November 1864 and 2 December 1864.

[7] In 'Climbing plants', p. 22, CD had reported that, according to Thomas Thomson and Joseph Dalton Hooker, some *Dalbergia* species ascend thick trees by twining. See also *Correspondence* vol. 12, letter from J. D. Hooker, 29 November 1864, and letter to J. D. Hooker, 4 December [1864]. On branch-climbers, see letter from Fritz Müller, [12 and 31 August, and 10 October] and nn. 2, 8, and 19.

[8] In Stephan Ladislaus Endlicher's *Genera plantarum*, the description of *Dalbergia* is immediately followed by that of *Trioptolemea*; both are placed in the tribe Dalbergieae (Endlicher 1836–42, pp. 1304–5).

[9] See letter to Fritz Müller, 20 September [1865].

[10] *Origin*. The first German edition was published in 1860 (Bronn trans. 1860).

[11] New Holland was the Dutch name for Australia (*EB*).

[12] In *Origin*, pp. 358–65, CD presented evidence suggesting that plants could migrate across large bodies of water by means of seeds or fruit that floated, or had been embedded in drift wood or icebergs, or carried by birds. To test his theory, CD had performed experiments between 1855 and 1856 in which he immersed seeds and fruits in salt water (*Correspondence* vols. 5 and 6); he discussed the theory with Hooker over several years (see letter to J. D. Hooker, 22 and 28 [October 1865] and nn. 12–14).

[13] Müller explored the region of the Itajaí Açu River in south-eastern Brazil between 1852 and 1856. In 1854, he acquired a homestead in the German settlement of Itajaí, at the mouth of the river. He began teaching mathematics at a provincial school in Destêrro (Florianópolis) in 1856, but retained his property in Itajaí (Möller ed. 1915–21, 3: 61–9).

[14] Louis Agassiz began his Brazilian expedition, which included a voyage on the Amazon River, in April 1865. While on the steamer travelling from New York to Rio De Janeiro, he delivered a lecture to

his students in which he expressed his conviction that the combination of animals in South America would give him the means of showing that the transmutation theory was 'wholly without foundation in facts' (J. L. R. Agassiz and Agassiz 1868, p. 33). Agassiz never published a formal description of the expedition's specimens, or an analysis of their variation or distribution (see Winsor 1991, pp. 66–76). His informal comments on the fish of South America are given in J. L. R. Agassiz and Agassiz 1868, pp. 216–27, 237–41, and 377–84.

From F. W. Farrar 6 November 1865

Harrow.
Nov 6. 1865.

Dear Sir,

I can say with perfect sincerity that your word of approval is to me the most pleasant result which my "Chapters on Language" have yet procured me.[1] That the theory which I have advocated appears to you to rest on secure grounds is to my mind a great confirmation of its reasonableness.

With regard to the "Origin of Species⟨"⟩ I am not vain enough to claim the right of attaching any great value to my own opinion.[2] Perhaps my Polygenist prejudices may have led me to it, although I am aware that your views are not absolutely *incompatible* with a belief in the *primordial* diversity of the present families of the human race,—since it is of course conceivable that there may have been different lines of development each terminating in a specific human type.[3] Still I suppose that you would hardly hold this view, & I see that Prof. Huxley in the little book which you name, & which I possess, distinctly gives in his adhesion to Monogenism.[4] Now I confess that, so far as I can see, History, & even Tradition, as far back as their primeval dawn, *prove* to us the existence of the several human races unchanged from their present physical characteristics; & if it be demonstrable that, under every possible variety of external influence & physical condition, the chief existing races have remained unaltered for say 5000 years—is not this a very strong argument for the Polygenist?—[5]

My state of mind however on this question is a mere *suspension of assent* & nothing would surprise me less than the discovery of some fresh palæontological fact, or the artificial production of some new species, which would practically decide the question in favour of your hypothesis.[6]

I am so ignorant on these subjects that I only hope that I have been saying nothing absurd. Pray pardon me if I have. Since in your former note you were so good as to mention some Ethnological papers of mine,[7] I have taken the liberty of enclosing one of my anonymous contributions to the Anthropological Review, which may possibly have a moment's interest for you, though I fear it is very crude.[8]

I hope that you will not take the trouble to acknowledge it, and that you will believe me to be, with very great respect, | Your's very faithfully, | Frederic W Farrar.

DAR 164: 35

1 For CD's favourable view of Farrar 1865, see letter to F. W. Farrar, 2 November [1865].
2 See letter to F. W. Farrar, 2 November [1865] and n. 6.
3 For CD's views on the origin and classification of human races, see *Correspondence* vol. 12, letter to A. R. Wallace, 28 [May 1864], and this volume, letter from Henry Denny, 23 January 1865, n. 3. CD did not publicly advocate monogenism, the view that humans originated as a single race, until 1871 (see *Descent* 2: 388). Many of the leading proponents of polygenism, however, were opposed to CD's transmutation theory (see *Correspondence* vol. 12, letter from A. R. Wallace, 10 May 1864 and n. 7).
4 See letter to F. W. Farrar, 2 November [1865] and n. 7. Thomas Henry Huxley had stated that there was 'no evidence whatever for saying, that mankind sprang originally from any more than a single pair' (T. H. Huxley 1863b, p. 117). Huxley went on to describe how distinct races and varieties could arise through the operation of natural selection (*ibid.*, pp. 121–3). See also letter to T. H. Huxley, 12 July [1865], n. 8.
5 Some of Farrar's objections to monogenism were expressed in papers delivered at the Ethnological and Anthropological Societies of London in 1864. Farrar rejected the alleged universality of cultural traditions, such as the belief in God, which was often appealed to as evidence in support of monogenism (Farrar 1864a and 1864b). He also argued that the fertility of the offspring of unions between different human races was inconclusive (Farrar 1864c). CD cited Farrar 1864b on the absence of belief in God among some human races in *Descent* 1: 65 n.
6 CD discussed the imperfection of the geological record in *Origin*, pp. 279–311. Huxley was the first to advance the view that CD's theory of transmutation could not be fully accepted until proved by the production of new species by means of artificial selection (see, for example, [T. H. Huxley] 1860b and T. H. Huxley 1863b, pp. 147–9). Huxley's suggestion was considered at length by CD and served as a partial impetus for some of his botanical experiments on cross and hybrid sterility (see, for example, *Correspondence* vol. 10, Appendix VI).
7 See letter to F. W. Farrar, 11 October [1865] and n. 3.
8 Farrar may refer to one of several anonymous articles that appeared in the 1864 and 1865 volumes of the *Anthropological Review*. The article 'Pott on myths of the origin of man and language' (*Anthropological Review* 2 (1864): 24–30), included a discussion of criticisms by German philologists of Friedrich Max Müller's theory of language. The article 'On the phenomena of hybridity' (*ibid.*, pp. 164–73), reviewed the work of the French anthropologist Paul Broca for its bearing on the monogenist theory of human descent. The May 1865 issue carried an article titled 'On the plurality of the human race' (*Anthropological Review* 3 (1865): 120–32).

From Asa Gray 6 November 1865

Cambridge. Mass.
Nov. 6. 1865.

My Dear Darwin

I am very glad to hear from you, and to see half your letter of the 19th Oct. in your own handwriting is a good sign.[1] I do hope you may get a comfortable winter, and bring out your next volume without breaking down.[2]

I am pleased that you approve my abstract of your Climber paper.—but *observe* it was only of the *1st* part of your elaborate article—[3] But as to *praise* you speak of, I am sure you pay me back, with interest.[4]

I lately sent to Silliman as much more—a large part, indeed, *extracts*, which I could not shorten—on the tendril-bearing part of your paper. But Dana sent me the proof, with all my long extracts omitted for want of room.[5] This reduced my article to incoherence, so I begged all to be laid over for the January number, when I hope to have room. I entertained our Social Scientific Club here with your article, and all were greatly interested.[6] As to Climbing Roses, they are the strong *summer* shoots, growing after flowering, which I find frequently running their heads into dark corners of the porch over my door, &c—[7]

That is very curious—but quite what I looked for—that dimorphous species self-fertilized, should act *like hybrids* (sterile or dwarfs, &c—)

You *must* publish these facts in some brief article—[8]

"Stephens" = Stevens, was a New Yorker—is dead, years ago.—[9] wrote most amusing & popular travels—in Egypt, as well— Central America was his first & freshest book.[10] But only *amusing*, as far as I recollect.

So Palmerston is gone—a fine specimen of a John Bull he was—a very typical specimen. The Yankees can't help admirng & liking him.—tho' not for any good he ever did us.[11] But as for his successor, he is a prig,—a juiceless stick.[12]

Don't you think Adams pays him back nicely for propo⟨sing⟩ that they should sit down and rejoice together over the abolition of slavery.[13] Just to see how the world has moved, turn back to Russell's letter/lecture to be read to Mr. Lincoln, on occasion of his proclamation of emancipation![14]

Good Bye, my dear good fellow, and recover health as fast as ever you can.

Yours affectionately | A. Gray

Postmark: NO 20 65
DAR 165: 149

1 The second half of CD's last letter to Asa Gray, 19 October [1865], was written by Emma Darwin.

2 In his letter of 19 October [1865], CD had told Gray that although his health was somewhat improved, he could still do 'no regular work'. CD had stopped work on *Variation* in April 1865 because of poor health (see letter to John Murray, 2 June [1865], n. 1, and 'Journal', Appendix II; see also Appendix IV).

3 Gray's abstract of 'Climbing plants' appeared in two parts (A. Gray 1865–6). The first part, covering the sections of the paper on twiners and leaf-climbers, appeared in the September 1865 issue of the *American Journal of Science and Arts.* The second part, covering tendril-bearers and the paper's concluding remarks, appeared in the January 1866 issue. CD's lightly annotated copy of A. Gray 1865–6 is in the Darwin Pamphlet Collection–CUL.

4 See letter to Asa Gray, 19 October [1865] and n. 3.

5 Benjamin Silliman Jr and James Dwight Dana were co-editors of the *American Journal of Science and Arts.*

6 Gray had helped to found the Cambridge Scientific Club in the 1840s. The group was composed largely of Harvard faculty, and met twice a month at a different member's house for a meal, and the presentation and discussion of a paper. Louis Agassiz was also a member of the club, and some of the meetings were occasions of debate between him and Gray. See Dupree 1959, pp. 121–2, 258, 285.

7 See letter to Asa Gray, 19 October [1865] and n. 4.

8 See letter to Asa Gray, 19 October [1865] and n. 10. See also letter from B. D. Walsh, 29 May 1865, n. 3.

9 John Lloyd Stephens had died in 1852 (*DAB*). See letter to Asa Gray, 19 October [1865].

10 Stephens's first book was *Incidents of travel in Egypt, Arabia, Petaea, and the Holy Land* (Stephens 1837). Gray also refers to *Incidents of travel in Central America, Chiapas, and Yucatan* (Stephens 1841).

11 Henry John Temple, third Viscount Palmerston, died on 18 October 1865; he had been prime minister since 1859 (*DNB*). Palmerston's policy of neutrality in the American Civil War is discussed in E. D. Adams 1925, 2: 37–74, and Jenkins 1974–80, 2: 66 and 167.

[12] John Russell was foreign secretary under Palmerston and succeeded him as prime minister (*DNB*).

[13] In his capacity as United States minister to Great Britain, Charles Francis Adams was involved in negotiations with the British government to obtain compensation for commercial losses inflicted by Confederate vessels that had been built or given shelter in British ports during the American Civil War. On 26 October 1865, the *New York Times* printed the diplomatic correspondence between Adams and Russell concerning the US claims. In his letter of 30 August 1865, Russell maintained that the British government had acted honourably as a neutral power, and could not be held responsible for damages caused by the southern vessels; he concluded by congratulating the US government for having resolved to abolish slavery, a practice for which the British had always held 'the deepest abhorrence': 'The efforts by which the United States Government and Congress have shaken off slavery have, therefore, the warmest sympathies of the people of these kingdoms'. In his reply, Adams reiterated the US claims and remarked that it had been his misfortune 'to observe in the process of so wonderful a revolution, a degree of coldness and apathy prevailing in many quarters' of the British public; he hoped, however, that 'the result arrived at [would] ultimately correct the hasty and harsh judgments that flowed from lack of faith and of confidence in [the US government's] fidelity to a righteous cause'. The correspondence between Adams and Russell also appeared in the *London Gazette* (supplement), 11 October 1865, pp. 4833–60.

[14] The 'letter' or 'lecture' has not been identified; however, Russell issued several statements following the decree made by the US president, Abraham Lincoln, on 22 September 1862, which gave freedom, effective from 1 January 1863, to slaves in the states that were currently at war with the Union. In a memorandum published by the Foreign Office and circulated to Cabinet ministers on 13 October 1862, Russell criticised the decree for 'want of consistency', and described it as 'an act of punishment and retaliation inflicted by a belligerent upon a hostile community', rather than an attempt to satisfy the 'claims of humanity' (see Bourne and Watt eds. 1986–7, 6: 94–5). See also Earl Russell to Earl Cowley, 'Despatch respecting the Civil War in North America', *Parliamentary Papers*, 72 (1863): 3–4, and Earl Russell to Lord Lyons, 17 January 1863, *ibid.*, pp. 51–2. Russell's response to the 'emancipation proclamation' is discussed in Jenkins 1974–80, 2: 174–5.

From George Henslow 6 November 1865

10 South Crescent | Bedford Sq: W.C.
Nov 6/65

My dear Sir,

I beg to thank you very much for yr kind & instructive letter:[1] there are one or two passages in it which,—with your permission I should like to read in connection with a few notes I purposing giving at the next Linnean Meeting.[2]

I should much like to read Sprengels account of Salvia, & am glad to find my surmise to have been corroborrated by yourself. (Would you kindly give me the reference to Sprengls paper?)[3] With regard to *Medicago*, perhaps the word *Irritability* is not exactly the true one. *Elasticity* also does not express the *facts* which are these. The stamens are at first *straight* & *horizontal* included within the alæ & Carina; the former of which have little processes overlying the stamens & pressing down upon them: this, they are able to do by being firmly locked with the carina.[4]

The instant any object, e.g. a pin, is directed down the *channel* of the Vexillum, it cannot help pressing upon these "processes" laterally. The stamens then spring up & the carina *spreads out* &, with the alæ, drop. But the stamens *not only* rise up but immeadiately become curved towards the Vexillum in the form of a $\frac{1}{2}$ circle;

so that there seems to me to be some other force required, besides mere *elasticity*, especially as this would imply the possibility of making them return to their original horizontal position: which they *will* not do, but *crack transversely* if it be attempted. Having however once assumed their '*curved*' '*rigid*' & '*vertical*' position it cannot be altered.[5]

I *have* read *Cohn's* paper & have been much interested in it, especially as I had been, *also*, examining some Centaureas,[6] & without knowing anything had been previously observed, *thought* I saw some peculiarity, but I found it was simply *elasticity*. I did not discover the independent irritability.[7] My own conclusion was this;—that as the style rises & stretches the anthers to their full extent by pushing against *their closed valvate extremities*, until it succeeds in forcing its way through,—so on the other hand, the instant the style *is* through, the tension of the filaments is relaxed & the anthers are drawn down again. So that two *opposing* forces are brought into play, the object being to enable the ring of hairs (if present) more effectually to sweep out the pollen, & so elevate on the *outside* of the two stigmatic ends which *now* & not till now, separate.

I presumed all this was intimately connected with intercrossing.[8]

Many thanks for your kind offer of giving me further information; which, I should like to avail myself off at another time.

Yours very truly | Geo. Henslow

P.S. I beg to thank you for the printed matter which I will return[9]

DAR 166: 151

[1] See letter to George Henslow, [2–5 November 1865].

[2] Henslow cited CD's observations of *Medicago lupulina* and other species in his paper delivered at the Linnean Society on 16 November 1865 (G. Henslow 1865; see letter from George Henslow, 1 November 1865 and n. 5).

[3] See letter from George Henslow, 1 November 1865. In the missing portion of his letter of [2–5 November 1865], CD evidently referred Henslow to the work of Christian Konrad Sprengel, who argued that floral structures were designed to bring about pollination by insects (Sprengel 1793); Sprengel described the anther movements of *Salvia pratensis*, which enabled pollen to be transferred from different flowers on the backs of bees, in *ibid.*, p. 61. CD scored this passage in his extensively annotated copy of the book, which is in the Darwin Library–CUL (see *Marginalia* 1: 774–85). CD referred to Sprengel's description of 'the admirable mechanical adaptations in [*Salvia*] for favouring or ensuring cross-fertilisation' in *Cross and self fertilisation*, p. 93 n. See also *Correspondence* vol. 3, letter to J. D. Hooker, [11–12 July 1845] and n. 13, and *Correspondence* vol. 12, letter to Friedrich Hildebrand, 25 June [1864] and n. 9.

[4] The carina is the structure formed by the two anterior petals of a papilionaceous flower. For the other botanical terms, see the letter from George Henslow, 1 November 1865 and n. 3.

[5] Henslow said that he had been unable to discover whether the curvature of the stamens was caused 'by distention of the cells on the convex side or a corresponding emptying of those on the concave' (G. Henslow 1865, pp. 327–8).

[6] Ferdinand Julius Cohn's paper on the contractile tissue of plants (Cohn 1860) was based principally on observations of the movements of stamens in *Centaurea*. CD's heavily annotated copy is in the Darwin Pamphlet Collection–CUL. On CD's interest in Cohn's work, see the letter from T. H. Huxley, 1 May 1865, n. 7.

[7] Cohn described how, on contact, the stamens of *Centaurea* first contracted, then began to extend themselves and acquire a curved condition similar to that which occurs in an irritated muscle; Cohn claimed that the movement of the plant was not due to mechanical elasticity, but to an 'expansive Kraft' (expansive force) and 'selbst fortgeleitete Kraft' (innate motor power) (Cohn 1860, pp. 4–6, 39–40, 47–8). A long abstract of the paper was published in English in the *Annals and Magazine of Natural History* 11 (1863): 188–202. CD cited this abstract in *Insectivorous plants*, pp. 256–7.

[8] Cohn observed that insects first lighted on the flowers, causing the contraction of the stamens and the consequent extrusion of pollen from the anthers; the insects then carried the pollen to other flowers, thereby effecting cross-pollination (Cohn 1860, pp. 44–5, 48).

[9] Henslow may refer to a copy of CD's letter to the *Gardeners' Chronicle*, [before 13 November 1858] (*Correspondence* vol. 7), on the agency of bees in the fertilisation of papilionaceous flowers. See letter from George Henslow, 1 November 1865 and n. 9. See also letter from George Henslow, 2 December 1865.

From Ernst Haeckel[1] 11 November 1865

Jena
11. November 1865.

Mein hochverehrter theurer Herr!

Nach längerer Zeit des Schweigens darf ich wohl wieder Ihre kostbare Zeit zum Lesen eines Briefes von mir in Anspruch nehmen.[2] Durch Herrn Huxley habe ich zu meinem grossen Bedauern erfahren, dass Sie leider wieder sehr krank gewesen sind.[3] Ich hoffe, dass es Ihnen wieder besser geht, und dass Sie der Wissenschaft und uns noch recht lange erhalten bleiben.

Leider ist mein Buch, welches Ihre Theorie ausführlich behandelt, immer noch nicht vollendet, trotzdem ich es schon vor einem halben Jahre fertig zu sehen hoffte.[4] Ich habe aber diesen ganzen Sommer mit Arbeiten am zoologischen Museum verlieren müssen;[5] da ich nämlich vor 6 Monaten hier zum ordentlichen Professor der Zoologie ernannt worden bin, nachdem ich eine Berufung nach Würzburg abgelehnt hatte.[6] Indessen hoffe ich das Buch bis nächsten Weihnachten vollenden zu können und werde es Ihnen sogleich nach Vollendung zusenden.

In diesem Winter hatte ich hier wieder öffentliche Vorlesungen "*über Darwins Theorie*", welche die besuchtesten von *allen* Vorlesungen sind, die an der hiesigen Universität gehalten werden.[7] Ich habe über 150 Zuhörer, und das grösste Auditorium unserer Universität ist gefüllt. Sie können daraus ersehen, welches ausserordentliche Interesse hier an unserer kleinen Universität für Sie und Ihre Theorie herrscht und wie lebhaft die Theilnahme an der grossen Reform der Wissenschaft ist, welche Sie begonnen haben, und welche durchführen zu helfen die grösste Aufgabe meines Lebens ist. Meine Zuhörerschaft ist sehr gemischt, es sind darunter ausser vielen Medicinern und Naturforschern auch viele Historiker und Studenten der Philosophie; selbst einige Theologen und Juristen fehlen nicht!

Für die Übersendung Ihres schönen Aufsatzes "on the climbing plants" sage ich Ihnen meinen herzlichsten Dank.[8] Ich habe ihn mit grossem Interesse gelesen.

Hoffentlich werden Sie mein Buch über die Geryoniden (mit 6 Tafeln), sowie einen Aufsatz "über fossile Medusen" und einen anderen über "Rhizopoden

Sarcode" richtig erhalten haben.[9] Ich hatte Herrn Engelmann in Leipzig beauftragt, Ihnen diese Arbeiten sogleich zu senden.[10] Unter den neuen Rhizopoden aus Nizza wird Sie der "Protogenes primordialis" (Fig 1, 2) vielleicht interessiren, als eines der allereinfachsten Geschöpfe, ein Organismus ohne Organe, durchaus homogen![11] Die Generatio aequivoca eines solchen lebenden Eiweiss-Klumpens liesse sich schon allenfalls denken, und damit wäre der Descendenz-Theorie über Ihren schwierigen Anfang hinweggeholfen.[12]

So oft ich Ihnen schreibe, mein theurer Herr, muss ich Ihnen den wärmsten Dank wiederholen für die unerschöpflische Quelle der Geistesthätigkeit und der lebendigen Anregung, welche Ihre classische Theorie für mich ist. Seitdem mir durch den Tod meines innigst geliebten Weibes alles Leben des Gemüthes abgeschnitten und getödtet ist, ist mir das Streben, mit dem geschärften Verstande die wahre Natur und den Zusammenhang der natürlichen Dinge zu erkennen, allein übrig geblieben.[13] Dieses Streben beschäftigt mich in meinem Einsiedlerleben Tag und Nacht, und dieses Streben verdanke ich nur Ihrem Wercke!

Mögen Sie uns noch lange in frischer Kraft erhalten bleiben, um die von Ihnen begonnene Reform der Biologie durchgeführt zu sehen und die Früchte Ihrer grossen Geistesthat selbst zu pflücken. Das ist der herzlichste Wunsch Ihres aufrichtig Ihnen ergebenen | Ernst Haeckel

DAR 166: 40

1 For a translation of this letter, see Appendix I.

2 Haeckel's last known letter to CD is dated 26 October 1864 (see *Correspondence* vol. 12); however, CD received a copy of Haeckel 1865b in May 1865 (see letter to J. D. Hooker, 4 May [1865]).

3 Haeckel refers to Thomas Henry Huxley. On CD's health, see the letter to J. D. Hooker, 22 and 28 [October 1865].

4 The book was eventually published as Haeckel 1866. In his letter of 9 [July 1864] (*Correspondence* vol. 12), Haeckel had claimed that in his next book he would show that the theory of descent threw 'light on every single chapter' of the history of creation and unified them 'in the most perfect harmony'. See also *ibid.*, letter from Ernst Haeckel, 26 October 1864. CD's annotated copy of Haeckel 1866 is in the Darwin Library–CUL (see *Marginalia* 1: 355–7).

5 Haeckel succeeded Carl Gegenbaur as director of the Zoological Museum in Jena in 1862, and began a programme of improvement and acquisition. In the summer of 1865, Haeckel successfully petitioned the Weimar state ministry to finance a complete reorganisation of the collection, a new hall to house acquisitions, and instruments for research. Haeckel was able to obtain government funding by establishing a separate Zoological Institute within the Museum. See Uschmann 1959, pp. 48–58.

6 Haeckel received a full professorship at Jena on 15 May 1865 (Uschmann 1959, p. 50). Five days previously, he had been offered a full professorship at Würzburg, but declined. The terms of the Würzburg position stipulated that control of the zoology museum and examinations remain with another, more senior, zoology professor (*ibid.*, p. 51).

7 The popularity of Haeckel's course on 'Darwins Theorie' is described in Uschmann 1959, p. 44. Since 1862, Haeckel had also given a separate annual lecture on CD's theory, and had devoted part of his zoology course to the topic (see *ibid.*, pp. 43–4, and *Correspondence* vol. 12, letter from Ernst Haeckel, 9 [July 1864]). On the reception of CD's theory in Germany, see letter from T. H. Huxley, 29 May 1865 and n. 3.

8 'Climbing plants'. No presentation list for the paper has been found.

[9] Haeckel refers to his monograph on the Geryonidae (now called Geryoniidae), a family of medusae (Haeckel 1865a), and to Haeckel 1865c and 1865d. A copy of Haeckel 1865a is listed in the 1875 catalogue of CD's library (DAR 240); however, it is not recorded in later catalogues and has not been found in the Darwin Library–CUL or at Down. A lightly annotated copy of Haeckel 1865d is in the Darwin Pamphlet Collection–CUL.

[10] The Leipzig firm Wilhelm Engelmann had published Haeckel 1865a, and was also the publisher of *Zeitschrift für wissenschaftsliche Zoologie*, in which Haeckel 1865c and 1865d appeared.

[11] Haeckel described *Protogenes primordialis* as the simplest type of Rhizopoda; a protoplasmic aggregate, without differentiation, organisation, or nucleus, it was able to surround and assimilate foreign bodies and to reproduce by simple division (Haeckel 1865d, pp. 360–3). In his later book, *Generelle Morphologie der Organismen* (Haeckel 1866, 2: XXII–XXIII), Haeckel created a new classification for the organism within a phylum he named the Monera, within the kingdom Protista, arguing that the Rhizopoda had possibly evolved from a *Protogenes* species. The organism is not recognised in modern taxonomic literature.

[12] *Generatio aequivoca* or equivocal generation: 'the (supposed) production of plants or animals without parents; spontaneous generation' (*OED*). Haeckel did not discuss *Primogenes primordialis* as a case of spontaneous generation (Haeckel 1865d); however, in *Generelle Morphologie der Organismen* (Haeckel 1866, 2: 174), he did speculate more generally on the possibility of spontaneous generation. For discussions of Haeckel's theories of protoplasm and the development of living organisms from inorganic matter, see Rehbock 1975 and Rupke 1976. On the Victorian debates over spontaneous generation, see Strick 2000.

[13] Haeckel's wife, Anne Sethe, had died on 16 February 1864; Haeckel discussed her death and its significance for his scientific work in his letters to CD of 9 [July 1864], 10 August 1864, and 26 October 1864 (*Correspondence* vol. 12).

From B. D. Walsh 12 November 1865

Rock Island. Illinois U.S.
Nov. 12, 1865

Chas. Darwin Esq.
Dear Sir,

I have not answered sooner your letter of July 9 because, as your health was so poor, I was afraid of worrying you by frequent scribbling.[1]

I received your Paper on Climbing Plants & was very much interested in it.[2] I had always supposed that Climbers had some unknown means of perceiving where there was a twig, or other support suitable for them to grasp, & went straight to their mark as a wolf goes straight at a deer. But this discovery of their sweeping circles & groping in the dark for support, like a blind Cyclops, is very astonishing. And yet, when we consider that plants have no eyes, it is evidently the only possible way, & like the other ways of Nature so beautifully simple. The gradations too that you trace between the different kinds of Climbers are of high philosophical interest. Don't be afraid that I shall ever find anything which you publish "not worth reading".[3] How you manage to get through such an amount of work is a mystery to me. If I did not know that you had been my Contemporary at college, I should suppose that you must be at least a hundred years old.[4] An ordinary life-time seems too brief for what you have accomplished.

I conclude that you are aware that there was a long notice of your Paper in Silliman's Journal by Asa Gray—the July number if I recollect right.[5] He speaks of it in very high terms.

The extract I sent you as from a work by *Gen. Paez* was, I find, from a work by *a son of Gen. Paez*.[6] My memory played me false.

I don't see—with my entomological ideas—how it can be possible for the *males* of *Daphnia* or any other genus to produce eggs. There may be dimorphous female forms in the genus, both of which produce eggs, but to call a form which oviposits a male seems to me simply an abuse of language. According to Dr. Minor, quoting from Wagner, both male & female *Daphnia* oviposit "without sexual influence", i.e. as I understand it without copulation or any other mode of impregnation.[7] Why then call one form a male & the other a female? If there is no impregnation, I don't see how there can exist any male sexual organs; for, if they had primordial existence, they would have long ago been obliterated by disuse. But probably this may be all very plain to you, though to me, who know but little of Natural History out of my own special department, it is unintelligible. Please, if not too much trouble, lighten my darkness. Perhaps the supposed males are really hermaphrodites.

I sent you a few days ago a Paper of mine on "the new Potato Bug", out of which you may possibly glean a useful fact or two.[8] I have another Paper just ready for the Press "On Phytophagic Species & Unity of Coloration", which will perhaps be of more interest to you. As soon as printed, I shall of course do myself the pleasure to forward you a copy.[9] By the way I was vexed at finding out some months ago, that your Post-office is so rigidly stringent, that they do not allow simple errors of the press to be corrected in MS. in the margin without charging letter postage;[10] & that I had in this way, through my thoughtless ignorance, put Mr. Bates to some trouble.[11] I shall know better for the future. Here every body does it, & everybody puts as many Pamphlets as they please under a single stamp, provided they don't exceed the just weight. Whereas I find your Post office requires a separate package & separate stamp for every Pamphlet, even if they each contain but four pages.[12]

I have been continuing this year my investigations into the Dimorphism of *Cynipidæ*, & though there are many mysteries still to solve, yet I think I have thrown some additional light upon the question.[13] I find that (unless I am mistaken) the agamous dimorphous ♀ form *aciculata* o.s. produced galls out of which came many specimens of the same agamous female form, (A single ♀ of course out of a single gall, this gall being monothalamous,) & also one ♀ specimen of the copulative ♀ form *spongifica*. Assuming the truth of the above—and to be quite certain of the truth I have been repeating my experiments this autumn—two consequences follow:—1st. that *aciculata* & *spongifica* are necessarily the same species, (as I had concluded from other evidence,) since one gives birth to the other, & 2nd. that an agamous ♀ form may generate the same agamous ♀ form for at least two successive generations & probably for more. My present idea is, that the supposed agamous *Cynipidæ* generate only agamous ♀ ♀ for an indefinite series of years, say 4 or 5 or 8 or 10, & that then comes forth a brood ♂ ♀ of the copulative dimorphous form. There is an analogy to this in *Aphis*, some species of which produce a long series of agamous apterous ♀ generations all through the summer & then in the autumn the winged copulative ♂ ♂ ♀ ♀, while others produce the copulative & the agamous forms promiscuously all through the season.

You may perhaps like to know the *rationale* of my experiments. They proceed upon a postulation, which I have ascertained by long observation & which you must take my word for, viz. that *Cynips* scarcely ever fly at all but breed from year to year on the same tree. A tree that is infested by a particular species one year, will always have more or less galls of the same species on it the following year; & in the case of those particular kinds of gall (such as *spongifica*, *ficus* & *poda*) which hang on the tree more than 12 months, the particular boughs which bear the old galls are very generally, as I have repeatedly noticed, those which bear the new galls.

First, then, I select Black Oak Trees, which are isolated from others, & which I know have borne no *spongifica* galls the preceding year. Then in the autumn I place upon each of these trees 8 or 10 lively *aciculata*, sometimes as high up as I can reach on the trunk, sometimes on the overhanging boughs, taking notes in my Journal of the particular trees & where I placed the insects. Of three trees experimented on thus last autumn, one bore no galls at all, one, a rather small tree, where I had put the insects on the trunk, was covered all over with the galls this spring, so that I had to climb to get them, & the third—a very large tree with overhanging boughs, where I had put all the Cynips on certain particular boughs—had about a dozen galls on these particular boughs, which formed perhaps $\frac{1}{20}$ part of the whole tree, & none whatever on any of the other boughs. Of course, if these galls had been made, not by my bred *Cynips*, but by wandering individuals, the chances are some 19 to 1 against their selecting the same particular boughs as I did.

I carefully preserved alive the dimorphous ♀ agamous forms raised in each of these cases, & have placed them this autumn on isolated oaks again, so as to find out what they will produce next year. It is tedious having to wait so long for the fruits of one's labor, but it seems to me the only possible way of solving the enigma. I have tried enclosing the insects on the trees in gauze bags, & the hateful boys will not let them alone.

But I shall tire you to death. Adieu, & may your health soon improve. | Yours ever very truly, | Benj. D. Walsh

P.S. Has Mr. Lubbock arrived at any conclusions on this subject?[14]

Field Museum of Natural History, Chicago (Riley bequest of 1948)

1 See letter to B. D. Walsh, 9 July [1865].

2 CD had sent Walsh his paper 'Climbing plants' earlier in the year (see letter to B. D. Walsh, 9 July [1865]). No presentation list for the paper has been found.

3 See letter to B. D. Walsh, 9 July [1865].

4 See letter from B. D. Walsh, 1 March 1865 and n. 2.

[5] The first part of Asa Gray's review of 'Climbing plants' appeared in the September 1865 issue of the *American Journal of Science and Arts* (A. Gray 1865–6). The journal was also known as 'Silliman's Journal' after its founder, Benjamin Silliman. See letter from Asa Gray, 6 November 1865 and n. 3.

[6] In his letter of 29 May 1865, Walsh had discussed a passage from *Wild scenes in South America, or life in the llanos of Venezuela* by Ramón Páez (Paez 1863).

[7] Walsh refers to the review of Wagner 1865 by William Chester Minor (Minor 1865). See letter from B. D. Walsh, 29 May 1865 and n. 14.

[8] Walsh 1865. Walsh's article 'The new potato bug, and its natural history' described a beetle, *Doryphora lineata*, that had recently migrated from the Rocky Mountain region into the state of Illinois. Walsh suggested that the insect fed on wild species of *Solanum*, and had migrated eastward as cultivated varieties of *Solanum*, such as domestic potatoes, were introduced into the Plains region. For a discussion of Walsh's research on the potato beetle, see Sorensen 1995, pp. 121–5.

[9] The second part of Walsh's article 'On phytophagic varieties and phytophagic species' was published in the December 1865 issue of *Proceedings of the Entomological Society of Philadelphia* (Walsh 1864–5). CD's heavily annotated copy is in the Darwin Pamphlet Collection–CUL. Walsh had sent CD the first part of the paper earlier in 1865 (see letter from B. D. Walsh, 1 March 1865 and n. 14); for CD's comments on it, see the letter to B. D. Walsh, 27 March [1865] and nn. 3 and 5. See also letter from B. D. Walsh, 29 May 1865.

[10] For periodicals carrying the newspaper stamp, British postal regulations prohibited any writing or other marks on the publication, except for the name and address of the recipient; an additional postage of 2*d.* was charged for a breach of the rule. Each publication had to carry a separate stamp. Offprints of articles, and periodicals published at intervals greater than thirty days, would usually travel by book post. A book packet could contain any number of separate publications, but no letters or 'communication of the nature of a letter'; packets containing such communications were charged additional postage (*British postal guide*, 1 July 1865, pp. 10–14).

[11] Henry Walter Bates. Walsh had sent copies of some of his papers to Bates earlier in the year (see letter from H. W. Bates, 22 March 1865).

[12] See n. 10, above.

[13] Walsh had sent CD an article on dimorphism in *Cynips* in 1864 (see *Correspondence* vol. 12, letter from B. D. Walsh, 29 April – 19 May 1864, and letters to B. D. Walsh, 21 October [1864] and 4 December [1864]). Walsh argued that the gall insect *Cynips quercus aciculata*, hitherto regarded as a distinct species, was a dimorphic female form of *C. q. spongifica* (Walsh 1864a, pp. 447–8). See also letter from B. D. Walsh, 1 March 1865, and letter to B. D. Walsh, 27 March [1865]. No further article by Walsh on the subject has been found.

[14] CD had forwarded Walsh 1864a to John Lubbock, who wrote a brief notice of the paper in the January 1865 issue of *Natural History Review*, pp. 138–41. Lubbock thought the evidence for dimorphism in *Cynips* was inconclusive (see *Correspondence* vol. 12, letter from John Lubbock, 3 November 1864, and Lubbock 1865, p. 141)

From Hermann Kindt 13 November 1865

Bampton, Penrith
13. Nov. 1865.

Dear Sir,

I herewith send you the facsimile reproduction of your handwriting, and beg you will be kind enough to let me know if you like the manner in wh: it has been executed.[1] At the same time accept my warmest thanks for your courteous attention. I am sure that the paragraph as well as the specimen of your handwriting are a great ornament to our publication, and will be received with the greatest pleasure by your numerous friends and admirers. I can at least vouch for my own countrymen.

If you can possibly send me for the same purpose of a facsimile reproduction an interesting note or letter of your celebrated Grandfather, I shall feel greatly obliged to you.[2]

I beg to enclose a biographical notice intended for that purpose, which was chiefly extracted from Miss Seward's "Life of Dr. Erasmus Darwin".[3] I copied and compiled it some time ago but cannot remember the biographical compilations I made use of for that purpose.

With renewed thanks for your kindness, believe me to remain | Your's very faithfully | Hermann Kindt

Charles Darwin Esq. M.A., F.R.S., etc.

DAR 169: 17

[1] Kindt had requested a page of handwriting from CD for facsimile reproduction in the *Autographic Mirror* (*L'Autograph Cosmopolite*; see letters from Hermann Kindt, 17 October 1865, and 23 October 1865 and n. 2).

[2] A facsimile reproduction of a letter from CD's paternal grandfather, Erasmus Darwin, to his maternal grandfather, Josiah Wedgwood I, was published in the *Autographic Mirror* (*L'Autograph Cosmopolite*) 4 (1866): no. 70.

[3] Kindt refers to Seward 1804. The biographical sketch of Erasmus Darwin appeared in the *Autographic Mirror (L'Autograph Cosmopolite)* 4 (1866): 23.

To Eliza Meteyard 16 November [1865]

Down Bromley | Kent
Nov. 16

My dear Madam

The Editor of the Autographic Mirror has applied to me for a specimen of the handwriting of Dr Erasmus Darwin & I sh^{d} be glad to oblige him.[1] It ought to be a moderately long note signed with his name & if possible possessing some sort of interest or as little dull as possible.

Will you have the kindness to see if you can find such a note among the letters & send it me or even two or three to choose from, & they shall be returned to you if required.[2]

Hoping that you will excuse my giving you this trouble I beg leave to remain | My dear Madam | yours very faithfully | Charles Darwin

LS
Postmark: NO 16 65
Fitzwilliam Museum, Cambridge

[1] See letter from Hermann Kindt, 13 November 1865 and n. 2.

[2] CD had sent Meteyard letters from his paternal grandfather, Erasmus Darwin, to his maternal grandfather, Josiah Wedgwood I, for use in Meteyard 1865–6 (see letter from Eliza Meteyard, 25 April 1865 and n. 4, and Supplement, letter to Hurst and Blackett, 15 November [1863] and n. 3). The letter from Erasmus Darwin to Josiah Wedgwood I, 27 April 1766, was reproduced in the *Autographic Mirror* (*L'Autograph Cosmopolite*); this letter is quoted in Meteyard 1865–6, 1: 436.

From Eliza Meteyard 17 November 1865

Wildwood | North End | Hampstead. NW.
Nov: 17. 1865

Dear Sir,

I send you nineteen of the letters you so kindly lent me: and hope amongst the number, one may be found suitable for the pages of the Autographic Mirror.[1] If not, and Mrs Darwin[2] will kindly inform me, I will send the rest, although dipping into them for the Wedgwood as I go on.[3]

I heartily apologize for retaining them so long, but I conscientiously feel that no one can reverence them more, or take greater care; and this perhaps makes me take unwarrantable liberties as to time.[4] But your goodness I am sure pardons me. The circumstances under which my book has had to be written has much harrassed me—and with hundreds of pages to copy, and no helping hand—I necessarily take liberty where I can. But as soon as the last page of Wedgwood is off to the printers—and we begin to print though slowly on Dec. 1—I will set to, at once, and make what notes I shall further require for the Life of Thomas Wedgwood[5]—and then return them speedily—with very sincere thanks for your indulgence.

I sincerely hope your health is a little better than it was.

With compliments to Mrs Darwin, I am, | Dear Sir | With grateful respect | your's faithfully | Eliza Meteyard.

C. Darwin Esq.

DAR 171: 161

1 See letter to Eliza Meteyard, 16 November [1865] and n. 2.

2 Emma Darwin.

3 Meteyard refers either to the second volume of her life of Josiah Wedgwood I (Meteyard 1865–6), which was published in September 1866 (*Publishers' Circular* 1866), or to Meteyard 1871 (see n. 5, below). See also letter from Eliza Meteyard, 25 April 1865.

4 CD had sent the letters in November 1863 (see *Correspondence* vol. 11, letter to J. D. Hooker, [22–3 November 1863] and n. 11, and this volume, Supplement, letter to Hurst and Blackett, 15 November [1863]).

5 Meteyard's study of Josiah Wedgwood's son Thomas Wedgwood, and his circle, was published under the title *A group of Englishmen* (Meteyard 1871).

From Edward Cresy to Emma Darwin 20 November 1865

Metropolitan Board of Works | Spring Gardens
20 Nov 65

My dear M[rs] Darwin,

Permit me to awaken your feminine sympathies in behalf of a very admirable young lady & dear friend of ours Miss Elizabeth Garrett who has, after encountering an amount of opposition which few men would have had the courage to encounter, succeeded in obtaining the diploma of the Apothecaries Company, & has started in practice—[1]

Your brother in law Mr Erasmus Darwin is Chairman of the Council of the Bedford College for Girls[2] & Miss Garrett is a candidate for their Professorship of physiology applications for which are to go in on Wednesday next— I have no doubt that if a properly qualified lady can be obtained that the council would be disposed to consider her as possessing many advantages for instructing girls, especially in that particular branch—[3] I cannot of course ask you to urge the claims of one who is a stranger to you—but if you could say to Mr Erasmus Darwin that you believe that my recommendation would not be lightly given and that I have had the opportunity of watching Miss Garretts career closely & of frequently observing & testing her scientific acquirements and know them to be of a high order—& also that her industry & zeal are beyond all praise, I think possibly that even at second hand such testimony might do her good service— She is frequently at our house & my wife[4] & I both entertain the greatest regard for her—

I may add what I know will interest you although it cannot help her in the matter now under consideration, viz that the very special career to which she has devoted herself has nothing impaired the charm of her manner or her social converse she is neither masculine nor pedantic & except you knew her intimately you would only recognise a well bred English Lady— I hope you will be able to give me a more favorable account of Mr Darwin than the last.[5] pray remember us both most kindly to him—

Yours very truly | E Cresy

Mrs Darwin

DAR 161: 247

[1] Elizabeth Garrett received a licence to practise medicine from the Society of Apothecaries in September 1865, and began a private consultancy in London shortly after (Manton 1965, pp. 115–18). The Society of Apothecaries was the only medical organisation whose charter did not prohibit the admittance of women. Although Garrett was accepted as an apprentice by the society in 1861, she later threatened legal action when the society's Court of Examiners refused to let her sit for the licensing examination. Garrett had already experienced difficulties obtaining the clinical hours and lecture courses that were required for the medical certificate, because many universities and hospital schools refused to admit her as a student (*ibid.*, pp. 112–68).

[2] CD's brother, Erasmus Alvey Darwin, had served as chairman of the council of the Ladies' College, Bedford Square (Bedford College) since it was founded in 1849. Emma's sister-in-law Frances Wedgwood was also a member of the council (Tuke 1939, pp. 36–42).

[3] Garrett had been advised to offer her services as lecturer in physiology by a trustee of Bedford College; however, on deliberation, the council considered the subject unsuitable for women students, and the position was not created until 1882 (Tuke 1939, p. 104).

[4] Mary Louis Cresy.

[5] See letter to Edward Cresy, 19 October [1865].

From E. A. Darwin [20 November 1865?][1]

Monday.

Dear Charles

John Brent[2] is come & I will send it down by Hope or Effie—[3] I enclose the address that you may know where it comes from. I hope your journey has been prosperous & that you wont be frightened at coming again[4]

E D

DAR 105: B39

[1] The year is conjectured on the basis of the publication of J. Brent 1865; the day and month are conjectured on the assumption that the letter was written on the day of CD's return to Down from London in November 1865 (see nn. 2 and 4, below).

[2] Possibly a reference to *Village bells, Lady Gwendoline and other poems*, by the novelist and antiquarian John Brent (J. Brent 1865). The book was published between 1 and 15 July 1865 (*Publishers' Circular* 1865, p. 369).

[3] Hope Elizabeth and Katherine Euphemia Wedgwood, the daughters of Frances Emma Elizabeth and Hensleigh Wedgwood, made occasional visits to Down.

[4] According to his 'Journal' (Appendix II), CD 'staid ten days' in London with his brother Erasmus, beginning 8 November 1865; however, Emma Darwin recorded in her diary (DAR 242) that she and CD returned to Down on 20 November.

From James Shaw 20 November 1865

Tynron Parish School | Dumfriesshire.

Nov[r]. 20— 1865.

Sir.

Here in the solitude of a remote glen your noble work has reached me and is another proof to me that the Age of Romance has not perished but that Truth is indeed stranger than Fiction.[1] You have given articulation to a thousand things lying dumb within me. When the *Vestiges* was attacked by friends and foes I used to skirmish in its defence[2] (*vide* my letter pub. by Lewis then Ed. of the Leader)[3] but never at least within the present century did I expect so scientific an exposition of the grandest general facts—of the deepest truth—yet presented to the human mind to contemplate. Never before have I conceived so clearly the wonderful relation of the individual to the august whole. Those dim ages when nature was silently strewing on her barren shores the dust of future continents and making out of her dead organisms quarries from which Ionic & Doric columns should be wrought are no longer in imagination spectral gleams but wear rather all the interest of history. I am a Man, said the Roman, and everything human has an interest for me.[4] I a humble student of that Nature with which we are all so strangely knit and everything natural has an interest for me. The comparative anatomy of brain or pelvis—a thousand things hard to understand—hard to attract—have now by a light which never before shone on sea or land assumed a meaning and an interest which never attached to them before. Newton's ocean of wonder on whose shore he gathered shells seems no longer a dark unnavigable sea crested with howling breakers.[5] Here and there through its awful bosom the strong vision with which you have supplied us enables us to see at least that there are other islands whose conditions are explicable by our own. It was a grand moment in the history of discovery when Lord Rosse was able to stand at one end of his magnificent tube and command the belt of Orion to swim forward millions of leagues nearer him as it were through unfathomable Space.[6] Grander still was that hour when using the plumb-line of generalization you enabled us to dip back into the far Past and bring it home to our hearths and our hearts. Wisdom says the inspired man dwelt with the Creator ere he laid the foundations of the world. By wisdom, toil, patience, perseverance the race has arrived at a point when a foremost man here and there looking back from the amazing pinnacle, can construct a chart of the lands over which Life has travelled and of the epochs through which it has progressed, and then spreading his wings in flight can conduct us through those unvisited regions until we stand like archangels at the gates of morning and at the Creation of rock of river and of hill, & of the tenants which therein are.

Some strictures were lately passed on your hypothesis by the Duke of Argyll.[7] I endeavoured to point out where his Lordship was at fault.[8] Of some of my objections to his criticism he took notice but to other of my objections he was silent. Out of that communication there is only one suggestion which I think is worth transmitting to you, but if the smallest scintillation of mine is acceptable for the light which you have shed on me you are welcome.

His Lordship (*vide* Good Words 1865) is puzzled to understand why there is so much beauty in the world—beauty in the shell—beauty in the flower—exquisite beauty, wonderful patterns of colour, in the plumes of humming birds. Then he proceeds to account for it not on your hypothesis but on the hypothesis that beauty affords a direct pleasure to its creator, remarking that your hypothesis is not yet wide enough to treat this subject. Now I have no objection to the Creator being delighted with beauty and making it to please himself, but not being taken in to his confidence at the creation of it, I cannot give an opinion on the subject.[9]

But to return to observation. Does his Grace deny all sense or love of beauty in the brutes? (To this query he returns no answer) Comte says the religion of brutes is the reverence gratitude and obedience which they display towards their masters.[10] If dogs be credited with religion why should birds be denied all aesthetical faculty? Birds like benighted Africans are fond of glittering pieces. Witness how larks in France are caught by twirling for them. The gorgeous Birds of Paradise are extremely attentive to their plumage and exceedingly sensitive concerning its cleaness &c &c His Grace points to the savage as an example of the width of range of the love of ornament,[11] and Carlyle suggests that rather than the desire of comfort the love of ornament was the first step towards clothes.[12] Strange if a passion diffused throughout all humanity should have no counterpart in other regions of life. Beauty like song is developed in the male chiefly at the breeding season. If the female give preference to the male of gaudiest hue or richest plumage it will be a chance for the preservation of such. The question propounded concerning the beautiful in Nature is I grant wider than this answer—but I think this answer ought not to be forgot in favour of your hypothesis.

I am, | Sir, | Yours most respectfully, | Ja[s]. Shaw.

Charles Darwin Esq[re]. F.RS.

I had yr address as given in a letter to the Athenaeum but have laid it aside therefore I have taken the liberty of sending this on to your publisher.[13] J.S.

DAR 177: 149

1 Shaw refers to *Origin*.

2 First published anonymously, *The vestiges of the natural history of creation* ([Chambers] 1844) presented a continuous, progressive view of development in accordance with natural laws and divine purpose. The work gave rise to extensive public debate, and successive editions were published in the 1840s and 1850s responding to criticisms. On the controversial reception of *Vestiges* and the speculation about its author, see J. Secord 2000. For CD's assessment of the book, see *Correspondence* vols. 3 and 4, *Correspondence* vol. 5, letter to T. H. Huxley, 2 September [1854], and the 'historical sketch' added to the third and subsequent editions of *Origin*.

3 George Henry Lewes was one of the founders and editors of the *Leader*, a radical weekly journal published in London (*DNB*; see also J. Secord 2000, pp. 483–4). Shaw's letter appeared in the 24 September 1853 issue of the journal, in response to a lengthy review of the tenth edition of *Vestiges* by Lewes (Lewes 1853). Shaw defended the view that 'the progression (of animal life)' was 'the result of *an aspiration towards new and superior fields of existence*' (see R. Wallace ed. 1899, pp. lii–liv).

4 Shaw refers to a famous line from the play *The self-tormentor*, by the Roman dramatist Terence: 'homo sum: humani nil a me alienum puto' (77).

5 Shaw alludes to the remarks of Isaac Newton, which were published in his memoirs (Brewster ed. 1855, 2: 407):

> I do not know what I may appear to the world, but to myself I seem to have been only like a boy playing on the sea-shore, and diverting myself in now and then finding a smoother pebble or a prettier shell than ordinary, whilst the great ocean of truth lay undiscovered before me.

6 William Parsons, third earl of Rosse, designed and constructed large reflecting telescopes, and made detailed observations of nebulae (*DNB*).

7 In his article 'The reign of law' in the religious weekly *Good Words*, George Douglas Campbell, eighth duke of Argyll, had criticised CD's theory of natural selection for failing to explain adequately the beauty and variety in nature (G. D. Campbell 1865, p. 231):

> The evidence is indeed abundant, that ornament and variety are provided for in nature for themselves and by themselves, separate from all other use whatever. Any theory on the origin of species which is too narrow to hold this fact, must be taken back for enlargement and repair.

Campbell had made the same criticism in his address to the Royal Society of Edinburgh on 5 December 1864 (G. D. Campbell 1864). For a discussion of Campbell's address and CD's response, see the letter from Charles Lyell, 16 January 1865 and nn. 6–11, and the letter to Charles Lyell, 22 January [1865] and nn. 5–14.

8 No published reply by Shaw to G. D. Campbell 1865 has been found; however, Campbell is listed as a correspondent of Shaw's (R. Wallace ed. 1899, p. xxiv).

9 Shaw refers to G. D. Campbell 1865, pp. 227–32; see also n. 7, above. In an undated autobiographical note, Shaw remarked that he had written this letter to CD without having read the discussion of sexual selection in *Origin* (R. Wallace ed. 1899, pp. lv–lvi).

10 In his most extended discussion of the origins of religion, the French philosopher and founder of positivism, Auguste Comte, drew an analogy between the love and devotion felt by domestic animals for their human masters, and the worship of gods and angels by humans (Comte 1851–4, 1: 496).

Shaw is described as being 'for a great many years a pronounced Positivist' (R. Wallace ed. 1899, xxii). On the reception of Comte's writings on religion in Britain, see T. R. Wright 1986.

11 Campbell referred to the 'habits of the rudest savage', who covered 'with elaborate carving the handle of his war-club', as illustrating the universal love of ornament (G. D. Campbell 1865, p. 230).

12 In *Sartor Resartus*, Thomas Carlyle wrote: 'The first purpose of Clothes, as our Professor imagines, was not warmth or decency, but ornament' ([Carlyle] 1838, p. 37).

13 Shaw refers to the *Athenæum* and to CD's publisher, John Murray.

From Hermann Kindt 24 November 1865

Bampton, Penrith
24. Nov. 1865.

Dear Sir,

I am obliged for your so kindly sending the three letters of Dr. Darwin, two of which I have already returned to Miss Meteyard.[1] I have kept the one relating to the French gentleman, which seemed to me the most interesting of the three.[2] A long and interesting letter of Mr. Wedgwood (Josiah) appeared in the II. volume of our publication, and I beg, therefore, to thank you for your kind intention.[3] I am glad you like the reproduction of your own handwriting and only regret the two misprints.[4] I do not correct the proofs on account of the great distance, etc.[5]

With best wishes for your health, believe me to remain, | ever your's faithfully | Hermann Kindt

Charles Darwin, M.A. Esq.

DAR 169: 18

1 Kindt refers to CD's grandfather Erasmus Darwin, and to Eliza Meteyard. See letter from Hermann Kindt, 13 November 1865 and n. 2, and letter from Eliza Meteyard, 16 November [1865].

2 The letter from Erasmus Darwin to Josiah Wedgwood I, 27 April 1766, was reproduced in the *Autographic Mirror* (*L'Autograph Cosmopolite*) 4 (1866): no. 70. The letter described Darwin's visit from a French nobleman who offered to sell the 'secret of making the finest old China as cheep as [Wedgwood's] pots' for £12,000.

3 A facsimile reproduction of a letter from Josiah Wedgwood I appeared in the *Autographic Mirror* (*L'Autograph Cosmopolite*) 1 (1864), no. 88–89. The letter, dated 31 January 1790, gave Wedgwood's view of the French Revolution.

4 The biographical sketch of CD appeared on pp. 82–3 of the *Autographic Mirror* (*L'Autograph Cosmopolite*) 3 (1865).

5 The *Autographic Mirror* (*L'Autograph Cosmopolite*) was published from offices in the Strand, London; in 1865, it was issued weekly.

From E. A. Darwin to Emma Darwin 25 [November 1865][1]

25th.

Dear Emma.

I have not yet received the Photo's & if I do wish to have any of them I will

EAD them on the back.[2] My present impression is that they are too hideous to be borne.[3] I think you had better write to the man & say that C will try again when he comes up to town after Christmas, & you may quote your feminine feelings that Mr Crecy alludes to as a reason.[4] (That was all a mistake that we were having a Professorship if you should have occasion to write to Mr Crecy but I read his letter to Council)[5]

I am very sorry to hear that C has been so poorly & I am afraid it was a mistake his going before Dr B J had wound him up.[6]

The last account I had of Susan spoke of her having had a good deal of her faintness.[7] Catherine goes there today.[8]

yours affec. | E D

DAR 105: B119–20

[1] The date is established by the relationship between this letter and the letter from Edward Cresy to Emma Darwin, 20 November 1865.

[2] CD evidently sat for a photographer during his stay in London, from 8 to 20 November 1865. This may have been Ernest Edwards, who supplied photographs for the series *Portraits of men of eminence in literature, science, and art, with biographical memoirs* (Reeve and Walford eds.1863–7). CD had agreed to appear in the series (see letter to Edward Walford, 22 [January–April 1865?] and n. 2). An entry in CD's Classed account book (Down House MS) records a payment of £1 for 'E. Edwards Photo' on 2 March 1866. According to Emma Darwin's diary (DAR 242), CD did not visit London again until 22 April 1866.

[3] The photographs of CD by Edwards appear to have been taken at two different sittings. It is likely that the portrait used in Edward Walford's series was taken in 1866, during CD's visit to London from 22 April to 1 May (see Reeve and Walford eds. 1863–7, 5: facing 49, and Browne 1998, pp. 269–71). Another photograph of CD by Edwards, possibly taken at the November 1865 sitting, is reproduced as the frontispiece to this volume, and in Browne 1998, p. 256. Photographs from both sittings were sold commercially as cartes de visite by the London Stereoscopic Company and the firm of Edwards and Bult. CD evidently purchased copies of one or more of these photographs for his own use. A second payment, of £3 8*s*. 6*d*. was made to Edwards on 5 September 1866 (CD's Classed account book (Down House MS)).

[4] See letter from Edward Cresy to Emma Darwin, 20 November 1865.

[5] See letter from Edward Cresy to Emma Darwin, 20 November 1865 and nn. 2 and 3.

[6] In his letter to J. D. Hooker, 22 and 28 [October 1865], CD wrote that he intended to visit London for a week in order to consult the physician Henry Bence Jones. According to Emma Darwin's diary (DAR 242), CD and Emma had stayed with Erasmus in London from 8 to 20 November; CD became feverish on the 20th and did not improve until the 27th (see also 'Journal', Appendix II). CD had been consulting Jones since July, and had reported an improvement in his health under the rigorous diet prescribed by Jones (see letter to J. D. Hooker, 27 [or 28 September 1865], letter to T. H. Huxley, 4 October [1865], and letter to W. D. Fox, 25–6 October [1865] and n. 2). CD's Account book–cash account (Down House MS) records payments to Bence Jones of £12 12*s*. on 28 November and 9 December 1865.

[7] Susan Elizabeth Darwin, the elder sister of CD and Erasmus, had been ill (see letter to W. D. Fox, 25–6 October [1865]).

[8] CD's younger sister, Emily Catherine, lived in Shrewsbury with her husband Charles Langton; Susan Elizabeth Darwin lived at The Mount, Shrewsbury (Freeman 1978).

To James Shaw 30 November 1865

Down, Bromley,
30th Nov. 1865.

Dear Sir,—

Illness has prevented me sooner thanking you for your letter on the *Origin*, written with so much fervour and kindness.[1] I am much gratified to hear that you have been defending my views, and in your country (Scotland), those who are bold enough to take this side of the question are few and far between.[2] I am so weak, I must write briefly. I have reflected much on the question of beauty.[3] It is a very complicated one. I quite agree with what you say on the beauty of birds, and the same view may be extended to butterflies and some other beings.[4] I think I can show that the beauty of flowers and of many kinds of fruit is solely to attract, in the former case, insects for their intercrossing, and in the latter case, to attract birds for the dissemination of the seed.[5]

Ch. Darwin

R. Wallace ed. 1899, p. lvi.

[1] See letter from James Shaw, 20 November 1865.

[2] Shaw had defended CD's theory of natural selection against criticisms made by George Campbell, eighth duke of Argyll, in the journal *Good Words* (G. D. Campbell 1865; see letter from James Shaw, 20 November 1865 and nn. 7 and 8). On the reception of the theory of natural selection in Scotland, see *Correspondence* vol. 10, Appendix V, *Correspondence* vol. 12, letter to J. D. Hooker, 5 April [1864] and n. 10, and Burkhardt 1974, pp. 67–71.

[3] For CD's early interest in the subject of beauty in relation to natural selection, see the letter to Charles Lyell, 22 January [1865] and n. 10.

[4] In *Origin*, pp. 88–90 and 199, CD suggested that many beautiful characteristics, such as the ornamental plumage of birds, could arise through sexual selection. CD expanded his discussion of beauty in *Origin* 4th ed., pp. 238–41. On the subject of beauty in birds, see also the letter from Charles Lyell, 16 January 1865, n. 11, and *Descent* 2: 108–24. On the operation of sexual selection in butterflies, moths, and other insects, see *Descent* 1: 386–423.

[5] In *Orchids*, pp. 346–60, CD argued that the unusual or beautiful forms of orchid flowers facilitated pollination by insects. See also *Origin* 4th ed., pp. 239–40, on the colour of flowers and fruit attracting insects and birds.

To Robert Burn 2 December [1865][1]

Down. | Bromley. | Kent. S.E.
Dec 2.

Dear Sir

I should be much obliged if you would enter my son Francis Darwin at Trinity for next October. I enclose his master the Rev. Dr Wrigley's certificate[2] & a cheque for entrance & caution money for £20.[3]

If it is possible I sh^d^ much wish my son to have rooms in college & sh^d^ therefore be very much obliged if you w^d^ put his name down for the first chance.

I beg leave to remain | Dear Sir | yours very faithfully | Charles R. Darwin

To Rev^d^. R. Burn

LS(A)
DAR 221.5: 4 (photocopy)

[1] The year is established by Francis Darwin's matriculation at Trinity College, Cambridge, in October 1866 (*Alum. Cantab.*).
[2] Alfred Wrigley was headmaster of Clapham Grammar School. Francis Darwin attended the school from the summer of 1860 to the summer of 1866 (CD's Classed account book (Down House MS)).
[3] Caution money: 'money deposited as a security for good conduct, [especially] by a student on entering a college' (*OED*). A payment to Burn of £20 for 'entrance for Trinity' is recorded in CD's Account book–cash account (Down House MS) on 2 December 1865.

From George Henslow 2 December 1865

10 South Crescent | Bedford Sq; | W.C.
Dec 2/65

My dear Sir,

I have been very remiss in not writing to thank you for your last interesting letter;[1] but I have been so much engaged since then that I have not had even time to write: in fact I have been engaged in writing a resumé of your deeply interesting paper on tendrils &c. for the "Popular Science Review", & had only a short time to do it in.—[2] It has so greatly interested me that I quite long to see the plants themselves; & therefore I am going to make the request for you to allow me to run down to Bromley & give myself the pleasure of calling on you: I am afraid I must however wait until January,[3] as being engaged in my school (of wh. I beg to enclose a Prospectus)[4] I cannot get away till then: but if you are at home at that period I would I assure you, esteem it a great favour to be allowed to examine the plants in situ.

Very many thanks for your offer of Sprengel's work:[5] I will see if it is in the Linnæan Library or British Museum if not perhaps I may be allowed to borrow yours: but I do not like to put you to any inconvenience

I read your note on Medicago lupulina in conjunction with mine on *M. sativa* &c at the Linnæan (last meeting)[6] It seemed to create interest

I should feel very much obliged if you will allow me to keep the printed paper on the *Kidney beans*[7] & to accept your offer as it will be very useful to me in Botanical teaching; for I am glad to get any illustrations pictures &c. or printed matter, for my "*Instructional Herbarium*"; originated by my father & now proving invaluable to me: i.e. a Herbarium with the addition of drawings of dissections of some 1 or more plant of every order, with morphological monstrosities Diagrams &c . . . interspersed for the purpose of teaching Botany in a School or in private lessons, which I am now doing,—having a class every Saturday Evening; as well as in my own school.—[8]

Have you ever examined *Crocus* to see if it requires intercrossing; I have often suspected it, & mean—as far as I can in London—to try a few experiments this spring. The method pursued by bees is admirably adapted for the purpose which only requires proving.[9]

Hoping you are in better health, | Believe me | Dear Sir, | Yours very Sincerely | Geo. Henslow

DAR 166: 152, 152/1

1 CD's letter to Henslow has not been found.
2 Henslow's summary of 'Climbing plants' appeared in the January 1866 issue of *Popular Science Review*, pp. 55–65.
3 According to Emma Darwin's diary (DAR 242), George Henslow did not visit Down until 2 April 1866.
4 Henslow was headmaster of the Grammar School, Store Street, London (*Alum. Cantab.*). The prospectus has not been found in the Darwin Archive–CUL.
5 Sprengel 1793. See letter from George Henslow, 6 November 1865 and n. 3.
6 G. Henslow 1865. See letter from George Henslow, 6 November 1865 and n. 2.
7 CD had sent Henslow his paper 'On the agency of bees in the fertilization of papilionaceous flowers, and on the crossing of kidney beans'. See letters from George Henslow, 1 November 1865 and n. 9, and 6 November 1865.
8 George Henslow's father, John Stevens Henslow, had actively promoted the use of drawings and diagrams, in conjunction with herbaria, in the teaching of botany (see Russell-Gebbett 1977, pp. 42–66, and Walters and Stow 2001, pp. 62, 240–2). Copies of his teaching materials, developed for schoolchildren at Hitcham, Suffolk, where he was rector, were specially commissioned by the Department of Science and Art and exhibited in the South Kensington Museum. He also wrote a pamphlet on the use of illustrations for elementary instruction in botany (J. S. Henslow 1858). George Henslow had assisted his father in teaching at Hitcham, and was called upon in 1864 by Rugby school to prepare herbaria and diagrams for a new botany course, the syllabus of which was designed by Joseph Dalton Hooker (see Russell-Gebbett 1977, p. 67). On the use of illustrations in botanical education in early Victorian Britain, see A. Secord 2002.
9 No record that CD experimented with *Crocus* has been found; however, in a letter to the plant hybridiser William Herbert, 26 June 1839 (*Correspondence* vol. 2), CD stated his belief in the necessity of occasional intercrossing between individuals, and asked whether insects, 'by using force with their jaws', could carry pollen from one *Crocus* flower to another of the same species. George Henslow had expressed interest in the floral mechanisms that favoured intercrossing in the Leguminosae, noting that *Crocus* afforded an example of cross-pollination by means of bees (see letter from George Henslow, 1 November 1865 and nn. 6 and 8). No record has been found of the results of experiments by Henslow on *Crocus*.

From William Pollard Pattison 4 December 1865

28 Guildford St: | Russell Sq:, London.[1]
4th. Dec: 1865.

Dear Sir,

I am about to take a liberty which I would not venture to do but for the kind way in which scientific men allow what might be their exclusive wealth to be drawn upon by all earnest enquirers. I must frankly say, however, that the object of my present enquiry is a practical one rather than the obtaining of knowledge for itself. But being on a point which not more than 4 or 5 can give *any* information & yourself the most valuable I have no option but to ask your kind permission to submit it to you.

In the course of my profession as an Actuary I have the following case:

The succession of certain people to a considerable property depends, with other contingencies, on issue being born to a gentleman of pure race (aged 35) to a lady (aged 35) who is a hybred—the offspring of an Englishman and an Indian lady. They have been married 21 years without issue & there have been, I believe, no indications of a power of fecundity. If the circumstances applied to two Europeans, not blood relations, a fair estimate might be formed of the probability of issue; and what I wish to ascertain is whether the failure of issue is functional & therefore impossible, or whether otherwise it is more or less probable than if both were Europeans.

Though I would highly prize your opinion, I will only ask you to name the books (kindly indicating in what parts) I should find any trustworthy information, or valuable opinion.[2] I do not ask for your opinion, only because you might not wish to give it & yet at the same time would not wish to decline giving it.

I am sometimes in the neighbourhood of Down, and was half tempted to call rather than write this long letter: but a letter can be answered in the leisure hour of the day, and my call might have been in the most busy one.

Believe me, Dear Sir, | Yours faithfully & obediently | W. P. Pattison.

Charles Darwin, Esqre | Down.

DAR 174: 29

1 Pattison apparently wrote 'Guildford' in error for 'Guilford' (*Post Office London directory*).

2 CD had discussed hybrid sterility in *Origin*, pp. 245–72, and continued extensive research on the subject following its publication (see *Correspondence* vol. 10, Appendix VI). Although his own experiments focused on plants, CD also collected information on animals, and was interested in analogous cases in humans. In a letter to W. B. Tegetmeier, 27 [December 1862] (*Correspondence* vol. 10), CD inquired about '*individual* fowls or pigeons which are sterile together', adding that this was a 'parallel case to those recorded of a man *not* impotent long living with a woman who remained childless; the husband died & the woman married again & had plenty of children'. CD suggested that in such cases the man and woman were 'dissimilar in their sexual organisation'. See also *Correspondence* vol. 12, letter to T. C. Eyton, 29 December [1864?]. In the course of his botanical research, CD debated whether the sterility of species when crossed was a specially selected quality, or was incidental upon other acquired differences (see letter to M. E. Wichura, 3 February [1865] and nn. 8 and 9). In *Descent* 1: 220–3, CD briefly considered the subject with respect to human races, drawing comparisons with the intercrossing of varieties and species in plants and animals. CD stated that the evidence for hybrid sterility in humans was inconclusive; he also noted that fertility was easily affected by changed conditions of life and close interbreeding, which were themselves governed by 'highly complex laws'. The subject of hybrid sterility in humans was of particular interest in debates about the unity of the human species and the origin of the human races. For a contemporary discussion of the topic, see Farrar 1864c. See also letter from F. W. Farrar, 6 November 1865 and n. 5.

From Thomas Gold Appleton 5 December [1865][1]

10 Commonwealth Avenue. [Boston, Mass.]

Dear Sir,

A gentleman Col Mr Charles C. Fuller[2] has just returned to us from California, bringing with him a very curious fish which you may not have seen. It seems that the mountain lakes there very often suddenly dry up & the homeless fish are forced to wander in quest of new water. Nature to that end has furnished them with the legs you will see upon the specimen the Col above mentioned has been good enough to give me, for the purpose of sending it to you.

These fish have been met en route as I am confidently told. At least their migration is the general belief. If you or your friends should desire other specimens, I can procure another of which I know.—

The fish may be familiar to you and of no value to illustrate the object of your recent work, but to me it seems a most remarkable corroboration of your idea.[3]

If these legs do not illustrate the energy of the principle of the "Struggle for life" I know nothing of the matter.[4]

My sister wrote that you and Mrs Darwin were in London[5] Hoping that this & the fish may find you there, and with the kindest regards to Mrs Darwin & ⟨t⟩he children I am | very sincerely | T. G. Appleton

Decr 5th.

DAR 159: 112

1 The year is established by the relationship between this letter and the letter from J. E. Gray, 28 February 1866 (*Correspondence* vol. 14; see n. 3, below).
2 Charles C. Fuller has not been further identified.
3 CD sent the specimen to John Edward Gray of the British Museum, who initially identified it as a larval form of a batrachian, noting its similarity to the axolotl (a larval salamander from Mexico, *Ambyostoma mexicanum*), and the 'mud devil' or 'hellbender' (*Cryptobranchus alleganiensus*, a North American salamander). See *Correspondence* vol. 14, letter from J. E. Gray, 28 February 1866, and letter to T. G. Appleton, 2 March [1866]. In *Origin*, pp. 107 and 330, and *Descent* 1: 204 and 212, CD described the *Lepidosiren* (lung-fish) as a transitional form between the fishes and amphibians.
4 For CD's use of the expressions 'struggle for life' and 'struggle for existence' see, for example, *Origin*, pp. 60–3. Appleton had written favourably of *Origin* in his letter to CD of 24 April [1862] (*Correspondence* vol. 10).
5 Appleton's sister was Mary Mackintosh. CD and Emma Darwin had been in London from 8 to 20 November (Emma Darwin's diary (DAR 242)).

To Ernst Haeckel 6 December [1865]

Down Bromley | Kent
Dec 6

My dear Sir

I am much obliged for your last letter of Nov. 11., for your letters always give me pleasure.[1] I am not surprized at the delay in the publication of your book, more especially as you are appointed Professor & have been doing other work.[2] I shall feel much interest in seeing this book when it appears. I most sincerely wish that you could work with better spirits; but time in the long run will do something for you.[3]

With your lectures & various writings, no one I think will do so much as you in spreading & perfecting sound views on species in Germany.[4]

Every now & then I find some good young worker taking the same side in England; I have just found this is the case with one of our best rising paleontologists Dr Duncan.[5] I occasionally hear from Max Müller in Desterro & he seems to be grandly in earnest on the subject.[6]

You tell me that you have sent me a book with plates on Medusæ & on fossil Medusæ & on some Rhizopods, but I have not received these.[7]

I have received only a description of new "Craspedoter medusen" from Nice, & the Monats bericht of Berlin on the generation of the Geryoniden &c[8]

This latter paper interested me & surprized me much for I have often speculated whether any such case did occur in Nature & thought it possible, but never expected to see it proved.[9] I suppose the paper with plates is on this same subject.[10]

I am sorry to say I can give but a poor account of my health: since April I have been able to do no scientific work, nor do I see any probability of any near approach to such happy days.[11]

With the most sincere respect & good wishes believe me my dear Sir yours very sincerely | Ch. Darwin

Endorsement: '1865'
LS
Ernst-Haeckel-Haus (Bestand A-Abt. 1: 1–52/7)

1 Letter from Ernst Haeckel, 11 November 1865.

2 In his letter of 11 November 1865, Haeckel wrote that completion of his book *Generelle Morphologie der Organismen* (Haeckel 1866) had been delayed owing to the increased responsibilities of his new position as professor at University of Jena.

3 Haeckel's wife, Anne Sethe, had died on 16 February 1864 (see letter from Ernst Haeckel, 11 November 1865 and n. 13).

4 See letter from Ernst Haeckel, 11 November 1865 and n. 7. On Haeckel's support for CD's views, see also *Correspondence* vol. 12, letter to Ernst Haeckel, 9 March 1864 and nn. 3 and 4.

5 In a paper read to the Geological Society of London on 8 March 1865, Peter Martin Duncan remarked that natural selection acted 'with varying intensity as the great excitant' in the dispersal of both stable and variable species (Duncan 1865, p. 361), and cited passages from *Origin*, p. 351, which suggest that natural selection operates 'only so far as it profits the individual in its complex struggle for life', so that 'the degree of modification in different species will be no uniform quantity'. Finally, Duncan referred to CD's discussion of the geographic dispersion of species during the glacial period in chapter 11 of *Origin*, claiming that it testified to 'the identity of species over vast areas of latitude and longitude with or without gaps' (Duncan 1865, p. 362). Duncan may have corresponded with CD on British fossil corals (see *Correspondence* vol. 9, letter to P. M. Duncan?, 18 July [1861]). On Duncan's support for Darwinian theory, see A. Desmond 1982, pp. 184–5.

6 CD wrote 'Max' in error; he refers to Fritz Müller, who lived in Desterro, Brazil, and who had begun to correspond with CD on climbing plants (see, for example, letter to Fritz Müller, 10 August [1865]). In his letter of 26 October 1864 (*Correspondence* vol. 12), Haeckel had mentioned Müller and his recent book, *Für Darwin* (Müller 1864), which presented facts on the development of Crustacea favourable to CD's theory of transmutation.

7 CD refers to Haeckel 1865a, Haeckel 1865c, and Haeckel 1865d. See letter from Ernst Haeckel, 11 November 1865 and nn. 9 and 11.

8 Haeckel 1864 and Haeckel 1865b. Haeckel 1864 described new species of 'Craspedoter medusen' (Hydromedusae). CD's annotated copies are in the Darwin Library–CUL. See *Correspondence* vol. 12, letter from Ernst Haeckel, 9 [July 1864] and n. 15.

[9] Haeckel 1865b, p. 88, described a form of asexual reproduction in *Geryonia hastata* (now *Geryonia proboscidalis*), a six-tentacled medusa. Haeckel claimed to have observed a new eight-tentacled species, *Cunina rhododactyla* (now *Pegantha rubiginosa*), produced by budding inside the stomach cavity of the parent. Both species could also reproduce sexually. CD referred to Haeckel 1865b in his chapter on pangenesis in *Variation* 2: 384 n.; however, CD described the budding as producing not a new species, but 'a widely different form of medusa'. See also letter to J. D. Hooker, 4 May [1865] and n. 10.

[10] Haeckel 1865a was a more general study of the history, morphology, and manner of reproduction of the genera within the family Geryonidae (now Geryoniidae).

[11] On CD's health, see the letter from E. A. Darwin to Emma Darwin, 25 [November 1865] and n. 6.

To Henry Wentworth Acland 8 December [1865][1]

Down. | Bromley. | Kent. S.E.
Dec 8

My dear Professor Acland

I am much obliged for your kind note & present.[2] I remember perfectly & with much pleasure our short acquaintance at Oxford.[3]

I have read your Oration with great interest, & much in it about Hervey &c was new to me.[4] I believe we entirely agree that purpose or design is one of the surest & simplest roads to discovery in Natural History.[5]

I have been the more interested in your Oration from having a few years ago corresponded a good deal on the subject with Professor Asa Gray but I confess I finished in hopeless confusion of mind.[6] On the one hand it grates against one's common sense to look at this world with all its inhabitants especially man as originating without express design. On the other hand I cannot believe that any one structure is expressly designed, in the common meaning of the word. Asa Gray, who believes in Nat. Selection, believes that the initial variations are designed, but he could not maintain that the variations of domestic animals, such as those by which the Pouter pigeon has been formed, were expressly designed; nor did he dispute that the variations under domestication & under nature are of the same order & follow the same laws.[7] So that looking at the subject from two opposite points of view, I am driven to two opposite conclusions.

I send by this post a little book on climbing plants which w^{d} only be worth your looking at as shewing how much power of movement, both spontaneous & from various stimuli, there is in plants of all kinds.[8]

With very sincere thanks believe me yours very faithfully

My health continues so bad that I have done no work for the last 8 months.

Ch. Darwin

LS(A)
Bodleian Library (MS Acland d. 81: 63–4)

[1] The year is established by the reference to Acland 1865 (see n. 4, below).

[2] The letter from Acland has not been found. The gift was probably a copy of Acland 1865 (see n. 4, below).

[3] In a letter to Jeffries Wyman, 3 December [1860] (*Correspondence* vol. 8), CD wrote that he had seen Acland 'several years ago ... & was charmed with him'. No other record of their meeting has been found. CD's only known visit to Oxford, where Acland was regius professor of medicine, was for the 1847 meeting of the British Association for the Advancement of Science (see *Correspondence* vol. 4).

[4] In 1865, Acland delivered the Harveian Oration, an annual address given in commemoration of William Harvey and other benefactors of the Royal College of Physicians (Acland 1865; *DNB*). CD's lightly annotated copy of Acland 1865 is in the Darwin Pamphlet Collection–CUL.

[5] Using the work of Harvey on the circulation of the blood, Acland defended the principle of design in nature as a guiding assumption of scientific discovery, concluding: 'All science goes to prove the existence of Order; and in Order there is nothing that we *know* to exclude Design' (Acland 1865, p. 51). Acland also criticised at length the claims of the French philosopher, Auguste Comte, that the results of modern science provided evidence against belief in the operation of final causes, and that natural productions such as the lens of the eye were inferior to the works of human artifice, and best explained as the result of chance (*ibid.*, pp. 5–7, 20–1, 26–8, 39–40). Arguing against Comte, Acland cited a passage from *Origin*, p. 186, on the development of the eye by natural selection (Acland 1865, p. 25). He also quoted a passage from *Orchids*, p. 351, in which CD remarked on the superiority of nature's 'contrivances and beautiful adaptations' to those 'which the most fertile imagination of the most imaginative man could suggest with unlimited time at his disposal' (Acland 1865, p. 40). On the relationship between CD's work and the tradition of natural theology in Britain, see Brooke 1991, pp. 219–25, 275–317.

[6] CD had discussed design in nature at length with Asa Gray (see especially *Correspondence* vol. 8, letters to Asa Gray, 22 May [1860], 3 July [1860], and 26 November [1860]). In 1860, Gray wrote a series of articles that presented CD's theory of natural selection as compatible with belief in the operation of divine providence, suggesting that new species could develop from variations that were themselves designed ([A. Gray] 1860, pp. 413–14). CD financed the publication of a pamphlet version of the articles under the title *Natural selection not inconsistent with natural theology* (A. Gray 1861; see *Correspondence* vol. 9, Appendix III). See also n. 7, below. CD's annotated copies of [A. Gray] 1860 and A. Gray 1861 are in the Darwin Pamphlet Collection–CUL.

[7] The source of Gray's views on variation in domestic animals has not been identified; a number of Gray's letters to CD in 1860 and 1861 are missing or incomplete. In his letter to Gray of 26 November [1860] (*Correspondence* vol. 8), CD stated that he could not believe, as Gray's theory of designed variability seemed to suggest, that variations in domestic animals, such as fan-tail pigeons, had been brought about 'in order to gratify the caprice of a few men'. In a letter to Gray of 5 June [1861] (*Correspondence* vol. 9), he wrote that he had come to differ more from Gray on the subject of design: 'from studying lately domestic variations & seeing what an enormous field of undesigned variability there is ready for natural selection to appropriate for any purpose useful to each creature'. See also *Correspondence* vol. 9, letter to Charles Lyell, [1 August 1861], and *Correspondence* vol. 11, letter to Asa Gray, 4 August [1863]. CD expressed his disagreement with Gray's view that variation had been 'led along certain beneficial lines' in *Variation* 2: 432.

[8] 'Climbing plants'.

To Fritz Müller 9 December [1865][1]

Down. | Bromley. | Kent. S.E.
Dec 9

My dear Sir

I have received your interesting letter of Oct. 10th. with its new facts on branch-tendrils.[2] If the Linn. Soc. publishes your paper, as I am sure it ought to do, I will append a note with some of these new facts.[3]

I forwarded immediately your M.S. to Prof. Max Schultze but I did not read it, for German hand writing utterly puzzles me, & I am so weak I am capable of no exertion: I took the liberty however of asking him to send me a copy if separate ones are printed;[4] & I reminded him about the spunge paper.[5]

I knew of the difference in the spicula, but your difficulty had not occurred to me; from analogy I should rather expect that spunges have existed with spicula of the 2 kinds, & that the one had ultimately preponderated over the other.[6]

You will have received before this my book on Orchids & I wish I had known that you w^{d} have preferred the English edition. Should the German ed. fail to reach you I will send an English one.[7]

That is a curious observation of your daughter about the movement of the apex of the stem of Linum & w^{d} I think be worth following out; I suspect many plants move a little, following the sun; but all do not for I have watched some pretty carefully.[8]

I can give you no Zoological news for I live the life of the most secluded hermit.

I occasionally hear from Ernst Haeckel who seems as determined as you are to work out the subject of the change of species. You will have seen his curious paper on certain Medusæ reproducing themselves by seminal generation at two periods of growth.[9]

With sincere respect | Believe me my dear Sir | Yours very truly | Ch. Darwin

Endorsement: 'Received Jan. 25.— 1866.—'
LS
British Library (Loan MS 10, no. 4)

[1] The year is established by the relationship between this letter and the letter from Fritz Müller, 10 October 1865.

[2] See letter from Fritz Müller, 10 October 1865 and n. 10.

[3] CD added the information on branch tendrils contained in Müller's letter of 10 October 1865 to the paper he sent to the Linnean Society. See letter from Fritz Müller, [12 and 31 August, and 10 October 1865] and n. 1.

[4] Müller had sent CD the manuscript of a paper on the wood of twining plants; Müller's paper, which CD forwarded to Max Johann Sigismund Schultze, was published in *Botanische Zeitung* (Müller 1866). See letter from Fritz Müller, 10 October 1865 and nn. 6 and 7. CD's letter to Schultze has not been found.

[5] CD refers to Müller 1865c, a paper published in the journal edited by Schultze, *Archiv für mikroskopische Anatomie*. Müller had asked Schultze to send a copy directly to CD (see letter from Fritz Müller, 10 October 1865 and n. 8).

[6] In his letter to CD of 10 October 1865, Müller had discussed the problem of explaining, according to Darwinian theory, the development of two types of sponges, one with calcareous spicules, the other with siliceous spicules. Müller thought it unlikely that either form was the ancestor of the other. In his paper (Müller 1865c, p. 351), Müller proposed that the sponge *Darwinella aurea* represented the ancestral form from which both types could have developed, because it possessed spicules composed entirely of organic fibres with no mineral component.

[7] In his letter of 10 October 1865, Müller had requested the original English edition of *Orchids*. Before receiving Müller's letter, however, CD had already sent him the German translation (Bronn trans. 1862). See letter to Fritz Müller, 20 September [1865]).

[8] CD refers to the observations of Rosa Müller. See letter from Fritz Müller, 10 October 1865 and n. 12.

[9] Haeckel 1865b described two species of medusae that he claimed were both capable of asexual and sexual reproduction (see letter to Ernst Haeckel, 6 December [1865] and n. 9). In his paper, Haeckel had referred to Müller 1859, which discussed different forms of a related species, *Liriope catharinensis* (now placed in synonymy with *Liriope tetraphylla*).

From Roland Trimen 13 December 1865

Colonial Office, | Cape Town.
13th. December, 1865.

My dear Mr. Darwin,

I am very greatly obliged for your kindness in sending me copies of the *Bonatea* paper. They reached me yesterday morning. I am very glad that the Linnean Society has at length published the paper, and do not doubt that I owe its having done so to your recommendation.[1] Fitch has copied my sketches with his usual accuracy.[2] The only thing that does not clearly appear in the lithograph is the aspect of the caudicles of pollinia *in situ* in figures A & B. The caudicles are rather conspicuous in nature, owing to their bright-yellow colour, and the line of their inverting membrane is rather marked, immediately within the reverted edge of the horns of rostellum, as shown in Fig. C.

I rejoice to observe that the address on the cover of the packet is in your own writing, as it leads me to think that your health must be better again.

I have done nothing in Orchid observation lately, or I should have written to you.[3]

Yesterday, I looked through a series of drawings of Natal Orchids made by Mr. Sanderson for Dr. Harvey.[4] They are very characteristic, but in most cases lack the careful detail so essential for studying the mode of fertilisation.

I regret to see Dr. Lindley's death announced in the English newspapers: botanical Science in England can ill spare such an able and devoted leader.[5]

Since I last wrote to you, I have paid a hurried visit to Mauritius. I was of course greatly interested in the Island and its productions; & I was fortunate in finding such friendly & hospitable naturalists as Sir Henry Barkly and Mr. Newton there.[6] My time was too short to do much, and was principally devoted to the collection of Lepidoptera. I think I have pretty clearly established that the Butterflies (at least) of Mauritius are all but one natives of Africa or Madagascar, & mostly of both Island & Continent. I am preparing a short paper on the subject.[7]

I was longing to visit Madagascar when in so close proximity to it, but could not arrange to do so.

I suppose you have heard of the very curious Raptorial Bird sent to Mr. Gurney by Andersson.[8] I have not seen it (there was but a single specimen), but it appears to combine the different characters of most of the chief groups of Raptores with the wide gape of a Goatsucker. Its flying by twilight & feeding on *bats* are also remarkable facts.[9] Andersson says that he only saw this solitary specimen during all his travels.

Sir H. Barkly & M^r. Newton have both written to me about the great *Dodo* discovery in Mauritius: there seems to be no doubt about the remains being really those of *Didus ineptus*.[10]

With kind regards | Believe me yours very truly | Roland Trimen

DAR 178: 185

[1] CD had communicated Trimen's paper on *Bonatea speciosa*, an orchid found in the Cape region of South Africa, to the Linnean Society, where it was read on 1 December 1864 (Trimen 1864; see *Correspondence* vol. 12, letter to Richard Kippist, 24 November [1864]). For CD's interest in *B. speciosa*, see also *Correspondence* vol. 12, letter to Roland Trimen, 25 November 1864 and nn. 1–3, and *Orchids* pp. 302–5. CD's annotated copy of Trimen 1864 is in the Darwin Pamphlet Collection–CUL. It is cited in 'Fertilization of orchids', p. 157 (*Collected papers* 1: 154), and *Orchids* 2d ed., pp. 76–7.

[2] Walter Hood Fitch was employed regularly as a lithographer for the *Journal of the Linnean Society* (*Botany*). His lithograph of *Bonatea speciosa*, produced from Trimen's sketches, appeared in the *Journal of the Linnean Society* (*Botany*) 9 (1867): tab. I.

[3] CD began to correspond with Trimen in January 1863, after Trimen sent notes and drawings of South African orchids (see *Correspondence* vol. 11, letter to Roland Trimen, 31 January [1863]).

[4] John Sanderson, a merchant and botanical collector based in Durban, in the Natal Province of South Africa, contributed specimens and drawings to William Henry Harvey's works on the flora of the Cape region (Gunn and Codd 1981).

[5] John Lindley died on 1 November 1865. His death was reported in the 4 November 1865 issue of *Gardeners' Chronicle*; an obituary notice appeared in the 11 November and 18 November issues, pp. 1058–9 and 1082–3. His death was also announced in *The Times*, 4 November 1865, p. 5.

[6] Trimen spent three weeks on Mauritius in July 1865. He collected butterflies with Henry Barkly, the governor of the island, and received notes on the distribution of birds on Mauritius and neighbouring islands from the ornithologist Edward Newton, acting auditor-general of the colony (see Trimen 1866, pp. 329, 335, 342 n.).

[7] In an 1866 paper on the butterflies of Mauritius, Trimen concluded that, with the exception of two native species, nearly all of the island's butterflies had dispersed from Africa, via Madagascar (Trimen 1866, p. 344).

[8] John Henry Gurney, an expert on raptors, had published several papers on the birds of south-western Africa based on notes from his friend, the explorer Charles John Andersson (see, for example, Gurney 1864). CD had been introduced to Andersson by Francis Galton, who had organised an expedition with Andersson to Lake Ngami, South Africa, in 1850 (*DSAB*). CD had requested information from Andersson on the cattle and dogs kept by native peoples of the west coast of South Africa (see *Correspondence* vol. 5, letter to E. L. Layard, 9 December 1855).

[9] Gurney described the raptorial bird at the 14 November 1865 meeting of the Zoological Society of London (Gurney 1865). The specimen, later identified as *Machaerhamphus alcinus* (bat hawk), had been shot by a servant of Andersson's. The bird's stomach contained a bat, and on this basis the bird was believed to be nocturnal or semi-nocturnal in habit. The specimen was deposited in the Norwich Museum, of which Gurney was a leading patron (Gurney 1865, *Modern English biography*).

[10] A large number of bones of the dodo (*Didus ineptus*), an extinct flightless bird, had been found on Mauritius in October 1865. Two largely complete skeletons were sent to England: one to Richard Owen, who concluded that the dodo most nearly resembled birds of the family Columbidae (pigeons) (R. Owen 1866, pp. 50–2, 72–9). The other skeleton was sent to Alfred Newton, who announced the discovery in *Ibis* (A. Newton 1866). Newton had recently described bones of another species, *Didus nazarenus*, that had been found in November 1864 by his brother, Edward, on the neighbouring Rodriguez Island (see A. Newton 1865a and 1865b, and E. Newton 1865).

To B. D. Walsh 19 December [1865]

Down. | *Bromley.* | *Kent. S.E.*
Dec. 19.

My dear Sir

I am much obliged for your interesting letter of Nov. 12—[1] I hope you will meet with the success which you well deserve in solving the very curious problem of the Cynips.[2]

I presume that you expect that the sexual brood, whenever it appears, would be more locomotive, & thus spread the species. On the other hand, the new gall which has appeared in England recently has spread very rapidly, & yet only females have been found.[3] I received your paper on the potatoe bug & it seems to me uncommonly well done.[4]

Sir J. Lubbock & M[r] Busk called here the other day[5] & neither knew or believed in the male Daphnia laying eggs.[6] The former would be almost sure to have heard of it. He believes in Wagner's case of the breeding larva of the fly.[7] I sh[d] not be very much astonished at the Daphnia case, for certain male & female Medusæ *whilst sexually mature* throw off reproductive buds & if these buds were encased in a shell, they might be as undistinguishable from true eggs as the ovules & buds in Aphis.[8]

It is curious about the post office that I some months ago was expressing much indignation at your government being so particular about writing in, & sending, single pamphlets. There are no such rules within England & it seems that they apply only to the transit from one country to the other.[9]

I have done no work since April owing to my health, but have just begun some easy jobs, such as counting seeds of experimental Primulas; amongst others of seedlings from John Scott's Primulas, & these afford widely different results from what he gives.[10] I mention this because I see that you quote him.[11]

With every good wish pray believe me my dear Sir | yours very sincerely | Ch. Darwin

LS
Postmark: DE 20 65
Field Museum of Natural History, Chicago (Walsh 6)

[1] See letter from B. D. Walsh, 12 November 1865.

[2] In his paper on dimorphism in *Cynips*, Walsh had argued that there were two female forms of *Cynips quercus spongifica*, one agamous, the other sexual (Walsh 1864a). In his letter of 12 November 1865, he discussed his ongoing experiments at length, and suggested that the sexual or brood form was produced from the agamous form only at intervals of four, five, eight, or ten years. Walsh also remarked that *Cynips* species rarely fly, and breed from year to year on the same tree. Walsh planned to complete a second paper on *Cynips* in the autumn of 1866 (see *Correspondence* vol. 14, letter from B. D. Walsh, 17 July 1866).

[3] In later correspondence, CD and Walsh discussed an unspecified form of oak-gall that had spread rapidly in England (*Correspondence* vol. 14, letter to B. D. Walsh, 20 August [1866], and letter to B. D. Walsh, 9 August [1867], *Calendar* no. 5603). The steady increase of 'the oak gall' is also mentioned in

letters to *The Times*, 17 August 1866, p. 10, and 18 August 1866, p. 12, and in the letter from Alfred Newton, 15 March 1874, *Calendar* no. 9364.

4 CD refers to Walsh 1865. See letter from B. D. Walsh, 12 November 1865 and n. 8.

5 According to Emma Darwin's diary (DAR 242), John Lubbock and George Busk came to lunch at Down on 17 December. Lubbock had succeeded to the baronetcy on the death of his father, John William Lubbock, in June 1865 (*DNB*).

6 See letter from B. D. Walsh, 29 May 1865 and n. 14. See also letter from B. D. Walsh, 12 November 1865.

7 In his letter to CD of 1 March 1865, Walsh had expressed doubts about Nikolai Petrovich Wagner's description of paedogenesis in the larvae of the gall-midge, *Miastor metraloas*. CD suggested that such cases be regarded as forms of gemmation (see letter to B. D. Walsh, 27 March [1865]; see also letter from B. D. Walsh, 29 May 1865). In the concluding paragraph to his 1866 paper on willow-galls, Walsh cited subsequent research that confirmed Wagner's description, adding: 'Mr. Darwin writes me word that [Wagner's observations] are believed by the distinguished English naturalist, Sir J. Lubbock' (Walsh 1866, p. 288).

8 In *Variation* 2: 384 n., CD referred to a paper by Ernst Haeckel on a species of medusa that, while sexually mature, produced a different form by budding (Haeckel 1865b; see letter to Ernst Haeckel, 6 December [1865] and n. 9).

9 See letter from B. D. Walsh, 12 November 1865 and n. 10.

10 On John Scott's research on *Primula*, and CD's efforts to corroborate Scott's findings, see the letter from John Scott, 20 January 1865 and n. 3, and the letter to Asa Gray, 19 October [1865] and n. 7. On CD's differing results, see the letter from B. D. Walsh, 29 May 1865, n. 3. CD's research on intra-specific and hybrid sterility is also discussed in the letter to M. E. Wichura, 3 February [1865].

11 CD had sent Walsh a copy of Scott 1864b; Walsh praised the paper in his letter of 29 May 1865, noting the 'remarkable fact' of intra-specific sterility in *Primula* as demonstrated by Scott.

To J. D. Hooker 22 December [1865]

Down. | *Bromley.* | *Kent. S.E.*
Dec 22—

My dear Hooker

I take pride & credit to myself at not having written to you for so long a time, as you must still be very busy; but I want much to hear how you are.[1] Yet I sh^{d} not have written now, had not Haast urged me to help him for Royal Soc.[2] Can you send me the certificate to sign? As I cannot go to the Soc. to sign it. I was wonderfully well in London & had long talks with Lyell & Huxley;[3] but I then caught a cold which threw me back a whole month. I am now getting stronger & am actually able to write about an hour on most days.[4]

I have been summing up the facts on the sterility & dwarfness &c of the illegitimate seedlings of Lythrum & Primula. I can prove that they are in all respects like hybrids, & this seems to me extremely curious, for I have in fact made hybrids within the limits of the same undoubted species.[5] I have at the same time tested the characters of seedlings which inherited all the external characters from John Scott's sterile Primrose & red Cowslip; but functionally they behaved very differently, & I begin to doubt his accuracy.[6]

I must ask a question or two. I have just read in the Annals Karsten's paper against the Parthenogenesis of Cœlobigyne. Do you believe in him? I really want to know. I hardly can believe, as the seedlings have all turned out females.[7] It is a shame that some of you Kew men do not examine some flowers with a lens, mark them with a thread, & enclose them in a gauze bag. 2[ndly] Do you still wish me to subscribe (you asked me just before y[r] illness) to the Palestine explo. fund?[8] If so tell me how much & to whom. I know nothing of the subject.

I was glad to see y[r] letter in Gard. Chron. about Naudin.[9] Here is a really curious thing, considering that Brewster is President & Balfour secretary, I have been elected Hon. member of Royal Soc. of Edin.[10] And this leads me to a 3rd question. Does the Berlin Acad. of Sciences send their proceedings to Hon. members? I want to know, to ascertain whether I am a Member. I suppose not for I think it w[d] have made some impression on me, yet I distinctly remember receiving some diploma signed by Ehrenberg.[11] I have been so careless I have lost several diplomas & now I want to know what Socs I belong to, as I observe every tacks their tittles to their names in the Catalogue of the Royal Soc.[12]

Nothing in this letter requires any immediate answer, excepting indeed Haast's Cert. if you can send it & about your own health. Is there any chance of your being able to spend a Sunday & w[d] M[rs] Hooker come with you?[13] It w[d] give us extreme pleasure. I sh[d] be able to have 2 or 3 half hours conversation with you—

yours affectionately | Ch Darwin

Endorsement: '/65'
L(S)
DAR 115: 278, 278b

[1] CD's last letter to Hooker was that of 22 and 28 [October 1865]. After a period of illness following his father's death, Hooker had returned to the Royal Botanic Gardens, Kew, on 26 October; he officially became director of Kew on 1 November (see letter from J. D. Hooker, [3 November 1865] and nn. 2 and 3).

[2] Julius von Haast had asked CD to support his candidacy for fellowship of the Royal Society of London (see letter from Julius von Haast, 27 September 1865 and n. 6).

[3] Charles Lyell and Thomas Henry Huxley. According to Emma Darwin's diary (DAR 242), CD was in London from 8 to 20 November.

[4] In his 'Journal' (Appendix II), CD wrote that while at Erasmus Alvey Darwin's house in London he 'fell ill again with cold; but got better in the beginning of December'. In his letter to Asa Gray of 19 October [1865], he said that he was able to do no 'regular work', by which he seems to have meant writing. On CD's health in late 1865, see the letter from E. A. Darwin to Emma Darwin, 25 [November 1865] and n. 6.

[5] CD described same-form crosses between dimorphic and trimorphic individuals as 'illegitimate unions' (see, for example, 'Three forms of *Lythrum salicaria*', pp. 186–7; *Collected papers* 2: 120). As a result of crossing experiments with dimorphic species of *Primula* and with the trimorphic *Lythrum salicaria*, CD determined that the offspring of illegitimate unions were in their degree of sterility and dwarfness identical to the offspring of hybrids. CD presented a summary of his findings in *Origin* 4th ed., pp. 322–3, and *Variation* 2: 181–4, and gave more detailed accounts in 'Illegitimate offspring of dimorphic and trimorphic plants' and *Forms of flowers*, pp. 188–243. CD's notes on *Lythrum* for 1865 are in DAR 109: B30–5, 40, and 55–86; his notes on *Primula* for 1865 are in DAR 108. His list of illegitimate seedlings is in DAR 109: B51–4.

[6] On CD's efforts to test John Scott's experimental results, see the letter to B. D. Walsh, 19 December [1865] and n. 10.

[7] *Coelebogyne ilicifolia* had first been described by the Kew gardener John Smith (1798–1888), who had observed only female forms during ten years of cultivation (see J. Smith 1839; see also letter from J. D. Hooker, 24 December 1865). Subsequent observations had confirmed that the plant was parthenogenetic (see, for example, Daubeny 1860a, p. 12, and letter from J. D. Hooker, [23] December 1865). In an article in the *Annals and Magazine of Natural History* for 1861, Hermann Karsten concluded that parthenogenesis never occurs in plants. He claimed to have observed the attachment of the pollen-tube to the embryo sac in the ovule of *C. ilicifolia* and the resulting fertilisation (Karsten 1861, pp. 93, 200–2). Karsten also claimed to have observed hermaphrodite flowers on the plant that had been missed by other botanists (Karsten 1861, pp. 88–93). In 1862, CD had received a letter from Scott discussing Karsten's observations, and remarking that only females of the plant had been produced (*Correspondence* vol. 10, letter from John Scott, 17 December [1862]; see also letter from John Scott, 15 November [1862], and letter to John Scott, 11 December [1862]). *Coelebogyne ilicifolia* is listed as a synonym of *Alchornea ilicifolia* (family Euphorbiaceae) (*Index Kewensis*). It is on CD's 1862 list of hothouse plants (see *Correspondence* vol. 11, Appendix VI).

[8] See letter from J. D. Hooker, 13 July 1865 and n. 22, and letter to J. D. Hooker, [29 July 1865] and n. 5.

[9] Hooker's letter on Charles Victor Naudin appeared in the 2 December 1865 issue of *Gardeners' Chronicle*, p. 1133. It was written in defence of Naudin, who had been accused of claiming priority for the hypothesis of the 'derivative origin of species' (see *Gardeners' Chronicle*, 4 November 1865, pp. 1033–4). The accusation was originally made in an editorial note attached to a paper by Naudin in the *Natural History Review* for October 1865 (Naudin 1865b). The note alleged that Naudin 'failed to perceive the points of Mr. Darwin's hypothesis, and hence suppose[d] that he had forestalled its author' (Naudin 1865b, p. 568). The *Gardeners' Chronicle* for 18 November 1865, pp. 1081–2, carried a letter from Naudin denying having made any claim to priority. Naudin also wrote a letter to the same effect to Hooker, who remarked (*Gardeners' Chronicle*, p. 1133):

> What M. Naudin justly claims, is the merit of having independently convinced himself of the truth of the hypothesis by a series of laborious and valuable physiological experiments instituted for the purpose of testing it, and which in my opinion do him great credit, and enable him justly to take the highest rank as a working apostle of the doctrine of the derivative origin of species.

Naudin 1865b was a portion of a much longer paper on hybridism (Naudin 1865a) that Naudin had sent to CD (see letter from C. V. Naudin, 18 June 1865 and n. 4). On the relationship between CD's and Naudin's work on hybridism and heredity, see the letter to M. E. Wichura, 3 February [1865] and n. 6.

[10] CD was elected an honorary fellow at the 4 December 1865 meeting of the Royal Society of Edinburgh (*Proceedings of the Royal Society of Edinburgh* 5 (1862–6): 518). David Brewster and John Hutton Balfour were opposed to CD's transmutation theory. Brewster had written a highly unfavourable review of *Origin*, in which he criticised CD for 'poisoning the foundations of science, and disturbing the serenity of the Christian world' (Brewster 1862, p. 3; see also *Correspondence* vol. 10, letter from T. H. Huxley, 13 January 1862, and letter to T. H. Huxley, 22 January [1862]). Balfour had expressed his opposition to CD's theory as applied to humans in his letter of 14 January 1862 (*Correspondence* vol. 10). See also letter from John Scott, 20 January 1865, n. 5. For the diploma, see Appendix III.

[11] CD was elected a corresponding member of the Königliche-Preussiche Akademie der Wissenschaften in 1863 (see *Correspondence* vol. 11, letter to the secretary, Royal Prussian Academy of Sciences, Berlin, 16 March 1863). The diploma, signed by Christian Gottfried Ehrenberg, one of the Academy's secretaries, is in DAR 229: 50a, with CD's other extant honorary diplomas; the text is reproduced in *Correspondence* vol. 11, Appendix III. CD became a fellow in 1878 (Freeman 1978).

[12] CD refers to the *List of the Royal Society*, which was published annually by the society and contained the names of the current fellowship. For a list of the societies of which CD was an honorary member, see Freeman 1978.

[13] CD refers to Frances Harriet Hooker. According to Emma Darwin's diary (DAR 242), Hooker next visited Down on 24 March 1866.

From J. D. Hooker [23] December 1865[1]

Kew
Saturday 22 Dec | /65.

My dear Darwin

When I read your letter & came to the part that said you could work an hour a day, my heart leaped—[2] I am so glad, & like the boys I say 'Oh my how jolly—'

No one believes Karsten—[3] he has been asked to come forward & would not: I heard the subject discussed by Decaisne & Naudin when in Paris,[4] & by Braun & others at Vienna in 1858 I think Karsten is a very *bad* observer— Braun you know I think it was who found a pollen tube in the ovary, but attributed it to some other plant.—[5] I have often *selected* a plant of Cœleb. to operate upon, but one & another thing always put it off! I think Brauns paper is in Ann. Sc. Nat. but am not sure.[6]

Your observation on the undoubted offspring of species simulating hybrids is a surprizing fact, & I cannot understand it— have you ever had 2 or more such pseudo-hybrids from one capsule or plant; or produced the same pseudo-hybrid twice.[7]

John Scotts observations were always too much for me— but I do not like to be churlish in my estimate of success in a line of enquiry I have not followed—& could only be guided to a judgment by appearances.[8] Then too I was suspicious of results obtained by snatches at a time when he should have been doing something else & was always in dread of being caught.[9]

I would let Palestine alone just now— we have shipped an explorer off, & if at all successful more funds will be wanted & then will be the time.[10]

Anderson has given Scott a capital app^{t}. at Calcutta[11] I hope he has squared with you.[12]

I am diverted at your quandary about membership of societies—as it was only yesterday that I got a hint that I had not had the civility to acknowledge one that I had forgotten all about— Latterly I have taken to send my titles as they come to Kippist who inserts them in the Linnæan list—[13] I do not like to parade them in the Royal, & it is not polite to the said Academies to ignore them altogether The Linnæan list seemed to me a modest way of embalming them & if I had only kept it up, it would have been a means of reference useful to myself. The Berlin, S^{t}. Petersburgh & Austrian Academys' are real honors, also the Bavarian.—[14] The French Institute is no longer worth having quoad *Correspondence* but *membership* is the greatest continental honor that can be passed on a foreigner.[15]

I *am* surprized at your election to Edinb^{h}. R. S.[16]—the Edinburgh people being as prejudiced partial & unfair as the Parisian.

The Berlin Academy does *not* send proceedings to Foreign correspondents or members I believe.

By all means sign Haasts paper when drawn up,[17] I shall—*but* I am in a bit of a fix here Haast asked me to draw up & sign his paper, being no Geologist I transferred it to Murchison, who being then on Council transferred it to Ramsay, who will do it.—[18] I however felt very strongly that if Haast is F.R.S. before Hector it will be a bad thing.[19] Because 1. Hectors claims are far higher as previously Geologist to American Expedition; 2. Hector is Director Genl. of Geolog. Survey of all N. Z. & high above Haast who is only provincial Geologist to Canterbury. 3. There is an ugly attack made on Haast in the Colony, for appropriating some other mans facts & theories *bodily*.[20] Considering all these facts, & as a friend to both parties, & knowing that modesty alone kept Hector from putting his name forward—I wrote consulted confidentially Ramsay & Lyell,[21] & both agreeing, I wrote to Hector the other day saying that I would have him proposed, though there will not be time to get his answer or approval.— I was also influenced by the fear that Haast was not elected (& I hear he would not be) the first year & Hector comes forward the next with Haasts name still up, the collision would be awkward Hector I think would get in at once—[22] Haast I think not, there being a more deserving Geologist on the list; but one not superior to Hector.

I am really extremely well though still a little stiff in the joints—[23] I have taken to Gardening i.e. laying out walks lawns shrubberies & planting trees &c, with enthusiasm[24]

I shall look to go down to Bromley at an early Sunday— Frances cannot till the bairns go back to school.[25]

Ever Yrs affec | J D Hooker

I send a quip to amuse the boys.[26]

I do not think I quite understand the Hybrid affair.[27]

DAR 102: 47–50

1 The date '22' was written in error; in 1865, 22 December was a Friday.

2 See letter to J. D. Hooker, 22 December [1865].

3 In an 1861 paper, Hermann Karsten claimed to have observed fertilisation in *Coelebogyne ilicifolia*, a species previously considered parthenogenetic (see letter to J. D. Hooker, 22 December [1865] and n. 7).

4 Prior to Karsten's publication on *Coelebogyne ilicifolia*, Charles Victor Naudin and Joseph Decaisne had undertaken experiments indicating that parthenogenetic reproduction occurred in species of *Cannabis*, *Mercurialis*, and possibly *Bryonia* (see Naudin 1856 and Radlkofer 1857, pp. 250–1). Hooker had visited Naudin and Decaisne in Paris in 1863 (see *Correspondence* vol. 11, letter from J. D. Hooker, 24 January 1863).

5 Alexander Carl Heinrich Braun had supported the view that *Coelebogyne ilicifolia* was parthenogenetic, reporting that a pollen tube once seen by another observer on the stigma of *C. ilicifolia* had probably fallen accidentally from a different plant (Braun 1856, p. 324). Hooker visited Vienna in 1856 (*Correspondence* vol. 6, letter to J. D. Hooker, 28 September [1856]).

6 A French translation of Braun 1856 appeared in the *Annales des Sciences Naturelles* (*Botanique*) 7 (1857): 229–46.

7 See letter to J. D. Hooker, 22 December [1865] and n. 5.

8 See letter to B. D. Walsh, 19 December [1865] and n. 10, and letter to J. D. Hooker, 22 December [1865]. CD had guided John Scott in his Primulaceae crossing experiments, and had expressed enthusiasm for Scott's research on this and other subjects in letters to Hooker (see letter from John Scott, 20 January 1865, nn. 3, 4, and 6). In a letter to CD of 16 September 1864 (*Correspondence* vol. 12), Hooker said he had found Scott's 1864 paper on *Primula* 'obscure' and had set it aside 'waiting leisure'.

9 Scott's superiors at the Royal Botanic Gardens, Edinburgh, had claimed that he had neglected his regular duties as a gardener in order to pursue botanical researches; Scott denied the charges, however (see letter from John Scott, 21 July 1865 and n. 6).

[10] At the September 1865 meeting of the British Association for the Advancement of Science, a grant of £100 was awarded by the general committee to help finance an expedition, led by Charles William Wilson of the Royal Engineers. The party undertook topographic measurements for the preparation of maps of the region between Jerusalem and Damascus (*Report of the thirty-sixth meeting of the British Association for the Advancement of Science; held at Nottingham in August 1866*, Transactions of the sections, p. 110). On the Palestine Exploration Fund, see the letter from J. D. Hooker, 13 July 1865 and n. 22, and the letter to J. D. Hooker, [29 July 1865] and n. 5.

[11] On Scott's previous position at a *Cinchona* plantation near Darjeeling, and his new appointment as curator of the Calcutta botanic garden under Thomas Anderson, see the letter from J. D. Hooker, [17 February 1865], and the letter from John Scott, 21 July 1865 and nn. 3 and 4.

[12] Scott had promised to repay the sum he had received from CD for his passage to India and sundry expenses (see letter from John Scott, 20 January 1865 and n. 6).

[13] See letter to J. D. Hooker, 22 December [1865]. As the Linnean Society's librarian, Richard Kippist was evidently responsible for updating the annually published *List of the Linnean Society of London*, which included the titles, appointments, and society memberships of each fellow. CD's entry for 1865 listed only his membership of the Royal and Geological Societies of London, and the Academia Caesaria Leopoldino-Carolina Naturae Curiosorum, a German academy of naturalists.

[14] CD was a corresponding member of the Königliche-Preussiche Akademie der Wissenschaften, Berlin; he became a fellow in 1878. He became a corresponding member of the Academia Scientiarum Imperialis Petropolitana, St Petersburg, in 1867. He became a corresponding member of the Kaiserliche Akademie der Wissenschaften, Vienna, in 1871, and an honorary foreign member in 1875. He became a foreign member of the Königliche-Bayerische Akademie der Wissenschaften, Munich, in 1878. See Freeman 1978, pp. 107–9.

[15] CD became a corresponding member of the Institut de France in 1878 (Freeman 1978, p. 108).

[16] See letter to J. D. Hooker, 22 December [1865] and n. 10.

[17] Julius von Haast had asked CD to support his candidacy for fellowship of the Royal Society of London (see letter from Julius von Haast, 27 September 1865 and n. 6).

[18] Hooker refers to Roderick Impey Murchison and Andrew Crombie Ramsay. Murchison was a member of the Royal Society council in 1865 (Royal Society, council minutes).

[19] Hooker refers to James Hector.

[20] Haast had been accused in the New Zealand press of plagiarising the work of other geologists (see letter to J. D. Hooker, 22 and 28 [October 1865] and n. 15).

[21] Charles Lyell.

[22] Hooker had written to Hector on 24 November 1865 of his decision to have Hector proposed for the fellowship at the same time as Haast (see Yaldwyn and Hobbes eds. 1998, pp. 63–4). Hector was elected a fellow of the Royal Society in 1866; Haast in 1867 (*DNB*).

[23] Hooker had suffered an attack of rheumatic fever (see letter from F. H. Hooker, [17 August 1865], and letter from J. D. Hooker, [26 September 1865]).

[24] On Hooker's involvement in landscaping at Kew, see R. Desmond 1995, pp. 226–31.

[25] See letter to J. D. Hooker, 22 December [1865] and n. 13. Hooker's wife was Frances Harriet Hooker.

[26] The enclosure has not been found.

[27] See n. 8, above.

From J. D. Hooker 24 December 1865

Kew
Dec 24/65.

Dear Darwin

I had a talk with Oliver about *Cœlebogyne*.[1] He says that Baillon found a stamen here & there on the female flowers;[2] & O. seems to disbelieve the "Virgin Mary plant", as its discoverer Allan Cunningham called it.[3] I examined many flowers without finding them, as did R. Brown,[4] Decaisne,[5] J. Smith,[6] A. Braun[7] & others.

The difficulty of enclosing the flowers in gauze[8] is that they grow quite down in the nests of the rigid leaves, & that not 1 in 100 ever ripens a seed. This, & the fact that every ovary is 3 celled & only 1 cell usually produces a ripe seed, renders the search for pollen tubes interminable. I tried a great many in vain.

Ever aff yrs | J D Hooker

I take for granted you know that Lyell is most good naturedly working to get me the Copley medal,[9] & after his funny & not-at-all-agreeable-to-me fashion, telling me all about it, & getting me to send him copies of my papers to send to Sir H. Holland,[10] who he wants to saddle with me. Of course I must not tell him so, but it is God's-truth, that not only shall I never think I deserve it if I get it; but that if I did deserve it, it would be *far* too dear at the cost of an after dinner Speech. These are things however which must take their courses

DAR 102: 51–2

CD ANNOTATION[11]
End of letter: 'coming [*illeg del*] | Gardening | Karsten | Lythrum x | Copley Medal' *pencil*

1 Hooker refers to Daniel Oliver. On the controversy over parthenogenesis in *Coelebogyne ilicifolia*, see letter to J. D. Hooker, 22 December [1865] and n. 7, and letter from J. D. Hooker, [23] December 1865 and nn. 3–5.

2 In a paper presented to the Société Botanique de France in Paris, Ernest-Henri Baillon reported having possibly observed an immature or underdeveloped stamen in the female flower of *Coelebogyne ilicifolia*; however, he was unable to confirm that the material extracted from the supposed anthers was in fact pollen grains (Baillon 1857, pp. 695–6).

3 In 1829, Allan Cunningham had sent three specimens of a plant, subsequently named *Coelebogyne ilicifolia*, to the Royal Botanic Gardens, Kew. The plants had been collected at Moreton Bay on the east coast of Australia (see J. Smith 1839, p. 509, and n. 6, below). The genus name derives from the Latin for 'celibate' and the Greek for 'female'. Cunningham later obtained specimens of the plant with male flowers, which he also sent to Kew (see Braun 1856, p. 327, and n. 7, below).

4 Robert Brown had collected specimens of *Coelebogyne ilicifolia* at Keppel Bay, on the east coast of Australia, in 1802 (J. Smith 1839, p. 509).

5 Joseph Decaisne. See letter from J. D. Hooker, [23] December 1865 and n. 4.

6 John Smith (1798–1888) had identified the specimens sent to Kew by Cunningham as a species of Euphorbiaceae, allied to *Sapium*, in a paper read to the Linnean Society (J. Smith 1839). See letter to J. D. Hooker, 22 December [1865], n. 7.

7 See letter from J. D. Hooker, [23] December 1865 and n. 5. In his 1856 paper on *Coelebogyne ilicifolia*, Alexander Carl Heinrich Braun described the male flower, based on specimens obtained from Kew, and concluded that the plant was dioecious and reproduced both sexually and parthenogenetically (Braun 1856, pp. 327–33). In a subsequent paper on polyembryony in *C. ilicifolia*, Braun commented that, as no contrary evidence had yet been provided, the prevailing view that the plant was parthenogenetic must stand (see Braun 1859, p. 117).

8 CD had suggested this procedure in his letter to Hooker of 22 December [1865].

9 Charles Lyell had written to the secretary of the Royal Society of London, William Sharpey, about the possibility of nominating Hooker for the Copley Medal in 1866; Lyell was concerned about the

poor representation of botany on the society's council, whose members voted on the awards (Charles Lyell to William Sharpey, 25 November 1865, American Philosophical Society (Lyell letters)). Hooker was not nominated for the Copley Medal in 1866. He was awarded the medal in 1887 (L. Huxley ed. 1918, 2: 307–9). On the small proportion of botanists on the Royal Society council, see also the letter to J. D. Hooker, 22 and 28 [October 1865] and n. 3.

10 Henry Holland had been chosen to serve on the council of the Royal Society for 1866 (Royal Society, council minutes, 1 November 1865).

11 CD's annotations are notes for his letter to Hooker of [31 December 1865].

To Julius von Haast 26 December [1865][1]

Down. | Bromley. | Kent. S.E.
Dec 26

My dear Dr Haast

I have received your last note & several printed papers,[2] some of which I have read with interest; others are still unread, lying in a large pile with other publications. These are all unread for I have been ill since April & for this same reason I have not sooner written to thank you. I am at last getting somewhat better ⟨I⟩ enclose my photograph which I had quite forgotten.[3]

I have written about your certificate for the Royal Soc. which I shall have much pleasure in signing & will do any thing I can in aid; but I hear there are so many applicants that I dare say you will have to wait a year or two.[4] I am much obliged for the N.Z. newspaper with your lecture which seemed to me very well written.[5] What magnificent scenery & glacial action you describe!

Mr S. Butler is now established in London as an artist.[6] He lately sent me a clever theological pamphlet.[7] I should have much liked to have seen him here & have heard N.Z. news, but the bad state of my health has rendered this impossible.

Believe me dear Dr Haast | yours sincerely | Charles Darwin

LS
Alexander Turnbull Library, Haast family papers, MS-Papers-0037-051

1 The year is established by the relationship between this letter and the letter from Julius von Haast, 27 September 1865.

2 See letter from Julius von Haast, 27 September 1865 and nn. 3 and 4.

3 See letter from Julius von Haast, 27 September 1865 and n. 13. CD probably sent the photograph taken in 1864 by his son William Erasmus Darwin. The photograph is reproduced as the frontispiece to *Correspondence* vol. 12.

4 Haast had requested that CD support his candidacy for the fellowship of the Royal Society of London (see letter from Julius von Haast, 27 September 1865 and n. 6, and letter from J. D. Hooker, [23] December 1865).

5 See letter from Julius von Haast, 27 September 1865 and n. 11.

6 See letter from Samuel Butler, 1 October 1865 and n. 2. Butler had lived in Canterbury, New Zealand, from 1858 to 1864 (*DNB*).

7 [Butler] 1865. See letter to Samuel Butler, 30 September [1865] and nn. 2 and 3.

From J. T. Moggridge 27 December [1865][1]

Maison Fontana | Mentone
Dec. 27

Dear Sir

I have forwarded a tin full of Ophryses[2] by my Brother to England for you;[3] & as he hoped to take but about 60 hours in the transit, perhaps they may arrive with you in tolerable condition.

⟨ ⟩ each parcel carefully ⟨ ⟩ I fear much that the damp may efface much of the writing.— My chief object was to distinguish those which came from spots where last year only *Oph. insectifera* a. *genuina* Rch. grew;[4] from those which came from habitats of *pink-winged* Varieties or from places unknown to me—[5]

Believe me yrs. very truly | J Traherne Moggridge.

DAR 171: 210

[1] The year is established by the relationship between this letter and the letter to J. T. Moggridge, 13 October [1865].

[2] See letter to J. T. Moggridge, 13 October [1865] and nn. 2, 3, and 5. In order to determine whether the late spider ophrys (*Ophrys arachnites*) and the bee ophrys (*O. apifera*) were crossing and self-fertile forms of the same species, CD had requested that Moggridge mark a half-dozen specimens of *O. arachnites*, noting the soil and situations in which the plants grew, and dig them up when he left Mentone. CD intended to cultivate the specimens and observe whether they retained the same character. In his letter of 14 October [1865], Moggridge said he would send CD 'an instalment [of orchids] coming into flower by rail or private hand'.

[3] Moggridge's brother has not been identified.

[4] Moggridge wrote 'a.' as an abbreviation of 'aranifera'; '*Ophrys aranifera genuina*' was described in Reichenbach 1851, pp. 88 and 178.

[5] In the text accompanying plate XLIV of *Flora of Mentone* (Moggridge 1865–8), Moggridge identified, as a variety of *Ophrys insectifera*, a form with 'whitish or pink-tinted petals and sepals' that flowered in March.

From Robert Monsey Rolfe 28 December 1865

40, Upper Brook Street. | *W.*
28 Decr 1865

My dear Mr Darwin—

I was quite mortified to find from Mrs Darwins[1] letter that I left Holwood[2] in the Autumn without sending to the annual cheque of which you so kindly undertake the distribution—[3] The fact is that it quite escaped my memory— I now enclose a cheque the rect of which you will perhaps acknowledge to me at Holwood where I hope to be this Evening—

I am afraid from your not writing yourself that you have not been very comfortable, But Mrs Darwin did not mention your health so we hope for the best—

With best wishes of the Season Believe me, | yours very truly | Cranworth

DAR 161: 233

1 Emma Darwin.
2 Holwood House, Rolfe's family seat, was $1\frac{1}{2}$ miles from Down (Freeman 1978).
3 Rolfe was a regular subscriber to the Down Friendly Club and the Down Coal and Clothing Club, of which CD was treasurer (Freeman 1978). The Down Friendly Club was a local savings and insurance society; the Coal and Clothing Club supplied parishioners with cheap coal and clothes in exchange for regular savings. In previous years, CD had distributed Rolfe's contributions to a day school and Sunday school in Down (CD's Account book–cash account (Down House MS)). In his Down Coal and Clothing Club account book (Down House MS), CD recorded a contribution of £2 2*s.* from Rolfe in 1865. On CD's involvement in local charities, see J. R. Moore 1985, p. 466.

To J. D. Hooker [31 December 1865]

Down. | *Bromley.* | *Kent. S.E.*
Sunday Evening

My dear Hooker

I write a line to say, firstly, to say how rejoiced I am that there is a good chance of seeing you here for a Sunday:[1] do keep to so good an intention— it will be a real joy to me.— Secondly to thank for for information about Karsten & the case of parthenogenesis, about which I became very curious.[2] It is jolly to hear of you working in the garden & very good for mind & body.

I will explain about my *so-called* hybrids of Lythrum when we meet.—[3]

I had heard & talked with Sir H. H. & Lyell about the C. medal[4] (which I knew you would not care very much about, tho' that is no reason for your friends not wishing it most earnestly) & it was very foolish in Lyell, who, I believe originated the idea this year, speaking to you about it. If I had my way it sh[d]. not be mooted in Council *this* year with so very few naturalists of any kind (deserving the name) on the Council.—[5]

Yours affectionately | C. Darwin

As for your thinking that you do not deserve the C. Medal, that I declare is mere insanity.—[6]

Endorsement: 'Dec 31/65.'
DAR 115: 279

1 See letter from J. D. Hooker, [23] December 1865. Hooker's next recorded visit to Down was 24 March 1866 (Emma Darwin's diary (DAR 242)).
2 CD refers to Hermann Karsten. See letter to J. D. Hooker, 22 December [1865] and n. 7, and letters from J. D. Hooker, [23] December 1865 and 24 December 1865.
3 See letter to J. D. Hooker, 22 December [1865] and n. 5, and letter from J. D. Hooker, [23] December 1865.
4 Charles Lyell had contacted Henry Holland about nominating Hooker for the Royal Society of London's Copley Medal in 1866. See letter from J. D. Hooker, 24 December 1865 and nn. 9 and 10. CD had met Lyell, and evidently Holland, on his visit to London in November (see letter to J. D. Hooker, 22 December [1865]).
5 For a list of the members elected to the Royal Society council for 1866, see *Proceedings of the Royal Society of London* 14 (1865): 513. The British Museum zoologist John Edward Gray was the only professional naturalist on the council; however, several other members had made contributions to natural history.

Holland often reviewed natural historical works; Francis Galton and Paul Edmund de Strzelecki, had both gained reputations as explorers; Warington Wilkinson Smyth was a mineralogist and surveyor; and the president of the Royal Society, Edward Sabine, had been credited with a number of natural historical discoveries in his early career and was a fellow of the Linnean Society (*DNB*).

6 See letter from J. D. Hooker, 24 December 1865.

SUPPLEMENT, 1822–1864

This supplement includes all the letters from or before the year 1864 that have been located or re-dated since the publication of *Correspondence* volumes 7–12; an earlier supplement was included in volume 7. It also includes letters with a wide date range that seem to have been written before 1865. The letters are arranged in chronological order according to the usual practice of the edition. A record of CD's alterations, information about the correspondents and other individuals mentioned in the letters, and bibliographical details, are included in the Manuscript alterations and comments, the Name register, and the Bibliography for volume 13.

To Dear Friend[1] 1 January 1822

January 1st. 1822

My Dear friend

I think it is a most abominable thing, the minute Erasmus went out of the room, they began abusing him for being out of temper, you must know that he is not well, and that he hath got the rheumatism, and very pale, and particularly Maryan[2] of couse you know how peevish Maryane is when she is unwell[3] but just the reverse I think Erasmus very good humered to day he came back from enquiring about his ticket, on purpose, because he thought I should like it, the reason was because he was going to draw for Cam bell Bible,[4] and I must now conclude

so therefore your

Note this day I received this ye cabinet[5]

AL
Ursula Mommens, Memorandum book, Darwin Archive, CUL

[1] This is the first of six entries written in a 'Memorandum book' comprising four sheets folded into a gather and sewn together in book form. The entries are in the style of letters addressed to an unnamed friend and are dated between 1 and 12 January 1822, shortly before CD's thirteenth birthday. As they were written straight into the memorandum book, it is clear that they were never sent through the post, but were either to an imaginary recipient, or intended to be read by someone in the household, possibly CD's youngest sister, Emily Catherine Darwin (known as Catherine; see this volume, Supplement, letters to Dear Friend, 2 January 1822, n. 3, 12 January 1822, n. 6; see also Browne 1995, pp. 11–12, 14). Catherine is known to have been close to CD as a child, and the absence of any reference to her in the letters is best explained by her being the intended recipient. Another suggested recipient is John Price (Browne 1995, p. 28). CD and Price were both pupils at the Royal Free Grammar School, Shrewsbury, but would have been home on holiday during the period when

the letters were written. The school year comprised two terms with two seven-week holidays in June and July and December and January (Oldham 1952, p. 158).

2 CD refers to his brother, Erasmus Alvey Darwin, his sister Marianne Darwin, and probably to one or more of his other sisters, Caroline Sarah Darwin, Susan Elizabeth Darwin, and Emily Catherine Darwin. According to CD's later reminiscences, Erasmus suffered ill health from childhood (*Autobiography*, p. 42).

3 The words 'a big lie' are interlined at this point in a different hand.

4 The reference is probably to an edition of the English translation of the New Testament by George Campbell (G. Campbell trans. 1789; G. Campbell *et al.* trans. 1818).

5 The words 'Note . . . cabinet' are written between the first and second entries in the 'Memorandum book' and are separated from them by faint lines. This suggests that CD kept the memorandum book in a new box, possibly the same cabinet in which he later kept mineral samples (see *Correspondence* vol. 1, letter from E. A. Darwin, 14 November 1822).

To Dear Friend[1] 2 January 1822

My Dear friend

I think Erasmus is quite good humered to day as he was yesterday and they have not abused at all,[2] I hope you will help me in looking out and washing the fossils out of the plate closet, as soon as you can conveniently,[3] I hope you received yesterdays letter—

I remain, *you know what*

A PUg,[4] January 2^d^ *1822*

AL
Ursula Mommens, Memorandum book, Darwin Archive, CUL

1 The friend has not been identified. See this volume, Supplement, letter to Dear Friend, 1 January 1822, n. 1.

2 See this volume, Supplement, letter to Dear Friend, 1 January 1822 and n. 2.

3 CD had developed an early interest in collecting 'chiefly seals, franks &[c] but also pebbles and minerals' (*Correspondence* vol. 2, Appendix III; see also *Autobiography*, p. 45). An interest in fossils in particular was shared with his sister Catherine, and his brother, Erasmus (see *Correspondence* vol. 1, letters from E. A. Darwin, 25 [October 1822] and [10 January 1825], and letter from E. C. Darwin, [*c.* June 1823]). The plate closet, named after its original purpose of storing silverplate ware, was used by Erasmus to store scientific books (see *ibid.*, letter from E. A. Darwin, [8 March 1825]).

4 An abbreviation of 'affectionate pug'; see this volume, Supplement, letter to Dear Friend, 12 January 1822.

To Dear Friend[1] 3 January 1822

January ^the^ 3 1822

My dear friend

I hope you like that inquisitive man Monseur Beodoes,[2] when I was a boy I used to think him a most impertinent man but you may differ in your opinion, he always used to be asking about my father and other things; In you last letter you told me you knew, who, my sisters at Shrewsbury were talking of, I think Papa[3] you will inform me in return of post

I remain—yours Tobe Case[4]

P.S. Mr Bayly[5] was formerly a *devlish* boor but I dare say he is not *so to you*

AL

Ursula Mommens, Memorandum book, Darwin Archive, CUL

[1] The friend has not been identified. See this volume, Supplement, letter to Dear Friend, 1 January 1822, n. 1.

[2] 'Monseur Beodoes' has not been identified. The reference may be to the poet and physician Thomas Lovell Beddoes, whose father, the physician Thomas Beddoes, had been a close friend of the Wedgwood and Darwin families (B. Wedgwood and Wedgwood 1980). 'Beddoes' was, however, a fairly common name in early-nineteenth-century Shropshire.

[3] CD refers to two or more of his sisters, Marianne, Caroline Sarah, Susan Elizabeth, and Emily Catherine Darwin, and to his father, Robert Waring Darwin.

[4] The form of the signature is unexplained. CD had previously attended a school run by George Augustus Case in Shrewsbury (*Autobiography*, p. 22 and n. 1).

[5] CD possibly refers to William Bayley, of the banking firm Rocke, Eyton, Campbell, Leighton and Bayley in Market Square, Shrewsbury, who had been instructed to act in a financial transaction on Robert Waring Darwin's behalf in January 1820 (DAR 227.6: 339 and 340, *Commercial directory for Shropshire*). CD's sister Emily Catherine wrote in her letter of 15 January [1826] that a Major Bayley desired '*most particularly*' to be remembered to CD (*Correspondence* vol. 1).

To Dear Friend[1] 4 January 1822

My Dear friend,

You must known that I am very sorry about shakespeare when Maryane, told me to put it up,[2] and I like Mariane excessively, she is so very good to M^iss^ Jone,[3] I will tell you a little accident that happened to me of course you know M^iss^ Jones is in bed, and when I went there with Mariane last night, while she stoopped there I went in town and bought the cakes, and I came back sooner than Mariane was ready, and that foolish old man shoed in the bedroom then there was a bit of Kick-hoop, not that I went quite in, oh no, quite the reverse

Your affectionate

Justice brust with a nose as big as your fist

PS I hope you were not much alarmed at M^iss^ Clare[4] accident

January 4, 1822

AL

Ursula Mommens, Memorandum book, Darwin Archive, CUL

[1] The friend has not been identified. See this volume, Supplement, letter to Dear Friend, 1 January 1822, n. 1.

2 CD refers to his sister Marianne Darwin.

3 Possibly Martha Jones, a housemaid, who appears in Robert Waring Darwin's accounts as receiving money between 1817 and 1822 (DAR 227.5: 82, f. 51 and contents page). The local poor and sick whom Erasmus Alvey Darwin visited in 1826 included Mrs O. Jones (*Correspondence* vol. 1, letter from Susan Darwin, [27 March 1826]).

4 Possibly Clare Leighton, a friend of the family (see *Correspondence* vol. 1, letter from Caroline and Susan Darwin, 2 [January 1826], and letter from Catherine Darwin, 15 January [1826]).

To Dear Friend[1] 4 January 1822

My Dear friend

you must know that after my Georgraphy, she[2] said I should go down to ask for Richards[3] poney, just as I was going, she said she must ask me not a very decent question, that was whether I wash all over every morning no then she said it was quite disgustin then she asked me if I did every other morning, and I said no then she said how often I did, and I said once a week, then she said of cour you wash your feet every day, and I said no, then she begun saying how very disgusting and went on that way a good while, then she said I ought to do it, I said I would wash my neck and shoulders, then she said you had better do it all over then I said upon my word I would not, then she told me, and made me promise I would not tell, then I said, why I only wash my fett once a month at school, which I confess is nasty, but I cannot help it, for we have nothing to do it with,[4] so then Caroline pretended to be quite sick, and left the room, so then I went and told erasmus,[5] and he bust out in laughing and said I had better tell he to come and wash them her self, besides that she said she did not like sitting by me or Erasmus for we smelt of not washing all over, there we sat arguing away for a good while.

I remain your affectionately

Justice *[burst]* w[ith] a nose as big as your fist

January the 2, 1822 The second letter today[6]

AL

Ursula Mommens, Memorandum book, Darwin Archive, CUL

1 The friend has not been identified. See this volume, Supplement, letter to Dear Friend, 1 January 1822, n. 1.

2 CD refers to his sister Caroline Sarah Darwin. Caroline had taught CD at home before he started attending school and was presumably supervising his studies during the school holiday (see *Autobiography*, p. 22, and this volume, Supplement, letter to Dear Friend, 1 January 1822, n. 1).

3 Richard or Richards has not been further identified.

4 On the sanitary arrangements at Shrewsbury School in the early nineteenth century, see Oldham 1952, p. 155.

5 Erasmus Alvey Darwin.

6 Since this letter appears in the book immediately after the letter to Dear Friend, 4 January 1822 (this volume, Supplement), it has been given a clear date of 4 January. The written date, 2 January, is presumably a mistake.

To Dear Friend[1] 12 January 1822

My Dear friend

I hope you had a pleasant ride yesterday on Domben[2] I had a very pleasant walk at first we were very dull but afterward we were quite merry till Curnell Burg Leighton[3] stoped us and walked with us in the quarry I have had such fun with nany and May[4] about some stuff* that I thought very shocking it is in paling *[skin]* I thought it rogue[5] I believ⟨e⟩ it cost eighteen pence a saucer of it—and that *[Maryan]* called very dear

I remain your Pug puller

*that stuff in the washroom

January 12th. 1822—

remember next summer to make two cave one for warlike instruments, the othe⟨r⟩ for relicks. Note spoon, old spear knife squirt if it can be found, and the name cut on the ash tree over the seat in the bank by the nut tree I belife that is all ove[r] the lief a plan of a machine

x a pin drove in the bark on which a piece of wood can turn up and down, aa the piece of wood, bb two pieces of string going down *[over]* the bank, cc two weights insted of springs, dd two pieces of coulored rag or ribbon e the tree on whi⟨ch⟩ I sit, g the bank, f *[one word illeg]* when the red rag is pulled up come to me, when the other lie still.[6]

Papa told me that Fridy perhaps I should go to Walcut to see all the beutiful things there[7] but he told me not to set my mind upon it so therefore your affectionate pug whom you gave a good cough.

Remember this time ask Downs, the drums shall beat, the fifes shall play— and it is all for the sake of——[8]

I went to Athelton where I say Miss reynols and I like her very much I think her very pretty but I do not know she was a friend of Mariane and Caroline[9] he[r] mothe[r] is half a quaker and I like her very murch, I believe Miss Reynols is going to be marriend to a clergyman, and has been under trial 1 2 3 4 or 5 year before Mrs Reynols would let her may him, and he has given some living, but I am not shure about this[10] I like M^r^ and M^rs^ W Cluide[11] very much indeed particularly Mr Clude but I dont thing Erasmus does for your must know he was with me indeed I dont think Erasmus liked it all together but I am *[oky]* I did, I think Miss and young M^aster^ Clude much mor[e] *[talkative]*

AL
Ursula Mommens, Memorandum book, Darwin Archive, CUL

1 The friend has not been identified. See this volume, Supplement, letter to Dear Friend, 1 January 1822, n. 1.

2 CD refers to the Darwin family horse, Dobbin (see *Correspondence* vol. 2, Appendix III).

3 Burgh Leighton, lieutenant-colonel in the Fourth Light Dragoons lived at Quarry Place, Shrewsbury (*Burke's peerage*, *Commercial directory for Shropshire*). The Leightons were family friends of the Darwins (see this volume, Supplement, first letter to Dear Friend, 4 January 1822 and n. 4).

4 CD probably refers to his sister, Marianne Darwin. The nanny has not been further identified.

5 Presumably a misspelling of 'rouge'.

6 In a separate series of notes probably written at a similar time, and certainly after 1821, CD recorded 'Our language and signs', including 'One pull,— Come | Two pulls,— Lie still' (CUL, Ursula Mommens deposit). These may have been written for Emily Catherine Darwin (Browne 1995, p. 14).

7 Walcot is a township about six miles from Shrewsbury. Walcot Park was the home of Edward Clive, first earl of Powis, and Henrietta Antonia Clive, countess of Powis (*Burke's peerage*). Both the earl and the countess were patients of CD's father, Robert Waring Darwin, who gave medical advice to the countess on 5 January 1822, and visited the earl at Walcot on 25 January 1822 (DAR 227.5: 67 f. 260). CD later remembered often accompanying his father on visits to patients (*Correspondence* vol. 2, Appendix III). Walcot Park had been rebuilt and furnished by the earl's father, Robert Clive, also known as 'Clive of India', one of the richest men in the country, and enlarged by the earl to house treasures acquired during his governorship of Calcutta (*DNB*, *Treasures from India*).

8 Possibly a reference to William James Downes, a fellow pupil of CD's, who left Shrewsbury School in 1823 (*Shrewsbury School register*). See also *Correspondence* vol. 1, letter from E. A. Darwin, 5 [March 1823]. The fife and drum were the traditional instruments used for signalling in English infantry regiments, and also for medieval mumming (Grove 1980).

9 Marianne Darwin and Caroline Sarah Darwin.

10 CD probably refers to Arleston, a hamlet in the township of Wellington, Shropshire, about eleven miles east of Shrewsbury (*Post Office directory of Gloucestershire, with Bath, Bristol, Herefordshire, and Shropshire* 1856). Hannah Reynolds and her daughter, Susanna, lived at Ketley Bank, near Arleston. Hannah Reynolds was both niece and daughter-in-law of the Quaker ironmaster and philanthropist Richard Reynolds, a friend of CD's grandfather, Josiah Wedgwood I (*DNB*). Susanna Reynolds married John Bartlett; Bartlett was appointed perpetual curate of Buildwas, Shropshire, in 1822 (*Alum. Cantab.*). See *Two Shropshire ironmasters*.

11 William Pemberton Cludde of Orleton Hall, near Wellington, Shropshire, and his wife, Anne Maria (*Burke's landed gentry*, s.v. Herbert of Orleton).

To John Lort Stokes [after 31 May 1837][1]

Are there masses of coral or beds of shells some yards above high water mark, on the coast fronting the barrier reef?[2]

Incomplete
Stokes 1846, 1: 331

[1] The date is established by the connection between this fragment and CD's reading of his paper 'Elevation and subsidence in the Pacific and Indian Oceans' at the Geological Society of London on 31 May 1837 (see n. 2, below).

[2] This question was one of several written down by CD for Stokes and is quoted by Stokes in his *Discoveries in Australia* (Stokes 1846, 1: 331); CD's original list of queries has not been found. Stokes was a shipmate of CD's during the *Beagle* voyage of 1831 to 1836, serving as mate and assistant surveyor (see *Correspondence* vol. 1). From 1837 to 1843 the *Beagle* was employed in surveying coastal areas of Australia under the successive commands of John Clements Wickham and Stokes, sailing from Plymouth on 5 July 1837 (Stokes 1846). On 31 May 1837, Stokes attended a meeting of the Geological Society at which CD read his paper 'Elevation and subsidence in the Pacific and Indian Oceans' (*Collected papers* 1: 46–9; see also Stokes 1846, 1: 331). In this paper CD proposed a theory of the development of coral reefs. For the development of CD's theories on coral formation and subsidence, see *Correspondence* vol. 1, Appendix V, and Stoddart 1976. Stokes wrote (Stokes 1846, 1: 331) that his interest in the paper was greatly enhanced:

> by a series of queries, kindly furnished by Mr. Darwin, and drawn up with a view to confirm or invalidate his views, his purpose being to elicit truth from a combination of well attested facts, and by inducing the research of others to further the objects of science.

In June 1839 Stokes discovered a raised beach of coral and shells on the coast of Cape Upstart that he argued, following CD's theory, supported the view that the north-east coast of Australia had not been subject to subsidence (Stokes 1846, 1: 332–3). CD published a full account of his theory in 1842 (*Coral reefs*), and acknowledged Stokes for supplying information on Australian coral (*ibid.* p. 86).

To John Richardson [24 July 1837][1]

My dear Sir

Will you excuse me troubling you with some questions, which I am very anxious to get answered, and do not know what work to refer to.— How far north do woods of any extent occur? Did I understand you to say that *forest trees* grew over ground, which at the depth of a few feet was perpetually frozen?—[2] I suppose this as much as to say, that there may exist extensive woods, where the *mean annual* temperature is the below the freezing point.— What trees grow in such cold climates, & do they attain any size? Do plains covered with bushes, occur in any part of the extreme north. My object in these questions, is to be enabled to compare the mere *quantity* of vegetation, in parts of South America, where large animals formerly did live, and likewise in Africa where large animals are now living, with the quantity growing in climates far north, and extremely cold.[3]

If you would have the kindness to answer me briefly these questions, I should be greatly obliged, but I really I ought to apologise for asking you to take so much trouble.

In what part of your works could I find (if at all introduced) the account of the frozen sandstone: you were going to look for me, when I was at your house, but something interrupted us.—[4] My government petition for assistance is in statu quo, and I think it will thus remain, for a long time. It was presented to the Chancellor of the Exchequer who did not receive it unfavourably, but said until Parliament met, he could give no answer.—[5] I am working away at my journal,[6] but get on very slowly:—building a pyramid is an insignificant task to writing a book; I had no idea what a hard working wretch an author, even on the humblest scale, mu⟨st⟩ be.—

Pray present my be⟨st⟩ respects to M^rs. Richardson,[7] an⟨d⟩ | Believe me dear Sir | Yours most truly | Chas. Darwin

Monday Evening
36 Grt Marlborough St^t.—

Can you give me any idea, how far from shore the sea becomes converted into a *mass resting on the bottom* of ice on any very *shallow* coast? Do you suppose such a *mass* ever extends any number of miles to seaward?— Do you suppose in such case, that the ice during the winter becomes cooled down many degrees below the freezing point? If such is the case the *mean annual* temperature, of the stratum a few feet *beneath* the *bottom* of the sea near to such a frozen coast, may possibly fall below the freezing point; although during summer the ice might break away; precisely in the same manner as the ground a few feet deep is constantly frozen, though during each summer the surface is thawed.—[8] Will you have the kindness to consider this question & tell me what you think.

Postmark: 25 JY 25 | 1837
Scott Polar Research Institute (MS 1503/16/1)

[1] The date is established by the postmark and the reference to Monday evening. In 1837, 24 July was a Monday.

[2] CD evidently refers to a recent conversation with Richardson. The letter refers to a visit by CD to Richardson's house; CD's plan to make a visit in May or June 1837 was announced in his letter to Caroline Darwin, [19 May – 16 June 1837] (*Correspondence* vol. 2). CD was trying to estimate the quantity of vegetation necessary for the survival of large quadrupeds in different environments (see nn. 3 and 8, below). In *Journal of researches*, p. 102 n., CD quoted some of Richardson's data on forest growth and the depth of frozen soil in the Arctic, citing the author's 'Zoological Remarks' (pp. 475–522 in Back 1836). Richardson completed publication of his four-volume *Fauna Boreali-Americana; or, the zoology of the northern parts of British America* (J. Richardson 1829–37) in 1837, having studied the area as surgeon and naturalist on John Franklin's first two polar expeditions.

[3] To determine whether the large South American quadrupeds could have survived in an area with little vegetation, CD considered extant mammals in Africa, as well as the evidence from carcasses preserved in the frozen ground of the Arctic where the quantity of vegetation was comparable (*Journal of researches*, pp. 98–103). CD concluded that 'luxuriant vegetation' was not necessary to the survival of large mammals in any of these regions, including the plains near Bahia Blanca and the Rio Negro, where he had found fossils of quadrupeds during the *Beagle* expedition (*ibid.*, pp. 103–5; see also pp. 293–5).

[4] In an account of the topography and geology of the North American Arctic, Richardson described sand near the Mackenzie river that had 'frozen into a hard sandstone' (Franklin 1828, Appendix I,

p. xix). CD's query may have related to understanding how mammal carcasses had been preserved in frozen mud at the bottom of shallow Arctic seas (see *Journal of researches*, pp. 295–8), or to the means of survival of large quadrupeds in the extreme climate of the South American pampas (see *ibid.*, pp. 102–4, 293–6; see also nn. 2 and 3 above, and n. 8, below). Richardson was cited by CD for information on ice and the geology of the North American Arctic in *Journal of researches*, pp. 619–20, and in the letter to Edward Sabine, 16 March [1857] (*Correspondence* vol. 6). For bibliographies of Richardson's works, see Curvey and Johnson 1969, and Huntley *et al.* 1972.

5 With the support of the presidents of the Linnean, Zoological, and Geological Societies of London, CD was seeking government assistance to pay for approximately 150 engravings to illustrate the zoological collections of the *Beagle* voyage (*Zoology*; see *Correspondence* vol. 2, letter to Francis Beaufort, 16 June 1837 and n. 2). The Treasury had provided similar assistance to Richardson for J. Richardson 1829–37 (see *ibid.*, Minutes of the Board of the Treasury, 25 August 1837). A preliminary application was submitted on CD's behalf to Thomas Spring Rice, Chancellor of the Exchequer, by their mutual friend, John Stevens Henslow (see *ibid.*, letter from J. S. Henslow to the Chancellor of the Exchequer, T. S. Rice, 21 June 1837). CD's own formal request for £1000 followed on 3 August 1837 (see *ibid.*, letter to the Chancellor of the Exchequer, T. S. Rice, 3 August 1837) and was granted by mid-August 1837 (see *ibid.*, letter to J. S. Henslow, 16 August [1837]). See also Browne 1995, pp. 367–70.

6 In 1837, CD worked on *Journal of researches* from 13 March to the end of September (see *Correspondence* vol. 2, Appendix II).

7 Mary Richardson.

8 In *Journal of researches*, CD questioned whether the stratum on the bottom of shallow coastal seas could remain frozen and thereby preserve carcasses of large quadrupeds. For CD's conclusions, which do not include data from Richardson, see *ibid.*, pp. 297–8, 625–6. No answer to this letter has been found; however, see this volume, Supplement, letter to John Richardson, [11 August 1837] and n. 3.

To John Richardson [11 August 1837][1]

My dear Sir

I am ashamed to trouble you again.[2] But in your last letter you said you probably would pass through town about the fifteenth of this month.—[3] I have this morning received intelligence (though not officially), that the Chancellor of the Exchequer had ordered a 1000£ for the publication of the Zoology of the Beagle's Voyage.—[4]

I am particularly anxious to ask your advice upon one or two points, now that the plan is more settled.—[5] If you could spare the time I should be most grateful for half an hour's conversation.

If you would let me know I would call upon you, at any place you might fix, at any hour.— Perhaps you would send me a line telling me your plans.—

Excuse this letter—for I am writing against time, as I hear the Postman's Bell.—[6] I was very much obliged by the trouble you took in writing me so long account of the climate of North America—[7] I cannot succeed in burying the Mammoth in the ice quite to my satisfaction[8]

Most truly yours| Chas. Darwin

36 | Great Marlborough St.
Friday.—

Scott Polar Research Institute (MS 1503/16/2)

[1] The date is established by the references to the probable acceptance of CD's formal application to the Chancellor of the Exchequer for funds to assist with the expense of illustrating *Zoology*, sent on 3 August 1837 (see *Correspondence* vol. 2, letter to the Chancellor of the Exchequer, T. S. Rice, 3 August 1837), and to Richardson's expected visit to London 'about the fifteenth'. CD learned that his application had been approved in an interview with the Chancellor of the Exchequer, Thomas Spring Rice, on 16 August (see *ibid.*, letter to J. S. Henslow, 16 August [1837]). In 1837, the Friday preceding 15 August was 11 August.

[2] That summer, CD had already paid a visit and written at least once to Richardson (see this volume, Supplement, letter to John Richardson, [24 July 1837]).

[3] The letter from Richardson has not been found.

[4] See n. 1, above; see also this volume, Supplement, letter to John Richardson, [24 July 1837] and n. 5.

[5] Richardson was a former recipient of a Treasury grant to assist the publication of a scientific work (see this volume, Supplement, letter to John Richardson, [24 July 1837], n. 5).

[6] Until 1846, in London and some provincial towns, letters for the last post were handed to a 'bellman' in the street; this form of collection was discontinued in the wake of Rowland Hill's postal reforms (Daunton 1985, pp. 7, 8, 240).

[7] This letter has not been found, but see this volume, Supplement, letter to John Richardson, [24 July 1837] and nn. 2–4 and 8. CD cited information on climate from Richardson's zoological appendix to Back 1836 in *Journal of researches*, pp. 102, 274.

[8] CD is referring to *Elephas primigenius* (see, for example, *Journal of researches*, pp. 150–1). For the topic of large animals preserved in ice, see this volume, Supplement, letter to John Richardson, [24 July 1837], nn. 3, 4, and 8. By October or November 1837, the *Journal of researches* was at proof stage (*Correspondence* vol. 2, Appendix II) and the problem of the entombment of large animals in ice had been resolved in CD's mind (*Journal of researches*, pp. 297–8, 625–6).

To Richard Owen [15 December 1837 – 9 June 1838][1]

My dear Owen

I send you a box with Isle of Wight fossils from Fox, with a letter from him.—[2] He will tell you their history.

These fossils appear to me a glorious fact. if any man a hundred years ago had said the time will come, when a naturalist looking at a heap of strata will be able to prophesy that bones of pig-like animals of many kinds are probably entombed there.—[3] Would he not in the language of the Persians have been called the grandfather of all liars?—[4]

I forget whether I told you that Londsdale has some one or two teeth from the Isle of Wight, which he wanted you to see, in case you undertook the Eocene quadrupeds.—[5]

Balliere[6] sends a parcel either on Saturday or at beginning of next week.— I have sent him a copy of 1st Iss^{e} for the Institute.[7]

Yours most truly | Chas. Darwin

Friday

Balliere charges about 1^{s} " 6^{d}. for carriage

DAR 185: 115

[1] The date is established by the relationship between this letter and the letter to W. D. Fox, [15 June 1838] (*Correspondence* vol. 2), in which CD asked William Darwin Fox when he would like his fossils returned. That letter recorded CD's receipt of the fossils three days earlier and the previous delay in their description by Richard Owen (see n. 2, below). It is likely that the fossils were despatched by Fox to CD some time after their first mention in the correspondence (see *Correspondence* vol. 2, letter to W. D. Fox, [11 December 1837]). In 1837, the Friday after 11 December was 15 December; in 1838, the Friday before 12 June was 9 June.

[2] The letter from Fox has not been found. Owen described the fossils, found by Fox in freshwater beds at Binstead and Seafield quarries on the Isle of Wight, in a paper delivered on 7 November 1838 to the Geological Society of London, and compared them with earlier finds from the same area and from the Paris basin (R. Owen 1838; see also *Proceedings of the Geological Society of London* 3 (1838): 1–3).

[3] R. Owen 1838 discussed the similarities between the anatomical features of the fossils (teeth of four species of *Palaeotherium* and two species of *Anoplotherium*, and a right ramus of the lower jaw of *Chaeropotamus*) and those of members of the hog tribe. C. Lyell 1855, pp. 210–11, also briefly described the fossils found by Fox, referring to *Palaeotherium* as resembling a tapir.

[4] CD also used this phrase in the letter to Charles Whitley, 23 July 1834 (*Correspondence* vol. 1): 'A Persian writer could alone do justice to it, & if he succeded he would in England, be called the "grandfather of all liars".'

[5] In his paper on quadrupeds of the Eocene (see n. 2, above), Owen described the structure of the teeth and the form of the jaw in a fossil obtained from the Binstead quarries and deposited in the museum of the Geological Society, of which William Lonsdale was curator and librarian. Owen concluded that the fossil should be attributed to a new species of *Dichobunes* (R. Owen 1838, pp. 44–5).

[6] Hippolyte Baillière.

[7] *Zoology* first appeared as a series of nineteen numbers, constituting five parts. The first number contained Owen's description of parts of the fossil mammal *Toxodon platensis*, found by CD about 120 miles north-west of Montevideo. It was communicated to the journal *L'Institut* by the French zoologist Henri Milne-Edwards, and a brief anonymous review appeared in *L'Institut*, 1st section, 6 (1838): 110–11.

To John George Children 22 February [1838][1]

36 Grt. Marlbro' St

Dear Sir

During the last six months, I have had occasion several times to consult the collection of Birds in the British Museum.—[2] With respect to the subject of your letter, I have much pleasure in informing you, that Mr George Gray has uniformly acted towards me with courtesy.[3]

He not only has always opened as many cases, as I required with readiness, but has likewise *repeatedly* given me much useful information, often suggesting books of reference and other means by which I could obtain a knowledge of the subject I had in view.—[4]

I have the pleasure to remain | Dear Sir | Very faithfully yours | Chas. Darwin

February 22^d^.—

British Museum Officers Reports 20 (1838): 5314

[1] The year is established by the relationship between this letter and the letter from J. G. Children, 21 February 1838 (*Correspondence* vol. 7, Supplement).

[2] In late 1837, CD was working on a scheme for the publication of *Zoology*; *Zoology* included a volume on birds (see *Correspondence* vol. 2, Appendix II; see also *Correspondence* vol. 2, letter to John Gould, [February 1838]). CD made extensive use of the ornithological collections of the British Museum (Brandon-Jones 1996).

[3] CD refers to Children's letter of 21 February 1838 (*Correspondence* vol. 7, Supplement), in which he requested a character reference for George Robert Gray. Gray was Children's assistant in the zoology department of the museum, and had been the subject of a complaint by Edward Blyth, who was then preparing the sections on mammals, birds, and reptiles for a new edition of Georges Cuvier's *Animal kingdom* (Cuvier 1840).

[4] CD's opinion of Gray's good character was one of fifteen responses to Children's enquiry into Blyth's accusation of Gray for gross incivility and discourtesy. The background to this episode is described in Brandon-Jones 1996.

To John Stevens Henslow 3 November 1838[1]

Saturday, 3 Novr 1838

My dear Henslow

I am preparing an appendix to my Journal, in which I mean to add a few remarks on some of the subjects, which I have there discussed.—[2] You may recollect how often I have talked over the marvellous fact of the species of birds being different, in those different islands of the Galapagos.—[3] Lately I have gained some curious facts, bearing on the same points, regarding the lizards & tortoises of those same islands;[4] & now I want to know whether you can tell me anything about the plants.— Pray understand, I do not want you to take any trouble in giving me names &c &c.— all I want is to know whether in casting your eye over my plants, how many cases (for you told me of some one or two) there are of *near* species, of the same genus,;—one species coming from one island, & the other from a second island.—[5]

If there are any number of these cases you can mention to what natural families they belong. (or the genera if you happen to know them).— Also how near the species are;—whether they require comparison to be distinguished; or whether, merely in any natural classification, they would follow one after another. I have already given you trouble enough about this same Galap: Arch: but all I now want is for you to answer me, as far as your present knowledge goes; & in doing this, I apprehend it will not take up more than a hour, & it will (if such be the fact with the plants) support my case of the birds & Tortoises in a *glorious* manner.— [6]

Many thanks for your letter, which I received some time.—[7] Leonard Jenyns[8] will see about the Savings Bank.—

Will you have the kindness to write pretty soon, as I believe (but am not sure) I shall go to press with the appendix immediately,—

Ever yours most truly | Chas. Darwin

I have written to you on a torn piece of Paper without perceiv⟨ing⟩ it.

DAR 249: 83 (photocopy)

[1] A summary of this letter, taken from a sale catalogue, was printed in *Correspondence* vol. 7, Supplement. It was dated [before 27 October 1838]. Since then a facsimile of the original has been acquired by CUL; the text is reprinted here, with a corrected date.

[2] CD's supplementary remarks appear as the Addenda to the first edition of *Journal of researches*, pp. 609–29. They cover a range of topics, but principally geology.

[3] In an earlier letter to Henslow, CD mentioned the birds of the Galápagos, suspecting them to be 'very curious' (see *Correspondence* vol. 1, letter to J. S. Henslow, [28–9] January 1836 and n. 2). Observations by CD on the diversity of mocking-birds and finches and their differences on the different islands appeared in *Journal of researches*, pp. 461–2 and 475. A fuller account of Galápagos birds, including illustrations of finches, was given in *Journal of researches* 2d ed., pp. 378–81, 394–5. For comment on the status of Galápagos finches in CD's exposition of his species theory, see *Correspondence* vol. 12, letter from [C. P.], 29 April 1864 and n. 3.

[4] In the Addenda to *Journal of researches*, p. 628, CD recognised that his own research had not been extensive enough to prove that each island had species of lizard and tortoise distinct from those of other islands. However, CD quoted Gabriel Bibron's identification of different species of tortoise from different islands, and of two species of the marine iguana, *Amblyrhynchus*, within the archipelago (*ibid.*, p. 628).

[5] Henslow's reply has not been found, but an extract from it is quoted in *Journal of researches*, p. 629: 'there are several instances of distinct species of the same genus, sent from one island only: that is, whilst the genus is common to two or three islands, the species are often different in the different islands. In some cases the species seem to run very close to each other, but are, I believe, *distinct*.' Henslow qualified his reply by saying that he had not yet examined the material in detail. The growing evidence of divergent populations of both animals and plants in the Galápagos caused CD to regret 'not having procured a perfect series in every order of nature from the several islands' and to explain that he had not anticipated while on the *Beagle* 'that islands in sight of each other should be characterized by peculiar faunas' (*ibid.*, p. 629).

[6] Henslow arranged the collection of CD's Galápagos plants, now in the herbarium of the Department of Plant Sciences, Cambridge University, and described a small part of it (see J. S. Henslow 1837, 1838). CD continued for several years to encourage Henslow to describe the plants (see *Correspondence* vol. 2, letters to J. S. Henslow, [November 1839] and [22 January 1843]). Joseph Dalton Hooker eventually identified many of them (see *Correspondence* vols. 3 and 4), and observations arising from those identifications were added to *Journal of researches* 2d ed., pp. 392–3, 395–8. For CD's notes on his collection of plants during the *Beagle* voyage, see D. M. Porter 1987.

[7] The letter from Henslow has not been found.

[8] Leonard Jenyns was Henslow's brother-in-law and responsible for the descriptions of fish in *Zoology*.

From [Mr Edwards?][1] [before end of 1839?][2]

at pointer bitch belonging to Chas Wilkins Esq. Hay, Breconshire[3]—put to a pointer dog threw one setter puppy.[4] The other puppies were pointers with marks of the highest breed, fine sterns &c

The breed had been in the family for generations and had been sold at high prices. No cross was remembered and it must have been far back

A Mem, incomplete
DAR 163: 4

CD ANNOTATION
Bottom of page: 'Mr. Edwards | Surgeon' *ink*; '16'[5] *brown crayon, circled brown crayon*

[1] The name of the correspondent is suggested by CD's endorsement: 'Mr. Edwards | Surgeon.' He has not been further identified.

[2] The date is conjectured from the date at which the Wilkins family adopted the name de Winton (see n. 3, below).

[3] Probably Charles Lorenzo Wilkins, son of Walter Wilkins of Hay Castle, Brecon, Wales. The family adopted the name de Winton in 1839. (*Burke's landed gentry* 1846 s.v. de Winton.)

[4] CD was soliciting information on the transmission of characteristics between, and instances of reversion in, crossbred animals from at least 1839, and this memorandum may have been written in answer to his printed questionnaire 'Questions about the breeding of animals', probably circulated between 1 January and 6 May 1839 (see *Correspondence* vol. 2, Appendix V). The information was probably for the chapters on variation under domestication in CD's manuscript on species (see *Correspondence* vol. 6, Appendix III).

[5] The annotation may refer to a numbered portfolio of notes, possibly on hybridism or heredity. CD kept a number of such portfolios, amassed over a long period of time and since dispersed. The likely contents of some portfolios, but not one numbered '16', were reconstructed when some of CD's papers were catalogued in 1932 (see DAR 220: 13).

To John Maurice Herbert[1] [January 1839 – September 1842][2]

[London]

[In his undated letter, Darwin gives his correspondent two names of bird-preservers for a friend and notes, 'I should recommend your friend to call herself and explain in which styles she wants them mounted.' Darwin then replies to an unknown remark made by his correspondent in an earlier letter in a particularly animated manner: 'As for Birds of Paradise from the West Indies, tell that to the marines, as we used to say on board the Beagle.'[3]]

Sotheby's New York (13 June 1991)

[1] Herbert is identified as the correspondent in the sale catalogue.

[2] The date range supplied in the sale catalogue is apparently based upon the London address. CD lived at 12 Upper Gower Street, London, from January 1839 to September 1842, when he moved to Down, Kent.

[3] 'Tell that to the marines': said of a far-fetched story. The saying, attributed to King Charles II of England, was prompted by an account of flying fish given by Samuel Pepys. The courtiers reacted with incredulity until an officer of the marine regiment confirmed the existence of flying fish. The king replied: 'Henceforward ere ever we cast doubts upon a tale that lacks likelihood, we will first "Tell it to the Marines"' (*Brewer*, p. 1160).

To Benjamin Silliman 26 February 1840[1]

[London]

Mr. Darwin presents his compliments to Prof. Silliman and trusts that Prof. Silliman will excuse the liberty Mr. D. has taken in enclosing a letter on the subject of whirlwinds to be forwarded by any opportunity to Mr. Redfield whose address is unknown to Mr. D.[2] Mr. Darwin hopes that Prof. Silliman received some time since a copy of his 'Journal of Researches during the voyage of the Beagle'[3] which he took the liberty of sending him as a mark of respect.

Gallery of History (4 December 1996)

[1] The date is given in the sale catalogue.

[2] William C. Redfield was a saddler, harness maker, and self-taught meteorologist. His articles on storms and hurricanes in the *American Journal of Science and Arts* (Redfield 1831 and 1834) became classics (*DAB*, *DSB*). The letter to Redfield, dated 24 February [1840], is printed in *Correspondence* vol. 2. See also *ibid.*, letter to W. C. Redfield, [22 December 1840].

[3] *Journal of researches* was first published in 1839 as volume 3 of *Narrative*, with the title *Journal and remarks*. See Freeman 1977, pp. 31–2. Silliman acknowledged receipt of the volume in his letter to CD of 7 April 1840 (*Correspondence* vol. 2).

To Richard Owen [March? 1840][1]

12, Upper Gower St.
Sunday

I send you a proof title-page in place of the temporary title, now affixed to no. I of Fossil Mammalia.[2] Please to observe that it is the general title to whole work,[3] & matches the general one to Mammalia (which was given in last no. of birds).[4] I hope you will approve of it. I have put your honorary titles only briefly in this general title, as they are more fully given in the title devoted to your department.[5] I do not at all know, whether I have put your honorary titles properly. I only just put them down to fill up, what appeared to be the proper space in the page.— Will you be so kind as to return, after having made any alterations you choose, the proof to me.—

I hope soon to receive proofs of the remaining sheets.—[6] I fear after all this publication of the number at the proper time is beyond hope.—[7]

Believe me | Very truly yours | Chas. Darwin

PS. Since I wrote this, Mr Stewart, the Printer[8] has called & he tells me, you have not sent any list of plates or contents.[9] I trust & hope for the uniformity of the work you will have the kindness to do so, even if they be brief ones.

Incomplete
Christie's, New York (29 October 1993)

[1] The date is conjectured from the relationship between this letter and the letter to Richard Owen, 24 [February 1840] (*Correspondence* vol. 2), containing CD's request that Owen submit his manuscript of the final number of *Fossil Mammalia* by the end of the first week of March. The reference in this letter to the 'proper date', which according to the schedule of publication of *Zoology* was originally 1 March 1840, and the absence of any reference to the postponement mentioned in the letter to Owen of 24 [February 1840], suggest that the letter may have been written earlier, on Sunday 16 February. However, the printer's request for lists of plates and contents, which require page references and are therefore usually set up in type last, provides a strong reason for setting the date later than 24 February. The first Sunday after 24 February was 1 March. The final number of *Fossil Mammalia* was published in April 1840 (Freeman 1977).

[2] *Fossil Mammalia*, by Richard Owen, contained four of the nineteen numbers that made up CD's *Zoology*. The first number of *Fossil Mammalia* appeared in February 1838 with a temporary title page. In common with other parts of *Zoology*, the final number of *Fossil Mammalia* was issued with a title page (to which CD here refers), other preliminaries, and an index, so that the separate numbers might be bound into a single volume. For details of the complex publishing history of the parts and numbers of *Zoology*, see Freeman 1977, pp. 26–31.

[3] The 'whole work' to which CD refers is the 1840 publication of all the separate numbers of *Fossil Mammalia* in a single volume as Part I of *Zoology*. The title page gives the general title, *The zoology of the voyage of H.M.S. Beagle*, with CD as editor, above the title and author of the part.

[4] *Mammalia*, by George Robert Waterhouse, is Part II of *Zoology*; it was completed in 1839, before Part I. The title page of *Mammalia*, issued in September 1839 with the last number of the part, omitted the name of its author. A corrected title page was issued with the fourth number of *Birds* in November 1839. See Freeman 1977, p. 29.

[5] On the published title page of *Fossil Mammalia*, Owen's titles are given as 'professor of anatomy and physiology to the Royal College of Surgeons in London, and corresponding member of the Institute of France, etc. etc.' On a second title page, which does not mention *Zoology*, Owen's professorship at the Royal College of Surgeons is repeated and followed by additional titles. CD's proofs of these pages have not been found.

[6] CD may refer to the remaining proofs of Owen's text of the final number of *Fossil Mammalia* (see n. 1, above).

[7] The original scheme, published in a prospectus of 1837, was to issue successive numbers of *Zoology* 'on the first day of every alternate month'; the number preceding Owen's final number appeared in January 1840 (Freeman 1977, pp. 27, 29). Therefore, the 'proper time' of publication of the last number of *Fossil Mammalia* was 1 March; it was not, however, published until 1 April (see *Correspondence* vol. 2, letter to Leonard Jenyns, [27 February 1840]).

[8] Stewart and Murray were the printers of *Zoology* and *Geology of the 'Beagle'*. Their offices were at 15 Old Bailey and Green Arbour Court, London (*Post Office London directory*).

[9] The plates for *Fossil Mammalia* were the work of George Scharf (see *Correspondence* vol. 2, letter to Richard Owen, 24 [February 1840]). The last number of each part was issued with the part's title page, half-title, preliminaries, index, and list of plates (see Freeman 1977, p. 27).

From David Milne 28 March 1840

10 York Place, Edinburgh,
28th *March* 1840.

Sir,—

I have been so much occupied professionally till within the last few days, that I have not been able to read and duly study your paper on South American volcanic action to enable me to acknowledge it.[1] I assure you that it is appreciated by me most highly, and I have perused it a *second* time with increasing benefit. There are views in it as well as facts most interesting to the subject of my own inquiries. There is one passage in your paper on which I would venture to offer a remark. On page 618 you observe that the disturbance appears to emanate *not* from one point, but from many points, ranged in a band, otherwise the fact of the linear and unequal extension of earthquakes would be unintelligible.[2] The Scotch earthquakes are felt over a greater extent of country in a N.E. or S.W. direction than in any other direction. They all emanate from one point, which I have ascertained to be N.W. of Comrie village in Perthshire. Fraserburgh, in the N.E. point of Aberdeenshire, is about 150 miles from Comrie, and the shocks were more sensibly felt there than at Stirling, which is about 25 miles S.S.E. from Comrie. The reason of this, I consider to be, that a chain of hills runs across the country through Comrie to Fraserburgh, consisting chiefly of gneiss, granite, and mica slate; whilst between Comrie and

Stirling there are various secondary formations. The former transmits the vibration with tenfold more facility than the latter.

Is it not possible that the linear and irregular extension of the shocks in South America may be owing to a similar cause? Von Hoff makes a similar remark with regard to the earthquakes of Germany, which follow the line of the basaltic hills that stretch across the country.[3] In your most interesting volume, which forms part of the notes on the *Beagle's* voyage, you speak of the connection subsisting between earthquakes and the atmospheric disturbances which *follow*, and you express your belief in *these*. But I am somewhat inclined to go further. I think there are strong reasons for suspecting a connection between earthquakes and meteoric change *preceding* them.[4] One of these may be the diminished atmospheric pressure as noticed by you, and in support of my view I have some very remarkable proofs drawn from our own country. But another of these influences I suspect to be *water*, a circumstance which I see from your book is matter of belief also in South America.[5]

I am much obliged to you for warning me against too ready a belief in the connection between earthquakes felt on opposite sides of the equator, especially when the interjacent countries evince no indication of them; and on this account I am beginning to be sceptical as to any connection between our Scotch earthquakes and those in Italy, for both in Calabria and in Savoy the volcanic fires were raging when Perthshire was excited.[6]

Another phenomenon has lately come to my knowledge, which I am anxious to make known to you. On the 16th November jets of smoke or steam were observed to issue from the side of a hill in Glen Almond, eight miles north-east from the focus of the Comrie shocks, and stones were by these jets raised out of their beds. On the 15th January last, about 200 yards from the same place, a flame was seen by a different person, who had not heard of the former eruption, which alarmed him so—he was within 150 yards of it—that he ran away for two miles without stopping. I have caused precognition, to use a Scotch lawyer's phrase, to be taken regarding this matter. The result is, that such a flame has been seen occasionally for the last twenty or thirty years arising from the same spot.

Do you know of any analogous cases where flames rise from the ground not connected with craters or any volcanic disturbance? The formation is clay or mica slate; the latter, I rather think.

I wrote to Dr Daubeny about it. He thinks it must be carburetted hydrogen, but can suggest no cause for it, or any instance precisely similar. However, he speaks of flames issuing from the calcareous mountains of Albania; and the rocks of Pietra Mala in Italy are also calcareous.[7] These gases may arise from bituminous matter, but of which none can exist in Glen Almond. I should feel very much obliged if you could suggest to me any case analogous to this, and any points which should be specially attended to in my investigations.—

I am, dear Sir, yours very truly, David Milne.

Milne Home 1891, pp. 69–72.

[1] Milne was an advocate at the Scottish bar and a fellow of the Royal Society. He had become interested in earthquakes as a result of earthquake shocks at Comrie, Perthshire, in 1839, and had written to CD in early 1840 (Milne Home 1891, p. 63; this letter has not been found). CD replied in a letter that should now be dated 20 [February 1840] (it was published in *Correspondence* vol. 2 with the date [19 March 1840] and in Milne Home 1891, pp. 67–9, with the date 20 March 1840). In the letter, CD offered to send Milne, in a week or ten days, a copy of his article 'Volcanic phenomena and the formation of mountain chains', which he was expecting to be published 'immediately' in the *Transactions of the Geological Society of London.* The paper appears in the third part of volume 5 (2d series) of the *Transactions*, and a page following the table of contents of the second part of the volume states: 'The Third and concluding Part of the Volume being nearly ready, will be published in February'.

[2] See 'Volcanic phenomena and the formation of mountain chains', p. 618 (*Collected papers* 1: 70).

[3] The reference is to Karl Ernst Adolf von Hoff, and possibly to Hoff 1822–41 (see, for example, Hoff 1822–41, 4: 61).

[4] Milne refers to *Journal of researches*, pp. 430–3. CD had reservations about the theory that earthquakes were caused by a lowering of atmospheric pressure, which also tended to bring rain, since the weather that accompanied some earthquakes was itself abnormal and thus seemed more likely to have been a consequence of the earthquake than vice versa. Milne referred to CD's account of the mercury in the barometer falling before the Valparaiso earthquake, and quoted his conclusion that there must be some connection between subterranean and atmospheric disturbances, in Milne 1841–4 (see *Edinburgh New Philosophical Journal* 31 (1841): 299 and *Journal of researches*, p. 433).

[5] Milne may be referring to his belief that water filtering into underground fissures and turning into steam could generate electricity, and thus account for the electrical phenomena associated with earthquakes (see Milne 1841–4, *Edinburgh New Philosophical Journal* 31 (1841): 307). CD noted in *Journal of researches*, p. 431, as part of a discussion of the association between earthquakes and the weather, that the inhabitants of northern Chile were firmly convinced of a connection between earthquakes and rain.

[6] See *Correspondence* vol. 2, letter to David Milne, [19 March 1840], which should now be dated 20 [February 1840] (see n. 1, above). Milne discussed the evidence for the coincidence of earthquakes around the world in Milne 1841–4 (*Edinburgh New Philosophical Journal* 36 (1843–4): 362–373).

[7] The letter from Milne to Charles Giles Bridle Daubeny, and Daubeny's reply, are published in Milne Home 1891, pp. 63–7.

To William Buckland [November 1840 – 17 February 1841][1]

12 Upper Gower St
Wednesday

My dear Sir

I received yesterday the enclosed paper from Mr. Murchison with his pencil marks on it:—[2] I cannot think of any information to give you relating to Dr. Du Gard.—

He is no relation to any of my family.—[3]

I have to thank you for your kindness in sending me long since some maps of the country near Glen Roy, which I daresay will be very useful, if I ever visit that country again—[4] I should much like to hear your's & Agassiz's opinion on the parallel roads, though I believe I know its outline—[5] I cannot give up the sea, after thinking over many points of minor detail in that country, though, I am very sure, if your theory had occurred to me, during the first two days of my examination, I should have given up their marine and *ordinary* lacustrine origin at once[6]

But I have no business to trouble you with my notions.

Believe me | dear Sir | Yours very faithfully | Charles Darwin

Oxford University Museum of Natural History, Buckland papers, Glaciation /4 (iv)

[1] The date is established by the reference to Thomas Dugard (see n. 3, below), and by the reference to Buckland's and Louis Agassiz's opinion on the parallel roads of Glen Roy (see n. 5, below). In 1841, 17 February was a Wednesday.

[2] CD refers to Roderick Impey Murchison. The paper sent by Murchison has not been identified.

[3] Dugard, physician to the Shropshire Infirmary, and honorary fellow of the Geological Society, died on 19 January 1840 (*Gentleman's Magazine* n.s. 14 (1840): 556). Buckland, who was president of the Geological Society, included a brief obituary notice of him in his anniversary address to the society on 19 February 1841 (*Proceedings of the Geological Society of London* 3 (1838–42): 523). Buckland had probably written to CD to ask for information about Dugard to include in his notice. There had been some scandal about Dugard in 1831 and 1832 that is mentioned in CD's sisters' letters to CD (see *Correspondence* vol. 1).

[4] CD visited Glen Roy in June 1838. His paper 'Parallel roads of Glen Roy' appeared in 1839. On CD's theory about the parallel roads, in the context of other nineteenth-century theories about their origin, see Rudwick 1974; see also *Correspondence* vol. 9, Appendix IX. The maps that Buckland sent have not been found in the Darwin Archive–CUL.

[5] Buckland and Agassiz visited Glen Roy together in 1840 (Buckland 1840, p. 333). Agassiz's account of his theory of the glacial origin of the parallel roads of Glen Roy, delivered at the Geological Society on 4 November 1840, was summarised in the *Proceedings of the Geological Society of London* (J. L. R. Agassiz 1840, p. 332; see also Rudwick 1974); Buckland's remarks at the society on 4 or 18 November were also reported (Buckland 1840, p. 333). Buckland and Agassiz had concluded that the parallel roads of Glen Roy were beaches formed as the water of a lake formed by a glacial dam at the foot of the Glen subsided. CD's theory was that they were beaches of a sea that once covered that area of Scotland (see n. 4, above).

[6] In the letter to Robert Chambers, [June 1848] (*Correspondence* vol. 4), CD wrote of his visit to Glen Roy: 'during my first two days I was a convert to the lacustrine theory' (that is, the theory that the parallel roads were remnants of the shores of ancient lakes, though the lake barrier was not at that time thought to have been a glacier); but he also explained that a number of features of the topography of Glen Roy had convinced him that his marine theory was correct. See also *ibid.*, letter to Robert Chambers, 11 September 1847, and Rudwick 1974, p. 117. CD did not abandon his marine theory until 1861, when Thomas Francis Jamieson, with the support of CD and Charles Lyell, visited Glen Roy. Jamieson published his conclusions in Jamieson 1863. See Rudwick 1974 and *Correspondence* vol. 9.

From Richard Owen [17 September 1841?][1]

6, Park Cottages, Park Village East.
Friday.

Dear Darwin,

In the possible event of your being from home I write to say that I shall have a holiday musical evening tonight, and that if you & M^rs. Darwin will join us—we have tea at $\frac{1}{2}$7—it will add to our pleasure very much.[2]

Ever your's, | R^d. Owen.

Royal College of Surgeons of England, T. M. Stone/J. H. Watson Collection

[1] The date is conjectured from the address, and from Owen's reference to a 'holiday musical evening' (see n. 2, below). Owen is recorded in the *Royal blue book* as residing at this address from 1840 to 1842.

[2] On 8 September 1841, Owen wrote to William Clift that he hoped that Clift would come to dinner 'at No. 6 Park Cottages' on 12 September, following Owen's return from holiday on 11 September. He promised: 'a long and most delightful tour through Cornwall will be re-performed, with innumerable and most rare adventures by land, rock, mine, and sea; and the characters of divers amiable and hospitable friends will be portrayed' (R. S. Owen 1894, 1: 187). It is probable that Owen performed other holiday entertainments during this period. Owen's biographer records: 'About this time Owen and his wife saw a good deal of Mr. and Mrs. Darwin' (*ibid.*, p. 188). In 1841, the first Friday after 12 September was 17 September.

To Francis Boott [3 February 1842][1]

12 Upper Gower St
Thursday

My dear Sir

I have called on you, to solicit your vote & interest at the Athenæum Club (if you ever attend the evening meetings)[2] in favour of my brother, Erasmus Darwin, at the ballot on next Monday (7th).[3] If you can so far oblige me, I shall esteem it a great kindness.[4]

Believe me my dear Sir | Yours very faithfully | Charles Darwin

Sherman Bull (private collection)

[1] The date is established by the date of the vote on Erasmus Alvey Darwin's candidacy for membership of the Athenaeum Club, Monday 7 February 1842 (see *Correspondence* vol. 2, letter to J. F. Royle, [2 February 1842] and n. 1, and n. 4, below). The preceding Thursday was 3 February.

[2] The balloting on candidates for membership of the Athenaeum Club was held between 9 and 10:30 P.M. (Ward 1926).

[3] See n. 1, above.

[4] Erasmus became a member of the Athenaeum Club in 1842 (Waugh [1888]).

To Edward Cresy 26 [December 1843 – April 1846 or September 1855 – October 1860?][1]

Down, Bromley, Kent.
26

My dear Sir,

In answer to 2 or 3 letters of mine I heard only this morning from Dr. Hooker, who desires me to thank you *particularly* for offer of plant, but they do not want any.[2] He says he has been so overwhelmed with work, that he really could not write before.

In Haste, | Yours sincerely, | C. Darwin.

Copy
DAR 143: 311

[1] The date is conjectured by the address, the reference to Joseph Dalton Hooker, and the salutation. CD moved to Down in September 1842 and used the Down, Bromley, Kent address until April 1846 and then again from after August 1855 until 1869. He also used this address in the second half of 1853. Hooker returned from his Antarctic expedition in September 1843 (R. Desmond 1999, p. 85); CD appears to have written to him for the first time in November (see *Correspondence* vol. 2, letter to J. D. Hooker, [13 or 20 November 1843]). CD started addressing Cresy as 'Dear Cresy' in his letters rather than 'My dear sir' in November 1860.

[2] The letters referred to have not been found, and the plant has not been identified. 'Dr. Hooker' is J. D. Hooker, assistant director of the Royal Botanic Gardens, Kew, from 1855 to 1865.

To Henry Denny [27 July – 10 August 1844][1]

Down Bromley Kent
Saturday.

Dear Sir

I herewith send you the four little packets of Lice.—[2] I remember thinking that one, though minute, was an odd form.—

I sincerely wish I could aid you in obtaining fresh materials: should I have any opportunity, I will not fail to remember to ask anyone, whom I may meet & think likely to have specimens: but I live so retired in the country, that I now see few naturalists. Have you ever applied to Dr. A. Smith, (Fort Pitt Chatham) the S. African Traveller, if you like to write to him, you can use my name & say, I was sure that he would be glad to assist every naturalist.[3]

I am sorry to say I shall not be at York,[4] otherwise you might depend on my aid, feeble though it would be.

With my best wishes for success in your laborious pursuit.[5] Pray believe me | dear Sir | yours very faithfully | C. Darwin

P.S I have kept no copy of the extract of 'White's Regular Gradation in Man'—a foolish book with some odd facts—published, I think, at Manchester. You wd easily find the passage to which I refer.— I *think* it was in the latter part of the volume.—[6]

The next time I write to Dr Hooker I will ask him, whether he has Lice of Penguins.—[7]

I find I can make the bottle safe, so will send it by Post at same time with this note.— The bottle shd be opened, as it is not quite full— the spirits are quite strong enough

C. D

Yale University, Harvey Cushing John Hay Whitney Medical Library

[1] The date is established by CD's offer in this letter to write to Joseph Dalton Hooker and ask whether he had specimens of penguin lice, and by the fact that CD responded to Denny's queries about the specimens sent with this letter in his letter to Denny of 12 August [1844] (*Correspondence* vol. 3). In 1844, the last Saturday before 12 August was 10 August, the latest possible date for this letter. CD wrote to Hooker mentioning Denny's request for penguin lice in a letter originally published in *Correspondence* vol. 3 with the date [25 July – 29 August 1844]. The nearest extant letter from CD to Hooker before this one is dated 22 July [1844] (*Correspondence* vol. 3), and does not mention Denny or penguin lice.

In 1844, the first Saturday after 22 July was 27 July, the earliest possible date for this letter. The letter to Hooker mentioning Denny's request for penguin lice, which was written on a Thursday, should therefore be dated [1–29 August 1844], 1 August being the first Thursday after 27 July.

2 For identification of these packets see *Correspondence* vol. 3, letter to Henry Denny, 12 August [1844]. Denny had requested specimens of lice from CD in January 1844 (see *ibid.*, letter to Henry Denny, 20 January [1844]), but CD was unable to send any until July, when he had sorted his collections (see *ibid.*, letter to Henry Denny, 3 June [1844] and n. 2).

3 Andrew Smith was an army surgeon and an authority on South African zoology (*DNB*). CD had become acquainted with him in June 1836 at the Cape of Good Hope (see *Correspondence* vol. 1, letter to J. S. Henslow, 9 July 1836, and Appendix I).

4 The British Association for the Advancement of Science met in York from 26 September to 2 October 1844 (*Report of the fourteenth meeting of the British Association for the Advancement of Science, held at York in September 1844*, p. xxx).

5 Denny, at the request of the Natural History Section of the British Association for the Advancement of Science, had undertaken to investigate and illustrate exotic species of Anoplura (*Report of the fourteenth meeting of the British Association for the Advancement of Science, held at York in September 1844*, p. 392). He had already published an account of the British Anoplura (Denny 1842).

6 CD read C. White 1799 in May or June 1844 (see CD's reading notebooks, *Correspondence* vol. 4, Appendix IV, 119: 14a). The book was compiled from papers read to the Literary and Philosophical Society of Manchester in 1795, and published in London. Lice are mentioned on page 79:

> Mr. Long takes notice of a fact, which seems to have escaped the observation of naturalists:—that the lice which infest the bodies of negroes are blacker, and generally larger, than those which are found on white people.

7 See *Correspondence* vol. 3, letter to J. D. Hooker, [25 July – 29 August 1844] (for the correction to the date of this letter, see n. 1, above).

To William Benjamin Carpenter 6 December [1844]

Down near Bromley Kent
Dec 6th.

Sir

I hope you will excuse the liberty I take in addressing you, but I trust that our several common acquaintances may be some excuse & Mr Lyell[1] has been so good as to say that I might use his name as an introduction to you. My object in writing is to ask you to confer a favour on me by examining a specimen of calcareous rock. The great deposit of the Pampas, in which so many strange mammiferous remains are embedded, in its southern parts, passes into an indurated, calcareous, tuff-like stone. Comparing a specimen of this, with a specimen from a post-pleiocene calcareous bed from the coast of Chile, which has resulted from the attrition of recent shells, of which many are embedded in it, I was struck with the general resemblance between them, and especially in some minute, rounded, included fragments.—[2]

I am most anxious to know, whether these minute fragments, which are very indistinct, are really pieces of shells, like those in the Chilian stone, because the origin of the Pampæan formation has been lately disputed & attributed to a debacle, & is in many points of view of considerable interest.—[3] Having seen some of your curious researches on the microscopical structure of shells,[4] it has occurred to me, that perhaps you would be willing to confer this great favour on me;[5] and would be so kind, if you think the little fragments in their matrix would stand cutting & polishing, to have it done at my expence.— Should you have time to spare to look at my specimen, I would send it (carriage free) either to anywhere in London, or perhaps better, direct to Ripley, if you would tell me by what conveyance is best—

Hoping that you will excuse the liberty which I am taking Pray | Believe me, dear Sir | Yours faithfully | Charles Darwin

Postmark: DE 7 1844
DAR 185: 114

[1] Charles Lyell.

[2] CD had been working on *South America*, the third part of the geology of the *Beagle* voyage, since 27 July 1844 (see *Correspondence* vol. 3, Appendix II). He described the Pampean formations in chapter 4, giving Carpenter's conclusions on pp. 76–7. As a result of Carpenter's examination of the specimens, CD concluded that it would be hazardous to conjecture that the Pampas rocks were, like those from Coquimbo, Chile, the result of attrition and decay of shell fragments. See also n. 5, below.

[3] Alcide Charles Victor Dessalines d'Orbigny, in volume 3, part 3, of his *Voyage dans l'Amérique Méridionale*, had argued that the Pampas was formed as the result of the sudden deposition of a large quantity of mud (Orbigny 1835–47, 3 (pt 3): 82–4). Part 3 of volume 3 was published in 1842. CD, who followed Lyell's uniformitarian theory of changes by 'causes now in action' (C. Lyell 1830–3, 1: 311), would have been interested in evidence of formations laid down by slow sedimentation, attested by the presence of sea-shells or their calcareous remains. See *South America*, p. 98.

[4] Carpenter had read a paper 'On the microscopic structure of shells' at the York meeting of the British Association for the Advancement of Science, 26 September to 2 October 1844 (*Report of the fourteenth meeting of the British Association for the Advancement of Science, held at York in September 1844*, pp. 1–24).

[5] See *Correspondence* vol. 3, letter to W. B. Carpenter, [11 or 18 December 1844], and letters from W. B. Carpenter, 21 December 1844, 2 January [1845], and 5 May 1845, and *Correspondence* vol. 7, Supplement, letter to W. B. Carpenter, 25 December [1844].

To Hugh Falconer [1845?–7 or 1857–64][1]

Friday

My dear Falconer

Instead of your coming here I should prefer coming to you at 2½.— I **know** I may speak truth to you; I prefer it, because if I feel my head swimming I can go away at once;[2] but if you were here I could not bear to send you away

Yours affect^ly^. | C. Darwin

If I do not come punctually I shall be unable from head.

Copy
DAR 144: 21

[1] The date range is conjectured from those years when CD and Falconer were on visiting terms and when Falconer was resident in London. CD visited Falconer from at least 1845, and possibly from as early as 1843; they met in London in June 1847 (see *Correspondence* vol. 4, letter to J. D. Hooker, [10 June 1847]); Falconer left for Calcutta in 1848 and returned to settle in London early in 1857. CD visited him in London in 1857 and 1858 but they did not meet again until at least November 1859 (see *Correspondence* vol. 6, letter to J. D. Dana, 5 April [1857]; *Correspondence* vol. 7, letter from Hugh Falconer, 25 October and 12 November [1859]). They met again in London in August 1860 and May

1862 (*Correspondence* vol. 8, letter to Charles Lyell, 28 August [1860]; *Correspondence* vol. 10, letter to J. D. Hooker, 9 May [1862]). In October 1862 CD wrote to Falconer: 'I had hoped to have called on you on Monday evening, but was quite knocked up' (*Correspondence* vol. 10, letter to Hugh Falconer, 1 October [1862]). They may have met when CD was in London in February 1863 (see *Correspondence* vol. 11, letter to Hugh Falconer, 20 [January 1863]). No further meeting are known to have taken place before Falconer's death early in February 1865. CD was not in London in January 1865.

2 CD complained of swimming of the head from mid-1848 to spring 1849, in 1854, February 1859, and September 1863 (see *Correspondence* vol. 4, letter to J. S. Henslow, 6 May 1849, *Correspondence* vol. 5, letter to Edward Sabine, 28 June [1854], *Correspondence* vol. 7, letter to W. D. Fox, [12 February 1859], and Emma Darwin's diary (DAR 242)).

From the principal inhabitants of Down[1] to the secretary of the Post Office[2] [1845–51?][3]

To the Secretary of the Post-office.

We the undersigned the principal inhabitants of the village of Down in Kent & its neighbourhood, respectfully request your attention to the arrangement for the delivery our letters. Although but a small place we receive an average from 50 to 60 letters & newspapers &c. daily.— A letter posted by general post in London we receive not until after 9 oclock the next morning, which is almost too late to act on that day. But our chief grievance is that a letter written in any part of Great Britain (except a few places in the South), say on a Monday, is not delivered here till past 9 on the Wednesday morning; such letters lying 23 hours at Bromley, only six miles distant.[4] Our hopes for many months have been raised by constant reports of some improvement; but the present result & only change is that the Down Postman delivers all letters in Farnborugh, & our delivery has been delayed a full half-hour.—[5]

By a change granted us a few years ago, our letters leave this place at one oclock, by which means they get to London in time to be sent out by the general post of the same day, & are delivered within London on that night; we most earnestly pray that this arrangement may not be disturbed.[6]

Hoping that you will consider our case favourably | Sir | Your obedient servants

A draft[7]
DAR 96: 6

1 There were fourteen members of the gentry listed under Down in the *Post Office directory of the six home counties* for 1845, and six in 1851, including, at both dates, CD himself and John William Lubbock.

2 William Leader Maberley was secretary to the Post Office from 1836 to 1854 (*DNB*).

3 The beginning of the date range is suggested by known changes to the postal arrangements for Down and Farnborough in 1845 (see nn. 5 and 6, below). The end of the date range is suggested by the fact that, at the time of writing, letters posted to Down from outside London, and letters posted within London, were evidently arriving at Bromley at different times, the post from outside London being delivered too late for the morning dispatch to Down. 'London and all letters' were delivered to Bromley at the same time in 1851 (*Post Office directory of the six home counties* 1845; see n. 4, below).

[4] From 1851 at the latest both the general post and the London post were being delivered to Bromley by the same mail. It appears that in 1845 London mail was being delivered separately. According to the *Post Office directory of the six home counties* for 1845 and 1851, letters were arriving at Down at 8:00 A.M. during both these years. (*Post Office directory of the six home counties* 1845, 1851.)

[5] A foot-messenger was appointed in March 1845 to take letters to and from Bromley to Farnborough and Down (*Postmaster general's minutes*, vol. 81, minute 3376). CD is first known to have used Farnborough in his letterhead in 1846 (see *Correspondence* vol. 3, letter to J. D. Hooker, [May 1846]). Rowland Hill, as secretary to the postmaster-general from 1846, made repeated attempts to reform the postal service; these were consistently rejected by Maberley (*DNB*).

[6] An application was made in 1843 by the inhabitants of Down for mail to be collected early enough to reach Bromley in time for the twopenny post dispatch to London (*Postmaster general's minutes*, vol. 71, minute 2087). This may not have been put into immediate effect, as, according to the *Post Office directory of the six home counties* 1845, mail was dispatched from Down at 2 P.M. It was dispatched at 1 P.M. in 1851.

[7] No final version of this petition has been found, and no reference to its receipt has been found in the *Postmaster general's minutes* between 1839 and 1865.

To George Robert Waterhouse [after 22 May 1845][1]

Down, Bromley, Kent
Sunday

[thanks him for his description of the niata ox][2]

I am delighted to hear that you are grappling with the Galapagos insects:[3] the more I go into the Fauna, the more peculiar it is: out of 17 land shells 16 are new species confined to the group![4]

[. . .] of you, but it will not cost you five minutes: it is to tell me by return of Post, whether any Entomologists, besides yourself, Walker,[5] & A. White,[6] have described any number of my insects:[7] who described some flys?[8]

I want to know, because in my Preface I give an *outline* of the scientific publications of the Voyage.[9]

Ever my dear Waterhouse | yours truly | C. Darwin

[refers to a cargo of bones from 'a little north of the St. of Magellan',[10] & says that if he does not hear from White about the spiders[11] in eight days it will be too late.]

Incomplete[12]
Bloomsbury Book Auctions (1990)

[1] The date is established by the relationship between this letter and the letter from G. R. Waterhouse, 21[–22] May 1845 (*Correspondence* vol. 3).

[2] Waterhouse's unpublished description of a niata ox skull is mentioned in *Journal of researches* 2d ed., p. 146 n.

[3] Waterhouse was preparing Waterhouse 1845, 'Descriptions of coleopterous insects collected by Charles Darwin, Esq., in the Galapagos Islands'. See *Correspondence* vol. 3, letters from G. R. Waterhouse, 21[–22] May 1845, and [*c.* June 1845] (first letter) and n. 1.

[4] See *Journal of researches* 2d ed., pp. 390–1.

[5] Francis Walker was a widely travelled entomologist (*Entomologist* 7 (1874): 260–4). For Walker's descriptions of CD's insects, see Walker 1838, 1839, and 1842–3.

[6] Adam White was a naturalist employed in the zoological department of the British Museum (*DNB*). See A. White 1841.

[7] Others who described CD's insects were Frederick William Hope (Hope 1837), and Charles Cardale Babington (C. C. Babington 1838). For CD's notes on his collection of insects during the *Beagle* voyage, see R. D. Keynes ed. 2000; see also DAR 29.

[8] Walker sent specimens of Diptera to the Dublin entomologist Alexander Henry Haliday in 1837 and 1838, but Haliday did not publish on them (K. G. V. Smith 1987, pp. 30–4).

[9] The preface to the second edition of *Journal of researches* included a paragraph detailing CD's separate publications on the *Beagle* voyage, and adding:

> Mr. Bell, I hope, will describe the crustacea, and Mr. Sowerby the shells. Messrs. Waterhouse, Walker, Newman, and White, have published several able papers on the Insects which were collected, and I trust that many others will hereafter follow.

CD referred to Thomas Bell, George Brettingham Sowerby and Edward Newman.

[10] On 4 June 1845, CD reported to Hooker that he had just heard from Bartholomew James Sulivan that he had made 'a grand collection of fossil quadrupeds from the R. Gallegos in the southern part of Patagonia' (see *Correspondence* vol. 3, letter to J. D. Hooker, [4 June 1845], and letter from B. J. Sulivan, 13 January – 12 February 1845). The fossils are mentioned in a footnote to *Journal of researches* 2d ed., p. 172.

[11] White had described 'New or little known Arachnida' collected by CD (A. White 1841). No letter from White has been found in CD's correspondence of 1845. CD may be referring to information to be included in the second edition of *Journal of researches*, on which he was working between 25 April and 26 August 1845 (see *Correspondence* vol. 3, Appendix II). A. White 1841 is cited in new material about spiders on p. 35 n. The second edition was originally issued in monthly parts as nos. XXII, XXIII, and XXIV of the Colonial and Home Library published by John Murray. Part XXII, in which the spiders are described, was advertised as published on 28 June 1845 (see Freeman 1977, p. 35).

[12] The original letter is complete and is described in the sale catalogue as being three pages long.

To Smith, Elder & Co. 6 June [1846][1]

Down Farnborough Kent
June 6th.—

Dear Sir

I shall be ready in a week's time to send the M.S. of my Geolog. Obser. on S. America to press,[2] & therefore wish you to be so good as to draw up an agreement & make arrangements. The volume on the type hitherto used, would take up as I roughly calculate, 350 or 360 pages: there are 21 woodcuts—4 *large* copper-plates of engraved shells & one very large coloured section & a map.[3] Hence the expence will be considerable & I must use every possible economy.[4] I have already contracted through the Vice-Sec. of Geolog. Soc for **all** the engraving, printing off & paper for Plates, & cutting of wood-blocks, & these I will pay at once myself, so shall not draw on your funds for this expence.[5] You are aware, that I shall give you no trouble going through the press, & that all you will have to do is to superintend sales &c & advance for printing, Paper, & Binding & advertisements.— Under these circumstances, I trust to your liberal kindness (which I have always experienced) to draw up as easy an agreement for me as you can fairly do towards yourselves; if it would make any *considerable reduction* in the per-centage commission charges &c

I would undertake to pay the whole expence on the day of Publication; if not, I presume, at the balance at the end of the year.—[6] Please turn this in your mind & let me hear **soon**: you had better perhaps, at once, send me a *single* copy of an Agreement for my Approval.

I sh[d] like the paper to be of same price as before (has not the quality improved of late for same price?) namely 16s. 6d per ream.— With respect to type I find that a page of the Geological Journal (of which I send specimen, & please return it sometime) contains on an average 566 words, whereas my Coral & Volcanic vol's contain only 381 words.; so that by using this type there would be a **very great** reduction in paper, printing off & cold pressing.[7] Therefore with reluctance, I have *resolved* to sacrifice uniformity, keeping the *outsides* uniform, & use the type of the Geolog. Journal for text & notes.—

Being on the Council,[8] I find M[r]. Taylor's[9] charge for *750* copies, per sheet is £3" 8s" 0d, but *this includes small type & woodcuts*; which are very numerous in the Journal.— M[essrs] Stewart & Murray[10] for my works charges for *500* copies *not including small type & woodcuts* £2.9s. so that there will be a *great* saving in using the smaller type in setting up.—

I have asked the Vice-Secretary to make some more exact enquiries what M[r] Taylor's charges are, for 500 copies, & not including small type & woodcuts. I sh[d]. be sorry not to print with M[essrs] Stewart & Murray, to whose attention & kindness, I am indebted, but I have no choice, & am compelled to be *rigidly economical.* If they think it worth while to send me in an estimate of their charge for sheet of exactly same size, if the estimate be nearly the same, with M[r] Taylor, I will certainly send my M. S. to them & shall much prefer doing so.:— I sh[d]. be glad of an answer *soon.*—

With respect to the number of copies to be struck off, though I know, if more than 250, I must pay for 500, yet seeing that hitherto, only 216 copies of Coral Reef & 143 of Volcanic isl[d] have been sold, I really doubt whether it be worth printing off more than 350 copies,— I sh[d]. like to hear your opinion on this.— I suspect it would answer for you, (as well as me) if you could offer all three Geological Parts bound together at a much reduced price: I am determined to try this Part much cheaper than the others.—[11]

I am sorry to have troubled you with so long a letter.

Dear Sir | Your's very faithfully | C. Darwin

C. C. Kohler (dealer)

[1] The year is established by the publication date of *South America*, the third volume of the *Geology of the 'Beagle'*, late in 1846 (Freeman 1977, p. 57).

[2] The manuscript of one chapter of *South America*, the last, is in DAR 1.

[3] *South America* contains 286 pages (including preliminaries), four fold-out plates of shells, a coloured fold-out section and a fold-out map. There are twenty-four woodcuts.

[4] A Treasury grant of £1000 had been intended to defray the cost of the *Zoology* and of the three parts of the *Geology of the 'Beagle'*; however, it was exhausted by the publication cost of the *Zoology* and of *Coral reefs*, the first part of the *Geology*, published in May 1842 (Freeman 1977, pp. 26, 57). The account

submitted for the final number of the *Zoology*, *Reptiles* no. 2, which was published in October 1843, claimed only a portion of the costs, because the grant had been exhausted (see *Correspondence* vol. 2, letter to A. Y. Spearman, 9 October 1843 and n. 1). In the preface to *South America*, CD noted that, in addition to the Treasury grant, a further sum had been advanced by Smith, Elder and by himself to publish the *Geology* and the *Zoology*. For correspondence about the grant, see *Correspondence* vol. 2; for the publication of *South America*, see *Correspondence* vol. 3.

5 No correspondence with David Thomas Ansted, the assistant secretary of the Geological Society of London, about this contract has been found.

6 According to CD's Classed accounts (Down House MS), CD paid Smith, Elder £61 8*s.* on 10 August 1847. Payments are also recorded of £10 2*s.* on 4 September 1846 to Lovell Reeve & Co. for a lithograph (the large coloured section); of £4 10*s.* on 15 December 1845 and £10 on 19 August 1846 to George Brettingham Sowerby for work with and engraving shells, respectively; and of £6 7*s.* on 5 July 1846 for wood-cuts.

7 CD refers to the *Quarterly Journal of the Geological Society of London.* The typeface of *South America* is smaller than that of *Coral reefs* or *Volcanic islands* (the first two parts of the *Geology of the 'Beagle'*), and the same size as that of the *Quarterly Journal.*

8 In 1846, CD was a member of the Council of the Geological Society, as he had been in 1845 (*Journal of the Geological Society of London* 1 (1845) and 2 (1846)). In 1843 and 1844 he had been vice-president (*Proceedings of the Geological Society of London* 4 (1843–5): 64, 349), and in 1839 and 1840 he had been secretary (*Proceedings of the Geological Society of London* 3 (1838–42): 60, 268).

9 Richard and John E. Taylor, of Red Lion Court, Fleet Street, London, published the *Quarterly Journal of the Geological Society of London.*

10 Stewart and Murray were printers with premises at 1 Green Arbour Court, Old Bailey, London (*Post Office London directory* 1846). They had printed *Zoology*, *Coral reefs*, and *Volcanic islands*, and went on to print *South America.*

11 The three parts were not reissued in one volume until 1851; the price was 10*s.* 6*d.* The volume was made up of the original sheets bound together of *Coral reefs* (1838), original price 15*s.*; *Volcanic islands* (1844), original price 10*s.* 6*d.*; and *South America* (1846), original price 12*s.* See Freeman 1977, pp. 58–9.

To Charles Lyell [1847 or 1848][1]

Down. Bromley.
Tuesday.

My dear Lyell.

I happen to be in a great bustle—but just send one line in answer to your note[2] to say that in Fitz Roys volume Vol II p 359 there is a very good (i.e. very faithful) representation of a glacier entering the sea, *viewed from a distance.*—[3] I am sure I have somewhere seen a nearer view.— There is a distant one, but very poor in Scoresby.[4]

I think Gaimards Great Expedition to the North is published (perhaps in the Athenaeum Club)[5] & that would be sure to have drawings, taken for very geological object, as the naturalist Martens has written, as you know, on subject.—[6]

I am not well acquainted with Parry Franklin, or Ross'[7] works.

Farewell for present. | C. Darwin.

Copy
DAR 146: 329

1 The year is conjectured from the estimated publication date of the volumes of Gaimard ed. [1843–8] to which Charles Frédéric Martins contributed (see n. 6, below).

[2] The letter from Lyell has not been found.

[3] CD refers to *Narrative* vol. 2, written by Robert FitzRoy. The plate facing page 359 is captioned, 'Mount Sarmiento (from Warp Bay)'. Mount Sarmiento is in Tierra del Fuego.

[4] Scoresby 1820. There is a lightly annotated copy in the Darwin Library–Down (see *Marginalia* 1: 747). CD probably refers to plate 5 in the second volume, 'View of the east coast of the island of Jan Meyen'.

[5] CD refers to Gaimard ed. [1843–8]. The Athenaeum was a London club of which Lyell and CD were members. A principal object of its library was to buy works on the fine arts, natural history, and antiquities that were too expensive for most private persons ([Cowell] 1975, p. 66).

[6] Gaimard ed. [1843–8] is divided into several multi-volume parts. Martins contributed a paper on glaciers to each of the two volumes in the part titled 'Géographie physique, géographie botanique, botanique et physiologie': to the first he contributed 'Observations sur les glaciers du Spitzberg comparés à ceux de la Suisse et de la Norvège' (pp. 138–92), and to the second 'Mémoire sur les températures de la Mer Glaciale, à la surface, à de grandes profondeurs, et dans le voisinage des glaciers du Spitzberg' (pp. 279–352). The second volume may have been published in 1847 (*NUC*). The first two sections of the 'Mémoire' were also printed in the *Annales de Chimie et de Physique* 3d ser. 24 (1848): 220–56, and an abstract appeared in *Comptes rendus hebdomadaires des séances de l'Académie des Science* 26 (1848): 333–5.

[7] CD refers to William Edward Parry, John Franklin, and either James Clark Ross or John Ross. The works to which he alludes are probably Parry 1824 and 1826, Franklin 1823 and 1828, and J. Ross 1835 or J. C. Ross 1847. CD recorded having read Franklin 1823 on 20 September 1845, and Robert Brown's appendix to Parry 1824 on 28 December 1847 (*Correspondence* vol. 4, Appendix IV, 119: 16b and 20a).

To J. L. Stokes 2 January [1847][1]

Down Farnborough | Kent
Jan 2[d]

Dear Stokes

Your magnificent books arrived quite safely yesterday.—[2]

You once gave me hopes of seeing you here. I have asked some of our mutual friends here on Saturday the 16th to stay over Sunday; will you give us the pleasure of joining them.[3] I hope Falconer & Forbes will come.[4] Your best way here is by the Croydon Train which leaves London Bridge at *3°.15'* & stop at Sydenham Station & take a Fly (above 9 Miles Eheu)[5] with Forbes & Falconer, if, as I expect, one or both come. Please send me a line & I hope it won't be inconvenient to you to come.

I expect Hooker & Waterhouse[6] here.

Very truly Yours | C. Darwin

University of Akron

[1] The year is established by the reference to Stokes's publication (Stokes 1846) and by the invitation to visit Down House on 16 January; see n. 3 below.

[2] CD refers to the two volumes of Stokes 1846. The volumes are not in the Darwin Library–CUL or the Darwin Library–Down. See also *Correspondence* vol. 3, letter to J. L. Stokes, [*c.* 26 November 1846] and n. 2.

[3] Joseph Dalton Hooker was also invited for Saturday 16 January 1847, and Charles and Mary Lyell may have come the same day or shortly after; other guests, 'the old set', were evidently invited

for the Sunday (*Correspondence* vol. 3, letter to J. D. Hooker, [December 1846 – January 1847], and *Correspondence* vol. 4, letter to Charles Lyell, [23 January 1847]).

4 Hugh Falconer and Edward Forbes.

5 Sydenham was on the South-eastern line. The *Railway Chronicle* travelling chart, *circa* 1846, comments that the South-eastern line, 'instead of pursuing the old coach-road, through Rochester and Canterbury, as was proposed, was compelled by the benighted, and now repentant, opposition of the inhabitants of North Kent to take a circuitous route through the Weald of Kent'. *Eheu*: 'Alas!' (ancient Greek).

6 George Robert Waterhouse.

To John Russell [10 March 1847][1]

To The Right Honourable Lord John Russell, M.P.,
First Lord of the Treasury, &c. &c.

We, the undersigned,[2] members of the British Society for the Advancement of Science, and of various scientific societies, respectfully submit to the consideration of your Lordship, that a strong feeling pervades the naturalists of our country, that the promotion of the science of Natural History is very inadequately provided for by the present constitution of the Board of Trustees of the British Museum.[3]

So long as that institution consisted only of a library, with limited collections of antiquities and natural curiosities, it was easily managed by a body consisting of public functionaries, men of rank and general attainments, and the representatives of the chief donors of the Museum.[4] When, however, the establishment was so enlarged as to become a national deposit of the productions of nature, both recent and fossil, together with vast additions to the books, manuscripts, prints, statues, &c., supplemental trustees were elected to aid in the direction.[5] We rejoice that eminent men of letters were thus associated in the Trust;[6] but even so amended, the Board, with the best intentions, we believe, has found it very difficult satisfactorily to perform the numerous and various duties devolving upon it; for although the "ex officio" Trustees, the Presidents of the Royal Society and of the College of Physicians,[7] are necessarily persons of weight in science; and although we gratefully witnessed the election of the eminent astronomer, Sir John Herschel,[8] we would earnestly represent to your Lordship, that the qualifications of these gifted individuals do not necessarily include an interest in, or the ability to judge of, many of those measures which may best promote Natural History; and, consequently, that there is no effective provision (in the absence of other men of science) for the proper guidance of the Natural History department, or for having at the Board, trustees who can explain to their associates the desiderata of naturalists, and estimate the value of new specimens, either offered to or purchased by the nation.

Fully acknowledging, that in their accomplishments and high characters the present Trustees offer the best sureties for the satisfactory execution of any duties connected with their own pursuits, we still think that, with the best disposition (and they have already done much good service) these distinguished men are unable adequately to direct the vast and rapidly increasing Natural History departments of the Museum;[9] and we can even well suppose that they would themselves be happy to be relieved from the heavy responsibility which must be attached to the application of the large sum annually voted by Parliament for the support of natural science.[10]

Now, in the event of a knowledge of Natural History being in future recognized among the grounds for election to the Trusteeship of the British Museum, we should have reason to anticipate that the sum allotted to this subject would be applied so efficiently and regularly to its extension and improvement, as would best secure the progress of science, and yield most interest and instruction to the public.

Deeply impressed with these sentiments, we beg to suggest, for the consideration of your Lordship, that steps should be taken to effect such an improvement in the constitution of the Trust, as shall render the management of the Natural History departments of the British Museum, as far as possible, independent of the other divisions; and on this point we would beg to refer your Lordship to the original plan of Sir Hans Sloane.[11]

In offering this suggestion, we do not contemplate a separation of the Natural History collections from the other departments of the British Museum, as we well know that the cultivation of natural science cannot be efficiently carried on without reference to an extensive library.

What we chiefly desire to see is, the formation of such a responsible system of management as may satisfy the public and ourselves, that in this great national establishment, the interests of all branches of natural science will be thoroughly protected and advanced, and that the halls devoted to it shall be so enriched with well selected and well classified objects of contemplation and comparison, as shall not merely gratify the curiosity and excite the wonder of the multitude, but shall prove of real use to the researches of the student and the man of science.[12]

Memorial
British Library, *Parliamentary Papers* 1847 (vol. 34): 253–6.

[1] According to a note at the head of the text, this petition was presented on 10 March 1847.

[2] CD's name appears thirtieth in a list of fifty-seven signatories, in the form 'C. Darwin, F.R.S. F.G.S. &c.' The first four signatories were Roderick Impey Murchison, president of the British Association for the Advancement of Science, William Willoughby Cole, president of the Linnean Society, William Buckland, reader in geology and professor of mineralogy at the University of Oxford, and Henry Thomas de la Beche, president of the Geological Society of London. Among the remainder were leading figures from all fields of natural history including Richard Owen, Charles Lyell, William Jackson Hooker, Edward Sabine, Joseph Dalton Hooker, Thomas Bell, William Spence, Charles Cardale Babington, Nathaniel Wallich, John Lindley, Adam Sedgwick, and Charles Giles Bridle Daubeny.

[3] The original board of forty-one trustees of the British Museum was composed of three principal trustees (the archbishop of Canterbury, the lord chancellor and the speaker of the House of Commons); seventeen official trustees, drawn from major state office-holders and all appointed *ex officio*; six family trustees representing descendants of Hans Sloane, Robert and Edward Harley, and Robert Bruce Cotton, whose collections formed the original museum holdings, and fifteen additional trustees who

were elected by the twenty-six permanent members of the Board. The *ex officio* trustees included the presidents of the Royal Society of London and the Royal College of Physicians of London. By 1847 the original board had been augmented by the appointment, between 1805 and 1824, of representatives of the Towneley, Elgin, and Knight families following major donations, and by the addition of the president of the Society of Antiquaries and the president of the Royal Academy in 1824, and, from 1832, of a trustee appointed by the sovereign (*An act to enable his majesty, his heirs and successors to appoint a trustee of the British Museum* [23d June 1832]). See P. R. Harris 1998, pp. 2, 11 and n. For contemporary concern over the perceived lack of representation of men of science, see Stearn 1981, p. 16.

4 For the original composition of the Board of Trustees, see n. 3 above.

5 For details of the expansion of the museum's collections in the late eighteenth and early nineteenth centuries, see P. R. Harris 1998, pp. 29–45, 75, 134, and 144. The Royal Society's natural history collection was transferred to the British Museum in 1781, and the Hatchett, Mordaunt, and Cracherode collections of minerals in 1799 (Stearn 1981, p. 20). The principal acquisitions of the early nineteenth century were the King's Library in 1823, which increased the collection of printed books by over fifty per cent, and the natural history collections of Joseph Banks in 1827 (P. R. Harris 1998, pp. 31–2). For changes to the composition of the Board of Trustees, see n. 3, above. Although no additional trusteeships had been created since 1832, following a House of Commons Select Committee investigation in 1836, any of the fifteen elected trustees who did not attend meetings for a period of more than twelve months was expected to resign, and the committee recommended that eminent men in the sciences, literature, and art should be better represented on the board (*ibid.*, pp. 11, 105–6).

6 In addition to John Frederick William Herschel (see n. 8, below), the most recently elected members of the Board of Trustees of the British Museum were the constitutional historian Henry Hallam (1837), the antiquary and diplomat William Richard Hamilton (1838), and the diplomat George Leveson Gower Granville (1841). Robert Peel, known for his patronage of men of science, had been a trustee since 1833. See *Statutes and rules for the British Museum* and Rupke 1994, p. 53.

7 For the *ex officio* trustees, see n. 3, above. In 1847, the president of the Royal Society was Spencer Joshua Alwyne Compton, and the president of the Royal College of Physicians was John Ayrton Paris.

8 Herschel was elected to the Board of Trustees of the British Museum in August 1843 (*Herschel calendar*, no. 5571).

9 The natural history department of the British Museum was effectively created in 1835 when the museum's various botanical collections were brought together and placed in the care of Robert Brown. Following the recommendations of a House of Commons Select Committee in 1837, the natural history department was formally split into three sections, botanical, zoological, and mineralogical and geological, each with its own keeper. See Stearn 1981, pp. 22–4.

10 For details of government expenditure on the British Museum in this period, see *History of the king's works* 6: 403–21.

11 For Sloane's will, see Faulkner 1829, 1: 363–73. Under its terms, Sloane left his collections for the public use and stipulated that they should remain in London, and be kept together, since they tended to 'the manifestation of the glory of God, the confutation of atheism and its consequences, the use and improvement of physic, and other arts and sciences, and benefit of mankind' (*ibid.*, p. 365). See also MacGregor ed. 1994. The head of the department of printed books, Anthony Panizzi, had been campaigning for the natural history collections to be moved to make more space available for the library (Stearn 1981, pp. 23–4).

12 Largely as a result of this petition, a Royal Commission was appointed in June 1847 to investigate the constitution and administration of the museum; Buckland, one of the signatories to the petition (see n. 2, above) was appointed a trustee in the same year (*Statutes and rules for the British Museum*). The commission reported in 1850 and recommended that the various collections should not be dispersed and that the building should be enlarged. It also recommended changes to the organisation of the natural history departments (P. R. Harris 1998, pp. 165–73). No serious reorganisation took place, however, until the establishment of the British Museum (Natural History) and the transfer in 1880 of

the natural history collections to a separate site in Kensington (*ibid.*, pp. 241–2; see also Rupke 1994). For CD's involvement with the continued campaign on behalf of the natural history collections, see *Correspondence* vol. 7, Appendix VI.

To Samuel Pickworth Woodward 10 July 1848

Down Farnborough Kent
July 10th./48

My dear Sir

I have much pleasure in stating that from your long connection with the Museum of the Geological Society & from your general knowledge of Geology and Palæontology, in my opinion, you would be enabled to render very material service in taking care of & in arranging the specimens in the Fossil department of the British Museum.[1]

With my wishes for your success,—I beg to remain | Dear Sir | Yours faithfully | C. Darwin

S. P. Woodward Esqre

British Museum Central Archives: Staff Applications and Testimonials (S. P. Woodward; CE33/712/3)

[1] Woodward was sub-curator of the Geological Society of London from 1839 to 1845 and assisted CD by preparing specimens from the collections for him to study while he was working on *Volcanic islands* (see *Correspondence* vol. 2, letters to S. P. Woodward, [12 November 1842] and 29 [September 1843]). Woodward's application was successful; he was appointed first-class assistant in the department of geology and mineralogy in the British Museum in 1848, and in that capacity provided CD with information and specimens for his work on fossil cirripedes (*Correspondence* vol. 4, letters to S. P. Woodward, 21 March [1850] and [April 1850 – January 1851]; *DNB*; *Fossil Cirripedia* (1851)). For Woodward's career at the British Museum, see Stearn 1981, pp. 229–30.

To Peter Lund Simmonds[1] 25 February [1849]

Down Farnborough Kent
Feb 25th

Sir

The state of my health prevents me replying at length to your letter.[2] I saw **very** little of the country near New Bay & was not at the Chupat;[3] what I did see was poor & water generally absent: there wd be brush-wood in the valleys in all probability. About the year 1790 there was a Spanish Settlement there; but every soul except one was murdered by the Indians.— Indians now perhaps tamer, but never to be trusted; between 1820 & 1830, they attacked every separate farm near the Rio Negro: further south at the Str of Magellan they are much tamer: coal probably not to be found; that in Straits said to be only lignite.— I know nothing about the Fish on the coast. Cattle cd be easily driven from the R. Negro to the Chupat. Anyone settling there, wd I apprehend be under the government of

B. Ayres, about as bad as could be.[4] Interior absolutely unknown; probably as bad as that seen by us on the S. Cruz.[5] The climate fine & dry in summer: no severe frosts or snow in winter.— Wheat does excellently in the *valley* of the R. Negro: the plains everywhere sterile. Port Desire was the best spot which we saw on the Patagonian coast; but the country is poor; it was once settled by the Spaniards, but long since deserted. Altogether I have a very poor opinion of the whole coast of Patagonia south of R. Negro & I sh^{d} sincerely pity any emigrants there: on the other hand if such a coast existed, under British Rule, as in Australia, I have no doubt it w^{d} be thinly settled soon & would feed sheep.— If I had to choose, I w^{d} *incomparably* prefer the Falkland Islands to Patagonia— there you w^{d}. have cattle & horses & pigs—*peat* & brushwood for fire—plenty of fish as is asserted. There are no severe frosts & the snow does not lie long. On other hand *no* timber, country looks desolate & is very stormy. I have reason to believe that though twice there we happened to be very unfortunate in our weather—[6] An emigrant w^{d}. be there under British Government & free of Indians.—[7] Wild cattle thrive there & most vegetables: I rather think wheat has lately succeeded. I wish I could hav⟨e⟩ given you fuller information.

Sir | Your obet. sevt. | C. Darwin

Postmark: ⟨ ⟩ 26 1849
Quentin Keynes (private collection)

[1] This letter was published in *Correspondence* vol. 4 in the form in which it appeared in a sale catalogue and is republished here transcribed from the original text.

[2] Simmonds's letter has not been found. CD recorded in his 'Journal' for 1849 'Jan 1st to March. 10th. Health very bad with much sickness & failure of power'. This was the end of a period of ill health that had lasted from July 1848. See *Correspondence* vol. 4, Appendix I.

[3] CD refers to the Chubut river, Patagonia (see *Correspondence* vol. 1, letter from Edward Lumb, 13 November 1833 and n. 1).

[4] Simmonds may have asked CD for information on the suitability of Patagonia for colonial settlers, either in his capacity as editor of *Simmonds's Colonial Magazine and Foreign Miscellany* or as proprietor of a colonial agency office. He published the magazine, which regularly carried articles on likely places of settlement, until March 1849, and actively promoted emigration in his editorials as a cure for social ills. See *Simmond's Colonial Magazine and Foreign Miscellany*, January 1849 and April 1849; *Post Office London directory*, 1848 and 1849. For CD's opinion of the government in Buenos Aires, see *Correspondence* vol. 1, letter to Edward Lumb, 30 March 1834.

[5] CD visited the Santa Cruz river in April 1834 during the *Beagle* voyage (see *Correspondence* vol. 1, letter to Catherine Darwin, 20–9 July 1834, and Appendix I); for his unfavourable impressions of Patagonia in general, see *Correspondence* vol. 1, letter to Catherine Darwin, 6 April 1834, and *Journal of researches*, chapters 10 and 11, where he commented that the 'curse of sterility' was on the land (*ibid.*, p. 215).

[6] CD visited the Falkland Islands twice during his voyage on board the *Beagle*. He first arrived in March 1833 after a succession of gales (see *Correspondence* vol. 1, letter to Caroline Darwin, 30 March – 12 April 1833); he spent four days exploring the island in March 1834, during the whole of which time 'it blew a gale of wind with hail & snow' (*Correspondence* vol. 1, letter to Catherine Darwin, 6 April 1834).

[7] British sovereignty over the Falkland Islands had been re-established early in 1833, shortly before CD's first visit, following a brief occupation by Argentina (see *Correspondence* vol. 1, letter to Caroline Darwin, 30 March – 12 April 1833 and n. 6.)

To Emily Harriet Stanhope 9 August [1849][1]

Down
Augt. 9th.

Dear Lady Mahon[2]

M^r. Darwin is very much obliged for the honour of your Ladyships kind invitation for Thursday, & will have much pleasure in accepting it.[3]

I beg to remain | Your Ladyships | Truly obliged | Charles Darwin

Centre for Kentish Studies, Maidstone, Kent (U1590 C481/18)

1 The year is established by the relationship between this letter and the letter to Charles Lyell, [2 September 1849] (*Correspondence* vol. 4).
2 Emily Harriet Stanhope was the wife of Philip Henry Stanhope, Viscount Mahon (*Complete peerage*).
3 In 1849, the Thursday following 9 August was 16 August. In his letter to Charles Lyell, [2 September 1849], CD described the dinner that took place at Viscount Mahon's ancestral home at Chevening, Kent (*Correspondence* vol. 4). Chevening is about three miles south-east of Down.

To James Scott Bowerbank 25 February [1850][1]

Down Farnborough Kent
Feb 25th.

My dear Sir

I have found a **few** sponges in my collection: some dry & some in Spirits.

I do not suppose they will be of any use to you—but on the same principle that I like to look at all old rubbishy barnacles, you may like to look at & have any rubbishy Sponges.—[2] I will leave them directed to you at Geolog. Soc. on Thursday next.—[3]

I believe I have a few more specimens but cannot now lay my hands on them.

Yours sincerely | C. Darwin

I return your Plumularia on which the beautiful & most interesting Scalpellum ornatum was attached.—[4]

Postmark: FE 26 18⟨50⟩
DAR 185: 104

1 The year is established by the oval cancellation stamp on the cover containing the code '58'. This code was in use by the chief Post Office in London between May 1849 and September 1850 (Parmenter 1981, catalogue no. 58A).
2 Bowerbank was an expert on sponges and published the first part of *A monograph of the British Spongidæ* in 1864 (Bowerbank 1864–82); CD is not mentioned in the list of acknowledgments. For the sponge specimens CD collected on the *Beagle* voyage, see R. D. Keynes ed. 2000.
3 In 1850, 25 February was a Monday and the following Thursday was 28 February. CD is not known to have visited London on that date, and may instead have sent the specimens by the weekly carrier, which operated a service to London on Thursdays (*Post Office directory of the six home counties* 1851). However, in his letter to J. de C. Sowerby, 3 March [1850], CD refers to delayed plans to visit London in order to deliver and collect cirripede specimens (*Correspondence* vol. 4).
4 *Plumularia* is a genus of feathery, branching colonies of hydrozoan polyps that have a protective horny covering. *Scapellum ornatum*, a barnacle from Algoa Bay, South Africa, is described in *Living Cirripedia*

(1851), p. 244. Both were sent to CD by Bowerbank in 1849 (see *Correspondence* vol. 4, letter to J. S. Bowerbank, 24 February [1849]). In a letter to Bowerbank of 25 June [1851], CD requested the return of the *Scalpellum* in order to re-examine the peduncle (*Correspondence* vol. 5).

To Adam Sedgwick 11 October [1850][1]

Down Farnborough Kent
Oct. 11th.

My dear Professor Sedgwick

Having sent yesterday to the Geolog. Soc. for parcels, I was surprised & delighted at receiving your Discourse[2] this morning.

I thank you cordially for this act of kindness & for your remembrance of me.— By an odd chance about a fortnight since, seeing your Book advertised, I was speculating how I could borrow it.— It has wonderfully grown since I read it as a first Edition.[3]

I have got to decide upon what school to send my eldest Boy of eleven, & I have been thinking of sending him not to a purely classical school; so that I shall read your ideas on Education with a practical end in view.—[4]

I most sincerely hope that your health is pretty good: mine is much better, thanks to the inestimable Water Cure,[5] than it has been for several years, but I see that I shall never have a sound stomach & therefore never be really strong again.

Accept my sincere thanks for your kindness & believe me, | Your's truly obliged, | Charles Darwin

Photocopy
Gerald M. Friedman (private collection)

1 The year is established by the reference to the fifth edition of Sedgwick's *Discourse on the studies of the University of Cambridge* (Sedgwick 1850; see n. 2, below).

2 Sedgwick 1850. CD's annotated copy, inscribed 'From the Author', is in the Darwin Library–CUL; see *Marginalia* 1: 749–52. CD recorded reading this edition on 12 November 1850 (*Correspondence* vol. 4, Appendix IV, 119: 22a).

3 Sedgwick 1833. The fifth edition (Sedgwick 1850) was expanded by a preface of 442 pages, longer than the original work, which, as reproduced in the fifth edition, runs to 322 pages. There is no record of the date on which CD read the first edition, although it was recommended to him by his sister Caroline Darwin in 1834 (*Correspondence* vol. 1, letter from Caroline Darwin, 9–28 March [1834]).

4 CD refers to his eldest son, William Erasmus Darwin. William was still 10 years old at the time this letter was written, reaching his 11th birthday on 27 December 1850; he entered Rugby School in 1852. On the question of William's schooling, see *Correspondence* vol. 4, letter to W. D. Fox, 10 October [1850] and n. 2. Sedgwick, in the first edition of *Discourse on the studies of the University of Cambridge* (Sedgwick 1833), promoted the teaching of the natural sciences, alongside the traditional subjects of mathematics and classics, believing that a better understanding of the natural world would serve to reinforce a belief in divine providence. The preface to the fifth edition (Sedgwick 1850) expands on the same theme, containing a lengthy refutation of the theory of transmutation of species proposed in *Vestiges of the natural history of creation* ([Chambers] 1844), but welcoming the recently established natural sciences tripos at Cambridge University.

5 CD had first tried the water cure in 1849 at James Manby Gully's hydropathic establishment at Malvern in Worcestershire (see *Correspondence* vol. 4, letter to Susan Darwin, [19 March 1849]). He

visited Malvern again from 11 to 18 June 1850 (*ibid.*, Appendix I) and had also been practising the cure at home (see *ibid.*, letter to J. D. Hooker, 13 June [1850], and letter to W. D. Fox, 4 September [1850] and n. 2). See also Browne 1990.

To Edward Adolphus Seymour Seymour[1] May 1851[2]

[London]

[Testimonial on behalf of Dr. Joseph Dalton Hooker ... praising Hooker's achievements 'during his voyage to the Antarctic Regions under Captain Sir James Clarke Ross,[3] and again during a three years' expedition made on behalf of the Royal Gardens at Kew to the North-Eastern districts of India',[4] his botanical knowledge, his researches into Geology, Zoology and Meteorology, and hoping to secure 'to Her Majesty's Botanical Establishment at Kew his future services in the following up the investigations so ably commenced'.[5]]

[...] the accuracy of observation and high scientific views shown in the publication of the results of his first voyage have secured to him an European reputation. We therefore gladly avail ourselves of the occasion of his return from his second Expedition to represent to your Lordship the claims we conceive he has acquired on the Country to which he is an honor [...][6]

Memorial incomplete[7]
Julian Browning (Catalogue 10, 'Historical Manuscripts and Documents')

[1] According to the sale catalogue, the testimonial was sent to Seymour in his capacity as Chief Commissioner of Her Majesty's Woods and Forests (see n. 5, below).

[2] The date is given by the sale catalogue.

[3] Joseph Dalton Hooker served as assistant surgeon and botanist on board HMS *Erebus*, commanded by James Clark Ross, on the Antarctic expedition of 1839 to 1843, following which he published *Flora Antarctica* (Hooker 1844–7); see also *Correspondence* vol. 2.

[4] Hooker collected plants for the Royal Botanic Gardens, Kew, in India and the Indian and Tibetan Himalayas, between 1848 and 1850, returning to Britain in March 1851 (Allan 1967, p. 177; see also *Correspondence* vol. 4).

[5] The testimonial was submitted to Seymour in support of an attempt by Hooker's father, William Jackson Hooker, to obtain a government salary for his son to continue to work on specimens brought back from India (see R. Desmond 1999, pp. 200–1). Seymour, as first commissioner of woods and works (sometimes known as the commissioner of woods and forests), controlled the government department that had contributed funds to the Indian expedition (*History of the king's works* 6: 242; Allan 1967, p. 180). Representations were also made on Hooker's behalf by the British Association for the Advancement of Science (R. Desmond 1999, p. 183; *Report of the twenty-first meeting of the British Association for the Advancement of Science held at Ipswich in July 1851*, p. xxx). In 1852, the Treasury awarded Hooker an annuity of £400 a year for three years in order to produce a flora of India (Hooker and Thomson 1855; see also Allan 1967, p. 181).

[6] Besides CD, the signatories named in the catalogue entry include Robert Brown, Charles Giles Bridle Daubeny, John Stevens Henslow, John Lindley, John Forbes Royle, Edward Forbes, Nathaniel Wallich, George Bentham, William Hopkins, Charles Lyell, Roderick Impey Murchison, William John Hamilton, Leonard Horner, Frederick McCoy, Adam Sedgwick, Philip de Malpas Grey-Egerton, Edward Sabine, Thomas Hill, and Henry Thomas de la Beche. Thomas Hill has not been further identified.

[7] The original memorial is complete and is described in the sale catalogue as being four pages long.

To J. S. Henslow [1852–60][1]

57 Queen Anne St
Thursday morning

My dear Henslow

The enclosed has been forwarded to me from Down: it is quite correct.—[2]

In Haste | Ever yours | C. Darwin

Harvard Medical Library in the Francis A. Countway Library of Medicine (H MS c3.3)

[1] The date range is established by the address. CD visited his brother, Erasmus Alvey Darwin, at 57 Queen Anne Street, London, from 1852, when Erasmus moved there (see *Correspondence* vol. 5, letter to W. D. Fox, 24 [October 1852]). In spring 1860 the house number was changed to '6' (see *Post Office London directory* 1861 and *Correspondence* vol. 8, letter to a Bookseller, 13 [May 1860]).

[2] The enclosure has not been identified.

To J. D. Hooker [9 or 16 February 1854][1]

Down
Thursday Afternoon

My dear Hooker

I do not know what to say— I have within these few minutes received your Book,[2] & after admiring some of the illustrations, I fell on the Dedication to me.—[3] My dear Hooker, you have gratified me, more than I can speak, but really it is altogether too great a compliment to me.— Only fancy that I should live to have a Book, dedicated to me,—& such a book, so beautiful & magnificent.— I have not even read the Title-Page nor looked at all the illustrations, & now Emma[4] has seized on the Book. Most of the illustrations seem to me quite beautiful: there is a capital woodcut, showing the great plain in an admirable manner,—[5] it reminds me of the Pampas from the Andes.— Your book will delightfully last me for a long time, & I shall begin this evening, & lay aside the Salt-Lake Book.—[6] What a sum of money the whole must have cost! I had no idea whatever that it would be so splendid,—And that I should have such a book dedicated to me.— Well, my dear Hooker, I thank you from my heart, & M^rs. Hooker, as *joint-author*:[7] Emma says she feels a good share of the honour, & she desires me to give her best love to M^rs. Hooker.

I wish I felt more worthy of the honour, but I am so far worthy, that I am & have long been, my dear Hooker, your very affectionate friend | C. Darwin

I see that the Dedication has put it out of my head even to thank you for this beautiful present of the Book itself—

Oliver N. Hooker (private collection)

[1] The date is established by the publication date of Hooker 1854 (see n. 2, below), and by the relationship between this letter and the letter to Charles Lyell, 18 February [1854] (*Correspondence* vol. 5), in which there is a reference to the dedication in Hooker 1854. In 1854, 9 and 16 February were Thursdays.

[2] Hooker's *Himalayan journals* (Hooker 1854) was published in mid-February 1854 (*Publishers' Circular* 1854). An annotated copy is in the Darwin Library–CUL (see *Marginalia* 1: 392–3). CD recorded it under the heading 'Books read' in his reading notebooks on 7 March 1854 (see *Correspondence* vol. 4, Appendix IV, 128: 8).

[3] The dedication reads: 'To Charles Darwin, F.R.S., &c. These Volumes are Dedicated, by his affectionate friend, J. D. Hooker. Kew, Jan. 12th, 1854' (Hooker 1854, 1: [v]).

[4] Emma Darwin.

[5] CD may refer to a fold-out lithograph, 'Tibet and Cholamoo Lakes from Donkia Pass', in volume 2 of Hooker 1854, facing page 124.

[6] CD recorded Stansbury 1852, *An expedition to the valley of the Great Salt Lake of Utah*, under the heading 'Books read' in his reading notebooks on 23 March 1854 (see *Correspondence* 4, Appendix IV, 128: 8).

[7] In his *Life and letters of Sir Joseph Dalton Hooker* (L. Huxley 1918, 2: 190), Leonard Huxley wrote of Frances Harriet Hooker: 'As a good writer of English, she constantly aided him in his writing and correction of proofs, where he relied greatly on her judgment.'

From [John Innes][1] [after 8 February – August 1855][2]

Hore. W S. | St Clements Rectory | Oxford—[3]
No communication from Ainslie about water—[4]
Probably I may have mentioned the pointer and setter.

Either Mr. Tournay of Saltwood[5] or Capt Johnson of Newington[6] had a pointer bitch who appeared perfectly purely bred. being put to a thorough bred pointer dog she produced a litter of puppies which all appeared pure pointers but one which was a black and white brindled setter, with feathered legs and stern, appearing a thorough bred setter— I mention this one instance because I remember the setter well but I was told that in several litters this result occurred and that the setter was always distinctly indicated when pupped. I do not know where Capt Johnson is if alive— Mr. Tournay I believe still lives at Saltwood— Lynedock Douglas Esq Hythe[7] and Capt Douglas of the Down's Rifles[8] would remember the circumstance and so probably would Genl. and Col. King at Hythe[9] and others.

Nothing is more common than for crossed dogs to take after one parent. The breeds commonly crossed are terriers with bull and pointers with fox hounds. there are often some of the first offspring which have but little sign of cross in their formation but continuing to breed with pointers and terriers the marks of the former cross frequently reappear in individuals

Incomplete
DAR 163: 5

CD ANNOTATIONS
1.1 Hore . . . water— 2.1] *crossed pencil*
5.5 individuals] 'But ultimately absorbed into pure breed.—' *pencil*
Top of letter: '16'[10] *brown crayon, circled brown crayon*

[1] The attribution to Innes is based on the handwriting, and on the content of the letter, which suggests that the correspondent was well known to CD, and was familiar with Down and with the Hythe area of Kent. John Innes's mother, Mary Innes, was buried at Hythe (see *Correspondence* vol. 7, letter to John Innes, 4 March [1859], n. 2).

[2] The date range is established by the date at which Donald Douglas became a captain in the Royal North Down Rifles (see n. 8, below), and by the month by which William Strong Hore had left Oxford (see n. 3, below).

[3] Hore was rector of St Clement's, Oxford, from 1850 to 1855 (*Clergy list*). Innes may be replying to a request from CD to supply Hore's address; as a fellow clergyman Innes would have had access to that information from the *Clergy list*. In October 1855, CD wrote to William Darwin Fox, 'I had a letter some 2 months ago from Hore, who is settled, an old Batchelor, in Devonshire' (*Correspondence* vol. 5, letter to W. D. Fox, 14 October [1855]). Fox, Hore, and CD were all at Cambridge together (*Correspondence* vol. 1, letter to W. D. Fox, 12 [June 1828], n. 3; *Alum. Cantab.*).

[4] Innes refers to Robert Ainslie, resident at Pond House, later Tromer Lodge, Down, from 1845 to 1858, and with whom CD and other residents of Down had a series of disagreements (*Correspondence* vol. 3, letter to Susan Darwin, 3[–4] September 1845, and *Correspondence* vol. 7, letter to W. E. Darwin, 14 [May 1858] and n. 5).

[5] Thomas Tournay of Brockhill House, Saltwood, Kent (*Post Office directory of the six home counties* 1851, 1855).

[6] Newington is $2\frac{1}{2}$ miles north-east of Hythe, Kent (*Post Office directory of the six home counties* 1851). There is no Captain Johnson listed in any of the Post Office directories for Kent from the 1840s or 1850s.

[7] Lynedock Douglas of High Street, Hythe, Kent (*Post Office directory of the six home counties* 1855).

[8] Donald Douglas was a captain in the Royal North Down Rifles from 8 February 1855 (*Army list* 1855).

[9] Major-general Richard Thomas King of Swiss Cottage, Hythe, and Colonel William James King of West End Road, Hythe (*Hart's army list* 1856; *Post Office directory of the six home counties* 1855). By April 1857, CD had received information from 'Col. King near Hythe' about the markings of wild and domesticated pigeons, and he is acknowledged in *Variation* for supplying CD with specimens of rock-pigeon (*Correspondence* vol. 6, letter to Laurence Edmonston, 19 April [1857] and n. 5; *Variation* 1: 184).

[10] The annotation may refer to a numbered portfolio of notes, possibly on hybridism or heredity (see this volume, Supplement, letter from [Mr Edwards?], [before end of 1839?] and nn. 4 and 5.) CD was actively soliciting information on the transmission of characteristics between, and instances of reversion in, crossbred animals from at least 1839 (see *Correspondence* vol. 2, Appendix V). In 1855 he specifically requested information on mongrels from Fox (*Correspondence* vol. 5, letter to W. D. Fox, 27 [June 1855]; see also letter to W. D. Fox, 22 [July 1855]), probably for the chapters on variation under domestication in his manuscript on species (see *Correspondence* vol. 6, Appendix III).

To William Henry Benson 7 December [1855][1]

Down Bromley Kent.
Dec: 7th

My dear Sir,

I thank you cordially for the very kind manner with which you have answered my note,[2] & for the very great trouble you have taken in answering it so fully & so very clearly.

The facts which you give are the very thing which I wanted.

The range of shells over the land seems to me in many cases much more confined than plants: on the other hand what a curious case that of Madeira & St Helena shells![3] I collected several semi-fossil shells at St Helena, which were described by

Mr Sowerby:[4] I cannot remember whether any are left, but if they would interest you I would look: I know I gave away many but forget whether I gave all. I am particularly obliged to you for mentioning the case of Bulimus Bengalensis, which is so curious; I am very glad to hear of it, for it is so exactly parallel to what, as is so well known, sometimes occurs with plants.— I presume such cases are not common with land Mollusca.[5] The whole subject of the distribution of pulmoniferous[6] Mollusca seems to me very interesting. Am I not right in believing that many fresh-water species have very wide ranges?[7] Though I ask this, I am not so unreasonable as to expect you again to take the trouble to answer me in full.[8]

Thank you *much* for saying that you will keep in mind my very strong wish for information in regard to distribution of alpine or semi-alpine shells.

You kindly enquire about my health: though I appear a strong man, I am far from being so, & shall never be so again; but I contrive to do some Nat. History work every day in my life.

Dr Gully did me *much* good.[9]

I sincerely hope that you profited by the water cure, & are now stronger.—

With many thanks pray believe me in haste | yours sincerely | CH. Darwin

Copy
DAR 143: 89

[1] The year is established by the relationship between this letter and the memorandum from W. H. Benson, 5 December 1855 (see *Correspondence* vol. 5).

[2] CD refers to an earlier letter to Benson that has not been found, and to the memorandum from W. H. Benson, 5 December 1855 (*Correspondence* vol. 5). CD had asked Benson whether any alpine shells were common to the heights of Ceylon (now Sri Lanka) and India (see *Correspondence* vol. 5, letter to E. L. Layard, 9 December 1855).

[3] See *Correspondence* vol. 5, memorandum from W. H. Benson, 5 December 1855 and n. 14. See also *Origin*, p. 391.

[4] See *Journal of researches*, p. 582; see also p. 426. George Brettingham Sowerby described extinct land-shells from St Helena in the appendix to *Volcanic islands*, pp. 155–8.

[5] Benson had indicated that *Bulimus bengalensis* had a widespread, but patchy, distribution (*Correspondence* vol. 5, memorandum from W. H. Benson, 5 December 1855). In *Origin*, pp. 365–76, CD discussed the occurrence of the same species, particularly of plants, on the summits of distant mountain ranges, and argued that this was due not to creation in more than one location, but to the stranding of these species, which once occupied continuous but lower terrains, on mountains, when the temparature rose after the glacial period. He did not mention land-shells in this discussion, but in *Natural selection*, the longer manuscript of which *Origin* was an abstract, CD commented (p. 555):

> land-shells have either been so frequently created, or as I should infer so easily modified, that there do not appear to be many species in common even on mountain-summits as near to each other as the Alps & Pyrenees, or, as I am informed by Mr. Benson, on the Nilghiri & heights of Ceylon.

[6] Pulmoniferous: 'bearing or having lungs (or lung-like organs); pulmonate; . . . belonging to the group Pulmonifera (=Pulmonata) of gastropod molluscs' (*OED*).

[7] CD discusses the distribution of freshwater shells in *Origin*, pp. 385–6.

[8] No reply from Benson has been found.

[9] James Manby Gully was the proprietor of a hydropathic establishment in Malvern, Worcestershire,

at which CD was treated in 1849. In 1851 he took his eldest daughter, Anne Elizabeth Darwin, there for treatment of an illness that proved fatal. See *Correspondence* vols. 4 and 5.

To ? 7 December [1855–7?][1]

Dec. 7th

Dear Sir

Will you be so kind as to take the trouble to inform me the address of Mr.—(?) Mason, who I believe is at Madeira,[2] & *whether he intends visiting the Azores.* I want him to collect for me 2 or 3 trifling specimens[3] & I do not know how to write to him.—

Dear Sir | Yours faithfully | Ch. Darwin

DAR 249: 93 (photocopy)

[1] The years are conjectured from the reference to a Mr Mason; Nathaniel Haslope Mason collected plants in Madeira between 1855 and 1857 (R. Desmond 1994).
[2] Nathaniel Haslope Mason; see n. 1, above.
[3] The specimens may have been bats. In the letter to Charles Lyell, 5 [October 1860] (*Correspondence* vol. 8), CD wrote: 'The subject of Bats is washed out of my head: I got Mr Mason to collect in Madeira & the species turned out to be European. . . . I failed to get the Azores species.' CD was evidently also seeking bat specimens from the Azores in May 1855 (see *Correspondence* vol. 5, letter from T. C. Hunt, 2 July 1855). Bats are discussed in a section on the inhabitants of oceanic islands in the second chapter on geographical distribution (chapter 12) in *Origin* (pp. 394–5); Madeira and the Azores are not mentioned, however. See also *Origin*, pp. 104–7.

To George Gulliver 27 December [1855]

Down Bromley Kent
Decr 27th

My dear Sir

I write one line to say that I have this morning received safely your most kind present of Hewson's work.[1] I have just looked at the Table, & am astounded at the number of the measurements.[2]

With very sincere thanks | yours truly obliged | Ch. Darwin.

Postmark: DE28 1855
Josh B. Rosenblum (private collection)

[1] In his letter to Gulliver of 18 December [1855] (*Correspondence* vol. 5), CD had accepted Gulliver's offer of *The works of William Hewson* (Gulliver ed. 1846). CD's annotated copy is in the Darwin Library–CUL (see *Marginalia* 1: 378).
[2] Gulliver had tabulated comparative measurements of red blood cells of various species of mammals, birds, reptiles, amphibians, and fish for his edition of the works of the physiologist William Hewson (Gulliver ed. 1846, pp. 237–44). CD was interested in obtaining measurements of corpuscles of pigeons and dogs (see *Correspondence* vol. 5, letter to George Gulliver, 18 December [1855]; see also *Correspondence* vol. 6, letter from George Gulliver, 20 January [1856]).

To John Davy 3 January [1856][1]

Down Farnborough Kent[2]
Jan. 3^{d}.

My dear Sir

I am much obliged for your very kind note,[3] & am delighted to hear that you have been prosecuting your Researches.—[4]

I am sure I am much pleased & flattered at the form you are so good as to put your communication.[5] I heard, on enquiry, the other day that your Paper would soon be published.[6] There has been inexcusable delay in regard to it: but, I may mention, that we had the other day at the council a discussion,[7] how we could expedite the Reports of Referees, & I found the subject involved in rather more difficulty than I had thought.—

Pray believe me | my dear Sir | Yours sincerely | Charles Darwin

David Schulson Autographs, New York (1991)

[1] The year is established by the reference to J. Davy 1855 (see n. 6, below).
[2] CD's address after September 1855 was 'Down, Bromley, Kent', not 'Down, Farnborough, Kent', but strong evidence supports dating the letter 1856 and interpreting the misaddress as a lapse.
[3] The letter from Davy has not been found.
[4] From November 1855 to January 1856 Davy was carrying out experiments on the ova of salmon that were a continuation of experiments originally suggested by CD and first reported on in J. Davy 1855 (J. Davy 1856; *Correspondence* vol. 5, letter from John Davy, 30 January 1855). On 10 January 1856, Davy sent CD a letter detailing the results; this was later published in the *Proceedings of the Royal Society of London* (J. Davy 1856; see also *Correspondence* vol. 6, letter from John Davy, 10 January 1856).
[5] Both J. Davy 1855 and J. Davy 1856 were written in the form of letters to CD, and forwarded by CD to the Royal Society of London. See *Correspondence* vol. 5, letter from John Davy, 21 March 1855, and letters to John Davy, 25 March [1855] and 26 March [1855], and *Correspondence* vol. 6, letter from John Davy, 10 January 1856.
[6] CD refers to J. Davy 1855. The paper was read before the Royal Society on 26 April 1855. A summary appeared in the *Proceedings of the Royal Society of London* 7 (1854–5): 362, and the full version was published in the *Philosophical Transactions of the Royal Society of London* 146 (1856): 21–9.
[7] CD had attended the council meeting of the Royal Society on 20 December 1855 (*Royal Society council minutes*).

To Miles Joseph Berkeley 18 March [1856][1]

Down Bromley Kent
March 18th

My dear Sir

I am very much obliged for your note of the 7th & kind permission to incorporate your facts: I have now sent a little notice to the Linnean Soc.—[2]

Many thanks, also, for your remarks on the Peas.— I send a few of the Peas, which produce the black or intensely purple pods.—[3]

Pray believe me | Yours truly obliged | Ch. Darwin

Joseph R. Sakmyster (dealer)

[1] The year is established by the relationship between this letter and the letter from M. J. Berkeley, 7 March 1856 (see *Correspondence* vol. 6).

[2] Berkeley had experimented on the immersion in sea-water of seeds of different kinds of plants to determine how long they could remain immersed and still germinate. See *Correspondence* vol. 6, letter to M. J. Berkeley, 29 February [1856] and n. 3, and letter from M. J. Berkeley, 7 March 1856. The notice CD refers to is 'On the action of sea-water on the germination of seeds' (read 6 May 1856; *Journal of the Linnean Society (Botany)* 1 (1857): 130–40, *Collected papers* 1: 264–73). CD referred to Berkeley several times in his account of their joint experiments; see also *Origin*, pp. 358–60. CD's notes for and manuscript of this paper are in DAR 27.1 E and F. See also *Correspondence* vol. 5, letter to *Gardeners' Chronicle*, 21 November [1855].

[3] In his letter to Berkeley of 29 February [1856] (*Correspondence* vol. 6), CD mentioned a neighbour's 'curious pea with black pods'; Berkeley had suggested the difference might be due to 'mere variation and not to impregnation', and asked for a sample (letter from M. J. Berkeley, 7 March 1856).

To John William Lubbock 28 [June 1856 – January 1865?][1]

Down.—
28th

Dear Sir John Lubbock

I am very much obliged for your invitation to dinner on the 30th & I shd. have very much liked to have come, but my health has been so indifferent of late that I have resolved not to dine out for the next few weeks or months till I get better.—

With many thanks for your kindness, I am, dear Sir John Lubbock | Your's truly obliged | Ch. Darwin

The Royal Society (LUB: D26)

[1] The date range is conjectured from the letter's position in file in which it is contained at the Royal Society of London, and from Lubbock's death in 1865. The letter is at the end of its file, following the letter to J. W. Lubbock, 27 May [1856] (*Correspondence* vol. 6). Lubbock died 20 June 1865 (*DNB*); in his letter to CD, 22 and 26 March 1865, his son, John Lubbock, reported that his father was 'far from well'. John William Lubbock lived at High Elms in Down (*Post Office directory of the six home counties*).

From James Dwight Dana 8 September 1856

New Haven,
Sept. 8, 1856.

My dear Sir:—

I received your most welcome letter a few days before the meeting of our Scientific Association:[1] and as I should meet Prof. Agassiz[2] there, who could best answer your queries respecting the Mammoth Cave Animals, I concluded to defer my answer till my return. Here I am, back again, at last and I seat myself for a few words with you, socially and Scientifically.—

First as to the Mammoth Cave.— Professor Agassiz told me that the family to which the Fishes belong—the Cyprinodonts—was rather strikingly American.[3] With regard to the Insects, Dr John L. LeConte an Excellent Entomologist says that the genera of beetles are not American, but the same that occur in Caverns in Europe & elsewhere.[4] The genus of fly Anthomyia is common in Europe.[5] The Crustacean, *Astacus pellucidus*, belongs to that subdivision of the genus, (Cambarus, as it has been called), which is peculiarly American. Cambarus is made a distinct genus by some writers: the only difference is in the number of branchiæ: Cambarus has 17, on a side or one less than Astacus.— The Crustacean genus *Triura*, has not been found any where except at the Mammoth Cave. You may have seen some notice of the species of the Cave in the Amer. Jour. Sci., vol. xi, p. 127 (1851).—[6] Of the spiders I cannot speak definitely.— I would add respecting the genus Cambarus, that its Species are very numerous and widely spread over North America. Agassiz has collected a large amount of information on the peculiarities of the North American Fauna, but he has not yet embodied them in any work or article. One of the most interesting of our peculiar tribes, as you undoubtedly know, is that of the Gar-pikes, of which there are several genera & near two dozen known species occurring over the Continent between Cuba & the northern Lakes—and not represented elsewhere over the globe.— It is not to the point in view, yet I may mention here a fact of geological interest brought out by Agassiz at our Assoc. meeting a fortnight since. There were some young individuals, alive, shown, which had the tail of the Ancient Ganoids— That is, the vertebræ were actually continued to the extremity of the upper lobe—[7] This upper lobe, as here drawn, drops off as the animal grows & the fish then is of the modern type of form.[8]

It was a striking illustration of the principle discussed by Agassiz—that the earliest species in Geological history often have relations to the early or embryonic forms of existing species.—[9]

With respect to Genera represented by both northern & Southern species and not by tropical:—[10]

There are several which may be as correctly considered northern as southern, or the reverse.

Lithodes is a cold water genus, found about the South Extremity of S. America, and also in the N. Atlantic & N. Pacific. There are 2 known northern species, & 4 Southern. Still I should doubt the propriety of calling it a Southern genus. There is also another closely related genus described by Adam White[11] called *Echidnocerus* found near the mouth of the Columbia R.— I have not seen the species

(The Genus Eurypodius of Southern S. America is represented in Puget's Sound, by the Genus *Oregonia* (D.) differing but little.)

Cancer (Platycarcinus of Edw.) occurs about Britain, Eastern North America (northern part of the U.S.), also along the coast of Chili, and about New Zealand & Tasmania; but not in the Tropics. There are 5 species in the N. Temperate Zone, and as many in the South Temperate.

Atelecyclus is found about Britain (3 species), on the coast of Chili (1 sp.), and along Japan (1) *Peltarion* is a closely related genus found at the Falkland's.

Bernhardus, of the Paguridea, is represented quite largely in the North and South temperate zones and not in the warmer tropical zone, as far as known.

Homarus has two species in the North Atlantic, (one American & one European,) and one near Cape Town, in the South Atlantic.

Crangon is one of the *northern* genera not yet found south.

Portunus is eminently north temperate: but like *Cancer*, has species both in Britain, and in New Zealand & Tasmania; in no other parts of the globe as far as ascertained do these two genera occur together (see p. 1578 of my work)—[12] we may *say* that it is a northern genus, as the larger part of the known species are in the north. Yet it is possible that if there had been as much coastline in the South, there would have been as many Southern species. In such comparisons the amount of coast line—that is, the extent of possible habitats—is an important element.

I do not know that the facts here stated meet precisely the question you have under consideration: and if not, do not hesitate to put more enquiries, and I will do my best to answer them.— In my Chapter on the subject of Geog. Distribution I should not have used the word Kingdom as I have done.[13] Some other term for the great sections of the globe would have been better.

You may have noticed that in citing the chapter in the Amer. Jour. Sci., I changed the name *Nova Scotia Province*, to *Acadian Province*[14] I have always looked on my review of the Distrib. of Crustacea, as only a beginning, which the future would extend & correct—a first effort, as regards Crustacea, in a field that will yet afford great results. For want of a knowledge of the difference of temperature between the waters of the coasts of Europe & N. America Milne Edwards made poor work in his "Crustacés" (1834) in a chapter on the Distribution of Species;[15] and for a like reason, naturalists generally had failed to appreciate the wide difference, in water & land distribution on the same coasts. It is something to know that the plants and birds may have equatorial heat, when the fishes of the neighboring waters are in the temperate zone: & this at least is clear from my survey of the subject.—

As to fossil trees in the South Shetland's— Dr James Eights says (Transactions of the Albany Institute vol. ii, p. 64)[16]—(after describing the rocks of the Islands as basalt & a conglomerate, passing into a tufa (if I may judge from his description) and farther into "a fine argillaceous substance with an imperfect slaty structure and a Spanish-brown aspect" often containing zeolites especially near "the incumbent amygdaloid")— "The only appearance of an organized remain that I any where saw, was a fragment of Carbonized wood, imbedded in this conglomerate. It was in a vertical position, about $2\frac{1}{2}$ ft. in length and 4 inches in diameter Its color is black, exhibiting a fine ligneous structure; the concentric circles are distinctly visible on its superior end; it occasionally gives sparks with steel and effervesces slightly in nitric acid."—[17]

The part of your letter which afforded special pleasure was that alluding to your home group—[18] I cannot speak of mine as "large", yet I may as "good &

happy"—Fanny the oldest 10 years—Edward, 6 (nearly 7) and James, 3.—[19] I was also pleased to hear of Mr Lubbock's happiness.[20] A long life to them full of joys.— Please thank him from me for the pamphlets which he has sent. He is doing excellent work among the Entomostracas,[21] and dear *little* things they are.

—You query as to my whereabouts. My Professorship is here at New Haven where I have been living since the return of the Expl. Expedition after the first two years—in other words, since 1844—the year in which I was married at this place.

I have been interested recently in the "Embryogeny" of N. America, and before many months have past will send you a pamphlet on the subject.— That title sounds queer, yet is not inappropriate: I entitle the article "On the plan of development in the Geological History of N. America"—[22] I have not space or time for explanations now and therefore must ask you to wait a little before you set me down as a mere speculator.

With warm wishes for your welfare, I remain | Sincerely your friend | James D. Dana

Ch. Darwin Esq.

DAR 205.3: 269 (Letters), DAR 162: 38

CD ANNOTATIONS
0.1 New Haven . . . Cave.— 2.1] *crossed pencil*
2.2 Cyprinodonts] *underl pencil*
2.2 rather strikingly American] *underl pencil*
2.7 peculiarly American] *underl pencil*
3.1 It was . . . species.— 3.3] *crossed pencil*
4.1 With respect . . . reverse 5.2] 'See that I am correct about Crustacea N & S'[23] *ink*
8.1 Platycarcinus] *underl pencil*
14.1 I do not know . . . subject.— 15.11] *crossed pencil*
15.1 You may . . . same coasts 15.9] 'Silicified Tree Shetlands' *ink*
16.1 (Transactions . . . p. 64) 16.2] *cross in margin pencil*
16.7 a fragment . . . structure; 16.9] *double scored pencil*
16.8 vertical position] *underl pencil*
17.1 The part . . . Ch. Darwin Esq. 21.1] *crossed pencil*
Top of first page: 'Dana' *pencil*; '19'[24] *brown crayon*

[1] Letter to J. D. Dana, 14 July [1856] (*Correspondence* vol. 6). The meeting of the American Association for the Advancement of Science was held in Albany, New York, from 20 to 28 August 1856.

[2] Louis Agassiz.

[3] See letter to J. D. Dana, 14 July [1856] and n. 5. The information on fish from the Mammoth Cave, Kentucky, had already been given in J. L. R. Agassiz 1851. CD referred to Dana and Agassiz in his discussion of blind cave animals in *Origin*, pp. 137–9.

[4] John Lawrence Le Conte was recognised in the United States and elsewhere as the leading American entomologist (*DAB*).

[5] This sentence was added in the margin.

[6] For the work of Agassiz and others on the classification of the blind crayfishes (*Triura*), see Winsor 1991, pp. 107–15. Dana refers to J. L. R. Agassiz 1851.

[7] This fact was eventually published in J. L. R. Agassiz 1857.

[8] This letter was published to here, including the drawing, in *Correspondence* vol. 6. The remaining eight pages in DAR 162: 38 were subsequently discovered to be part of this letter.

[9] Agassiz's contribution to the theory that the development of a class of organisms as recorded in geological evidence was recapitulated in the development of the embryo was first published in J. L. R. Agassiz 1848. For CD's assessment of Agassiz's view, see *Origin*, pp. 338, 449–50. See also Lurie 1960, pp. 162–3, 285–91.

[10] In the letter to J. D. Dana, 14 July [1856] (*Correspondence* vol. 6), CD had asked for information on the geographical distribution of northern and southern species of Crustacea. See also *Origin*, pp. 374–81, and *Natural selection*, pp. 556–7.

[11] Adam White was a naturalist employed in the zoological department of the British Museum (*DNB*).

[12] Dana refers to his *Crustacea* (Dana 1852), volume 13 of the reports of the United States exploring expedition between 1838 and 1842. The page reference given is part of a discussion of the geographical distribution of Crustacea. This section of Dana 1852 is cited in *Natural selection*, p. 556; see also *Origin*, pp. 372, 376. The parts of Dana 1852 on the classification and geographical distribution of Crustacea were reprinted separately as Dana 1853a. There is an annotated, inscribed copy of Dana 1853a in the Darwin Library–CUL (see *Marginalia* 1: 178–9).

[13] See Dana 1852 and 1853a, p. 1554; Dana divided the world into the Occidental, European, Oriental, Arctic, and Antarctic kingdoms. See also *Correspondence* vol. 5, letter from J. D. Dana, [before 6 December 1855].

[14] Abstracts of Dana's report on the geographical distribution of Crustacea (Dana 1852) were published in the *American Journal of Science and Arts* (Dana 1853b and Dana 1854–5). For the reference to the Acadian province, see Dana 1853b, p. 325, and compare Dana 1852 and 1853a, p. 1564.

[15] The reference is to Henri Milne-Edwards and to Milne-Edwards 1834–40, 3: 555–91, 'De la distribution géographique des crustacés'. CD's annotated copy of Milne-Edwards 1834–40 is in the Darwin Library–CUL (see *Marginalia* 1: 579–81). CD had been strongly influenced by Milne-Edwards's work on classification in his own studies on cirripedes (see *Correspondence* vol. 4, Appendix II), and dedicated *Living Cirripedia* (1854) to him.

[16] In his letter to J. D. Dana, 14 July [1856] (*Correspondence* vol. 6), CD had asked whether any author had described fossil trees in the South Shetland Islands off the coast of Antarctica, and suggested 'Mr Eights whose writings I have never seen'. James Eights was an American naturalist who had accompanied a voyage of discovery to the South Shetland Islands and published a number of papers on the specimens he brought back (*American medical biography*, Stewart 1990). The work Dana refers to is Eights 1833–52.

[17] For CD's interest in this fossil wood, see CD annotations, and *Natural selection*, p. 579, n. 3.

[18] CD had written, 'As for myself I live a very quiet & retired life, with a large set of very happy & good children round me' (*Correspondence* vol. 6, letter to J. D. Dana, 14 July [1856]).

[19] Dana and his wife, Henrietta Frances, eventually had six children, of whom four survived. Edward Salisbury Dana became the leading mineralogist of the United States. (*DAB*, *ANB*.)

[20] CD had referred to John Lubbock's marriage to Ellen Frances Hordern in his letter to J. D. Dana, 14 July [1856] (*Correspondence* vol. 6).

[21] Entomostraca was the name used to refer to all crustaceans other than Malacostraca (Leftwich 1973).

[22] Dana's paper 'On the plan of development in the geological history of North America' was published in the *American Journal of Science and Arts* (Dana 1856); it was also published together with Dana 1855 as a pamphlet, 'On American geological history' (New Haven: Ezekial Hayes, 1856). There is an annotated copy of the pamphlet in the Darwin Pamphlet Collection–CUL. CD wrote on the back, '(Very poor Pamphlet)'.

[23] See *Natural selection*, p. 556.

[24] The annotation refers to a portfolio of notes on the distribution of animals. CD kept a number of such portfolios, amassed over a long period of time and since dispersed. The likely contents of some portfolios were reconstructed when some of CD's papers were catalogued in 1932 (see DAR 220: 13).

To J. M. Herbert 18 November [1856][1]

Down Bromley Kent
Nov 18

Dear Herbert,

It would be both a duty & a pleasure to me to give any suggestion to Lieut. Blakiston.[2] But I am at present thus situated: my wife is out of health & expects her confinement in a few weeks, & I cannot *possibly* receive anyone here or leave home.—[3]

The middle of January I shall be in London[4] and would have much pleasure in meeting Lieutenant Blakiston. Or if that be too late, I would write anything which might suggest itself to me; but to do this I sh^{d} wish to know to what branches of Nat. History, Lieut. Blakiston means to attend.

As far as general instructions are concerned, I could not improve those given by me in the "Admiralty Manual of Scientific Instruction."—[5]

I was very glad to hear of old friend Watkins.[6]

Farewell, my dear old friend | Yours most sincerely | Ch. Darwin

If you think fit you could forward this to Lieut. Blakiston.—

Sotheby's New York (June 1991); Remember When Antiquities (catalogue 26, 1992); Gerard A. J. Stodolski Inc. (1995)[7]

1 The year is established by the reference to Emma Darwin's pregnancy (see n. 3, below) and the allusion to Thomas Wright Blakiston's expedition in 1857 (see n. 2, below).

2 Blakiston evidently sought an interview with CD in preparation for his part in the scientific expedition led by John Palliser to explore British North America between Ontario, Canada, and the Rocky Mountains in 1857 (*DNB*). No letter from Herbert containing this request has been found.

3 Charles Waring Darwin, the Darwins' tenth and last child, was born on 6 December 1856 (*Correspondence* vol. 6, Appendix II).

4 CD was in London from 13 to 16 January 1857 (Emma Darwin's diary (DAR 242)).

5 Herschel ed. 1849; there was a second edition in 1851. CD had contributed the section on geology. CD offered some additional ideas for the geologist on Palliser's expedition in his letter to William Sharpey, 24 January [1857] (*Correspondence* vol. 6).

6 Frederick Watkins was a friend of CD's from Shrewsbury School and a member of the Cambridge dining club dubbed the Glutton Club by Herbert (*Shrewsbury School register*, *Correspondence* vol. 1, letter from Frederick Watkins, [18 September 1831] and n. 1).

7 The letter has been reconstructed from three sale catalogue entries. The letter as transcribed is believed to be complete.

From George Howard Darwin [28 November 1856][1]

[Clapham Grammar School]
Friday. | morning.

You must excuse bad writing in this page because I had an abominable quill pen

My dear Papa

You need not have sent me to School Yesterday because today M^{r}. P Says I am to stop up in the sick room all day.[2] Last night I left all my goodies in the sick room so I thought of course they would all be safe but this morning low and behold they are all gone except some ginger Bread and a little bit of cake. They took the letter into M^{r}. P. last night.

I like the Book I am reading now now very much it is called Ben Sylvester's word,[3] it is all about robbers & murdering that is to say a good deal about them. When once I get in steady work again, the time will soon fly. We have no chapel now and on Sunday some boys are going to Westminster Abbey and goodness knows where, all sorts of Churchs. M^{rs}. Morrey gave me a great peice of Toffey and I never eat a bit of it myself, and I only had one peice of ginger Bread.[4] I forgot both the pencills and the nibbs for my pens, so I shall have to buy some. I forgot to put to put away my painting things so I must tell you where to put them.

Pencil in pencil case in box
compass in compass case in ditto
pen and nibbs ditto
gold paint ditto
paintbox in the setting board cupboard on the middle shelf that is just below the setting boards.
pallate ditto
mug ditto
painting hankerchief ditto.
LOCK UP THE BOOK

I expect you have put my things in the wrong places but now you will be able to put them right. I wish I had stoped.

Unfortunately I could not get an inside place in an omnibus so I had a cold drive. I got up this morning with a bit of a headache but it soon went of.

The first bit of time will pass awfully slow. I have begun those thick trowsers, so I dont know what to say about flannel draws Mama will settle that I hope shall come to London to meet H. Hemmings.[5] I suppose Brooks came home very late last night.[6] Will you or Mama *write* to me *very soon*

Good bye I remain | your affec son | G H Darwin

Nucklebones come from sheep's ankles I suppose all four legs I have no pen

DAR 251: 2222

[1] The date is established by the fact that George, aged 11, started attending Clapham Grammar School in August 1856 (J. R. Moore 1977, p. 54), by the reference to Mrs Morrey, who left Down at the end of 1856 (see n. 4, below), and by references in Emma Darwin's diary (DAR 242). The only Thursday in 1856 on which Emma recorded in her diary George's returning to school was 27 November.

[2] The reference is to Charles Pritchard, headmaster of Clapham Grammar School (*DNB*). George had been ill while at home: for 25 November 1856 Emma Darwin's diary (DAR 242) records, 'G. very poorly'.

[3] Yonge 1856.

[4] Sarah Morrey had been cook to Sarah Elizabeth Wedgwood, CD's and Emma's aunt, who had died 6 November 1856 (Freeman 1978). The Darwin children, except for William, had visited Sarah's

former home in Down, Petleys, on 22 November for a farewell tea party with the servants, of whom they were fond (*Correspondence* vol. 6, letter to W. E. Darwin, 25 [November 1856]). Henrietta Emma Darwin later wrote, 'Mrs Morrey's gingerbread was like no other we have ever tasted before or since' (*Emma Darwin* (1915), 2: 106). Petleys was auctioned and the servants departed in mid-December 1856 (*Correspondence* vol. 6, letter to W. E. Darwin, 25 [November 1856]).

5 Henry Hemmings had been a servant of Sarah Elizabeth Wedgwood (Freeman 1978). After Sarah's death he moved to Staffordshire, where he remained in contact with the Darwins and Wedgwoods. In the letter to W. E. Darwin, 10 [December 1856] (*Correspondence* vol. 6), CD wrote that Hemmings would be in London in the latter part of December. In a letter probably written on 17 December 1856, George wrote to William that he had met Hemmings unexpectedly that day, presumably in Clapham, and added: 'On Monday I am going to London & on Tuesday you will come & I beleive you are to stop a night or two and there will be Hemmings who I know you will be glad enough to see & we will have plenty of larks and skramages with H.H.' (DAR 210.6: 13).

6 William Brooks was a servant at Down (F. Darwin 1920); he may have driven George to meet the omnibus. George's departure had been delayed for an hour by a nose-bleed (letter from Emma Darwin to W. E. Darwin, [1 December 1856], DAR 219.1: 13).

To [W. E. Darwin][1] [1857?][2]

Will you tell Mr Linton,[3] that I am sorry to have troubled him with so many messages; but I had always thought that canaries & goldfinches, when paired, produced fewer young-birds than the pure parents; & I was particularly curious to know the cause of this.— I am sure that fewness of young is generally the rule, as in common mule from Horse & ass,—& from pheasant & Fowl.— But I shall be very much obliged for any facts about the eggs either of hybrids or of first crosses.—[4]

Mem
DAR 210.6: 187

1 The letter is thought to be to William Erasmus Darwin because it was found among other letters to William from CD. See also n. 2, below.

2 The date is conjectured by the reference to Mr Linton and by the relevance of the subject matter to CD's work on *Natural selection*. The note may have been written while William and Henry Linton's son Sydney were fellow pupils at Rugby, between 1855 and 1857 (see n. 3, below); this is the only time there is a recorded connection between William and the Lintons. There were no Lintons at Cambridge while William was an undergraduate there (*Alum. Cantab.*). For CD's work on *Natural selection*, see n. 4, below.

3 Henry Linton was a contemporary of Hensleigh Wedgwood's at Rugby School (*Rugby School register*). William was at Rugby from 1852 to 1857 (*Rugby School register*, *Correspondence* vol. 6, letter to W. E. Darwin, 29 [October 1857]). Henry Linton's son Sydney was at Rugby from August 1855 until 1860 (*Rugby School register*, *Alum. Oxon.*).

4 The fertility of crosses between horses and asses, pheasants and fowls, and canaries and goldfinches, and the fertility of their offspring, are discussed in *Natural selection*, pp. 438–9 (see also pp. 431–2, 438). Bernard Peirce Brent is cited for the information on canaries; there is a letter from him to CD on the subject dated 23 October 1857 (*Correspondence* vol. 6). This section of *Natural selection* was completed in December 1857 (*Correspondence* vol. 6, Appendix III).

To Bernard Peirce Brent 7 February [1857][1]

Down Bromley Kent
Feb. 7th

My dear Sir

I am extremely sorry not to hear better news of your affairs— Your case is a *most* cruel one: it seems extra hard that you $sh^{d.}$ not have even your ancient Family relics.— I most sincerely hope that your London Lawyers may see daylight without ruinous expence.[2]

Very many thanks for the offer of the Cock, but as I have a Hen, & as the Bird is not quite full-grown, I think it is not worth sending. If ever you have occasion to write, I $sh^{d.}$ like to hear whether these silk Fowls can fly as well as other Fowls.—[3]

Do not trouble yourself at present to send the C. Gardeners.—[4]

I certainly $sh^{d.}$ be *extremely* glad of a German old fashioned Pouter: by a strange coincidence I was writing to $M^{r.}$ Tegetmeier this morning, & had asked him to apply to Carstang for one; but I have struck out the passage.[5] If you will be so kind as to enquire, I $sh^{d.}$ be very much obliged, & Hammond could send the Bird by enclosed address; sending the Bird alive in a Basket between 12 & 1 on Thursday to the Nags Head. He $c^{d.}$ send account to me direct.—[6] I suppose it $w^{d.}$ be best to specify that I would give only a few shillings for it. Will Hammond tell you truly whether it is a really a common German or Dutch Pouter?—[7]

I am ashamed to ask you in the midst of all your troubles to attend to these trifles; & I am sure that it is uncommonly kind of you to think at all about me & my work under your circumstances. It is more, I fear, than I $sh^{d.}$ do if placed as you are.

I am beginning to write my Book on Variation; but it will take me some two years to complete.—[8]

Let me repeat again my wishes for your success, & believe me | My dear Sir | Yours sincerely | Ch. Darwin

Richard Brent (private collection)

1 The year is established by the relationship between this letter and the letter to W. B. Tegetmeier, 6 February [1857] (*Correspondence* vol. 6); see n. 5, below.

2 Brent's legal problems have not been identified; however, see *Correspondence* vol. 12, letter from B. P. Brent, 2 September 1864.

3 Brent provided information to CD on poultry, and had also given him specimens (see *Correspondence* vol. 6, letter to W. D. Fox, 3 January [1856], and letter to W. B. Tegetmeier, 20 March [1856]). See *Variation* 1: 270 for CD's interest in the flight of silk fowls.

4 Brent was a regular contributor to the *Cottage Gardener*.

5 For CD's request to William Bernhard Tegetmeier for German pouter pigeons, see *Correspondence* vol. 6, letter to W. B. Tegetmeier, 6 February [1857]; the passage of the letter that CD struck out is in Manuscript alterations and comments. CD also refers to Philip Castang, a game dealer, who is listed in CD's Address book (Down House MS) under 'Pigeons', at Ship Tavern Passage, Leadenhall Market, an area of London associated with meat and poultry markets; see also *Post Office London directory* 1861.

6 Hammond is listed under 'Pigeons' as 'dealer' in CD's Address book (Down House MS), at 37 Kent Street, Borough; however, Hammond is not listed in the *Post Office London directory* and has not been

further identified. The address that CD enclosed would have been that of George Snow, Nag's Head Inn, 102 Borough High Street, Southwark. Snow was the Down carrier who made weekly trips to and from the Nag's Head public house, in south London (Freeman 1978). No mention of the purchase of a pouter pigeon about this time has been found in CD's Account books (Down House MS).

7 CD had previously ordered some German pouter pigeons but was not satisfied that they had been purely bred (see *Correspondence* vol. 6, letter to W. B. Tegetmeier, 11 February [1857]).

8 CD had begun writing his species book, or what he sometimes called his 'Book on Variation', in 1856, but he interrupted his work in 1858 to write an abstract of it, *Origin* (see *Correspondence* vols. 6 and 7, Appendix II). The first two chapters of the species book became *Variation*, and the remaining chapters were published posthumously as *Natural selection* (see *Natural selection*, p. 1).

To Hugh Falconer [7 March 1857][1]

Saturday

My dear Falconer

I have been thinking over the remarkable paragraph you read to me out of your paper;[2] and it occurs to me (though hardly worth mentioning to you) that what Owen might say in answer (and I think truly) is, that, according to *his* belief all the Purbeck Mammals are Marsupials and that it is not right to compare the teeth in two such distinct groups.[3] He might maintain that for the Placentata, or for his still more restricted group of Gyroencephala (or some such name)[4] that the law of the more general type of teeth held for the older mammals. I cannot remember whether the law was enunciated for all mammals; but if so, I should think in my ignorance that he might uphold it for a more restricted group.[5]

My dear Falconer | Yours most truly | C. Darwin

Copy
DAR 144: 26

1 The date is established by the reference to Falconer 1857, and the reference to CD and Falconer having met (see n. 2, below). In 1857, 7 March was a Saturday.

2 The reference is to the manuscript of Falconer 1857; the paragraph referred to probably corresponds to that on page 276 of the published paper (see n. 3, below). Falconer's paper was read before the Geological Society of London on 11 March 1857; it was published later in the year. According to Emma Darwin's diary (DAR 242), CD visited London from 4 to 7 March 1857; he apparently saw Falconer during his visit (see *Correspondence* vol. 6, letter to J. D. Dana, 5 April [1857] and n. 8).

3 In his published paper (Falconer 1857, p. 276), Falconer suggested that *Plagiaulax*, a fossil marsupial found in Purbeck, Dorset, furnished a 'crucial test' of the generalisation made by Richard Owen, William Benjamin Carpenter, and others, that the degree of specialisation within a group such as the Mammalia increased over geological time. In particular, he noted, these naturalists argued that the earliest Eocene mammals usually possessed the full complement of teeth, while forms characteristic of later times were 'remarkable for the special suppression of these organs'. In opposition to this view, Falconer argued that, while *Plagiaulax* was 'the oldest well-ascertained herbivorous mammal yet discovered', it represented 'the most specialized exception . . . from the rule to be met with in the whole range of the Marsupialia, fossil or recent'. CD evidently thought that Owen might distinguish between placental mammals, to which his rule definitely applied, and marsupials (see n. 5, below).

4 CD refers to Owen's classification of the Mammalia, in which the Gyrencephala included the placental mammals with more highly developed brains, except for humans (R. Owen 1857, p. 19). Owen's classification was announced in a paper read before the Linnean Society on 17 February and 21 April

1857; CD's letter to J. D. Dana, 5 April [1857] (*Correspondence* vol. 6), suggests that he had doubts about classifications based upon a single feature, in this case, brain structure.

5 In R. Owen 1852, cited in Falconer 1857, p. 276, Owen wrote (pp. 904–5):

> Examples of the typical dentition are exceptions in the actual creation; but it was the rule in the forms of Mammalia first introduced into this planet. . . . The *Chæropotamus* . . . and other ancient (eocene and miocene) tertiary mammalian genera presented the forty-four teeth, in number and kind according to that which is here propounded as the typical or normal dentition of the placental Mammalia.

R. Owen 1850, also cited by Falconer, is similarly ambiguous about whether the progressive specialisation of the typical mammalian dentition applies to marsupials or not: see p. 495.

To William Sharpey 9 April [1857][1]

Down Bromley Kent
Ap. 9th

My dear Dr. Sharpey

It is very puzzling trying to mark the Catalogues;[2] especially the entomological one. There are lots of good Books, but I fancy most are too special; I have marked a few of more general interest.—

I think we ought to have all Agassizs works.—[3] You will be surprised at two which I have marked doubtfully, viz Low's Domestic animals,[4] for I believe that the time is coming when all records of domestic variation will be admitted as most very important.— Macgillivrays British Birds (I have it myself) is, also, a special Book, but quite above the common run—[5]

I am delighted to hear of progress in arrangement of Library.—[6] With respect to candidates; Davidson from his Palæontological work & Sorby from that on cleavage & sea-bottom have the highest possible claims.—[7] Mr Beckles I think has, also, *very strong* claims.[8] I do not myself know Warrington Smyth's writings, but I have always heard him spoken of with much respect;—[9] I shd. think he might wait, if you are pressed.— Andrew Smith, whose Zoology of S. Africa I know well, has very strong & paramount claims for admission. in my opinion.—[10]

General Sabine in letter to Sir C. Lyell, forwarded to me, says I am on Government Grant Committee, of which I have never heard;[11] & both wish me to attend at next meeting in respect to Mr Beckles grant.—[12] This I cannot do, therefore I have written separate note, if you will be so kind as to read it to Committee.—[13]

Your's very sincerely | Ch. Darwin

Quentin Keynes

1 The year is established by the reference to the meeting of the Government Grant Committee of the Royal Society of London at which a grant to Samuel Husband Beckles was approved. See n. 12, below.

2 CD had evidently been asked by Sharpey, the secretary of the Royal Society, to recommend, probably from publishers' catalogues, books to be purchased for the Royal Society Library. See also n. 6, below.

3 CD refers to the zoologist Louis Agassiz.

4 Low 1845. There is an annotated copy in the Darwin Library–CUL (see *Marginalia* 1: 506–10).

[5] Macgillivray 1837–52. CD's heavily annotated copy is in the Darwin Library–CUL (see *Marginalia* 1: 551–7).

[6] The Royal Society Library was being augmented and rearranged following the removal of the society to Burlington House (Royal Society, council minutes, 5 March 1857).

[7] Thomas William St Clair Davidson and Henry Clifton Sorby were both elected fellows of the Royal Society in June 1857 (*Record of the Royal Society of London*, p. 476). Sorby pioneered the use of the polarising microscope in geology and established the cause of slaty cleavage in sedimentary rock (*DSB*). For CD's favourable assessment of Sorby 1856, see *Correspondence* vol. 6, letter to J. D. Hooker, 8 September [1856].

[8] Beckles was not elected a fellow of the Royal Society until 1859 (*Record of the Royal Society of London*, p. 477).

[9] CD refers to Warington Wilkinson Smyth. Smyth was elected a fellow of the Royal Society in 1858 (*Record of the Royal Society of London*, p. 476).

[10] CD refers to Andrew Smith. Smith was elected a fellow of the Royal Society in 1857 (*Record of the Royal Society of London*, p. 476). There is an annotated copy of A. Smith 1849 in the Darwin Library–CUL (see *Marginalia* 1: 759).

[11] The letter from Edward Sabine to Charles Lyell has not been found. Sabine was treasurer of the Royal Society. Although the grant voted by Parliament to be distributed by the Royal Society was only formally established as an annual payment in 1856, the Government Grant Committee of the Royal Society had been in existence since 1850, and CD had served as a member of the sub-committee that recommended a grant to Richard Owen in that year (MacLeod 1971b). In January 1857, CD had solicited a grant from the committee on behalf of William Freeman Daniell (*Correspondence* vol. 6, letter to J. D. Hooker, 17 January [1857] and n. 2). See also n. 12, below.

[12] The Government Grant Committee of the Royal Society recommended at a meeting held on 3 July 1857 that the Council approve an award to Beckles of £150 'for further prosecuting the search for fossil remains in the Purbeck Strata' (Royal Society, Council minutes, 4 July 1857). Lyell had encouraged Beckles in this work and had reported the initial finds to a meeting of the Philosophical Club of the Royal Society on 18 January 1857 (Bonney 1919, p. 134; see also *Correspondence* vol. 6, letter from Charles Lyell, [16 January 1857] and n. 2.)

[13] CD's note has not been found.

CD memorandum July 1857

July. 1857.—

I agreed to receive 2 per cent on the £250 from Ansted & Ransome; & in lieu of principal share which Co^{y}. may divide—the final winding up to be deferred for 4 years, i.e in 1861[1]

Mem
DAR 210.10: 23

[1] CD invested £250 in fifty shares in the Patent Siliceous Stone Company, owned by David Thomas Ansted and Frederick Ransome, in 1852. He made a loan to the company of £250 in 1853, and two separate loans, also each for £250, to Ransome in 1854 and to Ansted in 1855. CD by this memorandum recorded his intention to write off his investment in the company, and to reduce to 2 per cent the rate of interest on his loans to Ansted and Ransome, originally set at $3\frac{1}{2}$ per cent. Both men made interest payments at this rate between 1857 and 1861, but there is no evidence that any of the principal was ever repaid. See CD's Investment book (Down House MS), p. 59; and *Correspondence* vol. 5, letter to J. S. Henslow, 8 March [1853] and n. 1. For the later history of CD's dealings with

the company, see *Correspondence* vol. 12, letter from Frederick Ransome, 7 March 1864 and nn. 1 and 2.

To Charles Spence Bate 29 November [1857]

Down Bromley Kent
Nov. 29th

My dear Sir

Mr Lubbock[1] told me yesterday of a fact, observed, I believe, by a friend of yours, which interests me particularly; & I shd. be *extremely* much obliged if you could get me a little more information on the subject, namely the act of cross impregnation in some Balanus.—[2]

['I am really very curious' about 'the act of deep impregnation in some Balanus', including whether the 'recipient was a willing agent or adulterer, or whether it was a case of rape by act' if the specimens were under water all the time, and who was the observer.][3]

Incomplete[4]
Postmark: [30 November and 1 December 1857][5]
Sotheby's London (21 July 1992)

1 John Lubbock.
2 Bate's friend, Richard Bishop, provided answers to CD's questions (see *Correspondence* vol. 6, letter from Richard Bishop to C. S. Bate, 3 December 1857). In *Living Cirripedia* (1854), p. 102, CD had noted that in several specimens of the hermaphroditic barnacle *Balanus balanoides*, the penis was truncated and imperforate, yet there were perfectly developed larvae, indicating that cross-fertilisation must have occurred. He had not, however, observed the act of impregnation.
3 Bishop, in his letter to Bate, noted that the recipient was 'exerting its cirri, with more than usual energy, & gave evidence of the intruder being a welcome guest' (see *Correspondence* vol. 6, letter from Richard Bishop to C. S. Bate, 3 December 1857).
4 The original letter is complete and is described in the sale catalogue as being four pages long. The bracketed section is from the catalogue description. See also Bishop's reply (*Correspondence* vol. 6, letter from Richard Bishop to C. S. Bate, 3 December 1857), from which it is possible to reconstruct CD's queries.
5 The postmark is given in the sale catalogue.

From J. Burningham[1] [10 September 1858?][2]

Sir

I have examined the ants nest o⟨ ⟩ ⟨ ⟩ high ground but did not find any yellow ones— thare whare a grate many eggs b⟨u⟩t saw a very few ants either in the nest or on the outside the ceasd to carry eggs in about a week after you left[3] and soon after deserted that Part altogeather I mean the ⟨*2 or 3 words missing*⟩ is not ⟨*1 word missing*⟩ a⟨n⟩t to ⟨be⟩ seen those ⟨*1 word missing*⟩ yellow ⟨ ⟩s under the ⟨ ⟩ dont seem to have been disturbed the ants on the high ground have not collected any rubbish for a nest.

I know ⟨a⟩n Hor⟨ne⟩ts nest near the Caves[4] in an hollow tree part of it is quite open

Yours Obediently | J Burningham

Endorsement: 'Mr ⟨ ⟩ | Sep ⟨ ⟩ 10 ⟨ ⟩5⟨ ⟩'
DAR 160: 186 (fragile)

[1] Possibly James Burningham of Gatten, just north of Shanklin, Isle of Wight (Census returns 1851 (Isle of Wight Record Office)).
[2] The date is conjectured from the endorsement, and from CD's 1858 visit to the Isle of Wight (see n. 3, below).
[3] CD made notes on ants carrying pupae during a visit to the Isle of Wight from 16 July to 13 August 1858, when he stayed in both Sandown and Shanklin (DAR 205.11: 94; see also *Correspondence* vol. 7, Appendix II). However, he also made notes on the yellow ant, *Formica flava*, during visits to Hartfield, Sussex, between 11 and 14 July 1858 and between 30 July and 1 August 1860 (DAR 205.11: 90–3, 99–102), and during a visit to Moor Park, Surrey, on 23 and 24 July 1859 (DAR 205.11: 96–7). Some of these notes form the basis for his discussions of the 'slave-making instinct' in *Origin*, pp. 219–24.
[4] There are caves in the area south of Shanklin and in particular around Ventnor, Isle of Wight.

To [J. D. Hooker] [before 6 October 1858][1]

Is there any Dog in Spain closely like our English Pointer, in *shape* & size, & *habits*,—namely in pointing, backing[2] & not giving tongue.— Might I be permitted to quote Mr Borrow's answer to the query?[3] Has the improved English pointer been introduced into Spain?

C. Darwin

Incomplete
Tim Lewens (private collection)

[1] The recipient and date are established by the relationship between this letter and the letter to J. D. Hooker, 6 October [1858] (*Correspondence* vol. 7).
[2] Back: 'to follow the lead of a dog that "points," by falling into the same perfectly stiff or semi-cataleptic state' (*OED*).
[3] George Henry Borrow was a well-known writer who had lived and travelled in Spain for several years. In a biography of Borrow in which the above letter was reproduced, the author noted that CD sent the inquiry through Hooker, who passed it on to Borrow's friend, Thomas Brightwell (Shorter 1913, p. 318). CD cited Borrow's reply in *Variation* 1: 42, but Borrow's letter to CD has not been found (see *Correspondence* vol. 7, letter to J. D. Hooker, 6 October [1858] and n. 8).

To Hugh Falconer 15 October [1858][1]

Down,
Oct. 15th.

My dear Falconer

I mistook your former note[2] and thought you intended to let me know when you returned to London. Unfortunately I cannot come up till Tuesday.[3] Will this be too late? If it be, will you kindly send me one line by return of post, that I may not have my journey for nothing. If I do *not* hear, you may expect to see me on Tuesday morning, about 11·15 to ·30 A.M. As my health is so indifferent I am never certain of my movements;[4] but if I do not appear by 11.45 you will understand the cause; for *certainly* I will come if I possibly can. Shall you be in London on Wednesday if I should fail on Tuesday?

With hearty thanks | Yours most truly | C. Darwin

DAR 144: 34

[1] The year is established by the relationship between this letter and the letter to J. D. Hooker, 20 [October 1858] (*Correspondence* vol. 7). See n. 3, below.

[2] Falconer's letter has not been found.

[3] In 1858, the Tuesday after 15 October was 19 October. In his letter to J. D. Hooker of 20 [October 1858] (*Correspondence* vol. 7), CD referred to meeting Falconer in London on the previous day. CD wished to consult Falconer on the identification of skeletal remains of breeds of pigeon for his 'skeleton & Pigeon M.S.'; he had been working on the manuscript since August, and it later formed the basis for a discussion in *Variation* (*Variation* 1: 162–7; see *Correspondence* vol. 7, letter to T. C. Eyton, 4 October [1858] and n. 4, and Appendix II).

[4] CD's stomach was troubling him, and he was planning to visit a hydropathic establishment at Moor Park, which he did between 25 October and 1 November, to take the water cure (see *Correspondence* vol. 7, letter to W. E. Darwin, 15 [October 1858], letter to J. D. Hooker, 20 [October 1858], and Appendix II).

To J. D. Hooker [14 November 1858][1]

There are 84 *genera* common to Europe & marked x = extra-tropical or chiefly so.— Could there be **any** (certainly few I presume) genera added to these, equally extratropical, but found only in Asia or North America, for they would in fact come into same category.

These 84 genera (or more, if there are N. Asiatic or N. American genera) contain how many species in Australia? And of such species in these 84 genera, how many are endemic to *Australia* & how many *common to Europe* (& N. Asia or N. America)?

This would be **very interesting** to me, as it would show how many species had probably been there manufactured since glacial Epoch.—[2]

Of the species included in the 84 genera how many are annual? Is it a large proportion compared to the proportion in Europe? This bears on passage of the Tropics: but perhaps it would be best to take only the species which are terrestrial.[3]

In your list of species common to (Victoria & Tasmania) & Europe. Are those marked * **lowland** Indian plants;[4] I suppose so.— It would greatly add to interest of list to me, if you would append ⊙ for annual & some mark for aquatic & sub-aquatic. Was it not very difficult to eliminate the introduced plants.?

Mem
DAR 50: E55–6

JDH ANNOTATIONS
1.2 Could there . . . category 1.4] 'Haloragis, Microtis, &c see list on small paper' *ink*
2.1 These 84 . . . Australia? 2.2] 'I have added the numbers' *ink*
2.2 And of . . . *Australia* 2.3] 'I have added the number (approx) to each genus' *ink*
2.3 & how many . . . *Europe*)?] 'these are in the European lists of *species*!' *ink*
4.1 Is it a large proportion 4.2] *underl*; 'A great proportion of *both* are herbaceous weeds, but few are true annuals,— Consult Loudon[5] in which all will be found—' *ink*
5.2 **lowland**] 'Yes mostly.' *ink*
5.4 the introduced plants.?] *underl*; 'No I have very few dubious & have marked all that are *possibly* introduced' *ink*

CD note:
It would be interesting to see whether same genera or *orders* have varied into species *in* S. America & Australia. | What does Temperate India mean | Is not the *extra-tropical [*interl*] resemblance of *Cape* & SW Australia due to Glacial epoch as shown by character of Abyssinian Flora & on W. African Hills?[6]

[1] This is an enclosure that was sent with the letter to J. D. Hooker, 14 November [1858] (*Correspondence* vol. 7). It is a series of questions and comments on lists of European species found in south-west Australia and Tasmania, and European genera found in Australia; the lists were sent to CD with the letter from J. D. Hooker, 12 November 1858 (*Correspondence* vol. 7). These lists have not been found in the Darwin Archive–CUL; however, copies that CD had made are in DAR 50: E61–74. Hooker's replies to CD's questions were made on the same sheet as the questions and have been transcribed as annotations.

[2] CD discussed the possibility that plants from Europe had crossed the equator during the glacial period and colonised parts of the southern hemisphere in *Origin*, pp. 374–82; see also *Natural selection*, pp. 534–54.

[3] CD was told by Hugh Falconer that the damp heat of the tropics was destructive to perennials from a temperate climate (*Origin*, p. 378). In addition, CD made a distinction in *Origin* between the dispersal of terrestrial and aquatic plants (see *Origin*, pp. 386–8).

[4] According to CD, the plants from temperate regions that had crossed the equator would ascend the mountains as warmth returned, and be exterminated in the lowlands (*Origin*, p. 378). Hooker had discussed the possibility of plants passing through India and other countries when dispersing from Europe to Australia (see *Correspondence* vol. 7, letter from J. D. Hooker, 12 November 1858.)

[5] Hooker refers to Loudon 1841; there is an annotated copy in the Darwin Library–CUL (see *Marginalia* 1: 504–6).

[6] CD discusses the flora of Abyssinia and north-west Africa in *Natural selection*, pp. 551, 552 and n. 1; see also *Origin*, p. 375. In his letter to C. F. J. Bunbury, 21 April [1856] (*Correspondence* vol. 6), CD referred to Abyssinia as a 'channel of communication' during the glacial period. See also *Correspondence* vol. 7, letter to J. D. Hooker, 27 [November 1858], and letter from J. D. Hooker, 22 December 1858.

To James Paget 19 December [1858][1]

Down Bromley Kent
Dec. 19th

Dear Paget

I am heartily obliged for your present, but hardly feel worthy of it. I was so very much interested by many things which you told me, that you may be sure I shall read your Lectures;[2] but I shall not be able very soon, as I have several big borrowed books, which must be returned soon. Will you be so very kind as to remember me, if anything occurs to you, in regard to inheritance at *corresponding or rather earlier* ages; & in regard to constitution & complexion.—[3]

I wish I could give you any facts on your Chronometry of Life;[4] I am sure I have often met with striking facts; but I have disregarded such facts & *deviations* alone would have struck me.

You know of course that the same bird in state of nature further S. or N. lays eggs at different times; & not rarely 2 broods in the S. & only one in N. The degree to which they sit, varies under different temperatures; but I do not mean by this that eggs hatch at different periods.— Some Batrachians of same species are oviparous or ovoviviparous under different climates. The periodical shedding of wool in sheep is no doubt affected by tropical climate. The appearance of second teeth has been greatly affected & accelerated in our domestic quadrupeds. Certain breeds of fowls, acquire their perfect plumage slower than others, so that after the down they are apt to be almost naked.

I see Youatt does not seem to doubt that period of breeding has been accelerated with the general early maturity of our improved cattle.[5] Tessier gives 321 days as longest period & 240 as shortest period of gestation in cattle.[6]

I do not suppose these rough remarks will be of any interest to you, but I send them for mere chance

Pray believe me, with very many thanks | Yours sincerely | Ch. Darwin

Cocks can be distinguished from Hens, earlier in some breeds than others.

Wellcome Library for the History and Understanding of Medicine, MS 5703/28

1 The year is established by the relationship between this letter and the letter to James Paget, [15 October – 19 November 1859] (*Correspondence* vol. 7).

2 It is not known what present Paget had given CD. The discussion that CD refers to may have taken place at the 16 December 1858 meeting of the Philosophical Club of the Royal Society of London, which CD and Paget both attended (see Royal Society, Philosophical Club minutes). The 'Lectures' were probably Paget's *Lectures on surgical pathology* (Paget 1853). There is a heavily annotated copy of the first volume of Paget 1853 in the Darwin Library–CUL (see *Marginalia* 1: 658–61); it is not inscribed, so it is unlikely to have been the 'present' that CD refers to. Volume 1 of the two-volume work dealt with hypertrophy, atrophy, repair, inflammation, mortification, and specific diseases; volume 2 dealt with tumours. CD read the first three chapters and the last chapter of the first volume in early 1859 (see CD's reading notebooks (*Correspondence* vol. 4, Appendix IV, 128: 24), and his annotations in the back of the volume at CUL).

3 See *Correspondence* vol. 7, letter to James Paget, [15 October – 19 November 1859].

4 Paget delivered his lecture 'On the chronometry of life' (Paget 1859) at the Royal Institution of Great Britain on 8 April 1859. The lecture was intended 'to illustrate the law that the processes of organic life are regulated with a regard to time, as exact as that which is observed by them in respect of size and weight and quantity of material employed in them' (Paget 1859, p. 117).

5 A heavily annotated copy of William Youatt's *Cattle: their breeds, management, and diseases* (Youatt 1834) is in the Darwin Library–CUL (see *Marginalia* 1: 885–8). Youatt discusses 'The proper age for breeding' in Youatt 1834, pp. 526–7; see also *Variation* 1: 87.

6 CD probably refers to the note in Youatt 1834, p. 527, which mentions the observations of Alexandre Henri Tessier. He also cited the note in *Variation* 1: 87 and n. 58.

From William Charles Linnaeus Martin [1859–61][1]

Comments on Mr. Darwin's Grand Theory= (touching *Misuse*)

Mr. Darwin p. 338—says—"Agassiz insists that ancient animals resemble to a certain extent the *embryos of recent animals* of the same classes;—or that the geological succession of extinct forms, is in some degree *parallel to the embryological development of recent forms*. I must follow Pictet & Huxley in thinking that the truth of this doctrine is very far from *proved*— Yet I *fully expect* to *see* it *hereafter confirmed*, at least in regard to subordinate groups which have branched off from each other in comparatively recent times For the doctrine of Agassiz accords well with the theory of Natural Selection".—[2]

Now I cannot help thinking, from the little knowledge I have, that there is a glimpse of truth in the theory of Agassiz—& well indeed does it accord with the views of Mr. Darwin, who supposes improvement, by Nat. Selection in the multitudinous assemblage of Organic Beings.— The tendency is to progress by development,—*Natural selection* always benefitting the living creature.— The *forms*, which stop short in the progress of divergence & adaptation to the struggle of Life, become rarer & rarer & at last extinct.—

Among the surviving wrecks of ancient Life he reckons the Blind=Fish, *Amblyopsis*—& the *Proteus anguinus* (which is confessedly a sort of semideveloped Tadpole,)—as is the Axolotl,[3]—& some others— Still they do well for their destined abode, & mode of life—whatever hereafter may become of them.— Were the caves of Carniola[4] destroyed by any agency—& the deep waters in its natural dungeons dried up, or poured out over the land & so into the rivers, with all the Proteus train, what would be the result— would the survivors of the catastrophe put on new forms—or would all perish— if the latter & circumstances favoured not fossil preservation, they would pass away, & leave no trace hereafter of their once having existed— Has this not often been the case?—

—At p: 188—Mr Darwin speaks of that marvellous organ the *Eye*— & he supposes it to be under the jurisdiction of natural selection.— "We must suppose that there is a power always intently watching each slight accidental alteration, which under varied circumstances may in any ways or in any degree, tend to produce a *distincter image*— We must suppose each new state of the instrument to be multiplied by the million & each to be preserved till a better be produced, & then the old ones to be destroyed—" p: 189.—[5]

Look now at the beautiful eyes of the frog— look at those of the Tadpole— Look at the beaming eyes of the gazelle— reflect upon them in the embryo.—

But is there not an *arrest* in development balanced by & subervient to mingled contingences— The eye of the mole is not that of the eagle,—but what upon the Principle of Natural Selection is to prevent it so becoming;— Has the eye of the degenerated for *want of use*—[6]

It seems strange to us that carrying out his theory of Progress, according to the laws of Natural Selection Mr Darwin should entertain the views he appears to do, in his Chapter on *Use & disuse.*—[7] Might he not have have used the Term, "Non development from an embryotic or semiembryotic state", the contingencies of existence not needing the operation of the selective power inherent in the nature of organic beings?—

After reading so much on the benefits conferred by Natural Selection on organic beings, & their consequent advancement in those qualities which tend to render them more & more capable of maintaining the Struggle of Life, it occasions some surprise to learn, that under certain circumstances deterioration & unfitness are the result of certain casual circumstances.— For example,—that the non use of wings, once fully developed, entails a progressive reduction in the wings of succeeding generations until at length they become mere rudimentary appendages—[8] Per contra—Is it not that the rudimentary winged birds are in state of progression,—& nearer the unknown primeval source, than the fully winged birds— And if natural selection has attended more to the legs than the *wings*, is it not because the *latter* were rather neglected & left under arrest, the whole force for the benefit of the race being directed to the other limbs—?—

Birds tenanting oceanic islands are usually great flyers.— But the Dodo & Solitaire feeding upon the shore had no need of developed wings, & the force of Nature was elsewhere directed.— The non=use of a bird's wing is the result not the cause of a rudimentary condition & on M^r^. Darwin's principles a gradual organic development ought to take place till the bird is capable of flight.— Is it not the strife for wings that we see in the Apteryx the Cassowary, the Rhea, the Emeu &c

How many wingless birds have passed away— New Zealand & Madagascar present us with their fossil or semifossil remains, & but for these who would have dreamed of their having existed?— Does not this shew that wingless birds were the earliest or among the earliest of the descendants of the unknown primeval root?— M^r^. Darwin says, "As Professor Owen has remarked, there is no greater anomaly in nature than a bird which cannot fly."[9] It is bold to differ from so great a philosopher—but we venture with submission to think the contrary— It is because we usually see birds with wings & are accustomed to consider birds as feathered flying creatures, that we come to settle it as fact that birds are essentially winged—(albeit in some the wings are fitted only for aquatic progression).—

But were we accustomed to see Birds without wings, or only with rudimentary wings (& great is the number of such) should not think a bird of flight an "anomalous" creature— Among mammalia do not the Bats & Galeopitheci startle us?— However they acquired flying membranes (or as in the case of certain squirrels & Phalangers) ample parachutes, certain it, that as these organs are beneficial, so they do not deteriorate,—perhaps improve.[10] And when we contemplate a Bird without wings, instead of attributing their loss to Disuse may we not rather on Mr Darwins theory believe that something has produced an arrest, & that the time has not yet arrived for their full development— With respect to the malformed creatures of domestication,—as upright or Penguin Ducks,—otter sheep,—two legged Pigs, & the like[11]—we cannot bring them into the argument— they are artificial not natural beings—although they may propagate their breeds.—as do Lop-eared Rabbits—Fowls destitute of a tail, or with reversed feathers, or with silky plumage & a black periosteum.—[12]

Turning for a moment to Insects—some of which as *Scarabæus*, *Ateuchus* &^{c}, want the anterior tarsi,—while others, with undivided *Elytra* have the membranous wings in a more rudimentary condition, is it necessary for us to have recourse to non use as a cause?=[13] Is non=use so influential?— it may affect the Individual—but can it affect a whole race— And after all, what induced non=use in the beginning?— Deficiency of development,—to be modified & bettered by the agency of Natural selection—at all events as soon a maze of contingencies call for the operation of this principle Of course we follow M^r. Darwin's Theory.—

With respect to the eyes of moles & such like underground creatures—fitted for their habits,—natural selection is not likely until, circumstances gradually operate to do more than make them efficient delvers.— In the case of the Cave rat we verily believe that if a nest of blind young were reared in a room their eyes would be day seeing or twilight=seeing—& in a few generations as good as the eyes of rabbits or hares.—[14]

From what has been said—& it is impossibble here to enter at large into the subject, it will be seen that we hesitate to accecpt the theory as to the permanent influence of non use, in the *establishment* of Imperfection.— It may cause imperfection, nay it does so, for our rabbits cramped up in small hutches, wherein they have been born & bred are very small & feeble in the hinder limbs— But turn them out into the coppice or upon the heath, there let them breed & then see if their young are not as able with their hinder extremities as any of the old true wild stock around them.—[15] We must be pardoned then, if we cannot subscribe to the opinion that *disuse* unites the elytra of beetles into one shard,—curtails their wings & *necessarilily* renders them creepers under stones—[16] Are we not changing cause & effect?— We must not forget that the females of many moths are wingless—as is the female of the glow-worm—that the Gnats which dance in the air are males,—plumed, but not armed with a keen suctorial instrument, as are the females— That the male crickets & grasshoppers alone are stridulous,—the females wanting the necessary apparatus— Has disuse removed the wings from the moths in question—the piercer from the male Culex,—the stridulous apparatus from the female Cricket— If this theory of disuse or Non=use be sound, surely it ought to apply in such cases—no less than in others.— Can *Deterioration* be consistent with the *Great Principle of Natural Selection*.—

Let us pass over these points, & again revert to Wingless or rather Brevipennate Birds— Had we *only seen such*—& we suspect (on M^r. Darwin's system) that such were the primeval creatures out of which Birds were fashioned by almost insensible stages,—should we not regard— should we *not have regarded* (supposing our existence in those remote periods of times,) a winged bird hovering in the air as a terrible monster.—

Does it follow that all birds have advanced *pari passu*?— May not many an arrests or stoppages have occurred from a combination of contingences, leaving vast numbers of birds, not advanced up to *the mark* in *alar* development, whose brevipennate descendants, surviving the lapse of Time remain to tell a story of strange mutations & undreamed of phases on the surface of our planet, from the Silurian epoch, even to the present— But alas we we know not the language—no nor can we decipher the mystic hieroglyphics,—the lithoglyphs of a world gone by.—

But it is time to conclude a paper perchance unworthy of consideration= Fearing indeed that this may be the case we will not draw it out to a tedious length—at the same time, we have ventured upon a sort of imperfect & very crude diagram, explanatory of our views, Birds alone being its subject— What bears upon one great division of Organic Being, bears, so we venture to opine, upon every other.—

W. C. L. Martin.—

[Enclosure 1]

As may easily be perceived, this is a very rude "*Ebauche*" a mere rough sketch, (on rough paper), intended to convey an imperfect, & as yet, not maturely studied idea of the development of the leading forms of birds divaricating from a primeval, extinct, & of course unknown root.— For its roughness & imperfection I must apologize,—indeed I ought to have recopied it—but I am not in health—& every trifle is a labour— Should you however, deem this rough sketch not altogether worthless, I will copy it upon more sightly paper, with such improvements & alterations as further reflexion may suggest.—

I throw the terrestrial brevipennate Birds on one side,—the brevipennate & finwinged aquatic birds on the other side,—as diverse branches from the great stem— Birds are not related to each other simply because they are brevipennate—

I suppose each circle (or most of the circles) to represent a group which in the wear of time has proved its stability & contended successfully against contingences in the great struggle of Life— I suppose them to have given off successors, still better fitted for this struggle according to the alteration of influencing conditions, & taking the place of their effete predecessors, and so the progression of change & development as I concieve to go on—group succeeding group—& branching into families & genera, often multitudinous & complex in their affinities—till at last we come to the artificial families & genera of modern ornithologists, whose systems are contradictory & based upon grounds often untenable—seldom philosophical— Thus we have the fanciful systems of the Trinarians, the Quinarians, the upholders of representation[17]—& so on to Prince C. Lucien Bonaparte, Gray, & others,—strenuous maintainers of their own systems, & each system at variance with every other.—[18]

Between the rude circles I suppose a long but variable period to intervene—a period mostly perhaps to be measured by ages—geological ages,—for the modifications must be slow— Between these circles I have placed nothing—but the circles represent forms of longer or shorter duration— The small dots, like flower stamens,

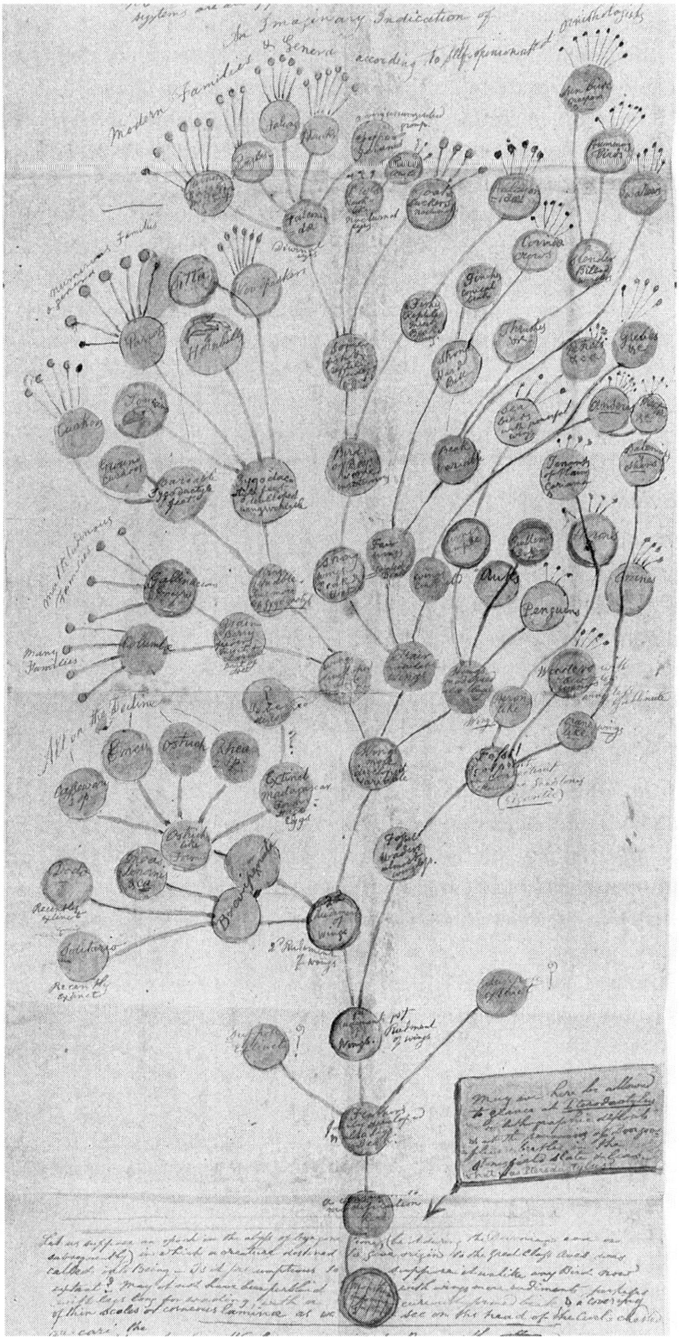

Letter from William Charles Linnaeus Martin, [1859–61], Enclosure 2

convey an idea of the ornithological multitude of families & genera—which please modern naturalists & look well on paper, with the appendix of their names to the genera they would fain establish— Great is the rage for genus making— I have known the male & female of an Ortyx (American Quail) made not only into two species, but each into the type of a distinct genus—(Vigors)[19]—& similar blunders are constantly occurring—

It will be easily percieved that I have no great respect for any ornithological system as yet promulgated— Not one with which I am acquainted but is a baseless fabric—built up by caprice, & fleeting opinion— In my rude sketch I aim at none—& follow none— My object is to elucidate the great Theory of natural selection, & to shew that the rudimentary condition for wings does not arise from *Disuse* but is one of Partial or arrested Development perhaps necessitating at no distant date the extinction of the survivors of a wingless multitude some of whose relics have been recently brought to light;—while those of others will perchance never be recovered.—

[Enclosure 2]

No two Ornithologists maintain a similar system; their systems are all opposed to each other,—& capriciousness reigns throughout.—

An Imaginary Indication of Modern Families & Genera according to self-opinionated Ornithologists

[Transcription][20]

Monstrum informe ignotum[21]

a considerable modification here May we here be allowed to glance at *Pterodactylus* in lithographic deposits[22] & at the remains of Dragonflies & beetles in the stonesfield slate & lias. What was Pterodactylus?

–Feathers fairly developed also legs & beak
—suppose extinct?
—1st Rudiments of Wings
——2^{d} Rudiment of Wings
———Brevipennate
————Solitario[23] Recently extinct
————Dodo Recently extinct
————Moas. Dinornis & **ca**
————Ostrich like Forms
—————Cassowary 2 sp. }
—————Emeu } All on the Decline
—————Ostrich }
—————Rhea 2 sp. }
————Extinct Madagascar Forms—Huge Eggs
————To be discovered?

——Wings more developed Variable
———wings developed Legs ⟨variab⟩le
——— Grain & Berry feeders. Insects & Larvæ part of diet
———— Columbæ
————— Many Families
———— Gallinaceous Groups
————— Multitudinous Families
——— Wings variable—Toes more legs zygodactyle
———— Variable Zygodactyle feet
————— Cuckoos
————— Ground cuckoos
———— Zygodactyle feet developed wings variable
————— Toucans
————— Parrots
—————— Numerous Families & genera
————— Hornbills
————— Woodpeckers
—————— Sitta
———/Fair/ available wings
————strong wings, Beak & Feet
—————Birds of Prey. some insectivorous
——————Some fish & reptile feeders Insects also
———————Falconidæ diurnal eyes
————————abnormal Birds of the Falcon group
————————Eagles
————————Falcons
————————Hawks
————————Gypogeranus a very circumscribed group.[24]
———————Owls Hawk-like Nocturnal eyes
————————Eared owls
———————Goat-suckers nocturnal
——————Fishes Reptiles Insects Beak strong
———————Halcyonidæ
————Fair wings Pointed Beak
—————Strong Hard Bill
——————Finches Conical beak
——————Corvidæ Crows
—————Beak variable
——————Thrushes &c
———————Slender Billed warblers
————————Sunbirds Creepers
————————Humming Birds
——————Swallows

———wings contracted
————Grebes &c
————*Anseres*
—————Phaeton &c
———Wings modified also legs
————wings ample
————Sea birds with powerful wings
—————Petrels &c
————Auks
—————Guillemot Puffin
————Penguins
——Fossil Waders almost wingless
———Fossil footprints Connecticut Permian Sandstones (Gigantic)
———Heron-like *Wings*
————Herons
—————Tenants of Plains Cariama
———Waders with Powerless wings wingless Gallinule
———*Crane-like. wings*
————Cranes
—————*Balæniceps & others*
—suppose extinct?

Let us suppose an epoch in the abyss of bygone time, (be it during the Devonian era or subsequently) in which a creature destined to give origin to the great Class Aves was called into Being— Is it presumptuous to suppose it unlike any Bird now extant? May it not have been purblind with wings mere rudiments, perhaps with legs long for wading, with a curiously formed beak, & a covering of thin scales or corneous laminæ as we see on the head of the Curl-crested aracari, the hackle feathers of Sonnerats Junglefowl & on the back of the young ostrich. After a long lapse of time may we not suppose these corneous laminæ to be resolved into a shaft & barbules more or less close, & rigid, & so on by degrees till true plumage is acquired?

We have no definite Proof of the existence of *Birds* or *Mammalia* during the long long travel of Time, whilst the Devonian & Silurian strata were in process of formation

Nevertheless, *Red* & *warm-blooded* vertebrata may even then have existed, few, strange, undreamed of by the Naturalist—perhaps so unlike any thing which we now see, as to startle imagination could they revisit our planet.

Devonian & Silurian Transition Formations.

Devonian. Plants—Corals, Crinoidea, mollusca shelled—Cephalopoda

Crustaceans—Trilobites. Fishes &^{c}—occasional layers of coal, with *[one word illeg]* impressions.—

Silurian. Corals—Mollusca—Trilobites, Crinoideæ, Spirifers, Orthocerata

Fishes &^{c} Strata of marine origin—

Volcanic & Hypogene Primary Rocks We here lose all positive proof of the existence of Organic Beings, but it would *be rash* to *fix* the *first creation* of *Plants* & *animals* at the precise point where our retrospective knowledge happens to stop.—

DAR 171: 56/1–15

CD ANNOTATIONS
7.3 Has the eye ... *use?*— 7.4] *two crosses in margin, pencil*
7.4 degenerated] *cross added before, pencil*
9.7 Per contra ... progression,— 9.8] *cross in margin, pencil*
10.3 The non=use ... condition 10.4] *cross in margin, pencil*
11.3 Does not ... root?— 11.5] *cross in margin, pencil*
11.7 but we venture ... winged— 11.10] *cross in margin, pencil*; 'Whole shape of Birds, *[one word illeg]* hollow Bones &c' *pencil*
13.6 Deficiency ... selection 13.7] 'I agree doubtful | Eyes in *[*earlier*]*' *pencil*
15.4 wherein ... limbs 15.5] *cross in margin, pencil*; *cross over* 'small', *pencil*
15.15 Has disuse ... question] *two crosses in margin, pencil*
16.2 Had we ... monster.— 16.6] *cross in margin, pencil*
17.8 by.—] *cross after, pencil*
Verso of enclosure: 'W. C. L. Martin *abstracted* Diagram' *pencil*

[1] The date range is established by the references to *Origin* or *Origin* 2d ed., published in 1859 and 1860, respectively; the page references given by Martin could not be to the third edition, which was published in 1861.
[2] *Origin*, p. 338; the emphasis was inserted by Martin. CD referred to Louis Agassiz, François Jules Pictet de la Rive, and Thomas Henry Huxley.
[3] Agassiz's views on the North American 'blind fish' (now called the cavefish, *Ambyopsis*), and the European *Proteus*, a blind cave salamander that retains gills and tail fin throughout its life, are discussed in *Origin*, p. 139. CD also mentioned the blindness of *Proteus* in *Journal of researches* 2d ed., p. 52. Martin also refers to the axolotl (*Ambyostoma*), a salamander from Mexico that retains its gills and tail fin when an adult. See also *Natural selection*, pp. 295–7.
[4] Carniola is a mountainous region, once part of the Austrian crown lands; in the southern part (the Karst Mountains) there are a great number of caves and subterranean streams (*EB*). It is now in Slovenia. CD mentioned the caves of Carniola in *Origin* 3d ed., p. 154, rather than the caves of Styria mentioned in the first edition (see, for example, *Origin*, p. 137). CD also mentioned the caves in Carniola in an addition to *Origin* 3d ed., p. 155 (see also Peckham ed. 1959, p. 284). Andrew Murray had provided CD with information and references on the Carniola caves (see *Correspondence* vol. 8, letter from Andrew Murray, 3 May 1860, and *Origin* 3d ed., p. 156).
[5] *Origin*, p. 189. The emphasis was inserted by Martin, who in the first sentence omitted the words 'in the transparent layers; and carefully selecting each alteration' between 'accidental alteration' and 'which under'.
[6] See *Origin*, pp. 137, 473; see also CD annotations.
[7] Martin refers to the section 'Effects of use and disuse' in chapter 5, Laws of variation (*Origin*, pp. 134–9).
[8] *Origin*, p. 134. See also *Natural selection*, p. 290.
[9] *Origin*, p. 134. The reference is to Richard Owen.
[10] The flight of squirrels, bats, and *Galeopithecus*, the flying lemur, is mentioned in *Origin*, pp. 180–1.
[11] Martin discusses the penguin duck and the otter sheep in W. C. L. Martin [1849a] and W. C. L. Martin [1849b]. The penguin duck, according to Martin, waddled 'in an upright position like the penguin, owing to the unnatural position of its legs'. He added: 'Such varieties are of no value; they are the result of accidental malformation transmitted from generation to generation' (W. C. L. Martin [1849b], p. 589). CD discusses the penguin duck in *Variation* 1: 277, 279–80, 282, and

had kept them himself (*Variation* 1: 279; see also *Correspondence* vol. 10, letter to *Journal of Horticulture*, [before 27 December 1862]). The otter sheep was a North American breed of sheep, with a long body and short legs: Martin said that it was 'evidently an example of hereditary malformation' (W. C. L. Martin [1849a], p. 73). CD discusses it in *Variation* 1: 100, where he calls it a semi-monstrous breed. CD mentions a race of two-legged pigs in *Variation* 2: 4.

12 Lop-eared rabbits are discussed in *Variation* 1: 105–6 and 2: 301; rumpless fowls, frizzled or caffre fowls 'with the feathers curling backwards', and silk and sooty fowls, both of which breeds have the skin and periosteum black, are listed in *Variation* 1: 230, and discussed on the following pages. Martin himself discussed the 'Persian or rumpless fowl', the 'frizzled or Friesland fowl', the 'silky fowl', and the 'black fowl' in W. C. L. Martin [1849b], pp. 513–14. He saw the first three as interesting but not necessarily useful products of domestication.

13 *Scarabaeus* and *Ateuchus* are genera of beetles. CD suggested that the absence of anterior tarsi in *Ateuchus* was due to disuse, and discussed the wingless beetles of Madeira, in *Origin*, pp. 135–6. See also *Natural selection*, pp. 291–4.

14 The eyes of moles and other burrowing or cave-dwelling creatures are discussed in *Origin*, pp. 137–9. See also this volume, Supplement, second letter from W. C. L. Martin, [1859–61]. CD added to his discussion of vision in the cave-rat (*Neotoma*) in *Origin* 3d ed, pp. 154–5; see also *Correspondence* vol. 8, letter to Benjamin Silliman Jr, 4 December [1860].

15 CD discussed changes in limb length and weight in domesticated rabbits in *Variation* 1: 124.

16 See *Origin*, pp. 135–6.

17 The quinarian system of classification, proposed by William Sharp Macleay and his followers, arranged organisms into groups and subgroups, each of which contained five forms whose affinities to one another were such that they could be depicted in the form of a circle. Macleay believed that such a regular scheme existed in the natural world and was evidence of God's plan of creation. See Winsor 1976, pp. 83–7, and Ospovat 1981, pp. 101–13; see also Stevens 1994 and McOuat 1996. For CD's discussion of classification in *Origin*, see *Origin*, pp. 411–35. The trinarians and upholders of representation have not been identified.

18 Martin refers to Charles Lucien Bonaparte and probably to John Edward Gray. Bonaparte worked extensively on the classification of animals: he divided the birds into two subclasses, based on whether or not they were monogamous and whether the chicks were capable of feeding themselves from birth (*EB* s.v. Ornithology (Taxonomy); *DSB*). Gray was keeper of the zoological collections at the British Museum from 1840, and compiled or edited the catalogues of the department of zoology. His later work was criticised because of what was thought to be the unnecessary number of genera and species that he introduced. (*DNB*.) For more on systematics in the eighteenth and nineteenth centuries, see Stevens 1994 and McOuat 1996.

19 In 1830, Nicholas Aylward Vigors had discussed his observations on the genus *Ortyx* before the Zoological Society of London (*Proceedings of the Zoological Society of London* 1 (1831): 2–4). The *Proceedings* reported: 'To these nine described species, he added two others apparently new to science, and which he characterized under the names of *Ort. neoxenus* and *affinis*; stating at the same time his doubts whether both might not be the females or young males of the imperfectly known species *Ort. Sonninii* or *cristatus*.'

20 Martin's sketch, a genealogical tree, has been transcribed starting at its base, or root: each indented item or block of indented items represents a branch or branches stemming from the nearest less-indented item above it. The sketch is reproduced on p. 403 at 27 per cent of its original size.

21 'Monstrum informe ignotum': formless, unknown monster.

22 The *Pterodactylus* was named and its true nature described by Georges Cuvier in 1809 (*EB*). On the discovery and subsequent history of work on the pterodactyl, see Seeley 1901, pp. 27–36. See also A. Desmond 1982.

23 Martins refers to the solitaire.

24 *Gypogeranus* is a synonym of *Sagittarius*. *Sagittarius serpentarius* is the secretarybird (see G. R. Gray 1841 and *Birds of the world* 2: 206–15).

From W. C. L. Martin [1859–61][1]

& that their loss, or semiobliteration has resulted first from a combination of circumstances, bringing Disuse into play as an *active* agent,= 2dly by a perpetuation of the misfortune by inheritance,—till what was first accidental becomes normal— Now if the darkness of habitual residential places,— (burrows deep caves & the like), produces this effect in certain cases, why not in all where the like causes prevail?— But such is not the case.[2]

Look for example at your little Tuco=Tuco—(*Ctenomys Brasiliensis*)—[3] this subterranean creature is not blind, although it often loses its sight by disease—(as the English soldiers in Egypt lost theirs from the irritation of impalpable sand)—nor is our mole,—although, if I remember rightly its optic nerve is either rudimentary or wanting, a fibril of the 5th pair appearing to supply its place— neither is the Hamster blind= But the Diplostoma which lives under ground, on roots, in the plains of the Missouri, appears to be so,—its eyes being covered by hair—[4]

In the Bathyergus, of the Cape, the eyes are uncovered, though very minute—[5]

In the Mole-rat of Asia minor—& the steppes of Tartary—(*Asphalax Typhlus*, Desmarest) the specks representing eyes are concealed beneath the skin—this subterreous creature is truly sightless—[6]

If we turn from these rodents to certain burrow dwelling Insectivora we shall find that eyes, though often very minute are not wanting— we may instance our shrew mouse (*Sorex*)—the Desman (*Mygale* G. Cuv:) of Muscovy, (also Pyrennees) in the *Talposorex* of N: America in the Cape *Chrysochloris*, & the Condylura of N. Amer. &^{c}.—[7]

Now may we not rather attribute the non=developement or even nullity of eyes in these & other instances, to *non=necessity* for them, their presence in perfection being of no service, than to Disuse, which latter seems to me to say that eyes *were once* in perfection—whereas *I think* they have been *stinted* or *arrested* in *their development*, nature never wasting her resources,—& therefore stopping where her end is accomplished.— I am not however dogmatic on this point— Much may be said on both sides.—

But let us return to your Tucotuco (so interestingly described in your Journal[8] a work more delightful to me than any work of the kind I have ever read, & a model for travelling naturalists)— Let us, I say return to your Tucotuco— We learn (& I remember examining the specimen dissected by Mr Reid)[9] that its thigh bone is destitute of the Ligamentum teres,— accounting for its queer movements.— Now this ligament is wanting in very few mammals,—(*but I am not aware of its absence in any birds*) among which are four of very different character—viz the Orang, & the Elephant—The Seal tribe & the 3=toed Sloth. (I have mislaid my notes on this subject, & therefore hesitate to name any others)—

What are we to say respecting this *absence*, in such diverse examples.— now Animals as ponderous & lourd in motion as the Elephant, as good climbers as the orang & as efficient burrowers as the Tucotuco, do possess this ligament (so needful in man) thoroughly developed— Surely we cannot attribute these respective deficiencies to *Disuse*— Is it not because there is here no need of this ligament, or perhaps that it would be worse than unserviceable that Nature has not given it?— Is there not a negation in nature, when a priori, we should not anticipate (scrutiny alone teaching us) such a structural peculiarity.— Consider the double cæcum of Birds (present & single in man the Orang & one or two more)— Does not anatomy reveal strange facts? structural facts which we cannot, as I think attribute to disuse—& totality of anatomical structure must be the basis of our speculations— Can *Disuse* thoroughly alter the great points of anatomy— or take away the cells of the camel's stomach in that of the ox?— In this latter instance has natural selection obtained & perpetuated them?—that is in the Camel— most probably—

But then what shall we say to the Ligamentum teres?— Has Natural Selection adopted it in the majority of mammals, & neglected it in the few;—or has Disuse, or the non=necessity of it operated so as to prevent its development= Is Natural selection confined only to adoption? or does it involve rejection?—

AL incomplete
DAR 47: 211–13

CD ANNOTATIONS
1.1 & that . . . the case 1.6] 'Ch VII' *red crayon*
1.1 & that . . . pair appearing 2.5] *crossed pencil*
6.1 Now . . . sides.— 6.7] 'Ch VI W. C. L. Martin' *red crayon*
9.1 Has Natural . . . rejection?— 9.4] *scored red crayon*

1 The date range is established by the references to *Origin* or *Origin* 2d ed., published in 1859 and 1860, respectively; the page references given by Martin could not be to the third edition, which was published in 1861.

2 CD discussed the effects of use and disuse in *Origin*, pp. 134–9.

3 *Ctenomys*, the tuco-tuco (Nowak 1999), a South-American burrowing rodent, is described in *Origin*, p. 137, as an example of blindness 'probably due to gradual reduction from disuse, but aided perhaps by natural selection'. CD also discussed the tuco-tuco in *Journal of researches*, pp. 58–60.

4 Martin refers to the sight of several burrowing animals, including the genus then called *Diplostoma.*

5 Martin refers to *Bathyergus*, the dune mole-rat (Nowak 1999, 2: 1640–1).

6 Martin refers to what was then called *Asphalax* or *Aspalax*, and is now called *Spalax*, the Ukrainian blind mole-rat; this genus includes a species from eastern Ukraine and south-eastern Russia (Nowak 1999, 2: 1426–7). CD referred to the blindness of 'Aspalax' in *Journal of researches* 2d ed., p. 52. See also Keynes ed. 2000, pp. 165–6, 231.

7 Martin refers to several insectivores by their contemporary generic names, including the Russian desman (now *Desman*), the Pyrennean desman (now *Galemys*), and the *Condylura*, or star-nosed mole (Nowak 1999, 1: 230–3, 242–3).

8 *Journal of researches*, pp. 58–60.

9 The dissection of the tuco-tuco by Mr Reid is described in *Journal of researches*, pp. 59–60; in a letter to Jeffries Wyman, 3 October 1860 (*Correspondence* vol. 8), CD said that the dissection of the eyes had been carried out by 'a clever young Surgeon'. He was presumably the James Reid who contributed 'Notes on several quadrupeds in Mr. Darwin's collection' to the *Proceedings of the Zoological Society of London* 5 (1837): 4, but these notes did not mention *Ctenomys brasiliensis*. Reid has not been further identified.

From Charles Lyell 4 October 1859[1]

Oct. 4th 1859

They who never doubted about Creation can by assuming this, explain very plausibly why endemic species in the Galapagos sh^{d}. bear the stamp of the S. Ameri^{n} type.[2] The creative power foreknows not only what range of temperature they must be prepared to withstand & what organic beings they will be in contact with for the next century but what colonists will arrive in the next 100,000 years, & as most of these will be immigrants from S. America, the same motives wh. led to the endowment of the last created mollusk, fish, reptile bird or mammal with peculiar attributes on the mainland, will lead to the originating of some modification of like types for those Galapagan species which will have to struggle with the new Colonists, so that even were there no inorganic conditions, common to the Galapagos & S. America, or common to Africa & other regions, which is very improbable, there w^{d}. still be reasons why the new insular species sh^{d}. be gifted with S American characters. Without these they will not last.

This same foreknowledge must alike govern the plans of Nature or this unknown First Cause even if creating nothing directly he endowed a single primeval egg with the germs of all that was to be afterwards developed.

Lamarck's monads coming daily into being supply a perpetual crop of the simplest forms, so that there is no lack of these last while others are in the lapse of ages improving & rising in grade. Any system wh. begins with fewer & simpler types & ends with Man requires Lamarck's progression. If it be said that some forms are always degenerating or passing from complex to simple, ought not data to be shown in confirmation & could the simple be so perfectly adapted for their place.[3] However long it took to develop an orang into a Boscheman[4] or this to a Newton, the period from the newer Pliocene without Man to the Post Pliocene with him is geologically a moment, & is not Man's appearance in the organic what a cataclysm w^{d}. be in the inorganic world. What c^{d}. Selection do in adding new powers, attributes & forces, were not a creative power assumed as part of the reproductive or variety making natural law.

When once a principle of improvement is added to "Selection" when the Marsupial are supposed to have supervened on a less intelligent set of Vertebrata the Elephant to supersede the marsupial the Negro the Elephant & the White man the Negro some power must be appealed to as mysterious & undefinable as "Creation"

If the present inorganic causes can be held to be the same as those wh. acted in the past we can interpret the past & foresee the future changes but not so if "selection" be coupled with "progression" In part of a geological epoch a Man may replace the Orang & a being as much higher may supersede Man.

In the Principles I assumed to escape the dilemma that the intellectual & moral world of Man was as diff^{t}. as were the animal & vegetable worlds from the spiritual Man as an animal not higher or more perfect than a Gorilla.[5]

According to Darwin Lamarcks plan must be adopted to get rid of the intermediate variety between Man & the Apes by the dominant race soon exterminating them.[6]

It will be necessary to regard man as one of the mammalia, his instincts & reason as differing only in degree & as the human embryo may become a rational & intellectual being in a given number of years so may the first created embryo in a given number of million of years, be developed into a Man. He has merely had to pass thro' more metamorphoses.

This theory cannot I fear compare scientifically to the verae causae to which you & I can now refer the inorganic changes.[7] It resembles more the inventions of the old cosmogonists for altho' it assumes that all this machinery is now going on & at work & that the changes are not brought about by extinct or unknown causes, yet the power which can superinduce the intelligence of the Elephant into a planet which had nothing above that of a fish at an earlier period & then produce Man in his highest state is not a comprehensible vera causa—or one the mode of working of which we can hope to understand?

What Huxley calls the "laws of morphology" & which he says must be assumed to account for there being no passages between the great classes of organic beings are I presume a part of this mysterious & unknown Creative Power.[8] If this power has caused the successive phenomena to appear by the ordinary laws of generation, the proving this is no doubt a grand step, but it would be better to put more broadly & fairly how little it explains, and how much of the mysterious intervention of some other & higher & what we call creative power is required— I do not mean to say that ordinary generative succession does not throw great light in your hands on a vast number of phenomena which you have so well accounted for but it struck me that some sentences here & there to show the subordinate influences which you must feel that such machinery can alone supply would have been useful, not in the style of the Bridgewater Treatise,[9] "as per contract", but as modestly limiting the pretensions of "selection" when all the wonders of organic Nature & the plan of Nature are to be accounted for, & their origin in some degree traced to it's source.

But it is more easy to recommend than to pen such sentences. In rewriting & recasting my books I have to make up my mind whether I am to embrace not only your views but also "progression" & whether Man coming last is or is not a reason for inferring all the prior steps of development.

Any hints as to the subject would be welcome perhaps I had better evade it as much as possible but I wish to break ground because my protest against the "finality men" or the extent to which men such as Agassiz & Owen often assume that the first appearance of each being as at present ascertained, is really the date of it's origin (I was very nearly saying creation, it is growth by natural selection I mean) my continued protest, I say has been skilfully misinterpreted by some antagonists as a positive denial on my part of progression.[10] I have purposely been very reserved on the point, & so have you, but I incline to believe in progress provided (as you w^{d} require) they give us as many ages *before* as since the Lower Silurian epoch.

In a letter of Gaudins some months ago he told me of Heer of Zurich having made out a fine Miocene Flora in Iceland & many species common to Swiss Miocene beds which throws light on your migration theory—[11] I suspect that when this Miocene vegetation flourished so far north there may have been more species in the world not only than in the subsequent Glacial Period but also than in our ⟨own⟩ epoch—[12]

p—315. Ch. X. old copy

Each form^{n} marks only an "occasional scene" this is quite my idea, and Agassiz has no idea how true this is to a practical geologist.[13]

327. old copy

"The forms which are beaten &c inheriting some inferiority in common"—[14] This inferiority seems to me too often put without guarding the expression. A species may last 100,000 years or more having lungs fitted for a tropical & nearly uniform temperature & able to resist the then existing assemblage of enemies & to flourish on the food then at hand. The climate becomes extreme & other species then immigrate being superior merely in reference to the new conditions. But these invaders though victorious may not endure half as long as the older & extinct ones did. If the white man in Liberia gives way before the black or the mulatto, it is only because of *inferiority*[15] in that climate.

What you say in Ch. X of intermediate form between extinct ones & why not exactly intermediate is very good & original.[16] The Summary of the 10th Ch is very striking.[17]

Ch. XI I have been always in the habit of considering the dis⟨si⟩milarity of African & American species as the necessary resu⟨lt⟩ of "Creation" adapting new species to the preexisting ones— Granting this unknown & if you please miraculous power acting as steadily as does extinction and all the consequences are the same as those of "natural selection." It is as necessary to give species certain endowments to do battle against living enemies, & to hold a place amidst contemporary organisms as to give them for aquatic aerial terrestrial or subterranean conditions.

But if once the variety-making power is accepted as a vera causa & as the mode & form under which the creative force manifests itself on earth then of course the difference of the plants & animals in the Old & New World are referable to a better understood mode of creation, which was in no small degree guessed at by Lamarck & others & wh. you are establishing by the facts of domestication breeding &c. I was afraid you w^{d}. have had to shake my faith in specific centres more than you do. It is not exactly that we have a single pair as progenitors, but the manufacturing goes on in a limited area as in Porto Santo, which if joined to the main Island of Madeira or to the Continent would be a centre of 60 or more species of land-shells.—[18]

Are there no intermediate or transitional forms between the primrose & the cowslip? when they are from the seed of one plant.[19]

Did not the Lamarckians appeal to Man's Coccygean vertebræ in proof of his origin from a tailed ancestor.[20] Which came first the tailed or tailless monkeys?

Grove once suggested to me that rudimentary organs may be quite as often the germs of organs about to be enjoyed by descendants as the remains of organs became abortive. What say you to this?[21]

Before an Apteryx can grow into a flying bird must there not be thousands of generations with rudimentary wings? In that case instead of the rudiment being genealogical, it becomes as Agassiz might say prophetic.[22]

I expect your book to form an era in geological literature. If people c^d. favor the Vestiges,[23] why sh^d they be shocked by y^r views. Subversive of old ideas they no doubt are, & like every great truth hitherto unfolded by Geol^y. they will be unpopular to a vast majority. In reality the out & out progressionists are quite as subversive of received dogmas & the popularity they enjoy is under false pretences of supposed agreement with endowed opinions with which properly understood they do not by any means accord.

Contemporary copy in hand of Mary Elizabeth Lyell and AL incomplete
DAR 170: 81; Kinnordy MS, Charles Lyell's notebooks 241, pp. 75–90

CD ANNOTATIONS
13.1 In a . . . epoch— 13.6] '20 Glacial'[24] *brown crayon*
18.1 What you . . . preexisting ones 19.3] *crossed pencil*
19.1 Ch. XI . . . acting 19.4] *scored pencil*; 'No' *pencil*

1 This letter continues Lyell's remarks in his letter of 3 October 1859 (*Correspondence* vol. 7) on the proof-sheets of the second half (chapters 8 to 14) of *Origin* (see also *Correspondence* vol. 7, letter to Charles Lyell, 30 September [1859], and letter from Charles Lyell, 3 October 1859, n. 2). Most of the text is taken from a contemporary copy in Lyell's notebooks. However, there is a damaged scrap of part of the original in DAR 170: 81 (see nn. 12, 15, and 17, below).

2 In *Origin*, pp. 397–9, CD pointed out that the species of the Galápagos Islands resembled (but were not identical with) species in South America, and were dissimilar from those of the Cape Verde Islands, despite the fact that the terrain and climate of the Galápagos were more like that of the Cape Verde Islands than like that of South America. He argued that this could not be accounted for 'on the ordinary view of independent creation', but was to be explained by the fact that the Galápagos were more likely to receive colonists from South America, and the Cape Verde Islands from Africa. For CD's response to Lyell's point, see *Correspondence* vol. 7, letter to Charles Lyell, 11 October [1859].

3 In his discussion of Jean Baptiste de Lamarck's theory of the transmutation of species (C. Lyell 1853, p. 567–77), Lyell explained Lamarck's hypothesis that 'Nature' was continually creating monads, the 'elementary rudiments of animal and vegetable existence', corresponding to 'what the ancients termed *spontaneous generation*' (*ibid.*, p. 574). This continual creation was necessary, in Lamarck's theory, to explain the wide range of simple and complex organisms presently existing, in the context of the progressive development of organic life that could be inferred from geological observations (*ibid.*, p. 573); otherwise, Lyell asked, 'Why have the majority of existing creatures remained stationary throughout this long succession of epochs, while others have made such prodigious advances?' (*ibid.*, p. 574). For CD's response, see *Correspondence* vol. 7, letter to Charles Lyell, 11 October [1859]; for further comments by Lyell in his scientific journal, see L. G. Wilson ed. 1970, pp. 294–5.

4 Bosch-man or bushman: 'a member of an aboriginal people of Southern Africa' (*OED*). Lyell discussed Lamarck's account of how the orang-utan developed into the human species in C. Lyell 1853, pp. 575–7.

[5] In C. Lyell 1853, pp. 148–9, Lyell argued that the human race was distinguished from other animals by its intellectual and moral qualities, and that the physical organisation of humans did not alone make them pre-eminent. He continued: 'the sudden passage from an irrational to a rational animal, is a phenomenon of a distinct kind from the passage from the more simple to the more perfect forms of animal organization and instinct. To pretend that such a step, or rather leap, can be part of a regular series of changes in the animal world, is to strain analogy beyond all reasonable bounds.' Lyell seems in this letter to doubt further the power of natural selection alone to yield progressive advances in intelligence throughout the animal kingdom. CD replied that he entirely rejected as 'quite unnecessary any subsequent addition "of new powers, & attributes & forces"; or of any "principle of improvement" ', except in so far as every character that was naturally selected or preserved was in some way an advantage or improvement, 'otherwise it would not have been selected' (*Correspondence* vol. 7, letter to Charles Lyell, 11 October [1859]). See also L. G. Wilson ed. 1970, pp. 249–51, for earlier remarks on this subject in Lyell's scientific journals, and Bartholomew 1973.

[6] In his account of Lamarck's theory about how the orang-utan developed into the human race, Lyell described how the new race of humans, inspired by the 'desire of ruling', drove out their nearest competitors, confining them to unfavourable habitats where their development was checked, while humans themselves, lived in large groups and developed new desires and the industries needed to satisfy them, 'perfecting their means and faculties' (C. Lyell 1853, p. 576).

[7] *Verae causae*: true conditions. Lyell refers to his theory, accepted by CD, of the uniformity of the laws of inorganic nature (see C. Lyell 1853, pp. 71–2, 181).

[8] Thomas Henry Huxley at this time supported Karl Ernst von Baer's theory of embryological types; see, for example, T. H. Huxley 1856, p. 189 (M. Foster and Lankaster 1898–1903, 1: 306), where Huxley asserts: 'morphology demonstrates that the innumerable varieties of the forms of living beings are modelled upon a very small number of common plans or types'. See also Di Gregorio 1984, p. 30.

[9] The Bridgewater Treatises were eight works of natural theology published between 1833 and 1836 under the terms of a bequest made to the Royal Society of London by Francis Henry Egerton, the eighth earl of Bridgewater. Each author was to write on 'the Power, Wisdom and Goodness of God, as manifested in the Creation'. (*EB*.)

[10] By 'finality men', Lyell possibly means teleologists, that is, those who believed that each species had been created to fulfil a particular purpose (cf. *OED*, and *Correspondence* vol. 7, letter to J. D. Hooker, 23 [December 1859]); however, there are thought to have been elements of teleology in Lyell's own thinking (see Bartholomew 1973, p. 287). Louis Agassiz believed in progressive development, that is, that ever more complex species were successively and independently brought into being by a creator or superior intelligence (Lurie 1960, p. 85). Lyell disagreed with this view on the ground that the fossil record, which had appeared to support such a progression, was imperfect (see Bartholomew 1973, p. 274; see also *Origin*, p. 302); in the light of his uniformitarian theory of the earth's development, he argued that organisms as physiologically complex as those of the present day may have and probably had existed in past ages despite the lack or paucity of fossil evidence for them, and that the fossil succession could not be used as evidence for progression (see Bartholomew 1973). Richard Owen also supported the view of a progressive succession of fossil types (Rupke 1994, p. 238). On Lyell's manoeuvrings on the question of progression and transmutation, see Bartholomew 1973.

[11] Lyell refers to a letter from Charles Théophile Gaudin dated 18 May 1859 (Lyell papers, University of Edinburgh library). Gaudin translated Oswald Heer's book on the climate and vegetation of the Tertiary period (Heer 1860) into French (Heer 1861). There are copies of both works in the Darwin Library–CUL; Heer 1860 is annotated (see *Marginalia* 1: 364–5). The Miocene flora of Iceland was discussed in Heer 1860, pp. 115–20. These pages are annotated in CD's copy of this work, and a note on one of the back pages on which an abstract is pinned reads: 'Read from P 115 to end'. On page 120, CD wrote in the margin: 'Tertiary Vegetation of Iceland, like rest of Europe a decidedly American character.' For CD's discussion of the migration of plants between the Old and New Worlds during the worldwide glacial period, see *Origin*, pp. 365–72.

[12] The text of this paragraph was taken from the part of the original letter in DAR 170: 81.

[13] The published statement was, 'Each formation, on this view, does not mark a new and complete act of creation, but only an occasional scene, taken almost at hazard, in a slowly changing drama' (*Origin*, p. 315). See n. 10, above.

[14] The published statement was, 'The forms which are beaten and which yield their places to the new and victorious forms, will generally be allied in groups, from inheriting some inferiority in common' (*Origin*, p. 327). For CD's response to Lyell, see *Correspondence* vol. 7, letter to Charles Lyell, 11 October [1859].

[15] The phrase 'is only . . . *inferiority*' is taken from the part of the original letter in DAR 170: 81. In the contemporary copy the text reads: 'it is only because he is *inferior* in that climate'.

[16] *Origin*, pp. 334–5.

[17] This paragraph, taken from the part of the original letter in DAR 170: 81, is not included in the contemporary copy. The published summary is in *Origin*, pp. 341–5.

[18] Porto Santo is a smaller island about twenty-five miles north-east of Madeira in the eastern Atlantic Ocean (*EB*). Lyell visited Madeira and Porto Santo in 1853 and 1854 (L. G. Wilson ed. 1970, p. xxxviii). It had been discovered in 1833 that of seventy-one species of land shell from the Madeiran group, only two were common to both Madeira and Porto Santo (*ibid.*). See *Origin*, pp. 352–6; in *Origin*, pp. 402–3, CD discussed land shells on Madeira and Porto Santo, citing Lyell.

[19] For CD's response, see *Correspondence* vol. 7, letter to Charles Lyell, 11 October [1859].

[20] The reference has not been identified. CD replied: 'I believe os coccyx gives attachment to certain muscles, but I cannot doubt that it is a rudimentary tail' (*Correspondence* vol. 7, letter to Charles Lyell, 11 October [1859]). CD discussed rudimentary organs in *Origin*, pp. 450–6.

[21] The reference is to William Robert Grove. For CD's response, see *Correspondence* vol. 7, letter to Charles Lyell, 11 October [1859].

[22] On Agassiz's theory of prophetic types, see Lurie 1960, p. 290. In Agassiz's view, the existence of such features indicated that the whole plan of creation was premeditated.

[23] Lyell refers to *Vestiges of the natural history of creation* ([Chambers] 1844), an anonymous evolutionary work.

[24] The annotation refers to a portfolio of notes on geographical distribution. CD kept a number of such portfolios, amassed over a long period of time and since dispersed. The likely contents of some portfolios were reconstructed when some of CD's papers were catalogued in 1932 (see DAR 220: 113).

To Elizabeth Drysdale [22 or 29 October 1859][1]

Wells Terrace | Ilkley Otley | Yorkshire
Saturday

My dear Lady Drysdale

I am extremely much obliged for your kindness about the Hornet's combs.[2] Will you be so good as to send them packed in a little Box with cotton-wool (in 3 or 4 weeks' time) per Railway addressed as follows—

C. Darwin
care of G. Snow
Nag's Head
Southwark
London.[3]

I staid in the establishment for a fortnight and then took a House and all my family are here.[4] We shall stay for 4 or 5 weeks more, so that I may give the treatment a fair trial. Please tell Dr Lane[5] that it has done me much good, though I was unfortunate in spraining my ankle[6]

[Moor] Park[7] would spoil one for almost any place. We thank you very truly for your kind invitation; but our plans are necessarily at present very uncertain depending on how long we stay here.—[8]

I heartily wish I could hear of Dr. Lane being fixed to his satisfaction.[9] It is wonderful how these northern Establishments seem to answer.—

My wife begs to send her very kind remembrances to yourself & all your party, & pray believe me | Dear Lady Drysda⟨le⟩ | Yours sincerely obliged | Ch. Darwin

I began by accident this with one sheet slipped into another.—[10]

John Wilson (Autographs) Ltd, Catalogue 88; Clive Farahar & Sophie Dupré, Catalogue 55; B & L Rootenberg (May 1991)[11]

[1] The date is established by the dates the Darwins were in Ilkley and the reference to four or five weeks' stay. CD left for Ilkey, in Yorkshire, where he was planning to visit the hydropathic establishment, on 2 October 1859, and left there for London on 9 December (*Correspondence* vol. 7, Appendix II). In his letter to J. D. Hooker, 15 October [1859] (*Correspondence* vol. 7), CD wrote that his family were arriving on Monday, which would have been 17 October, and staying for three or four weeks. Emma left with Henrietta and Elizabeth on 24 November (Emma Darwin's diary (DAR 242)). In mid-October CD himself seems to have been planning to leave before December (*Correspondence* vol. 7, letter to John Murray, 15 October [1859]). CD's reference to staying for four or five weeks more thus suggests that the letter was written before November. There were two Saturdays in October after Emma's arrival: 22 and 29 October.

[2] CD had been interested in the cell-making instincts of bees and wasps since at least 1857, when he began a correspondence with George Robert Waterhouse on the subject (see *Correspondence* vols. 6 and 7). His notes on the subject are in DAR 48B, and are dated between 1858 and 1860; he discussed the cell-making instinct of the hive bee in *Origin*, pp. 224–35. Earlier in 1859, CD thanked John Stevens Henslow for specimens, including the comb of a hornet, and asked Henslow to send more fresh hornet combs next year (*Correspondence* vol. 7, letter to J. S. Henslow, 10 January [1859]).

[3] George Snow of Down, Kent, operated a weekly carrier service between Down and London (*Post Office directory of the six home counties* 1859). The Nag's Head Inn, 102 Borough High Street, Southwark, was his point of departure in London (*Post Office London directory* 1859, *Correspondence* vol. 7, letter to Richard Owen, 10 December [1859], n. 5).

[4] CD initially stayed at Ilkley Wells House, a hydropathic establishment (see Metcalfe 1906, pp. 107–8, and *Correspondence* vol. 7, letter to W. D. Fox, [6 October 1859]). Emma and the children (Henrietta, Elizabeth, and Francis, and probably Leonard and Horace), arrived on 17 October (Emma Darwin's diary (DAR 242); letter from Emma Darwin to William Erasmus Darwin, [24 or 31 October 1859], DAR 210.6: 50).

[5] Edward Wickstead Lane was the son-in-law of Elizabeth Drysdale (*Emma Darwin* (1904), p. 184). See n. 7, below.

[6] CD had sprained his ankle on 9 October 1859 (see *Correspondence* vol. 7, letter to W. E. Darwin, [14 October 1859]).

[7] Moor Park, Surrey, was the hydropathic establishment owned by Lane at which the Darwins had became acquainted with Drysdale. They visited it in 1857 and 1858 (see *Correspondence* vols. 6 and 7), and CD had been there three times in 1859 (*Correspondence* vol. 7, Appendix II).

[8] See n. 1, above. No letter from Drysdale containing this invitation has been found.

[9] Lane was trying to acquire a new establishment. In 1860, he moved to a hydropathic establishment at Sudbrook Park, Richmond, Surrey (Metcalfe 1906, p. 57; see also *Correspondence* vol 8, letter to Charles Lyell, 25 [June 1860]).

[10] Two pages of this four-page letter are blank on the verso.

[11] The letter has been reconstructed from three sale catalogue entries. The transcription given is believed to be substantially complete.

From Charles Lyell 22 October 1859

Mildenhall
22^d. Oct^r '59

Respecting the doctrines of Pallas?[1] about the various races of dog having come from several distinct wild species (or sub-species)[2]

I wish you would enlarge a little in a letter on that point because I know that when I reperuse your pages I shall still feel (tho' I expect to appreciate it better than now) that there is an uncomfortable indefiniteness in the doctrine & that it somewhat invalidates the specific centre theory & leaves us very much "dans le vague". Is it meant that Man found several canine species just on the verge of passing from permanent varieties into species & before they had diverged from some anterior ancestral type so far as to make frequent interbreeding impossible they were blended so as to produce a wider range of variety than would otherwise have come about. Or did Nature mingle the several supposed wild stocks of dog.[3]

I can understand all of them coming from the wolf, especially as Bell says the period of gestation is the same,[4] altho' Sir H. Holland in his last review in Edinburgh assumes that period to be different[5] & when I told him that Bell says it is the same to a day, he could not tell me where he learnt the contrary but said he thought it was in Cuvier.[6]

If I find myself perplexed with the distance between the European, Negro, Hottentot & Australian races & am told that they probably may have sprung from several indigenous stocks or species settled in remote & isolated regions which preceded all those now living & which by blending together gave origin to the diversity now observed I feel instead of this being an explanation of my difficulty that it simply obscures all my ideas. Of course according to your system there was some common ancestor of all these races, as of the greyhound, pug, shepherd's dog &c. But how does Pallas notion of several wild species being the ancestors in common of these races help us & does it not make the doctrine of specific centres less defensible?

Hooker asked me whether I was confounded by your migration of many European plants from N. to S. across the intermediate tropical zone by aid of mountain chains during glacial period.[7] I told him I was by no means staggered by it. Indeed the necessity of expecting a vast number of difficulties of this kind was fully appreciated by me long ago. I rather marvel at the extent to which the non-intermixture of provinces holds good, assuming single birthplace of species.

I presume that Pallas interpreted by the Darwinian key merely meant that some canine species, the wolf or one near to it was in a very varying state before Man meddled with the several races of said single species. Man took advantage of it & selected & made new & stronger or diff^t races but they all kept true to their ancestors' period of gestation—whereas the Jackall, fox & other canine species which branch off at a much earlier period have diverged so much farther as to have distinct periods of gestation from each other & from the dog.

If this be all then it sh^d be distinctly declared that neither the dog nor Man are any more derived from several aboriginal or wild species than other plants or animals having a wide range of races.

Whatever you yield in regard to the dog you will have to concede to every variable species of plant or animal (wild or cultivated) Man included.

If I understand you the first egg or seed (or the first few types of plants or animals whatever their number) may have had powers of adaptation imparted to them which were not of necessity *progressive*[8] The first types or stocks may or may not have been very simple. If simple some of their descendants may have continued for ever equally simple. If by so continuing they were most in harmony with surrounding conditions organic & inorganic. As these latter changed, they w^d almost certainly require modification, but not always transmutation to higher grades

But unless there was an inherent tho' for ages perhaps latent principle of improvement how evolve the Elephant & finally Man out of a Lepidosyren or even the much higher Ornithorhyncus. Selection or breeding is merely giving the organism a free & fair chance of developing the inherent capacity. It is simply putting the acorn in the fittest soil.

E. de Beaumont thinks that of the successive catastrophes which have suddenly convulsed & revolutionized the earth that of the Alps was the latest or nearly so & the greatest.[9]

The appearance of Man viewed as progressive development views it, is in the organic world, I presume, the grandest & latest catastrophe. Is it not a greater event than the originating of the first types if none of them were in the beginning higher than those at present discovered in the Silurian system? I care not for the term "Creation" but I want something higher than "selection", unless the latter divinity be supposed to have produced the primeval egg or seed or germ of the first great branches of the organic kingdom—in some ante-Cambrian epoch— To that power I must refer all the wonders of successive groups of species, more especially every step gained in organization & intelligence, generation being a very subordinate part of the machinery, tho' the proving it to be the proximate way by which the new creations are brought like individuals successively on the stage is a vast step if secured among those humble advances which we may be permitted to make.

I perceive that Agassiz looks on Bryozoa as degraded mollusca,[10] & Hooker on the Apetala as degraded Monopetalous or polypetalous plants.[11] This idea of occasional simplification is no doubt a gain at least to me it is so—& makes your system more intelligible answering the question why in spite of higher forms coming in, there is so large a stock of the simpler & lower ones. But they (if it be true the whole organic world is advancing) I cannot understand & seem to require some arbitrary hypothesis.

C. L

Contemporary copy in hand of Mary Elizabeth Lyell
Kinnordy MS, Charles Lyell's notebook 242, pp. 15–24

[1] The question-mark is in the text. Mary Elizabeth Lyell, who made the copy of Lyell's letter from which this text is transcribed, may have wondered whether her spelling of the name was correct.

[2] Lyell refers to Pyotr Simon Pallas and Pallas 1780. See *Origin*, pp. 253–4; see also *Variation* 1: 16 and n. 3.

[3] CD replied to Lyell's queries in his letters to Lyell of 25 October [1859] and 31 [October 1859] (*Correspondence* vol. 7). See also Lyell's related queries in the letter from Charles Lyell, 4 October 1859 (this volume, Supplement), and CD's response to that letter (*Correspondence* vol. 7, letter to Charles Lyell, 11 October [1859]). These letters are part of CD and Lyell's discussions on the proof-sheets of *Origin*. For CD and Lyell's continuing discussion on the origins of dogs and humans, see *Correspondence* vol. 7.

[4] Lyell refers to Thomas Bell and T. Bell 1837, pp. 197–8.

[5] Lyell refers to Henry Holland and [Holland] 1859, p. 249.

[6] Lyell refers to Georges Cuvier. In *Le règne animal* (Cuvier 1817), Cuvier gives the gestational period of dogs (*ibid*, p. 153), but not that of wolves.

[7] Lyell refers to Joseph Dalton Hooker. See *Origin*, pp. 365–82.

[8] See *Correspondence* vol. 7, letter to Charles Lyell, 11 October [1859] and n. 7.

[9] Lyell refers to Léonce Elie de Beaumont and probably to Elie de Beaumont 1831.

[10] The reference is to Louis Agassiz. For Bryozoa as degraded molluscs, see J. L. R. Agassiz 1857–62, 1: 186, and J. L. R. Agassiz and Gould 1848, p. 193.

[11] The reference has not been identified.

To Charles Lyell 4 February [1860][1]

Down.
Feb. 4th.

My dear Lyell.

With respect to antiquity of man,[2] I really think p 356, 357, 358 of my Journal are worth your looking at & perhaps p 355 to see nature of country.—[3] Mr Gill's remark on deflexion of streams not applicable in these cases.[4] It is a queer thing but I had forgotten the whole case as completely as if I had never written it.

Yours— | C Darwin.

P. S. Very many thanks for your pleasant little note[5] just received, though I am sorry you shd. have had trouble to write it.— I have just got Haldeman from Royal—[6] it is perplexing to know what he thinks.—

I am extreemly glad to hear about Ramsay.[7]

I have had letter from Bronn: to my surprise he seems slightly stagered.[8] If he could see how life itself could arise he says he wd. be converted to Nat. Selection! Says he has himself published forebodings of the theory of Nat. Selection. in a work on the productions of Islands.—[9] I wd. send letter but I suppose you wd. not care to see it.[10]

Copy
DAR 146: 229

[1] The year is established by the relationship between this letter and the letter to Charles Lyell, 12 [February 1860] (*Correspondence* vol. 8).

[2] On 9 January 1860, Lyell wrote to George Ticknor that he had been 'much occupied' with the 'antiquity of man as implied by the flint hatchets of Amiens, undoubtedly contemporaneous with the

mammoth, and also the human skeletons of certain caves near Liége', which he believed to be of a similar age (K. M. Lyell ed. 1881, 2: 330). In his letter of 14 January [1860] (*Correspondence* vol. 8), CD had written to Lyell that he would be very interested in reading Lyell's discussion of humans (see also *Correspondence* vol. 8, letter to Charles Lyell, 4 [January 1860]). Lyell was working on a new edition of his *Elements of geology* (*Manual of elementary geology*; C. Lyell 1855) until 1861 (see K. M. Lyell ed. 1881, 2: 329, and *Correspondence* vol. 9, letter to Charles Lyell, 20 July [1861] and n. 7); by late in 1861 he had evidently decided to incorporate the material on human antiquity into a new work instead (see *Correspondence* vol. 9, letter to Asa Gray, 11 December [1861] and n. 15; see also this volume, Appendix V, the revised preface to the third edition of *Antiquity of man*, C. Lyell 1863c). *Antiquity of man* was published in 1863 (C. Lyell 1863a).

3 *Journal of researches* 2d ed. The first three pages CD cites are concerned with evidence in northern Chile and Peru that humans had inhabited South America 'for an immensely long period'.

4 Gill, identified by CD as a civil engineer, found a former river-bed in Peru in which a ridge had been uplifted, deflecting the water to form a new channel; the river's former course turned into a desert (*Journal of researches* 2d ed., p. 359). CD thought that the abandonment of habitations found in locations miles from any water was probably due to a change in climate rather than to deflection of streams (*Journal of researches* 2d ed, p. 358). Gill has not been further identified.

5 Lyell's letter has not been found.

6 CD refers to Haldeman 1843–4 and to the Royal Society of London. Lyell had reminded CD of Samuel Steman Haldeman's paper in June 1859 (see *Correspondence* vol. 7, letter from Charles Lyell to T. H. Huxley, 17 June 1859, and letter to Charles Lyell, 21 June [1859] and n. 2). CD referred to Haldeman's paper in the historical sketch that he was preparing in January and February 1860 for the second American edition of *Origin*. The sketch cited naturalists who had preceded CD in espousing favourable views of the transmutation of species; see *Correspondence* vol. 8, Appendix IV. CD had apparently first read Haldeman 1843–4 in 1844 (see *Correspondence* vol. 4, Appendix IV, *119: 19v.).

7 It is not known what news CD had received about Andrew Crombie Ramsay. Ramsay's health had broken down, and at the end of December 1859 it had been arranged that his lectures at the Royal School of Mines be given by Joseph Beete Jukes, while he himself went to stay with a friend in Cheltenham. He was ordered to take several months' rest, and went abroad in early April 1860. See Geikie 1895, p. 261.

8 The reference is to Heinrich Georg Bronn; his letter has not been found. CD sent it to Lyell on 12 February 1860, asking him to return it to Erasmus Alvey Darwin's house in London (see *Correspondence* vol. 8, letter to Charles Lyell, 12 [February 1860]). CD sent Bronn a presentation copy of *Origin* (see *Correspondence* vol. 8, Appendix III); Bronn published a review of it (Bronn 1860a), and the first German translation (Bronn trans. 1860). See also *Correspondence* vol. 8, letter to T. H. Huxley, 2 [February 1860] and n. 2. For CD's reaction to Bronn's review, see *Correspondence* vol. 8, letter to Charles Lyell, 18 [and 19 February 1860] and nn. 2–5.

9 Bronn 1860b. There is an annotated copy in the Darwin Pamphlet Collection–CUL.

10 See n. 8, above.

To William Rathbone Greg 21 March [1860?][1]

[Down]

[Darwin refers to remarks made by his correspondent on the subject of *On the Origin of Species*.[2] He writes: 'I am very glad indeed to read remarks made of a man who possesses such varied and odd knowledge'.[3] Darwin continues his letter with discussion of a manuscript on proportions of the sexes at birth that W. R. Greg has sent along for review[4] but returns to commentary of *On the Origin of Species*. He notes his own 'remarks on mankind are superficial, and given merely as some sort of standard for comparison with the lower animals'. He closes his letter remarking, 'I am quite delighted that my book interests you enough to lead you to read it with such care [. . .]']

Ch. Darwin

Incomplete[5]
Sotheby's New York (December 1996)

[1] The date 21 March [1860?] is taken from the sale catalogue.
[2] *Origin.* The letter from Greg has not been found.
[3] Greg wrote for a number of periodicals on politics and economics, as well as on 'speculative and theological subjects' (*CDEL*).
[4] The manuscript has not been identified. Greg had earlier approached CD to write reviews: see *Correspondence* vol. 6, letter to F. M. Wedgwood, 18 [August 1856 – January 1858].
[5] The original letter is complete and is described in the sale catalogue as being six pages long.

From Thomas Stewardson 16 April 1860

Hall of the Academy of Natural Sciences of Philadelphia.
April 16th *18*60

To Charles Darwin F.R.S *Sir,*

I have the honour to inform you, that at a meeting held March 27th 1860 *you were elected a Correspondent of this Society.*[1]

To this announcement of the confidence of the Academy, I beg leave to add that it is with sentiments of the highest respect, I subscribe myself,

Your obedient servant, | Tho.s Stewardson M.D. | *Corresponding Secretary,* | *Of the Academy of Natural Sciences of Philadelphia.*

Extracts from the Constitution and By-Laws of the Academy.

Chapter 3, Article 7. No pecuniary contributions shall be required from Correspondents, but Correspondents residing in the United States shall be charged a diploma fee of five dollars.

[Enclosure]

The Academy of
Natural Sciences
of Philadelphia
has elected
Charles Darwin
London.
a Correspondent *of the Association this* twenty seventh *day of* March *18*60.

Isaac Lea *President*

Tho.s Stewardson *Corresponding* } *Secretary* — Robert Bridges } *Vice Presidents*
Howard Rand[2] M.D. *Recording* } *Secretary* — Wm S Vaux[3] } *Vice Presidents*

LS
DAR 229: 3, 230: 4

[1] CD acknowledged receipt of the diploma in his letter to Stewardson of 8 May 1860 (*Correspondence* vol. 8). For CD's assessment of the significance of the honour, see his letter to Charles Lyell of 8 [May 1860] (*ibid.*).
[2] Benjamin Howard Rand.
[3] William Sanson Vaux.

To M. T. Masters 25 April [1860][1]

Down Bromley Kent
Ap. 25

My dear Sir

I beg you not to speak of "apologies for intruding", when you are doing me the greatest kindness in giving me valuable information.[2]

I am heartily glad to hear of the papers in Linn: Journal, & then I shall see about peloria: I hope you will give any information which you may possess on the power of reproduction in peloria & other monsters.[3] Of course you will naturally compare monstrous to normal structures: as far as I have seen, such comparisons have generally to be instituted between members of *distinct groups*, which to my mind greatly destroys the value of such parallelism in relation to the origin of species.[4]

Unless your Father[5] actually experimentised on *mixed* pollen it is curious how he discovered the truth, which Gärtner has shown without doubt, namely that the stigma does select its own kind of pollen out of others; & more than that, for if the plant's own pollen be put on stigma within a certain number (forgotten by me) of hours after foreign pollen, all influence from that foreign pollen is completely eliminated.[6] But then no facts are known showing that this holds with the pollen of different *varieties*: on the contrary some facts lead to the suspicion that the pollen of a *distinct variety* has a prepotent effect over a plant's own pollen.—[7] In the case of Hollyhocks, I have suspected from some facts that each variety preferred its own pollen; & I have in vain been searching for information, from those who have raised many hollyhocks whether they found it necessary, in order to get the varieties pretty true by seed, to keep the several varieties in different parts of the garden. Has your Father raised with care, many seedling Hollyhocks.?[8]

I observe that your Father is a strong believer in the rule that when variation has once commenced it goes on.[9]

If at any time you were staying with your Father it would be a grand thing to get his reasons for this belief.—

That is a curious case about the yellow Hyacinths.[10] The laws of inheritance seem to be determined to puzzle everyone.

With sincere thanks | My dear Sir | Yours sincerely | C. Darwin

Shrewsbury School Archives, SR/Darwin box 1

[1] The year is established by the relationship between this letter and the letter from William Masters, 8 May 1860 (*Correspondence* vol. 8); see n. 8, below.

[2] The letter from Masters from which CD quotes has not been found. CD had invited Masters to let him know if anything occurred to him in support of or opposing CD's theories in his letter to Masters of 13 April [1860] (*Correspondence* vol. 8).

[3] In the missing letter, Masters may have discussed the work on peloric orchids that he later published in *Journal of the Linnean Society (Botany)* (Masters 1864). He also published an article on peloria in the *Natural History Review* in 1863 (Masters 1863c). Masters may also have mentioned his work on prolification, a type of floral monstrosity; two of these papers were published in *Transactions of the Linnean Society*, and one in *Journal of the Linnean Society (Botany)*, but they did not concern what CD had called peloria in *Origin*, p. 145 (see Masters 1861, 1862b, and 1863a, and *Correspondence* vol. 10, letter to M. T. Masters, 24 July [1862] and n. 4).

[4] Since many monstrosities resembled other species in distinct and fairly remote groups, CD suspected that the resemblances were accidental, and he doubted that the monstrous forms could develop into new species (see *Natural selection*, pp. 318–21, and *Correspondence* vol. 8, letter to Charles Lyell, 18 [and 19 February 1860]; see also *Origin*, pp. 44, 150–70). Masters was developing a specialty in plant teratology, or the study of malformations and abnormal growth (see n. 3, above, and Masters 1860). For CD's interest in peloria, see *Correspondence* vol. 4, letter to J. D. Hooker, [12 June 1847], *Correspondence* vol. 8, letter to M. T. Masters, 13 April [1860], and *Origin*, p. 145. CD would soon seek to establish whether peloria in flowers resulted in changes in fertility (see, for example, *Correspondence* vol. 9, letters to *Journal of Horticulture*, [before 18 June 1861] and [before 9 July 1861], and *Correspondence* vol. 10, letter to M. T. Masters, 8 July [1862]).

[5] William Masters.

[6] See *Origin*, p. 98. CD refers to the experiments described by Karl Friedrich von Gärtner in Gärtner 1849, pp. 35, 38–46, 46–58. CD's heavily annotated copy of this work is in the Darwin Library–CUL (see *Marginalia* 1: 248–98). There is also an abstract of Gärtner 1849 in DAR 116. The reference to William Masters's observations confirming Gärtner's experiments has not been identified, but it may have been included in the missing letter from M. T. Masters (see n. 2, above). See also CD's most recent letter to M. T. Masters, in which he included notes for William Masters on his own experiments in crossing different varieties of sweet peas (*Correspondence* vol. 8, letter to M. T. Masters, 13 April [1860] and *CD note*).

[7] See *Origin*, p. 99. See also *Correspondence* vol. 7, letter to Asa Gray, 11 August [1858], and *Correspondence* vol. 8, first letter to Andrew Murray, 28 April [1860].

[8] For CD's queries on hollyhocks, see *Correspondence* vol. 5, letter to J. D. Hooker, 14 [August 1855], and letter to J. S. Henslow, 10 November [1855]; see also *Correspondence* vol. 2, letter from William Herbert to J. S. Henslow, 5 April 1839. William Masters replied to CD's query in his letter of 8 May 1860 (*Correspondence* vol. 8), and evidently provided further information on the cultivation of hollyhocks (see *Variation* 2: 107–8).

[9] Masters may have included information from his father in his missing letter (see nn. 2 and 6, above); however, see also the letter from William Masters, [after 7 April 1860] (*Correspondence* vol. 8).

[10] In his missing letter (see n. 2, above), Masters may have reported his father's finding that yellow hyacinths, unlike other coloured varieties, reproduce their colour. Since the blue varieties were not nearly so true in their transmission of colour to offspring plants, William Masters remarked, 'we see that a garden variety may acquire a more permanent habit than a natural species' (see *Variation* 2: 20). No letter in which William Masters made this remark to CD has been found.

From J. D. Hooker [11 May – 3 December 1860][1]

reducing individuals here & destroying **genera** elsewhere— Whether man reduces the bulk of organic & increases that of inorganic matter or no there cannot be a doubt that his influence is devoted to reducing existing species & hence improving our means of classification, which is only another expression for, further sundering allied forms—[2] The number of new varieties he creates are as nothing to the destruction of old & the reduction of the battle field for the remainder.— the struggle is hence all the harder in the space he leaves to native vegetation.[3] Man & his intellect are all part of the scheme & must not be shirked. Still the question makes me giddy from its complication.

I think little of his objection on the score of varieties of the lineal descendent of A. becoming confounded with those of ⟨B⟩[4] no doubt such cases occu⟨r⟩ I have somewhere I think alluded to them as very probably frequent. ⟨I should⟩ now put into this category all those anomalous plants which hover between two otherwise very remote species in genera or families.[5] It struck me that such cases were well explained by your divergent series—[6] Such cases *if decided ones* should be rare— so they are— There is however the perpetual difficulty of deciding between reversion of character & persistence of character. Let us suppose that Dicots. were developed through & after Monocots, & regard Conifers as anomalous Dicots nodding towards Cryptogams & we have a case in point,— the Conifers are either Dicots that vary towards Cryptogs. or (as I should incline to suppose) a sub series of Dicots. developed parallely to the main Monocot branch, but in which alone ⟨ ⟩ the Cryptog. characters **persisted**[7] Watson must think that subsequent variation may as easily be towards[8] the old type as away from it— this is conceivable of course, but I maintain it is not so in nature— it is opposed to my idea of centrifugal variation.—[9] Now variation is not centrifugal because of any repulsive or repellent power, but simply because in fact it is a million chances to one that identity of form once lost will be returned to.[10] Each character we estimate is a compound of lesser characters each too minute to estimate in the gross, but missed if with-drawn. A petal if further varying after it has once varied has a thousand characters of form color consistence nervation &c &c &c to chuse amongst (so to speak) besides those its grandparents bore. After all, experience is our best guide, & we do not find in the human race any reversion so strong as would lead us to confound a man with his ancestor, a Yankee with an Englishman— Whoever saw a Grandfathers portrait that would really pass for his Grandsons—& if we do not find reversion amongst individuals so close in kin & in time, how can we expect them in organisms that have reached the specific term of divergent development—

True you may have a 10th removed lineal descendent (who supposing the name of Wedgwood to be lost to genealogists—except by a stray portrait) may so resemble the Wedgwood portrait that an Ethnologist would call him an anomalous Darwin representing the lost Wedgwood type.—but the chances are a hundred to one in favor of both Wedgwood & Darwin physiognomy being wiped out long before the 10th. generation—[11] How often do we consider 2 people strikingly alike till we see them together & then consider them wholly unlike; this is because **one** minute character common to both was **alone** carried away, in the mind, but which, though no doubt present, is not perceived when *again* sought for amongst the thousand other characters of the 2 faces.

Incomplete
DAR 205.5: 217 (Letters), DAR 47: 214

CD ANNOTATIONS
1.1 reducing individuals . . . frequent. 2.3] *crossed brown crayon*
2.4 now put . . . they are— 2.7] *scored brown crayon*
2.7 There is . . . **persisted** 2.13] '11·'[12] *brown crayon, circled brown crayon*
2.7 There is . . . with-drawn. 2.20] *crossed brown crayon*
2.23 & we . . . Grandsons— 2.25] *scored brown crayon*
2.25 & if . . . faces. 3.10] *crossed brown crayon*

[1] The date range is established by the relationship between this letter, the letter from H. C. Watson, 10 May 1860 (*Correspondence* vol. 8), and the letter to J. D. Hooker, 4 December [1860] (*ibid.*).

[2] Hooker advocated uniting species in classification: 'It is strange that local naturalists cannot see that the discovery of a form uniting two others they had previously thought distinct, is much more important than that of a totally new species, inasmuch as the correction of an error is a greater boon to science than is a step in advance' (Hooker and Thomson 1855, p. 36). For further information on the 'splitting' and 'lumping' schools in taxonomy, their effect on the number of species, and the involvement of Hooker and Hewett Cottrell Watson in the debates, see, for example, the letter to J. D. Hooker, 17 April [1865] and n. 12.

[3] According to Hooker, human activity represented 'a new enemy to scarce old forms [of plants], and a strong ally to those already common'; he recorded the destruction of local genera in Australia, New Zealand, and South Africa in the wake of human activity (Hooker 1859, pp. civ–cv).

[4] Hooker apparently refers to Watson's hypothesis that variation could cause the descendants of two different plants ultimately to become morphologically indistinguishable (see *Correspondence* vol. 8, letter from H. C. Watson, 10 May 1860). CD had earlier communicated Watson's arguments in favour of 'convergence' to Charles Lyell and Hooker (*ibid.*, letter to T. H. Huxley, 11 January [1860] and n. 5). These were contained in the letter from H. C. Watson, [3? January 1860] (*ibid.*), and included Watson's view that his principle of 'convergence' explained the occurrence of characters in common between two species where CD's theory of divergence did not. CD discussed Watson's views and the idea of 'convergence' in the revised American edition of *Origin* (*Origin* US ed., pp. 116–17*; see *Correspondence* vol. 8, Appendix IV, pp. 576–7) and in *Origin* 3d ed., p. 141 (see n. 6, below).

[5] Hooker's interest in plants that were difficult to classify owing to their similarity to widely differing taxa is exemplified by his research on the anomalous *Welwitschia mirabilis* (Hooker 1863).

[6] CD, with the aid of a diagram, discussed the operation of his principle of divergence on the species of a large genus in *Origin*, pp. 111–26. CD addressed Watson's objection directly in a new passage added to *Origin* 3d ed., p. 141:

> I will only say that if two species of two closely allied genera produced a number of new and divergent species, I can believe that these new forms might sometimes approach each other so closely that they would for convenience sake be classed in the same new genus, and thus two genera would converge into one; but from the strength of the principle of inheritance, it seems hardly credible that the two groups of new species would not at least form two sections of the supposed new single genus.

[7] For a summary of the main divisions of the contemporary classification of the vegetable kingdom, see *Correspondence* vol. 7, letter from J. D. Hooker, [9 March 1859]. In contrast to the prevailing view, Hooker maintained that conifers were advanced dicotyledons that retained some cryptogamic characters (see, for example, *Correspondence* vol. 7, letter from J. D. Hooker, [26 December 1858] and n. 3).

8 The letter text from this point was previously published in *Correspondence* vol. 10, as the letter from J. D. Hooker, [after 26 March 1862?]. It appears here with new footnotes.

9 Watson's view of transmutation included both centrifugal variation and centripetal variation (the former generating divergence and the latter what Watson called convergence). See *Correspondence* vol. 8, letter from H. C. Watson, [3? January 1860]. See also this volume, Supplement, letter from H. C. Watson to J. D. Hooker, 4 January 1861 and n. 9, and *Correspondence* vol. 9, letter to J. D. Hooker, 15 January [1861].

10 Hooker had introduced CD to his ideas of centrifugal and centripetal forces of variation in *Correspondence* vol. 3, letter from J. D. Hooker, 1 February 1846; the terms also occur in Hooker 1844–7, p. 315. CD had criticised Hooker's use of the word 'force' and warned that 'centrifugal' variation implied a single centre (*Correspondence* vol. 3, letter to J. D. Hooker, [16 April 1846]).

11 Hooker may have been aware of the Wedgwood family joke that the Darwins were 'more Wedgwood than the Wedgwoods', since CD was the son of Susannah Wedgwood, and had married his cousin, Emma Wedgwood (B. Wedgwood and Wedgwood 1980, p. 269).

12 The annotation probably refers to the numbered portfolio in which CD kept his notes on divergence (see the 1932 catalogue of CD's papers, DAR 220: 13). A quantity of notes marked '11' are now collected in the Darwin Archive–CUL (DAR 205.5).

To J. D. Hooker 19 [June 1860][1]

Down
19th

My dear Hooker

Have pity on me & let me write once again on Orchids for I am in a transport of admiration at most simple contrivance, & which I shd. so like you to admire.[2] How I wish I was a Botanist.[3] Such brief books as I have, say structure of Gymnadenia like Orchis, but I find the stigma consists of two most curious *lateral* horns;[4] I must think these horns are the stigma, from what follows— push *thick* needle down mouth of nectary & the two pollen masses adhere by thin long naked strap-shaped sticky glands, & then when needle withdrawn by elasticity become nearly parallel to needle. In doing this not a grain of pollen is left on the horn-like stigma— But here comes pretty fact, hold needle in same relative position to parts of flower & push into nectary of **another** flower, & lots of grains of pollen are almost inevitably left on the two humid stigmas. In fact G. conopsea would hardly ever get a grain of pollen of its *own* flower![5] I was led to observe this by finding that the 13 pollen-masses on the proboscis of Acontia luctuosa belonged to this orchis.—[6]

I see Bentham puts this plant into genus Orchis, but surely such a difference in stigma, & naked & curious strap-formed glands must be good generic character.[7] If, as is likely, you do not care or are too busy to look at this, as it seems to me *really beautiful* relation of part, ask Mr Bentham, whether he will.—[8] I see the stigmatic horns rise from spot where stigma lies in true Orchis.

C. D.

Etty decidedly rather better.[9]

(I vow I will not bore you again.)

DAR 261.10: 69 (English Heritage MS 88206052)

[1] The date is established by the relationship between this letter and the letter to J. D. Hooker, 26 [June 1860] (*Correspondence* vol. 8).

[2] Orchids had been a frequent topic in CD's correspondence with Hooker since April 1860 (see *Correspondence* vol. 8; see also his questions on orchids to the readers of *Gardeners' Chronicle* (*Correspondence* vol. 8, letter to *Gardeners' Chronicle*, [4–5 June 1860]).

[3] In his letters to Hooker, CD had long been scornful of his own botanical capability (see, for example, *Correspondence* vol. 3, letter to J. D. Hooker, [3 September 1846]). In later life, CD claimed that, while at Cambridge, he attended John Stevens Henslow's botanical lectures and excursions as an amateur and 'did not study botany' (*Autobiography*, p. 60).

[4] CD was evidently observing the reproductive organs of either *Gymnadenia conopsea*, which he described in *Orchids*, pp. 79–82, or *G. albida*, which he described in *Orchids*, pp. 82–3. One of the books to which CD probably referred is Bentham 1858 (see n. 7, below); there is a lightly annotated copy in the Darwin Library–CUL (see *Marginalia* 1: 51).

[5] CD's experiment imitated the insertion of the proboscis of a moth or butterfly into the flower. In the introduction to *Orchids*, CD stated that a principal object of the book was to show that flowers in that family have adapted to assist cross-fertilisation, thereby challenging earlier texts, for example Sprengel 1793, that described orchids as being generally self-fertilised (*Orchids*, pp. 1–3). In the conclusion, CD wrote, 'Nature thus tells us, in the most emphatic manner, that she abhors perpetual self-fertilisation' (*Orchids*, p. 359).

[6] The moth *Acontia luctuosa* was also listed in *Orchids*, p. 36, with twenty-two other lepidopterous species, as having been observed with pollinia of *Orchis pyramidalis* adhering to its proboscis.

[7] The reference is to Bentham 1858, in which the fragrant orchid, *Gymnadenia conopsea*, is classified as *Orchis conopsea*, and *G. albida* as *Habernaria* (now *Habenaria*) *albida* (Bentham 1858, pp. 511–12). The morphological differences between *Orchis* and *Gymnadenia*, noted in Bentham 1858, p. 511, did not include their stigmas or the sticky glands of their pollinia. See also n. 4, above.

[8] George Bentham had been compiling *Genera plantarum* with Hooker since 1857 (Allan 1967, p. 213). From that year, Bentham went regularly to Kew to collaborate with Hooker (Bentham 1883, p. 305).

[9] On Henrietta Emma Darwin's recurrent ill health between 1857 and 1861, see Bowlby 1990, pp. 329–30, 354, 359–60. In April 1860, she contracted either typhus or typhoid fever (*Correspondence* vol. 8, letter to J. D. Hooker, 11 May [1860] and n. 5).

To Charles Giles Bridle Daubeny 16 July [1860][1]

Hartfield | Tonbridge Wells[2]
16 July

My dear Dr. Daubeny

Henslow has sent me part of a note of yours & his answer.—[3] You are quite right that there is nothing direct in my Book on the final cause or manner of origination of the sexes.[4] It seems to me one of the profoundest mysteries in nature. Especially since the facts of parthenogenesis have been clearly made out.[5] I have heard Huxley remark that the old saying was "it was a wise child who knew who his Father was", but now the saying might be, "it was a wise child who knew whether he had a Father."—[6]

I have often speculated on your view,[7] & I have somewhere got a case of a trifling variation being propagated by gemmation with anim⟨als⟩ (in a coral) just as we see almost every shade of difference propagated by grafts & buds with plants.[8]

I daresay that you are quite right that one of the final causes of sexuality is the origination of varieties;[9] but the number of cases which can be collected of "sporting" plants ie plants which have varied by gemmation has made me rather doubtful on this head.—[10] I am inclined to lay **more** stress on sexuality (p. 103 Origin of Species 1st edit) being the means of keeping forms constant ie of checking variation. This is a subject on which I have worked much; but have said hardly anything in the Origin.—[11] The whole subject of sexuality seems to me profoundly interesting & mysterious; & the more it is stirred up the better.

Pray believe me | My dear Dr. Daubeny | Yours very sincerely | Charles Darwin

Magdalen College, Oxford (MS 400, fol. 118)

[1] The year is established by the date on which Daubeny read his paper in Oxford (see n. 3, below) and from CD's letter to J. S. Henslow, 16 July [1860] (*Correspondence* vol. 8), which refers to CD's sending the present letter.

[2] From 10 July to 2 August, CD joined Emma and Henrietta Darwin at The Ridge, Hartfield, East Sussex, the home of Emma's sister, Sarah Elizabeth Wedgwood (see *Correspondence* vol. 8, including Appendix II).

[3] Daubeny's note and John Stevens Henslow's answer have not been found; however, see the letter to J. S. Henslow, 16 July [1860] (*Correspondence* vol. 8). The correspondence between Daubeny and Henslow was evidently prompted by a paper read by Daubeny at the meeting of the British Association for the Advancement of Science in Oxford, in a session chaired by Henslow on 28 June 1860 (see *Athenæum*, 7 July 1860, p. 25). CD did not attend the meeting (see *Correspondence* vol. 8, letter to J. D. Hooker, 26 [June 1860]). In writing the present letter to Daubeny, CD probably relied on information from Henslow on the content of Daubeny's paper, as he did not receive the published text (Daubeny 1860a) until later (see this volume, Supplement, letter to Charles Daubeny, 1 August [1860] and n. 2). CD may also have read a report of Daubeny's paper in the press (see, for example, *Athenæum*, 7 July 1860, pp. 25–6, reproduced in *Correspondence* vol. 8, Appendix VI).

[4] According to the report of his paper in the *Athenæum* (see n. 3, above), Daubeny questioned the final cause of sexes in plants in view of the allegedly identical results of sexual and asexual reproduction. Daubeny cited various possible uses of sexual reproduction in plants, such as the provision of fruit for food, the gratification of the human senses, and the spatial dissemination of species, but contended that CD's theory of natural selection, subject to its acceptance, offered a more plausible and universal final cause than these. CD's own earlier interest in the final cause of sexual difference is reflected in his *Notebooks*: 'My theory gives great final cause ... of *sexes* ...: for otherwise, there would be as many species, as individuals. . . . [I]f my theory be true then the formation of sexes rigidly necessary' (Notebook E, 48–9; see also *Notebooks*, Notebook C, 236, Notebook D, 167). In *Origin*, however, CD confined himself to more general observations on the origin of sexual reproduction, such as, for example, 'that it is a general law of nature ... that no organic being self-fertilises itself for an eternity of generations' (*Origin*, p. 97). In his 'Provisional hypothesis of pangenesis', CD again considered the final cause of sexual reproduction (see, for example, *Variation* 2: 362). For recent discussions of the teleology implied in the nineteenth-century use of the term 'final cause', and of the teleology implied in CD's writings in particular, see England 2001 and Sloan 2001.

[5] During the previous decade, several publications on parthenogenesis had appeared, for example R. Owen 1849 and Siebold 1857, annotated copies of which are in the Darwin Library–CUL (see *Marginalia* 1: 656–7, 756–7). These were followed by Thomas Henry Huxley's paper on parthenogenesis in aphids, which challenged Richard Owen's explanation of the phenomenon (T. H. Huxley 1858; see also *Correspondence* vol. 6, letter to T. H. Huxley, [before 12 November 1857]). In the first edition of *Origin* there is no reference to parthenogenesis, but CD added a single reference in the second edition, published on 7 January 1860, in the phrase 'with the exception of the curious and not well-understood cases of parthenogenesis' (*Origin* 2d ed., p. 96).

[6] This quotation has not been found in Huxley's writings. It appears to derive from 'It is a wise father that knows his own child', Shakespeare, *Merchant of Venice*, 2.2.72–3 (S. Wells and Taylor eds. 1988).

Huxley's work on parthenogenesis in aphids prompted him to question the necessity of the male in reproduction (T. H. Huxley 1858, p. 220).

[7] Daubeny held that sexual reproduction in plants provided a mechanism for variation, whereas asexual reproduction produced 'a mere counterpart' of the parental stock (see *Athenæum*, 7 July 1860, pp. 25–6; see also Daubeny 1860a).

[8] Gemmation in this sense means 'the formation of a new individual by the protrusion and complete or partial separation of a part of the parent' (*OED*). CD's 'case of a trifling variation' propagated by the vegetative reproduction of an animal in coral has not been identified. For CD's later discussion of asexual reproduction and variation, see *Variation* 1: 374 and 2: 358–9, 371–3.

[9] See n. 7, above.

[10] CD's point about variation in plants raised by gemmation was cited by Daubeny in a footnote, with a reference to this letter (Daubeny 1860a, p. 23 n.). 'Sports' are new varieties that arise from mutations, chiefly in buds, and are then propagated by vegetative reproduction. CD gave numerous examples of sports in *Variation* 1: 373–411. As examples of variation independent of sexual reproduction, sports were a challenge to Daubeny's view; see also *Correspondence* vol. 8, letter to J. S. Henslow, 16 July [1860].

[11] CD alludes to the sentence 'Intercrossing plays a very important part in nature in keeping the individuals of the same species, or of the same variety, true and uniform in character' in *Origin*, p. 103. CD maintained this view in all editions of *Origin*. There are further details of his work on the role of crossing in the origination of species in *Natural selection*, p. 271, where he states: 'Intercrossing will prevent or retard the process of natural selection; but here we are involved in much doubt.'

To C. G. B. Daubeny 1 August [1860][1]

(Down Bromley Kent)
Aug 1

My dear Sir

I thank you sincerely for your present of your interesting pamphlet & more especially for the *extremely* kind & liberal manner in which you allude to my work.—[2] I can wish & hope for nothing better.— Permit me to add that I am convinced from my own mental experience, that he who goes as far as you, *if he keeps the subject in mind*, will ultimately go much further.[3]

I quite agree with your view of the importance of sexual generation, as far more freely admitting variation, than gemmation;[4] but I cannot avoid the suspicion that some more important or at least some other final cause lies hidden.— When an animal is propagated by any process analogous to gemmation, (ie *not* by sexual generation) I doubt whether it even passes through embryological metamorphoses. On other hand we have reason to *suspect* that there is always metamorphosis (often concealed) in case of sexual generation.[5]

But this is a mere loose & crude speculation, which I have sometimes thought of following up.— I daresay it w^d. break down.—

With respect to varieties *never* being infertile together pray look to my abstract of Verbascum & Tobacco cases.—[6]

At p. 30 you give some grand facts, some of which are unknown to me & will aid me capitally.— [7]

With many thanks personally for myself, & for subject-sake, as it is a grand thing for a man in your position deliberately to consider it.

I remain | My dear Sir | Yours truly obliged | Charles Darwin

Magdalen College, Oxford (MS 400, fol. 119)

1 The year is established by the reference to Daubeny 1860a (see n. 2, below).

2 An annotated and inscribed copy of Daubeny 1860a is in the Darwin Pamphlet Collection–CUL (see also this volume, Supplement, letter to C. G. B. Daubeny, 16 July [1860] and n. 3). The paper contains information from CD's letter to Daubeny of 16 July [1860] about variation in plants derived from vegetative reproduction (see Daubeny 1860a, p. 23 n.). Daubeny also deprecated the interference of theologians in questions of science and referred to CD's transmutation theory as 'the principle so ingeniously put forward in Mr. Darwin's work' (Daubeny 1860a, pp. 26 n., 29). Daubeny 1860a is summarised in Daubeny 1860b.

3 CD probably alludes to Daubeny's reservations about the Darwinian hypothesis; Daubeny thought that it failed to explain the origins of the imagination, moral sense, and mental powers of humans (Daubeny 1860a, pp. 23–6).

4 See Daubeny 1860a, pp. 20–1, and this volume, Supplement, letter to C. G. B. Daubeny, 16 July [1860] and nn. 7, 8, and 10.

5 CD wrote 'Embryology' in the margin of his copy of Daubeny 1860a, p. 6, where Daubeny questions the distinction between vegetative multiplication and sexual reproduction. CD considered the significance of metamorphosis in reproduction, and questioned the origin of sexual reproduction, in *Variation* 2: 361–3. This, and other unresolved questions, provided the impetus for CD's provisional hypothesis of pangenesis (*Variation* 2: 357–404).

6 Daubeny and others objected that CD's theory could not be considered proven until a new species arose under domestication (Daubeny 1860a, pp. 28–9; see also T. H. Huxley 1860a, [T. H. Huxley] 1860b, and *Correspondence* vol. 10, including Appendix VI). CD refers Daubeny to the examples of partial infertility in the progenies of Karl Friedrich von Gärtner's crosses in *Verbascum* and Joseph Gottlieb Kölreuter's crosses in tobacco (see *Origin*, pp. 270–1).

7 On the back cover of his copy of Daubeny 1860a, CD wrote, 'p 30 excellent facts new to me'. On page 30, Daubeny gave examples of monstrous plants that, in their abnormal state, resembled plants of quite different genera. These examples, with references to Alphonse de Candolle, Maxwell Tylden Masters, and Alfred Moquin-Tandon, were marked by CD (Daubeny 1860a, p. 30). CD did not subsequently cite any of the facts given in Daubeny 1860a, but did cite other examples of monstrosities from the botanical work of Candolle, Masters, and Moquin-Tandon in *Variation* 1: 365–6 and nn., and 2: 58 and n., 59 n., 61 n., 245–7, and 254 n.

To A. D. Bartlett 24 August [1860][1]

Down Bromley Kent
Augt. 24th

Dear Sir

I have directed a copy of my "Origin of Species" to be sent to your address to the Soc^{y}. Rooms in Hanover Sq^{re}, & I hope that you will do me the favour to accept it.—[2] If you will read the Ch. on Hybridism at p. 264 you will see why I am curious about the embryos in eggs from *first* crosses.—[3] I was very glad to see a Donkey with a wild ass in the Gardens for I infer from this that you intend rearing a Hybrid; if so I hope that you will look carefully for stripes on shoulder *& legs* **in the foal**; you will see why I am so curious on this head, if you will read the little discussion in the Origin from p. 163–167.—[4]

I will let you hear about the Moscow Rabbits, after I have heard from the young Lady who bought them, whether she consents to their being sent to the Gardens.—[5]

I was much interested by the facts you kindly communicated to me[6] & remain Dear Sir | Yours very faithfully | Charles Darwin

If you should hear from Hunt anything about the Records of the gestation of the Canidæ—or about the parents of Hybrid Jackal, perhaps you will be so kind as to inform me.—[7]

The LuEsther T. Mertz Library Archives at The New York Botanical Garden (Charles Finney Cox Collection)

1 The year is established by the dates of publication of the first three editions of *Origin* (Freeman 1977). The discussions and pages to which CD refers are found in the first and second editions of *Origin*, published in November 1859 and January 1860 respectively. The third edition of *Origin* was published in April 1861 with different pagination.

2 Bartlett was superintendent of the gardens of the Zoological Society of London (*Modern English biography*); the offices of the society were at 11 Hanover Square (Scherren 1905, p. 99).

3 CD refers to part of the section entitled 'Causes of the sterility of first crosses and of hybrids' (*Origin*, pp. 263–7). On page 264, CD considered two mechanisms of sterility in first crosses, that is, crosses between distinct species; his curiosity related to abortion following fertilisation, a mechanism that he thought had not been 'sufficiently attended to'. Bartlett, whose assistance CD acknowledged frequently in *Variation* and *Descent*, was not cited on this subject in CD's works.

4 CD evidently visited the gardens of the Zoological Society while in London on or after 21 August 1860 (Emma Darwin's diary (DAR 242)). The wild ass is likely to have been *Asinus hemionus* (the Kiang or Tibetan wild ass) or *A. hemippus*, specimens of both species having been acquired in 1859 by the Zoological Society (Sclater 1862, pp. 163–4). CD considered the presence of stripes in horses and their relatives as a reversion to a primitive character (*Origin*, pp. 163–7; see also *Correspondence* vols. 6–8 and *Variation* 2: 351).

5 In *Variation* 1: 106–7, CD reported receiving two live rabbits from Moscow and commented on their colouring; no indication has been found in the *Proceedings of the Zoological Society of London* of their receipt at the society's gardens in Regent's Park. These rabbits appeared to be closely related to a species that Bartlett had provisionally named *Lepus nigripes* (Bartlett 1857; see also *Variation* 1: 108–10).

6 No letter from Bartlett in 1860 has been found.

7 James Hunt was head keeper at the gardens of the Zoological Society until 1859 (Scherren 1905, p. 104). No information from him on these subjects has been found. However, details of the gestation periods of members of the Canidae, and of a sterile female from a jackal crossed with a domestic dog at the zoological gardens, appeared in *Variation* 1: 29–30 and 32–3.

To a bookseller 9 September [1860][1]

Down Bromley Kent
Sept. 9th

Dear Sir

If you can get from Trübner[2] the "Atlantic Monthly" for last *August*,[3] please send me a copy by Post here; but it is not worth sending to America for.—

Please order for me *two* copies of the Atlantic Monthly for next *October*[4] & send here.—

Dear Sir | yours faithfully | C. Darwin

Christophe Stickel Autographs

1 The year is established by the relationship between this letter and the letter to Asa Gray, 10 September [1860] (*Correspondence* vol. 8).

[2] CD refers to Trübner and Co. of London.

[3] The August 1860 issue of the *Atlantic Monthly* contained the second, and final, part of an anonymous review, 'Darwin on the origin of species' ([A. Gray] 1860, pp. 229–39). There is an annotated copy of each part in the Darwin Pamphlet Collection–CUL.

[4] The October 1860 issue of the *Atlantic Monthly* contained 'Darwin and his reviewers', the anonymous sequel to the two-part review ([A. Gray] 1860, pp. 406–25; see n. 3 above). There is an annotated copy in the Darwin Pamphlet Collection–CUL. For the attribution of the review and its sequel to Asa Gray, see *Correspondence* vol. 8, letter to J. D. Hooker, 19 [July 1860] and n. 2. In 1861, Trübner published all three parts of [A. Gray] 1860 as a pamphlet on which Gray's name appeared as author (A. Gray 1861). There are two copies in the Darwin Pamphlet Collection–CUL; one is lightly annotated.

From Hewett Cottrell Watson to George Gordon 19 September 1860

Thames Ditton
19 Sep^{t} 60

My dear Sir

I took the liberty of giving your address to Mr. Darwin, anxiously inquiring how he could procure some living examples of Goodyeræ.—[1] His name would doubtless be known to you, & more favourably than any introduction by me. But you will perhaps still allow me to express sincere thanks for your kindness in so promptly responding to his application.[2] He seems highly pleased with the successfulness of his application.—[3]

Mr. D. takes immensity of pains in putting his theoretic views to the test of observation & experiment; &, as a true seeker of truth, deserves all help & encouragement, whether or not we adopt his theoretic views.— The more I cogitate on those views, the more I feel that there must be some fallacy or insufficiency which vitiates ideas partly sound & important. 'Natural Selection' is a fact, & a vera causa. Can it operate to the extent Darwin contends?—or, is it not also checked and counterbalance by some other provision or prevision?[4]

Yours, Re^{d} Sir, | Very truly | Hewett C. Watson

The Rev^{d} | G. Gordon

Elgin Museum: Gordon Archive (60.15)

[1] The letter from Watson to CD has not been found. Watson had already assisted CD in the procurement of orchids (see, for example, *Correspondence* vol. 8, letter to A. G. More, 24 June 1860). *Goodyera* was of special interest to CD for his study of the comparative anatomy of the rostellum in British orchid species (see *Correspondence* vol. 8, letter to George Gordon, 11 September [1860]; see also *Orchids*, pp. 112–16).

[2] For CD's application to Gordon for specimens, see *Correspondence* vol. 8, letter to George Gordon, 11 September [1860].

[3] Gordon sent the requested specimens of *Goodyera repens* to CD (see *Correspondence* vol. 8, letter to George Gordon, 17 September [1860]). CD acknowledged Gordon for supplying specimens of 'this rare Highland Orchid' in *Orchids*, p. 112 n.

[4] For Watson's views on forces that counterbalance the proliferation of species through natural selection, see, for example, *Correspondence* vol. 8, letters from H. C. Watson, [3? January 1860] and 10 May 1860.

To David Forbes 11 December [1860][1]

[Down]

[in reply to various contributions and criticisms by Forbes, and with Darwin's request for information on a point he wanted to check for the new edition of *Origin of Species*, which was completed a few weeks later:[2] 'Your letter pleased me exceedingly.[3] When I saw how much error you detected in d'Orbigny, I said to myself "good heavens my turn will come next"';[4] encouraging Forbes to submit his observations on Chile to the Geological Society and offering advice on publication;[5] replying to various matters raised by Forbes including 'I am sorry for my blunder about [Andesite]',[6] asking Forbes if he had done any work on plutonic rocks, commenting on some observation of the Norwegian geologist B. M. Keilhau: '... it seems to me very mysterious & indicating some great unknown class of facts';[7] asking for help in sorting out a point in the first edition of *Origin of Species*: 'I know how busy you are & must be: but it would be a *great favour*, if you would write literally one line to me to say whether the(?) great mound of detritus in Val del Jesu(?) has any connection with ice-action; because I am preparing new [sic] edition of Origin of Species, & if you say "no" I will strike out 3 lines: if you say "yes" I will leave them as they now stand' [...],[8] and commenting that features he saw near the hot springs of Cauquenes in the Cordillera (in 1834), he was now inclined to attribute to glacial (rather than fluvial), action;[9] asking Forbes if he would undertake some observations on mules.]

[...] it trouble you to make a few observations on mules, as by enclosed paper, & some time write me the results?[10]

With every good wish for your success & prosperity, & with cordial thanks for your letter and kind expressions. Believe me | My dear Sir | yours very sincerely | C. Darwin

Incomplete[11]
Bonhams (12 September 2000)

1 The date is given in the sale catalogue; the year is confirmed by the relationship between this letter and the publication of the third edition of *Origin* (see n. 2, below).

2 In his Journal, CD noted that he worked in December 1860 and January 1861 on the third edition of *Origin*, which was published in April 1861 (see *Correspondence* vol. 9, Appendix II). The new edition included information from Forbes about glacial action on the rocks of the Andes (see *Correspondence* vol. 8, letter from David Forbes, [November? 1860] (now dated [after 11 December 1860]; see n. 3, below), and *Origin* 3d ed., pp. 403–4).

3 Forbes's letter has not been found. The letter from David Forbes, [November? 1860] (*Correspondence* vol. 8), which should now be dated [after 11 December 1860], is a reply to this letter.

4 CD refers to D. Forbes 1860, in which Forbes challenged many of Alcide Charles Victor Dessalines d'Orbigny's findings on the geology of South America (Orbigny 1835–47; see *Correspondence* vol. 8, letter to Charles Lyell, 4 December [1860]). For a summary of the differences between Forbes and Orbigny, see *Proceedings of the Geological Society of London* 33 (1877): 44–6.

5 D. Forbes 1860 was read to the Geological Society of London on 21 November 1860. It concerned the geology of Bolivia and southern Peru, and was intended as the first part of a memoir that would include two further parts on the geology of Chile and the Argentine (D. Forbes 1860, p. 62). However, Forbes read no further paper on the geology of South America to the Geological Society.

[6] CD refers to andesite, the plutonic rock that CD named having first found it in the vicinity of the Beagle Channel (*South America*, pp. 154–5 *et seq.*). Forbes had revised CD's nomenclature, preferring diorite on account of its similarity to European diorites (D. Forbes 1860, p. 29).

[7] CD had earlier been impressed by Baltazar Mathias Keilhau's theory of the origin of rocks by heat-induced crystallisation (see Keilhau 1838–40 and 1844, and *Correspondence* vol. 3, letter to Adolf von Morlot, 10 October [1844] and n. 7).

[8] CD described the geological features of the Yeso valley, in Chile, in *South America*, p. 178. This may have been the source for the passage about the 'vast mound of detritus' in an Andean valley in *Origin*, p. 373. After receiving Forbes's reply (*Correspondence* vol. 8, letter from David Forbes [November? 1860], now redated [after 11 December 1860]), CD restated his conviction that the mound was due to glacial action (*Origin* 3d ed., pp. 403–4).

[9] See *South America*, pp. 64–7. While a small number of boulders near Cauquenes in Chile had previously been described by Adolphe Théodore Brongniart as erratics deposited by glaciers, CD argued that river action and uplift could have displaced them from their original strata (*South America*, p. 66 and n.).

[10] The enclosure listing questions for Forbes has not been found. By mules, CD probably means hybrids between goats and sheep. For Forbes's reply, see *Correspondence* vol. 8, letter from David Forbes, [November? 1860], now redated [after 11 December 1860]. CD cited Forbes on the sheep of Chile in *Variation* 1: 95, and considered the question of the possible relatedness of sheep to goats in *Variation* 1: 94, 98–9, and 102.

[11] The original letter is complete and is described in the sale catalogue as being five pages long.

From H. C. Watson to J. D. Hooker 4 January 1861

Thames Ditton S.W.
19 Jnry. 61

My dear Hooker,

Better late than never; and where there is ample reason for lateness, there can be none for complaint. Many thanks for your letter de convergentibus &c.—[1] How much outside the botanical circles,—or, at any rate the great circle,—not even to have heard that you had been beyond seas again![2] Was there ever a botanist who had seen so much of the earth & its vestments?[3]

How vastly I should have enjoyed the run over Syria with you! And yet if I had chanced to be away from England those eight weeks, inconveniences to myself & others would have been so incurred, that might hardly have been removed in as many years. I seem always called on to give time & thought to things no wise botanical, while wishing to give them to matters that are botanical, or connected with botany.[4]

I can hardly conceive you & Mr. Bentham employed more serviceably than on a 'Genera Plantarum'.[5] So many qualifications & advantages concentrate in you two, towards such an undertaking, that the accomplishment of your task ought to date as a new formation in the series of botanical strata—& doubtless will do so.—Can a *list* of species under each genus be compassed, if only single names?[6]

You have not yet grasped the whole idea of convergence, as I conceive of it whether rightly or erroneously.[7] No wonder. If Darwin had sought to explain his notions of divergence & selection in compass of a letter sheet, he might have got further or fuller than I could do with its (presumed) counterpoise, & yet done it most imperfectly.

Darwin watches minutesimal variations; & he supposes that these may & do accumulate into any extent of diversity in long time.— I suspect that we should all have said, that is pure fancy or pure absurdity, if geology had not shown the fact, that enormous differences do exist between the present & the past, & must have come about somehow.[8]

I watch minutesimal variations also; & I suppose that these may & do accumulate into any degree of similarity in long time. But, seeing that we know not the future, and cannot compare the present therewith, we lack corroboration. And we (i.e. you &c.) do think convergence to such a degree mere fancy or absurdity.

You & Darwin seem to me at present something like an Astronomer endeavouring to account for the planetary system by a centrifugal force alone, without taking into account a centripetal force, or recognising it only as occasional, anomalous, trifling.[9] Darwin's idea that similitudes must be always inherited from some predecessor, common ancestor of all that so resemble, if pushed to its logical consequences, would concentrate all organic characters in the dozen (or single) original types;—and that surely is a real as well as logical absurdity?[10]

Observe,—every the smallest possible change in a plant, which to that extent renders it more like another plant, (whether that other plant be the nearest, or the most remote otherwise) *is* a step in convergence. Advergence would perhaps be a better term. Can you point out divergences in any plants, which are certainly not advergences towards some other plants?

You will be ready to cry, 'Hold enough'. And I will not inflict on you a criticism of the various ideas in your letter;—some of which indeed I quite admit the force of; though I think that the arguments against $^{\text{ad}}_{\text{con}}$vergence are not proof against reply.—

The Indian specimens you kindly sent some time ago are still kept apart, awaiting some other introductions into my herbarium, before trying how closely I can concur with your labels of European identity.[11] Unlike you, I have to go through *all* the mechanical processes, such as washing over with poison, glueing to paper, & so forth,—sadly time-consuming.[12]

Writing with back to the fire, I still have to keep rubbing fingers, or the pen would fall from their cold benumbed trembling guidance—

Glad to have so favorable an account of your Father—[13]

Ever faith^{ly} | Hewett C. Watson

Dr J. D. Hooker

Royal Botanic Gardens, Kew, DC 105 (205)

1 The letter from Hooker to Watson has not been found.

2 Hooker had taken part in an expedition to Syria from mid-September to mid-November 1860 (L. Huxley ed. 1918, 1: 528; see also *Correspondence* vol. 8).

3 For details of Hooker's wide-ranging travels, see, for example, R. Desmond 1999.

4 Watson was active as a politician (*DNB*). By 1861, he travelled rarely and was engaged in philanthropic activity (Frank Egerton, personal communication).

[5] Hooker's collaboration with George Bentham on *Genera plantarum* is documented in Stearn 1956, and Hooker's work on the volume began in earnest in March 1860 (see *Correspondence* vol. 8, letter to J. D. Hooker, 12 March [1860] and n. 4). The aim of the work was to describe all known genera of flowering plants and gymnosperms from specimens in the herbarium at the Royal Botanic Gardens, Kew, and to include all known synonyms. The first part of the first volume was published in 1862 (Bentham and Hooker 1862–83).

[6] Bentham and Hooker 1862–83 did not contain lists of species. However, Hooker and Benjamin Daydon Jackson's *Index Kewensis*, financed by a bequest from CD and first published in 1893, did provide such a list.

[7] Watson's views on 'convergence' were summarised in a memorandum to CD (*Correspondence* vol. 8, letter from H. C. Watson, [3? January 1860]). Hooker wrote to CD on the subject in his letter of [11 May – 3 December 1860] (this volume, Supplement), and in the missing fragment of his letter of [26 November – 4 December 1860] (*Correspondence* vol. 8; see also *Correspondence* vol. 8, letter to J. D. Hooker, 4 December [1860] and n. 6). CD referred to Watson's criticism in the revised American edition of *Origin* and in the third and later editions of *Origin* (see, for example, *Origin* 3d ed., pp. 141–3).

[8] CD referred to the geological record in support of his transmutation theory in *Origin*, chapters 9 and 10.

[9] Hooker had previously described centripetal and centrifugal models of variation (*Correspondence* vol. 3, letter from J. D. Hooker, 1 February 1846); however, Watson and Hooker evidently understood the terms centripetal and centrifugal somewhat differently. For Watson, centripetal variation afforded a mechanism for the convergence of species (see also this volume, Supplement, letter from J. D. Hooker, [11 May – 3 December 1860]). Hooker used centripetal variation to designate variations that were prone to revert to a parental type. CD objected to Hooker's notion, arguing that *Gnaphalium* species tended to 'revert to more than one centre or type' (*Correspondence* vol. 3, letter to J. D. Hooker, [16 April 1846]). Hooker used centrifugal variation to designate variation that accompanied extensive geographical distribution, as in *Senecio* (Hooker 1844–7, p. 315), while Watson used it to signify the mechanism of divergence. CD disapproved of Hooker's use of the word centrifugal (*Correspondence* vol. 7, letter to Charles Lyell, 31 [October 1859] and n. 3).

[10] The correspondence between CD and Watson on 'convergence' is in *Correspondence* vol. 8 (letters from [H. C. Watson], [3? January 1860] and 10 May 1860, and letter to H. C. Watson, [5–11 January 1860]). See also this volume, Supplement, letter from J. D. Hooker, [11 May – 3 December 1860].

[11] Hooker had been on a botanical expedition to India from November 1847 to January 1851 (R. Desmond 1999, pp. 96–177). Hooker had presumably asked for comparisons between his Indian specimens and specimens in Watson's herbarium of British plants (*DNB*).

[12] These processes are stages in the preparation of herbarium specimens. For more on preparation of specimens, see Oliver 1881, pp. 269–73. Watson had private means and worked from home (*DNB*).

[13] Watson refers to William Jackson Hooker.

From Charles and Emma Darwin to W. E. Darwin [13 January 1861][1]

Down.—
Sunday morning

My dear William

There are 2 letters for you at Erasmuses (one marked to be forwarded), which the Boys took up as we did not then know your address—[2]

Poor George came back last night, having had teeth out & ever so many stopped.—[3] Consequently he did not call at Moores the Cutler (164. Oxford St) for his Skates—[4] Please be sure call for them & pay for them.— If not done have them sent to care of

Mitchell, Postman
Bromley.—[5]

Remember carrige will meet Train at Bromley which leaves London. B.[6] at 2º.30'—

Your affect | C. Darwin

Dear Wm,[7]

I expect Eva will come with you, so please call at the Macks & see. The girls stay till tomorrow week which is very nice.[8]

Georgy is frantic for caps if you have time 38 Broad St | Golden Sq[9]

DAR 185: 117

[1] The letter is dated by the reference to George Howard Darwin's dental treatment (see n. 3, below) and references in Emma Darwin's diary (DAR 242; see nn. 2 and 8, below). In 1861, 13 January was a Sunday.

[2] CD refers to his brother, Erasmus Alvey Darwin, who lived at 6 Queen Anne Street, London, and to Francis and George Howard Darwin, who, according to Emma Darwin's diary (DAR 242), went to London on 10 January 1861, returning to Down on 12 January.

[3] George underwent extensive dental treatment between December 1860 and February 1861 (see letter from G. H. Darwin to W. E. Darwin, [9 December 1860] (DAR 251: 2226), letters from Emma Darwin to W. E. Darwin, [30 January 1861] (DAR 219.1: 36), [12 February 1861] (DAR 219.1: 37), and [20 February 1861] (DAR 219.1: 38), and *Correspondence* vol. 9, letter to J. D. Hooker, 4 February [1861] and n. 3).

[4] CD refers to Sophia Matilda Morrell, cutler and jeweller of 164 Oxford Street, London (*Post Office London directory* 1861).

[5] Mr Mitchell (see also *Correspondence* vol. 9, letter to W. B. Tegetmeier, 22 March [1861]) has not been further identified.

[6] London Bridge station.

[7] Emma Darwin added this note to CD's letter.

[8] The references are to Eva Mackintosh, who arrived at Down House with William on 14 January 1861 (Emma Darwin's diary (DAR 242)); to the London home of her father, Robert James Mackintosh, at 2 Hyde Park Terrace (*Post Office London directory* 1861); and to Katherine Euphemia and Hope Elizabeth Wedgwood, who arrived at Down House on 11 January 1861, and left on 21 January (Emma Darwin's diary (DAR 242)).

[9] Emma was mistaken about the address. Eley Brothers, percussion-cap makers, occupied 38 Broad Street, Golden Square, London, until 1858; by 1861 they had moved to Calthorpe Place, Gray's Inn Road, and 38 Broad Street was being used by Oetzmann & Plumb, pianoforte makers (*Post Office London directory* 1858, 1861).

To Archibald Geikie 27 February [1861][1]

Down Bromley Kent
Feb. 27th

Dear Sir

You cannot know how much your note has pleased me.[2] I am naturally, & I hope not only selfishly, deeply interested in the progress of opinion on the mutability of species. I have long been convinced that species will be held to be all aboriginal creations as long as the geological record is looked at as nearly perfect.[3] Therefore you may believe with what high satisfaction I have read the remarks of one, who like yourself has large & practical knowledge of geology.[4]

He who can shake the general belief that we should be able to trace the mutation of species in our formations, if they (ie species) have changed, will I believe do high service to natural science; & this will be in your power.—

Again let me thank you for your note; though I was far from expecting that anyone would thank me for so trifling a present.—[5]

I have just finished correcting a new Edit. of the Origin with some additions, & I will do myself the pleasure of telling my Publisher to send you a copy, as soon as it is out.—[6]

Pray believe me | Dear Sir | Yours very faithfully | Charles Darwin

DAR 185: 131

[1] The year is established by the reference to the publication of the third edition of *Origin*; see n. 6, below.

[2] Geikie's letter has not been found.

[3] CD wrote about the imperfection of the geological record in chapter 9 of *Origin* (pp. 279–311).

[4] Geikie was a member of the Scottish branch of the Geological Survey of Great Britain (*DNB*). See also *Correspondence* vol. 9, letter to T. W. St C. Davidson, 30 April 1861 and n. 4.

[5] Geikie may have written to thank CD for the presentation copy of Asa Gray's pamphlet on *Origin* (*Natural selection not inconsistent with natural theology. A free examination of Darwin's treatise on the origin of species*; A. Gray 1861) that CD sent him in late February 1861 (see *Correspondence* vol. 9, Appendix III).

[6] The third edition of *Origin* was published in April 1861. Geikie's name is on CD's list of those sent presentation copies (see *Correspondence* vol. 9, Appendix VII); his name is not on CD's list of those sent presentation copies of the second edition (see *Correspondence* vol. 8, Appendix III).

To Richard Kippist 27 [February or March 1861?][1]

Down
27.

My dear Sir

I will send Carrier[2] for Books tomorrow Thursday Morning & he shall pay 6^{s} but please give a slip as receipt, for I do not like trusting Carriers.—

If you have in Library Agricolas Latin Work on Agriculture (or any Translation)[3] please send it.—

If you have **not** Agricola please send in lieu (for he quotes Agricola) 'Aldrovandi de Quadrupedibus digitatis 1637.'—[4]

Have you Castelnau voyage in S. America;[5] I want to see the Historical part, **not** the *pure* Zoological part.

Lastly (God forgive me) if at present in, please send Brown-Sequard's Journal de Phys: Vol. 2.[6] but I know Mr Sclater[7] had it out.—

Yours sincerely | C. Darwin

Morristown National Historical Park (Lloyd W. Smith MS 697)

[1] The date is conjectured by reference to the works CD wished to consult (see nn. 4–6, below), which are cited in *Variation* 1, chapters 3 and 4. In his journal, CD wrote that he finished chapter 3 of *Variation* and began chapter 4 on 20 March 1861 (see *Correspondence* vol. 9, Appendix II). In 1861, 27 February

and 27 March were both Wednesdays. Chapter 3 of *Variation* (1: 65–102) is on pigs, cattle, sheep, and goats; chapter 4 (1: 103–30) is on rabbits.

2 George Snow ran a carrier service between London and Down every Thursday (*Post Office directory of the six home counties* 1862). Kippist was the librarian of the Linnean Society (*DNB*); CD often arranged to borrow the books that he needed to consult for his research.

3 CD may refer to G. A. Agricola 1721, *A philosophical treatise of husbandry and gardening; being a new method of cultivating and increasing all sorts of trees, shrubs, and flowers*. However, it was originally written in High-Dutch, not Latin. It is mentioned, and marked 'read', in CD's list of books to be read in his reading notebooks (see *Correspondence* vol. 4, Appendix IV, *119: 19v.), but it is not cited in *Variation*.

4 Aldrovandus 1637. This work was cited in *Variation* 1: 104. G. Agricola 1614, a study of burrowing animals, was cited in Aldrovandus 1637, pp. 384 and 388, not G. A. Agricola 1721 (see n. 3, above).

5 Castelnau 1850–7, *Expédition dans les parties centrales de l'Amérique du sud . . . exécutée par ordre du gouvernement français pendant les années 1843 à 1847*. Francis de la Porte Castelnau is cited for information on cattle in *Variation* 1: 88, but no bibliographic reference is given.

6 The *Journal de la physiologie de l'homme et des animaux* was edited by Charles Edouard Brown-Séquard. Volume 2 contained the third and fourth part of an article on hybridity in hares and rabbits by Pierre Paul Broca (Broca 1858–9); the fourth part was cited in *Variation* 1: 105.

7 Philip Lutley Sclater.

To Williams & Norgate 4 March [1861][1]

Down Bromley Kent
March 4th

Dear Sir

I am not the less obliged to you for getting me Ferguson,[2] that on my telling an acquaintance that I could not purchase a copy he sent me his own as a present.—[3]

It is not much of a loss, but if you could dispose of copy sent, for a lesser price, I would return it you.— But this is not likely, if I do **not** hear I will understand in a few days that I must keep it or rather give it away.

Yours sincerely | C. Darwin

Staatsbibliothek zu Berlin – Preussischer Kulturbesitz

1 The year is established by the relationship between this letter, the letter from W. A. Wooler, 4 February 1861, and the letter to W. B. Tegetmeier, 25 February [1861] (*Correspondence* vol. 9); also, according to his journal, CD was working on chapter 7 of *Variation*, 'Fowls' (1: 225–75), from mid-March 1861 (see *Correspondence* vol. 9, Appendix II). Ferguson 1854 (see n. 2, below) is cited in *Variation* 1: 248 n.

2 Ferguson 1854 (*Ferguson's illustrated series of rare and prize poultry*); there is an annotated copy in the Darwin Library–CUL (see *Marginalia* 1: 227–9).

3 The acquaintance may have been William Alexander Wooler; see letter from W. A. Wooler, 4 February 1861 (*Correspondence* vol. 9).

To Daniel Oliver 23 March [1861][1]

Down Bromley Kent
March 23d

My dear Sir

You told me formerly that you did not much care about my troubling you: I hope to Heaven you keep of same mind.— Will you ask Mr. Fitch to make two little diagrams for me for woodcuts; as by enclosed paper.—[2] And will you kindly give him the specimens, viz two common Primroses, one "pin-headed, as the Florists say with stigma at mouth of corolla: & the other with stamens at the mouth.— And cut them longitudinally with sharp scissors Mr Fitch could keep a little account against me.— I *think* I shall publish an account of my observations & experiments on Primula; i.e. on its dimorphous condition.[3] By the way you will find the pin-headed & *non*-pin-headed in any bank of Primroses.— I find that with P. Sinensis, the short-stamened flowers fertilise themselves, whereas the long-stamened flowers will not set without man's or insects' aid.—

I am now crossing largely Cowslips & polyanthuses. And this leads me to beg a second favour, viz to send me 2 or 3 flowers (not whole trusses) of pin-headed & non-pin headed, of any species of Primula, except P. vulgaris, Sinensis & Auricula.— If you have any other species pray fold me a few flowers in *double* green leaf, & pollen would come neither too dry or too wet for measurement.— If you can send please send names of species.—[4]

Yours very truly | Ch. Darwin

[Enclosure]

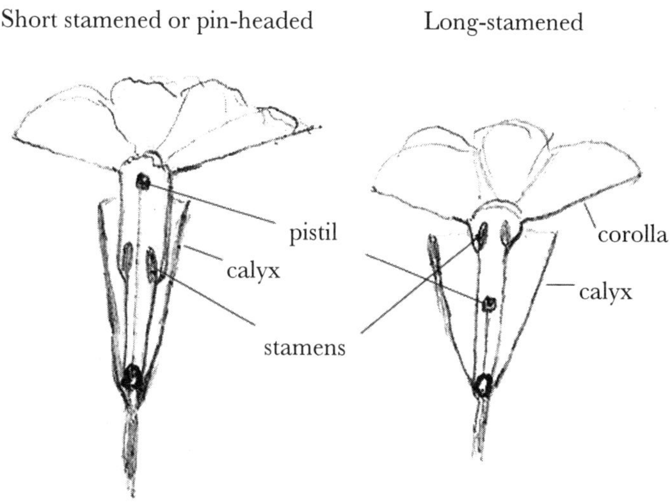

Points to be attended to: relative heights of pistil & stamens—of corolla & calyx; & of the *enlargement* of corolla, where anthers are situated.—

(Longitudinal Section of *Common Primrose*)— Clean, hard outlines for 2 woodcuts, matching in size & to stand near alongside each other.[5]

Corolla & calyx of same length in the two

DAR 261.10: 4 (English Heritage MS 88205988); diagram © Christie's Images

[1] This letter was published without its enclosure in *Correspondence* vol. 9. The year is established by the reference to CD's paper on *Primula* (see n. 3, below), which was read before the Linnean Society on 21 November 1861.

[2] Walter Hood Fitch was a freelance botanical artist who often worked at the Royal Botanic Gardens, Kew (R. Desmond 1995). The woodcuts he prepared were published in CD's paper on dimorphism in *Primula* (see n. 3, below).

[3] 'Dimorphic condition in *Primula*' was CD's first published study of dimorphism. CD stated that, although botanists recognised the existence of the long-styled and short-styled forms of *Primula*, they regarded the phenomenon as 'mere variability' (*ibid*, p. 78; *Collected papers* 2: 45). He illustrated that this recognition was not specific to the botanical community by referring to florists' use of the terms 'pin-eyed' and 'thumb-eyed' to describe what he believed to be two distinct forms of *Primula* ('Dimorphic condition in *Primula*', p. 77; *Collected papers* 2: 45).

[4] Oliver sent CD the flowers of a number of different *Primula* species ('Dimorphic condition in *Primula*', p. 81; *Collected papers* 2: 48).

[5] The diagram was published in 'Dimorphic condition in *Primula*', p. 78 (*Collected papers* 2: 46).

To B. P. Brent 1 April [1861][1]

Down Bromley Kent
April 1.

My dear Sir

Again I have to thank you for your *extreme* kindness, in sending me so much information.[2] I daresay I shall have to use & quote some of your remarks on combs: indeed I see some already which will be useful to me.—[3] Thank you much for your offer of the Pigeon; but your information will suffice, & I shall quote it, as I am glad to show variability in structure of tail.[4]

Many thanks for C. Gardeners, (returned by this Post), I see you have asked my questions.— The Angora seems very amiable & I presume not stupid.—[5] What a pleasant article you have written on Canaries & Gold-finches! I am too ignorant to attempt to say a word on Canaries.—[6]

It could not have been I, who told you about G. Bankiva having sometimes tuft of feathers on back of Head.— Mr Blyth has examined in India hundreds of specimens, & has mentioned every *trifling* point in which the wild birds vary.—[7] Indeed I cannot believe that G. Sonneratii has had anything to do with parentage of our fowls; because not only does the comb & curious hackles differ so much; nor solely because the hybrids for G. Sonneratii & the domestic fowl are very sterile; but chiefly because the voice is utterly & entirely different.—[8]

With cordial thanks | Yours very sincerely | C. Darwin

P.S Again I have to thank you for letter just received; but I have had not had time yet to consider it: but I can see that it will be very useful.—[9]

Richard Brent (private collection)

[1] The year is established by the references to articles in the 26 March 1861 issue of the *Cottage Gardener* (see nn. 5 and 6, below).

[2] The letter from Brent has not been found.

3 CD discussed fowl combs in *Variation* 1: 253–4. He cited Brent in note 50 (*ibid.*, p. 253): 'I am much indebted to Mr. Brent for an account, with sketches, of all the variations of the comb known to him, and likewise with respect to the tail . . .'

4 Brent may have offered CD an air-tumbler pigeon; in *Variation* 1: 159, CD discussed the 'remarkable' tail-feathers of several air-tumbler pigeons that Brent possessed.

5 CD returned the issue of the *Cottage Gardener and Country Gentleman* for 26 March 1861, for the notice of the Angora rabbit, see pp. 388–9. CD cited this discussion of the Angora in *Variation* 1: 106 n. 9, giving a different page number, and the name the journal acquired in April 1861 (CD wrote *Journal of Horticulture* for *Journal of Horticulture and Cottage Gardener*).

6 CD refers to B. P. Brent 1861, an article on goldfinches, on breeding goldfinches with canaries, and on the resulting hybrids; this article was part of a series Brent was writing on 'The canary and the British finches', appearing in consecutive issues of the *Cottage Gardener*. CD discussed canaries in *Variation* 1: 295, citing Brent; he also mentioned canaries several times in the second volume of *Variation*, occasionally citing Brent.

7 CD refers to the wild fowl *Gallus bankiva*, and to Edward Blyth. CD cited Blyth in his description of *G. bankiva* in *Variation* 1: 235–6. For Blyth's correspondence with CD on fowls, see *Correspondence* vols. 5 and 6.

8 For CD's description of *Gallus sonneratii* and for his argument that *G. bankiva* is the ancestor of all domestic fowl, see *Variation* 1: 233–9. See also *Correspondence* vol. 5, letters from Edward Blyth, [30 September or 7 October 1855] and 8 October 1855, *Correspondence* vol. 6, letter from Edward Blyth, 23 February 1856 and n. 21, and this volume, letter to W. B. Tegetmeier, 27 February [1865], n. 5.

9 The letter from Brent has not been found.

To W. B. Tegetmeier 8 May [1861][1]

Down. | Bromley. | Kent. S.E.
May 8th

My dear Sir

Your note is a mine of wealth to me:[2] I feared that you had been ill, because you said you had been in a former note[3] & I thought it might have recurred.

I did mean to count the little feather $\frac{1}{2}$ inch long, but Mr Gould[4] tells me it ought *not* to be counted, so that for the future we neither will count it: & I understand that your Game Bantam had 10 **not** counting this minute feather.—[5] I am sorry to say that I have not attended to Brain Development, so cannot aid you about Pigeon;[6] it would be most laborious work.

Can you really believe Pigeons ever fly a league a minute?— I am very glad to have the speed for long distances: I shall see your paper in due time in N. Hist. Review & that will suffice, so will not trouble you to send M.S.[7]

Thanks for Hackles; since writing I have examined G. Sonnertii & was convinced that what Mr. Brent wrote was without any foundation.[8] He is a very obliging kind man, but very crotchetty.— Thanks about identity of Polish Fowls.—[9]

Thanks about Malay tails (I have found same numbers in Shangais)—[10] a pretty man to trust is Ferguson with his 12 caudal feathers![11] I have some other questions to ask & remarks on your answers but it will be more convenient for you, if I write to match my former questions.

With sincere thanks | Yours very truly | C. Darwin

I hope in a week to put my M.S. in Copyist Hand.—[12] I have now examined 55 skulls.[13] I find great variability; so that few of the breeds seem *certainly* characterised by any difference.— I shall beg to be permitted to have two or three of your skulls engraved.—[14]

The LuEsther T. Mertz Library Archives at The New York Botanical Garden (Charles Finney Cox collection) (Tegetmeier 34)

[1] The year is established by the relationship between this letter and the letter from W. B. Tegetmeier, 4 May [1861] (*Correspondence* vol. 9).

[2] See *Correspondence* vol. 9, letter from W. B. Tegetmeier, 4 May [1861].

[3] This letter has not been found.

[4] John Gould.

[5] CD's remark relates to wing feathers; see *Variation* 1: 258.

[6] In his letter of 4 May [1861] (*Correspondence* vol. 9), Tegetmeier asked CD about cerebral development in newly hatched pigeons.

[7] See *Correspondence* vol. 9, letter from W. B. Tegetmeier, 4 May [1861]. No paper by Tegetmeier on racing pigeons appeared in the *Natural History Review*, and no such paper is listed in the *Royal Society catalogue of scientific papers*.

[8] Tegetmeier had sent hackles (neck feathers) from a 'Duckwing and Black red Game' with his letter of 4 May [1861] (*Correspondence* vol. 9). The written comment by Bernard Peirce Brent to which CD refers has not been identified, but may relate to CD's argument in *Variation* 1: 233–4 that *Gallus sonneratii* was not the ancestor of any modern domestic breed. CD commented: 'its hackles partially consist of highly peculiar, horny laminæ, transversely banded with three colours; and I have met with no authentic account of any such character having been observed in any domestic breed' (*Variation* 1: 233–4). See also this volume, Supplement, letter to B. P. Brent, 1 April [1861] and n. 8.

[9] Tegetmeier had written to CD that Polish fowl were not generally 'semi idiotic'; CD had thought that the modification of the form of the skull and consequently the brain of the Polish fowl might have affected its intellect (see *Correspondence* vol. 9, letter from W. B. Tegetmeier, 4 May [1861], and *Variation* 1: 264).

[10] CD first asked Tegetmeier about the number of tail feathers the Malay hen had in his letter to Tegetmeier of 25 February [1861] (*Correspondence* vol. 9), querying George Ferguson's statement on the subject (see n. 11, below); he repeated the question in his letter to Tegetmeier of 1 March [1861] (*ibid.*), and apparently in the now missing part of the enclosure to his letter to Tegetmeier of 14 April [1861] (*ibid.*; see *ibid.*, letter from W. B. Tegetmeier, 4 May [1861]). In *Variation* 1: 227, CD reported that the tails of both the Malay and the Shanghai breeds of fowl were 'generally formed of 16 feathers'.

[11] The reference is to George Ferguson, and presumably to Ferguson 1854, a copy of which CD had apparently acquired by March 1861 (see this volume, Supplement, letter to Williams & Norgate, 4 March [1861]). According to Ferguson 1854, Shanghai fowl have fourteen tail feathers (*ibid*, p. 15), and the Malay hen has ten (*ibid.*, p. 186). CD commented in *Variation* 1: 248 that Ferguson could not be generally trusted. See also *Correspondence* vol. 9, letter to W. B. Tegetmeier, 25 February [1861] and n. 6.

[12] CD finished his chapter on fowls for *Variation* on 16 May 1861 (*Correspondence* vol. 9, Appendix II).

[13] In *Variation* 1: 260, CD states that he had examined fifty-three skulls, nearly half of them supplied by Tegetmeier. Tegetmeier sent his collection of fowl skulls to CD in March 1861 (see *Correspondence* vol. 9, letters to W. B. Tegetmeier, 22 March [1861] and 28 March [1861]).

[14] Volume 1 of *Variation* has illustrations of two fowl skulls viewed from above on page 262, two longitudinal sections viewed laterally on page 263, a skull of a horned fowl on page 265, and two cervical vertebrae on page 267. The skull of the horned fowl is said to be 'in the possession of Mr. Tegetmeier'.

From H. C. Watson to George Gordon 27 June 1861

Thames Ditton | London— S.W.
27 June 1861.

Dear D^r. Gordon

I have not a botanical correspondent in Edinburgh now. Doubtless Prof. Balfour could readily get roots of Corallorhiza, through some student of his class, or by sending a gardener from the Bot^c. Garden.—[1] I will suggest this course to M^r. Darwin.[2]

I fear that Mr. D. gives much valuable time to minor matters. Instead of resting his theory on the broad general facts that tend to support it, & admitting much of difficulty & uncertainty in detail,—he seems to delight in singling out the difficulties, & trying to solve or explain them away. This adds weakness, rather than strength, to a theory that I think will turn out true in the main;—although the more I cogitate on it, the less do I see the necessity of tracing back all existent species to one or to one dozen originals.—[3]

My dear Sir | Yours Very truly | Hewett C. Watson

The rev^d. George Gordon D. D.

Elgin Museum: Gordon Archive (61.9)

[1] John Hutton Balfour was regius keeper of the Royal Botanic Garden, Edinburgh. CD had asked Gordon for help in obtaining a specimen of the orchid *Corallorhiza* (see *Correspondence* vol. 9, letter to George Gordon, 17 June [1861]).

[2] No letter from Watson to CD containing this suggestion has been found; see, however, *Correspondence* vol. 9, letter to George Gordon, 4 July [1861].

[3] Watson, a believer in the transmutation of species since 1845, had responded enthusiastically to *Origin* (see Watson 1845 and *Correspondence* vol. 7, letter from H. C. Watson, 21 November [1859]). Watson corresponded with CD on various aspects of CD's transmutation theory, and promoted the idea of a principle of convergence, which, he argued, was necessary to limit the large number of specific types that would proliferate under CD's principle of divergence (see, for example, this volume, Supplement, letter from H. C. Watson to J. D. Hooker, 4 January 1861, and *Correspondence* vol. 9, letter to H. C. Watson, [17 July 1861] and n. 3).

From George Gordon to John Hutton Balfour 1 July 1861

Manse of Birnie, Elgin
1 July 1861

My dear Sir,

Last year, at M^r. H. C. Watson's instance I sent a parcel of *Goodyera* to M^r. Darwin.[1] He (M^r. D.) has just applied to me for living specimens of *Corallorhiza*,[2] which I regret I cannot supply.

Our friend Mr Stables[3] has suggested that some of your pupils could procure them from Ravelrig bog.[4] "A few flowers in large bud" is what M^r. D. wants,[5] and his present address is Charles Darwin Esqr. (of Down, Kent) at Torquay, Devonshire".—[6]

I shall be happy to repay your pupil any expense he may be at in procuring & forwarding this plant;—and trusting you will excuse the liberty I have now taken I am | Yours truly | George Gordon

Prof^r Balfour &c &c

Royal Botanic Garden, Edinburgh

[1] See *Correspondence* vol. 8, letter to George Gordon, 17 September [1860]. Gordon's assistance in obtaining the specimen is acknowledged in *Orchids*, p. 112.
[2] See *Correspondence* vol. 9, letters to George Gordon, 17 June [1861] and 22 June [1861].
[3] Gordon probably refers to the Scottish naturalist and botanist, William Alexander Stables (R. Desmond 1994).
[4] The bog was probably located in the area of Ravelrig Hill, near Balerno, about seven miles south-west of Edinburgh (Ordnance Survey Map, Edinburgh (Midlothian), 6″: 1 mile scale, 1853). See this volume, Supplement, letter from H. C. Watson to George Gordon, 27 June 1861 and n. 1.
[5] See *Correspondence* vol. 9, letter to George Gordon, 22 June [1861].
[6] CD was in Torquay from 2 July to 27 August 1861 (see *Correspondence* vol. 9, letter to George Gordon, 22 June [1861] and n. 6, and Appendix II).

To Williams & Norgate 25 August [1861][1]

Aug 25th

Dear Sir

Please to send by Post to Down, Bromley Kent.

'Du Passe et de l'Avenir des Haras par F. Michel'.[2]

Also as soon as published 'The Three Barriers' published by Blackwood;[3] I saw it announced as out—

Dear Sir | Yours faithfully | Ch. Darwin

DAR 249: 119

[1] The year is established by the publication date of [Rorison] 1861; see n. 3, below.
[2] Michel 1861. CD's annotated copy is in the Darwin Library–CUL (see *Marginalia* 1: 241). CD cited the book in *Variation* 2: 203, 209.
[3] The publication of *The three barriers* ([Rorison] 1861) was announced in *Publishers' Circular*, 15 August 1861, p. 374. CD later referred to the book as 'a theological hash of old abuse of me' (see *Correspondence* vol. 9, letter to T. H. Huxley, 31 October [1861] and nn. 2 and 3).

To Neil Arnott 29 August [1861][1]

Down. | Bromley. | Kent. S.E.
Aug. 29th

My dear Sir

On my return home after a two months' absence in Devonshire,[2] on Tuesday night I found lying here the "Survey of Human Progress",[3] which you have very kindly sent me. I fear it has been lying here for some time. I am very glad to see that you have kept your intention of publishing on this grand & interesting subject & I hope in a few days to read your work.[4]

I thank you sincerely for sending me a copy & with much respect, I [. . .] remain My dear Sir | Yours sincerely | Charles Darwin

Remember When Antiquities (Catalogue 28)

[1] The year is established by the reference to CD's stay in Devon and by the date of publication of Arnott's book (see nn. 2 and 3, below).
[2] CD was in Torquay, Devon, during July and August 1861 (see *Correspondence* vol. 9, Appendix II).
[3] Arnott 1861 was published in August 1861 (*Publishers' Circular*, 15 August 1861, p. 373).
[4] After receiving a pamphlet from Arnott on human progress, CD had encouraged him to write a book on the subject (see *Correspondence* vol. 8, letter to Neil Arnott, 16 February [1860?] and n. 3).

From Thomas White Woodbury [9–22 October 1861][1]

I fancy the articles on "The Queen Bee" and on "Drone Influence" may not be without interest[2] Since writing the latter, I have met with another instance of the same kind, one of my drones (Ligurian) having hybridised a queen at a distance of a mile and a half[3] What renders it more remarkable, is that in both instances the city of Exeter lay between the lovers as well as several apiaries of common bees. One of these (Mr S. B. Fox's)[4] contained an unusual number of drones.

Yours | T. W. W.

Incomplete
Pinned to CD's unbound copy of *Journal of Horticulture*, 8 October 1861, p. 38 (Darwin Library–CUL)

CD ANNOTATION
End of letter: 'Woodbury | Mount Radford | Exeter' *ink*

[1] The date is established by the references to Woodbury's articles on bees (see nn. 2 and 3, below).
[2] 'Drone influence, to what distance does it extend?' and 'Age of the queen bee' appeared in the *Journal of Horticulture*, 8 October 1861, p. 39. CD's annotated copy of the issue, with Woodbury's letter pinned to the facing page, is in the Darwin Library–CUL.
[3] Woodbury communicated this observation, which he noted was made 9 October 1861, in the *Journal of Horticulture*, 22 October 1861, p. 77. CD cited Woodbury's observations in *Variation* 1: 299. CD used Woodbury's information on Ligurian drones to counter the view that 'common' bees were a case of 'inevitable close interbreeding' (*Variation* 2: 126).
[4] Sylvanus Bevan Fox. CD cited Fox as an authority on bee species in *Variation* 1: 298.

From Charles Lyell 22 October 1861

London
22 Oct. | 1861

My dear Darwin

The Swiss engineers said that they tunneled the ice barrier in the case of the valley of Bagnes because if once the water overflowed it would cause a debacle.[1]

Therefore I hold that ice could not have formed the blockages in Lochaber unless we can assume that in every case the water escaped over some 'col' either into a contiguous valley on the same watershed, or over into the eastern watershed—[2]

Whether the barrier be of ice or moraine-matter it cannot stand long eno' at one level to allow a shelf to form if the water issues at the ice-dam or moraine-dam— On the other hand if you suppose a continual increase of ice or moraine at the lower end of each glacier-lake you would have an ever-varying depth of water.

There is only one way to get a permanent level & that is by the water overflowing some "col" distinct from the lower or dammed-up extremity of the lake.

If you can cite any case in Lochaber where a shelf does not coincide with a "col" (did not Milne or some one say there was one?)[3] I cannot comprehend how it is to be reconciled with the ice or glacier-lake theory— I therefore do not suppose the "cols" were *land-straits*[4] but the places where the lakes were drained when forced to flow the wrong way—

Please to return this note that I may send it to Mr Jamieson & see if he agrees—[5]
ever sinly yrs | Cha Lyell

C. C. Kohler (dealer)

[1] The Val de Bagnes runs along the river Drance to the south-east of Martigny in the canton of Valais, Switzerland. Lyell refers to the ice and snow that commonly fell from the mountainsides and accumulated, impeding the flow of the Drance near Mauvoisin. In June 1818, after an exceptionally large accumulation of ice the previous winter, the lake that was formed above Mauvoisin burst through the ice barrier, causing devastation to the entire Val de Bagnes (Baedeker 1889, p. 292).

[2] For many years, the cause of the famous parallel terraces or 'roads' of Glen Roy in the district of Lochaber, southern Inverness-shire, Scotland, had been debated. For CD's theory of the marine origin of the parallel roads, published in 1839, see 'Parallel roads of Glen Roy'; see also Rudwick 1974. CD corresponded on the subject with Lyell and Thomas Francis Jamieson, after Jamieson's visit to Lochaber in August 1861 (see *Correspondence* vol. 9, Appendix IX). Jamieson became convinced that the roads represented the shores of former glacial lakes and that the level of these shores coincided with the level of various 'cols' or passes at the head of valleys (glens), indicating that the cols had once been points of outflow. Crucial to his hypothesis was the proposal that glacial barriers had at one time existed at the mouths of the glens. In a letter of 14 October 1861, Jamieson suggested to Lyell that the barriers of the Glen Roy lakes were 'mainly of ice' (*Correspondence* vol. 9, Appendix IX). After reading Jamieson's letter, CD had expressed doubts about the possibility of a 'mere barrier of ice' being able to account for the formation of shelves at Glen Roy (see *Correspondence* vol. 9, letter to Charles Lyell, 20 October [1861] and n. 1). Lyell, by analogy with the case of the Swiss ice barrier, evidently concluded that ice barriers in Lochaber would have given way under the force of the water or, if the ice were continually increasing, have resulted in a constantly changing depth of water. This argument was based on the supposition that the ice barrier would have been at a lower level than the col so that water would overflow the barrier rather than the col.

[3] David Milne had found a col at Glen Glaster, at the same level as one of the shelves, which had been missed by previous observers. He also mentioned two intermediate shelves, faintly discernable, which he took as evidence of the gradual lowering of a barrier (Milne 1847, pp. 398, 404).

[4] Land-strait: 'isthmus' (*OED*).

[5] Lyell evidently sent this letter to Jamieson, who, in his reply, agreed with Lyell's conclusion that cols had determined the location of the shelves and that water had overflowed the col rather than the barrier. He dismissed the conflicting evidence of intermediate shelves described by Milne, noting how faint these markings were (see *Correspondence* vol. 9, Appendix IX, letter from T. F. Jamieson to Charles Lyell, 27 October 1861).

To Daniel Oliver 3 November [1861][1]

Down. | Bromley. | Kent. S.E.
Nov 3^d

Dear Oliver

I thank you most sincerely for all the references.[2] Good Heavens how do you know such multitudinous references! It is a horrid bore, that I shall probably find all my recent work done & done better than I have.— Anyhow I have enjoyed the work. If you had not told me these references how nicely you might have reviewed. "disgraceful ignorance" "highly presumptuous" &c &c—[3]

I heartily thank you, though I expect to be heartily disgusted with the papers, when I get them. from Linn. Society.—[4]

Please thank Hooker for information about Dendrobium Chrysanthum;[5] & say that I sh^d. be very glad of a Cypripedium to have a look at the ducts of Labellum—[6]

Yours very sincerely | Charles Darwin

If Hooker sends any more orchids I sh^d like one more head of Eveylina carivata on account of its curious nectary.—[7]

DAR 261.10: 52 (English Heritage MS 88206035)

[1] The year is established by the relationship between this letter and the first letter to J. D. Hooker, 1 November [1861] (*Correspondence* vol. 9).

[2] See *Correspondence* vol. 9, letter from Daniel Oliver, [before 8 November 1861]; this letter should now be dated [before 3 November 1861]. The references were to works on orchid structure and pollination; CD was working on *Orchids* in the latter part of 1861 (*Correspondence* vol. 9, Appendix II).

[3] Oliver was an editor of the *Natural History Review*, which began publication in January 1861; he had written a number of botanical reviews. *Orchids* was reviewed by Joseph Dalton Hooker in the October 1862 issue of *Natural History Review*, pp. 371–6 (see *Correspondence* vol. 10, letter from J. D. Hooker, 7 November 1862).

[4] CD frequently borrowed books from the library of the Linnean Society; see, for example, *Correspondence* vol. 9, letter to Richard Kippist, 7 January [1861].

[5] CD had asked Hooker about the position of the labellum in *Dendrobium* in his first letter to Hooker of 1 November [1861]; see also the letter to John Lindley, 1 November [1861] (*Correspondence* vol. 9). Hooker's reply has not been found, but for CD's description of *D. chrysanthum*, see *Orchids*, pp. 172–8.

[6] CD had been investigating the ducts, or spiral vessels, of orchid flowers in order to determine homologies within the orchid family (see, for example, *Correspondence* vol. 9, letters to J. D. Hooker, 4 October [1861] and 27 October [1861]). In the first letter to J. D. Hooker of 1 November [1861] (*ibid.*), he wrote that he did not have a *Cypripedium* 'in spirits' to dissect for an examination of the ducts. For CD's discussion of the spiral vessels in orchids, including *Cypripedium*, and his accompanying diagram, see *ibid.*, letter to J. D. Hooker, 8 November [1861], and *Orchids*, pp. 290–305.

[7] CD mentioned the nectary of *Evelyna* in *Orchids*, p. 277; see also p. 295–6 n.

To Jean-Baptiste Pierre Guépin[1] 14 November 1861

Down. | Bromley. | Kent. S.E. England
Nov. 14 1861

Monsieur

J'éspère que vous excuserez la liberté que je prends à vous addresser cette lettre. J'ai lu un avis sur un papier dans les Annales de la Societé Linn: d'Angers, sur la fertilisation des Orchides, dont vous ètes l'auteur.[2] J'ai en vain essayé d'en obtenir un exemplaire à Londres, ce que je désire beaucoup, comme je suis occupé dans ce moment à préparer pour la publication un ouvrage sur ce sujet.[3]

Auriez vous la bonté de m'informer si vous avez pu indiquer l'éspèce des Orchides desquels les Abeilles avaient deplacé le pollen? Y avait il un, deux, ou plus de "pollen masses" collé sur une abeille? J'ai vu ces masses de pollen en grand nombre sur la trompe des Lepidoptères mais non pas sur les abeilles.[4]

Si par hazard vous possediez plusieurs specimens d'abeille avec le pollen attaché et que vous pourriez m'en épargner un, vous me rendriez un grand service. Si je pouvais éxaminer une de ces abeilles (que vous pourriez m'envoyer dans une lettre, enfermée dans une plume) je crois pouvoir déterminer de quelle éspèce d'orchide sont les "pollen masses". Surtout si vous pouviez m'informer quelles sont les éspèces d'orchides les plus fréquentes dans le voisinage d'Angers.[5]

Je suis avec beaucoup de respect et vous priant d'excuser cette liberté votre très humble serviteur | Charles Darwin

LS(A)
Archives Départmentales de Maine-et-Loire (24 J 1, pièce no. 32)

1 For a translation of this letter, see Appendix I.

2 Daniel Oliver had sent CD a list of references in French and German relating to orchids. Guépin 1853 was described on this list as dealing with 'fecundation of *Catasetum*' (see *Correspondence* vol. 9, letter from Daniel Oliver, [before 8 November 1861], and this volume, Supplement, letter to Daniel Oliver, 3 November [1861]).

3 CD refers to *Orchids*.

4 Frederick Bond is cited in *Orchids*, pp. 35–6, as having sent CD a large number of moths and butterflies with orchid pollinia attached to their proboscises. Virtually all the pollinia came from a single species, *Orchis pyramidalis*. See also *Correspondence* vol. 8, letter to J. D. Hooker, 17 June [1860] and n. 5.

5 Guépin published a flora of the area of Maine and Loire (Guépin 1830–54). Guépin had died in 1858, so the letter was never answered.

To W. E. Darwin 30 November [1861][1]

Down
Nov. 30th

My dear William

The consols were purchased under name of William Erasmus Darwin of Southampton & Banker was not introduced.[2]

Mamma[3] is in bed with bad Headach.— Miss. L.[4] is very bad with headach.— Lenny[5] has got a slight headach.— I am not very bright— The day is raining torrents, the children are ennuied—so I have not heart to write—

GoodBye | My dear old man | C. Darwin

DAR 210.6: 108

1 The year is established by the relationship between this letter and the letter to W. E. Darwin, 12 October [1861] (*Correspondence* vol. 9).

[2] 'Consols' is an abbreviation for consolidated annuities, a form of government stocks. CD purchased these in William's name to serve as a guaranteed deposit toward William's partnership in the bank of Atherley, Fall & Atherley (later the Southampton and Hampshire Bank; see *Correspondence* vol. 9, letter to W. E. Darwin, 12 October [1861] and n. 5).
[3] Emma Darwin.
[4] Camilla Ludwig, the Darwin children's governess.
[5] Leonard Darwin.

From Berthold Carl Seemann 24 April 1862[1]

[Enclosure][2]

"Mountains exceeding 2000 feet in elevation, situated principally in Western Veraguas, possess a vegetation which resembles in many respects that of the Mexican highlands;[3] one in which the forms of the torrid region are harmoniously blended with those of the temperate. Alders and Blackberries are found with Fuchsias and Salvias; the common Brake[4] grows in company with Lupins and Ageratums; Oaks and Palms are intermigled; fine large flowers are abundant. The genera represented are: Styrax, Rondeletia, Salvia, Lopezia, Fuchsia, Centradenia, Ageratum, Conostegia, Lupinus, Hypericum, Freziera, Galium, Smilax, Euphorbia, Rhopala,[5] Equisetum, Clematis, Chorisia,[6] Verbena, Condaminea, Inga, Solanum, etc. The Oaks, like most tropical ones, are scarcely higher than 30 feet, resembing neither in size nor in grandeur those which our heathen forefathers worshipped; their branches are smooth and devoid of that rugged appearance which renders those of the northern species so picturesque."

Extracted from Berthold Seemanns Botany of the Voyage of H.M.S. Herald. p. 67. (London. 1852–1857. Quarto. 484 pages, and 100 plates.).[7]

NB. The oaks here alluded to are: Quercus Seemanni, Lieb., Q. bumelioides. Liebm., and Q: Warscewiczii[8] Liebm.

DAR 50: E28

CD ANNOTATIONS
Top of enclosure: 'April 25—1862' *ink*
Bottom of enclosure: 'As in Himalaya, Fernando Po & here, temperate forms get mingled with Tropical, apparently from neighborhood of Temperate forms. so it would have been in Glacial Period' *ink*[9]

[1] For the text of the letter with which the copied passage was enclosed, see *Correspondence* vol. 10, letter from B. C. Seemann, 24 April 1862.
[2] The passage is taken from a description of the flora of the isthmus of Panama. The extract, copied by hand with minor alterations and spelling errors, is from Seemann 1852–7, p. 67.
[3] Veraguas is a province in west central Panama that extends across the entire isthmus from Mosquitos Gulf in the Caribbean Sea to the Pacific (*Columbia gazetteer of the world*).
[4] The printed version does not have 'common'.
[5] A synonym of *Roupala.*
[6] Now included within *Ceiba.*
[7] See n. 2, above.

[8] The correct spelling is *Q. warszewiczii* Liebm.

[9] In *Origin*, pp. 365–82, CD discussed the migration of plants during a former glacial period. He proposed that during such periods plants from temperate zones could have migrated towards the equator and, in many cases, become naturalised in tropical areas. He suggested that mountain ranges such as the Himalayas could have acted as 'lines of invasion' for temperate species (*ibid.*, p. 378); see also this volume, Supplement, letter to J. D. Hooker, [14 November 1858]. Joseph Dalton Hooker provided CD with information about temperate plants found on Clarence Peak, Fernando Po, off the coast of West Africa; these plants showed an affinity with the vegetation of Abyssinia, 1800 miles distant (see *Correspondence* vol. 10, letter from J. D. Hooker, [5 May 1862] and n. 4, and letter to J. D. Hooker, 9 May [1862] and n. 6). CD added Hooker's information, which he considered to be further proof of migration, to *Origin* 4th ed., pp. 445–50 (see also *Correspondence* vol. 10, Appendix VIII and n. 47).

To J. H. Balfour 15 June [1862?][1]

Down. | Bromley. | Kent. S.E.
June 15th

My dear Sir

I presume that I owe to your great kindness a fine specimen of Corallorhiza received this morning sound & safe.[2] I have been very glad to see it, but the wicked insects had done their work so effectually that every flower was fertilised.

I have planted the specimen in a pot & I have hopes that it may flower another year.—[3] Pray accept my sincere thanks for your kindness.—

As I am writing I will beg another favour, for the chance of your being able to grant it, through the Botanic Garden, viz seed of Corydalis claviculata.[4] I have searched everywhere for it. Of course you will not write; but if fortune favours me, pray send me some.—[5]

My dear Sir | Yours very sincerely | Ch. Darwin

P.S. Since the above was written your note has arrived—[6] again many thanks—

Yale University, Sterling Library

[1] The year is conjectured from the reference to *Corallorhiza* and to the seed of *Corydalis claviculata* (see nn. 2 and 5, below).

[2] CD had written to Balfour requesting a specimen of the rare orchid genus *Corallorhiza* in 1861 (see this volume, Supplement, letter from H. C. Watson to George Gordon, 27 June 1861, and *Correspondence* vol. 9, letter to George Gordon, 4 July [1861]). The letter to Balfour has not been found.

[3] No notes by CD on *Corallorhiza* have been found. The specimen would have arrived too late for CD to include observations on it in *Orchids*; the genus is not discussed in 'Fertilization of orchids' or *Orchids* 2d ed.

[4] Balfour was professor of botany at Edinburgh University and regius keeper of the Royal Botanic Garden, Edinburgh (*DNB*).

[5] CD received seeds of a *Corydalis*, probably *C. claviculata*, from Balfour in October 1864; by that time, however, CD had already obtained specimens and seeds from elsewhere, and had raised one specimen from seed (see *Correspondence* vol. 12, letter to J. H. Balfour, 21 October [1864] and n. 3). CD's notes on *C. claviculata*, written in September 1863, are in DAR 157.2: 26–8; the species is described in 'Climbing plants', pp. 45, 70–2.

[6] The letter from Balfour has not been found.

From John Murray [1 July – 23 August 1862][1]

Darwin on Orchids[2]

Dr.

1862				
June 30	To Printing 1500 N° Clowes[3]	89	1	9
	" 48 $\frac{1}{4}$ Rms. Sht $\frac{1}{2}$ Post[4]	51	17	4
	" Drawing & Engraving[5]	74	2	
	" Binding 1000 copies	30	4	2
	" Commission allowed. Agents	2	3	
	" Advertising	27	4	
	" Entering Stationers' Hall[6]		5	
		274	17	3
1862				
June 30	To Balance Deficiency	45	5	1

Cr.

1862				
May	By 1500 copies			
	5 Stationers' Hall[7]			
	12 allowed Author			
	30 Presented Reviews[8]			
	732 685 on hand June 30/62			
	768 Sold Viz			
	358 Trade 25 as 24 6/	103	4	
	410 D° " " 6/5[9]	126	8	2
	768	229	12	2
June 30	By Balance Deficiency	45	5	1
		274	17	3
1862				
June 30	By 685 on hand			

Document
DAR 171: 525

CD ANNOTATIONS

Bottom of page: 'Feb–27— 1866. Still about 600 copies on Hand & 30£ deficient.—'[10] *pencil*

Verso: '1875' *mauve crayon*; 'Orchis Book & 4th Edit of Origin.— '[11] *pencil*; 'Nov 1869' *blue crayon*; 'Orchis Book & 4th Edit of Origin' *pencil*

[1] The date range is established by the dates mentioned in the account and by the relationship between this document and the letter to John Murray, 24 August [1862] (*Correspondence* vol. 10).

[2] For a discussion of the publication of *Orchids*, see *Correspondence* vol. 10, Appendix IV.

[3] William Clowes & Sons, printers.

[4] Murray ordered 48¼ reams of sheet-and-a-half post paper. Sheet-and-a-half post was one and a half times the size of ordinary post; each sheet was about 500 mm by 600 mm, accommodating 12 leaves of an octavo size book (about 200 mm by 125 mm uncut) such as *Orchids* (Gaskell 1972, p. 224, and Nicholas Smith, personal communication).

[5] The illustrations for *Orchids* were made by George Brettingham Sowerby Jr.

[6] Under the 1842 Copyright Act, five shillings was the cost of making an entry in the Book of Registry of the Stationers' Company, whose headquarters were at Stationers' Hall, London. Registration did not affect copyright, but was 'a necessary preliminary to an action at law for copyright infringement' (Seville 1999, p. 235).

[7] Under the 1842 Copyright Act, four copies of any new work published in the United Kingdom and the British Dominions were to be delivered to an officer of the Stationers' Company for distribution on demand to the Bodleian Library in Oxford, the 'Public Library' (now the University Library) in Cambridge, the Faculty of Advocates at Edinburgh, and the library of Trinity College, Dublin. A copy was also to be delivered to the British Museum in London. Before the 1842 Act, the Stationers' Company had been responsible for handling the British Museum's copy as well as those of a number of other libraries legally entitled to copies. See Seville 1999, pp. 233, 262.

[8] For CD's presentation list for *Orchids*, see *Correspondence* vol. 10, Appendix IV. Sixty-five individuals and three institutions are listed. CD presumably paid for the extra copies himself: see *Correspondence* vol. 10, letter to John Murray, 24 August [1862] and n. 2. See also *ibid.*, letters to John Murray, 9 April [1862] and 2 May [1862].

[9] Trade purchasers received one free copy in every twenty-five, as was the general practice (Plant 1965, p. 405), and were charged 6*s.* or 6*s.* 5*d.* *Orchids* went on sale to the public priced at 9*s.* See Freeman 1977, p. 113.

[10] This information was conveyed to CD in the letter from John Murray, 24 February [1866] (*Correspondence* vol. 14).

[11] The fourth edition of *Origin* was published in 1866.

To William Pamplin 4 [July 1862][1]

Down Bromley Kent
4th

Dear Sir

The paper did not arrive this morning by our weekly Carrier, but I presume was sent too late yesterday (Thursday) for him.—[2]

I write now to ask whether you keep any white paper for mounting plants when dryed; if so would you be so kind as to send one sheet of what you recommend per post with price marked per ½ ream or quire.—

I quite forgot to speak about this, when I wrote for drying paper, & apologise for extra trouble thus caused.—[3]

Dear Sir | Yours faithfully | C. Darwin

Archives Department, University of Wales Bangor (Pamplin papers)

1 The month and year are established by the relationship between this letter, the letter to J. D. Hooker, 15 [May 1862], and the letter from J. D. Hooker, [17 May 1862] (*Correspondence* vol. 10). In 1862, the only month after May in which the 4th fell on a Friday was July.
2 George Snow, the Down village coal dealer, ran a carrier service to and from London every Thursday (*Post Office directory of the six home counties* 1862).
3 Joseph Dalton Hooker had recommended Pamplin, a botanical bookseller, as a source for paper for drying plants (*Correspondence* vol. 10, letter from J. D. Hooker, [17 May 1862]). CD evidently ordered the paper, but the letter has not been found. CD's Classed account book (Down House MS) records a payment, under the heading 'Science', of £2 9*s.* 6*d.* to Pamplin for botanical paper on 3 August 1862.

From George Chichester Oxenden 5 July [1862][1]

Dear Sir

I *carefully* showed to my *Foreman*, who goes with me on Snipe-shooting days, the Plants of "*Epipactis palustris*" and caused *him* to sit up with them last night—[2]

—He is a good Gardener, & certain to have made no mistake—

—I also taught him to approach the plants very cautiously—

—The day had been without rain, & the Evening was fine but cloudy—

—Nevertheless the absence of Insects in this *One* instance is *not conclusive*, since it may have been due to hygrometric causes—

—If I can find means to go *myself* I will make a *certainty* of it—

Sincerely | G C. Oxenden

Broome | July 5—

DAR 173: 64

1 The year is established by the relationship between this letter and the letter from G. C. Oxenden, 8 July 1862 (*Correspondence* vol. 10).
2 In June 1861, CD had asked Oxenden to observe the behaviour of insects visiting *Epipactis palustris*, and the resulting movements of the parts of the flowers (see *Correspondence* vol. 9, letter to A. G. More, 4 June 1861, and *Correspondence* vol. 10, letter from G. C. Oxenden, 8 July 1862). Oxenden's assistance in supplying CD with specimens and information is acknowledged in *Orchids*, pp. 31–2 n. CD's description of *E. palustris* and its pollination is in *Orchids*, pp. 94–102. In 'Fertilization of orchids', pp. 148–9 (*Collected papers* 2: 145–6), and *Orchids* 2d ed., pp. 99–100, CD added further observations, made by his son William, of insects visiting the species.

From G. C. Oxenden 13 July [1862][1]

Dear Sir

One or two chances yet remain, namely with

—Epipactis latifolia

—Epipactis purpurata—[2]

If you wish any work done with either of these, please write—& put the words "*to be forwarded*" upon the Envelope—as I am leaving home—

E. *latifolia* is a **very** pleasing handsome plant—

—I hope your domestic tribulations are disappearing [3]

Sincerely | G. C. Oxenden

Broome | July 13.

DAR 173: 65

[1] The year is established by the relationship between this letter and the letter from G. C. Oxenden, 4 June [1862] (*Correspondence* vol. 10).

[2] Oxenden had promised to send specimens of *Epipactis latifolia* and *E. purpurata* as well as *E. palustris* to CD (see *Correspondence* vol. 10, letter from G. C. Oxenden, 4 June [1862]). No further letter from Oxenden on *E. purpurata* has been found, but in *Orchids*, 2d ed., p. 102, CD cited Oxenden for the information that *E. purpurata* was frequented by 'swarms of wasps'; see also 'Fertilization of orchids', p. 150 (*Collected papers* 2: 147).

[3] In his letter of 21 June 1862, Oxenden had recommended a treatment for Leonard Darwin, who was ill with scarlet fever (see *Correspondence* vol. 10, letter from G. C. Oxenden, 21 June 1862 and n. 1).

To Friedrich Hildebrand 17 July [1862][1]

Down. | *Bromley.* | *Kent. S.E.*
July 17th

Dear Sir

I am much obliged for your kind note & offer of aid in the translation of my work on Orchids.[2] But I have heard from Mess. Schweizerbart of Stuttgart that the late lamented Prof. Bronn had finished the translation together with some additional notes which I had sent him.[3] How Mes. Schweizerbart intend to get the proof-sheets corrected, I do not know.

I have not yet received, but hope I shall soon, the work on the Distribution of Coniferous Trees, which you have kindly sent me;[4] I have no doubt it will interest me, as few subjects are so interesting as Geographical Distribution of Plants.—

Pray present my most respectful compliments to Prof. Treviranus: the notes which he spoke of appending to my little work, would have added *greatly* to its value.—[5]

I beg you to accept my sincere thanks for your very kind offer & I remain | Dear Sir | Yours truly obliged | Charles Darwin

Courtesy of Eilo Hildebrand (photocopy)

[1] The year is established by the relationship between this letter and the letter from Friedrich Hildebrand, 14 July 1862 (*Correspondence* vol. 10).

[2] In his letter of 14 July 1862 (*Correspondence* vol. 10), Hildebrand had offered to undertake a German translation of *Orchids*, unaware that one had recently been completed by Heinrich Georg Bronn (see n. 3, below).

[3] Bronn died on 5 July 1862, having recently completed work on the German translation of *Orchids* for the publishing firm E. Schweizerbart (Bronn trans. 1862; see also *Correspondence* vol. 10, letter from E. Schweizerbart'sche Verlagsbuchhandlung, 11 July 1862 and nn. 2 and 5). For the additional

material that CD had asked Bronn to translate and include in the German edition, see the enclosures to the letter to H. G. Bronn, 30 June [1862] (*Correspondence* vol. 10).

4 Hildebrand 1861; see *Correspondence* vol. 10, letter from Friedrich Hildebrand, 14 July 1862 and n. 2. CD's copy of Hildebrand 1861 is in the Darwin Library–Down (see *Marginalia* 1: 380).

5 CD had approached Ludolph Christian Treviranus, professor of botany at the University of Bonn, suggesting that Treviranus translate *Orchids* into German (see *Correspondence* vol. 10, letter from Friedrich Hildebrand, 14 July 1862 and n. 1). No further reference to the supplementary notes apparently proposed by Treviranus has been found; however, Treviranus published comments on *Orchids* in two papers in *Botanische Zeitung* in 1863 (Treviranus 1863a and 1863b; see also *Correspondence* vol. 10, letter from E. Schweizerbart'sche Verlagsbuchhandlung, 11 July 1862 and n. 6). Hildebrand was a colleague and former student of Treviranus at the University of Bonn (see Correns 1916).

To J. D. Hooker 24 [November 1862][1]

[Enclosure][2]

These three hermaphrodites coexist in about equal numbers; they differ greatly & *definitely* in length of pistil. Each form has two sorts of pollen; & *altogether* the three forms have three sorts of pollen & three sorts of pistils. The red lines show which pollen produces full fertility;[3] ie each stigma can be fertilised only by pollen of the stamens of *corresponding* height in the other two forms.— (pollen is attached to different parts of Bees' bodies).[4] Thus long-styled cannot be fertilised by *either* of its own two kinds of pollen, & only by the large green pollen of the tallest stamens of the mid-styled & short-styled. The long-styled could exist & seed if in company with either *one* of the two other forms; but then half its own anthers would be superfluous. Thus if long-styled & midstyled lived together without short-styled, the shorter stamens of both would be *utterly* useless. Hence we here have a triple marriage alliance between three Hermaphrodites! All this is certainly proved; but mid-styled is anomalous & can be *partially* fertilised by some of stamens of non-corresponding height; hence I will not publish, till more experiments are tried.—[5]

DAR 115: 279b

1 The date is established by CD's comment to Hooker in his letter of 24 [November 1862] (*Correspondence* vol. 10) that he was enclosing a diagram of *Lythrum salicaria*; see also *ibid.*, letter to J. D. Hooker, [10–]12 November [1862]. CD sent similar diagrams in 1862 to two other correspondents (see *ibid.*, letter to W. E. Darwin, [2–3 August 1862], and letter to Asa Gray, [3–]4 September [1862]). A similar diagram was published in CD's 'Three forms of *Lythrum salicaria*', p. 171 (*Collected papers* 2: 107), and *Forms of flowers*, p. 139.

2 The drawing (see p. 458) is an enclosure to the letter to J. D. Hooker, 24 [November 1862] (*Correspondence* vol. 10; see n. 1, above). It is reproduced here at 60 per cent of original size.

3 The red crayon lines of the original are represented here by dashed lines.

4 CD started crossing the three forms of *Lythrum salicaria* at the end of July 1862, and also began observing their pollination by bees; see, for example, *Correspondence* vol. 10, letters to Asa Gray, 28 July [1862] and n. 16, and [3–]4 September [1862]. Most of CD's notes on his experiments on and observations of *Lythrum* for 1862, 1863, and 1864 are in DAR 27.2B; his later work on *Lythrum* is in DAR 109.

5 In 1862, as CD was counting the seeds resulting from his crosses of *Lythrum salicaria*, he grew increasingly confident that each of the three forms (long-styled, mid-styled, and short-styled) were

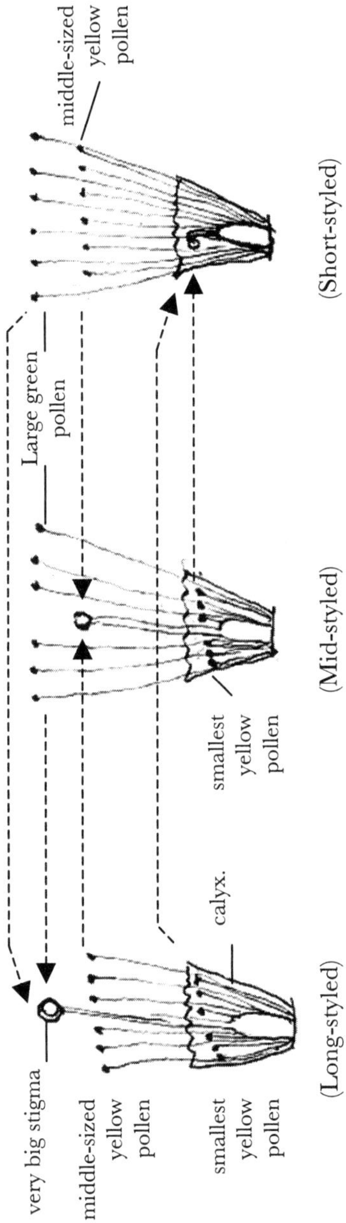

Enclosure to letter to J. D. Hooker, 24 [November 1862]

hermaphrodites, and together included three females and three sets of males (see *Correspondence* vol. 10, letter to J. D. Hooker, 6 October [1862], and letter to Asa Gray, 16 October [1862]). However, he also began to suspect additional differences in the pollen of the mid-styled form, and after making almost 100 crosses in 1862, he was determined to make more in 1863 (see *ibid.*, letter to J. D. Hooker, 27 [October 1862] and nn. 11 and 12, letter to W. E. Darwin, 4 [November 1862], and letter to Asa Gray, 6 November [1862]). See also 'Three forms of *Lythrum salicaria*', pp. 169, 186, 188 (*Collected papers* 2: 106, 119, and 122), and *Forms of flowers*, pp. 137–8, 162–4.

To [Camilla] Ludwig[1] 22 December 1862[2]

Down

In all money affairs it is better to have a memorandum in writing. I thought it so hard a case that your Father should be told that the £200 would be given him & then the offer partially retracted, that I, under these circumstances alone, offered to give the £200, which I have now done:[3] but I understand that if Miss Davies kept to her original intention, then, as I know perfectly well, you would repay me the £200;[4] under any other circumstances, I must freely & cordially give you the £200. This note is ill expressed but is written in a hurry.

My dear Miss Ludwig | yours very sincerely | Ch. Darwin

Incomplete[5]
R. M. Smythe (dealer)

[1] The recipient is identified as Camilla Ludwig, who was the Darwin children's governess at the time. In family letters, the appellation 'Miss Ludwig' or 'Camilla' refers to Camilla Ludwig, while 'Louisa' or 'Louisa L.' is used to refer to her younger sister (letter from Emma Darwin to W. E. Darwin, [27 May 1862] (DAR 219.1: 57)). See also n. 4, below.

[2] The date is given in the sale catalogue.

[3] On her return to Down in November 1862 after visiting her family in Hamburg, Ludwig reported that her family was in great poverty and that she had advanced her whole salary to them (letter from Emma Darwin to W. E. Darwin, [6 November 1862] (DAR 219.1: 64)). CD's Account book–banking account (Down House MS) records a payment to Miss Ludwig of £200 on 21 December 1862. Next to the entry CD wrote '(Class as Invested)' (square brackets in original). CD's Account book–classed accounts (Down House MS) records a payment of £10 to 'Ottocar Ludwig' on 4 July 1866. 'Ottocar' is probably the name of Camilla Ludwig's father, but he has not been further identified.

[4] CD's Account book–banking account (Down House MS) contains the following entry for 14 March 1863: 'Miss Davies Repayment of advance to C. L. (do not class) £200' (square brackets in original). Miss Davies has not been further identified.

[5] The original letter is complete and is described in the sale catalogue as being two pages long.

To John Scott 8 January [1863][1]

Down, Bromley,
Jan. 8th.

Dear Sir,—

What an indomitable worker you are and what a number of interesting experiments you are trying! I thank you for telling me about yourself; I felt respect for you before, and now this is sincerely increased.[2]

With your tastes, the fall in your circumstances must have been severely felt; but you have acted like a real man in making the best of things, and most truly do I wish you success in all your endeavours. I daresay a foreign appointment will be best in every way, and will be sure to offer a field for new observations; but I cannot avoid being sorry for it. Only imagine how grand it would be to see an insect at work at a Gongora, &c.[3]

Very sincere thanks for all your great trouble taken about the primulas, which I received this morning.[4] I have put them in my green-house, which is warmed at night; whether this is wise I know not. Your remarks on Gongora have interested me much, and I never saw a fresh specimen before. The labellum beats that of acropera. Do the tips of the basal inturned horns secrete any fluid? In one specimen, to the mouth of the stigmatis[5] cavity there adhered a large protuberant drop of *very slightly* viscid fluid, evidently in excess, but it supports your idea about acropera.[6] I have put specimens in spirits, (for I am too busy now) and some day will look at ovates. I am quite confused about their genera.[7] Can the plants be kept too dry and so cause stigma not to secrete enough? I can keep pod of acropera till whatever time I get one of other germs for comparison.[8] I shall be quite nervous when I first look at seeds. I must write briefly.

[. . .] Never trouble yourself to answer my letters unless inclined to do so, and pray believe me with every good wish,—Dear Sir, | Yours sincerely, | Charles Darwin.

Have you not been put to some little expense about primulas? If so, I beg you to have kindness to inform me.

What say you to cross a primrose (heteromorphic) with pollen of wild cowslip and of a highly cultivated polyanthus, and see which in, say, ten pods of each, yielded most seed.[9]

I am going immediately to build small hothouse, but whether my two rather ignorant men will succeed with the plants I know not.[10]

Transactions of the Hawick Archæological Society (1908): 67

1 The year is established by the relationship between this letter and the letter from John Scott, 6 January 1863 (*Correspondence* vol. 11).

2 For Scott's discussion of his various experiments with orchids, and of his personal circumstances, see his letter to CD of 6 January 1863 (*Correspondence* vol. 11).

3 Scott was hoping to move from his position as the foreman of the propagating department at the Edinburgh Botanic Garden to a foreign appointment (see *Correspondence* vol. 11, letter from John Scott, 6 January 1863). He and CD were interested in the pollination of orchids, including the tropical orchid *Gongora* (see *Correspondence* vol. 10, and *Correspondence* vol. 11, letter from John Scott, 6 January 1863).

4 In his letter of 6 January 1863 (*Correspondence* vol. 11), Scott wrote that he was sending CD three plants each of *Primula scotica* and *P. farinosa* for his observations and experiments on dimorphism.

5 Probably a mistranscription for 'stigmatic'.

6 Scott had sent CD *Gongora atropurpurea* specimens (see *Correspondence* vol. 11, letter from John Scott, 6 January 1863 and CD note). CD had examined the allied *Acropera luteola*, which he called the 'opprobrium' of his work on orchids, as all the parts seemed 'determinately contrived' so that the

plant should never be pollinated (*Orchids*, p. 203). However, Scott had succeeded in pollinating *Acropera loddigesii*, despite a stigmatic chamber as narrow as that of *A. luteola*; he suggested that when out of its native habitat the stigmatic orifice might grow too dry, losing the viscid fluid that would aid pollination (see *Correspondence* vol. 10, letter from John Scott, 11 November 1862, and letter to J. D. Hooker, 18 [November 1862]). CD discussed *Gongora* in 'Fertilization of orchids', p. 154 (*Collected papers* 2: 150) and *Orchids* 2d ed., pp. 168–70, 276.

7 'Ovates' is probably a mistranscription for 'ovules' and 'genera' a mistranscription for 'genitalia'; in his note describing Scott's *Gongora* specimens, CD mentioned that he would put the ovarium in spirits to examine it (see *Correspondence* vol. 11, letter from John Scott, 6 January 1863 and CD note).

8 'Germs' is probably a mistranscription for 'genera'. Scott had earlier offered to send CD his rapidly swelling *Acropera* ovary, and continued to report on its growth (see *Correspondence* vol. 10, letter from John Scott, 15 November [1862], and *Correspondence* vol. 11, letter from John Scott, 6 January 1863; see also *ibid.*, letter to John Scott, 20 [February 1863]).

9 Scott conducted crosses with species of Primulaceae on the advice of CD from 1862 (see *Correspondence* vols. 10–12, and Scott 1864). Following publication of 'Dimorphic condition of *Primula*', CD continued making crosses of *Primula* species. For his work comparing crosses of primroses with cowslips and with polyanthus, see 'Illegitimate offspring of dimorphic and trimorphic plants', 'Specific difference in *Primula*', *Forms of flowers*, and his notes in DAR 108.

10 The construction of CD's hothouse was completed by mid-February 1863 (see *Correspondence* vol. 11, Appendix VI). CD also refers to his servants William Brooks and Henry Lettington (see *Correspondence* vols. 11 and 12).

From John Crawfurd to E. A. Darwin 7 August 1863

The Athenæum
August 7th. 1863.

My dear Darwin/

I was requested by Archdeacon Sinclair to send the enclosed to your brother.[1] It contains a remarkable prophecy of your grandfather which long anticipated the opinion of Libig & well deserves attention.[2]

Will you forward with my kind regards | very sinly yours | J Crawfurd

Erasmus Darwin Esq

[Enclosure]

Nor did the Board confine their efforts to the Collection & diffusion of agricultural Knowledge;[3] they endeavoured also to arrange it—to connect husbandry with philosophy or as the Founder of the institution frequently expressed himself "to raise agriculture from the level of an art to the dignity of a science." They wished its process no longer to be carried on blindly or empirically, but to be formed into a regular system & conducted upon rational & experimental principles. The two branches of science from which the greatest aid might be expected were vegetable physiology & chemistry; the Board therefore endeavoured to prevail on men of eminence in these departments to employ their scientific attainments in arranging & combining theoretically the miscellaneous facts & loose notions of which agricultural knowledge at that time exclusively consisted. Among the most valuable results were Darwin's Phytologia & Davy's Lectures on Agricultural Chemistry.[4]

The Phytologia or Philosophy of Agriculture & Gardening was dedicated by the Author in the most cordial terms of friendship & esteem to Sir John Sinclair as a work "begun at his instigation, & forwarded by his encouragement." Dr Darwin entered warmly into Sir John's feeling of indignation that husbandry had till then been so much neglected by the learned, & that at a time when many branches of knowledge far inferior in importance had been carefully arranged & digested into sciences agriculture & gardening though of such great utility in producing the nutriment of mankind continued to be only arts. The work comprises three parts of which the first explains the physiology of vegetation; the second describes the economy of vegetation & the third applies the principles thus established to agricultural & horticultural productions. Among the casual discoveries which this ingenious work had the merit of bringing forward was the use of bone dust as a manure. Having mentioned phosphorus as an element under different forms, existing universally in vegetables, & not before sufficiently attended to, the Doctor specifies the different substances from which this essential food of plants may be obtained & which therefore might be advantageously employed as manure; he then proceeds— "and, lastly, the use of recent shells, or bones ground into powder, or of bone ashes may be deduced as they consist almost entirely of phosphorus & calcareous earth."[5] The fertilizing properties of this manure had been previously noticed by Hunter,[6] but they were first theoretically explained & brought forward with authority by D^r. Darwin. Perhaps no modern discovery has contributed so powerfully to improve the fertility & increase the produce of the soil, as the use of bone dust.[7] Nothing but experience would render credible the strong & lasting effects produced by even small quantities of this active Substance.—

Life of Sir John Sinclair Bar^t. | vol II. p. 83–5.

DAR 161: 237, 237/1

[1] The reference is to John Sinclair, archdeacon of Middlesex and vicar of Kensington. The enclosure was a handwritten extract from a book by Sinclair about the life of his father, John Sinclair (Sinclair 1837, 2: 83–5); it included a tribute to CD's grandfather, Erasmus Darwin.

[2] The enclosure discusses points in Erasmus Darwin's *Phytologia* (E. Darwin 1800) that the author thought predated Justus von Liebig's conclusions regarding agricultural fertiliser. Liebig had noted that fields required the replacement of essential minerals in order to maintain a constant state of fertility (Liebig 1840, p. 182). He recommended bone as a rich source of calcium and magnesium phosphates and further observed that the waste materials of the glue industry represented an untapped source of valuable minerals in a form particularly suitable for agricultural use (*ibid.*, p.185). For an analysis of Liebig's role in the development of agricultural chemistry, particularly in the analysis and development of artificial fertilisers, see Brock 1997, pp. 145–82.

[3] The reference is to the Board of Agriculture, founded in 1793 at the instigation of John Sinclair Sr, who became its first president (E. J. Russell 1966, pp. 54, 67).

[4] E. Darwin 1800 and H. Davy 1813. For a discussion of Erasmus Darwin's *Phytologia*, and his anticipation of later discoveries in agriculture and plant biology, see King-Hele 1999, pp. 333–8. Humphry Davy's *Elements of agricultural chemistry* is discussed in Russell 1966, pp. 67–76.

[5] The quotation is from E. Darwin 1800, p. 211.

6 Sinclair refers to a remark on the use of ash as a fertiliser in Hunter 1777, pp. 382–3.
7 In his biography of his grandfather, CD quoted this sentence, adding in a note that he was 'indebted to Dr Dowson's *Life of Erasmus Darwin*, for the reference to the *Life and Works of Sir J. Sinclair*' (*Erasmus Darwin*, p. 113); the quoted sentence appears in Dowson 1861, p. 56.

To Hurst & Blackett 15 November [1863][1]

Down. | Bromley. | Kent. S.E.
Nov 15

Gentlemen

I see that Miss Meteyard is preparing a life of my grandfather Josiah Wedgwood.[2] I have in my possession a good many letters to him from Dr Erasmus Darwin (also my grandfather) & if they would be likely to give any information, I shall be glad to send them to Miss Meteyard or to you.[3]

They are chiefly on semi-scientific subjects & probably would only be useful as shewing what subjects were occupying Mr Wedgwood's mind & the dates of them.

Perhaps you will forward this to Miss Meteyard

I am Gentleman | Yours faithfully | Charles Darwin

LS
Cleveland Health Sciences Library (Robert M. Stecher collection: FF8)

1 The year is established by the relationship between this letter and the letter to J. D. Hooker, [22–3 November 1863] (*Correspondence* vol. 11).
2 The novelist and journalist Eliza Meteyard had been researching the life of Josiah Wedgwood I since the 1850s (see Meteyard 1970, introduction, and B. Wedgwood and Wedgwood 1980, p. 271). The first volume of her *Life of Josiah Wedgwood* (Meteyard 1865–6) was published by Hurst and Blackett in April 1865.
3 In the first volume of her biography, Meteyard acknowledges the receipt of letters sent to her by CD (see Meteyard 1865–6, 1: xv, 436). The letters between Josiah Wedgwood and Erasmus Darwin are at Down House.

To Hermann Kindt 17 September 1864[1]

Dear Sir

I am most obliged for your kind & flattering letter.—[2] I have copied the last sentence of the "Origin", which I hope will do.—

In haste | yours faithfully | C. Darwin

[Enclosure]

Down. | Bromley. | Kent. S.E.

There is grandeur in this view of Life, with its several powers having been originally breathed by the Creator into a few forms or into one; & that whilst this planet has gone cycling on according to the fixed law of gravity, from so simple a beginning endless forms most beautiful & most wonderful have been, and are being evolved.—[3]

Charles Darwin
Sept. 17th. 1864.—

J. A. Stargardt, 11 and 12 June 2002; Local Studies Unit Archives, Manchester Central Library (Autograph Letters: Harland Collection, vol. 1, p. 67, MSf091 H 15)

[1] A brief description of this letter was published in *Correspondence* vol. 12; since then the full text has become available.
[2] See letter from Hermann Kindt, 16 September 1864 (*Correspondence* vol. 12).
[3] See *Origin* 2d ed., p. 490.

APPENDIX I
Translations of letters

From C. A. Royer[1] [April–June 1865][2]

'Election naturelle'[3] is virtually contained in the whole of these laws,[4] and especially in all that relates to the caste system, within which one must recognise races that are probably different and very unequal in their human development, at least at their origin.

This same principle of choice is the result of the prescriptions and recommendations given by the law makers to men of noble castes concerning the choice of their wives.[5] They are enjoined to avoid marrying women who show signs of ill-omen such as rheumy eyes or other deformities or infirmities.[6] Perhaps these are the precepts of a law that has modified the instincts, habits and physical characteristics of the Indo-Germanic race.

If such is the general spirit of these laws, a spirit that recurs throughout them all, the paragraphs in which it is most explicitly expressed are the following.

book III. par. 6 and following 10–14–19.[7]
book IX. par. 72–73 and 178[8]
book X. par. 5 up to 49 – 66 to 73.[9]

A Mem
DAR 80: B44

CD ANNOTATION
Top of memorandum: 'Castes of India are [apposite] to breeding like domestic animals' *pencil*

[1] For a transcription of this letter in the original French, see p. 104. There is an incomplete translation of this memorandum by Emma Darwin in DAR 80: B42–3.

[2] The date range is established from the relationship between this memorandum and the letters in which CD and Emma Darwin mention that CD is working on the translation of the second French edition of *Origin* (see n. 4, below).

[3] The phrase 'élection naturelle' appears to have been used as a translation of 'natural selection' first by Edouard Claparède (see *Correspondence* vol. 10, letter from Edouard Claparède, 6 September 1862). However, 'sélection' was already in use as a technical term in French zoology (*Grand Larousse de la langue française*). In the first French edition of *Origin*, Royer translated 'natural selection' as 'élection naturelle' (Royer trans. 1862). In the second French edition, 'sélection' was substituted for 'élection' at CD's request, although Royer expressed her disapproval of this change in a footnote (Royer trans. 1866, p. 95 n.; see also J. Harvey 1997, pp. 76–7). Royer's footnote referred to a paragraph that CD had inserted to defend the term natural selection against criticisms on various grounds, including its implication of conscious choice (*Origin* 3d ed., pp. 84–5). Royer's preference for the word 'élection',

which may be translated as 'choice', reflected her belief in the control of nature by an intelligent power (C. Blanckaert in Tort 1996, p. 3746).

4 The memorandum is evidently a note for the second French edition of *Origin* (Royer trans. 1866). In the first edition of her translation (Royer trans. 1862, p. 57), Royer inserted a note comparing animal breeding practices to the strict marriage laws governing Hindu castes. This memorandum appears to be an expansion of that earlier note, and refers to the Hindu 'Laws of Menu' or 'Manu' described in the Sanskrit text Manava-Dharma-Sastra, which was translated into French in 1833 (Loiseleur Deslongchamps trans. 1833). CD probably made corrections to the manuscript of Royer trans. 1866 between April and June 1865 (see letter to John Murray, 31 March [1865], letter to J. D. Hooker, 10 [April 1865], and letters from Emma Darwin to H. E. Darwin, [6 June 1865] and [21 June 1865] in DAR 219.9: 27, 32). See also 'Journal' (Appendix II). CD evidently rejected the additions proposed in this memorandum, as they do not appear in the published version of Royer trans. 1866; there is a copy of this work in the Darwin Library–Down.

5 Royer also discussed marriage selection and its effect on human evolution in the first French edition of *Origin* (see Royer trans. 1862, Préface). For a discussion of Royer's translations of *Origin* and CD's reaction to her editorial insertions, see J. Harvey 1997, pp. 62–9, 76–9. See also *Correspondence* vol. 10, letter to Asa Gray, 10–20 June [1862], letter to Armand de Quatrefages, 11 July [1862], letter from Edouard Claparède, 6 September 1862, and letter to J. D. Hooker, 11 September [1862]. In a letter from Emma Darwin to H. E. Darwin, [6 June 1865], in DAR 219.9: 27, Emma reported that CD was at work 'on the verdammte Mll^e^ Royer whose blunders are endless'.

6 The reference is to the Hindu 'Laws of Menu' (see n. 4, above). The text is also discussed in Lucas 1847–50; CD read Lucas 1847–50 in 1856 (*Correspondence* vol. 4, Appendix IV), and there is a heavily annotated copy of this work in the Darwin Library–CUL (see *Marginalia* 1: 513–23). CD refers to Lucas 1847–50 in an annotation to Emma's translation of this memorandum (see n. 1, above). 'Prosper Lucas about Beautiful men marrying beautiful in caste—'. The translation is also annotated 'Man' in red crayon, indicating that CD kept this material for possible use in *Descent*; however, he did not refer to it in any of his published works.

7 This section of book 3 of the 'Laws of Menu' discusses the rules of marriage, including caste restrictions for four named Hindu castes, and the signs of family or individual deformity or disease that one should avoid in a marriage (see W. Jones trans. 1869, pp. 40–1). CD wrote 'I have not looked' against this reference on Emma's translation of the memorandum (see DAR 80: B42–3).

8 In book 9 of the 'Laws of Menu', paragraphs 72 and 73 discuss conditions under which a marriage to a woman with an undetected physical defect, or a marriage contracted through deception, can be annulled; paragraph 178 designates a son 'begotten through lust' by a high-caste man on a low-caste woman as a 'living corpse' (see W. Jones trans. 1869, pp. 200–1, 211–12). On Emma's translation of the memorandum CD crossed through this and the subsequent citation (see n. 9, below) and added the word 'Nothing' (see DAR 80: B43).

9 These passages define the status of children born into various Hindu castes. The last passage cited (paragraphs 66 to 73) outlines the effect of the parents' good or bad conduct on the possibility of offspring from a mixed caste rising to a superior caste (see W. Jones trans. 1869, pp. 228–35). For CD's reaction to these passages, see n. 8, above.

From C. V. Naudin[1] 18 June 1865

Museum | d'Histoire Naturelle, | Culture. | Paris,
18 June 65.

Dear Mr Darwin,

M. Decaisne[2] and I have received the copies of your excellent work on climbing plants that you were kind enough to send us, and I want, in his name and in mine, to express my heartfelt thanks.[3]

In return, I hope soon to be in a position to give you a copy of my paper on hybridisation, which is in press at the moment.[4] At the same time, I plan to send you a short communication written by me, about ten years ago, on the tendrils of the Cucurbitaceae, of which perhaps you are not aware (Annales de Sc. Nat vol. . . .),[5] and in which I believe that I have shown that these organs are the veins of a leaf belonging to an abortive shoot and do not, as M. Payer has claimed, derive from the normally developed leaf.[6]

I am extremely busy at the moment, not only with my research on the Cucurbitaceae and hybridisation, but also in writing a general horticultural treatise applying to all the climates of France.[7] I am sorry to add that a lot of my time is taken up also with my struggle against a very painful facial neuralgia, which lets me rest neither by day nor by night.[8]

It would make me happy, dear Sir, to learn that your health is entirely restored. I wish it from the bottom of my heart, first for your sake, and secondly for the sake of Science, which has everything to gain from your ability to continue to serve it for a long time.

I received, with the utmost pleasure, the beautiful portrait which you sent me, and which I keep as a precious memento of you, and of your son, whose skill in photography it proves.[9]

Will you, dear Sir, accept with our collective thanks, the assurance of the sincere regard with which I have the honour to remain | your obedient servant and colleague, | Ch. Naudin | Member of the Institut de France| at the Muséum d'Histoire Naturelle.

DAR 172: 8

[1] For a transcription of this letter in its original French, see pp. 189–90.

[2] Joseph Decaisne.

[3] The reference is to 'Climbing plants'. No presentation list for 'Climbing plants' has been found.

[4] The article referred to is Naudin 1865a. CD's annotated presentation copy is in the Darwin Library–CUL (see *Marginalia* 1: 638–9).

[5] The reference is to Naudin 1855. No copy of this article has been found in the Darwin Pamphlet Collection, but the article had been mentioned to CD by Joseph Dalton Hooker (see *Correspondence* vol. 11, letter from J. D. Hooker, [31 July 1863] and n. 7). There was a long debate among a number of botanists, including Naudin and Jean-Baptiste Payer, in the Société Botanique from February to July 1857 on the origin of the tendril in the Cucurbitaceae (*Bulletin de la Société Botanique* 4 (1857): 109–111, 142–6, 322–4, 744–56, 787–8). CD mentions these discussions in a footnote to 'Climbing plants', p. 73. For some of CD's inquiries about tendrils, see *Correspondence* vol. 12, letter from Daniel Oliver, [before 31 March 1864] and nn. 3 and 5, and letter to Daniel Oliver, 11 March [1864] and nn. 3, 6, and 9.

[6] In Payer 1845, p. 164, the tendril is described as a modified leaf. During the 1857 discussions (see n. 5, above), Payer added that the tendril in the Cucurbitaceae was analogous to the stipule of the leaf in the Rosaceae (Guillard 1857, pp. 145–6). Naudin had attempted to show that the tendril was a modification of the first leaf of an aborted shoot; that is, a 'mixed organ' consisting of a branch at its base and a leaf at the top (Naudin 1855, p. 9). In 'Climbing plants', CD mentioned Naudin's view of the tendril as 'part leaf and part branch', referring to the 1857 discussions but not citing Naudin specifically. CD then cited Payer on the authority of Miles Joseph Berkeley (who referred to Payer as

'Payen' in his article in *Gardeners' Chronicle* 1864, p. 721) as providing the 'most probable' explanation, that the tendril was 'a separate portion of the leaf itself' ('Climbing plants', p. 73).

[7] The work referred to is Decaisne and Naudin 1862–71.

[8] Naudin's nervous disorder had left him totally deaf and in constant pain. He had already been forced to give up teaching, and eventually resigned from the Muséum in 1868 after two operations on his facial nerves failed to resolve the problem completely (letter from C. V. Naudin, 29 March 1868, *Calendar* no. 6068; see also J. Harvey 1997).

[9] At Naudin's request, CD had sent a photograph taken by his son William Erasmus Darwin in 1864 (see *Correspondence* vol. 12, letter to C. V. Naudin, 8 December [1864]; the photograph is reproduced as the frontispiece of *Correspondence* vol. 12). Naudin had sent a photograph of himself to CD in December 1864 (see *Correspondence* vol. 12, letter from C. V. Naudin, 6 December 1864 and n. 3). Although no photograph of Naudin has been found in the Darwin Archive–CUL, it is likely that the photograph reproduced in this volume is the same as the one CD received.

From Fritz Müller[1] 12 August 1865

Desterro, Brazil,
12. August 1865

My dear Sir,

A few days ago I received your essay on climbing plants,[2] and I hasten to express my thanks for this valuable gift. I have read it with the greatest interest and I am very happy that my attention has been directed to these remarkable plants which are so extraordinarily plentiful in our Flora. In a couple of days I collected in the surroundings of Desterro, after I had read your work, the following genera of climbing plants: *Spirally twining*: Dioscorea, Alströmeria (subgenus Bomarea,[3] Boussingaultia, Hebanthe,[4] (Goruplerina),[5] Aristolochia, Micania,[6] Manettia, Condylocarpon, Echites, Prestonia, Oxypetalum and a few other Asclepiads, Tournefortia, Ipomoea (numerous species), Calonyction, Cuscuta, Thunbergia (introduced), Mendozia, Cocculus (?), Davilla, Anchietia, Tetrapterys, Peixotoa, Brachypteris (?),[7] Thryallis, Dicella, Dalechampia, Mucuna, Caravillia;[8] *leaf climbers*: Clematis, Fumaria (introduced) Tropaeolum; *Tendril bearers*: Smilax, Mutisia, Strychnos, Haplolophium, Bignonia (numerous species), Cissus, Passiflora (6 species), Momordica, Sechium (cultivated) and various other Cucurbitaceae, Serjania, Paullinia, various Papilionaceae; *Hook-climbers*: Galium (Relbunium Endlicher)[9] Buttneria, Rubus; *Root-climbers*: Vanilla, various Aroids, Piper. I can add the genera Lygodium, Blumenbachia, Solandra, Combretum, Tragia, which I saw earlier in other parts of our province.[10] Since until now I have paid no particular attention to climbing plants, I am not at present able to give you any information on their habits.

Incomplete[11]
Möller 1915–21, 2: 72–3.

[1] For a transcription of this letter in the German of its published source, see p. 215. According to Alfred Möller, all Fritz Müller's letters to CD were written in English (see Möller ed. 1915–21, 2: 72 n.); most of them have not been found. Many of the letters were later sent by Francis Darwin to Möller, who translated them into German for his *Fritz Müller: Werke, Briefe und Leben* (Möller ed.

1915–21). Möller also found final drafts of some Müller letters among the Fritz Müller papers and included these in their original English form (*ibid.*, 2: 72 n). Where the original English versions are missing, the published version, usually appearing in German translation, has been used. See also n. 11, below.

2 'Climbing plants'. See letter to Fritz Müller, 10 August [1865], n. 4.

3 *Bomarea* is now considered to be a separate genus of the family Alstroemeriaceae.

4 *Hebanthe* is a synonym of *Pfaffia*.

5 This is probably a misspelling or incorrect transcription of *Gomphrena* (Amaranthaceae). Although *Gomphrena* has few climbers, there is one climbing species, *G. vaga*, which is found in the vicinity of Destêrro (Florianópolis).

6 A misspelling of *Mikania* (see letter from Fritz Müller, [12 and 31 August, and 10 October 1865] and n. 18).

7 A misspelling of *Brachypterys*.

8 Probably a misspelling or incorrect transcription of *Canavalia*.

9 Müller refers to *Galium*, subgenus *Relbunium*, as described in Endlicher 1836–42, p. 523. *Relbunium* is now considered to be a distinct genus.

10 Müller resided in Destêrro on Santa Catarina Island off the coast of the state of Santa Catarina, Brazil. Destêrro is now Florianópolis.

11 Part of the missing portion of this letter was submitted by CD, together with extracts from other letters from Müller to CD on climbing plants, to the Linnean Society. The resulting paper was read on 7 December 1865, and published in the 29 November 1866 issue of the *Journal of the Linnean Society (Botany)* (see letter from Fritz Müller, [12 and 31 August, and 10 October 1865], and Müller 1865b). When CD was preparing the letters for publication, he informed Müller that he had 'slightly modified the arrangement of some parts & altered only a few words' (letter to Fritz Müller, 17 October [1865]; see also letter to Fritz Müller, 9 December [1865]; for more on the modifications made to Müller's letters, see the letter to J. D. Hooker, 3 October [1865], and the letter to Daniel Oliver, 20 October [1865]).

From Fritz Müller[1] 31 August 1865

Desterro, Brazil,
31 August 1865

My dear Sir,

I hope you have received in good order a letter from me in which I thanked you for sending me your stimulating work about climbing plants.[2] Since then I have made some observations on this subject which I believe will be of some interest to you.

To the list of climbing plants which grow in the hedges and thickets around Desterro I can add the genera Lygodium, Valeriana, Tragia, Caulotretus. Three others I have not yet seen in bloom and do not know to which genera they belong.

Incomplete[3]
Möller 1915–21, 2: 73–4.

1 For a transcription of this letter in the German of its published source, see p. 227. See also letter from Fritz Müller, 12 August 1865, n. 1.

2 'Climbing plants'. See letter from Fritz Müller, 12 August 1865.

3 For the missing portion of the letter, see the letter from Fritz Müller, [12 and 31 August, and 10 October 1865]. See also letter from Fritz Müller, 12 August 1865, n. 11.

From Fritz Müller[1] 10 October 1865

Desterro, Brazil,
10 October 1865

My dear Sir,

Yesterday I received your kind letter of 10 August[2] and I feel very complimented by the interest you have taken in my little book about species;[3] I feel well rewarded if it has given you some pleasure. I thank you cordially for having sent me your photograph;[4] I am very happy to have it and hope soon to be able to send you mine but at the moment I have none. I accept with thanks your offer to send me your work on orchids, I would very much like to read it; since books always lose some of their interest in translation, I would prefer the English edition to the German.[5] When your letter arrived, I was just writing the last lines of a small paper on the wood of climbing plants. Assuming that it will be of some interest to you, I enclose the paper with this letter.[6] When you are finished with it, would you please send it to Prof. Max Schultze of the University of Bonn, or if you have a translator, the paper could perhaps be published in an English journal.[7] Prof. Max Schultze will send you in time a small paper about a remarkable sponge, which will be published in his Archiv für mikroskopische Anatomie.[8] When I started to study the sponges of our coast last year, I met with great difficulties in applying your theory to that class. There are, as you know, calcareous spicules in some sponges and siliceous ones in others; their forms are sometimes so similar that one could hardly suppose that they have evolved independently of one another; but on the other hand it seemed very improbable that siliceous spicules should have gradually transformed into calcareous ones, or vice versa. The most probable assumption seems to me that both derived from an ancestral form of spicule that was composed of an organic substance and that was later encrusted with a calcareous substance in some descendants and a siliceous one in others. This view of mine was confirmed by the discovery of a species, described in my paper, which has very beautiful horny spicules of considerable size. . . .[9] [missing text][10]

. . . My children had grown a single plant of Linum usitatissimum and my daughter Rosa[11] noticed on it that the tip of the stem, which was bent over slightly before the flower opened, had a peculiar revolving movement that followed the sun, or, since we live here in the southern hemisphere, that was counterclockwise.[12] I found this fact confirmed and observed that a revolution was completed in approximately eight hours; the movement was quite irregular; sometimes it stopped for an hour, then again it was so fast that the circle would have been completed in 2–3 hours had the speed remained constant. The following days were very windy so that I could not continue my observations.

Concerning Anelasma, you will have seen from my first letter that I, too, consider this form as a splendid link between the true Cirripedes and the Rhizocephala[13] or the Cirripedia suctoria (?), as they are called by Lilljeborg,[14] although they possess neither cirri nor any kind of "suctorial apparatus".

With the wish that this letter finds you in good health believe me Dear Sir with sincere respect | very truly yours | Fritz Müller.

Incomplete[15]
Möller 1915–21, 2: 74–6.

[1] See letter from Fritz Müller, 12 August 1865, n. 1. For a transcription of this letter in the German of its published source, see pp. 266–7.
[2] See letter to Fritz Müller, 10 August [1865].
[3] Müller refers to Müller 1864 (see letter to Fritz Müller, 10 August [1865], n. 2).
[4] CD had probably sent the photograph taken by his son William Erasmus Darwin in 1864. The photograph is reproduced as the frontispiece to *Correspondence* vol. 12.
[5] CD had offered Müller a copy of the German translation of *Orchids* (Bronn trans. 1862) in his letter to Müller of 10 August [1865]. In his letter of 20 September [1865], CD wrote that he had sent it.
[6] The reference is to the manuscript of an article later published in *Botanische Zeitung* (Müller 1866). CD's annotated copy of the published article, with a separate sheet of notes in his hand, is in the Darwin Pamphlet Collection–CUL.
[7] Max Johann Sigismund Schultze was the editor of the *Archiv fur mikroskopische Anatomie* and had been a friend of Müller's since 1846, when they were both students at the University of Greifswald (see Möller ed. 1915–21, 3: 25). CD later told Müller that he had immediately forwarded the manuscript of Müller's essay on the wood of twining plants to Schultze (see letter to Fritz Müller, 9 December [1865]). In a letter of 12 December 1865, Müller asked Schultze to try to get the essay published in a botanical journal (see Möller ed. 1915–21, 2: 74, 78).
[8] The reference is to Müller 1865c. An offprint of the article, with a note in CD's hand on the front cover, is in the Darwin Pamphlet Collection–CUL.
[9] Müller named the sponge *Darwinella aurea* (see F. Müller 1865c, p. 344). The genus *Darwinella* belongs to the order Dendroceratida, whose members are non-calcareous, and which is characterised by a skeleton composed entirely of spongin fibres (horny spicules) and lacking siliceous spicules (Bergquist 1978, p. 176).
[10] The letter is incomplete. CD extracted the sections of this letter on climbing plants and included them in an article that appeared in the *Journal of the Linnean Society* (*Botany*). This article is reproduced as the letter from Fritz Müller, [12 and 31 August, and 10 October 1865]; see also letter from Fritz Müller, 12 August 1865, n. 11.
[11] Rosa Müller.
[12] Müller later published the observation on *Linum usitatissimum* (common flax; Müller 1870, p. 137); CD cited the article in *Movement in plants*, p. 203. CD's annotated copy of Müller 1870 is in the Darwin Pamphlet Collection–CUL.
[13] Müller's observations on *Anelasma* were in a missing section of the letter from Fritz Müller, 12 August 1865; see, however, the letter to Fritz Müller, 20 September [1865] and nn. 9 and 10.
[14] The Swedish naturalist Wilhelm Lilljeborg coined the name 'Cirripedia Suctoria' to describe the genera *Liriope* and *Peltogaster*, from 'their sucking their nourishment from the animal to which they are attached' (see Lilljeborg 1859 and Lilljeborg 1860, p. 173). *Peltogaster* is now classified as a genus of Rhizocephala; *Liriope* is a genus of the order Isopoda, and is a hyperparasite (parasite upon a parasite) of some Rhizocephala.
[15] See n. 10, above.

From Fritz Müller[1] 5 November 1865

Desterro, Brazil,
5. November 1865.

My dear Sir,

I have to thank you for your kind letter of 20 September and also for your Orchid book and the two essays on *Linum* and *Lythrum* that I received in good order by the last French post.[2] I have read these various papers with the greatest interest. How remarkably complicated are the relations of the sexes in Lythrum salicaria! The only plant of that family which I have found up to now (a Nesaea, sub-genus Heimia), is monomorphic;[3] but there may perhaps be dimorphic species among the various *Cuphea* in our flora.[4] We are very rich in orchids here and I hope I will be able to see for myself some of the wonderfully perfect adaptations that you describe in your book.[5]

In the forests of the mainland here there are many climbing plants that ascend even the thickest trunks but I cannot remember whether there are twining forms among these; next month I plan to make an excursion to the Itajahy River and on this occasion I will direct my attention to that point.[6] Concerning *Dalbergia*, of which you state that it climbs thick trees, I have some doubts whether it is a true twiner or more likely a branch-climber like many other plants of the same family; that is, whether it climbs with the help of the swinging motion of its shoots or by means of bending to the side that by chance touches a trunk.[7] This latter movement together with the upward growth of the shoot would cause it to climb spirally like a true twiner. One of our branch-climbing dalbergias, which I found without leaves first and which now has young flower buds, appears to be a *Triptolemaea* and this genus Endlicher places immediately after *Dalbergia*.[8]

You ask if Natural History is not rendered exceptionally interesting by such views as we both hold.[9] To be sure! Ever since I read your book on the origin of species[10] and was converted to your views, many facts which I used to view with indifference have become quite remarkable; others which previously seemed meaningless oddities have acquired great significance, and thus the countenance of all Nature has been transformed. I will for that reason never be able to express fully my deep gratitude or the great obligation that I feel towards you.

I will mention just one fact that confused me a great deal until your book made sense of it. In the crevices in the granite rock on the east coast of our island, I found a species of *Apium* that, in its general appearance and still more in its scent, was extraordinarily similar to the wild celery that I collected years ago in the marshy meadows on the Baltic coast; this *Apium* was in the company of a *Tetragonia* resembling *Tetragonia expansa*, which in Germany is cultivated under the name of New Zealand spinach, and at the same spot the sandy coast was covered with a *Scaevola*, which in our flora, as far as I know, is the only genus of a family almost completely restricted to New Holland.[11] Now, through your book, I understand why coastal genera are more likely to have a so much wider distribution than other plants.[12]

I plan to remain in Desterro until 1870 or 1871 and then to return to the Itajahy River, where I lived from 1852–1856 and where I own a small homestead.[13] The landscape of our island is very beautiful; even travellers who have visited the islands of the Pacific, Java, etc. have said to me that our island was one of the most beautiful spots that they had ever seen. Unfortunately the vegetation has now lost much of its former splendour: the virgin forests have almost completely vanished, and many of our hills are now covered almost exclusively by low bushes of an unattractive *Dodonaea*.

Mr. L. Agassiz is at present researching the Amazon River; as I gather from a letter of his that is published in a Rio newspaper, he hopes that the geographic distribution of fishes in each river will furnish conclusive evidence against the theory of transmutation.[14] It seems impossible to me to acquire a fairly complete knowledge of the geographic distribution of fishes in this enormous river and its numerous tributaries in a couple of years, and I cannot imagine that any conclusive objection could be derived from incomplete information about this.

With the wish that your health will soon improve I am, dear Sir with the most sincere respect | Yours very truly | Fritz Müller.

Möller 1915–21, 2: 76–7

[1] For a transcription of this letter in the German of its printed source, see pp. 294–6.

[2] In his letter of 20 September [1865], CD said that he had sent Müller the German edition of *Orchids* (Bronn trans. 1862), and two papers on dimorphism, 'Two forms in species of *Linum*' and 'Three forms of *Lythrum salicaria*'.

[3] When CD described *Nesaea verticillata* (family Lythraceae) as trimorphic in *Forms of flowers*, p. 167, he added a note that Müller had observed a homostyled species of the genus in Santa Catarina, Brazil.

[4] In *Forms of flowers*, p. 168, CD described three species of *Cuphea* (family Lythraceae) as possessing traits of heterostyly, suggesting that they may have reverted from a heterostyled condition. CD's notes on *Cuphea* are in DAR 109: B2. Müller found only homostyled species of *Cuphea* in Brazil (see Müller 1868b, p. 113).

[5] Müller published papers on orchids, and is cited extensively in 'Fertilization of orchids', and *Orchids* 2d ed.

[6] CD had asked Müller for information on twining plants ascending thick trunks in his letter of 20 September [1865]. In 'Climbing plants', p. 21, CD had noted that English plants, with the exception of honeysuckle (*Lonicera periclymenum*), never twined around trees, while tropical climbers were able to ascend thick trees. In *Climbing plants* 2d ed., p. 37 n., CD added Müller's observations of root-climbers and tendril-bearers ascending thick trees in southern Brazil. See also *Correspondence* vol. 12, letter to J. D. Hooker, 26 November [1864], and letters from J. D. Hooker, 29 November 1864 and 2 December 1864.

[7] In 'Climbing plants', p. 22, CD had reported that, according to Thomas Thomson and Joseph Dalton Hooker, some *Dalbergia* species ascend thick trees by twining. See also *Correspondence* vol. 12, letter from J. D. Hooker, 29 November 1864, and letter to J. D. Hooker, 4 December [1864]. On branch-climbers, see letter from Fritz Müller, [12 and 31 August, and 10 October] and nn. 2, 8, and 19.

[8] In Stephan Ladislaus Endlicher's *Genera plantarum*, the description of *Dalbergia* is immediately followed by that of *Trioptolemea*; both are placed in the tribe Dalbergieae (Endlicher 1836–42, pp. 1304–5).

[9] See letter to Fritz Müller, 20 September [1865].

[10] *Origin*. The first German edition was published in 1860 (Bronn trans. 1860).

[11] New Holland was the Dutch name for Australia (*EB*).

[12] In *Origin*, pp. 358–65, CD presented evidence suggesting that plants could migrate across large bodies of water by means of seeds or fruit that floated, or had been embedded in drift wood or icebergs, or carried by birds. To test his theory, CD had performed experiments between 1855 and 1856 in which he immersed seeds and fruits in salt water (*Correspondence* vols. 5 and 6); he discussed the theory with Hooker over several years (see letter to J. D. Hooker, 22 and 28 [October 1865] and nn. 12–14).

[13] Müller explored the region of the Itajaí Açu River in south-eastern Brazil between 1852 and 1856. In 1854, he acquired a homestead in the German settlement of Itajaí, at the mouth of the river. He began teaching mathematics at a provincial school in Destêrro (Florianópolis) in 1856, but retained his property in Itajaí (Möller ed. 1915–21, 3: 61–9).

[14] Louis Agassiz began his Brazilian expedition, which included a voyage on the Amazon River, in April 1865. While on the steamer travelling from New York to Rio De Janeiro, he delivered a lecture to his students in which he expressed his conviction that the combination of animals in South America would give him the means of showing that the transmutation theory was 'wholly without foundation in facts' (J. L. R. Agassiz and Agassiz 1868, p. 33). Agassiz never published a formal description of the expedition's specimens, or an analysis of their variation or distribution (see Winsor 1991, pp. 66–76). His informal comments on the fish of South America are given in J. L. R. Agassiz and Agassiz 1868, pp. 216–27, 237–41, and 377–84.

From Ernst Haeckel[1] 11 November 1865

Jena
11 November 1865.

Most esteemed, dear Sir!

After a long period of silence, I hope I may take up your valuable time again with one of my letters.[2] To my great regret I heard from Mr. Huxley that unfortunately you have been very ill again.[3] I hope you are feeling better now and that you will remain with us and science for a long time yet.

Unfortunately my book, which deals in detail with your theory, is still unfinished, although I had hoped to complete it half a year ago.[4] However, I was forced to give up the whole summer in order to work at the zoological museum;[5] you see, I was made a full professor of zoology 6 months ago, after I turned down the offer of a chair at Würzburg.[6] Still, I hope to be able to complete the book by next Christmas and I will send it to you as soon as it is finished.

This winter I am giving public lectures "*on Darwin's theory*" again; these are the most popular of *all* lectures given at this university.[7] I have over 150 attendants and the largest lecture theatre of our university is packed. This gives you an idea of how exceptional the interest in you and your theory is here at our small university and how much support there is for the great reform of science that you have begun and that it is the greatest task of my life to realise. My audience is very mixed; in addition to many students of medicine and the sciences it includes numerous historians and students of philosophy; there are even a few theologians and lawyers!

My most heartfelt thanks to you for sending your beautiful essay "on the climbing plants".[8] I have read it with great interest.

I hope you received my book on Geryonidae (with 6 plates) in good order, as well as a paper "on fossil Medusae" and another one on "Rhizopodean protoplasm".[9] I instructed Mr Engelmann in Leipzig to send you these works without delay.[10] Among the new Rhizopoda of Nice the "Protogenes primordialis" (fig. 1, 2) might interest you as one of the simplest of all creatures, an organism without organs, homogeneous throughout![11] The *generatio aequivoca* of such a living gelatinous mass is at any rate conceivable, and this would help the theory of descent with the difficulty of explaining the first origin.[12]

Whenever I write to you, dear Sir, I must reiterate my warmest thanks for the unfailing inspiration and the vigorous stimulus that your classical theory gives me.

Ever since the death of my most dearly beloved wife put an end to and destroyed my emotional life, all I have left is the aspiration to employ the keenest intellect in understanding the true nature and continuity of natural things.[13] This aspiration occupies me in my hermit-like existence day and night, and I owe this aspiration solely to your work!

May you long remain in vigorous health, so as to witness the reform of biology that you began, and to enjoy the fruits of your great intellectual achievement. This is the most heartfelt wish of your sincerely devoted | Ernst Haeckel

DAR 166: 40

[1] For a transcription of this letter in its original German, see pp. 302–3.

[2] Haeckel's last known letter to CD is dated 26 October 1864 (see *Correspondence* vol. 12); however, CD received a copy of Haeckel 1865b in May 1865 (see letter to J. D. Hooker, 4 May [1865]).

[3] Haeckel refers to Thomas Henry Huxley. On CD's health, see the letter to J. D. Hooker, 22 and 28 [October 1865].

[4] The book was eventually published as Haeckel 1866. In his letter of 9 [July 1864] (*Correspondence* vol. 12), Haeckel had claimed that in his next book he would show that the theory of descent threw 'light on every single chapter' of the history of creation and unified them 'in the most perfect harmony'. See also *ibid.*, letter from Ernst Haeckel, 26 October 1864. CD's annotated copy of Haeckel 1866 is in the Darwin Library–CUL (see *Marginalia* 1: 355–7).

[5] Haeckel succeeded Carl Gegenbaur as director of the Zoological Museum in Jena in 1862, and began a programme of improvement and acquisition. In the summer of 1865, Haeckel successfully petitioned the Weimar state ministry to finance a complete reorganisation of the collection, a new hall to house acquisitions, and instruments for research. Haeckel was able to obtain government funding by establishing a separate Zoological Institute within the Museum. See Uschmann 1959, pp. 48–58.

[6] Haeckel received a full professorship at Jena on 15 May 1865 (Uschmann 1959, p. 50). Five days previously, he had been offered a full professorship at Würzburg, but declined. The terms of the Würzburg position stipulated that control of the zoology museum and examinations remain with another, more senior, zoology professor (*ibid.*, p. 51).

[7] The popularity of Haeckel's course on 'Darwins Theorie' is described in Uschmann 1959, p. 44. Since 1862, Haeckel had also given a separate annual lecture on CD's theory, and had devoted part of his zoology course to the topic (see *ibid.*, pp. 43–4, and *Correspondence* vol. 12, letter from Ernst Haeckel, 9 [July 1864]). On the reception of CD's theory in Germany, see letter from T. H. Huxley, 29 May 1865 and n. 3.

[8] 'Climbing plants'. No presentation list for the paper has been found.

[9] Haeckel refers to his monograph on the Geryonidae (now called Geryoniidae), a family of medusae (Haeckel 1865a), and to Haeckel 1865c and 1865d. A copy of Haeckel 1865a is listed in the 1875 catalogue of CD's library (DAR 240); however, it is not recorded in later catalogues and has not been found in the Darwin Library–CUL or at Down. A lightly annotated copy of Haeckel 1865d is in the Darwin Pamphlet Collection–CUL.

[10] The Leipzig firm Wilhelm Engelmann had published Haeckel 1865a, and was also the publisher of *Zeitschrift für wissenschaftliche Zoologie*, in which Haeckel 1865c and 1865d appeared.

[11] Haeckel described *Protogenes primordialis* as the simplest type of Rhizopoda; a protoplasmic aggregate, without differentiation, organisation, or nucleus, it was able to surround and assimilate foreign bodies and to reproduce by simple division (Haeckel 1865d, pp. 360–3). In his later book, *Generelle Morphologie der Organismen* (Haeckel 1866, 2: XXII–XXIII), Haeckel created a new classification for the organism within a phylum he named the Monera, within the kingdom Protista, arguing that the Rhizopoda had possibly evolved from a *Protogenes* species. The organism is not recognised in modern taxonomic literature.

12 *Generatio aequivoca* or equivocal generation: 'the (supposed) production of plants or animals without parents; spontaneous generation' (*OED*). Haeckel did not discuss *Primogenes primordialis* as a case of spontaneous generation (Haeckel 1865d); however, in *Generelle Morphologie der Organismen* (Haeckel 1866, 2: 174), he did speculate more generally on the possibility of spontaneous generation. For discussions of Haeckel's theories of protoplasm and the development of living organisms from inorganic matter, see Rehbock 1975 and Rupke 1976. On the Victorian debates over spontaneous generation, see Strick 2000.

13 Haeckel's wife, Anne Sethe, had died on 16 February 1864; Haeckel discussed her death and its significance for his scientific work in his letters to CD of 9 [July 1864], 10 August 1864, and 26 October 1864 (*Correspondence* vol. 12).

To J.-B. P. Guépin[1] 14 November 1861

Down. | *Bromley.* | *Kent. S.E.* England
Nov. 14 1861

Sir

I hope you will forgive the liberty I am taking in writing to you. I read a notice of a paper in the Annales de la Societe Linn: d' Angers, regarding the fertilisation of Orchids, of which you are the author.[2] I have unsuccessfully tried to obtain a copy in London, something I am very keen to do, as I am at present preparing a book for publication on this subject.[3]

Would you be so kind as to let me know whether you have been able to show from which species of Orchids these Bees had been gathering nectar? Were there one, two, or more 'pollen masses' stuck to a bee? I have seen these pollen masses in great numbers on the proboscis of lepidoptera but not on bees.[4]

If by chance you had in your possession several specimens of bees with the pollen attached and you were able to spare one of them, you would be doing me a great service. If I could study one of these bees (which you could send me in a letter, wrapped in a feather), I think I would be able to identify the species of orchid from which the 'pollen masses' come. Especially if you were able to enlighten me as to which are the most common species of orchids in the Angers area.[5]

I am with much respect and begging you to forgive this liberty your very humble servant | Charles Darwin

LS(A)
Archives Départmentales de Maine-et-Loire (24 J 1, pièce no. 32)

1 For a transcription of this letter in the original French, see pp. 449–50.

2 Daniel Oliver had sent CD a list of references in French and German relating to orchids. Guépin 1853 was described on this list as dealing with 'fecundation of *Catasetum*' (see *Correspondence* vol. 9, letter from Daniel Oliver, [before 8 November 1861], and this volume, Supplement, letter to Daniel Oliver, 3 November [1861]).

3 CD refers to *Orchids*.

4 Frederick Bond is cited in *Orchids*, pp. 35–6, as having sent CD a large number of moths and butterflies with orchid pollinia attached to their proboscises. Virtually all the pollinia came from a single species, *Orchis pyramidalis*. See also *Correspondence* vol. 8, letter to J. D. Hooker, 17 June [1860] and n. 5.

5 Guépin published a flora of the area of Maine and Loire (Guépin 1830–54). Guépin had died in 1858, so the letter was never answered.

APPENDIX II
Chronology 1865

This appendix contains a transcription of Darwin's 'Journal' for the year 1865. Darwin commenced his 'Journal' in August 1838 and continued to maintain it until December 1881. In this small notebook, measuring 3 inches by $4\frac{1}{2}$ inches, Darwin recorded the periods he was away from home, the progress and publication of his work, and important events in his family life.

The version published by Sir Gavin de Beer as 'Darwin's Journal' (de Beer ed. 1959) was edited before the original 'Journal' had been found and relied upon a transcription made by an unknown copyist. The original, now in the Darwin Archive in Cambridge University Library (DAR 158), reveals that the copyist did not clearly distinguish between the various types of entries it contains and that the transcription made was incomplete.

From 1845 onward, Darwin recorded all that pertained to his work (including his illnesses, since these accounted for time lost from work) on the left-hand pages of the 'Journal', while the periods he was away from home, and family events, were noted on the right-hand pages. In order to show clearly Darwin's deliberate separation of the types of entries he made in his 'Journal', the transcription has the left- and right-hand pages labelled.

All alterations, interlineations, additions, and the use of a different ink or pencil have been noted. In addition, the editors have inserted additional information relevant to Darwin's correspondence throughout this transcription of the 'Journal' for 1865. These interpolations are enclosed in square brackets to distinguish them from Darwin's own entries, the source of the information being given in the footnotes.

[**Left**]
1865
Jan. 1. I continued on "Domestic Animals &c" Ch X[1] till April 22^{d}.[2] when (with exception of one subsequent week) I became ill & remained so & unable to do anything (except read Origin for 2^{d} French Edit.)[3] till early in December when I began counting Homomorphic seeds[4] On Dec 25 began again on Chapt X of Domestic Animals.

[**Right**]
[8–9 February. Sarah Elizabeth Wedgwood (1793–1880) visited][5]
[25–7 February. Edmund Langton and Katherine Elizabeth Sophy, Margaret Susan, and Lucy Caroline Wedgwood visited.][6]
[4–6 March. Joseph Dalton Hooker visited.][7]
[8 April. George Howard Darwin home from Cambridge.][8]
[13 April. William Erasmus Darwin arrives home.][9]
[13–24 April. Elizabeth, Francis, and Leonard Darwin home from school.][10]
[14 April. John Lubbock comes to lunch.][11]
Fell ill on April 22[d]—got a little better in the middle of September—
[3 May. William Jenner (physician) visited.][12]
[20 May. John Chapman (physician) visited.][13]
[28 May. John Chapman visited.][14]
[29 May – 22 June. Henrietta Emma Darwin in Wales.][15]
[1 November. Henry Bence Jones (physician) visited.][16]
On Nov. 8th went to Erasmus[17] & staid ten days & fell ill again with cold; but got better in the beginning of December
[17 December. John Lubbock and George Busk to lunch.][18]

1 'Ch X' written in ink above 'Animals'. When *Variation* was published in 1868, the first eight chapters were on domestic mammals, birds, fish, and insects.
2 CD began to revise the earlier chapters of *Variation* after 16 November 1864 (CD's 'Journal', *Correspondence* vol. 12, Appendix II).
3 Royer trans. 1866.
4 See letter to B. D. Walsh, 19 December [1865] and n. 10.
5 Emma Darwin's diary (DAR 242). See also letter to J. D. Hooker, 9 February [1865] and n. 16.
6 Emma Darwin's diary (DAR 242). See also letter to J. D. Hooker, 15 [February 1865] and n. 3.
7 Emma Darwin's diary (DAR 242).
8 Emma Darwin's diary (DAR 242). See also letter to J. D. Hooker, 17 April [1865], n. 1.
9 Emma Darwin's diary (DAR 242). See also letter to J. D. Hooker, 17 April [1865], n. 1.
10 Emma Darwin's diary (DAR 242). See also letter to J. D. Hooker, 17 April [1865], n. 1.
11 See letter to J. D. Hooker, 17 April [1865].
12 CD's Account book–cash account (Down House MS).
13 Emma Darwin's diary (DAR 242).
14 Emma Darwin's diary (DAR 242).
15 Emma Darwin's diary (DAR 242). See also letter to J. D. Hooker, [17 June 1865] and n. 3.
16 Emma Darwin's diary (DAR 242).
17 CD's brother, Erasmus Alvey Darwin, lived at 6 Queen Anne Street, London (*Post Office London directory* 1865).
18 Emma Darwin's diary (DAR 242). See also letter to B. D. Walsh, 19 December [1865].

APPENDIX III
Diplomas presented to Charles Darwin

In 1865, Darwin received the following diplomas from the Royal Swedish Academy of Sciences and the Royal Society of Edinburgh. Although not letters in the conventional sense, they represent significant communication between Darwin and scientific societies, and the citations in such diplomas often provide valuable indications of those aspects of Darwin's work that were considered worthy of honour. In view of this, they have been included here, together with translations.

From Royal Swedish Academy of Sciences 8 November 1865

REGIA
SCIENTIARUM ACADEMIA
SVECICA
VIRUM
INGENIO ET DOCTRINA INCLYTUM
DOMINUM CELEBERRIMUM
CAROLUM DARWIN
SOCIUM EXTRANEUM
SOLENNI SUFFRAGIO ELECTUM
IN COMMUNITATEM STUDIORUM
LITTERIS HIS PATENTIBUS
QUA PAR EST OBSERVANTIA
INVITAT.

HOLMIÆ DIE VIII NOVEMB. MDCCCLXV
M. Thulstrup | *P. t. Præses.*[1]
P. F. Wahlberg. | *Secr. Perp.*[2]

[Translation]

By these letters patent, and with due deference, the Royal Swedish Academy of Sciences invites the most famous Master Charles Darwin, a man renowned for his genius and learning, to join its learned society as a Foreign Fellow elected by formal vote.

At Stockholm, 8th November 1865

M. Thulstrup | *President*

P. F. Wahlberg | *Permanent Secretary*

DAR 229: 13

[1] Carl Magnus Thulstrup.
[2] Peter Fredrik Wahlberg.

From Royal Society of Edinburgh 4 December 1865

Q. F. FQ. S.
Rogantibus Sociis referente Praeside
Societas Regia Edinensis
virum ingenii lumine clarissimum
Carolum Darwin.
honoris cause suffragiis suis collegam sibi cooptavit adscivitque.

In cujus rei fidem sigillo Societatis hoc Diploma muniri jussit:

Act: Edinb: Prid. Non. Dec^{is}: *an: Sal: Hum: MDCCCLXV*
David Brewster *Praeses.*
Joannes Hutton Balfour M.D. *a Secretis.*

[Translation]

Q. F. FQ. S.

At the petition of the Fellows and on the proposal of the President, the Royal Society of Edinburgh, by its votes, elected and admitted as an honorary colleague Charles Darwin, *a man most famous for the brilliance of his genius.*

As guarantee of this it instructed that this diploma be furnished with the seal of the Society.

Executed at Edinburgh, 4th December *AD 1865.*
David Brewster *President*
John Hutton Balfour MD *Secretary*

DAR 229: 14

APPENDIX IV
Note on Darwin's health

After a period of severe illness lasting some twelve months, Darwin's health began to improve at the end of March 1864 under the care of William Jenner. In November and December, however, his health grew worse again. He reported in a letter to W. D. Fox, 30 November [1864] (*Correspondence* vol. 12), that he was 'very weak & continually knocked up, but able most days, to do from 2 to 3 hours work'. Accounts of his health are slightly less frequent in his letters in the early months of 1865; he remarked on several occasions to Joseph Dalton Hooker that he had become discouraged with Jenner's treatment (see, for example, letters to J. D. Hooker, 7 January [1865] and 6 April [1865]). In his 'Journal' (Appendix II), Darwin wrote that he fell ill again on 22 April and was unable to 'do anything' (except some reading for the second French edition of *Origin*) until early December. Emma Darwin's diary (DAR 242) records that he experienced almost daily attacks of vomiting from 21 to 30 April. In a letter to Hooker of 4 May [1865], Darwin wrote that Jenner had been to see him and was 'evidently perplexed' at his case, adding that the physician struck him as 'a more able & sensible man, than he did before', and that he would 'consult no one else'. On 16 May, however, having suffered more attacks of vomiting on the preceding two days, Darwin wrote to John Chapman, requesting a professional visit. A prominent London publisher, Chapman had studied medicine in London and Paris in the early 1840s; he had returned to medical training in 1855, and eventually obtained a licence to practice in London from the Royal College of Physicians in 1857 (Poynter 1950). In 1863, he began using ice to treat epilepsy, and developed a theory of 'neuro-dynamic medicine' according to which many diseases were treatable through applications of heat or cold to the spine over long periods (see Chapman 1863a and 1863b). Darwin had received a copy of Chapman's 1864 book on sea-sickness, which described the use of ice for the cure of dyspeptic conditions. Chapman maintained that the ice acted as a sedative to the spinal cord, reducing the amount of blood in it and therefore lessening its 'automatic or excito-motor power' (Chapman 1864, pp. 5–6). Emma Darwin's diary (DAR 242) records that Chapman visited Down on 20 May 1865. On that day, Darwin wrote an account of the various symptoms of the illness that had extended over most of his adult life. The note was found together with letters from Darwin to Chapman, and was evidently written for Chapman's use in consultation.[1]

Age 56–57.— For 25 years extreme spasmodic daily & nightly flatulence: occasional vomiting; on two occasions prolonged during months.[2] Extreme secretion of saliva with flatulence Vomiting preceded by shivering, hysterical crying dying sensations or half-faint. & copious very palid urine. Now vomiting & every paroxys[m] of flatulence preceded by singing of ears, rocking, treading on air & vision. focus & black dots—[3] All fatigues, specially reading, brings on these Head symptoms ?? nervousness when E. leaves me.—[4]

(What I vomit intensely acid, slimy (sometimes bitter) corrodes teeth.)[5]

Doctors puzzled, say suppressed gout[6] Family gouty.—[7] No organic mischief, Jenner & Brinton.—[8]

Tongue crimson in morning ulcerated— stomach constricted dragging. Feet coldish.— Pulse 58 to 62—or slower & like thread. Appetite good—not thin. Evacuation regular & good. Urine scanty (because do not drink) often much pinkish sediment when cold— seldom headach or nausea.—

Cannot walk above $\frac{1}{2}$ mile— always tired— conversation or excitement tires me most.—[9]

Heavy sleep— bad day.

Eczema—(now constant) lumbago— fundament— rash.—[10]

Always been temperate— now wine comforts me much— could not take any formerly.[11] Physic no good—[12] Chalk & Magnesia.—[13] Water-cure & Douche— Last time at Malvern could not stand it—[14]

I fancy that when much sickness my stomach is cold—at least water is very little warmed.

I feel nearly sure that the air is generated somewhere lower down than the stomach & as soon as it regurgitates into the stomach the discomfort comes on—

Does not throw up the food.[15]

Instruction— How soon any effect? long long continue treatment?[16]

Endorsement: 'Charles Darwin | 20 May 1865'

University of Virginia Library, Albert and Shirley Small Special Collections Library, Darwin Evolution Collection (3314)

MANUSCRIPT ALTERATIONS

1.2 Extreme . . . flatulence 1.3] *added, left margin*
1.3 hysterical crying] *interl*
1.4 sensations] 'a' *over* 'i'
1.4 or half faint] *interl*
1.6 focus & black dots] *interl in Emma Darwin's hand*
1.6 All fatigues, . . . leaves me.— 1.7] *interl*
1.7 ??] *reversed in MS*
2.1 (What . . . teeth.)] *square brackets in MS*
3.1 puzzled,] *interl*
3.1 Family gouty.—] *interl*
4.1 Feet coldish. 4.2] *interl*

7.1 (now constant)] *interl*
10.1 I feel nearly sure ... food. 11.1] *added in Emma Darwin's hand*

[1] A previous transcription of the note, published in Colp 1977, pp. 83–4, contains a number of errors. Colp's transcription has also appeared in whole or in part in Bowlby 1990, pp. 6–7, A. Desmond and Moore 1991, p. 531, and R. Porter and Rousseau 1998, p. 163. A wide range of modern medical explanations have been given for CD's symptoms; for references to the extensive literature on CD's health, see Colp 1977 and 1998, and Bowlby 1990.

[2] On CD's early stomach troubles, see *Correspondence* vol. 2, letter to Caroline Wedgwood, [May 1838], and letter to Robert FitzRoy, [20 February 1840]. CD's health diary (Down House MS), which he kept between 1 July 1849 and 16 January 1855, describes almost daily occurrences of flatulence (see Colp 1977, pp. 46–7). CD first mentioned attacks of 'periodical vomiting' in a letter to W. D. Fox, [7 June 1840] (*Correspondence* vol. 2). He suffered from persistent sickness in 1849, describing 'incessant vomiting' in his letter to Richard Owen, [24 February 1849], and 'vomiting every week' in his letter to J. D. Hooker, 28 March 1849 (*Correspondence* vol. 4). Throughout the winter of 1863 and spring of 1864, he was sick almost daily (see *Correspondence* vol. 12, letter from Emma Darwin to W. D. Fox, [6 May 1864]). According to Emma Darwin's diary (DAR 242), such regular attacks had occurred again in the last week of April 1865, and the third week of May, just before CD's decision to consult John Chapman.

[3] Shivering or trembling, faintness, and black spots had been noted in journal entries and correspondence during periods of sickness in 1848, 1852, and 1859 (see Colp 1977, pp. 38, 47, 64). Fainting and 'rocking' had been recorded in Emma Darwin's diary (DAR 242) on several occasions in 1864 and 1865. 'Bad hysteria & sickness' were recorded in Emma Darwin's diary for 18 October 1864. No other record of 'hysterical crying' or 'dying sensations' has been found.

[4] On CD's reliance on Emma Darwin's companionship and care see, for example, *Correspondence* vol. 4, letter to Emma Darwin, [27–8 May 1848]. See also Browne 1995, pp. 428–9. On his reading difficulties, see letters to J. D. Hooker, 1 June [1865] and 27 [or 28 September 1865]. Emma or another member of the household regularly read to CD at intervals during the day and in the evening (see *LL* 1: 113, 118–19, 121–5, and A. Desmond and Moore 1991, pp. 359, 529).

[5] In his letter to Chapman of 16 May [1865], CD stated that his sickness was 'always caused by acid & morbid secretions'. CD had sometimes noted the acidity of his vomit in his health diary (Down House MS; see Colp 1977, p. 47).

[6] CD's condition had been diagnosed as 'suppressed gout' by Henry Holland in 1849 (*Correspondence* vol. 4, letter to W. D. Fox, 6 February [1849]; see also *Correspondence* vol. 7, letter to Adam Sedgwick, 24 August [1859]). According to some medical literature of the period, stomach disorders such as dyspepsia and flatulence could be caused by the same accumulation of toxic substances that, in outward attacks of gout, caused pain and inflammation of the joints (see, for example, Holland 1855, p. 233, and Garrod 1863, pp. 263–4). The diagnosis of suppressed gout may also have been made more recently by William Brinton, William Jenner, and George Busk (see letter to J. D. Hooker, [7 January 1865], and letter from George Busk, 28 April 1865). For a discussion of the diagnosis, see Colp 1977, pp. 109–11, and R. Porter and Rousseau 1998, pp. 155–6, 162–4, 170–1.

[7] CD believed that his father had suffered from gout (see *Correspondence* vol. 1, letter to W. D. Fox, [25–9 January 1829], and *Correspondence* vol. 2, letter to Sarah Elizabeth Wedgwood, [28 August 1837]). His grandfather, Erasmus Darwin, had described his own attacks of gout in his Commonplace book, p. 89 (Darwin Library–Down; see King-Hele 1999, pp. 161–2). Erasmus also wrote a letter to CD's father, in which he claimed that CD's great-grandfather, Charles Howard, had died of gout (see *Autobiography*, p. 224). CD discussed gout as an inherited disease in *Variation* 2: 7, 77–8. On Victorian views of hereditary illness, see Olby 1993.

[8] In November and December 1863, CD had consulted the stomach specialist, Brinton, who had prescribed mild acid solutions to aid digestion (*Correspondence* vol. 11, Emma Darwin to W. D. Fox, 8 December [1863]). In his letter to J. D. Hooker, 10 [November 1863] (*Correspondence* vol. 11), CD

wrote that Brinton did not believe his brain or heart to be 'primarily affected'. In March 1864, CD began to consult Jenner, who prescribed alkali and purgative substances in conjunction with dietary restrictions (see *Correspondence* vol. 12, letter to J. D. Hooker, 13 April [1864], letter from William Jenner to [William Walmisley Baxter?], [after 7 May 1864?], and letter from William Jenner, 14 August 1864). In his letter to J. D. Hooker, 26[–7] March [1864] (*Correspondence* vol. 12), CD remarked that Jenner had found 'no organic mischief'.

9 CD often remarked that excitement, conversation with visitors, or any departure from his normal routines made him exhausted or sick, and he frequently excused himself from social occasions on these grounds (see, for example, *Correspondence* vol. 2, letter to J. S. Henslow, 14 October [1837], *Correspondence* vol. 7, letter to Robert Monsey Rolfe, 10 November [1858], and *Correspondence* vol. 12, letter to F. T. Buckland, 15 December [1864]).

10 CD's various skin conditions are discussed in Colp 1977, pp. 31–2, 47, 98. In his letter to J. D. Hooker, 5 March [1863] (*Correspondence* vol. 11), CD remarked: 'A good severe fit of Eczema would do me good & I have a touch this morning & consequently feel a little alive'. See also *Correspondence* vol. 12, letter from Emma Darwin to J. D. Hooker, 17 March [1864].

11 On CD's consumption of alcohol, see *LL* 1: 118, and Colp 1977, p. 103.

12 CD tried a variety of medications over the course of his illness, including calomel (mercurous chloride), logwood, cinnamon, potassium bicarbonate, croton, aloe, bitters, bismuth nitrate, mineral acids and alkalies (see Colp 1977, pp. 12, 22, 37, 45–6, 65, 76, 78–80). Most recently, he had taken phosphate of iron as a tonic, and purgatives derived from *Colchicum autumnale* and *Podophyllum peltatum* (May apple) (see *Correspondence* vol. 12). A book of prescriptions used by the Darwin family, some of which are in CD's hand, is in Down House ('Receipts and memoranda book'), and has been transcribed in Colp 1977, pp. 147–67.

13 CD began taking small, frequent doses of chalk, magnesia, and other antacids in March 1864 (see Emma Darwin's diary, DAR 242, and n. 8, above). He reported that the treatment, prescribed by Jenner, had checked his chronic vomiting (*Correspondence* vol. 12, letter to J. D. Hooker, 13 April [1864]).

14 CD had undergone hydropathic treatment on occasion for many years and had at first been enthusiastic about the treatment (see, for example, *Correspondence* vol. 4, letter to W. D. Fox, 24 [March 1849], and *Correspondence* vol. 7, letter to W. D. Fox, 13 November [1858]). He first visited the establishment of James Manby Gully at Great Malvern, Worcestershire, where he and his family spent three months in March 1849 (see *Correspondence* vol. 4). He also took regular treatments in a specially built bath at Down for several years (see *Correspondence* vol. 4, letter to J. D. Hooker, 12 October 1849, and Colp 1977, pp. 43–6). He underwent hydropathic treatments at Moor Park, under Edward Wickstead Lane, between 1857 and 1859, and at Ilkley, under Edmund Smith, in 1859 (see *Correspondence* vols. 6 and 7). He also stayed at Lane's new establishment in Sudbrook Park, Surrey, at the end of June 1860 (*Correspondence* vol. 8). CD's last hydropathic treatment was taken at Malvern Wells, under James Smith Ayerst, in September and October 1863 (see *Correspondence* vol. 11, letter from Emma Darwin to W. D. Fox, 8 December [1863], and Appendix II). On CD's use of hydrotherapy, see Browne 1990.

15 The section, 'I feel nearly … food', is in Emma Darwin's hand. In a letter to J. D. Hooker, [20–]22 February [1864] (*Correspondence* vol. 12), CD remarked that his vomiting usually occurred two to three hours after eating, and that he seldom threw up food.

16 According to Emma Darwin's diary (DAR 242), CD began the ice treatment on 20 May 1865. In his letter to Chapman of 7 June 1865, he reported that the ice had failed to stop either flatulence or sickness. By the second week of July, he had evidently given up the treatment (see letter from Charles and Emma Darwin to J. D. Hooker, [10 July 1865]).

APPENDIX V
The Lyell–Lubbock dispute

In May 1865 a dispute arose between John Lubbock and Charles Lyell when Lubbock, in his book *Prehistoric times*, accused Lyell of plagiarism. The dispute caused great dismay among many of their mutual scientific friends, some of whom took immediate action to mediate a solution. Charles Darwin had close ties with both men and both sought his advice, but Darwin's correspondence during this period reveals his reluctance to become directly involved in the dispute.

In the concluding paragraphs of *Origin*, Darwin had predicted that a 'revolution in natural history' would result when his views on the origin of species were 'generally admitted' (*Origin*, p. 484). He went on to suggest that new areas of research would emerge, and claimed simply, 'Light will be thrown on the origin of man and his history' (*Origin*, p. 488). During the first half of the nineteenth century, the amount of empirical evidence about early humans was increasing, but no unifying hypothesis about human origins had found acceptance at the time *Origin* was published. In 1836, Jacques Boucher de Perthes had found flint implements in the gravel-beds of the Somme Valley and subsequently had argued strongly that the human race was older than had hitherto been supposed, but his views were generally derided.[1]

In 1859, Lyell visited several sites in France where flints that appeared to have been worked had been found, and at the 1859 meeting of the British Association for the Advancement of Science, he announced his belief that these were indeed implements of early humans (C. Lyell 1859). In September 1860 he visited sites in both France and Germany (K. M. Lyell ed. 1881, 2: 336). In April 1860, Lubbock travelled with Joseph Prestwich, Douglas Strutt Galton, and George Busk to France, to visit the Somme Valley. They met Boucher de Perthes and examined flint implements he had discovered that were contemporaneous with the remains of extinct species such as the mammoth (*Correspondence* vol. 8, letter to Charles Lyell, 4 May [1860] and n. 3; Hutchinson 1914, 1: 51). This was the first of many visits Lubbock would make to sites in various European locations in search of evidence regarding the age of the human species. The visits by both Lyell and Lubbock reflected the growing interest, following the publication of *Origin*, in discovering evidence to establish the age of the human race.

In 1861, Lubbock joined Thomas Henry Huxley, Busk, and several other supporters of Darwin in editing the *Natural History Review*, and wrote 'The kjökkenmöddings:

recent geologico-archæological researches in Denmark' (Lubbock 1861) for the October 1861 issue. The article reported on his recent visit, with Busk, to study the kjökkenmöddings (kitchen-middens) of ancient Danish settlements. Lubbock reviewed the literature on the topic and noted that Charles Adolphe Morlot had summarised, in French, earlier reports written in Danish (Morlot 1859, Forchhammer *et al.* 1851–5); Lubbock cited Morlot as the source of many of the 'details' for his article (Lubbock 1861, p. 494). Meanwhile, Lubbock continued his work on ancient settlements and produced three more articles for the *Natural History Review* based on visits to Switzerland, France, and Scotland (Lubbock 1862a, 1862b, and 1863a). In the July 1864 issue of *Natural History Review*, Lubbock produced a final article on 'Cave-men' (Lubbock 1864) that summarised recent evidence for the existence of humans coeval with the quaternary mammals. These five articles later formed the basis of Lubbock's book, *Prehistoric times* (Lubbock 1865).

By 1860, Lyell had begun work on a sixth edition of *Elements of geology* (C. Lyell 1865). At the same time, he was working on a new book, to be named *Antiquity of man*, which would present all the recent material available pertaining to the antiquity of humans. In 1865, he wrote that the section on kitchen-middens that appeared in this new book had been completed and set in type for *Elements of geology* in 1860 and then re-set in 1861 for *Antiquity of man* (see below, 'Textual changes made to C. Lyell 1863c'). On 6 February 1863, *Antiquity of man* (C. Lyell 1863a) was published. The second chapter dealt with Danish kjökkenmöddings and began with a note citing the work of Morlot as the source for information on the topic. Lyell also added the following note on page 11:

> *Mr. John Lubbock published, after these sheets were written, an able paper on the Danish 'shell-mounds' in the October Number of the Natural History Review, 1861, p. 489, in which he has described the results of a recent visit to Denmark, made by him in company with Messrs. Busk, Prestwich, and Galton.

In February 1863, Lubbock received a letter from Lyell, evidently in response to an earlier letter from Lubbock, discussing various aspects of Lyell's book. Lyell requested that Lubbock give him 'the benefit of errata' that he may have seen, and then mentioned:[2]

> I have struck out Galton & Prestwich at p. 11 who will be surprisd [*sic*] to learn that they were in Denmark.

Lyell had been mistaken when he wrote in the first note that Galton and Prestwich accompanied Lubbock and Busk during the trip to Denmark to study the kjökkenmöddings. Lubbock had apparently not made any other comment on the note and did not pursue the matter further at this time. His correspondence with Lyell on scientific topics continued to be very cordial; for example, he provided Lyell with information about the level of the Nile when Lyell was preparing his presidential address for the British Association meeting at Bath in 1864 (C. Lyell 1864).[3]

By November 1863 a third edition of *Antiquity of man* had already appeared, an indication of its popularity, which was no doubt partly inspired by the controversies associated with it.[4] One area of controversy centred around claims by Hugh Falconer that Lyell had failed to credit him and Joseph Prestwich properly for their work in the Brixham cave explorations of 1858 and 1859.[5] Another controversy arose when Richard Owen, writing in the *Athenæum*, accused Lyell of misrepresenting his position in the debate over the comparative anatomy of human and ape brains.[6] Many of Lyell's supporters were privately critical of several aspects of the book. Throughout the first half of 1863, Darwin discussed the book in correspondence with Joseph Dalton Hooker, Asa Gray, and Huxley but he never spoke out publicly about any controversial aspect.

Darwin's chief complaint about the book was more personal. He confided to Hooker that he was 'deeply disappointed' that Lyell had avoided taking a clear position on the transmutation of species.[7] Later, he wrote to Lyell himself, expressing considerable dissatisfaction that Lyell had not 'given judgment & spoken fairly out' what he thought about 'the derivation of Species'.[8] Darwin continued to feel aggrieved about Lyell's failure to support him. In April 1863, in a letter to the *Athenæum*, he discussed a passage in Lyell's book that touched on the theory of transmutation; he also wrote to Lyell telling him about the letter to the *Athenæum*.[9]

In the same letter, Darwin touched on an area of public controversy when he expressed support for Lyell's recent response to Falconer's accusation, which had just appeared in the *Athenæum*. Darwin had not advised Falconer personally, but had tried, indirectly, to influence him. He told Hooker:[10]

> Do see Falconer & see whether you can at all influence him, by saying what ill appearance Reclamations always have, & that the future historians of Science alone ought to settle such points.— It is wretched to see men fighting so for a little fame.—

During this initial period of controversy, Lubbock also urged Falconer to tone down his attack on Lyell and agreed, on Hooker's advice, to soften a passage in the manuscript of his own review of *Antiquity of man* in which he had originally argued that Lyell had done 'an injustice' to Falconer and Prestwich.[11] In the same review Lubbock expressed publicly what Darwin had said privately; that is, that Lyell should have been more explicitly supportive of the theory of descent with modification, given that 'the whole tenor of his argument' supported Darwin's theory ([Lubbock] 1863b, p. 213).

In May 1864, Lubbock received a letter from Falconer, who reiterated his version of the events leading to his own accusations against Lyell. Falconer accused Lyell of making use of information received 'in conversation in friendly & intimate intercourse' with him, and referring to it 'in language such as men use when they speak of their own original researches'. He then added:[12]

> Very many other parts of the book were submitted to me when passing through the press for correction but all the parts which had any reference to

> myself were studiously withheld from a fine sense of delicacy on the part of the author and so with Prestwich also

Lubbock himself still did not pursue any grievance against Lyell until the spring of 1865.[13]

In the course of preparing his own book, *Prehistoric times* (Lubbock 1865), for publication, however, Lubbock sought to clarify the relationship between his previous work on human origins and Lyell's *Antiquity of man*. In particular, he took exception to the wording of the note on p. 11 of C. Lyell 1863c, which implied that Lubbock 1861 had been written *after* the chapter on shell-mounds in Lyell's book. Lubbock would later argue that the close similarity of certain passages in C. Lyell 1863c and Lubbock 1861 (and consequently in Lubbock 1865), combined with the wording of Lyell's note, could lead some readers to the conclusion that Lubbock had copied passages from Lyell without acknowledgment. Sometime between the end of February and the beginning of March 1865, Lubbock wrote the note which would later appear at the end of the preface of *Prehistoric times* (Lubbock 1865, p. x).[14]

> Note – In his celebrated work on the 'Antiquity of Man,' Sir Charles Lyell has made much use of my earlier articles in the 'Natural History Review,' frequently, indeed, extracting whole sentences verbatim, or nearly so. But as he has in these cases omitted to mention the source from which his quotations were derived, my readers might naturally think that I had taken very unjustifiable liberties with the work of the eminent geologist. A reference to the respective dates will, however, protect me from any such inference. The statement made by Sir Charles Lyell, in a note to page 11 of his work, that my article on the Danish Shell-mounds was published *after his sheets were written*, is an inadvertence, regretted, I have reason to believe, as much by its author as it is by me.

Evidently, he then showed the note to Huxley and asked for his opinion on the matter. Huxley wrote, 'I have read over & weighed carefully what you have written and I think it is all perfectly justifiable & proper', but went on to advise caution, suggesting that Lubbock inform Lyell of his complaint and seek an explanation from him, or at least show him the note before it appeared in print.[15] Lubbock wrote to Lyell a few days later, explaining his position and citing passages in Lubbock 1861 and C. Lyell 1863c that were almost identical. He did not, as Huxley had suggested, send Lyell the text of the proposed note, but concluded:[16]

> If I allow this to pass unnoticed, I lay myself open to the misapprehension of having transcribed verbatim, & without acknowledgment from your work.
>
> I should be glad if you would authorise me to say that the reverse of this, through an inadvertence, I suppose, on your part, has occurred.

Lyell's answer to Lubbock has not been found, but in rough notes for a response he wrote:[17]

> You can say that the early chapters [of *Antiquity*] were written many months & for some time in print before I availed myself of the information obtained

from your paper to make a few additions & corrections but that I was unable to recast the first chapters which I must have done in order to do justice to all the new information which you had laid before the public

Lubbock evidently was unsatisfied with this response and proceeded with his plan to insert the note at the end of the preface. There is no evidence to suggest that Lubbock warned Lyell that the note would, in fact, appear.

When Lyell received a copy of Lubbock's book, published in mid-May 1865, he immediately wrote to express his dissatisfaction with the note Lubbock had inserted at the end of the preface and went on to say that he intended to make a copy of his letter to show to friends.[18] In addition to the perceived accusation of plagiarism against him by Lubbock, Lyell had also recently received a similar accusation from Andrew Crombie Ramsay in a note to an article published in the April 1865 issue of the *Philosophical Magazine*.[19]

Lubbock responded, expressing surprise at Lyell's reaction, reiterating his own position, and adding that he felt that several of their mutual friends supported his position.[20] Lyell wrote back quickly suggesting that had the friends seen Lyell's initial explanation of the facts their reaction would have been different.[21] As he had promised, Lyell sent copies of all three letters to a number of friends, including Darwin.[22]

Just before he received these letters from Lyell, Darwin had discussed the matter in person with Lubbock, and Emma Darwin wrote to Henrietta Emma Darwin, 'whereas after talking to John, he thought him not wrong, after seeing all the letters, he thinks he was quite wrong not to allude to Sir C's explanation of the matter'.[23] Hooker, who had also been sent copies of the letters, wrote to Darwin to ask what he thought of the affair (letter from J. D. Hooker, [2 June 1865]). Hooker, for his part, could see little evidence of plagiarism by Lyell, thought Lubbock had 'taken up the matter in a very evil spirit', and described Lubbock's note as 'rude & insulting' and, in part, hardly intelligible. Darwin responded that, while he thought Lubbock should have given Lyell's explanation in print, he disagreed with Hooker's assessment of Lubbock's note, saying it was not insulting and 'only too intelligible'. Moreover, he reiterated his admiration for Lubbock's book (letter to J. D. Hooker, [4 June 1865]). A week later he sent Lubbock a letter, praising both the book and Lubbock himself in glowing terms, while avoiding any mention of the note in the preface (letter to John Lubbock, 11 June [1865]). No correspondence with Lyell on the topic of the dispute has been found, but Lyell sent Darwin the corrected proofs of the revisions he made to his preface and note (see below, 'Textual changes made to C. Lyell 1863c'). By this time, Darwin clearly wished to avoid direct involvement in the dispute. When Hooker pressed him for an opinion (letter from J. D. Hooker, 13 July 1865), Darwin wrote back (letter to J. D. Hooker, [29 July 1865]):

Lyells corrected pages came when I was extra miserable; I read them & threw them away & now to my surprise find that I have no clear recollection about them,—only a feeling that I was disgusted with everything in world—

Another indication of Darwin's wish to avoid involvement is the fact that, although he corresponded with Huxley in June and July and had seen Huxley's letter to Hooker about the affair,[24] he does not seem to have mentioned the dispute in his letters to Huxley.

Lyell and Lubbock had no direct communication after the end of May 1865, each appealing to friends to resolve the dispute. Lubbock continued to seek advice from Huxley, Hooker, and other X-club friends[25] and, as mentioned above, discussed the matter in person with Darwin. Lyell wrote to Darwin, Hooker, and Huxley and also showed the correspondence to Busk.[26] In the end, it was Huxley who advised both parties on a course of action to resolve the dispute. He encouraged Lyell to cancel the offending note and make some acknowledgment of Lubbock in his preface.[27] Hooker also encouraged Lyell to follow Huxley's advice, and told Huxley, 'Lyell has nothing to do but make a clean breast of it— then, & not till then, can one deal with Lubbock'.[28] Lyell quickly agreed to Huxley's proposal, although he decided to change, rather than cancel the note, and within a few days told Huxley that he had sent the printer his manuscript of the addition to the preface. Huxley received proofs of the amended preface the next day (see below, 'Textual changes made to C. Lyell 1863c', for the revisions Lyell made to the note and the preface).[29]

Lubbock reluctantly agreed to delete his own note. In his last letter to Huxley dealing with the affair, he revealed that while he was satisfied with Lyell's new note, 'as far as it goes', he still intended to keep a note in his own book, concluding, 'I must of course still say something.'[30] However, two weeks later, in his last letter to Hooker on the matter, Lubbock's tone was resigned. He pleaded with Hooker to try to understand his point of view but concluded:[31]

> I hope I have now heard the end of it. It is a hateful business, & sometimes I almost wish that I had run the risk of misinterpretation.

Thus, in print-runs after the end of June 1865, Lubbock had cancelled his note at the end of the preface to Lubbock 1865 and Lyell had re-written a portion of the preface of C. Lyell 1863c and reworded the note on p. 11.

Unlike the earlier controversies of 1863 where the disputants had quarrelled openly on the pages of the *Athenæum*, this controversy was debated and settled privately among a small group of advisers who were friends of both interested parties. Only one known review of Lubbock 1865 draws attention to Lubbock's note; this was in the October 1865 issue of the *Anthropological Review*. The unnamed reviewer quoted the note verbatim and went on to say that, in effect, it accused Lyell of taking 'unjustifiable liberties' with Lubbock's work. The reviewer argued that since the disputed material was not original work (Lubbock had based much of his 1861 article on earlier Danish studies) it therefore did not 'justify so severe an attack on Sir Charles Lyell'.[32]

Darwin's analysis of the situation was succinct. In his letter to Hooker of [4 June 1865] he warned that no one could do much to heal the breach, concluding: 'Time alone could do it if it ever can be done.' In the end, the dispute was resolved, but the breach between the two men seems to have persisted. There was never a return to the friendly correspondence that had previously existed; in fact, no further correspondence between Lubbock and Lyell has been found.

Textual changes made to C. Lyell 1863c

Lyell revised both the preface and the note on page 11 of the third edition of *Antiquity of man* (C. Lyell 1863c; see letter from J. D. Hooker, [15 June 1865] and n. 13). The third edition had originally appeared in November 1863. In spite of Lyell's 1865 revisions, the title page of copies appearing after August 1865 was never altered and there is no indication that the print-run occurred in 1865 or that any changes had been made. Very few copies of this revised version of C. Lyell 1863c were produced.[33] The original and revised versions of the end of the preface and of the note on page 11 are included below.

Preface, C. Lyell 1863c, pp. vii–viii (original version of the last section, printed in November 1863)

In conclusion, I wish it to be understood that the present work makes no pretensions to give a complete analysis of all even the most important memoirs lately published on the coexistence of man with many species of extinct animals. I have thought it best, for the sake of brevity, and sufficient for the object of my work, to treat of those instances which I had been able to verify by personal examination of the localities referred to; or if in some cases I have departed from this rule, it was only where my intimacy with the original observers, and my opportunities of freely communicating with them, and of seeing their collections of fossils and works of art, made me feel competent to test the value of the evidence appealed to.

53 Harley Street:
November 1863

Preface, C. Lyell 1863c, pp. vii–ix (revised version of last section, printed in August 1865, but dated 1863 on the title page)

In conclusion, I will take the opportunity of stating that the second chapter of this work, treating of the Danish peat-mosses and kitchen-middens, as well as the Swiss lake-dwellings, was originally written in 1860 for the sixth edition of the 'Elements of Geology',[34] and the printed proofs were transferred early in the autumn of the next year to be set up in another form for the present work, which I had then determined to get out before the 'Elements.' In the shape which this chapter then assumed it remained in type for two years, being unchanged in substance and in the sequence of the arguments, and receiving no more additions than was consistent with the paging remaining undisturbed.

I mention this fact as an apology for not having availed myself more largely of several valuable contributions to our knowledge both of Danish and Swiss antiquities of the stone and bronze periods, which were published in the interval between the autumn of 1861 and February 1863. In this long interval my thoughts had been entirely absorbed in the composition and printing of the chapters on the glacial phenomena, and those relating to theories of the origin of species. My account of the Danish peat-mosses and shell-mounds had been derived chiefly from the admirable digest given by my friend M. Morlot of the labours of the Danish and Swiss archæologists and naturalists, which he had kindly sent me in English in MS. before its appearance in print; first in French, dated Berne, Sept. 1859, in the 'Mémoires de la Société Vaudoise des Sciences Naturelles, Tome VI.;' and afterwards in English, in a translation for the Smithsonian Institution, Washington, 1861.[35]

The distinguished zoologist M. Claparède had also conversed with me in 1859 on the researches of the best Danish writers, especially Steenstrup and Forchhammer, whom he had lately seen, and of whose papers, with others written in Swedish, he gave me an abstract for my use, in a letter dated December 1859. He referred me chiefly to 'Oversigt over det Konglike Danske Videnskabernes Selskabs Forhandlingen.' It was impossible for me, with the aid of such able investigators, to overlook any of the most striking discoveries and conclusions which had been made before 1860; but I gladly took advantage of the later numbers of Keller's 'Pfahlbauten,' and of Mr. Lubbock's 'Memoir on the Danish Kjökkenmöddings,' printed in the October number of the 'Natural History Review' for 1861, to improve the wording, and occasionally the subject-matter, of certain passages for which M. Morlot had already supplied the principal data. I had no space, without disturbing my type, for entering on a single new field of enquiry, or any new deductions furnished by Messrs. Keller, Lubbock, or other writers. Had I attempted to do justice to them, or to any authors of later date than the summer of 1860, I must have expanded the plan of my whole book, and seriously delayed the publication of the first edition, as well as of the subsequent issues.

Note on page 11, C. Lyell 1863c (original version)

*Mr. John Lubbock published, after these sheets were written, an able paper on the Danish 'shell-mounds' in the October Number of the Natural History Review, 1861, p. 489, in which he has described the results of a recent visit to Denmark, made by him in company with Mr. Busk.

Note on page 11, C. Lyell 1863c (revised version)

*Mr. John Lubbock published, in the October Number of the Natural History Review, 1861, p. 489, an able paper on the Danish 'shell-mounds,' in which he has described the results of a recent visit to Denmark, made by him in company with Mr. Busk.— See above, Preface, p. vii.

[1] Oldroyd 1980, p. 299. For more on Victorian debates over human antiquity, see Grayson 1983, Stocking 1987, and Van Riper 1993.

[2] Letter from Charles Lyell to John Lubbock, 20 February 1863 (British Library, Add. MSS 49640).

[3] Letters from Charles Lyell to John Lubbock, 22 February 1864 and 24 February 1864 (British Library, Add. MSS 49640).

[4] For an overview of CD's correspondence relating to the appearance of C. Lyell 1863a, see the introduction to *Correspondence* vol. 11, pp. xv–xvii. For a comparison of the first three editions of *Antiquity of man*, see Grayson 1985.

[5] For two interpretations of Hugh Falconer's attack on Charles Lyell, see Bynum 1984 and L. G. Wilson 1996.

[6] Owen's complaints about C. Lyell 1863a are discussed in Bynum 1984, pp. 154–9.

[7] See *Correspondence* vol. 11, letter to J. D. Hooker, 24[–5] February [1863]. On Lyell's unwillingness to commit himself to CD's theory of transmutation, see Bartholomew 1973.

[8] See *Correspondence* vol. 11, letter to Charles Lyell, 6 March [1863] and n. 7.

[9] See *Correspondence* vol. 11, letter to *Athenæum*, 18 April [1863], and letter to Charles Lyell, 18 April [1863].

[10] *Correspondence* vol. 11, letter to J. D. Hooker, 17 March [1863].

[11] See *Correspondence* vol. 11, letter from J. D. Hooker, [24 March 1863] and n. 9. In his published review, Lubbock wrote that though Lyell had failed to give 'due prominence' to Falconer and Prestwich, his omission was 'unintentional' ([Lubbock] 1863b, p. 214).

[12] Letter from Hugh Falconer to John Lubbock, 24 May [1864] (British Library, Add. MSS 49640). Another portion of this letter is quoted in L. G. Wilson 1996.

[13] For two interpretations of the dispute that arose between Lyell and Lubbock, see Bynum 1984 and L. G. Wilson 2002.

[14] As no manuscript copy of the note still exists, the published version has been reproduced. It is likely that the published version of the note had been toned down. Huxley told Hooker: 'It was as much as I could do to get him [Lubbock] to write to Lyell for an explanation before coming out with a preface to which what you have seen is milk & water' (see enclosure to letter from J. D. Hooker, [15 June 1865]). Later, Lubbock claimed that he had relied on an unnamed friend to compose the note that appeared at the end of the preface to Lubbock 1865. He told Hooker, 'I did not trust myself to write the note. It was written for me by a mutual friend of ours' (letter from John Lubbock to J. D. Hooker, 23 June 1865, in Royal Botanic Gardens, Kew, Letters to J. D. Hooker, vol. 14, doc. 183–4).

[15] Letter from T. H. Huxley, 7 March 1865 (British Library, Add. MSS 49641).

[16] Letter from John Lubbock to Charles Lyell, 13 March 1865 (University of Edinburgh, Lyell 1, Gen. 113: 3644–5).

[17] Rough notes for letter from Charles Lyell to John Lubbock, undated (University of Edinburgh, Lyell 1, Gen. 113: 3649).

[18] For the announcement of the publication of Lubbock 1865, see *Publishers' Circular*, 1 June 1865. For Lyell's response to Lubbock's note, see the first enclosure (letter from Charles Lyell to John Lubbock, 25 May 1865) to the letter from Charles Lyell to J. D. Hooker, [31 May 1865].

[19] See letter from J. D. Hooker, 2 May 1865 and n. 10.

[20] See the second enclosure (letter from John Lubbock to Charles Lyell, 29 May 1865) to the letter from Charles Lyell to J. D. Hooker, [31 May 1865].

[21] See the third enclosure (letter from Charles Lyell to John Lubbock, 30 May 1865) to the letter from Charles Lyell to J. D. Hooker, [31 May 1865].

[22] See letter from Charles Lyell to J. D. Hooker, [31 May 1865] and n. 1.

[23] Letter from Emma Darwin to Henrietta Emma Darwin, [1 June 1865] (DAR 219.9: 28).

[24] See the enclosure to the letter from J. D. Hooker, [15 June 1865].

[25] Lubbock wrote to Huxley, 'I should be quite satisfied with anything you & Tyndall & Busk settle' (letter from John Lubbock to T. H. Huxley, 7 June 1865, Imperial College, Huxley papers 6: 110). The X-club was a small group of friends interested in science who dined together monthly. For more on the X-club, see Barton 1998.

[26] Lyell mentioned he had shown the correspondence to Busk in a letter to Huxley, 5 June 1865 (Imperial College, Huxley papers 6: 104).
[27] Rough draft of a letter from T. H. Huxley to Charles Lyell, [3 June 1865] (Imperial College, Huxley papers 6: 102).
[28] Letter from J. D. Hooker to T. H. Huxley, 6 June 1865 (Imperial College, Huxley papers 3: 109).
[29] Letters from Charles Lyell to T. H. Huxley, 7 June 1865, and 8 June 1865 (Imperial College, Huxley papers 6: 108, 111).
[30] Letter from John Lubbock to T. H. Huxley, 9 June 1865 (Imperial College, Huxley papers 6: 110).
[31] Letter from John Lubbock to J. D. Hooker, 23 June 1865 (Royal Botanic Gardens, Kew, Letters to J. D. Hooker, vol. 14, doc. 183–4).
[32] See *Anthropological Review* 3 (1865): 340.
[33] Huxley wrote to Lyell to tell him that he had arranged for proofs of the revised preface and note to go to the printer. The instructions for the printer read: 'Print 100 copies' (rough draft of letter from T. H. Huxley to Charles Lyell, 11 June 1865, Imperial College, Huxley papers 6: 116).
[34] C. Lyell 1865.
[35] See letter from Charles Lyell to J. D. Hooker, [31 May 1865] and n. 8.

MANUSCRIPT ALTERATIONS AND COMMENTS

The alteration notes and comments are keyed to the letter texts by paragraph and line numbers. The precise section of the letter text to which the note applies precedes the square bracket. The changes recorded are those made to the manuscript by CD; changes of hand in letters written partly by CD and partly by amanuenses are also recorded. Readers should consult the Note on editorial policy in the front matter for details of editorial practice and intent. The following terms are used in the notes as here defined:

del	deleted
illeg	illegible
interl	interlined, i.e., inserted between existing text lines
omitted	omitted by the editors to clarify the transcription
over	written over, i.e., superimposed

To Thomas Henry Huxley 4 January [1865]
2.1 Oh] *after del double quote*
3.6 that] *interl*
3.7 for writing] *after del* 'you'

To Joseph Dalton Hooker 7 January [1865]
0.2 Jan 7th.] *added pencil*
1.1 Gärtner] 'G' *over* 'I'
1.2 to the Translation] *interl*
2.4 (But . . . the former.—) 2.6] *square brackets in MS, added red crayon*
2.5 12] *after del* 'about'
2.6 See P.S.] *added in left margin, red crayon*
2.6 P.S.] *after del* 'note'
3.4 there are 3.5] *interl*
7.1 *P.S.*] *added red crayon*
7.1 & following] *interl*

To Thomas Campbell Eyton 9 January [1865?]
2.2 Hounds] *over illeg*

To Henry Denny 17 January [1865]
2.4 Polynesian] *interl*

To J. D. Hooker 19 January [1865]
1.2 have done] *above del* 'did'
2.1 one of] *interl*
2.11 important] *after del* 'position of'
3.3 curious is] 'is' *interl*
7.1 a note of] *interl*

To Charles Lyell 22 January [1865]
5.9 that of 5.10] *interl in CD's hand*
9.1 If ever . . . Selection. 9.2] *in CD's hand*

To Alfred Russel Wallace 29 January [1865]
3.1 Do . . . you? 4.3] *in CD's hand*
3.1 hard] *interl in CD's hand*
4.1 killed &] *interl in CD's hand*
4.3 & run wild.] *interl in CD's hand*

To Abraham Dee Bartlett 30 January [1865]
0.2 Jan. 30th . . . You have 1.1] *in CD's hand*
3.2 tipped] 'ped' *added in CD's hand*

To J. D. Hooker 2 February [1865]
2.2 bad] 'b' *over* 'v'
2.5 hope] *above del* 'hype'

To [Robertson Munro?] [3 February 1865 or 1866?]
0.1 Feb 3rd] *added pencil, left margin*
1.2 perhaps] *interl*
1.3 have [been]] *pencil, above* 'am' *del pencil*

1.3 read your . . . inaccuracy of 1.4] *pencil, above* 'hear [*del pencil*] that Mr John Scott has been inaccurate'
1.4 & I fear] 'my first reac' *added above, pencil*
1.4 recording] *after del* 'inventing'
1.5 This would be a detestable crime] *interl*
1.5 cannot] *above del* 'cannot'
1.7 the crosses] 'the' *after del* 'his'
1.7 crosses] *above del* 'experiments'
1.7 in the day] *interl*
1.9 surprised] *after del* 'quite'
1.9 amounted to.] *above del* 'was.—'
1.11 for] *over* '&', *pencil*
1.11 for . . . consideration.— 1.12] *interl*
1.12 fairly] *below del* 'most truly'
1.12 Truth . . . consideration.—] 'truth must be followed, whatever may be the consequences' *added left margin, pencil*
1.15 whilst] *above del illeg*

To A. D. Bartlett 9 February [1865]
1.1 immediately] *interl in CD's hand*

To J. D. Hooker 9 February [1865]
1.9 Sic transit . . . vengeance.] *interl*
2.2 , but am much better today.] *interl*
2.4 Startin] 'n' *over* 'm'
2.8 the loss of nearly] *interl*
3.2 with] *interl*
3.7 by] *before omitted point*
3.9 love] *above del* 'love'

To Charles Lyell 21 February [1865]
3.2 work] 'k' *over* 'th'

To William Bernhard Tegetmeier 14 March [1865]
4.1 two] *interl in CD's hand*
9.1 P.S. . . . Turbit. 9.9] *added in CD's hand*
9.4 before they are cut] *interl*
9.5 *proofs*] 'p' *over* 'P'
9.5 So that . . . executed 9.6] *interl*
9.8 I sh^{d}] 'I' *over illeg*
9.8 *good*] *interl*

To J. D. Hooker 16 [March 1865]
1.3 advice &] *interl*

To Henry Denny 23 March 1865
3.3 had] 'h' *over illeg*
4.3 w^{d}.] *interl*

To Charles Lyell 25 March [1865]
3.1 domestic] *interl*
3.3 absolutely] *interl*

To Benjamin Dann Walsh 27 March [1865]
1.5 on your] 'on' *interl in CD's hand*
1.8 somewhat] *interl in CD's hand*
2.2 is] *in CD's hand above del* 'are'
2.5 at least . . . gall-making. 2.6] *interl in CD's hand*
2.11 the same genus of] *interl in CD's hand*
2.11 the 2 forms of 2.12] *interl in CD's hand*
2.12 posted to you] 'to' *interl in CD's hand*
3.5 to you,] *interl in CD's hand*
5.1 If . . . copy.—] *added in CD's hand*

To W. B. Tegetmeier 28 March [1865]
2.3 apparently] *interl*
2.7 (otherwise . . . big) 2.8] *interl*
2.10 soonish] *interl*
2.11 hear] *interl*

To John Murray 31 March [1865]
1.2 early] *interl*
1.3 to pledge] 'to' *interl*
2.1 want] *after del* 'have'
2.4 the wood] 'the' *interl*
3.2 (& *perhaps*] *parenthesis over stop*
3.2 of the 3.3] *interl*
3.4 of Field] *interl*
3.5 or trust . . . Sub-Editor. 3.6] *interl*
4.1 job] *over* '.—'
5.1 want] *interl*
5.1 or 6] *interl*
5.1 of Kew] *interl*
6.2 about] *interl*
6.3 number] *after del* 'Pigeons to 4'
10.2 now] *interl*
11.1 all] *after del* '12 more'
11.4 of] *over* '.—'
12.1 Aviary] 'i' over 'a'

To J. D. Hooker 6 April [1865]
2.3 Busk] *after del comma*

To W. B. Tegetmeier [7 April 1865]
2.1 (which please look at)] *interl*
2.3 likewise] *interl*
3.2 see] *above del illeg*
3.3 figures] *interl*
3.4 M^{r}. Wells . . . me.—] *added*
4.4 any] *interl*
4.4 M.S.] *interl*
4.4 by] *over* '—'

5.1 the] *interl*
8.1 (I enclose . . . *registered*.)] *square brackets in MS*
9.1 black] *interl*

To J. D. Hooker 10 [April 1865]
1.2 book] 'k' *over* 'ks'
1.5 parliament] *over illeg*
2.2 on] *over* 'of'
2.4 that] *interl*
4.3 that] *over* 'I'

To J. D. Hooker 13 April [1865]
1.2 not given] *above del* 'unknown'
1.2 & Bottle from Thwaites] *interl*
1.3 if so] *interl*
5.2 for nectar,] *interl*

To J. D. Hooker 17 April [1865]
3.8 one] *over* 'a'
3.13 Elements] *after del* 'Essa'
4.2 is] *over* 'I'

To Asa Gray 19 April [1865]
1.4 particular] *after del* 'to'
2.6 will] *interl*
2.6 shake] *altered from* 'shakes'
3.5 as] *over* '—'
3.8 "reciprocally] *interl*
4.2 Variation] 'iation' *added*
4.4 5 weeks.] *before omitted point*

To J. D. Hooker [1 May 1865]
1.3 published] *interl*
2.1 was] *above del* 'did'

To J. D. Hooker 4 May [1865]
3.1 apropos . . . Rafflesia] *interl*
3.2 the 2] 'the' *interl*
3.2 the same] *interl*
3.2 Gall-insect 3.3] *corrected from* 'Gall-insects'
3.3 being] *after del* 'laying their'
4.2 tadpoles] *after del* 'a'
4.3 being] *after del* 'ha'
9.1 edge of the] *interl*

From Emma and Charles Darwin 13 May 1865
1.1 yearly] *interl*
1.1 £50] '£' *above* '50' *in MS*
2.1 having been made] *added*

To T. H. Huxley 27 May [1865]
0.2 May 27th.—] *in CD's hand*

3.5 a very] 'a' *interl in CD's hand*
3.6 crude hypothesis] 'hypothesis' *interl in CD's hand*

To J. D. Hooker 1 June [1865]
1.3 very] *interl*
4.3 in Lubbock, . . . it] *interl*
5.2 book] *interl*
8.4 Order,] 'O' over *'o'*
9.3 & forgot] *interl*
9.5 an] *interl*

To W. B. Tegetmeier 2 June [1865]
2.1 weeks] *altered from* 'weekness'
7.1 young white] *interl*
7.2 red] *above del* 'black'
7.2 & *for*] *after del* 'with P'

To George Maw 4 June [1865]
1.4 W. H.] *interl in CD's hand*

To John Chapman 7 June 1865
6.1 and obliged] *added in CD's hand*

From Charles and Emma Darwin to J. D. Hooker [10 July 1865]
2.1 aloud] *interl*
2.1 & are] 'are' *over* 'am'
3.1 much . . . stronger 10.1] *in Emma Darwin's hand*

To T. H. Huxley 12 July [1865]
3.1 you have . . . trouble.—] *in CD's hand*
5.1 P.S. 2d. . . . Pangenesis.] *in CD's hand*

To T. H. Huxley [17 July 1865]
2.4 fully] *interl*
2.10 in the parent] *interl*

To J. D. Hooker [29 July 1865]
2.1 Lyells] *after del* 'L'
3.6 a striking] 'a' *interl*
3.8 we] *corrected from* 'were'
4.5 Politicks,] *interl*

To Fritz Müller 10 August [1865]
7.1 Could . . . one?—] *added in CD's hand*

To Fritz Müller 20 September [1865]
1.1 written . . . English 1.2] *interl in CD's hand*

To J. D. Hooker 27 [or 28 September 1865]
3.2 It is a . . . very good. 4.8] *in hand of amanuensis*
4.13 footnote] 'foot' *interl*

4.21 a poor] 'a' *interl*

To Jeffries Wyman 8 October [1865]
1.6 can] *in CD's hand above del* 'will'
1.8 self-twisting] *interl in CD's hand*
2.1 on the movement of plants] *interl in CD's hand*

To Fritz Müller 17 October [1865]
4.1 My health . . . nothing.] *added in CD's hand*

To Asa Gray 19 October [1865]
0.2 Oct. 19th . . . character. 2.7] *in CD's hand*
7.1 With . . . C. Darwin 8.1] *in CD's hand*

To Daniel Oliver 20 October [1865]
2.4 former] *in CD's hand above del* 'latter'

To J. D. Hooker 22 and 28 [October 1865]
0.3 (This . . . inclined.) N.B.)] *square brackets in original*
1.12 appears?] '?' *over* ':'
1.16 powers] *interl*
1.17 a nervous] 'a' *interl*
3.1 I have read . . . introduced.— 3.13] *In Emma Darwin's hand*
4.1 anonymous] *interl*
4.10 or ought not any longer] *interl*
7.2 please] *interl*

To William Darwin Fox 25–6 October [1865]
2.3 scanty] *in CD's hand above del* 'sharp'
3.4 thankful to say] *interl in CD's hand*
5.1 My dear . . . C. Darwin 5.2] *in CD's hand*

To Robert Burn 2 December [1865]
4.1 To . . . Burn] *added in CD's hand*

To Henry Wentworth Acland 8 December [1865]
6.1 My . . . months.] *in CD's hand*

To J. D. Hooker [31 December 1865]
3.2 friends not] 'not' *interl*

SUPPLEMENT

To Dear Friend 1 January 1822
1.1 the] *altered from* 'ther'
1.2 know] *altered from* 'known'
1.4 couse] *after del* 'a c'
3.1 ye cabinet] *interl*

To Dear Friend 3 January 1822
0.1 3] *altered from* '2'

To Dear Friend 4 January 1822
3.1 your] *over illeg*

To Dear Friend 4 January 1822
1.2 ask] *altered from* 'aske'
1.8 you had] *interl*
1.8 do it all over] *interl*

To Dear Friend 12 January 1822
1.1 I hope] *after del* 'Y'
1.5 and that . . . dear 1.6] *added*
6.1 Fridy] *above del* 'to day'
6.1 beutiful] 'u' *over* 'a'
8.2 was] *over* 'is'
8.3 half] *after del* 'a'

To John Richardson [24 July 1837]
1.4 was] *over* 'were'
1.7 covered] *after del* 'occur'
1.9 large] *after del illeg*
7.1 from shore] *circled, ink, and interl*
7.2 *resting on the bottom*] *interl*
7.6 frozen] *interl*
7.8 constantly] *after del* 'also'

To John Richardson [11 August 1837]
2.2 you] *over illeg*

To Richard Owen [15 December 1837 – 9 June 1838]
4.1 or at] *interl*

To John George Children 22 February [1838]
2.3 means] *above del* 'sources'
2.3 by] *over* 'of'

To John Stevens Henslow 3 November 1838
1.2 some] *after del* 'thos'
1.4 Lately] *after del* 'Since'
1.10 species] *interl*
2.1 natural] *interl*
2.6 answer] *above del* 'tell'
2.6 as far] *after del illeg*
2.7 a hour] 'a' *interl*
2.7 (if . . . plants) 2.8] *interl*
3.2 Savings] *after del illeg*

To Richard Owen [March? 1840]
1.6 at all] *interl*

2.2 publication] *after del* 'del'

To William Buckland [November 1840 – 17 February 1841]
3.7 and] *over* 'or'

To Henry Denny [27 July – 10 August 1844]
7.1 at same . . . this note 7.2] *interl*

From the principal inhabitants of Down to the Secretary of the Post Office [1845–51?]
0.1 To] *after del* 'Sir'
1.2 arrangement] *after del* 'circumstances'
1.3 an average] *interl*; 'an' *after del illeg*
1.4 daily.] *interl*
1.5 after] *interl*
1.7 (except . . . South)] *interl*
1.7 is] *above del* 'we do'
1.7 delivered here 1.8] *above del* 'receive'
1.9 only] *interl*
1.10 & only change] *interl*
2.1 By . . . disturbed. 2.4] *interl*
2.3 the same] 'the' *interl*
3.1 our case favourably] *altered from* 'favourably our case'

To Smith, Elder & Co. 6 June [1846]
1.4 21] '1' *over* '0'
1.5 very] *interl*
1.7 through . . . Soc] *interl*
1.8 , so . . . expence. 1.9] *interl*
1.11 advance] *interl above del* 'pay'
1.12 trust] *after del* 'hope'
1.13 for me] *interl*
1.13 can] *interl*
1.13 do] *above del* 'can'
1.14 commission] *interl*
1.16 balance at the] *interl*
2.1 like the] *interl*
2.1 to be] *interl*
2.2 16s. 6d] 's' *above* '16'; 'd' *above* '6'
2.8 Geolog.] *interl*
3.2 £3"8s"0d] '£' *above* '3'; 's' *above* '8'; 'd' *above* '0'
3.4 £2.9s] '£' *above* '2'; 's' *above* '9'
4.2 500] *after del* 'the simple'
4.6 estimate] *interl*
4.6 nearly] *interl*
4.7 to] *interl*
4.7 & . . . so.] *interl*
4.8 *soon.*] *before del interl* 'before a'
5.5 Geological] *interl*

To John Lort Stokes 2 January [1847]
2.4 Croydon] *interl*

To Peter Lund Simmonds 25 February [1849]
1.7 much] *interl*

To Adam Sedgwick 11 October [1850]
1.1 Soc. for] *interl*
5.1 my] *over illeg*

To J. D. Hooker [9 or 16 February 1854]
1.4 Only] 'O' *over* 'o'
1.7 the illustrations] 'the' *interl*
1.11 cost!] '!' *over full stop*

To ? 7 December [1855–7?]
1.1 the address] *after del* 'whether'
1.2 Madeira] *after del* 'the'

To John Maurice Herbert 18 November [1856]
2.4 attend.] *before del* 'to.'
3.1 by me 3.2] *interl*

To Bernard Peirce Brent 7 February [1857]
2.1 as the Bird] 'as' *interl*
4.1 be *extremely*] 'be' *interl*
4.3 I have] *interl*
4.4 Bird by] 'Bird' *interl*
4.5 on Thursday] *interl*
4.8 really] *over illeg*

To William Sharpey 9 April [1857]
2.3 admitted as most] *interl*
3.7 in my opinion.—] 'in' *over* '—'

CD memorandum July 1857
1.1 £250] '£' above '250'
1.2 winding] *after del* 'd'

To Charles Spence Bate 29 November [1857]
1.3 on] *over illeg*

To J. D. Hooker [14 November 1858]
1.2 (certainly . . . presume)] *interl*
1.3 only] *interl*
2.1 These] *after del* 'Of'; 'T' *over* 't'
4.2 in Europe] *after del illeg*

To James Paget 19 December [1858]
3.1 in state of nature] *interl*
3.4 of same species] *interl*
3.6 climate] 'e' *over* 'es'

To Maxwell Tylden Masters 25 April [1860]
3.2 namely] *interl*
3.6 the pollen of different 3.7] *interl*

To J. D. Hooker 19 [June 1860]
1.1 on Orchids] *interl*
1.2 How . . . Botanist. 1.3] *interl*
1.3 Such] *after del illeg*
1.3 brief] *after del illeg*
1.7 when needle withdrawn] *interl*
1.9 hold needle] 'needle' *after del* 'it'
1.9 relative] *interl*
1.9 to parts of flower] *interl*
1.10 almost inevitably] *interl*
1.12 observe] *interl*
1.13 proboscis of] *interl*
1.13 belonged] *after del* 'was'
5.1 (I vow . . . again.)] *square brackets in MS*

To Charles Daubeny 16 July [1860]
3.8 mysterious;] ';' *over* '—'

To Charles Daubeny 1 August [1860]
0.1 (Down Bromley Kent)] *square brackets in MS*
2.7 in case] *after del illeg*

To A. D. Bartlett 24 August [1860]
1.5 in the Gardens] *interl*

To William Erasmus Darwin [13 January 1861]
1.1 are] *over* 'is'
1.1 (one . . . forwarded)] *interl*
2.2 164.] *below del* '146'
5.1 Dear Wm, . . . Golden Sq 7.1] *added in Emma Darwin's hand*

To Archibald Geikie 27 February [1861]
2.2 (ie species)] *interl*

To Richard Kippist 27 [February or March 1861?]
1.1 Thursday Morning] *interl*
3.1 in lieu] *interl*

To Williams & Norgate 4 March [1861]
2.1 of a loss] 'of' *interl*

To Daniel Oliver 23 March [1861]
1.4 two] *interl*
1.5 And . . . scissors 1.6] *interl*
1.6 Mr Fitch] 'M' over 'I'
1.10 fertilise] *after del* 'self'
5.1 (Longitudinal] *after* 'Primrose' *pencil, del ink*
5.2 near] *after del* 'close'
Below diagram: 'Mark difference in shape of tube of corolla' *pencil, del ink*

To B. P. Brent 1 April [1861]
1.2 on combs 1.3] *interl*
3.1 have] *interl*
3.1 I] *interl above del* 'me'

To W. B. Tegetmeier 8 May [1861]
2.4 about Pigeon 2.5] *interl*
4.3 identity of] *interl*
5.2 I have] *after del illeg*
7.3 two or three] *after del* 'one or'

To Neil Arnott 29 August [1861]
1.2 here] *corrected from* 'there'

To Daniel Oliver 3 November [1861]
2.2 from Linn. Society.—] *added*

To Jean Baptiste Pierre Guépin 14 November 1861
0.1 England] *added in CD's hand*

BIOGRAPHICAL REGISTER

This list includes all correspondents and all persons mentioned in the letters and notes that the editors have been able to identify. Dates of letters to and from correspondents are given in chronological order. Letters to correspondents are listed in roman type; letters from correspondents in italic type; third-party letters are listed in italic type with the name of the recipient given in parentheses.

Abraham, Charles John (1814–1903). Clergyman and teacher. Fellow, King's College, Cambridge, 1837–50. Ordained deacon, 1837; priest, 1838. Master at Eton, 1839–49. Went to New Zealand in 1850. Worked in education and for George Augustus Selwyn, bishop of New Zealand. Became first bishop of Wellington in 1858. Encouraged Maori participation in church government; during the Maori war of 1860, urged fair treatment of the native population. Returned to England in 1868 and resigned his see in 1870. Took a leading part in founding Selwyn College, Cambridge, in 1882. (*DNB*, *DNZB*.)

Acland, Henry Wentworth (1815–1900). Physician. Lee's reader of anatomy at Christ Church, Oxford, 1845–58. Regius professor of medicine, University of Oxford, 1858–94. Appointed honorary physician to Edward, Prince of Wales, 1860. FRS 1847. (*DNB*, *Physicians*.)

8 December [1865]

Adams, Charles Francis (1807–86). American diplomat. Elected to the Massachusetts legislature as a Whig, 1840; led the state in opposition to slavery. Elected to Congress in 1858 and again in 1860 as a Republican. US minister to Great Britain, 1861–8. Published the papers of his father, the former president, John Quincey Adams (*DAB*). (*ANB*, *DAB*.)

Adhémar, Alphonse-Joseph (1797–1862). French mathematician and teacher. Author of several works on the application of mathematics to civil engineering and to perspective in art. (*DBF*.)

Agassiz, Alexander (1835–1910). Swiss-born zoologist, oceanographer, and mining engineer. Son of Louis Agassiz. Emigrated to the United States in 1849. Joined the US survey of Washington Territory boundaries in 1859. Superintendent, Calumet copper mine, Michigan, 1867; later president of the corporation. Curator of the Harvard Museum of Comparative Zoology, 1874. (*DAB*.)

Agassiz, Jean Louis Rodolphe (Louis) (1807–73). Swiss-born zoologist. Professor of natural history, Neuchâtel, 1832–46. Emigrated to the United States in 1846. Professor of natural history, Harvard University, 1848–73. Established the Museum of Comparative Zoology at Harvard in 1859. Foreign member, Royal Society of London, 1838. (*DAB*, *DSB*, *Record of the Royal Society of London*.)

Ainslie, Robert (1802/3–76). Independent Unitarian minister and religious writer. Resident at Pond House (later renamed Tromer Lodge), Down, 1845–58. Minister of New Court Chapel, Carey Street, London. Secretary of the London City Mission and of the Congregational Board of Education. Minister of Christ Church, Brighton, 1860–74. (Ainslie 1865; letter from O. A. Ainslie, 23 November 1880, *Calendar* no. 12842; *Correspondence* vol. 3, letter to Susan Darwin, 3[–4] September 1845; *Correspondence* vol. 7, letter to W. E. Darwin 14 [May 1858]; G. E. Evans 1897; *The Times*, 23 August 1876, p. 16.)

Albert Francis Charles Augustus Emmanuel, prince-consort of England (1819–61). Married Queen Victoria in 1840. A patron of the arts and sciences. Elected chancellor of Cambridge University, 1847. Instrumental in the inception and organisation of the Great Exhibition, 1849–51. FRS 1840. (*DNB*, *Record of the Royal Society of London*.)

Alice Maud Mary, grand duchess of Hesse-Darmstadt (1843–78). Second daughter and third child of Queen Victoria and Prince Albert. Married Prince Frederick William Louis of Hesse, later Louis IV, grand duke of Hesse-Darmstadt, in 1862. (*DNB*, Pakula 1996.)

Anderson, Elizabeth Garrett (1836–1917). Physician. One of the first women to practise medicine in Britain. Obtained a medical licence from the Society of Apothecaries, 1865. MD 1870, University of Paris. Senior physician, New Hospital for Women, 1866–92. Lecturer on medicine, London School of Medicine for Women, 1875–97; dean, 1883–1903. (*DNB*, Manton 1965.)

Anderson, Thomas (1832–70). Scottish physician and botanist. Entered the Bengal medical service in 1854. Superintendent, Calcutta botanic garden, 1861–8; conservator of forests, Bengal, 1864–9; retired because of ill health. Instituted experiments that led to the successful cultivation of *Cinchona* in India. (R. Desmond 1992, pp. 100–1; R. Desmond 1994; *DSB*; Stebbing 1922–6, 1: 515.)

Andersson, Charles John (Carl Johann) (1827–67). Swedish-born ornithologist, hunter, and explorer. Accompanied Francis Galton to South Africa in 1849. Organised a successful expedition to Lake Ngami and published an account of the expedition in 1856. Continued to explore the South African interior and to carry out ornithological studies. (*DSAB*.)

Ansted, David Thomas (1814–80). Geologist. Professor of geology, King's College, London, 1840–5. Assistant secretary to the Geological Society of London, 1844–7; editor of the society's journal from 1844. Lecturer on geology at the military college at Addiscombe, Surrey, and professor of geology at the College of Civil Engineers, Putney, from 1845. Consulting geologist and mining engineer from *circa* 1850. FRS 1844. (*DNB*, *Geological Magazine* n.s. 7 (1880): 336, *Record of the Royal Society of London*, Sarjeant 1980–96.)

Appleton, Thomas Gold (1812–84). Boston essayist, poet, and artist. Promoted the growth and improvement of Boston; a trustee of the Athenaeum, the Public Library, and the Museum of Fine Arts. (*DAB*.)
5 December [1865]

Argyll, duke of. *See* Campbell, George Douglas.

Arnott, Neil (1788–1874). Physician and natural philosopher. Surgeon with the East India Company in China, 1807–9, 1810–11. Practised as a physician in London, 1811–55. Gave lectures at the Philomathic Institution published as *Elements of physics* (1827). One of the founders of the University of London, 1836. FRS 1838. (*DNB.*)
29 August [1861]

Ayerst, James Smith (1824/5–84). Physician. Assistant surgeon, Royal Navy, 1847–55. Physician, Great Malvern, Worcestershire, 1856–7. Ran a hydropathic establishment at Old Well House, Malvern Wells, 1857–68, initially in conjunction with James Manby Gully. Practised homoeopathy and hygienics. Physician, Torquay, Devon, 1868–84. (*Medical directory* 1856–85; Metcalfe 1906, p. 94; *Navy list* 1847–55.)

Ayres, William Orville (1817–87). American physician and zoologist. Graduated from Yale in 1837 and began a career as a schoolteacher; returned to Yale to study medicine and qualified as a medical doctor in 1854. Practised medicine in San Francisco; professor of theory and practice of medicine and (from 1862) dean of the medical faculty at Tolland Medical College. An authority on the ichthyology of the United States's Pacific coast. Left California in the 1870s; practised medicine in New Haven, Connecticut, *circa* 1874–87. Lecturer on diseases of the nervous system at Yale Medical School, 1879–87. (Elliott 1979.)

Babington, Charles Cardale (1808–95). Botanist, entomologist, and archaeologist. Involved in natural history activities at Cambridge for more than forty years; an expert on plant taxonomy. A founding member of the Cambridge Entomological Society and the Cambridge Antiquarian Society. Editor of *Annals and Magazine of Natural History* from 1842. Chairman, Cambrian Archaeological Association, 1855–85. Professor of botany, Cambridge University, 1861–95. FRS 1851. (*DNB*, *DSB.*)
28 March 1865

Babington, Thomas H. Civil servant. Senior clerk at the Board of Trade; worked as Robert FitzRoy's assistant in the Meteorological Department. (*Post Office London directory* 1865, 'Official directory'; Mellersh 1968, pp. 288–9.)

Baer, Karl Ernst von (1792–1876). Estonian zoologist and embryologist. Professor of anatomy at Königsberg University, 1819; professor of zoology, 1826–34. Professor of zoology at the Academy of Sciences, St Petersburg, 1834–67. Demonstrated the existence of the mammalian egg, 1826. Propounded the influential view that embryological development proceeds from the general to the specific. Foreign member, Royal Society of London, 1854. (*DSB*, *NDB*, *Record of the Royal Society of London.*)

Baillière, Hippolyte (1808/9–67). Bookseller and publisher. Moved to London *circa* 1827 and established a business selling French medical and scientific texts. Collected books on behalf of the Royal College of Surgeons. (*Modern English biography.*)

Baillon, Ernest-Henri (1827–95). French botanist. Professor of medical natural history, and of hygiene and natural history applied to industry, Ecole Centrale des Arts et Manufactures, Paris, 1863–95. Editor of the botanical journal *Adansonia*, 1860–70. (*DBF.*)

Balfour, John Hutton (1808–84). Scottish physician and botanist. Professor of botany, Glasgow University, 1841–5. Professor of botany, University of Edinburgh, and regius keeper of the Royal Botanic Garden, Edinburgh, 1845–79. Founding member of the Botanical Society of Edinburgh, 1836. Founder of the Edinburgh Botanical Club, 1838. Co-editor of the *Edinburgh New Philosophical Journal*. FRS 1856. (*DNB*, *DSB.*)

15 June [1862?]

Ballance, Charles. Poultry fancier. (Tegetmeier 1867, p. 75.)

Baly, Joseph Sugar (1817–90). Physician and entomologist. Surgeon at Kentish Town and Camden Town dispensary, London; then at Warwick, 1868–90. Medical health officer, Leamington Priors, 1873–90. A leading authority on phytophagous Coleoptera. (*Modern English biography.*)

Banks, Joseph (1743–1820). Botanist. Travelled around the world on HMS *Endeavour*, 1768–71. President of the Royal Society, 1778–1820. Created baronet, 1781; privy councillor, 1797. FRS 1766. (R. Desmond 1994, *DNB.*)

Barkly, Henry (1815–98). Scottish-born colonial administrator. Served as governor of British Guiana, 1849–53; Jamaica, 1853–6; Victoria, 1856–63; Mauritius, 1863–70; and Cape Colony, 1870–77. Sent plants to the Royal Botanic Gardens, Kew. Knighted, 1853. FRS 1864. (Gunn and Codd 1981, *Record of the Royal Society of London.*)

Bartlett, Abraham Dee (1812–97). Zoologist. Taxidermist, 1834–52. Superintendent of the natural history department, Crystal Palace, 1852–9; of the Zoological Society's gardens, Regent's Park, 1859–97. (*Modern English biography.*)

24 August [1860], 30 January [1865], 9 February [1865], 14 February [1865]

Bartlett, John (1793/4–1861). Clergyman. Perpetual curate of Buildwas, Shropshire, 1822–61; vicar of Spratton, Northamptonshire, 1823–61. Married Susanna Hannah Reynolds. (*Alum Cantab.*, *Two Shropshire ironmasters.*)

Bate, Charles Spence (1819–89). Dentist and scientific writer. Practised dentistry in Swansea, 1841–51, then in Plymouth. Secretary of the Plymouth Institution, 1854–60; president, 1861–2 and 1869–70. Authority on the Crustacea. FRS 1861. (*DNB.*)

29 November [1857]

Bates, Henry Walter (1825–92). Entomologist. Undertook a joint expedition to the Amazon with Alfred Russel Wallace, 1848–50; continued to explore the area, after Wallace returned to England, until 1859. Provided the first comprehensive scientific explanation of the phenomenon subsequently known as Batesian mimicry. Published an account of his travels, *The naturalist on the River Amazons*, in 1863. Assistant secretary, Royal Geographical Society of London, 1864–92. President, Entomological Society of London, 1868, 1869, and 1878. FRS 1881.

(*DNB*, *DSB*.)

28 January 1865, 22 March 1865, 29 March 1865

Bates, Sarah Ann. Daughter of a Leicester butcher. As Sarah Ann Mason, married Henry Walter Bates in 1863. (*DSB* s.v. Bates, Henry Walter; Woodcock 1969, pp. 253–5.)

Bayley, William (*fl.* 1822). Banker. Member of the banking firm Rocke, Eyton, Campbell, Leighton and Bayley of Market Square, Shrewsbury. (*Alum. Oxon.*, p. 79, *Commercial directory for Shropshire*, *Shrewsbury School Register*, p. 71.)

Beckles, Samuel Husband (1814–90). Barrister and palaeontologist. Discovered in the Purbeck beds the oldest known mammalian fossils. FRS 1859. (*Modern English biography*, Sarjeant 1980–96.)

Beddoes, Thomas (1760–1808). Physician. Reader in chemistry, University of Oxford, 1788–92. Father of Thomas Lovell Beddoes. Wrote and edited several medical and other works. Friend of Josiah Wedgwood I and Robert Waring Darwin. (*DNB*, B. Wedgwood and Wedgwood 1980.)

Beddoes, Thomas Lovell (1803–49). Poet and physiologist. Son of Thomas Beddoes. Came to prominence as a poet on the publication in 1822 of *The Bride's Tragedy* while he was studying at Oxford University. (*DNB*.)

Bell, Charles (1774–1842). Anatomist and surgeon. Best known for his investigations of the nervous system and the expression of emotions in humans. Co-owner of and principal lecturer at the Great Windmill Street School of Anatomy, London, 1812–25. Surgeon at the Middlesex Hospital, 1812–36. Professor of surgery, Edinburgh University, 1836. Knighted, 1831. FRS 1826. (*DNB*, *DSB*, *Record of the Royal Society of London*.)

Bell, Thomas (1792–1880). Dental surgeon and zoologist. Dental surgeon at Guy's Hospital, London, 1817–61. Professor of zoology, King's College, London, 1836. President of the Ray Society, 1843–59. One of the secretaries of the Royal Society, 1848–53. President of the Linnean Society, 1853–61. Described the reptiles from the *Beagle* voyage. FRS 1828. (*Reptiles*, *DNB*.)

Bendyshe, Thomas (1827–86). Barrister. Fellow, King's College, Cambridge, 1848–86. Admitted at the Inner Temple, 1848; called to the bar, 1857. Expelled from the Conservative Club, London, for voting for John Stuart Mill at the Westminster election of 12 July 1865. Translated and edited Latin, German, and French anthropological works for the Anthropological Society of London. (*Alum. Cantab.*)

Bennet, Charles Augustus, 6th earl of Tankerville (1810–99). Styled Lord Ossulston, 1822–59. MP for North Northumberland, 1832–59. Became sixth earl in 1859. Queen Victoria's lord steward of the household, 1867–8. (*Burke's peerage* 1999, *Modern English biography*.)

Bennett, William (1804–73). Tea-dealer and botanist. Member of the Society of Friends. In 1851, retired to Brockham Lodge, Betchworth, Surrey. Father of Edward Trusted Bennett (1831–1908) and Alfred William Bennett (1833–1902). (R. Desmond 1994, *DNB* s.v. Bennett, Alfred William.)

Benson, William Henry (1803–70). Zoologist. Wrote extensively on the sea and land shells of India, Ceylon, Burma, and the Cape of Good Hope. (*IBN*, *Royal Society catalogue of scientific papers.*)
7 December [1855]

Bentham, George (1800–84). Botanist. Donated his botanical library and collections to the Royal Botanic Gardens, Kew, in 1854, and was provided with facilities there for his research from 1861. President of the Linnean Society of London, 1861–74. Published *Genera plantarum* (1862–83) with Joseph Dalton Hooker. FRS 1862. (*DNB*, *DSB.*)

Bentham, Sarah (d. 1881). Née Brydges. Wife of George Bentham. (Jackson 1906.)

Berkeley, Miles Joseph (1803–89). Clergyman and botanist. Perpetual curate of Apethorpe and Wood Newton, Northamptonshire, 1833–68. Vicar of Sibbertoft, Northamptonshire, from 1868. Editor of the *Journal of the Royal Horticultural Society*, 1866–77. An expert on British fungi; described fungi from CD's *Beagle* voyage. Royal Society Royal Medallist, 1863. FRS 1879. (*DNB*, *DSB.*)
18 March [1856]

Beyrich, August Heinrich Ernst (Ernst) (1815–96). German geologist and palaeontologist. One of the founding members of the German geological society, 1848. Professor of geology, University of Berlin, 1865. First director of the Museum für Naturkunde, 1873. Prussian state geologist, 1873. (*DBE*, *NDB*, Sarjeant 1980–96.)

Bishop, Richard. Of Plymouth. Provided CD with observations on *Balanus*. (*Correspondence* vol. 6, letter from Richard Bishop to C. S. Bate, 3 December 1857.)

Blainville, Henri Marie Ducrotay (Henri) de (1777–1850). French anatomist and zoologist. Appointed professor of comparative anatomy, Muséum d'Histoire Naturelle, Paris, 1832. Foreign member, Royal Society, 1832. (*DBF*, *DSB*, *Record of the Royal Society of London.*)

Blakiston, Thomas Wright (1832–91). Explorer and ornithologist. Second lieutenant in the Royal Artillery, 1851; second captain, 1858; resigned in 1862. Served in the Crimean campaign. Member of the scientific expedition for the exploration of British North America, between Ontario and the Rocky Mountains, 1857. Explored the Yang-tsze-Kiang river, China, 1861. Merchant at Hakodate, Japan, 1863–84. Eventually settled in New Mexico. Compiled, with H. Pryor, *A catalogue of the avifauna of Japan* (1884). (*DNB.*)

Blyth, Edward (1810–73). Zoologist. Pharmacist in Tooting, London. Wrote and edited zoological works under the pseudonym Zoophilus. Curator of the museum of the Asiatic Society of Bengal, Calcutta, India, 1841–62. Provided CD with information on the plants and animals of India. Returned to Britain in 1863, and continued to write on zoology and on the question of the origin of species. (*Correspondence* vols. 5–7, *DNB*, *DSB.*)

Bonaparte, Lucien Jules Laurent (Charles Lucien), Prince of Canino (1803–57). French ornithologist. Lived in America, 1822–8. In 1828, settled in Italy, where he was politically active. Attempted to enter France in 1848 but was expelled. Lived in England until he was allowed to settle in Paris in 1850. Devoted the rest of his life to work on taxonomy. (*DBF*, *DSB*.)

Bond, Frederick (1811–89). Entomologist and ornithologist. Elected member of the Entomological Society of London, 1841; of the Zoological Society of London, 1854. Collector of British Lepidoptera. (*Entomologist's Monthly Magazine* 25 (1888–9): 384, Gilbert 1977.)

Bonnet, Charles (1720–93). Swiss naturalist and philosopher. Although educated for the law, his main interest was natural history. Discovered parthenogenesis in aphids in 1746. Studied invertebrate regeneration, entomology, and plant physiology. Advocate of preformation. Wrote influential works on theoretical biology. (*DSB*.)

Boott, Francis (1792–1863). American physician and botanist. Resident in London from 1820. MD 1824. Lecturer on botany and materia medica, Webb Street School of Medicine, London, 1825. Conducted a successful medical practice in London, 1825–32. Secretary, Linnean Society of London, 1832–9; treasurer, 1856–61. (R. Desmond 1994, *DNB*.)

[3 February 1842]

Borrow, George Henry (1803–81). Traveller and writer. (*DNB*.)

Boucher de Crèvecoeur de Perthes, Jacques (Jacques Boucher de Perthes) (1788–1868). French customs official and archaeologist. Director of customs, Abbeville, France, from 1825. President, Société d'Emulation, Abbeville. Published on prehistoric archaeology and palaeontology. Discovered controversial evidence of early humans in the Somme River valley gravels. (*DSB*.)

Bowerbank, James Scott (1797–1877). London distiller and geologist. Had a special interest in London Clay fossils; devoted his later career to the study of sponges. Founder of the Palaeontographical Society, 1847. FRS 1842. (*DNB*, *Record of the Royal Society of London*, Sarjeant 1980–96.)

25 February [1850]

Bowman, William (1816–92). Surgeon. Author of works on physiology. Assistant surgeon, King's College Hospital, London, 1840–56; elected professor of physiology and of general and morbid anatomy, King's College, London, 1848; assistant surgeon at the Royal London Ophthalmic Hospital, Moorfields, 1846–51; surgeon, 1851–76. Created baronet, 1884. FRS 1841. (*DNB*, *DSB*.)

30 July [1865?]

Brace, Charles Loring (1826–90). American philanthropist, author, and ethnologist. After studying theology at Yale University, toured Europe, 1850–1. One of the founders of the New York City Children's Aid Society, 1853, and subsequently director. Published a small number of ethnological treatises. (*DAB*.)

Brady, George Stewardson (1832–1921). Zoologist and surgeon. Practised medicine in Sunderland, 1857–1892. Professor of natural history, Durham College of Science (College of Physical Sciences), Newcastle-upon-Tyne, 1875–1906. One of the secretaries of the Tyneside Naturalists' Field Club, 1863–9; president, 1875, 1892–3, 1906. Honorary curator of botany, museum of the Natural History

Brady, George Stewardson, cont.
Society of Northumberland, Durham, and Newcastle-upon-Tyne, 1866–74; vice-president, 1894–1921. Published on marine and freshwater Crustacea. Author of the reports on the Ostracoda and Copepoda of the *Challenger* expedition. FRS 1882. (*Medical directory* 1890–2, *Men and women of the time* 1899, *Natural History Transactions of Northumberland and Durham* 6 (1924–6): 1–7, *Proceedings of the Royal Society* Ser. B., 93 (1922): xx–xxiii, *Record of the Royal Society of London*, *Transactions of the Tyneside Naturalists' Field Club* 1863–4.)
19 March 1865

Braun, Alexander Carl Heinrich (1805–77). German botanist. Professor of botany, Freiburg, 1846–50. Professor of botany and director of the botanic garden, University of Berlin, 1851–77. Brother-in-law of Louis Agassiz. Deeply influenced by *Naturphilosophie*; studied plant morphology. Established the doctrine of spiral phyllotaxy. (*DSB*, *NDB*.)

Brehm, Christian Ludwig (1787–1864). German clergyman and ornithologist. His bird collection formed the basis of the Rothschild museum in Tring, Hertfordshire. (*NDB*.)

Brent, Bernard Peirce (1822–67). Bird-fancier and author. Studied pigeon breeding in France and Germany. (*CDEL*; family information.)
7 February [1857], 1 April [1861]

Brent, John (1808–82). Antiquary, poet, and novelist. Published on the archaeology and antiquities of Kent. Author of historical and romantic novels. Fellow, Society of Antiquaries. City treasurer of Canturbury. (*DNB*.)

Brewer, Thomas Mayo (1814–80). American ornithologist, journalist, and publisher. Practised medicine briefly after graduating from Harvard Medical School in 1838. Editor of the Boston *Atlas* and Washington correspondent for the paper during the 1840s and 1850s. Became a partner in the publishing firm Swan & Tileston, 1857; later head of Brewer & Tileston, a position he held until his retirment in 1875. Worked on birds' eggs and contributed to bird guides and lists. (*ANB*.)

Brewster, David (1781–1868). Scottish physicist who specialised in optics. Invented the kaleidoscope in 1816. Edited the *Edinburgh encyclopædia*, 1807–30. Assisted in the establishment of the British Association for the Advancement of Science in 1831. Principal of the United College of St Salvator and St Leonard, St Andrews University, 1838–59. Vice-chancellor, Edinburgh University, 1860–8. Knighted, 1832. FRS 1815. (*DNB*, *DSB*.)

Bridges, Robert (1806–82). Physician and botanist. Librarian, corresponding secretary, vice-president, and then president (1864) of the Academy of Natural Sciences of Philadelphia. Appointed professor of general and pharmaceutical chemistry, Philadelphia College of Pharmacy, 1842; professor of chemistry, Franklin Medical College, 1846–8. (*DAB*.)

Brightwell, Thomas (1787–1868). Solicitor in Norwich. Microscopist and botanist. Specialised in diatoms and protozoa. Fellow of the Linnean Society of London.

(R. Desmond 1994.)

Broca, Pierre Paul (Paul) (1824–80). French surgeon and anthropologist. MD, Paris, 1849. Assistant professor of the medical faculty in Paris and surgeon of the Central Bureau, 1853. Pioneer in the field of anthropology. Elected professor of pathology at Paris, 1867; of clinical surgery, 1868. Vice-president, French Academy of Medicine. (*DBF*, *DSB*.)

Brodie Innes, John. *See* Innes, John.

Brongniart, Adolphe Théodore (1801–76). French palaeobotanist, plant anatomist, and taxonomist. One of the founders of the *Annales des Sciences Naturelles*, 1824. Professor of botany, Muséum d'Histoire Naturelle, from 1833. (*DSB*.)

Bronn, Heinrich Georg (1800–62). German palaeontologist. Professor of natural science at Heidelberg University, 1833. Translated and superintended the first German editions of *Origin* (1860) and *Orchids* (1862). (*DSB*, *NDB*.)

Brooks, William (b. 1801/2). Servant at Down House, 1844–75. Employed as an outdoor servant, working in the garden, and with the cows, pigs, and horses; after 1853 listed under 'manservants' rather than 'Garden' in CD's Classed account books. Lived in a cottage near the cow-yard at Down House with his wife, Keziah (b. 1807/8). (CD's Classed account books (Down House MS); Census returns 1861 (Public Record Office RG9/462: 74–5); F. Darwin 1920, pp. 56–8; letter from Emma Darwin to William Erasmus Darwin, [3 February 1863] (DAR 219.1: 70).)

Brown, Robert (1773–1858). Scottish botanist. Naturalist to the expedition surveying the coast of Australia, 1801–5; published descriptions of the plants he collected. Librarian to the Linnean Society of London, 1806–22; to Sir Joseph Banks, 1810–20. Continued as curator of Banks's collections after his death in 1820 and negotiated their transfer to the British Museum in 1827. Keeper of the botanical collections, British Museum, 1827–58. FRS 1811. (*DNB*, *DSB*.)

Browning, Robert (1812–89). Poet. (*DNB*.)

Brown-Séquard, Charles Edouard (1817–94). French physiologist. Practised medicine in France and the United States before accepting a professorship at Virginia Medical College, Richmond, in 1854. Returned to Paris in 1855; moved to England in 1858. Physician, National Hospital for the Paralysed and Epileptics, 1860–3. Professor of physiology and pathology, Harvard Medical School, 1864–7. Professor of medicine, Collège de France, 1878–94. Conducted pioneering research in neurology and endocrinology. FRS 1860. (*DBF*, *DSB*, *Record of the Royal Society of London*.)

Bruno, Giordano (1548–1600). Italian philosopher. Known for his theories of an infinite universe and multiplicity of worlds. Arrested for heresy by the Venetian Inquisition in 1592; burned alive by the Roman Inquisition in 1600. (*DSB*.)

Bryant, Henry (1820–67). American physician and naturalist. MD 1843. Practised in France and Algeria and briefly in the United States before abandoning medicine because of ill health and turning to natural history, especially ornithology, after 1847. Served as surgeon in the Twentieth Regiment of Massachusetts

Bryant, Henry, cont.
Volunteers in the American Civil War, and took charge of Lincoln Hospital. Resigned his commission because of ill health before the end of the war and went abroad. Bought the La Fresnaye collection of birds, including 9000 specimens, in France, and presented it to the Boston Society of Natural History. (*Proceedings of the American Academy of Arts and Sciences* 7 (1867): 304–5.)

Buckland, Francis Trevelyan (1826–80). Naturalist, popular science-writer, and surgeon. Son of William Buckland. Trained and practised medicine at St George's hospital, London, 1848–53. Assistant surgeon in the Life Guards, 1854. Staff writer for the *Field*, 1856–65. Established an exhibit on pisciculture at the South Kensington Museum, *circa* 1865, and launched a weekly journal, *Land and Water*, in 1866. Inspector of salmon fisheries from 1867. (*DNB*.)
18 March 1865

Buckland, William (1784–1856). Geologist and clergyman. Professor of mineralogy, Oxford University, 1813; reader in geology, 1819–49. President of the Geological Society of London, 1824–5 and 1840–1. Dean of Westminster from 1845. FRS 1818. (*DNB*, *DSB*.)
[November 1840 – 17 February 1841]

Buckle, Henry Thomas (1821–62). Historian. Heir to a London shipowner. Travelled widely and published a number of volumes on the history of English and European civilisation. (*DNB*.)

Buckley, Arabella Burton (1840–1929). Popular scientific author, specialising in natural history. Secretary to Charles Lyell, 1864–75. Wrote in particular for young readers, and encouraged an active pursuit of natural history. A supporter of Darwin; she emphasised the importance of mutualism and dependence as forces of evolution. (*BDWS*, *CDEL*, Gates 1997, Tort 1996.)

Buffon, Georges Louis Leclerc, comte de (1707–88). French naturalist, philosopher, and mathematician. Keeper, Jardin du Roi, 1739–88. His theory of transmutation is outlined in *Histoire naturelle* (1749–1804). FRS 1739. (*DBF*, *DSB*, *Record of the Royal Society of London*.)

Burchell, William John (1781–1863). Explorer and naturalist. Schoolmaster and acting botanist to the East India Company, St Helena, South Africa, 1805–10. Collected plants in southern Africa, 1811–15, and Brazil, 1825–9. (R. Desmond 1994, *DNB*.)

Burn, Robert (1829–1904). Clergyman. Fellow of Trinity College, Cambridge, 1854–1904; tutor, 1856–72; dean, 1861–3. Ordained priest, 1862. (*Alum. Cantab.*)
2 December [1865]

Burningham, James (b. 1821/2). Police officer. Arrived in the Isle of Wight between 1846 and 1849; moved to Gatten, Shanklin, by 1851. (Census returns 1851 (Isle of Wight Record Office), Richard Smout, archivist, personal communication.)
[10 September 1858?]

Busk, Ellen. Daughter of Jacob Hans Busk. Married her cousin, George Busk, in

1843. (*DSB* s.v. Busk, George.)

Busk, George (1807–86). Russian-born naval surgeon and naturalist. Served on the hospital ship at Greenwich, 1832–55. Retired from medical practice in 1855. Member of several scientific societies. President of the Microscopical Society, 1848–9; of the Anthropological Institute, 1873–4. Zoological secretary of the Linnean Society of London, 1857–68. Hunterian Professor of comparative anatomy, Royal College of Surgeons of England, 1856–9; council member, 1863; member of board of examiners, 1868; president, 1871. Specialised in palaeontology and in the study of Bryozoa. FRS 1850. (*DNB*, *DSB*, Plarr 1930.)

20 February 1865, *28 April 1865*

Butler, Samuel (1774–1839). Educationalist and clergyman. Headmaster of Shrewsbury School, 1798–1836. CD attended the school, 1818–25. Bishop of Lichfield and Coventry, 1836; of Lichfield, 1836–9. (*DNB, Freeman 1978.*)

Butler, Samuel (1835–1902). Author and artist. Grandson of Samuel Butler (1774–1839). Emigrated to New Zealand in 1859; returned to Britain after publishing an account of his time farming in the colony, *A first year in Canterbury settlement* (1863). Published books on art, music, literature, and philosophy, including the novels *Erewhon* (1872) and *The way of all flesh* (1903); published a two-volume life of his grandfather, *Life of Samuel Butler, bishop of Lichfield* (1896). Became a critic of Darwinism from the 1870s. (*Autobiography*, pp. 167–219, *DNB*.)

30 September [1865], *1 October 1865*, 6 October [1865]

Butler, Thomas (1806–86). Clergyman. BA, St John's College, Cambridge, 1829. Rector of Langar with Barnston, Nottinghamshire, 1834–76. (*Alum. Cantab.*)

Butler, Timothy (1806–80). Leading portrait sculptor. Exhibited over 100 busts at the Royal Academy between 1828 and 1879. (*Dictionary of British sculptors*, *Modern English biography.*)

Campbell, Archibald (1805–74). Assistant surgeon, East India Company, 1827. Superintendent of the Darjeeling station, 1840–62, with responsibility for political relations between the British and the Sikkim government. Wrote many papers on Himalayan geography. Travelled with Joseph Dalton Hooker in Sikkim in 1849, and was imprisoned with Hooker by the Sikkim rajah. (R. Desmond 1994, Hooker 1854, L. Huxley ed. 1918.)

Campbell, George (1719–96). Scottish clergyman. Minister of religion in Aberdeen, 1757; principal, Marischal College, 1759; professor of divinity, 1771. Published a translation of the New Testament (1789), and works on philosophy and ecclesiastical history. (*DNB*.)

Campbell, George Douglas, 8th duke of Argyll (1823–1900). Scottish statesman and author of works on science, religion, and politics. A defender of the concept of design in nature. Chancellor of St Andrews University, 1851. President of the Royal Society of Edinburgh, 1861–4. Privy seal, 1853–5, 1859–60, 1860–6, and 1880–1; postmaster-general, 1855–8 and 1860; secretary of state for India, 1868–74. Succeeded to dukedom in 1847. FRS 1851. (*DNB*, *Proceedings of the Royal Society of London.*)

Campbell, John Francis (1821–85). Scottish writer on Highland folklore, geology, and meteorology. (*DNB*.)

Campbell, Walter Frederick (1798–1855). Scottish writer. MP for Argyleshire, 1822–32. Father of John Francis Campbell. (*Modern English biography*.)

Campden, Lord. *See* Noel, Charles George.

Candolle, Alphonse de (1806–93). Swiss botanist, lawyer, and politician. Active in the administration of the city of Geneva until 1860. Responsible for the introduction of postage stamps to Switzerland. Professor of botany and director of the botanic gardens, Geneva, from 1835. Concentrated on his own research after 1850. Foreign member, Royal Society of London, 1869. (*DSB*, *Record of the Royal Society of London*.)

Carlyle, Thomas (1795–1881). Essayist and historian. (*DNB*.)

Carpenter, Louisa. Married William Benjamin Carpenter in 1840. (*DNB* s.v. Carpenter, William Benjamin.)

Carpenter, William Benjamin (1813–85). Physician and naturalist. Fullerian Professor of physiology at the Royal Institution of Great Britain, 1844–56; professor of forensic medicine, University College, London; physiology lecturer, London Hospital, 1845–56. Registrar of the University of London, 1856–79. Founding member of the Marine Biological Association. FRS 1844. (*DNB*, *DSB*.)
6 December [1844]

Carpmael, Ernest (1844–1921). Barrister. Contemporary of George and Francis Darwin at Clapham Grammar School and Cambridge University. Son of William Carpmael. Matriculated, St John's College, Cambridge, 1863. Called to the bar, 1869. (*Alum. Cantab.*; Biographical notes, St John's College, Cambridge; *Eagle* (St John's College, Cambridge) 3: 356.)

Carpmael, William (1804–67). Engineer. Served on the council of the Institute of Civil Engineers. Member of the Institute of Mechanical Engineers. Member of the Metropolitan Board of Works, 1855–67. (*Minutes of Proceedings of the Institution of Civil Engineers* 30 (1869–70, pt 2): 430; *Proceedings of the Institution of Mechanical Engineers* (1868): 14–15.)

Carrière, Elie Abel (1818–96). French botanist and horticulturalist. Started work at the Muséum d'Histoire Naturelle at the age of 14 and rose to the rank of head perennial grower. Head of the botanic garden, Zaragossa, Spain, in the early 1860s; head of the nursery, Muséum d'Histoire Naturelle, from 1862; resigned in 1869. Editor of the *Revue Horticole* from 1862; editor-in-chief, 1869–96. Worked on peach trees, dimorphism, and hybridity. (Barnhart comp. 1965, *Revue Horticole* 1896, pp. 389–97, Tort 1996.)

Carter, Henry John (1813–95). Surgeon and naturalist. Studied medicine at University College, London, and Ecole de Médecine, Paris. Surgeon in the East India Company, 1842–62; stationed in Bombay from 1846. Retired to Devonshire in 1864. Noted for his papers on South Asian and Arabian geology, and for his extensive researches on sponges and foraminifera. Royal Society Royal Medallist, 1872. FRS 1859. (*Modern English biography*, *Proceedings of the Royal Society of London* 58 (1895): liv–lvii, Sarjeant 1980–96.)

Case, George Augustus (d. 1831). Unitarian minister of the High Street Chapel, Shrewsbury, from 1797. CD attended his school in Shrewsbury between the ages of 8 and 9. (*Autobiography*, *Christian Reformer* 18 (1832): 38.)

Caspary, Johann Xaver Robert (Robert) (1818–87). German botanist. Director, Bonn herbarium, 1856. Professor of botany and director of the botanic gardens at the University of Königsberg from 1858. Specialised in aquatic plants. (*ADB*.)

Castang, Philip. Dealer in live game, including pigeons, with a business address at 6 and 7 Ship Tavern Passage, Leadenhall Market, London. (CD's Address book (Down House MS), *Post Office London directory* 1861.)

Castelnau, Francis de la Porte, comte de (1812–80). Explorer. Travelled in North America, 1837–41. At the behest of the French government, made a voyage of exploration in South America, 1843–7. Consular official in various cities until 1877. (*DBF*.)

Castlereagh, Lord. *See* Stewart, Robert.

Chapman, John (1822–94). Physician, author, and publisher. Studied medicine in Paris and then at St George's Hospital, London, from 1844. Publisher and bookseller at 142 Strand until 1860. Editor and proprietor of the *Westminster Review*, 1851–94. MD, St Andrews, 1857; practised as a physician. Advocated the application of an ice-bag to the spine as a remedy particularly for seasickness and cholera. (*DNB*.)

16 May [1865], 7 June 1865

Chapuis, Félicien (1824–79). Belgian entomologist and physician. Studied medicine at the University of Liège and practised in Verviers. His entomological publications were mainly on the Coleoptera. Amateur pigeon fancier; wrote *Le pigeon voyageur belge* (1865). (*BNB*.)

Charles II (1630–85). Second son of King Charles I and Henrietta Marie of France. Escaped to France after Charles I's execution, 1649; returned to be defeated by Oliver Cromwell at Worcester, 1651; thereafter roamed Europe for eight years. King of England, Scotland, and Ireland from the restoration of the monarchy, 1660. (*DNB*.)

Children, John George (1777–1852). Naturalist. Librarian in the department of antiquities at the British Museum, 1816; employed in the zoological department, 1823–40. Secretary of the Royal Society, 1826–7 and 1830–7. FRS 1807. (*DNB*.)

22 February [1838]

Christy, Henry (1810–65). Ethnologist, archaeologist, geologist, and banker. Partner in the London banking house Christy & Co.; director of the London Joint-Stock Bank. Travelled for ethnological purposes in the East in 1850; in Scandinavia, 1852–3; in Canada, the United States, Cuba, and Mexico, 1856–7. Member of the Geological Society of London, 1858; conducted excavations of caves in the Vézère valley in southern France with Edouard Lartet in the 1860s. (*DNB*.)

Claparède, Jean Louis René Antoine Edouard (Edouard) (1832–71). Swiss naturalist and invertebrate zoologist. Professor of zoology and comparative anatomy, Academy of Geneva, 1862. Specialised in invertebrate anatomy, histology,

Claparède, Edouard, cont.
embryology, and evolution. One of the first Swiss naturalists to endorse CD's theory. (*Dictionnaire historique & biographique de la Suisse*, Gilbert 1977.)

Clift, William (1775–1849). Naturalist. Curator of the Hunterian Museum at the Royal College of Surgeons, 1793–1844. In 1835, Clift's daughter Caroline married Richard Owen, who succeeded Clift as curator in 1844. FRS 1823. (*DNB*; Dobson 1954, p. 103; *DSB*.)

Clive, Edward, 1st earl of Powis (1754–1839). Statesman. MP for Ludlow until 1794. Created Baron Clive of Walcot, 1794. Governor of Madras, 1798–1803. Married Henrietta Antonia Herbert, daughter of the earl of Powis and last of the Herbert family, in 1784. Granted the revived title of earl of Powis in 1804. (*DNB*.)

Clive, Henrietta Antonia, countess of Powis (after 1751–1830). Heiress of the last Herbert earl of Powis. Married Edward Clive in 1784. (*Burke's peerage* 1999, *DNB* s.v. Clive, Edward.)

Clive, Robert, Baron Clive (1725–74). Army officer, merchant, and colonial administrator. Father of Edward Clive. Led a number of successful military campaigns in India. Governor of the East India Company's territory in Bengal, 1757–60, 1765. MP for Shrewsbury, 1760–74. Created Baron Clive, 1762. (*DNB*.)

Clowes, William & Sons. Printers. William Clowes (1807–83), eldest son of William Clowes (1779–1847), joined his father's printing business in 1823; the name of the firm was changed to William Clowes & Sons in 1846. Printed the offical catalogue of the Great Exhibition of 1851. Introduced improvements in type-music printing. Printers to John Murray. (*DNB*.)

Cludde, Anne Maria (*fl.* 1781). Daughter of Edward Jeffreys of Shrewsbury. Married William Pemberton Cludde in 1781. (*Burke's landed gentry* 1952, s.v. Herbert of Orleton.)

Cludde, William Pemberton (d. 1829). Born William Pemberton; assumed the name Cludde on succeeding his uncle, Edward Cludde of Orleton, in 1785. Mayor of Shrewsbury, 1795; high sheriff of Shropshire, 1814. Owned the Shropshire manor of Wrockwardine from 1822. Married Anne Maria Jeffreys in 1781. (*Burke's landed gentry* 1952, s.v. Herbert of Orleton; *Victoria county history of Shropshire* 11: 311–12.)

Cobbold, Thomas Spencer (1828–86). Physician and zoologist. Studied medicine at Edinburgh University, graduating in 1851; appointed curator of the university anatomical museum. Moved to London in 1857; appointed botany lecturer at St Mary's Hospital. Obtained a similar position at the Middlesex Hospital in 1861, lecturing on zoology and comparative anatomy. Established a medical practice in London as a consultant in cases where internal parasites were suspected. Swiney Lecturer in geology at the British Museum, 1868–73. From 1873, professor of botany at the Royal Veterinary College, which shortly afterwards instituted a helminthology professorship for him. FRS 1864. (*DNB*.)

Cohen, Meyer. Father of Francis Palgrave. (*DNB* s.v. Francis Palgrave.)

Cohn, Ferdinand Julius (1828–98). German botanist and bacteriologist. Extraordinary professor, University of Breslau, 1859; professor, 1872. Founded the first institute for plant physiology, at Breslau, in 1866. In 1870, founded the journal *Beiträge zur Biologie der Pflanzen*, in which the earliest articles on bacteriology appeared. Worked on unicellular algae, and attempted to devise a system of classification for lower plants based on Darwinian transmutation theory. Foreign member, Royal Society, 1897. (*DSB*, *NDB*.)

Cole, William Willoughby, 3d earl of Enniskillen (1807–86). Geologist. Specialised in fossil fish. MP for Fermanagh, 1831–40. Succeeded to the peerage, 1840. FRS 1829. (*Modern English biography*, *Quarterly Journal of the Geological Society of London* 43 (1887): 38–9.)

Colenso, John William (1814–83). Clergyman, mathematician, and missionary. Fellow, St John's College, Cambridge, 1837; college tutor, 1842–6. Mathematical tutor, Harrow, 1839–42. Vicar of Forncett St Mary, Norfolk, 1846–53. Bishop of Natal, South Africa, 1853–83. Published *Critical examination of the Pentateuch* (1862–79), in which he argued that some passages in these books had been forged several centuries after their apparent date. There was an attempt to remove him from the bishopric in 1863, but his possession of the see was confirmed by the law courts in 1866. (*DNB*.)

Compton, Spencer Joshua Alwyne, 2d marquis of Northampton (1790–1851). MP for Northampton, 1812–20. President of the Royal Society, 1838–48. Succeeded his father as second marquis of Northampton, 1828. FRS 1830. (*DNB*, *Record of the Royal Society of London*.)

Comte, Isidore Auguste Marie François Xavier (Auguste) (1798–1857). French philosopher. Private secretary to Claude Henri de Rouvroy, comte de Saint-Simon, 1817–23. Founded the Association Polytechnique, a group devoted to education of the working classes, in 1830; and the Société Positiviste, devoted to the promulgation of the 'Cult of Humanity', in 1848. Adopted the term 'positivism' for his philosophy. (*DSB*.)

Cooke, Robert Francis (1816–91). Partner in John Murray's publishing company, 1837–91. (*Modern English biography*.)
18 April 1865 (W. B. Tegetmeier)

Cotton, Robert Bruce (1571–1631). Politician, historian, and antiquary. MP for Huntingdon, 1604; Old Sarum, 1624; Thetford, 1625; Castle Rising, 1628–9. Wrote political and historical treatises, and collected manuscripts and coins. His library was transferred to the nation in 1702 and deposited in the British Museum at its creation in 1753. Knighted, 1603; created baronet, 1611. (*DNB*.)

Crawfurd, John (1783–1868). Scottish-born physician, diplomat, and orientalist. Joined the East India Company medical service in 1803. Held several civil and political posts in Java, India, Siam, Cochin China, Singapore, and Burma. Returned to England in 1827; promoted the study of Indo-China and campaigned on behalf of the Singapore trading community against the East India Company. Published papers on ethnological and other subjects in various journals.

Crawfurd, John, cont.
President of the Ethnological Society of London, 1861–7. FRS 1818. (*DNB*, *Record of the Royal Society of London*, *Transactions of the Ethnological Society of London* n.s. 6 (1867): 6–7, Turnbull 1989, pp. 25–30.)
7 August 1863 (E. A. Darwin)

Cresy, Edward (1824–70). Surveyor and civil engineer. Son of Edward Cresy (1792–1858), the architect and civil engineer who advised CD about the purchase and improvement of Down House. Worked as an architectural draftsman in his father's office as a young man. Assisted his father in preparing his *Cyclopædia of civil engineering* in 1845. Assistant surveyor under the commissioners of sewers, 1849; afterwards engineer. Principal assistant clerk at the Metropolitan Board of Works, 1859. Architect to the fire brigade, 1866. Founder member of the Geologists' Association, 1858; president, 1864–5; vice-president, 1865–70. (*Annual Report of the Geologists' Association* 1859–70; Census returns 1861 (Public Record Office RG9/422: 118a); *DNB* s.v. Cresy, Edward (1792–1858); *Engineer* 30 (1870): 409.)
26 [December 1843 – April 1846 or September 1855 – October 1860?], *30 May 1865*, *9 June 1865*, 7 September [1865], *10 September 1865*, *18 October 1865*, 19 October [1865], *20 November 1865* (Emma Darwin)

Cresy, Mary Louis (b. 1820/1). Wife of Edward Cresy (1824–70). (Census returns 1861 (Public Record Office, RG9/422: 118a), family communication.)

Croll, James (1821–90). Scottish physical geologist. Worked for the Geological Survey of Scotland, 1867–81. Worked on oceanic circulation patterns, the mechanics of glaciers, and causes of the glacial epoch. FRS 1876. (*DNB*, Sarjeant 1980–96.)

Crotch, William Duppa (1831/2–1903). Entomologist and zoologist. Specialised in the Lepidoptera, Coleoptera, and Hemiptera. Travelled with his brother George Robert Crotch (1841–74) on several collecting expeditions. Married a Swede and settled in Scandinavia, where he studied lemmings. (*Entomologist's Monthly Magazine* 39 (1903): 256, Gilbert 1977.)
10 April 1865

Crüger, Hermann (1818–64). German pharmacist and botanist. Apothecary in Trinidad in the West Indies from 1841; government botanist and director of the botanic garden, Trinidad, from 1857. Collected plants in Jamaica, Trinidad, and Venezuela. (R. Desmond 1994, S[chlechtenda]l 1864.)

Cunningham, Allan (1791–1839). Botanist and explorer. Collected plants for the Royal Botanic Gardens, Kew, in Brazil, 1814–16; New South Wales, 1816–26 and 1827–30; New Zealand, 1826. Superintendent of the botanic garden, Sydney, 1836–8. (*Aust. dict. biog.*, R. Desmond 1994.)

Currey, Frederick (1819–81). Mycologist. Called to the bar, 1844. Secretary, Linnean Society of London, 1860–80; vice-president and treasurer, 1880–1. One of the editors of the *Natural History Review*. FRS 1858. (*Alum. Cantab.*, *DNB*.)

Cuvier, Jean Léopold Nicolas Frédéric (Georges) (1769–1832). French systematist, comparative anatomist, palaeontologist, and administrator. Professor of natural history, Collège de France, 1800–32; professor of comparative anatomy, Muséum d'Histoire Naturelle, 1802–32. Permanent secretary to the Académie des Sciences from 1803. Foreign member, Royal Society of London, 1806. (*DBF*, *DSB*.)

Dallas, William Sweetland (1824–90). Entomologist, author, and translator. Prepared lists of insects for the British Museum, 1847–58. Curator of the Yorkshire Philosophical Society's museum, 1858–68. Assistant secretary to the Geological Society of London, 1868–90. Translated Fritz Müller, *Für Darwin* (1869); prepared the index for *Variation* and the glossary for *Origin* 6th ed. Editor, *Annals and Magazine of Natural History*, 1868–90, *Popular Science Review*, 1877–80. (Freeman 1978; *Geological Magazine* n.s. decade 3, vol. 7 (1890): 333–6; *Modern English biography*, Sarjeant 1980–96.)

Dana, Edward Salisbury (1849–1935). American mineralogist. Son of James Dwight Dana. Curator of mineralogy, Yale University, 1874–1922; assistant professor of natural philosophy, 1879–90; professor of physics, 1890–1917. Published on mineralogy. (*DAB*, *Historical register of Yale University*.)

Dana, Henrietta Frances. Daughter of Benjamin Silliman. Married James Dwight Dana in 1844. (*DAB*.)

Dana, James Dwight (1813–95). American geologist and zoologist. Naturalist with the United States expedition to the Pacific, 1838–42; wrote reports on the geology, zoophytes, and Crustacea. One of the editors of the *American Journal of Science and Arts* from 1840. Professor of natural history, Yale University, 1856–64; professor of geology and mineralogy, 1864–90. Foreign member, Royal Society of London, 1884. (*DAB*, *DSB*.)
8 September 1856

Daniell, William Freeman (1818–65). Surgeon and botanist. Assistant surgeon to the British army in West Africa, 1841–53. Sent plants to the Royal Botanic Gardens, Kew. (R. Desmond 1994, *DNB*.)

Darwin, Anne Elizabeth (Annie) (1841–51). CD's daughter. (*Darwin pedigree*.)

Darwin, Caroline Sarah. *See* Wedgwood, Caroline Sarah.

Darwin, Charles Waring (1856–8). Youngest child of CD. Died of scarlet fever. (*Darwin pedigree*.)

Darwin, Edward Levett (1821–1901). CD's half first cousin. Son of Francis Sacheverel Darwin. Captain in the Second Derby militia from 1856. Lived in Buxton, Derbyshire. (*Army list* 1863, *Darwin pedigree*, Erasmus Darwin's Commonplace book (Down House MS).)

Darwin, Elizabeth (Bessy) (1847–1926). CD's daughter. (*Darwin pedigree*, Freeman 1978.)

Darwin, Emily Catherine (1810–66). CD's sister. Resided at The Mount, Shrewsbury, until she married Charles Langton in 1863. (*Darwin pedigree*.)

Darwin, Emma (1808–96). Youngest daughter of Josiah Wedgwood II. Married CD, her cousin, in 1839. (*Emma Darwin* (1904) and (1915).)
[13 January 1861] (W. E. Darwin), *[10 July 1865]* (J. D. Hooker)

Darwin, Erasmus (1731–1802). CD's grandfather. Physician, botanist, and poet. Advanced a theory of transmutation similar to that subsequently propounded by Jean Baptiste de Lamarck. FRS 1761. (*DNB*, *DSB*, King-Hele 1999.)

Darwin, Erasmus Alvey (1804–81). CD's brother. Attended Shrewsbury School, 1815–22. Matriculated at Christ's College, Cambridge, 1822; Edinburgh University, 1825–6. Qualified in medicine but never practised. Lived in London from 1829. (*Alum. Cantab.*, Freeman 1978.)
24 August [1865], 1 September [1865], [20 November 1865?], 25 [November 1865] (Emma Darwin)

Darwin, Francis (1848–1925). CD's son. Botanist. BA, Trinity College, Cambridge, 1870. Qualified as a physician but did not practise. Collaborated with CD on several botanical projects, 1875–82. Lecturer in botany, Cambridge University, 1884; reader, 1888–1904. Published *LL* and *ML*. President of the British Association for the Advancement of Science, 1908. Knighted, 1913. FRS 1882. (*DNB*, *DSB*.)

Darwin, George Howard (1845–1912). CD's son. Mathematician. BA, Trinity College, Cambridge, 1868; fellow, 1868–78. Called to the bar in 1872 but did not practise. Plumian Professor of astronomy and experimental philosophy, Cambridge University, 1883–1912. President of the British Association for the Advancement of Science, 1905. Knighted, 1905. FRS 1879. (*DNB*, *DSB*.)
[28 November 1856]

Darwin, Henrietta Emma (1843–1927). CD's daughter. Married Richard Buckley Litchfield (*Alum. Cantab.*) in 1871. Assisted CD with some of his work. Edited *Emma Darwin* (1904) and (1915). (*Burke's landed gentry* 1952, Freeman 1978.)

Darwin, Horace (1851–1928). CD's son. Civil engineer. BA, Trinity College, Cambridge, 1874. Apprenticed to an engineering firm in Kent; returned to Cambridge in 1875 to design and make scientific instruments. Founder and director of the Cambridge Scientific Instrument Company. Mayor of Cambridge, 1896–7. Knighted, 1918. FRS 1903. (*Alum. Cantab.*, *DNB*.)

Darwin, Leonard (1850–1943). CD's son. Military engineer. Attended the Royal Military Academy, Woolwich. Commissioned in the Royal Engineers, 1870; major, 1889. Served on several scientific expeditions, including those for the observation of the transit of Venus in 1874 and 1882. Instructor in chemistry and photography, School of Military Engineering, Chatham, 1877–82. Intelligence service, War Office, 1885–90. Liberal Unionist MP, Lichfield division of Staffordshire, 1892–5. President, Royal Geographical Society of London, 1908–11; Eugenics Education Society, 1911–28. Chairman, Bedford College, London University, 1913–20. (M. Keynes 1943, Sarjeant 1980–96, *WWW*.)

Darwin, Marianne. *See* Parker, Marianne.

Darwin, Reginald (b. 1818). Son of Francis Sacheverel Darwin. CD's half first cousin. (Freeman 1978).

Darwin, Robert Waring (1766–1848). CD's father. Physician. Had a large practice in Shrewsbury and resided at The Mount. Son of Erasmus Darwin and his first wife, Mary Howard. Married Susannah, daughter of Josiah Wedgwood I, in 1796. FRS 1788. (Freeman 1978.)

Darwin, Susan Elizabeth (1803–66). CD's sister. Lived at The Mount, Shrewsbury, the family home, until her death. (*Darwin pedigree*, Freeman 1978.)

Darwin, Susannah (1765–1817). CD's mother. Daughter of Josiah Wedgwood I. Married Robert Waring Darwin in 1796. (*Darwin pedigree.*)

Darwin, William Erasmus (1839–1914). CD's eldest son. Banker. BA, Christ's College, Cambridge, 1862. Partner in the Southampton and Hampshire Bank, Southampton, 1861. Chairman of the Southampton Water Company. Amateur photographer. (*Alum. Cantab.*, F. Darwin 1914.)
[1857?], [13 January 1861], 30 November [1861], *[late February–May 1865]*

Daubeny, Charles Giles Bridle (1795–1867). Chemist and botanist. Professor of chemistry, Oxford University, 1822–55; professor of botany, 1834–67; and rural economy, 1840. An active supporter of the British Association for the Advancement of Science; vice-president, 1847, president, 1856. FRS 1822. (*Alum. Oxon.*, *DNB*, *Record of the Royal Society of London.*)
16 July [1860], 1 August [1860]

Davidson, Thomas William St Clair (1817–85). Artist and palaeontologist. Fellow of the Geological Society of London. Expert on fossil brachiopods. FRS 1857. (*DNB*, Sarjeant 1980–96.)

Davis, Jefferson (1808–89). American statesman. As senator, announced the secession of Mississippi from the Union in 1861; became president of the Confederacy in the same year. Remained president until the fall of the Confederacy and his own capture by unionists in 1865. Advocated an independent Southern American nation. (*DAB.*)

Davis, John Edward (1815–77). Naval officer and hydrographer. Second master on the *Terror* during the Antarctic expedition, 1839–43. Drew the charts for the expedition. Surveyor to the North Atlantic telegraph expedition, 1862. Naval assistant (curator of original documents) in the Hydrographic Department of the Admiralty, 1865–76. Captain, 1870. (*Modern English biography*, *Navy list* 1864–77.)

Davy, Humphry (1778–1829). Chemist. Professor of chemistry at the Royal Institution, 1802–13. President of the Royal Society, 1820–7. FRS 1803. (*DNB*, *DSB.*)

Davy, John (1790–1868). Physiologist and anatomist. Brother of Humphry Davy. FRS 1814. (*DNB.*)
3 January [1856]

'Dear Friend'.
1 January 1822, 2 January 1822, 3 January 1822, 4 January 1822 (two letters), 12 January 1822

Decaisne, Joseph (1807–82). French botanist. Gardener at the Jardin des Plantes, Paris, 1824. Professor of agricultural statistics, Collège de France, 1848. Professor of plant cultivation, Muséum d'Histoire Naturelle, 1850. Foreign member, Royal Society of London, 1877. (*DBF*, *NBU.*)

De Filippi, Filippo (1814–67). Italian zoologist, embryologist, and geologist. Professor of zoology and director of the Museum of Zoology, Turin, 1847. Travelled as naturalist with a diplomatic and scientific mission to Persia in 1862; with a scientific voyage of global circumnavigation, 1865–7. Advocated belief in a limited transmutation of species, 1855. His lecture 'Man and the monkeys', delivered in Turin in January 1864, initiated the public debate on Darwin's work in Italy. (Corsi 1983, *DBI*, Pancaldi 1991.)

De la Beche, Henry Thomas (1796–1855). Geologist. First director of the Geological Survey of Great Britain, 1835–55. Established the Museum of Practical Geology and the School of Mines. Knighted, 1842. FRS 1819. (*DNB*, *DSB*.)

Denny, Henry (1803–71). Zoologist and botanist. Specialised in entomology, studying parasitic insects. Curator and assistant secretary, museum of the Literary and Philosophical Society of Leeds, 1826–71. Secretary to the West-Riding Geological and Polytechnic Society. (*DNB*, Gilbert 1977, *Proceedings of the Linnean Society of London* (1870–1): lxxxiv–lxxxv.)

[27 July – 10 August 1844], 17 January [1865], *23 January 1865*, 28 January [1865], 23 March [1865]

Disraeli, Benjamin (1804–81). Author and Conservative politician. Prime minister, 1867, 1874–6. Created earl of Beaconsfield, 1876. (*DNB*.)

Dixon, Edmund Saul (1809–93). Clergyman and poultry-fancier. Rector of Intwood with Keswick, Norfolk, 1842–93. Author of books on the history and management of poultry. Also published under the pseudonym Eugene Sebastian Delamer. (*Modern English biography*.)

Douglas, Donald (*fl.* 1850s). Army officer. Captain in the Royal North Down Rifles from 8 February 1855. (*Army list* 1855; see also this volume, Supplement, letter from [John Innes], [after 8 February – August 1855].)

Douglas, Lynedock (*fl.* 1850s). Resident at High Street, Hythe, Kent. (*Post Office directory of the six home counties* 1855; see also this volume, Supplement, letter from [John Innes], [after 8 February – August 1855].)

Dove, Heinrich Wilhelm (1803–79). German physicist and meteorologist. Extraordinary professor of physics at the University of Königsberg, 1828; professor of physics at the University of Berlin, 1845. Editor of *Repertorium der Physik*, 1837–49. Director of the Prussian Institute of Meterology from its foundation in 1849. Proposed that warm and cold air currents affected weather conditions in temperate climates. (*DBE*, *DSB*, *NDB*.)

Downes, William James (b. 1805). Schoolfellow of CD's. Left Shrewsbury School and entered Worcester College, Oxford, in 1823. (*Alum. Oxon.*, *Shrewsbury School register*.)

Drysdale, Elizabeth (1781/2–1882). Daughter of John Pew of Hillowton, Kirkcudbrightshire. Married William Copland of Colliston, Dumfries; widowed, 1808. Married Sir William Drysdale (1781–1847), for many years treasurer of the city of Edinburgh. Mother of John James Drysdale (1816–92), a leading homeopathic doctor. Mother-in-law of the hydropathic specialist Edward Wickstead Lane.

(*Emma Darwin* (1904), 2: 184; *Modern English biography* s.v. Drysdale, John James.)
[22 or 29 October 1859]

Du Bois-Reymond, Emil Heinrich (1818–96). German physiologist. A founder of the Physikalische Gesellschaft in Berlin, 1845. Instructor in anatomy, Berlin Academy of Art, 1848–53. Elected member of the Prussian Academy of Sciences, 1851; permanent secretary, 1876. Appointed professor of physiology at the University of Berlin, 1858. Worked on animal electricity, and in 1850 invented the nerve galvanometer to measure the electrical impulses in nerve and muscle tissue. (*ADB*, *DSB*, *NDB*.)

Dugard, Thomas (1777–1840). Physician. Physician to the Shropshire Infirmary, 1811–40. Honorary member of the Geological Society of London, 1807. (*Gentleman's Magazine* n.s. 14 (1840): 556, *Proceedings of the Geological Society of London* 3 (1838–42): 523.)

Duncan, Peter Martin (1821–91). Physician, zoologist, and geologist. Physician, Essex and Colchester Hospital, 1848–59; consultant physician, county asylum and Oldham Club. Practised at Blackheath from 1860. Professor of geology, King's College, London, 1870. Secretary of the Geological Society of London, 1864–70; president, 1876–8. Specialist on living and fossil corals and Mesozoic echinoids. FRS 1868. (*DNB*, *Medical directory* 1849–76.)

Dunlop, A. A. Landowner in Bengal. (Letters from E. P. Wright, 24 March 1865 and 31 March 1865; E. P. Wright 1864, p. 451.)

Earl, George Windsor (*c.* 1805–65). Merchant, travel writer, and colonial official. Wrote on the geography and ethnography of south-east Asia and northern Australia. Promoted British commercial interests in Singapore and the Malayan Archipelago. Commissioner for the crown lands, Port Essington, Australia, *circa* 1838–49. Assistant resident for the Straits Settlements in Singapore, 1858, and Penang, 1859–65. (Earl 1971.)

Edmonds, James 'Barney' (d. 1871). Menagerist. Ran Wombwell's Windsor Castle Menagerie, later re-named Edmonds' Menagerie. (Turner 1995.)

Edwards, Mr.
[before end of 1839?]

Edwards, Ernest (1837–1903). Photographer. Ran a photographic studio in London, specialising in portraits, 1864–9. Moved to Boston, and later New York, where he opened commercial firms in landscape photography and colour printing. (W. S. Johnson 1990, *Post Office London directory*, Pritchard 1994.)

Egerton, Francis Henry, 8th earl of Bridgewater (1756–1829). Scholar and patron of learning. His literary works, many on his notable ancestors, were mostly printed for private circulation. Bequeathed 8000 guineas in his will to commission works illustrating the 'goodness of God as manifested in the Creation'; the money to be divided among eight persons. The resulting essays have become known as the Bridgewater treatises. FRS 1781. (*DNB*, Topham 1993.)

Egerton, Philip de Malpas Grey-, 10th baronet (1806–81). Palaeontologist. Of Oulton Park, Cheshire. Tory MP for South Cheshire, 1835–68; for West

Egerton, Philip Grey-, cont.
Cheshire, 1868–81. Specialised in fossil fish. FRS 1831. (*DNB*, Sarjeant 1980–96.)

Ehrenberg, Christian Gottfried (1795–1876). German zoologist, comparative anatomist, and microscopist. Wrote extensively on infusoria and the development of coral reefs. Professor of medicine at Berlin University, 1839. Foreign member, Royal Society of London, 1837. (*DSB*, *NDB*.)

Eights, James (1798–82). American naturalist. Physician, Albany, New York. Accompanied Captain Edmund Fanning's voyage of discovery to the South Sea Islands in 1829, and published a number of papers on his discoveries. (*American medical biography*.)

Elie de Beaumont, Jean-Baptiste-Armand-Louis-Léonce (Léonce) (1798–1874). French mining engineer and geologist. Participated in the drawing-up of a geological map of France, 1825–41. Professor of geology, Ecole des Mines, 1827; professor, Collège de France, 1832. Foreign member, Royal Society of London, 1835. (*DBF*, *DSB*, *Record of the Royal Society of London*.)

Eliot, George. *See* Evans, Marian (or Mary Anne).

Endlicher, Stephan Ladislaus (1804–49). German botanist. (*NDB*.)

Engelmann, Georg (George) (1809–84). German-born physician and botanist. MD, University of Würzburg, 1831. A founder of the St Louis Academy of Science, Missouri, 1856, and member of numerous scientific societies. Played a principal role in the establishment of the botanical garden at St Louis. Made fundamental contributions to the classification and taxonomy of many plant families, especially grapes, cacti, and yuccas. Discovered disease-resistant grape species and the role of the pronuba moth in pollination of yuccas. Provided thousands of plant specimens to collections in Berlin and St Petersburg. (*ANB*, *DAB*.)

Engelmann, Wilhelm. German publishing firm, founded 1833, based in Leipzig. Publisher of *Zeitschrift für wissenschaftliche Zoologie* from 1848. (*NDB*, *Zeitschrift für wissenschaftliche Zoologie*.)

Erichson, Wilhelm Ferdinand (1808–49). German entomologist. Extraordinary professor and assistant at the zoological museum, University of Berlin. Edited, after 1841, the *Archiv für Naturgeschichte*. Worked on the Staphylinidae, devising a new classificatory system that he published in 1839–40. (Gilbert 1977, R. F. Smith *et al.* 1973, p. 129.)

Eudes-Deslongchamps, Jacques-Armand (1794–1867). French comparative anatomist and palaeontologist. Specialist in marine invertebrates and reptiles. Professor of natural history, Faculty of Sciences, Caen, from 1825; of zoology, from 1838. Founding member, Linnean Society of Normandy and Caen Museum of Natural History. (*DBF*, Sarjeant 1980–96, *Geological Magazine* (1867): 140–1.)

Evans, Marian (or Mary Anne) (1819–80). Novelist. Published under the name George Eliot. (*DNB* s.v. Cross, Mary Ann.)

Ewald, Julius Wilhelm (1811–91). German geologist and palaeontologist. (*DBE*, Sarjeant 1980–96.)

Eyton, Thomas Campbell (1809–80). Shropshire naturalist. Friend and Cambridge contemporary of CD. Author of several works on natural history. On coming into possession of the family estate at Eyton, Shropshire, in 1855, he built a museum for which he formed a collection of skins and skeletons of European birds. (*DNB*.)
9 January [1865?]

Falconer, Hugh (1808–65). Palaeontologist and botanist. Superintendent of the botanic garden, Saharanpur, India, 1832–42. Superintended the arrangement of Indian fossils for the British Museum in 1844. Superintendent of the Calcutta botanic garden and professor of botany, Calcutta Medical College, 1848–55. Retired owing to ill health and returned to Britain in 1855; pursued palaeontological research while travelling in southern Europe. Vice-president of the Royal Society of London and foreign secretary of the Geological Society of London, 1865. FRS 1845. (*DNB*, *DSB*.)
[1845?–7 or 1857–64], [7 March 1857], 15 October [1858], *3 January 1865* (E. A. Darwin), *5 January [1865]* (E. A. Darwin), 6 January [1865]

Farrar, Frederick William (1831–1903). Anglican clergyman and headmaster. Master at Harrow school, 1855–70. Canon of Westminster, 1876–95. Dean of Canterbury, 1895–1903. Promoted scientific education. Noted for his school stories, writings on language, and biographies of Christian figures. FRS 1866. (*DNB*.)
11 October [1865], 2 November [1865], *6 November 1865*

Ferguson, George (*fl.* 1850s). Author of works on fancy poultry. (*CDEL*.)

Ferguson, William (1820–87). Civil servant, botanist, and entomologist. Served in the Ceylon civil service, 1839–87. Author of a number of works on the botany of Ceylon. Collected plants, especially algae. (R. Desmond 1994, *DNB*.)

Fergusson, James (1808–86). Scottish writer on architecture and India merchant. Established a successful indigo factory in India, retiring after about ten years. Travelled widely in India and published many works on its architecture. General manager of the Crystal Palace Company, 1856–8. Member of the Royal Commission inquiring into the defences of Britain, 1857. After 1869, inspector of public buildings and monuments at the Office of Public Works. (*DNB*.)

Fielding, Henry (1707–54). Novelist, journalist, and barrister. (*DNB*.)

Filippi, Filippo de. *See* De Filippi, Filippo.

Fitch, Walter Hood (1817–92). Botanical artist at the Royal Botanic Gardens, Kew, from 1841. Produced over 10,000 published drawings. Illustrated Joseph Dalton Hooker's *Rhododendrons of Sikkim-Himalaya* (1849–51) and *Illustrations of Himalayan plants* (1855). (R. Desmond 1994, *Proceedings of the Linnean Society of London* (1891–2): 68.)

FitzRoy, Emily-Unah (d. 1856). Eldest daughter of Robert FitzRoy and Mary Henrietta O'Brien. (*Burke's peerage* 1865, s.v. Grafton, duke of; *Correspondence* vol. 6, letter to Syms Covington, 22 February 1857.)

FitzRoy, Fanny. Second daughter of Robert FitzRoy and Mary Henrietta O'Brien. (*Burke's peerage* 1865, s.v. Grafton, duke of.)

FitzRoy, George (1800–83). Army officer. Of Grafton Regis, Northamptonshire. Commissioned in the First Life Guards, 1818; lieutenant, 1821; captain, 1825. Lieutenant-colonel, Royal Buckinghamshire Militia. Brother of Robert FitzRoy. (*Burke's peerage* 1999, s.v. Grafton, duke of; *Hart's army list* 1846.)

FitzRoy, Katherine. Third daughter of Robert FitzRoy and Mary Henrietta O'Brien. (*Burke's peerage* 1865, s.v. Grafton, duke of.)

FitzRoy, Laura Maria Elizabeth (1858–1943). Daughter of Robert FitzRoy and his second wife, Maria Isabella Smyth. (*Burke's peerage* 1999, s.v. Grafton, duke of.)

FitzRoy, Maria Isabella (d. 1889). Daughter of John Henry Smyth of Heath Hall, Wakefield, Yorkshire. Second wife of Robert FitzRoy, whom she married in 1854. (*Burke's peerage* 1999, s.v. Grafton, duke of, *County families* 1864, s.v. FitzRoy, Robert.)

FitzRoy, Mary Henrietta (d. 1852). Second daughter of Major-general Edward James O'Brien. Married Robert FitzRoy in 1836. (*Burke's peerage* 1999.)

FitzRoy, Robert (1805–65). Naval officer, hydrographer, and meteorologist. Commander of HMS *Beagle*, 1828–36. Tory MP for Durham, 1841–3. Governor of New Zealand, 1843–5. Superintendent of the dockyard at Woolwich, 1848–50. Chief of the meteorological department at the Board of Trade, 1854; chief of the Meteorological Office from 1855. Rear-admiral, 1857; vice-admiral, 1863. FRS 1851. (*DNB*, *DSB*.)

FitzRoy, Robert O'Brien (1839–96). Naval officer. Son of Robert FitzRoy. Entered the navy in 1853; captain, 1872. Served in the China war, 1857–8. (*Modern English biography*.)

Flower, William Henry (1831–99). Anatomist and zoologist. Curator of the Hunterian Museum, Royal College of Surgeons of England, 1861–84; Hunterian Professor of comparative anatomy, Royal College of Surgeons, 1870–84. Director of the Natural History Museum, London, 1884–98. President of the Zoological Society of London, 1879–99. President of the British Association for the Advancement of Science, 1889. Knighted, 1892. FRS 1864. (*DNB*.)

Forbes, David (1828–76). Geologist and philologist. Supervised mining and metallurgical works in Norway, *circa* 1847–57. Travelled in South America, 1857–60, in search of ores of nickel and cobalt for the firm of Evans & Askins, nickel-smelters of Birmingham. After more years of travelling, he settled in England and became foreign secretary to the Iron and Steel Institute. FRS 1856. Brother of Edward Forbes. (*DNB*.)

11 December [1860]

Forbes, Edward (1815–54). Zoologist, botanist, and palaeontologist. Naturalist on board HMS *Beacon*, 1841–2. Appointed professor of botany, King's College, London, and curator of the museum of the Geological Society of London, 1842. Palaeontologist with the Geological Survey of Great Britain, 1844–54. Professor of natural history, Edinburgh University, 1854. FRS 1845. (*DNB*, *DSB*.)

Forchhammer, Johan Georg (1794–1865). Danish geologist and chemist. Became a lecturer in geology at the University of Copenhagen in 1821; professor of mineralogy and geology, 1831. Professor of chemistry and mineralogy, Polytechnic Institute, and manager of one of its two chemical laboratories. Secretary, Royal Danish Academy of Sciences, 1851–65. (*DBL*, *DSB*.)

Fox, Charles Woodd (1847–1908). Barrister. Son of William Darwin Fox. (*Alum. Oxon.*, *Repton School register*.)

Fox, Ellen Sophia (1820–87). Daughter of Basil George Woodd of Hillfield, Hampstead. Married William Darwin Fox in 1846. (*Darwin pedigree*.)

Fox, Gilbert Basil (b. 1864). Son of Ellen Sophia and William Darwin Fox. (*Darwin pedigree*.)

Fox, Harriet (1799–1842). Daughter of Sir Richard Fletcher. Married William Darwin Fox in 1834. (*Darwin pedigree*.)

Fox, Samuel William Darwin (b. 1841). Clergyman. Eldest son of Harriet and William Darwin Fox. Curate of St Paul's, Manningham, Yorkshire, 1865; of St Augustine's, Halifax, Yorkshire, 1867. Rector of St Peter Lymm, Oughtrington, Cheshire, 1874. Vicar of St Paul's, Maidstone, Kent, 1887. (*Alum. Oxon.*, *Crockford's clerical directory*, *Registers of Wadham College, Oxford*.)

Fox, Sylvanus Bevan (1825–1912). Dentist, beekeeper, and writer. Licentiate of dental surgery, Royal College of Surgeons of England, 1863. Dentist in Exeter, 1865 (or earlier) – 1908. Dental surgeon, Exeter Dental Hospital, 1880 (or earlier); consultant dental surgeon, 1889–1908. A contributor to the beekeeping section of the *Cottage Gardener*, 1860–1; of the *Journal of Horticulture*, 1861–73. Member and sometime councillor of the Devonish Association for the Advancement of Science and Art. Member of the Society of Friends. (Blaschko 1958, *Medical directory*, 1865–1911.)

Fox, William Darwin (1805–80). Clergyman. CD's second cousin. A friend of CD's at Cambridge; shared his enthusiasm for entomology. Maintained an active interest in natural history throughout his life and provided CD with much information. Rector of Delamere, Cheshire, 1838–73. Spent the last years of his life at Sandown, Isle of Wight. (*Alum. Cantab.*, *Autobiography*, *Correspondence*.)

6 January [1865], *[before 26 October 1865]*, 25–6 October [1865]

Frankland, Edward (1825–99). Chemist. Professor of chemistry, Putney College for Civil Engineering, 1850, and Owens College, Manchester, 1851–7. Lecturer in chemistry, St Bartholomew's Hospital, London, 1857–64. Professor of chemistry, Royal Institution, 1863–8, and Royal College of Chemistry, 1865. President of the Chemical Society, 1871–3; of the Institute of Chemistry, 1877–80. Knighted, 1897. FRS 1853. (*DNB*, *DSB*.)

Franklin, John (1786–1847). Naval officer and Arctic explorer. Lieutenant-governor of Van Diemen's Land (Tasmania), 1837–43. Leader of the 1845 expedition in search of a north-west passage during which all hands perished. Knighted, 1829. FRS 1823. (*DNB*.)

Frederick William, Emperor Frederick III (1831–88). Married Victoria, princess royal of Great Britain, in 1858. Became crown prince of Prussia in 1861, and crown prince of the German Empire in 1871. On the death of his father,

Frederick William, Emperor Frederick III, cont.
William I, in 1888, became Emperor Frederick III, three months before his death. (*Chambers biographical dictionary.*)

Gainsborough, earl of. *See* Noel, Charles George.

Galton, Douglas Strutt (1822–99). Military engineer and civil servant. Officer in the Royal Engineers. Joined the Ordnance Survey in 1846. Secretary, Railway Commission, 1847; Railway Department, Board of Trade, 1854. Assistant permanent under-secretary of state for war, 1862. Director of public works and buildings, Office of Works, 1869–75. Author of numerous government reports on sanitation, telegraphy, and railways. Associate, Institution of Civil Engineers, 1850. Member, British Association for the Advancement of Science, 1860; general secretary, 1871–95; president, 1895. Knighted, 1887. FRS 1859. (*DNB.*)

Galton, Francis (1822–1911). Traveller, statistician, and scientific writer. CD's cousin. Explored in south-western Africa, 1850–2. Carried out various researches on heredity. Founder of the eugenics movement. FRS 1860. (*DNB*, *DSB.*)

Garrett, Elizabeth. *See* Anderson, Elizabeth Garrett.

Garrod, Alfred Baring (1819–1907). Physician. Assistant physician to University College Hospital, 1847; physician and professor of therapeutics and clinical medicine, 1851–63. Gulstonian Lecturer, Royal College of Physicians, 1857; lecturer on materia medica, 1864; Lumleian Lecturer, 1883, and vice-president, 1888. Physician to King's College Hospital and professor of materia medica and therapeutics, King's College, London, 1863–74. Physician extraordinary to Queen Victoria, 1880. A specialist in the treatment of gout. Knighted 1887. FRS 1858. (*DNB.*)

Gärtner, Karl Friedrich von (1772–1850). German physician and botanist. Practised medicine in Calw, Germany, from 1796, but left medical practice in 1800 to pursue a career in botany. Travelled in England and Holland in 1802. Studied plant hybridisation from *circa* 1824. Elected a member of the Deutsche Akademie der Naturforscher Leopoldina, 1826. Ennobled, 1846. (*ADB*, *DBE*, *DSB.*)

Gaudin, Charles-Théophile (1822–66). Swiss palaeontologist. Tutor to the family of Lord Ashley in England from 1845 to 1851, when he returned to Switzerland and began his palaeontological studies. Collaborated in the publication of *Mémoires sur les animaux vertébrés trouvés dans le terrain sidérolithique du canton de Vaud* (1852). Tutor to the daughter of Mme de Rumine from 1854, spending many winters in Italy, where he contined his research. Member of the Sociétés Helvetique et Vaudoise de Sciences Naturelles. (*Dictionnaire biographique des Genevois et de Vaudois.*)

Gegenbaur, Carl (or Karl) (1826–1903). German anatomist and zoologist. A supporter of CD; emphasised the importance of comparative anatomy in evolutionary reconstruction. Professor extraordinarius of zoology, Jena, 1855–8; professor of anatomy and zoology, 1858–62; of zoology, 1862–73. Professor of anatomy and comparative anatomy, Heidelberg, 1873–1901. Elected to the Deutsche Akademie der Naturforscher Leopoldina, 1857. (*DBE*, *DSB*, *NDB.*)

Geikie, Archibald (1835–1924). Scottish geologist. Appointed member of the Scottish branch of the Geological Survey in 1855. Director of the Geological Survey of Scotland, 1867–82. Director-general of the Geological Survey of Great Britain, 1882–1901. Murchison Professor of geology and mineralogy, Edinburgh University, 1871–81. Knighted, 1891. FRS 1865. (*DNB*, *DSB*.)
27 February [1861]

Geoffroy Saint-Hilaire, Isidore (1805–61). French zoologist. Succeeded his father, Etienne Geoffroy Saint-Hilaire, as professor at the Muséum d'Histoire Naturelle in 1841. Continued his father's work in teratology. Became professor of zoology at the Sorbonne in 1850. (*DBF*, *DSB*.)

Gerstaecker, Carl Edouard Adolph (Adolph) (1828–95). German zoologist. Studied medicine at the University of Berlin and passed the state medical examination in 1852 but never practised. Habilitated in zoology, University of Berlin, 1856; became keeper of the entomological collection at the Berlin zoological museum, 1856. Docent in zoology at the agricultural institute, Berlin, 1864; extraordinary professor, 1874. Professor of zoology, University of Greifswald, 1876. Co-editor of the *Handbuch der Zoologie* (1863), for which he wrote the chapter on arthropods. (*ADB*, *NDB*.)

Gill, Theodore Nicholas (1837–1914). American zoologist. Taught at the Columbian College (now George Washington University) from 1860; professor of zoology, 1884–1910. Librarian, Smithsonian Institution, 1862–6. Assistant librarian, Library of Congress, 1866–74. President of the American Association for the Advancement of Science, 1897. (*DSB*.)

Girard, Charles Frédéric (1822–95). French-born zoologist and physician. Student and assistant of Louis Agassiz at the College de Neuchâtel, Switzerland, and from 1847 at Harvard University. Assistant to Spencer Fullerton Baird at the Smithsonian Institution, 1850–60. MD, Georgetown College, 1856. In France, 1860–3, returning with medical supplies for the Confederate Army. After the war settled in Paris, where he established a medical practice. (*ANB*.)

Gladstone, William Ewart (1809–98). Statesman and author. Chancellor of the Exchequer, 1852–5 and 1859–66. Prime minister, 1868–74. FRS 1881. (*DNB*.)

Glaisher, James (1809–1903). Astronomer and meteorologist. Assistant, Cambridge University observatory, 1833–5. In charge of the magnetic and meteorological department at the Royal Observatory, Greenwich, 1838–74. Secretary, Royal Meteorological Society, 1850–72. FRS 1849. (*DNB*.)

Goethe, Johann Wolfgang von (1749–1832). German poet and naturalist. (*DSB*, *NDB*.)

Goodsir, John (1814–67). Scottish surgeon and anatomist. Surgeon in Anstruther, Fifeshire, 1835–40. Conservator, museum of the Royal College of Surgeons, Edinburgh, 1841–3. Conservator, human and comparative anatomy museum, Edinburgh University, 1840; curator, anatomy and pathology museum, 1843; demonstrator in anatomy, 1844; professor of anatomy, 1846. FRS 1846. (*DNB*, *DSB*.)

Gordon, George (1801–93). Scottish botanist, geologist, and clergyman. Minister of Birnie, near Elgin, Morayshire, 1832–89. Specialised in the stratigraphy and fossil vertebrates of Elgin sandstone. Founded the Elgin and Morayshire Literary and Scientific Association. (R. Desmond 1994, Sarjeant 1980–96.)
1 July 1861 (J. H. Balfour)

Gosse, Philip Henry (1810–88). Zoologist, traveller, and writer. Collected zoological specimens in the West Indies for the British Museum, 1844–6. Made his living as a writer from 1847. Lived at St Marychurch, near Torquay, Devonshire, from 1852. Cultivated orchids. Studied marine invertebrates. FRS 1856. (*DNB*.)

Gould, John (1804–81). Ornithologist and artist. Taxidermist to the Zoological Society of London from 1827. Described the birds collected on the *Beagle* and *Sulphur* expeditions. FRS 1843. (*DNB*.)

Grant, Ulysses Simpson (1822–85). United States Army soldier; president of the United States. United States Military Academy, West Point, New York, 1839–43. Served in Fourth Infantry; captain, 1853. Resigned and held a variety of civilian posts, 1854–60. In 1861 joined the Union forces as colonel of Twenty-first Illinois Volunteers; major-general of Volunteers, 1862. In 1863 promoted to major-general in the regular army, then lieutenant-general with command of the armies of the United States. Played a decisive role in the ending of the American Civil War in 1865; created general in 1866. Nominated Republican Party candidate in 1868. President of United States, 1868–72 and 1872–7. (*DAB*.)

Granville, George Leveson Gower, 2d Earl Granville (1815–91). Statesman and diplomat. Under-secretary for foreign affairs, 1840–1. MP for Morpeth, 1837–40; for Lichfield, 1841–6; deputy lieutenant for Shropshire, 1846; vice-president, Board of Trade, 1848–51. Vice-president of the Royal Commission for the Great Exhibition. FRS 1853. (*Burke's peerage* 1999, *Men and women of the time* 1856.)

Gray, Asa (1810–88). American botanist. Fisher Professor of natural history, Harvard University, 1842–88. Wrote numerous botanical textbooks and works on North American flora. President of the American Academy of Arts and Sciences, 1863–73; of the American Association for the Advancement of Science, 1872. A regent of the Smithsonian Institution, 1874–88. Foreign member, Royal Society of London, 1873. (*DAB*, *DSB*.)
17 January 1865, 19 April [1865], *15 and 17 May 1865*, *24 July 1865*, 15 August [1865], 19 October [1865], *6 November 1865*

Gray, George Robert (1808–72). Zoologist. An expert on insects and birds. Assistant in the zoological department of the British Museum, 1831–72. Brother of John Edward Gray. FRS 1865. (*DNB*, Gilbert 1977.)

Gray, Jane Loring (1821–1909). Daughter of Charles Greely Loring, a Boston lawyer and politician, and Anna Pierce Brace. Married Asa Gray in 1848. Edited the *Letters of Asa Gray* (1893). (Barnhart comp. 1965; Dupree 1959, pp. 177–84.)

Gray, John Edward (1800–75). Botanist and zoologist. Assistant keeper of the zoological collections at the British Museum, 1824; keeper, 1840–74. President,

Botanical Society of London, 1836–56. FRS 1832. (R. Desmond 1994, *DNB*.)

27 January [1865]

Gray, Robert (1809–72). Clergyman. Bishop of Cape Town from 1847, and metropolitan of South Africa from 1853. (*DNB*.)

Greene, Joseph Reay (*fl.* 1857–77). Irish naturalist. Professor of natural history at Queen's College, Cork, 1858–77. One of the editors of the *Natural History Review*. Studied Hydrozoa. (L. Huxley ed. 1900, Praeger 1949.)

Greg, William Rathbone (1809–81). Essayist. Mill-owner, 1832–42. Author of the *Creed of Christendom* (1851). Wrote articles for the leading quarterlies, and books, mostly on politics and economics. Comptroller of the Stationery Office, 1864–77. (*DNB*, Tort 1996.)

21 March [1860?]

Grisebach, August Heinrich Rudolf (1814–79). German botanist. Travelled in the Balkan Peninsula and north-western Asia Minor, 1839–40, studying the flora of these regions. Professor of botany, Göttingen University, 1847. (*DSB*, *NDB*.)

Grote, George (1794–1871). Historian. One of the founders of University College, London; vice-chancellor of the college, 1862. FRS 1857. (*DNB*.)

Grove, William Robert (1811–96). Lawyer, judge, and natural philosopher. Professor of experimental philosophy, London Institution, 1847. An active member of the Royal Society of London; treasurer and chairman of the executive committee of the Philosophical Club, 1847; Royal Medallist, 1847. Member of the Royal Commission on the Law of Patents, 1864. Appointed to the bench, Court of Common Pleas, 1871. FRS 1840. (*DNB*, *DSB*.)

Guenée, Achille (1809–80). French lawyer. Resided in Châteaudun (Eure-et-Loir). Noted for his work on Lepidoptera. (Gilbert 1977.)

Guépin, Jean-Baptiste Pierre (1779–1858). French doctor and botanist. (Baillon *et al.* 1886, *Catalogue général de la librairie française* 1868.)

14 November 1861

Guizot, François Pierre Guillaume (François) (1787–1874). French historian and statesman. Professor of modern history at the Sorbonne, 1812–30. Suspended from teaching, 1822–8, because of his political writing. After 1814, held a number of offices in the interior ministry; elected to the chamber of deputies to represent Lisieux in Normandy, 1830, and made minister for the interior. Education minister, 1832–6; foreign minister from 1840 until the revolution of 1848. (*DBF*, *EB*.)

Gulliver, George (1804–82). Anatomist and physiologist. Surgeon to the Royal Horse Guards. Member of the council of the Royal College of Surgeons, 1852. Hunterian Professor of comparative anatomy and physiology, 1861. FRS 1839. (*DNB*.)

27 December [1855]

Gully, James Manby (1808–83). Physician. Practised medicine in London, 1830–42. Set up a hydropathic establishment in Great Malvern, Worcestershire, in 1842; a successful practitioner of hydropathy until his retirement in 1872. (*DNB*, *Modern English biography*.)

Gurney, John Henry (1819–90). Norfolk financier and ornithologist. MP, King's Lynn, Norfolk, 1854–65. (*Modern English biography*, Stenton 1976.)

Haast, John Francis Julius (Julius) von (1822–87). German-born explorer and geologist. Travelled to New Zealand in 1858 to report on the prospects for German emigration. Explored the western districts of Nelson province at the request of the provincial government in 1859. Appointed provincial geologist, 1861. Conducted the first geological survey of Canterbury province, 1861–8. Became a British national in 1861. Founded the Philosophical Institute of Canterbury in 1862, and the Canterbury Museum in 1863. Professor of geology, Canterbury College, 1876–87. Member of the senate of the University of New Zealand, 1879–87. Knighted, 1886. FRS 1867. (*DNB*, *DSB*, *DNZB*, H. F. von Haast 1948.)
27 September 1865, 26 December [1865]

Haeckel, Ernst Philipp August (Ernst) (1834–1919). German zoologist. MD, Berlin, 1857. Lecturer in comparative anatomy, University of Jena, 1861–2; professor extraordinarius of zoology, 1862–5; professor of zoology and director of the Zoological Institute, 1865–1909. Specialist in marine invertebrates. Leading populariser of evolutionary theory. His *Generelle Morphologie der Organismen* (1866) linked morphology to the study of the phylogenetic evolution of organisms. (*DSB*, *NDB*.)
11 November 1865, 6 December [1865]

Haldeman, Samuel Steman (1812–80). American zoologist and philologist. Assistant to Henry Darwin Rodgers on the geological survey of New Jersey, 1836–42. Professor of natural history, University of Pennsylvania, 1851–5; professor of natural science, Delaware College, 1853–8; professor of philology, University of Pennsylvania, 1868–80. (*DAB*.)

Haliday, Alexander Henry (1806–70). Irish naturalist. An expert on Hymenoptera. An editor of the *Natural History Review*, 1854–60. (Gilbert 1977, *Modern English biography*.)

Hallam, Henry (1777–1859). Lawyer and historian. Barrister; commissioner of stamps. Published works on constitutional and literary history. Treasurer, Statistical Society; vice-president, Society of Antiquaries. (*DNB*.)

Hamilton, William John (1805–67). Geologist. MP for Newport, Isle of Wight, 1841–7. Secretary of the Geological Society of London, 1832–54; president, 1854 and 1865. (*DNB*, Sarjeant 1980–96.)

Hamilton, William Richard (1777–1859). Antiquary and diplomat. Secretary to Lord Elgin, 1799–1809; supervised the transfer of Grecian antiquities to the British Museum in 1802. Under-secretary for foreign affairs, 1809–22. Published a translation of the Rosetta Stone inscriptions. One of the founders of the Royal Geographical Society, 1833. (*DNB*.)

Hardy, Charles (1803–85). Clergyman. Curate of the Subdeanery Church, Chichester, 1827–32. Vicar of North and South Hayling, Hampshire, 1832–80. (*Alum. Cantab.*, *Crockford's clerical directory*.)

Harley, Edward, 2d earl of Oxford (1689–1741). Antiquary. Son of Robert

Harley, 1st earl of Oxford. Added to his father's collection of books and manuscripts. After his death the manuscripts were sold to the British Museum. (*DNB*.)

Harley, Robert, 1st earl of Oxford (1661–1724). Statesman and antiquary. Leader of the Tory party. Lord Treasurer, 1711. Created Baron Harley, earl of Oxford and Mortimer, 1711. Purchased a number of important manuscript collections. Knighted, 1712. (*DNB*.)

Harris, Thaddeus William (1795–1856). American entomologist and librarian. Librarian of Harvard College, 1831–56. Gave instruction in natural history, 1837–42. Contributed articles on entomology and horticulture to scientific journals. Author of section 8, 'Insects', of the *Catalogue of the animals and plants of Massachusetts* (1833), also published as part 4 of Edward Hitchcock's *Report on the geology, mineralogy, botany, and zoology of Massachusetts* (1833), and of the *Report on the insects of Massachusetts injurious to vegetation* (1841). (*DAB*.)

Harvey, William (1578–1657). Physician. Royal physician to James I and Charles I. Fellow and benefactor of the Royal College of Physicians. Best known for his discovery of the circulation of the blood. (*DSB*.)

Harvey, William Henry (1811–66). Irish botanist. Colonial treasurer in Cape Town; collected plants in South Africa, 1836–42. Keeper of the herbarium, Trinity College, Dublin, from 1844; professor of botany, Royal Dublin Society, 1848–66; professor of botany, Trinity College, Dublin, 1856–66. Published works on South African plants, including *Flora Capensis* (1859–65) with Otto Wilhelm Sonder. Specialist in marine algae. FRS 1858. (R. Desmond 1994, *DNB*, *DSB*.)

Hawthorn, Miss. Friend of Frances Harriet Hooker (*Correspondence* vol. 11, letter from J. D. Hooker, 1 October 1863). Possibly Grace Hawthorn (b. 1824/5), a family friend (see *Correspondence* vol. 8, letter from J. S. Henslow to J. D. Hooker, 10 May 1860, and letter from Leonard Jenyns, 9 May 1868 (*Calendar* no. 6168)), and sister of Sarah Hawthorn (Census returns 1851 (Public Record Office HO107/1759/628a)), who married Leonard Jenyns in 1862 (*DNB* s.v. Blomefield, Leonard).

Hawthorn, Robert (d. 1874?). Clergyman. BA Cambridge, 1830; MA 1833. Vicar of Stapleford, 1845. (*Alum. Cantab.*, *Crockford's clerical directory*.)

Hearne, Samuel (1745–92). Explorer and colonial administrator in Canada. (*DNB*, *MDCB*.)

Hector, James (1834–1907). Scottish geologist. Surgeon and geologist on the government expedition to the western parts of British North America, 1857–60. Geologist to the provincial government of Otago, New Zealand, 1861–5. Director of the Geological Survey of New Zealand, 1865. Director of the meteorological department of the New Zealand Institute, of the Colonial Museum, and of the botanical garden, Wellington, 1866–1903. Knighted, 1887. FRS 1866. (*DNZB*.)

Heer, Oswald (1809–83). Swiss biogeographer, palaeontologist, and botanist. An expert on Tertiary flora. Lecturer in botany, University of Zürich, 1834–5; director of the botanic garden, 1834; associate professor, 1835–52; professor of botany and entomology, 1852–83. (*DSB*, *NDB*.)

Heine, Rudolph (b. 1799). German physician in Bitterfeld, near Leipzig. (Information received from Stadtarchiv, Bitterfeld.)
10 March 1865, 15 March [1865]

Hemmings, Henry. A servant of Sarah Elizabeth (Sarah) Wedgwood. (Freeman 1978.)

Hemsley, William Botting (1843–1924). Botanist. Gardener at the Royal Botanic Gardens, Kew, 1860; assistant in the herbarium, 1865–7, 1883–98; keeper, 1899–1908. FRS 1889. (R. Desmond 1994).

Henslow, Frances Harriet. *See* Hooker, Frances Harriet.

Henslow, George (1835–1925). Clergyman, teacher and botanist. BA, Christ's College, Cambridge, 1858. Headmaster at Hampton Lucy Grammar School, Warwick, 1861–4; at the Grammar School, Store Street, London, 1865–72. Lecturer in botany at St Bartholomew's Hospital, 1866–80. Younger son of John Stevens Henslow. (*Alum. Cantab.*, R. Desmond 1994.)
1 November 1865, [2–5 November 1865], *6 November 1865*, *2 December 1865*

Henslow, John Stevens (1796–1861). Clergyman, botanist, and mineralogist. CD's teacher and friend. Professor of mineralogy, Cambridge University, 1822–7; professor of botany, 1825–61. Extended and remodelled the Cambridge botanic garden. Curate of Little St Mary's Church, Cambridge, 1824–32; vicar of Cholsey-cum-Moulsford, Berkshire, 1832–7; rector of Hitcham, Suffolk, 1837–61. (*DNB*, *DSB*, *Historical register of the University of Cambridge*.)
3 November 1838, [1852–60]

Herbert, John Maurice (1808–82). Lawyer. BA, St John's College, Cambridge, 1830; fellow, 1832–40. Barrister, 1835. County court judge, South Wales, 1847–82. (*Alum. Cantab.*, *Modern English biography*.)
[January 1839 – September 1842], 18 November [1856]

Herschel, John Frederick William, 1st baronet (1792–1871). Astronomer, mathematician, chemist, and philosopher. Member of many learned societies. Carried out astronomical observations at the Cape of Good Hope, 1834–8. Master of the Royal Mint, 1850–5. Created baronet, 1838. FRS 1813. (*DNB*, *DSB*.)

Hewitt, Edward. Judge of poultry exhibitions. Resided in Sparkbrook, Birmingham. (Wingfield and Johnson 1856–7, p. 89.)

Hewson, William (1739–74). Surgeon and physiologist. FRS 1770. (*DNB*, *DSB*.)

Hildebrand, Friedrich Hermann Gustav (Friedrich) (1835–1915). German botanist. After studying mineralogy, geology, and agriculture at Berlin, he took up botany, studying at Bonn, then from 1855 to 1858 at Berlin, where he received his doctorate. Habilitated at Bonn, becoming privat dozent there, in 1859. Professor of botany, Freiburg im Breisgau, 1868–1907. Worked mainly on hybridity, dimorphism, and generation. (Correns 1916, Junker 1989, Tort 1996.)
17 July [1862]

Hill, Rowland (1795–1879). Inventor of the penny post. Established a school on his own plan at Hazelwood; it later moved to Bruce Castle, Tottenham. Invented the adhesive postage stamp, 1837, and established the penny post in 1840. Dismissed from the Post Office, 1842. Chairman of the Brighton Railway, 1843–6. Secretary to the postmaster-general, 1846; to the Post Office, 1854–64. Knighted, 1860. FRS 1857. (*DNB*.)

Hirst, Thomas Archer (1830–92). Mathematician. Lecturer in mathematics and natural philosophy, Queenwood College, Hampshire, 1853–6. Appointed mathematical master of University College School, London. Professor of physics, University College, London, 1865; of mathematics, 1866–70. Assistant registrar of the University of London, 1870–83. General secretary of the British Association for the Advancement of Science, 1866–70. Director of naval studies, Royal Naval College, Greenwich, 1873–83. From 1861 his research specialised in pure geometry. Took a prominent part in the founding of the London Mathematical Society in 1865; president, 1872–4. FRS 1861. (*DNB*.)

Hodgson, Brian Houghton (1800–94). Orientalist and ethnologist. In the service of the East India Company from 1816; assistant resident in Nepal, 1820; resident, 1833–43. Acquired an important collection of Sanskrit and Tibetan manuscripts. Wrote extensively on the geography, ethnography, and natural history of India and the Himalayas. Lived in Darjeeling, 1845–58; thereafter settled in England. FRS 1877. (*DNB*.)

Hoff, Karl Ernst Adolf von (1771–1837). German geologist, geographer, and civil servant. Entered the civil service in Gotha in 1791; director of the Ober-Consistorium, 1829. In charge of the astronomical observatory from 1826; became the supervisor of scientific collections in Gotha in 1832. Devoted his free time to geology and mineralogy. Published *Geschichte der durch Überlieferung nachgewiesen natürlich Veränderungen der Erdoberfläche* (1822), insisting, in opposition to the prevailing catastrophist views, that researchers should study forces in operation on the earth at the present time, and apply this knowledge to the earliest history of the earth, a principle later known as *Aktualismus*. (*ADB*, *DSB*, *NDB*.)

Hofmeister, Wilhelm Friedrich Benedikt (Wilhelm) (1824–77). German botanist. Joined the family publishing, music, and bookshop business in 1841. Honorary doctor of philosophy, University of Rostock (for his work on the embryology of flowering plants), 1851. Professor of botany and director of the botanic garden, University of Heidelberg, 1863–72; professor of botany, University of Tübingen, 1872–6. (*ADB*, *DSB*, Goebel 1926, *NDB*.)

Holland, Francis James (1828–1907). Clergyman. Son of Henry Holland. Vicar of St Dunstan's, Canterbury, 1853–61; minister of Quebec Chapel, London, 1861–82. Chaplain-in-ordinary to Queen Victoria, 1873. Canon of Canterbury from 1882. (*Alum. Cantab.*, *WWW*.)

Holland, Henry, 1st baronet (1788–1873). Physician. Cousin of the Darwins and Wedgwoods. Physician in ordinary to Prince Albert, 1840; to Queen Victoria, 1852. President of the Royal Institution of Great Britain, 1865–73. Created baronet, 1853. FRS 1815. (Caroe 1985, *DNB*, *Emma Darwin* (1904), *Physicians*, *Record of the Royal Society of London*.)
2 January 1865, *25 June [1865]*

Home, David Milne (1805–90). Scottish advocate and geologist. Born David Milne, he took the name Home in 1845. Studied earthquakes and the parallel roads of Glen Roy. Founder of the Scottish Meteorological Society. (*DNB* s.v. Milne, Sir David (1763–1845); *Modern English biography*; Sarjeant 1980–96.)
28 March 1840

Hooker, Brian Harvey Hodgson (1860–1932). Fifth child of Frances Harriet and Joseph Dalton Hooker. (Allan 1967 s.v. 'Hooker pedigree'.)

Hooker, Charles Paget (1855–1933). Physician and surgeon. Third child of Frances Harriet and Joseph Dalton Hooker. Trained at St Bartholomew's Hospital, London; made a licentiate of the Royal College of Physicians of London and the Royal College of Surgeons, Edinburgh, 1879, before being appointed to the staff of the Hertfordshire General Infirmary. Cottishall Cottage Hospital, Norfolk, 1880–5; Cirencester Cottage Hospital, Gloucestershire, 1885–1912. (Allan 1967, *Medical directory* 1881–1933, *Medical who's who* 1914.)

Hooker, Frances Harriet (1825–74). Daughter of John Stevens Henslow. Married Joseph Dalton Hooker in 1851. (Allan 1967; *DNB* s.v. Hooker, Joseph Dalton.)
[27 January 1865], [17 August 1865], [21 or 28 August 1865], 6 September [1865], 13 September [1865], 22 September [1865]

Hooker, Harriet Anne (1854–1945). Second child of Frances Harriet and Joseph Dalton Hooker. (Allan 1967 s.v. 'Hooker pedigree'.)

Hooker, Joseph Dalton (1817–1911). Botanist. Worked chiefly on taxonomy and plant geography. Son of William Jackson Hooker. Friend and confidant of CD. Accompanied James Clark Ross on his Antarctic expedition, 1839–43, and published the botanical results of the voyage. Appointed palaeobotanist to the Geological Survey of Great Britain, 1846. Travelled in the Himalayas, 1847–50. Assistant director, Royal Botanic Gardens, Kew, 1855–65; director, 1865–85. Knighted, 1877. FRS 1847. (*DNB*, *DSB*.)
[9 or 16 February 1854], [before 6 October 1858], [14 November 1858], *[11 May – 3 December 1860]*, 19 [June 1860], 24 [November 1862], *1 January 1865*, 7 January [1865], *[8–18 January 1865]*, 19 January [1865], *[20 January 1865]*, *[26 January 1865]*, 2 February [1865], *3 February 1865*, 9 February [1865], 15 [February 1865], *[17 February 1865]*, *[27 February 1865?]*, *[10 March 1865]*, 16 [March 1865], 6 April [1865], *[7–8 April 1865]*, 10 [April 1865], *12 April [1865]*, 13 April [1865], 17 April [1865], *[19 April 1865]*, [1 May 1865], *2 May 1865*, 4 May [1865], *[26 May 1865]*, 1 June [1865], *[2 June 1865]*, [4 June 1865], *[15 June 1865]*, [17 June 1865], *[after 17 June 1865]*, [10 July 1865], *13 July 1865*, [29 July 1865], 16 August [1865], *[26 September 1865]*, 27 [or 28 September 1865], 3 October [1865], *6 October 1865*, 22 and 28 [October 1865], *[3 November 1865]*, 22 December [1865], *[23] December 1865*, *24 December 1865*, [31 December 1865]

Hooker, Maria (1797–1872). Eldest daughter of Dawson Turner. Married William Jackson Hooker in 1815; acted as her husband's secretary. Mother of Joseph Dalton Hooker. (Allan 1967 s.v. 'Turner pedigree', R. Desmond 1994.)

Hooker, Maria Elizabeth (Minnie) (1857–63). Fourth child of Frances Harriet and Joseph Dalton Hooker. (Allan 1967 s.v. 'Hooker pedigree'.)

Hooker, William Henslow (1853–1942). Eldest child of Frances Harriet and Joseph Dalton Hooker. Civil servant, India Office, 1877–1904. Encouraged imperial ties between metropolitan institutions (particularly the Royal Botanic Gardens, Kew) and British East Africa, *circa* 1896–1906. (Allan 1967; *India list* 1904–5; *Zanzibar Gazette*, 5 February 1896, p. 6, and 28 November 1900, p. 5.)

Hooker, William Jackson (1785–1865). Botanist. Father of Joseph Dalton Hooker. Regius professor of botany, Glasgow University, 1820. Appointed first director of the Royal Botanic Gardens, Kew, 1841. Knighted, 1836. FRS 1812. (*DNB*, *DSB*.)

Hope, Frederick William (1797–1862). Entomologist and clergyman. Curate of Frodesley, Shropshire, 1823. President of the Entomological Society, 1835 and 1846. Founded a professorship of zoology at Oxford University in 1849, and in the same year donated his entomological and other collections. FRS 1834. (*DNB*, Gilbert 1977.)

Hopkins, William (1793–1866). Mathematician and geologist. Tutor in mathematics at Cambridge University. President of the Geological Society of London, 1851–3. Specialised in quantitative studies of geological and geophysical questions. FRS 1837. (*DNB*, *DSB*, *Record of the Royal Society of London*.)

Hordern, Ellen Frances. *See* Lubbock, Ellen Frances.

Hore, William Strong (1807–82). Clergyman and botanist. Curate of Stoke Damerel, Devon, 1841–7; of All Saints', Norwich, 1848–50. Rector of St Clement's, Oxford, 1850–5. Vicar of Shebbear with Sheepwash, Devon, 1855–82. (*Alum. Cantab.*, R. Desmond 1994.)

Horner, Leonard (1785–1864). Scottish geologist and educationalist. Founded the Edinburgh School of Arts in 1821. Warden of University College, London, 1828–31. Inspector of factories, 1833–56. A promoter of science-based education at all social levels. President of the Geological Society of London, 1846 and 1860–2. Father-in-law of Charles Lyell. FRS 1813. (*DNB*, *DSB*.)

Horsfield, Thomas (1773–1859). American naturalist. Served in the East Indies under the Dutch and British, 1799–1819. Keeper of the East India Company Museum, Leadenhall Street, London, 1820–59. (*DNB*.)

Houghton, William (1828–95). Clergyman and naturalist. BA Oxford, 1850; MA 1853. Wrote numerous articles on natural history for popular scientific and literary journals. (*Crockford's clerical directory*, *Wellesley index*.)

Howard, Charles (1707–71). Proctor in the ecclesiastical court at Lichfield. Father of Mary Howard, who married Erasmus Darwin; CD's great-grandfather. (King-Hele 1999.)

Howse, Richard (d. 1901). Schoolteacher, geologist, and museum curator. Schoolteacher in South Shields, County Durham, then Newcastle-upon-Tyne, until appointed as permanent museum curator to the Natural History Society of Northumberland in 1882. General curator, museum of the Natural History Society of Northumberland, Durham, and Newcastle-upon-Tyne, 1866–1901; editor

Howse, Richard, cont.
of the society's transactions. One of the secretaries of the Tyneside Naturalists' Field Club, 1876–1901; vice-president, 1861–4, 1866–8, 1870–1. (*Natural History Transactions of Northumberland and Durham* 1867–1903, *Post Office directory of Northumberland* 1858, *Post Office directory of the county of Durham* 1873, *Royal Society catalogue of scientific papers*, *Transactions of the Tyneside Naturalists' Field Club* 1863–4.)

Humboldt, Friedrich Wilhelm Heinrich Alexander (Alexander) von (1769–1859). Prussian naturalist, geographer, and traveller. Official in the Prussian mining service, 1792–6. Explored equatorial South America, Cuba, Mexico, and the United States, 1799–1804. Travelled in Siberia in 1829. Foreign member, Royal Society of London, 1815. (*DBE*, *DSB*, *NDB*.)

Hunt, James (1833–69). Speech therapist and anthropologist. Honorary secretary of the Ethnological Society of London, 1859–62. Founder and first president of the Anthropological Society of London, 1863. (*DSB*.)

Hunter, John (1728–93). Surgeon and anatomist. His collection of zoological specimens formed the basis of the Hunterian Museum of the Royal College of Surgeons. FRS 1767. (*DNB*, *DSB*.)

Hurst and Blackett. Publishing firm with offices at 13 Great Marlborough Street, London. (*Post Office London directory* 1865.)
15 November [1863]

Hutton, Frederick Wollaston (1836–1905). Geologist and army officer. Entered the India Mercantile Marine *circa* 1850–1. Studied at King's College London, 1854–5. Joined the Royal Welsh Fusiliers, serving in the Crimea and India, 1855–8. Completed his education at Sandhurst and Woolwich academies. Captain, 1862. Left the army in 1865, and emigrated to New Zealand in 1866. Employed by Auckland provincial government to survey coal deposits. Assistant geologist to the geological survey of New Zealand, 1871–3. Provincial geologist of Otago, 1873–6. Professor of natural science, University of Otago, 1877–80. Professor of biology, Canterbury College, 1880–93. Curator of Canterbury Museum from 1893. FRS 1892. (*DNZB*, Stenhouse 1990, G. M. Thomson 1884–5.)

Huxley, Ethel (1866–1941). Youngest child of Henrietta Anne and Thomas Henry Huxley. Married artist–author John Collier in 1889. (Clark 1968, p. 110, A. Desmond 1994–7, 1: 348.)

Huxley, Henrietta Anne (1825–1915). Born Henrietta Anne Heathorn. Emigrated to Australia in 1843. Met Thomas Henry Huxley in Sydney, Australia, in 1847, and married him in 1855. (A. Desmond 1994–7, Freeman 1978.)
1 January 1865

Huxley, Henry (1865–1946). Physician. Son of Henrietta Anne and Thomas Henry Huxley. Became a fashionable general practitioner in London. (Clark 1968.)

Huxley, Jessie Oriana (1858–1927). Daughter of Henrietta Anne and Thomas Henry Huxley. Married Fred Waller, architect, in 1877. Shared her mother's interest in Moravian principles of education, and published an article, 'Mental and physical training of children', in 1889. (Bibby 1959, Clark 1968, A. Desmond

1994–7, Waller 1889.)

Huxley, Leonard (1860–1933). Biographer, editor, and poet. Son of Henrietta Anne and Thomas Henry Huxley. Assistant master at Charterhouse, 1884–1901. Assistant editor, *Cornhill Magazine*, 1901–16; editor from 1916. Married Julia Frances Arnold, niece of Matthew Arnold, in 1885. (*DNB*, Clark 1968.)

Huxley, Marian (1859–87). Artist. Daughter of Henrietta Anne and Thomas Henry Huxley. Studied art at the Slade School, London. Exhibited at the Royal Academy, 1880–4. Married artist–author John Collier in 1879. Her sketch of CD, made when she was 18, hangs in the National Portrait Gallery. (Bibby 1959; *Bryan's dictionary of painters and engravers*; Clark 1968, p. 97 and *passim*; A. Desmond 1994–7; Petteys 1985.)

Huxley, Nettie (1863–1940). Singer and illustrator. Daughter of Henrietta Anne and Thomas Henry Huxley. Married Harold Roller, joint owner of a firm of picture restorers, in 1889, but spent most of her time travelling in Europe with her daughter, supporting herself as a singer. (Bibby 1959, pp. 15, 275, 283; Clark 1968, pp. 111, 252, and *passim.*)

Huxley, Noel (1856–60). Eldest child of Henrietta Anne and Thomas Henry Huxley. Died of scarlet fever. (L. Huxley ed. 1900.)

Huxley, Rachel (1862–1934). Daughter of Henrietta Anne and Thomas Henry Huxley. Married Alfred Eckersley, a civil engineer, in 1884 and lived in various countries until his death in 1895 in San Salvador. Returned to London, where she ran a laundry business until her marriage to Harold Shawcross, when she moved to Lancashire. (Clark 1968, pp. 98, 109, 129, 165, and *passim.*)

Huxley, Thomas Henry (1825–95). Zoologist. Assistant-surgeon on HMS *Rattlesnake*, 1846–50, during which time he investigated Hydrozoa and other marine invertebrates. Lecturer in natural history, Royal School of Mines, 1854; professor, 1857. Appointed naturalist to the Geological Survey of Great Britain, 1855. Hunterian Professor, Royal College of Surgeons of England, 1863–9. Fullerian Professor, Royal Institution of Great Britain, 1863–7. President of the Royal Society of London, 1883–5. FRS 1851. (Clark 1968, A. Desmond 1994–7, *DNB*, *DSB*, *Record of the Royal Society of London.*)

1 January 1865, 4 January [1865], *15 January 1865*, *1 May 1865*, 27 May [1865], *29 May 1865*, 30 May [1865], *1 June 1865*, 12 July [1865], *16 July 1865*, [17 July 1865], *2 October 1865*, 4 October [1865]

Innes, John (1817–94). Clergyman. Perpetual curate of Down, 1846–68; vicar, 1868–9. Left Down in 1862 after inheriting an entailed estate at Milton Brodie, near Forres, Scotland; changed his name to Brodie Innes in 1861 as required by the entail. Priest in charge of Milton Brodie Mission and general licentiate of the diocese of Moray, 1861. Chaplain to the Bishop of Moray, 1861–80 and 1886–94. (*Clergy list*, *County families* 1864, *Crockford's clerical directory*, Freeman 1978, J. R. Moore 1985.)

[after 8 February – August 1855]

Innes, John Brodie. *See* Innes, John.

Innes, Mary (1776?–1859). Mother of John Innes. (*County families* 1864, *The Times*, 9 March 1859, p. 1.)

Jackson, Benjamin Daydon (1846–1927). Botanist and botanical bibliographer. Editor of *Index Kewensis*, 1893–5; co-editor of its supplement, 1901–6. Biographer of George Bentham and translator of a biography of Linnaeus, upon whose work and collections he was an authority. Botanical secretary, Linnean Society of London, 1880–1902; general secretary, 1902–26. (R. Desmond 1994, *Proceedings of the Linnean Society of London* (1927–8): 119–23.)

Jamieson, Thomas Francis (1829–1913). Scottish agriculturalist and geologist. Factor on the Ellon estate, Aberdeenshire, for many years; later took the farm of Mains, Waterton. Appointed Fordyce Lecturer on agricultural research, University of Aberdeen, 1862. Carried out notable researches on Scottish Quaternary geology and geomorphology. Became a fellow of the Geological Society of London in 1862. (*Geological Magazine* 50 (1913): 332–3, Sarjeant 1980–96.)

Jeffrey, Francis (1773–1850). Scottish judge, critic, and Whig politician. A founder of the *Edinburgh Review*, 1802; editor, 1803–29. (*DNB*.)

Jenner, William, 1st baronet (1815–98). Physician. Professor of pathological anatomy at University College, London, 1849; Holme Professor of clinical medicine, 1860. Assistant, then full, physician to University College Hospital, 1849–76; consulting physician, 1879. Physician extraordinary to Queen Victoria, 1861; physician-in-ordinary, 1862. Established, through clinical studies, definitive evidence that typhus and typhoid were different diseases. FRS 1864. (*DNB*.)

Jenyns, Leonard (1800–93). Naturalist and clergyman. Brother-in-law of John Stevens Henslow. Vicar of Swaffham Bulbeck, Cambridgeshire, 1828–49. Settled near Bath in 1850. Founder and first president of the Bath Natural History and Antiquarian Field Club, 1855. Member of many scientific societies. Described the *Beagle* fish specimens. Adopted the name Blomefield in 1871. (R. Desmond 1994, *DNB*.)

Johnson, Samuel (1709–84). Lexicographer and man of letters. (*DNB*.)

Jones, Henry Bence (1814–73). Physician and chemist. Physician to St George's Hospital, 1846–62. Secretary of the Royal Institution from 1860. In his chemical studies, devoted himself especially to the application of chemistry to pathology and medicine. FRS 1846. (*DNB*.)

Jones, Martha (*fl.* 1820s). Housemaid. A creditor of Robert Waring Darwin, and possibly a member of his household. (DAR 227.5: 82, f. 51 and contents page; see also this volume, Supplement, first letter to Dear Friend, 4 January 1822.)

Jordan, Claude Thomas Alexis (Alexis) (1814–97). French botanist. Conducted field research, 1836–46, to complete and correct existing French floras. Assembled an important private herbarium. After giving up his botanical expeditions, worked in his own experimental gardens, trying to demonstrate the stability of species. A strong opponent of transmutation theory. (*DSB*, Tort 1996.)

Jowett, Benjamin (1817–93). Master of Balliol College and regius professor of Greek, Oxford University, 1855–93. Author of an article in *Essays and reviews*

(1860). (*DNB*.)

Jukes, Joseph Beete (1811–69). Geologist. Geological surveyor of Newfoundland, 1839–40. Naturalist aboard HMS *Fly* in the survey of the north-east coast of Australia, 1842–6. Geologist with the Geological Survey of Great Britain working in North Wales, 1846–50. Director of the Irish branch of the Geological Survey, 1850–69. Lecturer on geology at the Royal College of Science, Dublin. President of the Geological Society of Dublin, 1853–4. FRS 1853. (*DNB*, *DSB*, *Quarterly Journal of the Geological Society of London* 26 (1870): xxxii–xxxiv, *Record of the Royal Society of London*.)

Karsten, Hermann (1817–1908). German botanist and geologist. Studied medicine and natural science in Berlin and Greifswald; made two research voyages to the northern part of South America, having been recommended by Alexander von Humboldt, 1844–7 and 1848–56. Taught from 1856 at the agricultural institute, Berlin; founded a physiology laboratory, 1865. In 1868, became professor of plant physiology, University of Vienna, but left in 1872, because of student opposition. Thereafter, taught privately in Basel, then Berlin. Wrote on the flora and geology of South America, plant physiology, and pharmaceutical botany. (*ADB*, *NDB*.)

Keilhau, Baltazar Mathias (1797–1858). Norwegian geologist and petrologist. Lecturer, University of Christiania, 1821; professor, 1834. (*NBL*, Sarjeant 1980–96.)

Keller, Ferdinand (1800–81). Swiss archaeologist. Teacher at the Industrieschule in Zürich from 1831. Founder and first president of the Gesellschaft für vaterlandische Altertümer (later Antiquarische Gesellschaft), 1832. Honorary D. Phil., University of Zürich, 1847. Editor of *Anzeiger für schweize Altertumskunde*, 1869–79. Best known for his discovery of ancient lake dwellings at Obermeilen and for his numerous publications on the subject. (*ADB*, *HBLS*.)

Kindt, Hermann. Editor of the *Autographic Mirror* (*L'autograph cosmopolite*). Letters from Hermann Kindt, 17 October 1865 and 23 October 1865.)

17 September 1864, *17 October 1865*, *23 October 1865*, *13 November 1865*, *24 November 1865*

King, Richard Thomas (*fl.* 1802–59). Army officer. Second lieutenant, Royal Artillery, 1802; lieutenant-colonel, 1837; colonel, 1851; major-general, 1854. Resident of Hythe, Kent, *circa* 1845–59. (*Hart's army list* 1856; *Post Office directory of the six home counties* 1845–59.)

King, William James (*fl.* 1845–59). Army officer. Lieutenant-colonel, Royal Staffordshire Corps, 1846; colonel, 1854. (*Hart's army list* 1855; *Post Office directory of the six home counties* 1845–59.)

Kingsley, Charles (1819–75). Author and clergyman. Lecturer on English literature, Queen's College, London, 1848–9. Professor of modern history, Cambridge University, 1860–9. Rector of Eversley, Hampshire, 1844–75. Chaplain to the queen, 1859–75. (*Alum. Cantab.*, *DNB*.)

30 May 1865, *10 June 1865*, *14 June 1865*, [17 June 1865]

Kingsley, Frances Eliza (1814–91). Née Grenfell. Wife of Charles Kingsley. Acted as his amanuensis and edited his memoirs after his death. (*DNB* s.v. Kingsley, Charles; *The Times*, 14 December 1891, p. 7.)

Kippist, Richard (1812–82). Botanist. Librarian of the Linnean Society of London, 1842–80. Specialist in Australian plants. (R. Desmond 1994, *DNB*.)

27 [February or March 1861?], 18 January [1865], 8 February 1865, 4 June [1865]

Koenen, Adolf von (1837–1915). German geologist. Following two years of practical mining studies, studied geology at the University of Berlin; PhD 1865, habilitation 1867. Extraordinary professor of geology, University of Marburg, 1873; professor, 1878. Professor, University of Göttingen, 1881–1907. Worked on stratigraphy, palaeontology, geological cartography, and tectonics. (Sarjeant 1980–96, *Zeitschrift der Deutschen Geologischen Gesellschaft. Monatsberichte* 67 (1915): 229–68.)

Kölliker, Rudolf Albert von (1817–1905). Swiss anatomist and physiologist. Professor of physiology and comparative anatomy, University of Würzburg, 1847–64; professor of anatomy, 1849–97. Foreign member, Royal Society of London, 1860. (*DSB*.)

Kölreuter, Joseph Gottlieb (1733–1806). German botanist. Assistant keeper of the natural history collections, Imperial Academy of Sciences, St Petersburg, 1756–61. Professor of natural history and director of the gardens of the margrave of Baden, Karlsruhe, 1763–86. Carried out extensive hybridisation experiments on plants. (*ADB*, *DBE*, *DSB*, *NDB*, *Taxonomic literature*.)

Krohn, August David (1803–91). Russian-born zoologist, anatomist, and embryologist, working in Bonn; travelled extensively. (Blyakher 1982.)

Lacordaire, Jean Théodore (1801–70). French naturalist, entomologist, and traveller. Professor of zoology, Liège university, Belgium, from 1835; professor of comparative anatomy from 1838. Made numerous voyages to South America under the auspices of Georges Cuvier. Published mainly on entomology. (*DBF*.)

Lamarck, Jean Baptiste Pierre Antoine de Monet (Jean Baptiste) de (1744–1829). French naturalist. Held various botanical positions at the Jardin du Roi, 1788–93. Appointed professor of zoology, Muséum d'Histoire Naturelle, 1793. Believed in spontaneous generation and the progressive development of animal types; propounded a theory of transmutation. (*DSB*.)

Lane, Edward Wickstead (1823–89). Physician. Proprietor of a hydropathic establishment at Moor Park, near Farnham, Surrey, 1859 (or before), and at Sudbrook Park, near Ham, Surrey, 1860–79. Practised in Harley Street, London, 1879–89. Member of the Faculty of Advocates, the Botanical Society, and the Speculative Society, Edinburgh. Author of works on hydropathy. (Freeman 1978, *Medical directory* 1859–89, *Post Office directory of the six home counties* 1859–62.)

Langton, Charles (1801–86). Rector of Onibury, Shropshire, 1832–41. Left the Church of England in 1841. Resided at Maer, Staffordshire, 1841–7, and at Hartfield Grove, Hartfield, Sussex, 1847–63. Married Emma Darwin's sister, Charlotte Wedgwood, in 1832. After her death, married CD's sister, Emily Catherine

Darwin, in 1863. (*Alum. Oxon.*, *Emma Darwin* (1915), Freeman 1978.)

Langton, Edmund (1841–75). Son of Charles and Charlotte Langton. BA, Trinity College, Cambridge, 1864. Admitted at Lincoln's Inn, 1864. (*Alum. Cantab.*, *Emma Darwin* (1915).)

Langton, Emily Catherine. *See* Darwin, Emily Catherine.

Laugel, Antoine Auguste (Auguste) (1830–1914). French writer on a wide range of subjects, including science, philosophy, politics, history, and psychology. (*Dictionnaire universel des contemporains* 1893, *NUC*.)

Lea, Isaac (1792–1886). American naturalist and publisher. President, Academy of Natural Sciences of Philadelphia, 1858–63; American Association for the Advancement of Science, 1860. Did extensive work on freshwater molluscs, publishing a series of thirteen volumes on the genus *Unio*. (*ANB*, *DAB*.)

Lecky, William Edward Hartpole (1838–1903). Historian and essayist. Elected to the Athenaeum Club, 1867. Wrote on Irish history and politics. Privy councillor, 1897. (*DNB*.)

Le Conte, John Lawrence (1825–83). American entomologist. Son of John Eatton Le Conte. Trained as a physician but never practised. Published in many fields of natural history, with a particular interest in geographical distribution. Served with the army medical corps during the American Civil War. President, American Association for the Advancement of Science, 1874. Chief clerk to the United States Mint at Philadelphia from 1878. (*DAB*.)

Lee, Robert Edward (1807–70). United States Army soldier, later Confederate Army general. Commissioned as brevet second lieutenant of engineers, 1829. First lieutenant of engineers, 1836; captain, 1838; brevet colonel, 1848. Superintendent at United States Military Academy, West Point, New York, 1852–5. Lieutenant-colonel of Second Cavalry, 1855; colonel of First Cavalry, 1861; resigned in April 1861. Appointed commander of Virginia's armed forces, 1861. Unofficial military advisor to Confederate President Davis with rank of general. In command of the Army of Northern Virginia, 1862. General-in-chief of all Confederate armies, 1865. Applied for a pardon on 13 June 1865. Indicted for treason but never brought to trial. President of Washington College (now Washington and Lee University) Virginia, from 1865. (*ANB*, *DAB*.)

Leighton, Burgh (1760–1836). Army officer. Lieutenant-colonel, Fourth Light Dragoons. Brother of Sir Baldwin Leighton of Loton Park, Shrewsbury. Resident at Quarry Place, Shrewsbury. (*Burke's peerage* 1855; *Commercial directory of Shropshire*.)

Leighton, Clare. Daughter of Louisa Anne and Francis Knyvett Leighton. (*Burke's landed gentry* 1846.)

Leighton, William Allport (1805–89). Botanist, clergyman, and antiquary. Schoolfellow of CD's in Shrewsbury, 1817. Curate of St Giles's, Shrewsbury, 1846–8. Edited the *Transactions of the Shropshire Archaeological Society* for many years. Published *The flora of Shropshire* (1841). (*Clergy list* 1846–8, R. Desmond 1994, *DNB*, Freeman 1978.)

29 May 1865

Léon, Isidore. Sub-inspector of customs at Lorient (Morbihan), France. Member of and contributor of papers to the Société Botanique de France. (*Bulletin Société Botanique de France* 5 (1858): viii.)

Lepsius, Karl Richard (1810–84). German Egyptologist. Led a scientific expedition to Egypt, 1842–5. Professor of Egyptology at Berlin University, 1846. Co-director of the Berlin Egyptology Museum, 1855; director, 1865. President of the Roman archaeological institute in Berlin from 1867. Head of the royal library in Berlin, 1873–84. Author of *Denkmäler aus Aegypten und Aethiopien* (1849–58). (*DBE*, *NDB*.)

Leslie, Lewis James (b. 1806/7). Auctioneer at 45 Margaret Street, Cavendish Square, London. Father of Alfred Russel Wallace's fiancée, Marian Leslie. Resided at Rothsay Villa, Campden Hill, Kensington. (Census returns 1861 (Public Record Office RG9/18: 2); *Post Office London directory* 1865.)

Leslie, Marian (b. 1846/7). Betrothed to Alfred Russel Wallace, 1864–5. Daughter of Lewis James Leslie. (Census returns 1861 (Public Record Office RG9/18: 2); letter from A. R. Wallace, 20 January 1865; *Post Office London directory* 1865; A. R. Wallace 1905, 1: 409–11; Wallace's notes for A. R. Wallace 1905 (Wallace family papers, private collection).)

Lesquereux, Leo (1806–89). Swiss-born bryologist and palaeontologist. Moved to the United States in 1848; after brief periods assisting Louis Agassiz and Asa Gray, settled in Columbus, Ohio, where he assisted the bryologist William Starling Sullivant. Employed on geological surveys in the mid-western United States and Pennsylvania. Engaged by Agassiz to organise the palaeobotanical collections at the Museum of Comparative Zoology, Harvard University, 1867–72. (*DAB*, *DSB*, Rodgers 1968, Sarjeant 1980–96, Sarton 1942.)

Lettington, Henry (b. 1822/3). Gardener in the village of Down, Kent. Worked as a gardener at Down House, 1854–79, and occasionally in following years. Assisted CD with botanical experiments. Son-in-law of William Brooks, who was also employed by the Darwins. (CD's Classed account books (Down House MS); Census returns 1861 (Public Record Office RG9/462: 70); F. Darwin 1920, pp. 56–7; letter from Emma Darwin to George Darwin (DAR 210.3: 110); Recollections of CD by Francis Darwin (DAR 140.3: 90a).)

Lewes, Agnes (1822–1902). Eldest daughter of Swynfen Jervis. Married George Henry Lewes in 1841. (Ashton 1991.)

Lewes, George Henry (1817–78). Writer. Wrote articles on literary and philosophical subjects for the quarterly reviews. Editor, *Fortnightly Review*, 1865–6. Published on physiology and on the nervous system in the 1860s and 1870s. Common-law husband of Marian Evans (George Eliot), with whom he lived from 1854. (Ashton 1991, *DNB*.)

Lewis, George Cornewall, 2d baronet (1806–63). Statesman and author. Poor Law commissioner for England and Wales, 1839–47. MP for Herefordshire, 1847–52; for Radnor boroughs, 1855–63. Editor of the *Edinburgh Review*, 1852–5. Chancellor of the Exchequer, 1855–8. Succeeded to the baronetcy in 1855. Home secretary, 1859–61. (*DNB*, *Modern English biography*.)

Liebig, Justus von (1803–73). German organic chemist. Professor of chemistry, University of Giessen, 1825–51; University of Munich, 1852–73. Played a major role in the development of organic chemistry. Developed artificial agricultural fertilisers. President of the Bavarian Academy of Science, 1859–73. Foreign member, Royal Society of London, 1840. (*ADB*, *DBE*, *DSB*, *NDB*.)

Lilljeborg, Wilhelm (1816–1908). Swedish zoologist. Held academic posts in zoology at Lund University, 1844–53. Professor of zoology at Uppsala, 1854–82. Established a zoological laboratory at Uppsala in 1875. Specialised in biogeography and pioneered the systematics of the fauna of Sweden and Norway. (*SBL*.)

Lincoln, Abraham (1809–65). American lawyer and statesman. Republican president of the United States of America during the American Civil War, from 1861 until his assassination in 1865. (*DAB*.)

Lindley, John (1799–1865). Botanist and horticulturalist. Assistant in Joseph Banks's library and herbarium, 1818 or 1819. Garden assistant secretary, Horticultural Society of London, 1822–7; general assistant secretary, 1827–41; vice-secretary, 1841–58; honorary secretary, 1858–62. Lecturer on botany, Apothecaries' Company, 1836–53. Professor of botany, London University (later University College, London), 1829–60. Horticultural editor of the *Gardeners' Chronicle* from 1841. FRS 1828. (R. Desmond 1994, *DNB*, *DSB*.)

Linton, Henry (1803–87). Clergyman. Curate of Diddington, Huntingdonshire, 1827–35; vicar, 1836–56. Rector, St Peter-le-Bailey, Oxford, 1856–77. Wrote works on the New Testament. (*Alum. Cantab.*, *Rugby School register*.)

Linton, Sydney (1841–94). Clergyman. Vicar of Holy Trinity, Oxford, 1870–7; of St Philip, Heigham, 1877–84; bishop of Riverina, New South Wales, Australia, 1884–94. (*Aust. dict. biog.*, *Modern English biography*.)

Lockyer, Joseph Norman (1836–1920). Astronomer. Civil servant in the War Office, 1857–75; published papers on solar physics. Secretary to the Royal Commission on Scientific Instruction and the Advancement of Science, 1870–5; employee of the Science and Art Department at South Kensington from 1875; first director of the Solar Physics Observatory, and professor of astronomical physics, Royal College of Science, South Kensington, 1890–1911. Established the journal *Nature* in 1869. Knighted, 1897. FRS 1869. (*DNB*, *DSB*.)

Logan, William Edmond (1798–1875). Canadian stratigrapher. Director of the geological survey of Canada, 1842–70. Knighted, 1856. FRS 1851. (*DNB*, *DSB*.)

Lonsdale, William (1794–1871). Geologist. Served the Geological Society of London, 1829–42, first as curator and librarian, and after 1838 as assistant secretary and librarian. Suggested fossils were a reliable basis for estimating the age of geological strata. (*DNB*, *DSB*.)

Loring, Charles Greely (1794–1867). American lawyer and author. Father-in-law of Asa Gray. Elected to the Massachusetts State Senate, 1862. Member of the American Academy of Arts and Sciences and the Massachusetts Historical Society, and president of the Boston Union Club. Following the outbreak of the

Loring, Charles Greely, cont.
Civil War in the United States, his correspondence with Edwin Wilkins Field concerning American relations with Britain was published in his *Correspondence on the present relations between Great Britain and the United States* (1862). (O. F. Adams ed. 1897, *Appleton's cyclopaedia of American biography*, Chamberlain ed. 1898–1900, Dupree 1959.)

Lubbock, Ellen Frances (1834/5–1879). Daughter of Peter Hordern, clergyman, of Chorlton-cum-Hardy, Lancashire. Married John Lubbock in 1856. (*Burke's peerage* 1970; Census returns 1861 (Public Record Office RG9/462: 75).)
[27 August – 1 September 1865]

Lubbock, John, 4th baronet and 1st Baron Avebury (1834–1913). Banker, politician, and naturalist. Son of John William Lubbock and a neighbour of CD's in Down until 1861, when he moved to Chislehurst, Kent. Studied entomology and anthropology. A partner in the family bank from 1849. Liberal MP for Maidstone, Kent, 1870 and 1874; for London University, 1880–1900. Succeeded to the baronetcy in 1865. Created Baron Avebury, 1900. FRS 1858. (*DNB*, *DSB*, Hutchinson 1914, *Record of the Royal Society of London.*)
22 and 26 March 1865, 11 June [1865], *12 June [1865]*

Lubbock, John William, 3d baronet (1803–65). Astronomer, mathematician, and banker. A neighbour of CD's in Down. First vice-chancellor of London University, 1837–42. Partner in the family bank, 1825. Treasurer and vice-president of the Royal Society of London, 1830–5 and 1838–45. Succeeded to the baronetcy in 1840. FRS 1829. (*DNB*, *DSB*, *Record of the Royal Society of London.*)
28 [June 1856 – January 1865?]

Lubbock, Rolfe Arthur (1865–1909). Sixth child of John and Ellen Frances Lubbock. (Hutchinson 1914.)

Lucas, Prosper (1805–85). French physician and medical writer. Interested in heredity. Psychiatrist at the hospital at Bicêtre, then Sainte-Anne. (*NUC*, Tort 1996.)

Ludwig, Camilla. German governess. Governess to the Darwin family, 1860–3. Translated German works for CD. Married Reginald Saint Pattrick, vicar of Sellinge, Kent (*Alum. Oxon.*), in 1874. (CD's Classed account books (Down House MS); letter from R. S. Pattrick, 19 October 1881 (*Calendar* no. 13416).)
22 December 1862

Ludwig, Louisa. German schoolmistress. Acted as governess to the Darwin family for periods in 1862, 1864, and 1865. Sister of Camilla Ludwig. (CD's Classed account books (Down House MS); *Correspondence* vol. 10, letter to Camilla Ludwig, 26 August [1862].)

Lyell, Charles, 1st baronet (1797–1875). Scottish geologist. Uniformitarian geologist whose *Principles of geology* (1830–3), *Elements of geology* (1838), and *Antiquity of man* (1863) appeared in many editions. Professor of geology, King's College, London, 1831. President of the Geological Society of London, 1835–6 and 1849–50; of the British Association for the Advancement of Science, 1864. Travelled widely

and published accounts of his trips to the United States. CD's scientific mentor and friend. Knighted, 1848; created baronet, 1864. FRS 1826. (*DNB*, *DSB*.)
[1847 or 1848], *4 October 1859*, *22 October 1859*, 4 February [1860], *22 October 1861*, *16 January 1865*, 22 January [1865], 21 February [1865], 25 March [1865], *[31 May 1865]* (J. D. Hooker)

Lyell, Mary Elizabeth (1808–73). Eldest child of Leonard Horner. Married Charles Lyell in 1832. (Freeman 1978.)

Maberley, William Leader (1798–1885). Army officer, politician, and civil servant. Served in the army, 1815–81, reaching the rank of lieutenant-colonel. MP for Westbury, 1819–20; Northampton, 1820–30; Shaftesbury, 1831–2; Chatham, 1832–4. Secretary of the Post Office, 1836–54. Opposed the Post Office reforms that Rowland Hill proposed following Hill's appointment as secretary to the postmaster-general in 1846. At the Board of Audit, 1854–66. (*DNB*.)

McCormick, Robert (1800–90). Naval surgeon, explorer, and naturalist. Published accounts of his voyages. Surgeon on board HMS *Beagle*, 1831–2. Accompanied James Clark Ross's Antarctic expedition, 1839–43. (*Correspondence* vol. 1, *DNB*.)

McCoy, Frederick (1823–99). Naturalist and geologist. Employed in arranging the collection in the Woodwardian Museum, Cambridge University, 1846–54. Professor of mineralogy and geology, and curator of the museum, Queen's College, Belfast, 1850–4. Professor of natural science, University of Melbourne, 1855–99. Palaeontologist to the Australian Geological Survey, 1856. Director of the National Museum of Natural History and Geology, Melbourne, 1857–99. An ardent opponent of natural selection. FRS 1880. (*Aust. dict. biog.*, *DNB*.)

Mackintosh, Eva. Daughter of Robert James and Mary Mackintosh. (B. Wedgwood and Wedgwood 1980.)

Mackintosh, Mary. American-born wife of Emma Darwin's cousin, Robert James Mackintosh. Sister of Thomas Gold Appleton. (B. Wedgwood and Wedgwood 1980.)

Mackintosh, Robert James (1806–64). Emma Darwin's cousin. Barrister-at-law, Lincoln's Inn, 1833, (*Alum. Oxon.*, O'Leary 1989, B. Wedgwood and Wedgwood 1980.)

Macleay, William Sharp (1792–1865). Naturalist and diplomat. Originated the circular or quinary system of classification. Diplomat in France and then Havana; retired in 1837. In 1839, emigrated to Sydney, Australia, where he devoted himself to the study of Australian natural history. Made large collections of Australian insects and marine fauna, and supervised the cultivation of exotic plants in his private gardens. Trustee of the Australian Museum, 1841–62. (*Aust. dict. biog.*, *DNB*.)

McNab, James (1810–78). Scottish botanist. Collected plants in North America, 1834. Superintendent, Caledonian Horticultural Society, 1835. Curator of the Royal Botanic Garden, Edinburgh, from 1849. President of the Edinburgh Botanical Society, 1872. (R. Desmond 1994.)

Magnus, Heinrich Gustav (1802–70). German physicist and chemist. Extraordinary professor of 'Technologie', University of Berlin, 1834; professor of physics, 1845. In 1840, set up a laboratory in his home and gave several lecture courses. Founding member of the Deutsche physikalische Gesellschaft, 1845. In 1863, opened a physics laboratory in his new home, where he taught until his death. Worked on the chemistry of blood gases and respiration, and the physics of heat expansion; discovered the 'Magnus-Effekt' in ballistics. (*ADB*, *Archives Internationales d'Histoire des Sciences* 48 (1998): 211–12, *NDB*.)

Mann, Gustav (1836–1916). German-born botanist. Gardener, Royal Botanic Gardens, Kew, 1859. Botanical collector to the Niger expedition, 1859–62. Worked for the Indian Forest Service, 1863–91. (R. Desmond 1994, R. Desmond 1995.)

Mann, Horace (1844–69). American botanist. Collected plants in the Hawaiian Islands. Curator of the herbarium at Harvard University, 1866–9. (Dupree 1959, Messerli 1972.)

Marshall (or Martial), Mr. Surgeon on an English whaling vessel whom CD met in 1834 during the *Beagle* voyage. (*Correspondence* vol. 3, letter to Henry Denny, 3 June [1844], and this volume, letter to Henry Denny, 17 January [1865].)

Martin, William Charles Linnaeus (1798–1864). Writer on natural history. Resided in Lee, Kent. Superintendent of the museum of the Zoological Society of London, 1830–8. Wrote works on domestic animals. (*DNB*, *Gentleman's Magazine* 1 (1864): 536, *Modern English biography*.)
[1859–61] (two letters)

Martins, Charles Frédéric (1806–89). French botanist. Qualified as a doctor in Paris in 1834. Professor of botany, Faculté de Montpellier, 1846. Also published on geology and meteorology. (*Dictionnaire universel des contemporains*, *NBU*.)

Mason, Nathaniel Haslope (*fl.* 1850–70). Collected plants on Madeira, 1855–7. Published sets of Madeira ferns and woods. Fellow of the Linnean Society, 1856. (R. Desmond 1994.)

Mason, Sarah Ann. *See* Bates, Sarah Ann.

Masson, David (1822–1907). Biographer and editor. Studied for the ministry of the Church of Scotland, University of Edinburgh, 1839–42. Worked as editor and journalist in Scotland. Professor of English literature, University College, London, 1853–65. Editor, *Macmillan's Magazine*, 1858–67; *Reader*, spring–summer 1863. Professor of rhetoric and English literature, University of Edinburgh, 1865–1907. (*DNB*.)

Masters, Maxwell Tylden (1833–1907). Botanist, surgeon, and general medical practitioner. Sub-curator, Fielding Herbarium, University of Oxford, *circa* 1853–7. Lecturer on botany at St George's Hospital medical school, 1855–68. Editor of the *Gardeners' Chronicle*, 1865–1907. Active in the Royal Horticultural Society, succeeding Joseph Dalton Hooker as the chair of the scientific committee; secretary of the International Horticultural Congress, 1866. FRS 1870. (Clokie 1964, pp. 106, 208; R. Desmond 1994; *DNB*; *Medical directory* 1857–1908.)
25 April [1860], *7 February 1865*, *12 July 1865*, *September 1865*

Masters, William (1796–1874). Nurseryman in Canterbury. Founded the Canterbury Museum in 1823; honorary curator, 1823–46. Conducted hybridisation experiments on passion flowers. Designed the formal gardens at Walmer Castle. Father of Maxwell Tylden Masters. (R. Desmond 1994.)

Matthew, Patrick (1790–1874). Scottish gentleman farmer. Author of works on political and agricultural subjects. Advanced a theory of natural selection in the 1830s. (Dempster 1996, R. Desmond 1994.)

Maurice, John Frederick Denison (1805–72). Clergyman. Chaplain of Guy's Hospital, London, 1836–46. Professor of modern history and English literature, King's College, London, 1840–6; of ecclesiastical history, 1846–53. Chaplain of Lincoln's Inn, 1846–60. Perpetual curate of the chapel of St Peter's, Vere Street, London, 1860–6. Knightbridge professor of moral philosophy, Cambridge, 1866–72. Vicar of St Edward's, Cambridge, 1870–2. (*Alum. Cantab.*, *DNB.*)

Maury, Matthew Fontaine (1806–73). American naval officer, astronomer, hydrographer, and meteorologist. Superintendent of the Depot of Charts and Instruments and of the Naval Observatory, Washington, 1841. Published a series of wind and current charts with sailing instructions for all the oceans. His 1855 study, *The physical geography of the sea*, laid the foundation for the science of oceanography. Resigned in 1861 and was commissioned in the Confederate States Navy. Agent in Britain for the Confederate government from 1862. Professor of meteorology, Virginia Military Institute at Lexington, 1869–73. (*ANB*, *DAB*, *DSB.*)

Maw, George (1832–1912). Tile manufacturer, geologist, botanist, and antiquarian. Partner with his younger brother Arthur in the encaustic tile company, Maw & Co., of Brosley, Shropshire. Established a well-known garden at his residence at Benthall Hall, Shropshire; an expert on crocuses. Wrote on the geology of western England and North Wales. Travelled to Morocco and Algeria with Joseph Dalton Hooker in 1871 and independently in 1873, writing on the geology of these countries. (Benthall 1980; R. Desmond 1994; *Gardeners' Chronicle*, 12 February 1881, pp. 205–6, 208, 209; Sarjeant 1980–96.)
1 June 1865, 4 June [1865]

Maximillian (1832–67). Archduke of Austria; emperor of Mexico. (*EB.*)

Max Müller, Friedrich (1823–1900). German-born orientalist and philologist. Published an edition of the 'Rigveda', the most important of the sacred books of the Brahmans, 1849–73. Moved to Paris in 1845; settled in Oxford in 1848 after fleeing the revolution in France. Deputy Taylorian Professor of modern European languages, Oxford University, 1850–4; professor, 1854–68; professor of comparative philology, 1868–75. Curator of the Bodleian Library, Oxford, 1856–63 and 1881–94. (*DNB.*)

Meinert, Frederik Vilhelm August (1833–1912.) Danish entomologist. Graduated in theology, Christianshavn, 1857. Curator at the zoological museum, University of Copenhagen. Founder and editor, *Entomologiske Meddelelser*, 1887–93. (*DBL*; *Entomologist's Monthly Magazine* 39 (1903): 153, 175; Gilbert 1977.)

Mellersh, Arthur (1812–94). Naval officer. Midshipman and mate on HMS *Beagle*, 1825–36. Served off the coast of Syria, then in command of HMS *Rattler* in the Burma campaign in 1852. Served off the coast of China in the 1850s, suppressing piracy, and in the Caribbean and South America before retiring in 1864. (*Modern English biography*; *The Times*, 28 September 1894, p. 4.)

Meteyard, Eliza (1816–79). Author. Born in Liverpool; lived in Shrewsbury, 1818–29; in Thorpe, near Norwich, 1829–42; thereafter in London. Contributed fiction and social articles to numerous periodicals under the pen-name Silverpen. Her novels include *Struggles for fame* (1845), *Mainstone's housekeeper* (1860), and *Lady Herbert's gentlewoman* (1862). Published a number of works about the Wedgwoods including a two-volume life of Josiah Wedgwood I (1865–6), and *The Wedgwood handbook* (1875). (*DNB*.)

25 April 1865, 16 November [1865], *17 November 1865*

Mill, John Stuart (1806–73). Philosopher and political economist. Administrator, East India Company, 1823–58. (*DNB*, *DSB*.)

Miller, William Hallowes (1801–80). Mineralogist and crystallographer. Professor of mineralogy, Cambridge University, 1832–80. Foreign secretary, Royal Society of London, 1856–73; Royal Medallist, 1870. FRS 1838. (*DNB*, *DSB*.)

Milligan, Joseph (1807–84). Scottish-born surgeon and naturalist in Australia. Surgeon (afterwards surgeon-superintendent), Van Diemen's Land Co., Surrey Hills, Tasmania, 1830–42. Inspector of convict discipline, 1842–3. Superintendent of Aboriginals and medical officer, 1843–46 and 1847–55. Made a study of Aboriginal languages. Surveyed the territory for mineral resources. Chairman of the Douglas River Coal Mining Co., 1855. Secretary to the Royal Society of Van Diemen's Land, 1848–60. Returned to England in 1860; commissioner for Tasmania at the Great Exhibition, 1862. Fellow of the Linnean Society, 1850. (*Aust. dict. biog.*, R. Desmond 1994.)

Milne, David. *See* Home, David Milne.

Milne-Edwards, Henri (1800–85). French zoologist. Professor of hygiene and natural history, Ecole Centrale des Arts et Manufactures, 1832. Professor of entomology, Muséum d'Histoire Naturelle, 1841, with responsibility for the collections of crustaceans, myriapods, and arachnids as well as insects; professor of mammalogy, 1861. Foreign member, Royal Society of London, 1848. (*DSB*.)

Minor, William Chester (d. 1866). MD Yale University, 1863. Assistant and demonstrator in anatomy at Yale, 1862–4. Translated and reviewed German works on natural history. (*American Journal of Science and Arts* 34 (1862): 188–99, 39 (1865): 362–3, *Catalogue of the officers and graduates of Yale University*.)

Mitten, Annie. *See* Wallace, Annie.

Mitten, William (1819–1906). Pharmaceutical chemist and bryologist. Authority on mosses and liverworts. Corresponded with William Jackson Hooker. Resided at Hurstpierpont, Sussex. Described the liverworts for Joseph Dalton Hooker's *Flora Novae Zelandiae* (1853–5) and *Flora Tasmaniae* (1860). (*Journal of Botany* 44 (1906): 329–32.)

Moggridge, John Traherne (1842–74). Entomologist and botanist. Wintered in Mentone, France, and studied the flora of the area. (R. Desmond 1994, *Gardeners' Chronicle* n.s. 2 (1874): 723, Gilbert 1977.)

17 May [1865], 13 October [1865], *14 October [1865]*, *27 December [1865]*

Mohl, Hugo von (1805–72). German biologist. Professor of physiology, University of Bern, 1832–5; professor of botany, University of Tübingen, 1835–72. Known for his work on the microscopic anatomy of plants and for his study of the plant cell. Co-founder of *Botanische Zeitung*, 1843. Foreign member, Royal Society of London, 1868. (*DSB*, *Record of the Royal Society of London*.)

Moore, David (b. Muir) (1808–79). Scottish-born gardener and botanist. Trainee gardener, later foreman, on the estate of the earl of Camperdown, Scotland. Emigrated to Ireland, 1828. Foreman, Trinity College Garden, Dublin, 1829–34. Botanist to the Ordnance Survey of Ireland, 1834–8. Director, Botanic Garden, Glasnevin, 1838–79. Specialist in mosses and liverworts. PhD Zurich 1863. (*DNB*, R. Desmond 1994).

Moore, Thomas (1821–87). Gardener. Gardener, Botanic Garden at Regent's Park, London, 1844–7. Curator, Physic Garden, Chelsea, London, 1848–87. Published major works on ferns. Assistant editor, *Gardeners' Chronicle*, 1841–66; co-editor, 1866–81. Co-editor, *Gardener's Magazine of Botany*, 1850–1; *Garden Companion and Florist's Guide*, 1852; *Orchid Album*, 1881–7. Editor, *Florist and Pomologist*, 1868–74. (*Annals of Botany* 1888: 409–10, R. Desmond 1994, *DNB*, *Proceedings of the Linnean Society of London* (1886–7): 41–2.)

Moquin-Tandon, Horace Bénédict Alfred (Alfred) (1804–63). French botanist and naturalist. Professor of comparative physiology, Athénée de Marseille, 1829–33; professor of natural history, faculty of science, Toulouse, 1833–8; professor of botany, Jardin des Plantes, 1834–52; professor of botany, faculty of science, 1838–52; professor of natural history and director of the Jardin des Plantes, faculty of medicine, Paris, 1853–63. (*Dictionnaire universel des contemporains*, *NBU*, Tort 1996.)

More, Alexander Goodman (1830–95). Botanist and natural historian. Matriculated at Trinity College, Cambridge, 1850. Suffered ill health for many years and later settled in Ireland. Published his chief work, *Cybele Hibernica*, with David Moore, in 1866. Assistant, Natural History Museum, Dublin, 1867; curator, 1881–7. (*Alum. Cantab.*, R. Desmond 1994, *Journal of Botany* 33 (1895): 225–7.)

Morlot, Charles Adolphe (1820–67). Swiss stratigrapher and archaeologist. Appointed professor of geology and mineralogy, Lausanne, 1851. At one time, conservator of the Bern Archaeological Museum. Wrote on the Tertiary and Quaternary geology of Austria, Switzerland, and Denmark. Originator of the term 'Quaternary'. (*ADB*, Sarjeant 1980–96.)

Morrell, Sophia Matilda. Cutler and jeweller of 164 Oxford Street, London. (*Post Office London directory* 1861.)

Morrey, Sarah. Cook to Sarah Wedgwood at Petleys, Down, until 1856. (Census returns 1851 (Public Record Office HO 107/1606/1), Freeman 1978.)

Mueller, Ferdinand Jakob Heinrich (Ferdinand) von (1825–96). German-born explorer and botanist. Emigrated to Australia in 1847. Government botanist, Victoria, 1852. Botanist to the North West Australia Expedition, 1855–7. Director of the Botanical Gardens, Melbourne, 1857–73. President of the Australian Association for the Advancement of Science, 1890. FRS 1861. (*Aust. dict. biog.*, R. Desmond 1994.)

Müller, Johann (1828–96). Swiss botanist. Born in the canton of Aargau, he adopted the appellation 'Argoviensis' to distinguish himself from other botanists of the same name. Curator of the herbarium of Alphonse de Candolle, Geneva, 1851–74. Lectured at the Académie in Geneva from 1868. Professor extraordinarius of medical and pharmaceutical botany, University of Geneva, 1871–6; professor, 1876–89. Curator of the herbarium Delessert and director of the botanical garden of Geneva, 1874–96. In the 1860s, worked chiefly on lichens. (*ADB*.)

Müller, Johann Friedrich Theodor (Fritz) (1822–97). German naturalist. Emigrated to the German colony in Blumenau, Brazil, in 1852. Taught science at the Lyceum in Desterro (now Florianópolis), 1856–67. Appointed *Naturalista Viajante* of the National Museum, Rio de Janeiro, 1876–92. His anatomical studies on invertebrates and work on mimicry provided important support for CD's theories. (*ADB*, *DBE*, Möller ed. 1915–21, *NDB*.)

10 August [1865], *12 August 1865*, *[12 and 31 August, and 10 October 1865]*, *31 August 1865*, 20 September [1865], *10 October 1865*, 17 October [1865], *5 November 1865*, 9 December [1865]

Müller, Rosa (1854–79). Daughter of Johann Friedrich Theodor (Fritz) Müller. (Möller ed. 1915–21, pp. 60, 116).

Munro, Robertson (*fl.* 1868–1900). Scottish nurseryman and florist. Premises at 6 St Andrew Street (South), Edinburgh, and nurseries at Jock's Lodge and Abercorn. Published a paper on fertility in *Passiflora* (Munro 1868). Associate member of the Botanical Society of Edinburgh, 1870–1900. (*Post Office Edinburgh directory* 1870–1, 1890–1, *Transactions of the Botanical Society of Edinburgh* 21 (1900): 372.)

3 February [1865 or 1866?]

Murchison, Charles (1830–79). Physician. Assistant physician, King's College Hospital, London, 1856–60; Middlesex Hospital, 1860–6; full physician, Middlesex Hospital, 1866–71. Physician with the London Fever Hospital from 1861. A keen amateur geologist; edited Hugh Falconer's *Palaeontological memoirs* in 1868. FRS 1866. (*DNB*.)

Murchison, Roderick Impey, 1st baronet (1792–1871). Geologist and army officer. Served in the British army, 1807–14. Noted for his work on the Silurian system. President of the Geological Society of London, 1831–3 and 1841–3; of the British Association for the Advancement of Science, 1846–7; of the Royal Geographical Society of London, 1843–4, 1851, 1857–8, 1862–70. Director-general of the Geological Survey of Great Britain, 1855. Knighted, 1846; created baronet, 1866. FRS 1826. (*DNB*, *DSB*, *Journal of the Royal Geographical Society*.)

Murray, Andrew (1812–78). Lawyer, entomologist, and botanist. Practised law in Edinburgh until 1860. Assistant secretary to the Royal Horticultural Society,

1860–5. An expert on insects harmful to crops. In entomology, specialised in the Coleoptera; in botany, in the Coniferae. (*Entomologist's Monthly Magazine* 14 (1878): 215–16, Gilbert 1977, *Proceedings of the Royal Horticultural Society* 5 (1865): 1.)

Murray, John (1808–92). Publisher, and author of guide-books. CD's publisher from 1845. (*DNB*, Freeman 1978.)
[1 July – 23 August 1862], 31 March [1865], *1 April 1865*, 4 April [1865], 2 June [1865]

Nathusius, Hermann Engelhard von (1809–79). German livestock breeder. Studied zoology at the University of Berlin, 1828–30. Turned to agriculture, specialising in cattle and horse breeding, from 1830. Director, state economic board of Saxony, and advisor to the ministry of agriculture, 1869. Chairman, agricultural institute, Berlin, and lecturer in animal husbandry from 1870. (*DBE*, *NDB*.)

Naudin, Charles Victor (1815–99). French botanist. Joined the herbarium staff at the Muséum d'Histoire Naturelle and became professor of zoology at the Collège Chaptal, Paris, in 1846. Resigned his professorship almost immediately owing to a severe nervous disorder. Appointed aide-naturaliste at the Muséum d'Histoire Naturelle, 1854. Established a private experimental garden at Collioure in 1869, earning his living by selling seeds and specimens. First director of the state-run experimental garden at Antibes, 1878. Experimented widely on plants, particularly on acclimatisation and hybridity. Published a theory of transmutation based on hybridisation. (*DSB*, *Taxonomic literature*.)
18 June 1865

Newman, Edward (1801–76). Naturalist. Founder of the Entomological Club, the precursor of the Entomological Society, 1826. Natural history editor of the *Field*, 1858–76. (R. Desmond 1994, *DNB*, Gilbert 1977.)

Newton, Alfred (1829–1907). Zoologist and ornithologist. Travelled extensively throughout northern Europe and North America on ornithological expeditions, 1854–63. Professor of zoology and comparative anatomy, Cambridge University, 1866–1907. FRS 1870. (*DNB*.)
27 October 1865, 29 October [1865], *30 October 1865*

Newton, Edward (1832–97). Ornithologist and colonial official. Entered the colonial service, where he held a succession of appointments, including colonial secretary of Mauritius, 1868–78, and lieutenant-governor of Jamaica, 1878–83. Co-authored papers with his elder brother, Alfred Newton. Knighted 1887. (*Modern English biography*, A. F. R. Wollaston 1921.)

Newton, Isaac (1642–1727). Mathematician and natural philosopher. Advanced the theory of gravitation in *Principia mathematica* (1687) and of light in *Opticks* (1704). Lucasian Professor, Cambridge University, 1669–1701. Master of the mint, 1699. President of the Royal Society of London, 1703–27. Knighted, 1705. FRS 1672. (*DNB*, *DSB*.)

Nightingale, William Edward (1794–1874). Father of Florence Nightingale. Had estates at Lea Hurst, Derbyshire, and Embley Park, Hampshire. (*DNB* s.v. Florence Nightingale.)

Noel, Charles George, Viscount Campden and 2d earl of Gainsborough (1818–81). Military officer and politician. Captain of the Leicestershire Yeomanry. MP for Rutland, 1840–1. Lord-lieutenant and high sheriff of Rutland from 1848. Succeeded to the earldom, 1866. (*Alum. Cantab.*, *Burke's peerage* 1865–78.)

Norman, Ebenezer (b. 1835/6). Schoolmaster in Down from 1854. Employed as a copyist by CD. (Census returns 1861 (Public Record Office (RG9/462: 73)); Freeman 1978.)

O'Brien, Mary Henrietta. *See* FitzRoy, Mary Henrietta.

O'Callaghan, Patrick. BA Trinity College, Dublin, 1822; LL.B. & LL.D. 1864. Admitted *comitatis causa* to Oxford, 1865. (*Alum. Dublin.*, *Alum. Oxon.*)

Oliver, Daniel (1830–1916). Botanist. Assistant in the herbarium of the Royal Botanic Gardens, Kew, 1858; librarian, 1860–90; keeper, 1864–90. Professor of botany, University College, London, 1861–88. FRS 1863. (R. Desmond 1994, *List of the Linnean Society of London*, 1859–91.)

23 March [1861], 3 November [1861], 20 October [1865], *23 October 1865*, 24 October [1865]

Orbigny, Alcide Charles Victor Dessalines d' (1802–57). French palaeontologist. Travelled widely in South America, 1826–34, on a commission for the Muséum d'Histoire Naturelle. Professor of palaeontology, Muséum d'Histoire Naturelle, 1853. (*DSB*.)

Owen, Caroline Amelia (1801–73). Daughter of William Clift, conservator of the museum of the Royal College of Surgeons. Married Richard Owen in 1835. (Dobson 1954, R. S. Owen 1894.)

Owen, Richard (1804–92). Comparative anatomist. Assistant conservator of the Hunterian Museum, Royal College of Surgeons of England, 1827; Hunterian Professor of comparative anatomy and physiology, 1836–56. Superintendent of the natural history departments, British Museum, 1856–84; prime mover in establishing the Natural History Museum, South Kensington, in 1881. President of the British Association for the Advancement of Science, 1858. Described the *Beagle* fossil mammal specimens. Knighted, 1884. FRS 1834. (*DNB*, *DSB*.)

[15 December 1837 – 9 June 1838], [March? 1840], *[17 September 1841?]*

Oxenden, George Chichester (1797–1875). Author of satiric verse and parodies, and orchid-fancier. Son of Henry Oxenden, seventh baronet. Lived at the family seat at Broome Park, near Canterbury, Kent. (*Alum. Cantab.*, *BLC*, *Burke's peerage* 1895, *Correspondence* vol. 10.)

5 July [1862], 13 July [1862]

Páez, José Antonio (1790–1873). Soldier and statesman. General in the Venezuelan war of independence. Three times president of Venezuela. (*Diccionario de historia de Venezuela*.)

Páez, Ramón (*c.* 1810–94). Venezuelan diplomat, writer, botanist, geologist and painter. (*Diccionario de historia de Venezuela*.)

Paget, James, 1st baronet (1814–99). Surgeon. Assistant surgeon at St Bartholomew's Hospital, London, 1847; surgeon, 1861–71. Arris and Gale Professor of anatomy and surgery at the Royal College of Surgeons of England, 1847–52. Lectured on physiology in the medical school, St Bartholomew's, 1859–61; on surgery, 1865–9. Appointed surgeon-extraordinary to Queen Victoria, 1858; serjeant-surgeon, 1877. Created baronet, 1871. FRS 1851. (*DNB*.)

19 December [1858]

Paine, Thomas (1737–1809). Political writer. Author of *Common sense* (1776), *Rights of man* (1791–2), and *Age of reason* (1794–5). (*DNB*.)

Palgrave, Elizabeth. *See* Turner, Elizabeth.

Palgrave, Francis Turner (1824–97). Poet, critic, and educationalist. Assistant private secretary to William Ewart Gladstone (*DNB*), 1846. Joined the department of education in 1848; vice-principal, Kneller Hall, 1850–5; examiner and assistant secretary, 1855–84. Art critic for the *Saturday Review*. Professor of poetry, Oxford University, 1885–95. Brother of Robert Harry Inglis and William Gifford Palgrave. (*DNB*.)

Palgrave, Robert Harry Inglis (1827–1919). Banker. Educated at Charterhouse; entered the banking-house of Gurneys & Co. (later Barclay's), in which his grandfather Dawson Turner was a partner. Wrote on banking, statistics, and currency. Editor, the *Economist*, 1877–83. Brother of Francis Turner and William Gifford Palgrave. Knighted, 1909. FRS 1882. (*Men and women of the time* 1899, *The Times* 28 January 1919, p. 11.)

Palgrave, William Gifford (1826–88). Missionary, explorer, and diplomat. Jesuit missionary to Syria, 1853–61. Travelled through the Middle East disguised as a Syrian doctor and merchant in areas closed to Europeans, 1862–3. Published an account of his travels, *Narrative of a year's journey through Central and Eastern Arabia* (1865). Member of the British diplomatic corps, 1865–88. Fellow of the Royal Geographical Society, 1878. Brother of Francis Turner and Robert Harry Inglis Palgrave. (*DNB*.)

Pallas, Pyotr Simon (1741–1811). German naturalist and geographer. Travelled widely in the Russian empire. Foreign member, Royal Society of London, 1764. (*ADB*, *DSB*, *Record of the Royal Society of London*.)

Palliser, John (1807–87). Geographer and explorer. Undertook a hunting expedition in the western and north-western districts of North America in 1847, and in 1853 published a popular account of his experiences. Commanded the British North American Exploring Expedition, 1857–60, surveying the boundary territories of British North America between Lake Superior and the Pacific. (*DNB*.)

Pamplin, William (1806–99). Botanist, bookseller, and publisher. Son of the nurseryman, William Pamplin (1768–1844; R. Desmond 1994); assisted his father until 1839. Botanical bookseller in Soho, London, 1839–62. Retired to Wales and tried to establish the North Wales Central Botanic Gardens. Published the *Phytologist*, 1855–63. Associate of the Linnean Society, 1830. Member of the Botanical Society of London. (R. Desmond 1994, *Modern English biography*.)

4 [July 1862]

Panizzi, Anthony (1797–1879). Librarian. Italian exile in Britain. Assistant librarian of the British Museum, 1831; principal librarian, 1856–66. (*DNB*.)

Paris, John Ayrton (1785–1856). Physician. Practised in London and Cornwall. First secretary, Royal Geological Society of Cornwall. Lectured at the College of Physicians, 1819–26; president, Royal College of Physicians, 1844–56. Published a number of medical works. (*DNB*.)

Parker, Marianne (1798–1858). CD's eldest sister. Married Henry Parker in 1824. (*Darwin pedigree*.)

Parry, William Edward (1790–1855). Navel officer and explorer. Led three voyages between 1819 and 1825 in search of a north-west passage, and an unsuccessful expedition to the North Pole in 1827. Knighted, 1829. FRS 1821. (*DNB*.)

Parslow, Joseph (1811/12–98). CD's manservant at 12 Upper Gower Street, London, *circa* 1840–2, and butler at Down House until 1875. (Census returns 1861 (Public Record Office RG9/462: 74), Freeman 1978.)

Parsons, William, 3d earl of Rosse (1800–67). Astronomer. Constructed a telescope with a 72-inch disc at Birr Castle, Ireland, in 1845. Made important observations on nebulae. President of the Royal Society, 1848–54. Appointed chancellor of the University of Dublin, 1862. FRS 1831. (*DNB*, *DSB*.)

Pattison, William Pollard Actuary to the Commercial Union Assurance Company, 19 and 20 Cornhill, London. (*Post Office London directory* 1865.)
4 December 1865

Payer, Jean Baptiste (1818–60). French botanist. Professor of geology and mineralogy at Rennes, 1840. Professor of botany at the Ecole Normale, Paris, 1841. Professor of plant organography, Faculté des Sciences, Paris, 1852. Member of the Académie des Sciences, Paris, 1854. (*IBF* 814: 403–10.)

Peel, Robert, 2d baronet (1788–1850). Tory prime minister, 1834–5 and 1841–6. FRS 1822. (*DNB*.)

Pennell, Henry Cholmondeley (b. 1838). Civil servant, poet, and writer. Served as a clerk in the Admiralty, Whitehall, 1853–66. Appointed an Inspector of Fisheries, 1866. Author of poetry and books on angling. Editor of *Fisherman's Magazine and Review* 1864–5. (*British imperial calendar* 1853–66; *County families*, 1871, s.v. Pennell, Charles Henry; *Men and women of the time*.)

Pepys, Samuel (1633–1703). Diarist, naval administrator, and member of parliament. After serving in a variety of posts connected with the navy, appointed secretary of the Admiralty in 1686. President of the Royal Society 1684–6. (*DNB*.)

Perrault, Charles (1628–1703). French author. (France ed. 1995.)

Phillips, John (1800–74). Geologist. Keeper of the museum of the Yorkshire Philosophical Society, 1825–40. Assistant secretary, British Association for the Advancement of Science, 1832–62. Professor of geology, King's College, London, 1834–40. Palaeontologist to the Geological Survey of Great Britain, 1840–4. Deputy reader in geology, Oxford University, 1853; professor, 1860–74. President of the British Association for the Advancement of Science, 1865. FRS 1834. (*DNB*, *DSB*.)

Pictet de la Rive, François Jules (1809–72). Swiss zoologist and palaeontologist. Professor of zoology, University of Geneva, 1835. (Gilbert 1977, Sarjeant 1980–96.)

Pigeard, Charles. French naval officer. Naval attaché at the French Embassy, London. (*Post Office London directory* 1865, 'Court directory'.)

Pius IX (Giovanni Maria Mastai-Ferretti) (1792–1878). Italian pope, 1846–78. (*EB*.)

Post Office, secretary of the
[1845–51?]

Powell, Baden (1796–1860). Clergyman and writer. Wrote on optics and theological topics; active in university reform. Vicar of Plumstead, Kent, 1821–7. Savilian Professor of geometry at Oxford University, 1827–60. Author of an article on the study of the evidences of Christianity in *Essays and reviews* (1860). FRS 1824. (*DNB*, *DSB*.)

Prestwich, Joseph (1812–96). Geologist and businessman. Entered the family wine business in London in 1830; became proprietor in 1842. Professor of geology, Oxford University, 1874–88. President of the Geological Society of London, 1870–2. An expert on the Tertiary geology of Europe. Prominent in studies of human prehistory. Knighted, 1896. FRS 1853. (*DNB*, *DSB*.)

Price, John (1803–87). Scholar, school-teacher, and naturalist. Educated at Shrewsbury School, 1818–22, and St John's College, Cambridge. Assistant master, Shrewsbury, 1826–7. Headmaster of the junior department at Bristol College, then classics principal at the Liverpool High School, before settling in Chester. A founding member of the Chester Natural Science Society. (*Alum. Cantab.*, *Eagle* (St John's College, Cambridge) 15 (1888): 169–72, *Modern English biography*.)

Pritchard, Charles (1808–93). Clergyman and astronomer. Headmaster of Clapham Grammar School, 1834–62, where he established an observatory. Hulsean Lecturer, Cambridge University, 1867. Savilian Professor of astronomy, Oxford University, 1870–93. President of the Royal Astronomical Society, 1886. FRS 1840. (*Alum. Cantab.*, *DNB*, *DSB*.)

Quain, Richard (1816–98). Physician. MD, University College, London, 1842; fellow, 1843. Appointed to numerous medical commissions and societies. Physician extraordinary to Queen Victoria, 1890. Had a large, fashionable Harley Street practice. Created baronet, 1891. (*DNB*.)

Rae, William Fraser (1835–1905). Scottish-born author and journalist. Called to the bar, 1861, but took up journalism instead. Edited the *Reader*, 1863–4. Special correspondent in Canada and the United States for the *Daily News*; in later life wrote mainly on eighteenth-century English political history. (*DNB*, Sullivan ed. 1984.)

Ramsay, Andrew Crombie (1814–91). Geologist. Appointed to the Geological Survey of Great Britain, 1841; senior director for England and Wales, 1862; director-general, 1871–81. Professor of geology, University College, London, 1847–52; lecturer on geology at the Royal School of Mines, 1852–71. President

Ramsay, Andrew Crombie, cont.
of the Geological Society of London, 1862–4. Knighted, 1881. FRS 1862. (*DNB*, *DSB*.)

Rand, Benjamin Howard (1827–83). Physician and chemist. Lecturer on chemistry, Franklin Institute, 1850; professor of chemistry, Philadelphia Medical College, 1853. Secretary, Philadelphia Academy of Natural Sciences, 1852–64. Author of chemistry textbooks and contributor to medical journals. (*Appleton's cyclopædia of American biography*.)

Ransome, Frederick (1818–93). Inventor of an artificial sandstone. The material was used for building in Britain and in the colonies. (*Modern English biography*.)
6 March 1865, 9 March 1865

Ray Society
[before 7 January 1865], [14–18 January 1865]

Redfield, William C. (1789–1857). American meteorologist. Advocated improvements in steamboat navigation and railway construction. First president of the American Association for the Advancement of Science, 1848. (*DAB*.)

Reeve, Lovell Augustus (1814–65). Conchologist, publisher, and bookseller. Set up a shop in King William Street, Strand, for the sale of natural history specimens and the publication of conchological works; moved to Henrietta Street, Covent Garden, in 1848. Editor and proprietor of the *Literary Gazette*, 1850–6. (*DNB*, Sarjeant 1980–96.)

Lovell Reeve & Co. Publishers and booksellers at 8 King William Street, Strand, until 1848, then at 5 Henrietta Street, Covent Garden, London. (*DNB* s.v. Reeve, Lovell Augustus; *Post Office London directory* 1846, 1863.)

Reynolds, Hannah (*fl. c.* 1789–1822). Daughter of Joseph and Susannah Ball; her mother was a sister of Richard Reynolds. Married her cousin, Richard Reynold's son, William, an ironfounder at Ketley, Shropshire, in 1789. Mother of Susanna Hannah Reynolds. Moved to Broseley, Shropshire, but returned to Ketley after her husband's death in 1803. (*Two Shropshire ironmasters*; this volume, Supplement, letter to Dear Friend, 12 January 1822.)

Reynolds, Joshua (1723–92). Portrait-painter. (*DNB*.)

Reynolds, Richard (1735–1816). Ironfounder and Quaker philanthropist. Partner in the ironworks at Ketley, Shropshire, 1757–89, and Horshay, Shropshire, 1762–89. Bought the manor of Madeley, Shropshire, 1780–1. (*DNB*, *Two Shropshire ironmasters*, *Victoria county history of Shropshire* 11: 35–6.)

Reynolds, Susanna Hannah (b. 1799). Daughter of William Reynolds and Hannah Reynolds. Married John Bartlett. (*Two Shropshire ironmasters*.)

Rice, Thomas Spring. *See* Spring Rice, Thomas.

Richardson, John (1787–1865). Arctic explorer and naturalist. After serving as assistant surgeon and surgeon on various ships, acted as surgeon and naturalist on John Franklin's polar expeditions, 1819–22 and 1825–7. Surgeon to the Chatham division of marines, 1824–38. Appointed physician to the Royal Hospital, Haslar, 1838. Conducted a search expedition for John Franklin, 1847–9.

Retired to Grasmere, where he accomplished much literary work. Knighted, 1846. FRS 1825. (*DNB*; R. Johnson 1976.)
[24 July 1837], [11 August 1837]

Richardson, Mary (1807–45). As Mary Booth, became John Richardson's second wife in 1833. Niece of John Franklin, whom Richardson accompanied on two polar expeditions. (*DNB*.)

Richardson, Samuel (1689–1761). Novelist. (*DNB*.)

Rivers, T. Francis (1831–99). Nurseryman. Son of Thomas Rivers (1798–1877). Known for his experimental work on prolonging the season of various fruit trees. (R. Desmond 1994, *Gardeners' Chronicle* 3d ser. 26 (1899): 179.)

Rivers, Thomas (1798–1877). Nurseryman. Succeeded to the family business in Sawbridgeworth, Hertfordshire, in 1827. Specialised in the cultivation of roses and fruit. Author of works on rose and fruit culture; contributed extensively to gardening journals. A founder of the British Pomological Society, 1854. (*DNB*.)
6 January [1865], 6 July 1865

Rolfe, Laura (1787–1868). Married Robert Monsey Rolfe in 1845. (*DNB* s.v. Rolfe, Robert Monsey.)

Rolfe, Robert Monsey, 1st Baron Cranworth of Cranworth (1790–1868). Statesman and jurist. Whig MP for Penryn and Falmouth, 1832–9. Solicitor-general, 1834 and 1835–9. Created Baron Cranworth of Cranworth, 1850. Lord justice of appeal, 1851–2. Lord chancellor, 1852–8 and 1865–6. Holwood Park, his country residence, was a mile and a half north of Down House. (*DNB*, *Dod's parliamentary companion*, Freeman 1978.)
28 December 1865

Rolle, Friedrich (1827–87). German geologist, palaeontologist, and natural history dealer. Assistant at the Kaiserlich-königliche Hofmineralien-Cabinett, Vienna, 1857–9; associate, 1859–62. Returned to Bad Homburg in 1862. Author of *Ch. Darwin's Lehre von der Entstehung der Arten und ihre Anwendung auf die Schöpfungsgeschichte* (1863). (*ADB*, *BLKO*, G. P. R. Martin and Uschmann 1969, Sarjeant 1980–96.)
6 May [1865]

Römer, Ferdinand (1818–91). German geologist and paleontologist. Professor of mineralogy and geology at the University of Breslau from 1855. (*DBE*.)

Rose, Gustav (1798–1873). German mineralogist. Studied mining at Kömigshütte, 1816–20. Habilitated in mineralogy, University of Berlin, 1823; extraordinary professor, 1826; professor, 1839. Director, mineralogical institute, Berlin, 1856–73. Worked on the geometric and chemical properties of crystals and was instrumental in the discovery of isomorphism. (*ADB*, *DBE*.)

Ross, James Clark (1800–62). Naval officer and polar explorer. Joined the navy in 1812. Discovered the northern magnetic pole in 1831. Employed on the magnetic survey of the United Kingdom, 1838. Commander of an expedition to the Antarctic, 1839–43; and of a search expedition for John Franklin (*DNB*), 1848–9. Knighted, 1843. FRS 1828. (*DNB*.)

Ross, John (1777–1856). Naval officer and Arctic explorer. Entered the service

Ross, John, cont.
of the East India Company in 1794. Joined the Navy as a midshipman in 1799; commander, 1812. Leader of the expeditions in search of the north-west passage, 1818 and 1829–33. Consul in Stockholm, 1839–46. Led a private expedition to search for John Franklin, 1850–1. Knighted 1834. (*DNB*.)

Roth, Justus Ludwig Adolph (1818–92). German geologist. Studied pharmacy in Hamburg; chemistry, mineralogy, and geology at Berlin and Tübingen, then worked as an apothecary from 1844–8. Habilitated in geology, University of Berlin, 1861; extraordinary professor, 1867. Worked on volcanic and plutonic rocks. (*DBE*, Sarjeant 1980–96.)

Routh, Edward John (1831–1907). Mathematician. BA Cambridge, 1854; senior wrangler, James Clerk-Maxwell coming second. From 1862 until his retirement from coaching in 1888 he was the best-known mathematics tutor at Cambridge, having twenty-eight senior wranglers and forty-three Smith prizemen among his pupils. (*Alum. Cantab.*, *DNB*.)

Royer, Clémence Auguste (1830–1902). French author and economist. Studied natural science and philosophy in Switzerland. In Lausanne in 1859, founded a course on logic for women. Translated *Origin* into French in 1862. (*Dictionnaire universel des contemporains*, J. Harvey 1997.)
[April–June 1865]

Royle, John Forbes (1799–1858). Surgeon and naturalist in the service of the East India Company. Superintendent of the botanic garden in Saharanpur, India, 1823–31. Professor of materia medica, King's College, London, 1837. In charge of the East India House correspondence relating to vegetable productions, 1838. A founder of the Philosophical Club of the Royal Society of London, 1847. FRS 1837. (R. Desmond 1977, *DNB*.)

Russell, John, 1st Earl Russell (1792–1878). Statesman. Liberal prime minister, 1846–52 and 1865–6. Home secretary, 1835–9; secretary of state for foreign affairs, 1852–3; for the colonies, 1839 and 1855. Foreign secretary under Lord Palmerston, 1859–65. Created Earl Russell of Kingston Russell, 1861. In later life, occupied with literary work. President of the Royal Historical Society, 1872. FRS 1847. (*DNB*, *Record of the Royal Society of London*.)
[10 March 1847]

Rütimeyer, Karl Ludwig (Ludwig) (1825–95). Swiss palaeozoologist and geographer. Professor of zoology and comparative anatomy, University of Basel, 1855; rector, 1865; professor in the medical and philosophical faculties, 1874–93. Made important contributions to the natural history and evolutionary palaeontology of ungulate mammals. (*DSB*.)
3 January 1865

Sabine, Edward (1788–1883). Astronomer, geophysicist, and army officer. Entered the army in 1803; major-general, 1856; general, 1870. Astronomer to expeditions in search of a north-west passage, 1818 and 1819–20. Appointed one of three scientific advisers to the Admiralty, 1828. General secretary of the British

Association for the Advancement of Science, 1839–52 and 1853–9. Foreign secretary of the Royal Society of London, 1845; treasurer, 1850; president, 1861–71. Knighted, 1869. FRS 1818. (*DNB*, *DSB*.)

Sanderson, John (1820–81). Scottish journalist, trader, botanical collector, and horticulturalist. Worked as a journalist in the Scottish press before emigrating to Durban, South Africa, in 1850. Secretary of the Natal Times Company from 1851. Undertook trading expeditions to the interior of Natal. Representative in municipal and provincial government. President of the Natal Agricultural and Horticultural Society. Supplied Joseph Dalton Hooker and William Henry Harvey with specimens and drawings of South African plants. (Gunn and Codd 1981.)

Say, Thomas (1787–1834). American entomologist and conchologist. A founder of the Academy of Natural Sciences of Philadelphia, 1812. Curator of the American Philosophical Society, 1821–7; professor of natural history, University of Pennsylvania, 1822–8. Member of a utopian community at New Harmony, Indiana, where he carried out his scientific work, 1825–34. (*DAB*, *DSB*.)

Scharf, George (1788–1860). Bavarian-born draughtsman and lithographer. Lived in London. (*DNB*.)

Schaum, Hermann Rudolph (1819–65). German entomologist and doctor. Practised medicine in Stettin until 1846, when he turned to entomology. Habilitated in zoology at the University of Berlin, 1851; extraordinary professor, 1857. Entomological editor for the *Archiv für Naturgeschichte*, 1848–52. Travelled in Europe, America, and Egypt; taught entomology, natural history, and medical zoology. (*Berliner Entomologische Zeitschrift* 9 (1865): 397–406.)

Schleicher, August (1821–68). German linguist and philologist. Professor of philology, University of Prague, 1853–6; University of Jena, 1857–68. (*ADB*.)
9 February 1865

Schultze, Max Johann Sigismund (1825–74). Anatomist. MD, Greifswald, 1849; Privatdozent, 1850–4. Assistant professor of anatomy, Halle, 1854–9. Professor of anatomy and director of the anatomical institute, Bonn, 1859. Founder and editor of *Archiv für mikroskopische Anatomie*, 1865–74. Noted for his work in microscopy, the reform of cell theory, and descriptive and taxonomic studies of rhizopods and sponges. (*DSB*, *DBE*.)

Sclater, Philip Lutley (1829–1913). Lawyer and ornithologist. One of the founders of *Ibis*, 1858; editor, 1858–65 and 1878–1912. Secretary of the Zoological Society of London, 1860–1903. FRS 1861. (*DSB*, Scherren 1905.)

Scott, John (1836–80). Scottish botanist. Gardener at Chatsworth House, Derbyshire, before becoming foreman of the propagating department at the Royal Botanic Garden, Edinburgh, in 1859. Through CD's patronage emigrated to India in 1864, and worked briefly on a *Cinchona* plantation before taking a position as curator of the Calcutta botanic garden in 1865. Fellow of the Linnean Society of London, 1873. Carried out numerous botanical experiments and observations on CD's behalf. (R. Desmond 1994, Freeman 1978, Letter from John Scott,

Scott, John, cont.
21 July 1865, *Transactions of the Botanical Society* [*of Edinburgh*] 14 (1883): 160–1.)
8 January [1863], *20 January 1865*, *10 April 1865*, *21 July 1865*

Scott, Robert Henry (1833–1916). Irish mineralogist and meteorologist. Keeper of minerals and lecturer on mineralogy, Royal Dublin Society, 1862. Director of the Meteorological Department, 1867–1900. President of the Mineralogical Society, 1888–91. FRS 1870. (*Mineralogical Magazine* 18 (1917): 143–4.)

Sedgwick, Adam (1785–1873). Geologist and clergyman. Woodwardian Professor of geology, Cambridge University, 1818–73. Prebendary of Norwich Cathedral, 1834–73. President, Geological Society of London, 1829–31; British Association for the Advancement of Science, 1833. FRS 1821. (*DNB*, *DSB*, *Record of the Royal Society of London*.)
11 October [1850]

Seemann, Berthold Carl (1825–71). German-born traveller and botanist. Studied botany at the Royal Botanic Gardens, Kew, 1844–6. Naturalist to HMS *Herald*, 1847–51. Commissioned in 1860 to report on the Fiji Islands and published a botanical catalogue of the islands. Editor of *Bonplandia*, 1853–62; of the *Journal of Botany, British and Foreign*, 1863–71. Travelled in Venezuela in 1864; in Nicaragua, 1866–7. (R. Desmond 1994, *DNB*.)
24 April 1862

Sethe, Anna (1835–64). Cousin of Ernst Haeckel, whom she married in 1862. (*DSB* s.v. Haeckel, Ernst; Uschmann 1984, p. 317.)

Seymour, Edward Adolphus Seymour, 12th duke of Somerset (1804–85). Statesman. Liberal MP for Okehampton, 1830; for Totnes, 1834–55. Lord of the Treasury, 1835; secretary to the Board of Control, 1839; under-secretary, Home Department, 1841; commissioner of works, 1851–2; first lord of the Admiralty, 1859–66. (*DNB*.)
May 1851

Sharpey, William (1802–80). Scottish physiologist. MD, Edinburgh University, 1823. Joint lecturer on systematic anatomy, Edinburgh University, 1832. Professor of anatomy and physiology, University College, London, 1836–74. Examiner in anatomy, London University, from 1840. Secretary of the Royal Society of London, 1853–72. Member of the General Medical Council, 1861–76. FRS 1839. (*DNB*, *DSB*.)
9 April [1857]

Shaw, Charles (fl. 1808–65). Naval officer. Entered the Navy in 1808; lieutenant, 1827; employed in the Coast Guard, 1828–47. Listed as secretary of the Welsh Charity School, 55 Charing Cross, London, in 1865. (O'Byrne 1849, *Post Office London Directory* 1865).
3 October 1865

Shaw, George (1751–1813). Assistant lecturer in botany, Oxford University, 1786. A founder of the Linnean Society of London, 1788. Assistant keeper of the natural history collections in the British Museum, 1791; keeper, 1807–13. FRS 1789. (*Aust.*

dict. biog., *DNB*.)

Shaw, James (1826–96). Scottish writer and schoolmaster. Apprenticed as a pattern designer. Worked in the calico printing trade. Began training as a schoolmaster in 1855. After a succession of posts, became schoolmaster in Tynron, Dumfriesshire, 1862–96. Member of the Dumfriesshire and Galloway Natural History and Antiquarian Society. (R. Wallace ed. 1899.)
20 November 1865, 30 November 1865

Sherman, William Tecumseh (1820–91). American army officer. Union general in the American Civil War; general commanding the army, 1869. (*DAB*.)

Siebold, Karl Theodor Ernst von (1804–85). German zoologist and doctor. District physician, Heilsburg, 1831; Königsberg, 1834. Extraordinary professor of zoology, comparative anatomy, and veterinary medicine, Erlangen, 1840; Freiburg, 1845. Professor of physiology, Breslau, 1850. Professor of physiology and comparative anatomy, Munich, 1853; of zoology and comparative anatomy, 1855. Curator, Bavarian state zoological collection, 1855. Co-founder and editor of *Zeitschrift für wissenschaftliche Zoologie*, 1848. A supporter of Darwinian transmutation theory; did research on generation, especially parthenogenesis. (*ADB*, *DBE*, *DSB*.)

Silliman, Benjamin (1779–1864). American chemist, geologist, and mineralogist. Professor of chemistry and natural history, Yale University, 1802–53. Founder and first editor of the *American Journal of Science and Arts*, 1818. (*DAB*, *DSB*.)
26 February 1840

Silliman, Benjamin Jr (1816–85). American chemist. Teaching assistant to his father, Benjamin Silliman, 1837. An editor of the *American Journal of Science and Arts*, 1838–85. Professor of practical chemistry, Yale University, 1846; succeeded his father as professor of chemistry and natural history in 1853. Professor of chemistry in the medical department, University of Louisville, Kentucky, 1849–54. Consultant to the petroleum and mining industries. (*DAB*, *DSB*.)

Simmonds, Peter Lund (1814–97). Danish-born journalist and publisher. Originally named Peter Lund, he assumed the surname Simmonds on his adoption by a British naval lieutenant of that name. Midshipman on board the brig *Cygnet*, 1826. Sugar planter in Jamaica, 1831–4. Secretary to the founder of the Royal Agricultural Society from 1841. Founder of *Simmonds's Colonial Magazine and Foreign Miscellany*; editor, 1844–9. Newspaper and advertising agent in London, 1849–59. Owner and editor of the *Technologist*, 1860–6; of the *Journal of Applied Science*, 1870–81. Wrote for a wide variety of journals, and published several books on commercial agriculture. (R. Desmond 1994, *Men and women of the time* 1895, *Modern English biography*, *NUC*, *Post Office London directory* 1849–59, *Simmonds's Colonial Magazine and Foreign Miscellany* 1–16 (1844–9).)
25 February [1849]

Sims, Thomas (1826–1910). Photographer. Established a photographic business in Weston-super-Mare in 1847. Moved to London and opened a studio at 44 Upper Albany Street, Regents Park, 1854–5. Following a legal dispute about patents, he

Sims, Thomas, cont.
re-established his studio in Westbourne Grove, West London, in 1860. In 1868, moved to Tunbridge Wells, where he worked as a photographer until his death. Married Alfred Russel Wallace's sister, Frances. (Letter from A. R. Wallace, 31 January [1865]; Pritchard 1994, A. R. Wallace 1905, 1: 15, 263.)

Sinclair, John (1754–1835). Politician. First president of the Board of Agriculture. Father of John Sinclair (1797–1875). (*DNB.*)

Sinclair, John (1797–1875). Theologian. Ordained deacon, 1819; priest, 1820. Secretary, National Society, 1839. Vicar of Kensington, 1843–75; archdeacon of Middlesex, 1844–75. Author of sermons and works on the Church of England. (*DNB.*)

Skirving, Robert Scot. Of Edinburgh. Author of several tracts on the labour question in Scotland. (*BLC*, *NUC.*)

Sloane, Hans (1660–1753). Physician and collector. Founded the Chelsea Physic Garden, 1721. Bequeathed his collections to the nation in 1753, leading to the foundation of the British Museum in 1754. (*DNB.*)

Smith, Alexander (1832–65). Botanist. Son of John Smith (1798–1888). Curator, museum of the Royal Botanic Gardens, Kew, 1856–8; herbarium clerk from 1864. Interested in economic botany. (*Journal of Botany* 3 (1865): 199–200.)

Smith, Andrew (1797–1872). Army surgeon. Stationed in South Africa, 1821–37. Principal medical officer at Fort Pitt, Chatham, 1837; deputy inspector-general, 1845. Director-general, Army Medical Department, 1853–8. An authority on South African zoology. Knighted, 1859. FRS 1857. (*DNB.*)

Smith, Edmund (1804–64). Surgeon and hydropathic doctor. In practice in Huddersfield, 1847–9; in Sheffield, 1849–58. Proprietor of Ilkley Wells hydropathic establishment, Ilkley, Yorkshire, 1858–64. (*Medical directory* 1848–65; Metcalfe 1906, p. 107.)

Smith, Elder, & Company. Partnership of George Smith (1789–1846) and Alexander Elder (1790–1876). Publishers and East India agents. (*DNB*, *Modern English biography*.)

6 June [1846]

Smith, John (1798–1888). Scottish gardener and pteridologist. Gardener at the Royal Botanic Garden, Edinburgh, 1818; at the Royal Botanic Gardens, Kew, 1822. Curator of the Royal Botanic Gardens, Kew, 1842–64. (R. Desmond 1994, *Taxonomic literature.*)

Smith, John (1821–88). Scottish gardener. Gardener to the duke of Roxburgh; to the duke of Northumberland at Syon House, Middlesex, 1859–64. Curator, Royal Botanic Gardens, Kew, 1864–86. (R. Desmond 1994.)

Smyth, Henry (1816–91). Military officer. Commissioned in the Sixty-eighth Foot Regiment, 1833; lieutenant, 1836; captain, 1842; major, 1848; lieutenant-colonel, 1853; colonel, 1854; major-general, 1868; lieutenant-general, 1876; general, 1878. Brother of Maria Isabella FitzRoy, Robert FitzRoy's second wife. (*Army list* 1865–92, J. Foster 1874, *Modern English biography.*)

Smyth, John George (1815–69). Of Heath Hall, Wakefield, Yorkshire. MP for York, 1847–65. Justice of the peace and deputy lieutenant for the North and West Ridings of Yorkshire. Colonel of the Second West Yorkshire Militia. Brother of Maria Isabella FitzRoy, Robert FitzRoy's second wife. (*Alum. Cantab.*, *County families* 1864, J. Foster 1874.)

Smyth, Maria Isabella. *See* FitzRoy, Maria Isabella.

Smyth, Warington Wilkinson (1817–90). Geologist and mineralogist. Mining geologist, the Geological Survey, 1844. Lecturer in mining and mineralogy, School of Mines, 1851–81; lecturer in mining, 1881–90. Inspector of crown minerals, 1857. Honorary secretary, Geological Society of London, 1856–66; president, 1866–8. FRS 1858. (*DNB.*)

Snow, George (1820/1–85). Coal-dealer, Down, Kent. Operated a weekly carrier service between Down and London. (Census returns 1861 (Public Record Office RG9/462: 72); gravestone inscription, Down churchyard; *Post Office directory of the six home counties* 1862.)

Sorby, Henry Clifton (1826–1908). Geologist. Pioneered microscopic petrology. President of the Geological Society of London, 1878–80. Established the chair of geology at Sheffield University. FRS 1857. (*DNB*, *DSB.*)

Sowerby, George Brettingham (1788–1854). Conchologist and artist. Father of George Brettingham Sowerby, Jr. Described the *Beagle* fossil shells. (*DNB*, Sarjeant 1980.)

Sowerby, George Brettingham Jr (1812–84). Conchologist and illustrator. Assisted his father, George Brettingham Sowerby, in a business selling natural history specimens; succeeded to the business in 1854. Illustrated numerous works on shells. Fellow of the Linnean Society of London, 1844. (*DNB.*)

Spence, William (1783–1860). Entomologist. A founder of the Entomological Society of London, 1833; president, 1847. Author, with William Kirby, of *Introduction to entomology* (1815–26). FRS 1834. (*DNB*, Gilbert 1977.)

Spencer, Herbert (1820–1903). Philosopher. Employed as a civil engineer on the railways, 1837–41 and 1844–6. Became sub-editor of the *Pilot*, a newspaper devoted to the suffrage movement, in 1844. Sub-editor of the *Economist*, 1848–53. From 1852, author of papers on evolution and numerous works on philosophy and the social sciences. (*DNB*, *DSB.*)

22 April 1865

Spottiswoode, William (1825–83). Mathematician and physicist. Succeeded his father as queen's printer in 1846. Throughout his life pursued mathematical studies in which he supplied new proofs of known theorems and also did important original work; produced a series of memoirs on the contact of curves and surfaces. President of the mathematical section of the British Association for the Advancement of Science, 1865; of the Royal Society of London, 1878–83. FRS 1853. (*DNB.*)

Sprengel, Christian Konrad (1750–1816). German botanist. Rector of the Great Lutheran Town School, Spandau, where he taught languages and natural

Sprengel, Christian Konrad, cont.
science, 1780–94. Moved to Berlin, where he worked as a private tutor. Published his major work on the pollination of flowers in 1793. (*ADB*, *DSB*.)

Spring Rice, Thomas (1790–1866). Statesman. Secretary to the Treasury, 1830–4. Whig MP for Cambridge borough, 1832–9. Chancellor of the Exchequer, 1835–9. Created Baron Monteagle of Brandon in Kerry, 1839. FRS 1841. (*Alum. Cantab.*, *DNB*.)

Spruce, Richard (1817–93). Botanist and school-teacher. School-teacher at Haxby and at the Collegiate School, York. Collected plant specimens in the Pyrenees, 1845–6; in South America, 1849–64. PhD, Berlin University, 1864. Retired in poor health to Coneysthorpe, Yorkshire, where he worked on his plant collections. (R. Desmond 1994, *DNB*.)

Stables, William Alexander (1810–90). Naturalist and botanist. Factor for Cawdor, near Inverness. (R. Desmond 1994.)

Stainton, Henry Tibbats (1822–92). Entomologist. Founder and editor of the *Entomologist's Annual*, 1855–74, and of the *Entomologist's Weekly Intelligencer*, 1856–61. Secretary to the Ray Society, 1861–72; to the natural history section of the British Association for the Advancement of Science, 1864 and 1867–72. Co-founder, *Entomologists' Monthly Magazine*, 1864. FRS 1867. (*DNB*.)

Stanhope, Emily Harriet, Lady Mahon (1815–73). Married Philip Henry Stanhope, Viscount Mahon, in 1834. (*Complete peerage*.)
9 August [1849]

Stanhope, Philip Henry, 5th Earl Stanhope (1805–75). Historian and politician. Styled Viscount Mahon from 1816 until 1855 when he succeeded to the earldom. Tory MP for Wootton Bassett, 1830–2; for Hertford, 1832 and 1835–52. Under-secretary for foreign affairs, 1834–5; secretary to the board of control for India, 1845–6. Instrumental in the founding of the National Portrait Gallery and the Historical Manuscripts Commission. FRS 1827. (*DNB*.)

Stanley, Arthur Penrhyn (1815–81). Cleric, ecclesiastical historian, and travel writer. Fellow of University College, Oxford, 1838–50; secretary of the Oxford University commission, 1850–2. Canon of Canterbury, 1851. Professor of ecclesiastical history, Oxford University, 1856; canon of Christ Church, 1858. Dean of Westminster, 1864–81. (*DNB*.)

Stanley, Edward George Geoffrey Smith, 14th earl of Derby (1799–1869). Statesman. MP for North Lancashire, 1832–44. Colonial secretary, 1833–4 and 1841–4. Prime minister, 1852, 1858–9, and 1866–8. Appointed chancellor of Oxford University, 1852. Supported parliamentary reform. (*Burke's peerage* 1980, *DNB*.)

Startin, James (1806–72). Surgeon and skin specialist. Trained at St Bartholomew's, London, and practised at the General Hospital and Town Infirmary, Birmingham, before going to France to study at the Hôpital Saint-Louis and Montpellier, specialising in skin diseases. Became the senior surgeon at the Hospital for Diseases of the Skin, Blackfriars; also established a private consultancy at 3 Savile Row, Piccadilly. Member, Royal College of Surgeons of England, 1828; fellow, 1852. (*Medical directory* 1863, Plarr 1930.)

Steenstrup, Johannes Japetus Smith (Japetus) (1813–97). Danish zoologist. Professor of zoology and director of the Zoology Museum, University of Copenhagen, 1846–85. Foreign member, Royal Society of London, 1863. (*DBL*, *DSB*.)

Stephens, John Lloyd (1805–52). American lawyer, travel writer, and railroad executive. Practised law in New York, *circa* 1826–34. Wrote books based on travels in the Mediterranean, eastern Europe, Chiapas, Yucatan, and Central America. Vice-president, then president, of Panama Railroad from 1849. (*DAB*.)

Stewardson, Thomas (1807–68). American physician and naturalist. Lived in Philadelphia, Pennsylvania. For a time, physician to the Pennsylvania Hospital. (*CDEL*.)

16 April 1860

Stewart, Mr. Printer. Partner in Stewart and Murray. Related to Alexander Elder of Smith, Elder, & Company. (*DNB* s.v. Smith, George (1824–1901).)

Stewart, Robert, Viscount Castlereagh and 2d marquis of Londonderry (1769–1822). Statesman. Chief secretary of Ireland, 1798–1801. President, (East India) Board of Control, 1802–5. Secretary of state for the War and Colonial Department, 1805; secretary of state for war, 1807–9. Foreign secretary, 1812–22. Marquis of Londonderry, 1821. (*DNB*.)

Stewart and Murray. London printers, with premises at 15 Old Bailey and Green Arbour Court, London, 1838–47. Taken over by Smith, Elder, & Co., 1855. Printers of *Zoology* and *Geology of the 'Beagle'*. (B. Bell 1995, P. A. H. Brown 1982.)

Stokes, John Lort (1812–85). Naval officer. Midshipman on HMS *Beagle*, 1825–30; mate and assistant surveyor, 1831–6; lieutenant, 1837–41; commander, 1841–3. Captain on HMS *Acheron*, surveying off New Zealand, 1847–51; on half pay, 1851–60; employed surveying the Channel coasts, 1860–3; rear-admiral, 1864; vice-admiral, 1871; admiral, 1877. (*DNB*, *Navy list* 1847–64, O'Byrne 1849.)

[after 31 May 1837], 2 January [1847]

Strachey, Richard (1817–1908). Army officer. Served in the Bengal engineers; lieutenant, 1841; lieutenant-general, 1875. Carried out scientific expeditions in the Himalayas. Collected plants in Tibet, 1848. Served in government in India, 1862–71. Member of the Council of India, 1875–89. (*DNB*, Sarjeant 1980–96.)

Strauss, David Friedrich (1808–74). German theologian and writer. (*ADB*, *DBE*.)

Strzelecki, Paul Edmund de (1796–1873). Polish-born explorer. Led an expedition to the interior of New South Wales, Australia, 1839–40. Knighted, 1848. Became a British subject in 1850. FRS 1853. (*DNB*.)

Sulivan, Bartholomew James (1810–90). Naval officer and hydrographer. Lieutenant on HMS *Beagle*, 1831–6. Surveyed the Falkland Islands in HMS *Arrow*, 1838–9. Commander of HMS *Philomel*, 1842–6. Resided in the Falkland Islands, 1848–51. Commanded HMS *Lightning* in the Baltic, 1854–5. Naval officer in the marine department of the Board of Trade, 1856–65. Admiral, 1877. Knighted, 1869. (*DNB*.)

Sulivan, Bartholomew James, cont.

8 May [1865], 31 May [1865]

Sulivan, Catherine Sabine. Daughter of Sophia and Bartholomew James Sulivan. Married William Wallace Trench in 1870. (Baptism register, Holy Trinity Church, Bournemouth (10 September 1871): 121; Census returns 1881 (Public Record Office RG11/1194: 136); Marriage register, Holy Trinity Church, Bournemouth (1870): 2.)

Sulivan, Henry Norton. Son of Sophia and Bartholomew James Sulivan; edited his father's *Life and letters*. (Sulivan ed. 1896.)

Sulivan, James Young Falkland (1838/9–1901). Naval officer. Eldest son of Sophia and Bartholomew James Sulivan. Sub-lieutenant on HMS *Firefly* in the Mediterranean, 1864; lieutenant on HMS *Nassau*, surveying the Straits of Magellan, 1866–7; on HMS *Favorite* in North America and the West Indies, 1868–9; on HMS *Indus* at Devonport, 1870. Stationed in China, 1870–5; studied at the Royal Naval College, Greenwich, 1875–6; lieutenant and commander of HMS *Britomart* at Southampton, 1876–9; served in the Coast Guard, 1882–8. Commander, 1889. (*DNB* s.v. Sulivan, Bartholomew James, *Navy list* 1864–1901.)

Sulivan, Sophia (1809/10–90). Daughter of vice-admiral James Young, of Barton End, near Stroud, Gloucestershire. Married Bartholomew James Sulivan in 1837. (*County families* 1871, s.v. Sulivan, B. J.; *DNB*, s.v. Sulivan, B. J.; Sulivan ed. 1896, pp. xii, 395.)

Sulivan, Sophia Henrietta (b. 1837/8). Daughter of Sophia and Bartholomew James Sulivan. Married the surgeon Henry Bullock in 1882. (Census returns 1881 (Public Record Office RG11/1194: 136); Marriage register, Holy Trinity Church, Bournemouth, (1882): 176.)

Sulivan, Thomas Edward (d. 1873). Naval officer. Second son of Sophia and Bartholomew James Sulivan. Sub-lieutenant on HMS *Gannet*, 1865–6, and lieutenant on HMS *Duncan*, stationed in North America and the West Indies, 1866–7; Transport Service, 1867–8; surveyed the Mediterranean in HMS *Newport*, 1868–70; lieutenant on HMS *Pert*, stationed off the Cape of Good Hope, 1870, and the south-east coast of America, 1871–3. (*Navy list* 1865–73, Sulivan ed. 1896.)

Swinhoe, Robert (1836–77). Diplomat and ornithologist. Attached to the British consulate in Hong Kong, 1854; in Amoy, China, 1855. British vice-consul, Taiwan, 1860–5; consul, 1865–73. Acting consul, Amoy, 1865–71; Ning-po, 1871–3. Consul, Ning-po, 1873–5. Collected plants and animals in Eastern Asia. FRS 1876. (*Foreign Office list* 1877, P. B. Hall 1987.)

[before 1 October 1865?]

Tankerville, earl of *See* Bennet, Charles Augustus.

Taylor, Tom (1817–80). Playwright, classicist, lawyer, and public servant. Fellow of Trinity College, Cambridge, 1842–4; professor of English, London University, 1845. Called to the bar, 1846. Assistant secretary to the Board of Health; later, assistant secretary to the sanitary department of the Local Government Board,

1850–4; secretary, 1854–71. Editor of *Punch*, 1874–80. In thirty-five years, more than seventy of his plays appeared in principal London theatres. (*DNB*.)

Tegetmeier, William Bernhard (1816–1912). Editor, journalist, lecturer, and naturalist. Pigeon-fancier and expert on poultry. Pigeon and poultry editor of the *Field*, 1864–1907. Secretary of the Apiarian Society of London. (*Field*, 23 November 1912, p. 1070; E. W. Richardson 1916.)

8 May [1861], 27 February [1865], *[28 February – 5 March 1865]*, 6 [March 1865], *13 March 1865*, 14 March [1865], *27 March 1865*, 28 March [1865], *[29 March – 7 April 1865]*, 6 April [1865], [7 April 1865], *10 April 1865*, *17 April 1865* (John Murray and R. F. Cooke), 2 June [1865], *20 June 1865*

Temple, Emily Mary, 3d Viscountess Palmerston (1787–1869). Daughter of Peniston Lamb, first Viscount Melbourne. In 1805, married Peter Leopold, fifth Earl Cowper, who died in 1837. Married Henry John Temple, third Viscount Palmerston, in 1839, and often entertained the political élite in her several homes. Inherited her family's estates in Hertfordshire and Derbyshire in 1853. (Airlee 1922, *Modern English biography*.)

Temple, Henry John, 3d Viscount Palmerston (1784–1865). Statesman. Succeeded to the Irish peerage in 1802. Secretary, War Office, 1809–28. Foreign secretary, 1830–41 and 1846–51. Home secretary, 1852–5. Prime minister, 1855–8 and 1859–65. FRS 1853. (*DNB*.)

Tennyson, Alfred, 1st Baron Tennyson (1809–92). Poet. Poet laureate, 1850. Created Baron Tennyson, 1884. (*DNB*.)

Terence (Publius Terentius Afer) (d. 159?). Roman dramatist. (*Oxford classical dictionary*.)

Tessier, Alexandre Henri (1741–1837). French agronomist. Studied natural sciences at Paris. Member of the Société Royale Médecine, 1776; of the Académie des Sciences, 1783. Author of a number of works and papers on agriculture; publisher of the *Journal d'Agriculture*. (*NBU*.)

Thayer, Nathaniel (1808–83). American financier and philanthropist. Partner in the Boston investment banking house John E. Thayer & Brother. Benefactor of Harvard College; sponsored the construction of a herbarium, 1863–5. Overseer, Harvard College, 1866; member of the board of fellows, 1868–75. (*DAB*; Dupree 1959, pp. 237–8.)

Thomson, Thomas (1817–78). Naturalist. MD, Glasgow, 1839. Travelled to India as assistant surgeon to the East India Company. Curator of the Asiatic Society's museum, Calcutta, 1840. Accompanied Joseph Dalton Hooker to the Himalayas, 1849–51, and collaborated with him on various botanical publications. Superintendent of the Calcutta botanic garden and professor of botany at the Calcutta Medical College from 1854 until his return to England in 1860 or 1861. FRS 1855. (R. Desmond 1994, *DNB*.)

Thulstrup, Carl Magnus (1805–81). Swedish army officer and politician. Colonel, 1853. Had a significant parliamentary career from 1860. Member of numerous learned societies. Knighted, 1858. (*SMK*.)

Thwaites, George Henry Kendrick (1811–82). Botanist and entomologist. Superintendent of the Peradeniya botanic gardens, Ceylon, 1849; director, 1857–80. FRS 1865. (R. Desmond 1994, *DNB*.)

Thwaites, Henry John Bardwell (b. 1847). BA, Cambridge, 1868. Admitted at the Inner Temple, 1866; called to the bar, 1869. (*Alum. Cantab.*, *Men-at-the-bar*.)

Thwaites, John (1815–70). Draper. Chairman of the Metropolitan Board of Works, London, 1855–70. Knighted in recognition of the opening of the main drainage-works of London. (*Modern English biography*, *Men and women of the time* 1868.)

Ticknor, George (1791–1871). American educator and author. Professor of French, Spanish, and belles-lettres at Harvard University, 1819–35. Travelled widely in Europe. (*DAB*.)

Tournay, Thomas (*fl.* 1850s). Resident at Brockhill House, Saltwood, Kent. (*Post Office directory of the six home counties*, 1851, 1855; see also this volume, Supplement, letter from [John Innes], [after 8 February 1855 – August 1855].)

Travers, Henry Hammersley (1844–1928). Plant collector. Studied New Zealand flora with his father, William Thomas Locke Travers. Collected plants in Chatham Islands for Ferdinand von Mueller in 1864, and for the Botanic Garden, Wellington, 1870–1. (R. Desmond 1994.)

Treviranus, Ludolph Christian (1779–1864). German botanist. Professor of botany, University of Breslau, 1816–30; University of Bonn, 1830–64. (*ADB*, *DSB*.)

Trimen, Roland (1840–1916). Zoologist and entomologist. Emigrated to South Africa in 1858. Arranged the Lepidoptera at the South African Museum. Held civil-service positions in the Commission of Land and Public Works, the governor's office, and the colonial secretary's office; became part-time curator of the South African Museum in 1873; full-time curator, 1876. FRS 1883. (*DSAB*.)
13 December 1865

Tristram, Henry Baker (1822–1906). Clergyman and ornithologist. Secretary to the governor of Bermuda, 1847–9. Rector of Castle Eden, Durham, 1849–60. Canon residentiary of Durham, 1874. Formed extensive ornithological collections from travels in Algeria, Palestine, Egypt, and Japan. FRS 1868. (*DNB*.)

Trollope, Anthony (1815–82). Novelist and Post Office official. (*DNB*.)

Trübner, Johann Nicolaus (Nicholas) (1817–84). German-born publisher and oriental scholar. Foreign corresponding clerk for Longmans (publishers), 1843–51. In partnership with Thomas Delf and David Nutt, established a successful London publishing house specialising in oriental studies and translations of works in philology, philosophy, and religion. (*ADB*, *DNB*.)

Trübner & Co. Booksellers at Paternoster Row, London, specialising in oriental, American and European publications (*Post Office London directory*). Founded by Nicholas Trübner (*DNB*).

Turner, Dawson William (1815–85). Philanthropist and education writer. Son of Dawson Turner; uncle of Joseph Dalton Hooker. Ordained deacon, 1840. Headmaster of the Royal Institution School, Liverpool, 1851–85. Benefactor of

London hospitals. (*DNB* s.v. Turner, Dawson; *Modern English biography*.)

Turner, Elizabeth. Daughter of Mary Palgrave and Dawson Turner. Aunt of Joseph Dalton Hooker. Married the historian Francis Palgrave in 1823. (*DNB* s.v. Palgrave, Francis; L. Huxley ed. 1918.)

Tylor, Alfred (1824–84). Geologist. Educated at the School of the Society of Friends, Grove House, Tottenham. Entered the family brassfoundry business in 1840, but studied geology and anatomy in his spare time. Wrote on the Quaternary period. Brother of Edward Burnett Tylor. (*DNB*, Sarjeant 1980–96.)

Tylor, Edward Burnett (1832–1917). Anthropologist. Educated at the School of the Society of Friends, Grove House, Tottenham. Keeper of Oxford University Museum, 1883; reader in anthropology, Oxford University, 1883; professor, 1896; professor emeritus, 1909. President of the Anthropological Society, 1879–80, 1891–2. Brother of Alfred Tylor. Knighted, 1912. FRS 1871. (*DNB*, *Men and women of the time* 1899.)

Tyndall, John (1820–93). Irish physicist, lecturer, and populariser of science. Professor of natural philosophy, Royal Institution of Great Britain, 1853; superintendent, 1867–87. Scientific adviser to Trinity House and the Board of Trade, 1866–83. FRS 1852. (*DNB*, *DSB*.)

Unger, Franz (1800–70). Austrian botanist. Practised medicine in Stockerau and Kitzbühel, 1828–35. Professor of botany and zoology and director of the botanic garden, Johanneum, Graz, 1835–49. Professor of plant anatomy and physiology, University of Vienna, 1850–69. Considered one of the founders of palaeobotany; wrote in favour of species transmutation as early as 1852 and suggested hybridisation as a source of new species. (*ADB*, *DBE*, *DSB*.)

Vaucher, Jean-Pierre-Etienne (1763–1841). Swiss botanist and clergyman. Parish priest, Geneva, 1797–1822. Honorary professor of botany, Geneva University, 1798–1807; of ecclesiastical history, 1807–39; rector, 1818–21. Founding member of the Société de Physique et d'Histoire Naturelle de Genève. Wrote extensively on plant physiology. (*DSB*.)

Vaux, William Sanson (1811–82). American mineralogist and archaeologist. President, Zoological Society of Philadelphia; vice-president, Philadelphia Academy of Natural Sciences, 1864–73, 1876–82. Founder member, American Association for the Advancement of Science; treasurer, 1871–82. (*Appleton's cyclopædia of American biography*.)

Veitch, James (1815–69). Nurseryman. In partnership with his father, James Veitch (1792–1863), proprietor of a nursery in Chelsea, London, from 1853. (R. Desmond 1994.)

Verlot, Bernard (1836–97). French botanist and horticulturalist. Head of propagation, Muséum d'Histoire Naturelle. (*NUC*, Tort 1996, Verlot 1864.)

Victor Emmanuel II (1820–78). King of Sardinia and first king of Italy. (*EB*.)

Victoria (1819–1901). Queen of the United Kingdom of Great Britain and Ireland, and empress of India. Succeeded to the throne in 1837; designated empress of India, 1876. (*DNB*.)

Victoria Adelaide Mary Louise, Princess Royal of Great Britain and Empress of Germany (1840–1901). Eldest child of Queen Victoria and Prince Albert. Married Prince Frederick William of Prussia in 1858. (*DNB*.)

Vigors, Nicholas Aylward (1785–1840). Irish zoologist and politician. FRS 1826. (*DNB*.)

Vogt, Carl (1817–95). Swiss naturalist. Collaborated with Louis Agassiz on a treatise on central European freshwater fish. Professor of zoology, Giessen, 1847. Forced to leave Giessen for political reasons, 1848; settled in Geneva. Professor of geology, Geneva, 1852; director of the Institute of Zoology, 1872. (*ADB*, *DSB*.)

Wagner, Nikolai Petrovich (1829–1907). Russian zoologist, entomologist, and writer. Graduated from the University of Kazan, 1854. Professor of zoology, Kazan, 1860; St Petersburg University, 1871. Instrumental in the founding of the Solovki Biological Station (later relocated and renamed Murmansk Marine Biology Institute). Discovered the phenomenon of paedogenesis, or larval reproduction. (Gilbert 1977, *GSE*.)

Wahlberg, Peter (Pehr) Fredrik (1800–77). Swedish botanist and natural historian. Professor of pharmacy and natural history at the Karolinska Institute, 1828–62; of materia medica, 1836–62. Travelled to study botanical collections, natural history museums, and agricultural institutes in Europe. Secretary of the Royal Swedish Academy of Sciences, 1848–66; represented the academy at parliament, 1844–51. (*SMK*.)

Walford, Edward (1823–97). Compiler. Ordained priest, 1847. Edited *Once a Week*, 1859–65; *Gentleman's Magazine*, 1866–8. Engaged in the publication of several biographical and genealogical works including *County families of the United Kingdom* (1860) and *Windsor peerage* (1890). In later life, established a reputation as an antiquary. (*DNB*.)

22 [January–April 1865?]

Walker, Francis (1809–74). Entomologist. Specialised in chalcids, small parasitic Hymenoptera. Described the Chalcididae CD collected on the *Beagle* voyage. Catalogued a number of the insect collections in the British Museum. (*Entomologist's Monthly Magazine* 11 (1874): 140–1, Gilbert 1977.)

Wallace, Alfred Russel (1823–1913). Naturalist. Collector in the Amazon, 1848–52; in the Malay Archipelago, 1854–62. Independently formulated a theory of evolution by natural selection in 1858. Lecturer and author of works on protective coloration, mimicry, and zoogeography. President of the Land Nationalisation Society, 1881. Wrote widely on socialism, spiritualism, and vaccination. FRS 1893. (*DNB*, *DSB*.)

20 January 1865, 29 January [1865], *31 January [1865]*, 1 February [1865], *18 September 1865*, 22 September [1865], *2 October 1865*

Wallace, Annie (1848–1914). Eldest daughter of William Mitten. Married Alfred Russel Wallace in 1866. (Marchant ed. 1916, 2: 252–3.)

Wallich, Nathaniel (1786–1854). Danish-born botanist and surgeon. Surgeon at Serampore, India, 1807–13. Superintendent of the Calcutta botanic garden,

1817–46. Made collections of plants in India and Burma. FRS 1829. (R. Desmond 1994, *DNB*.)

Walsh, Benjamin Dann (1808–69). Entomologist, farmer, and timber merchant. Student at Trinity College, Cambridge, 1827–31; fellow, 1833. Emigrated to the United States, where he farmed in Henry County, Illinois, 1838–51; lumber merchant, Rock Island, Illinois, 1851–8. Retired from commerce in about 1858 and concentrated on entomology, making contributions to agricultural entomology. Suggested the use of natural enemies to control insect pests. Author of several papers in agricultural journals. Associate editor of the *Practical Entomologist*, 1865. Acting state entomologist, Illinois, 1867. Brother of John Henry Walsh. (*Alum. Cantab.*, *DAB*.)
1 March 1865, 27 March [1865], *29 May 1865*, 9 July [1865], *12 November 1865*, 19 December [1865]

Walsh, John Henry (1810–88). Sports writer. Wrote under the name Stonehenge. Practised medicine in London and Worcester before becoming editor of the *Coursing calendar* in 1856. Editor of *The Field*, 1857–88. Published many works on dogs, horses, guns, domestic economy, medicine, and cookery. Fellow of the Royal College of Surgeons, 1844. (*DNB*.)

Waterhouse, George Robert (1810–88). Naturalist. A founder of the Entomological Society of London, 1833. Curator of the Zoological Society of London, 1836–43. Assistant in the mineralogical and geological branch of the natural history department of the British Museum, 1843–50; keeper, 1851–6; keeper, geology department, 1857–80. Described CD's mammalian and entomological specimens from the *Beagle* voyage. (*DNB*, Gilbert 1977.)
[after 22 May 1845]

Watkins, Frederick (1808–88). Clergyman. Attended Shrewsbury School, 1823–6. A friend of CD's at Cambridge University. HM Inspector of Schools, 1841–73. Rector of Long Marston, Yorkshire, 1873–88. Archdeacon of York, 1874–88. (*Alum. Cantab.*)

Watson, Hewett Cottrell (1804–81). Botanist, phytogeographer, and phrenologist. Edited the *Phrenological Journal*, 1837–40. Collected plants in the Azores in 1842. Wrote extensively on the geographical distribution of plants. (*DNB*, *DSB*.)
19 September 1860 (George Gordon), *4 January 1861* (J. D. Hooker), *27 June 1861* (George Gordon)

Watters, Thomas. Diplomat. Joined the foriegn service as a student interpreter in 1863 and was sent to China. Acting consul in Taiwan between 1 March and 31 October 1866. (*Foreign office list* 1868.)

Weale, James Philip Mansel (1838 – after 1899). Orchidist, entomologist, and writer. Resided in South Africa, 1860–78. Curator of the Department of Practical Science Museum, 1864 (Weale 1865). CD communicated his papers on orchid pollination to the Linnean Society of London; his papers on South African insects appeared in the *Transactions of the Entomological Society*. (*Alum. Oxon.*; CD–Weale correspondence (American Philosophical Society; CUL; University of Virginia

Weale, James Philip Mansel, cont.
Library, Charlottesville Va.); Freeman 1978; *Orchids* 2d ed.; *Royal Society catalogue of scientific papers*; *The Times*, 26 September 1899, p. 9; Weale 1891, pp. vii–viii.)
6 May [1865]

Wedgwood, Caroline Sarah (1800–88). CD's sister. Married Josiah Wedgwood III, her cousin, in 1837. (*Darwin pedigree.*)

Wedgwood, Emma. *See* Darwin, Emma.

Wedgwood, Frances (Fanny Frank) (d. 1874). Daughter of the Reverend John Peploe Mosley of Rolleston, Staffordshire. Married Francis Wedgwood in 1832. (*Burke's peerage* 1980, Freeman 1978.)

Wedgwood, Frances Emma Elizabeth (Fanny) (1800–89). Second child of James Mackintosh and Catherine Allen. Married Hensleigh Wedgwood in 1832. (Freeman 1978, O'Leary 1989.)

Wedgwood, Francis (Frank) (1800–88). Master-potter. Partner in the Wedgwood pottery works at Etruria, Staffordshire, until 1876. Emma Darwin's brother. Married Frances Mosley in 1832. (*Alum. Cantab.*)

Wedgwood, Hensleigh (1803–91). Philologist and barrister. Emma Darwin's brother. Fellow, Christ's College, Cambridge, 1829–30. Metropolitan police magistrate at Lambeth, 1832–7; registrar of metropolitan carriages, 1838–49. An original member of the Philological Society, 1842. Published *A dictionary of English etymology* (1859–65). Married Frances Emma Elizabeth Mackintosh in 1832. (*DNB*, Freeman 1978.)

Wedgwood, Hope Elizabeth (Dot) (1844–1934). Daughter of Hensleigh and Frances Emma Elizabeth Wedgwood. Second wife of Godfrey Wedgwood. (Freeman 1978.)

Wedgwood, Josiah I (1730–95). Master-potter. Founded the Wedgwood pottery works at Etruria, Staffordshire. Grandfather of CD and Emma Darwin. Greatly interested in experimental chemistry. Contributed several papers on the measurement of high temperatures to the Royal Society of London's *Philosophical Transactions*. Actively associated with scientists and scientific societies. FRS 1783. (*DNB*, *DSB*.)

Wedgwood, Josiah III (1795–1880). Master-potter. Partner in the Wedgwood pottery works at Etruria, Staffordshire, until 1844, when he moved to Leith Hill Place, Surrey. Emma Darwin's brother. Married CD's sister Caroline, his cousin, in 1837. (Freeman 1978.)

Wedgwood, Katherine Elizabeth Sophy (Sophy) (1842–1911). Daughter of Caroline and Josiah Wedgwood III. CD's niece. (*Darwin pedigree*, Freeman 1978.)

Wedgwood, Katherine Euphemia (Effie) (1839–1931). Daughter of Hensleigh and Frances Emma Elizabeth Wedgwood. Married Thomas Henry Farrer in 1873. (*Burke's peerage* 1980, Freeman 1978.)

Wedgwood, Lucy Caroline (1846–1919). Daughter of Caroline Wedgwood and Josiah Wedgwood III. CD's niece. (*Darwin pedigree*, B. Wedgwood and Wedgwood 1980.)

[April–May 1865?], ?*[May 1865]* ([Emma Darwin?])

Wedgwood, Margaret Susan (1843–1937). Daughter of Caroline Wedgwood and Josiah Wedgwood III. CD's niece. Married Arthur Charles Vaughan Williams in 1869. Mother of Ralph Vaughan Williams (*DNB*). (*Emma Darwin* (1915), Freeman 1978.)

?*[May 1865]* ([Emma Darwin?])

Wedgwood, Sarah Elizabeth (Sarah) (1778–1856). CD and Emma Darwin's aunt. Resided at Camp Hill, Maer Heath, Staffordshire, 1827–47, then moved to Petleys, Down. (*Emma Darwin* (1915).)

Wedgwood, Sarah Elizabeth (Elizabeth) (1793–1880). Emma Darwin's sister. Resided at Maer Hall, Staffordshire, until 1847, then at The Ridge, Hartfield, Sussex, until 1862. Moved to London before settling in Down in 1868. (*Emma Darwin* (1915), Freeman 1978.)

Wedgwood, Susannah. *See* Darwin, Susannah.

Wedgwood, Thomas (1771–1805). Son of Josiah Wedgwood I. Published researches on heat and light, 1791–2. (*DNB*.)

Weir, Harrison William (1824–1906). Animal painter and illustrator. (*DNB*.)

Wells, Luke. Artist. Resided at 59, Great Percy Street, Pentonville, London (*Post Office London directory* 1865.)

Wells, William Charles (1757–1817). Physician. Born in South Carolina, of Scottish parents, and educated in Scotland. Elected assistant physician to St Thomas's Hospital, London, 1795; physician, 1800–17. Awarded the Rumford Medal of the Royal Society of London for his 'Essay on dew' (1814). FRS 1793. (*DNB*.)

Westwood, John Obadiah (1805–93). Entomologist and palaeographer. Founding member of the Entomological Society of London, 1833; honorary president, 1883. Hope Professor of invertebrate zoology, Oxford University, 1861–93. Entomological referee for the *Gardeners' Chronicle.* Royal Society Royal Medallist, 1855. (*DNB*, Gilbert 1977.)

White, Adam (1817–79). Naturalist. Official in the zoological department of the British Museum, 1835–63. Described some of the insect specimens collected on the *Beagle* voyage. (*DNB*, Gilbert 1977.)

White, Walter (1811–93). Miscellaneous writer. Cabinetmaker in Reading until 1834. Travelled in America, 1834–9. Attendant in the library of the Royal Society, 1844; assistant secretary and librarian, 1861–84. Author of many travel books. (*DNB*, *Modern English biography*.)

Wichura, Max Ernst (1817–66). German lawyer and botanist. Prussian government official in Breslau, 1859–66. Devoted his spare time to botany and made frequent excursions to study European flora. Published a number of papers, including one on dimorphism in *Scabiosa* and *Lythrum* (Ueber unvollkommene Diklinie, *Jahresberichte der schlesischen Gesellschaft für vaterländische Kultur* 36 (1858): 65–6), and a major work on hybridisation, *Die Bastardbefruchtung im Pflazenreich erläutert an den Bastarden der Weiden* (1865). (*ADB*.)

3 February [1865]

Wickham, John Clements (1798–1864). Naval officer and magistrate. First lieutenant on HMS *Beagle*, 1831–6; commander, surveying the Australian coast, 1837–41. Settled in Australia in 1842. Police magistrate in New South Wales, 1843–57; government resident, 1857. Left Australia and took his family to southern France in 1862. (*Aust. dict. biog.*, R. Desmond 1994.)

Wiegmann, Arend Friedrich August (1802–41). German naturalist. Extraordinary professor, University of Berlin. Founder and editor of the *Archiv für Naturgeschichte*, known as 'Wiegmann's Archiv', 1835–41. (*NUC.*)

Wilberforce, Samuel (1805–73). Clergyman. Rector of Brightstone, Isle of Wight, 1830–40; of Alverstoke, Hampshire, 1840–3. Chaplain to Prince Albert, 1841. Dean of Westminster, 1845. Bishop of Oxford, 1845–69. FRS 1845. (*DNB*, Sarjeant 1980–96.)

Wilkins, Charles Lorenzo (*fl.* 1839–66). Army officer. Third son of the reverend Walter Wilkins of Hay Castle, Brecon. Ensign, Sixteenth Foot, 1839; lieutenant, 1841; captain, 1848; major, 1859. Transferred to Seventy-sixth Foot by 1866. The family adopted the name de Winton in 1839. (*Burke's landed gentry* 1846, s.v. de Winton of Maesderwen; *Hart's army list* 1840–67.)

Wilkins, Walter (before 1819–51). Clergyman. Father of Charles Lorenzo Wilkins. Adopted the name de Winton in 1839. Resident at Hay Castle, Brecon. Vicar of Llangan and Bronllys, Brecon, and Boughrood, Radnorshire; prebendary of Brecon, 1850–51. (*Burke's landed gentry* 1868, s.v. de Winton; *Clergy list* 1841–51.)

Williams & Norgate. Booksellers and publishers specialising in foreign scientific literature, with premises at Covent Garden, London, and South Frederick Street, Edinburgh. A partnership between Edmund Sydney Williams (1817–91) and Frederick Norgate. Publishers of the *Natural History Review.* (*Modern English biography* s.v. Williams, Edmund Sidney.)
4 March [1861], 25 August [1861]

Williamson, Alexander (1819–70). Scottish gardener. Curator of the pleasure grounds at the Royal Botanic Gardens, Kew, 1848–66. (R. Desmond 1995.)

Wilson, Charles William (1836–1905). Military officer and surveyor. Entered the Royal Engineers in 1855, achieving the rank of major-general in 1894. Led a survey of Jerusalem, 1864–5. First director of the topographical department of the War Office, 1870. Head of the Ordnance Survey in Ireland, 1876–86. Knighted 1881. FRS 1874. (*DNB.*)

Wilson, Henry Bristow (1803–88). Theologian. Fellow, St John's College, Oxford, 1825–50. Rawlinsonian Professor of Anglo-Saxon, Oxford University, 1839–44. Vicar, Great Staughton, Huntingdonshire, 1850–88. Contributed an article to *Essays and reviews* (1860), as a result of which he was prosecuted for heterodoxy in the Court of Arches. His case was tried together with that of Rowland Williams; he was found guilty of three of the articles brought against him, but, as with Williams, the decision was reversed on appeal. (*DNB.*)

Wimmer, Christian Friedrich Heinrich (1803–68). German botanist and educationalist. (*ADB.*)

Wollaston, Thomas Vernon (1822–78). Entomologist and conchologist. Passed many winters in Madeira, where he collected insects and shells. Wrote a series of works on the Coleoptera. (*DNB*, Gilbert 1977.)

Wombwell, George (1777–1850). Menagerist. In 1805, established a travelling menagerie, known as Wombwell's Royal Menagerie from 1830, having received the patronage of Princess Victoria. Exhibited a vast number of animals and contributed to zoological knowledge. (Turner 1995.)

Wood, Charles Alexander (1810–90). Barrister. Commissioner of the Colonial Land Emigration Board, 1844–57; commissioner of emigration, 1859–74. Deputy chairman of Great Western Railway, 1866–90. Knighted, 1874. (*Modern English biography*.)

Wood, David Edward (1812–94). Soldier. Trained at the Royal Military Academy at Woolwich; commissioned in the Royal Artillery, 1829; lieutenant, 1831; captain, 1846; regimental colonel, 1860; general, 1877. Served with distinction in the Cape of Good Hope, 1842, the Crimea campaign, 1855–7, and in suppressing the Indian mutiny, 1857–9, where he served as brigadier-general commanding the field and horse artillery at the final seige of Lucknow. Knighted, 1859. (*DNB*.)

Wood, John. Proprietor of the Star Hotel, Market Place, Shiffnal, Shropshire. (*Post Office directory of Gloucestershire, with Bath, Bristol, Herefordshire, and Shropshire* 1863.)

Wood, Thomas (1804–72). Army officer and politician. Lieutenant, Grenadier Guards, 1827; lieutenant-colonel, 1854–8; lieutenant-general, 1864; colonel, 1868; general, 1872. MP for Middlesex, 1837–47. (*Harrow School register*, *Modern English biography*.)

Woodbury, Thomas White (1818–71). Journalist and beekeeper. From 1850 devoted himself exclusively to beekeeping. Introduced Ligurian bees to Britain in 1859 and developed the 'Woodbury hive', marketed by the London apiarian specialists George Neighbour & Sons. As 'A Devonshire Beekeeper', a regular contributor to the *Cottage Gardener*, the *Journal of Horticulture*, the *Gardeners' Chronicle*, and *The Times*. (R. H. Brown 1975, Ron Brown 1994, Dodd 1983, Fraser 1958, Neighbour 1865.)
[9–22 October 1861]

Woodward, Samuel Pickworth (1821–65). Naturalist. Sub-curator, Geological Society of London, 1839–45. Professor of geology and natural history at the Royal Agricultural College, Cirencester, 1845. First-class assistant in the department of geology and mineralogy, British Museum, 1848–65. (*DNB*, Sarjeant 1980–96.)
10 July 1848

Wooler, William Alexander (1813–91). Landowner, colliery-owner, and agriculturalist, of Halliwell House, Heighington, and Sadberge Hall, Darlington, County Durham. Started the conservative newspaper, the *North-Eastern Independent*, 1879. Founder of Darlington Conservative Association. Member of the Darlington Rural Sanitary Authority. Instrumental in founding the South Durham and North Yorkshire Chamber of Agriculture. (*North Eastern Daily Gazette*, 5 May

Wooler, William Alexander, cont.
1891; *North Star*, 5 May 1891; *Northern Echo*, 5 May 1891; *Post Office directory of the county of Durham* 1873.)

Worsaae, Jens Jacob (1821–85). Danish archaeologist. Inspector of ancient monuments, Denmark. Appointed professor of archaeology at the University of Copenhagen in 1854. Director, National Museum, Copenhagen, 1865–85. (*DSB.*)

Wright, Edward Perceval (1834–1910). Irish naturalist. One of the editors of the *Natural History Review*, 1854–65. Director of the university museum in Dublin, 1857. Lecturer in zoology, Trinity College, Dublin, 1858–68; professor of botany and keeper of the herbarium, 1869. Resigned the professorship in 1904 owing to ill health, but continued to superintend the herbarium. (*DNB.*)
24 March 1865, *31 March 1865*

Wyman, Jeffries (1814–74). American anatomist and ethnologist. Curator of the Lowell Institute, Boston, 1840; lecturer, 1840–1 and 1848–9. Travelled in Europe, 1841–3. Professor of anatomy and physiology, Hampden-Sydney Medical College, Virginia, 1843–8. Hersey Professor of anatomy, Harvard University, 1847–74. Professor and curator, Peabody Museum of American Archaeology and Ethnology, Harvard, 1866–74. (*DAB*, *DSB.*)
22 September 1865, 8 October [1865]

Yarrell, William (1784–1856). Zoologist. Newspaper agent and bookseller in London. Author of standard works on British birds and fishes. Member of the Zoological Society of London from its formation in 1826. (*DNB.*)

Youatt, William (1776–1847). Veterinary surgeon in London. Gave private lectures on veterinary science; from 1830 he delivered these lectures at London University. Founded the *Veterinarian* in 1828. Author of a series of handbooks on the breeds, management, and diseases of farm animals. (*DNB.*)

Zurhorst, Frederick William. Irish merchant and poultry breeder with premises at 17 St Andrew Street, Dublin. Leading breeder of White Cochin poultry. (*Thom's Irish almanac* 1865, letter from W. B. Tegetmeier, 10 April 1865.)

BIBLIOGRAPHY

The following bibliography contains all the books and papers referred to in this volume by author–date reference or by short title. Short titles are used for some standard reference works (e.g., *DNB*, *OED*), for CD's books, and for editions of his letters and manuscripts (e.g., *Descent*, *LL*, *Notebooks*). Works referred to by short titles are listed in alphabetical order according to the title; those given author–date references occur in alphabetical order according to the author's surname. Notes on manuscript sources are given at the end of the bibliography.

Acland, Henry Wentworth. 1865. *The Harveian oration, 1865*. London: Macmillan and Co.

Adams, Ephraim Douglass. 1925. *Great Britain and the American Civil War*. 2 vols. London: Longmans, Green & Co.

Adams, Oscar Fay, ed. 1897. *A dictionary of American authors*. Boston, Mass.: Houghton Mifflin & Co.

ADB: *Allgemeine deutsche Biographie*. Under the auspices of the Historical Commission of the Royal Academy of Sciences. 56 vols. Leipzig: Duncker & Humblot. 1875–1912.

Adhémar, Alphonse-Joseph. 1842. *Révolutions de la mer*. Paris: Carilian-Goeury et V. Dalmont.

Agassiz, George Russell, ed. 1913. *Letters and recollections of Alexander Agassiz, with a sketch of his life and work*. London: Constable & Co.

Agassiz, Jean Louis Rodolphe. 1840. On glaciers, and the evidence of their having once existed in Scotland, Ireland, and England. [Read 4 November 1840.] *Proceedings of the Geological Society of London* 3 (1838–42): 327–32.

——. 1848. On the differences between progressive, embryonic, and prophetic types in the succession of organized beings through the whole range of geological times. *Proceedings of the American Association for the Advancement of Science. First meeting, held at Philadelphia, September, 1848*, pp. 432–8.

——. 1850. *Lake Superior: its physical character, vegetation, and animals, compared with those of other and similar regions*. Boston: Gould, Kendall and Lincoln.

——. 1851. Observations on the blind fish of the Mammoth Cave. *American Journal of Science and Arts* 2d ser. 11: 127–8.

——. 1857. On some young gar pikes from Lake Ontario. *American Journal of Science and Arts* 2d ser. 23: 284.

——. 1857–62. *Contributions to the natural history of the United States of America*. 4 vols. Boston, Mass.: Little, Brown & Company. London: Trübner.

Agassiz, Jean Louis Rodolphe. 1863. *Methods of study in natural history*. Boston, Mass.: Ticknor & Fields.

Agassiz, Jean Louis Rodolphe and Agassiz, Elizabeth. 1868. *A journey in Brazil*. Boston: Ticknor and Fields.

Agassiz, Jean Louis Rodolphe and Gould, Augustus A. 1848. *Principles of zoölogy: touching the structure, development, distribution, and natural arrangement of the races of animals, living and extinct. Part 1. Comparative physiology*. Boston: Gould, Kendall and Lincoln.

Agricola, Georg Andreas. 1721. *A philosophical treatise of husbandry and gardening: being a new method of cultivating and increasing all sorts of trees, shrubs, and flowers*. Revised by Richard Bradley. London: P. Vaillant, W. Mears, and F. Clay.

Agricola, Georgius. 1614. *De animantibus subterraneis liber*. Wittenburg: impensis Zachariæ Schureri Bibliop.

Aïdé, Hamilton. 1862. *Carr of Carrlyon: a novel*. 3 vols. London: Smith, Elder and Co.

Ainslie, Robert. 1865. *Discourses delivered in Christ Church, New Road, Brighton*. London: Longmans, Green, Reader, and Dyer. Brighton: J. H. and C. Treacher.

Airlee, Mabell, Countess. 1922. *Lady Palmerston and her times*. 2 vols. London: Hodder and Stoughton.

Aldrovandus, Ulysses. 1637. *De quadrupedibus digitatis viviparis libri tres, et de quadrupedibus digitatis oviparis libri duo*. Berne: Marcus Antonius, Bibliop.

Alefeld, Friedrich Christoph Wilhelm. 1860. Ueber *Pisum*. *Botanische Zeitung* 18: 204–5.

Allan, Mea. 1967. *The Hookers of Kew, 1785–1911*. London: Michael Joseph.

Alter, Stephen. 1999. *Darwinism and the linguistic image: language, race, and natural theology in the nineteenth century*. Baltimore: The Johns Hopkins University Press.

Alum. Cantab.: *Alumni Cantabrigienses. A biographical list of all known students, graduates and holders of office at the University of Cambridge, from the earliest times to 1900*. Compiled by John Venn and J. A. Venn. 10 vols. Cambridge: Cambridge University Press. 1922–54.

Alum. Dublin.: *Alumni Dublinenses. A register of the students, graduates, professors and provosts of Trinity College in the University of Dublin (1593–1860)*. New edition with supplement. Edited by George Dames Burtchaell and Thomas Ulick Sadleir. Dublin: Alex. Thom & Co. 1935.

Alum. Oxon.: *Alumni Oxonienses: the members of the University of Oxford, 1500–1886: . . . with a record of their degrees. Being the matriculation register of the university*. Alphabetically arranged, revised, and annotated by Joseph Foster. 8 vols. London and Oxford: Parker & Co. 1887–91.

American medical biography: *Dictionary of American medical biography. Lives of eminent physicians of the United States and Canada, from the earliest times*. New York and London: D. Appleton and Company. 1928.

ANB: *American national biography*. General editors, John A. Garraty and Mark C. Carnes. 24 vols. New York and Oxford: Oxford University Press. 1999.

'Ancient glaciers of Caernarvonshire': Notes on the effects produced by the ancient glaciers of Caernarvonshire, and on the boulders transported by floating ice. By Charles Darwin. *Philosophical Magazine* 21 (1842): 180–8. [*Collected papers* 1: 163–71.]

Anderson, Katherine. 1999. The weather prophets: science and reputation in Victorian meteorology. *History of science* 37: 179 216.

Anon. 1865a. Fritz Mueller. Pour Darwin. [Review of *Für Darwin*, by Fritz Müller.] *Bibliothèque Universelle et Revue Suisse. Archives des Sciences Physiques et Naturelles* n.s. 22: 154–63.

——. 1865b. The Darwinian hypothesis supported by observations on Crustacea. *Annals and Magazine of Natural History* 3d ser. 15: 410–16.

Appel, Toby A. 1987. *The Cuvier–Geoffroy debate: French biology in the decades before Darwin.* New York and Oxford: Oxford University Press.

Appleton's cyclopædia of American biography. Edited by James Grant Wilson and John Fiske. 6 vols. New York: D. Appleton. 1887–9.

Army list: *The army list.* London: the compiler of the annual official army list; Her Majesty's Stationery Office. 1815–1900.

Arnott, Neil. 1861. *A survey of human progress, from the savage state to the highest civilisation yet attained. A progress as little perceived by the multitude in any age as is the slow growing of a tree by the children who play under its shade, but which is leading to a new condition of mankind on earth.* London: Longman, Green, Longman and Roberts.

Ashton, Rosemary. 1991. *G. H. Lewes: a life.* Oxford: Clarendon Press.

'"Auditory-sac" of Cirripedes': On the so-called 'auditory-sac' of Cirripedes. By Charles Darwin. *Natural History Review* n.s. 3 (1863): 115–16. [*Collected papers* 2: 85–7.]

Aust. dict. biog.: *Australian dictionary of biography.* Edited by Douglas Pike *et al.* 14 vols. [Melbourne]: Melbourne University Press. 1966–96.

Autobiography: *The autobiography of Charles Darwin, 1809–1882. With original omissions restored.* Edited with appendix and notes by Nora Barlow. London: Collins. 1958.

Babington, Anna Maria, ed. 1897. *Memorials journal and botanical correspondence of Charles Cardale Babington.* Cambridge: Macmillan and Bowes.

Babington, Charles Cardale. 1838. *Dytiscidæ darwinianæ*; or, descriptions of the species of *Dytiscidæ* collected by Charles Darwin . . . in South America and Australia, during his voyage in H.M.S. Beagle. [Read 4 June 1838.] *Transactions of the Entomological Society of London* 3 (1841–3): 1–17.

Back, George. 1836. *Narrative of the Arctic land expedition to the mouth of the great Fish River, and along the shores of the Arctic Ocean: in the years 1833, 1834, and 1835.* London: John Murray.

Baedeker, Karl. 1889. *Switzerland and the adjacent portions of Italy, Savoy, and the Tyrol. Handbook for travellers.* 13th edition. Leipzig: Karl Baedeker. London: Dulan & Co.

Baillon, Ernest-Henri. 1857. De l'hermaphroditisme accidentel chez les euphorbiacées. [Read 26 June 1857.] *Bulletin de la Société Botanique de France* 4: 692–9.

Baillon, Ernest-Henri, *et al.* 1876–92. *Dictionnaire de botanique.* 4 vols. Paris: Librairie Hachette & Cie.

Bairstow, Martin. 1993. *The East Lancashire Railway. A history of the railways around Bury and guide to the preserved line.* Halifax: Martin Bairstow.

Barlow, Nora. 1932. Robert FitzRoy and Charles Darwin. *Cornhill Magazine* n.s. 72 (1932): 493–510.

Barnhart, John Hendley, comp. 1965. *Biographical notes upon botanists . . . maintained in the New York Botanical Garden Library.* 3 vols. Boston, Mass.: G. K. Hall.

Barrett, Paul H. 1980. *Metaphysics, materialism, and the evolution of mind. Early writings of Charles Darwin.* With a commentary by Howard E. Gruber. Chicago: University of Chicago Press.

Barrow, Logie. 1986. *Independent spirits: spiritualism and English plebeians, 1850–1910.* London: Routledge & Kegan Paul.

Bartholomew, Michael J. 1973. Lyell and evolution: an account of Lyell's response to the prospect of an evolutionary ancestry for man. *British Journal for the History of Science* 6 (1972–3): 261–303.

——. 1975–6. The award of the Copley Medal to Charles Darwin. *Notes and Records of the Royal Society* 30: 209–18.

Bartlett, Abraham Dee. 1857. Description of a rabbit said to be found on the Himalayan mountains. *Proceedings of the Zoological Society of London* 25: 159–60.

Barton, Ruth. 1987. John Tyndall, pantheist. *Osiris* 3: 111–34.

——. 1998. 'Huxley, Lubbock, and Half a Dozen Others': professionals and gentlemen in the formation of the X Club, 1851–1864. *Isis* 89: 410–44.

Bates, Henry Walter. 1861. Contributions to an insect fauna of the Amazon valley. *Lepidoptera*: *Heliconidæ*. [Read 21 November 1861.] *Transactions of the Linnean Society of London* 23 (1860–2): 495–566.

——. 1861–6. Contributions to an insect fauna of the Amazon valley. *Coleoptera*: *Longicornes*. *Annals and Magazine of Natural History* 3d ser. 8 (1861): 40–52, 147–52, 212–19, 471–8; 9 (1862): 117–24, 396–405, 446–58; 12 (1863): 100–9, 275–88, 367–81; 13 (1864): 43–56, 144–64; 14 (1864): 11–24; 15 (1865): 213–25, 382–94; 16 (1865): 101–13, 167–82, 308–14; 17 (1866): 31–42, 191–202, 288–303, 367–73; 425–35.

——. 1863. *The naturalist on the river Amazons. A record of adventures, habits of animals, sketches of Brazilian and Indian life, and aspects of nature under the equator, during eleven years of travel.* 2 vols. London: John Murray.

——. 1892. *The naturalist on the River Amazons. A record of adventures, habits of animals, sketches of Brazilian and Indian life, and aspects of nature under the equator, during eleven years of travel. With a memoir of the author by Edward Clodd.* Reprint of the first edition. London: John Murray.

BDWS: *Biographical dictionary of women in science: pioneering lives from ancient times to the mid-20th century.* Edited by Marilyn Ogilvie and Joy Harvey. 2 vols. New York and London: Routledge. 2000.

Bean, William Jackson. 1970–88. *Trees and shrubs hardy in the British Isles.* 8th edition, fully revised by D. L. Clarke and George Taylor. 4 vols. and supplement. London: John Murray.

Bell, Bill. 1995. The secret history of Smith & Elder: the *Publishers' Circular* as a source for publishing history. In *A genius for letters; booksellers and bookselling from the 16th to the 20th century*, edited by Robin Myers and Michael Harris. Winchester: St Paul's Bibliographies. New Castle, Delaware: Oak Knoll Press.

Bell, Thomas. 1837. *A history of British quadrupeds, including the Cetacea*. London: John van Voorst.

Benthall, Paul. 1980. George Maw: a versatile Victorian. *National Trust Studies* (1980): 11–20.

Bentham, George. 1858. *Handbook of the British flora; a description of the flowering plants and ferns indigenous to, or naturalized in, the British Isles*. London: Lovell Reeve.

[——.] 1865. The ancient and modern floras of Montpellier. *Natural History Review* 5: 202–25.

——. 1883. On the joint and separate work of the authors of Bentham and Hooker's 'Genera plantarum'. [Read 19 April 1883.] *Journal of the Linnean Society* (*Botany*) 20 (1884): 305–8.

Bentham, George and Hooker, Joseph Dalton. 1862–83. *Genera plantarum. Ad exemplaria imprimis in herbariis Kewensibus servata definita*. 3 vols. London: A. Black [and others].

Bentham, George and Mueller, Ferdinand von. 1863–78. *Flora Australiensis: a description of the plants of the Australian territory*. 7 vols. London: Lovell Reeve and Company.

Bergquist, Patricia R. 1978. *Sponges*. London: Hutchinson.

Bevington, M. M. 1941. *The* Saturday Review, *1855–68*. New York: Columbia University Press.

Bibby, Cyril. 1959. *T. H. Huxley. Scientist, humanist and educator*. London: Watts.

Birds: Pt 3 of *The zoology of the voyage of H.M.S. Beagle*. By John Gould. Edited and superintended by Charles Darwin. London: Smith, Elder and Co. 1838–41.

Birds of the world: *Handbook of the birds of the world*. By Josep del Hoyo *et al.* 6 vols. to date. Barcelona: Lynx editions. 1991–.

Blainville, Henri Marie Ducrotay de. 1841–55. *Ostéographie, ou description iconographique comparée du squelette et du système dentaire des cinq classes d'animaux vertébrés récents et fossiles*. 3 vols. and atlas (2 vols.). Paris: Arthus Bertrand.

Blaschko, Mary D. 1958. *Sidcot school: register of old scholars, 1808–1958*. Winscombe, Somerset: Sidcot School.

BLC: *The British Library general catalogues of printed books to 1975*. 360 vols. and supplement (6 vols.). London: Clive Bingley; K. G. Saur. 1979–88.

BLKO: *Biographisches Lexikon des Kaiserthums Oesterreich, enthaltend die Lebensskizzen der denkwürdigen Personen, welche seit 1750 in den österreichischen Kronländern geboren wurden oder darin gelebt und gewirkt haben*. By Constant von Wurzbach. 60 vols. Vienna: L. C. Zamarski. 1856–91.

Blyakher, L. Y. 1982. *History of embryology in Russia from the middle of the eighteenth century to the middle of the nineteenth century*. Translated from the Russian by H. I. Youssef and B. A. Malek, with an introduction by Jane Maienschein. Washington, D.C.:

Al Ahram Center for Scientific Translations for the Smithsonian Institution and the National Science Foundation.

BNB: *Biographie nationale publiée par l'Académie Royale des Sciences, des Lettres et des Beaux-arts de Belgique.* 44 vols., including 16 supplements. Brussels: H. Thiry-Van Buggenhoudt [and others]. 1866–1986.

Bonnet, Charles. 1779–83. *Œuvres d'histoire naturelle et de philosophie.* 8 vols. Neuchatel: l'imprimerie de Samuel Fauche, Libraire du Roi.

Bonney, T. G. 1919. *Annals of the Philosophical Club of the Royal Society written from its minute books.* London: Macmillan.

Bourne, Kenneth and Watt, D. Cameron, eds. 1986–7. *British documents on foreign affairs: reports and papers from the Foreign Office confidential print. Part 1: from the mid-nineteenth century to the First World War. Series C: North America, 1837–1914.* 15 vols. [Frederick, Md.]: University Publications of America.

Bowerbank, James Scott. 1864–82. *A monograph of the British Spongidae.* 4 vols. London: Ray Society.

Bowlby, John. 1990. *Charles Darwin: a biography.* London: Hutchinson.

Brace, Charles Loring. 1863. *The races of the old world. A manual of ethnology.* London: John Murray.

Brady, George Stewardson. 1865. Report on the pelagic Entomostraca [dredged off the coast of Northumberland and Durham 1862–4]. *Natural History Transactions of Northumberland and Durham* 1 (1865–7): 29–40.

Brandon-Jones, Christine. 1996. Charles Darwin and the repugnant curators. *Annals of Science* 53: 501–10.

Braun, Alexander Carl Heinrich. 1856. Über Parthenogenesis bei Pflanzen. [Read 23 October 1856.] *Abhandlungen der Königlichen Akademie der Wissenschaften zu Berlin* (1857): 311–76.

——. 1859. Uber Polyembryonie und Keimung von Caelebogyne. Ein Nachtrag zu der Abhandlung über Parthenogenesis bei Pflanzen. [Read 3 March 1859.] *Abhandlungen der Königlichen Akademie der Wissenschaften zu Berlin* (1860): 109–263.

Brent, Bernard Peirce. 1861. The goldfinch. *Cottage Gardener* 25: 391–2.

Brent, John. 1865. *Village bells, Lady Gwendoline and other poems.* London: Hamilton, Adams, and Co.

Brewer: *Brewer's dictionary of phrase & fable.* Revised by Adrian Room. Millennium edition. London: Cassell. 1999.

Brewster, David, ed. 1855. *Memoirs of the life, writings and discoveries of Sir Isaac Newton.* 2 vols. Edinburgh: Thomas Constable and Co.

Brewster, David. 1862. The facts and fancies of Mr Darwin. *Good Words* (1862): 3–9.

British imperial calendar: *The British imperial calendar or general register of the United Kingdom of Great Britain and Ireland and its colonies.* London: Winchester & Son [and others]. 1812–1919.

British postal guide: *British postal guide; containing the chief public regulations of the Post Office.* London: Eyre and Spottiswoode. 1861–76.

Britten, James. 1917. In memory of Daniel Oliver (1830–1917). *Journal of Botany* 55: 89–95.

Broca, Paul. 1858–9. Mémoire sur l'hybridité en général, sur la distinction des espèces amimales et sur les métis obtenus par le croisement du lièvre et du lapin. *Journal de la Physiologie de l'Homme et des Animaux* 1: 432–71, 684–729; 2: 218–58, 345–96.

Brock, William Hodson. 1997. *Justus von Liebig: the chemical gatekeeper.* Cambridge: Cambridge University Press.

Brock, William Hodson and Macleod, Roy M. 1976. The scientists' declaration: reflexions on science and belief in the wake of *Essays and reviews* 1864–5. *British Journal for the History of Science* 9: 39–66.

Brogan, Hugh, ed. 1975. The Times *reports. The American Civil War. Extracts from* The Times, *1860–1865.* London: Times Books.

Bronn, Heinrich Georg. 1860a. Review of *Origin of species. Neues Jahrbuch für Mineralogie, Geognosie, Geologie und Petrefakten-Kunde* (1860): 112–16. [Translated in Hull 1973, pp. 120–4.]

——. 1860b. *Über den Stufengang des organischen Lebens von den Insel-Felsen des Ozeans bis auf die Festländer. Eine Fest-Rede mit erläuternden Beilagen.* Stuttgart: E. Schweizerbart'sche Verlagshandlung und Druckerei.

——, trans. 1860. *Charles Darwin, über die Entstehung der Arten im Thier- und Pflanzen-Reich durch natürliche Züchtung, oder Erhaltung der vervollkommneten Rassen im Kampfe um's Daseyn.* Stuttgart: E. Schweizerbart.

——, trans. 1862. *Charles Darwin, über die Einrichtungen zur Befruchtung Britischer und ausländischer Orchideen durch Insekten und über die günstigen Erfolge der Wechselbefruchtung.* Stuttgart: E. Schweizerbart.

Brooke, John Hedley. 1991. *Science and religion: some historical perspectives.* Cambridge: Cambridge University Press.

Brown, Philip A. H. 1982. *London publishers and printers* c. *1800–1870.* London: British Library.

Brown, R. H. 1975. *One thousand years of Devon beekeeping.* Devon: Devon Beekeepers Association.

Brown, Robert. 1831. On the organs and mode of fecundation in Orchideæ and Asclepiadeæ. [Read 1 and 15 November 1831.] *Transactions of the Linnean Society of London* 16 (1833): 685–745.

Brown, Robert. 1996. The complete edited manuscript of Francis Darwin's preliminary draft of the 'Reminiscences of my father's everyday life'. [Typescript.]

Brown, Ron. 1994. *Great masters of beekeeping.* Borrowbridge, Somerset: Bee Books New & Old.

Browne, Janet. 1983. *The secular ark. Studies in the history of biogeography.* New Haven, Conn., and London: Yale University Press.

——. 1990. Spas and sensibilities: Darwin at Malvern. In *The medical history of waters and spas*, edited by Roy S. Porter. *Medical History*, supp. 10. London: Wellcome Institute for the History of Medicine.

——. 1995. *Charles Darwin. Voyaging. Volume 1 of a biography.* New York: Alfred A. Knopf.

——. 1998. I could have retched all night. Darwin and his body. In *Science incarnate.*

Historical embodiments of natural knowledge, edited by Christopher Lawrence and Steven Shapin. Chicago and London: University of Chicago Press.

Bryan's dictionary of painters and engravers. New edition revised and enlarged under the supervision of George C. Williamson. 5 vols. London: George Bell & Sons. 1903–5.

Buckland, William. 1840. Memoir on the evidences of glaciers in Scotland and the North of England. [Read 4 and 18 November, and 2 December 1840.] *Proceedings of the Geological Society of London* 3 (1838–42): 332–7, 345–8.

Buckle, Henry Thomas. 1857–61. *History of civilization in England.* 2 vols. London: John W. Parker & Son.

Buffon, Georges Louis Leclerc, comte de, *et al.* 1749–82. *Histoire naturelle, générale et particulière, avec la description du cabinet du roi.* 44 vols. Paris: Imprimerie royale.

Bullard, Laura Curtis. 1856. *Christine; or, woman's trials and triumphs.* New York: Dewitt & Davenport.

Burke's landed gentry: *A genealogical and heraldic history of the commoners of Great Britain and Ireland enjoying territorial possessions or high official rank but unvisited with heritable honours. Burke's genealogical and heraldic history of the landed gentry.* By John Burke, *et al.* 1st–18th edition. London: Henry Colburn [and others]. 1833–1969.

Burke's peerage: *A genealogical and heraldic dictionary of the peerage and baronetage of the United Kingdom. Burke's peerage and baronetage.* 1st– edition. London: Henry Colburn [and others]. 1826–.

Burkhardt, Frederick H. 1974. England and Scotland: the learned societies. In *The comparative reception of Darwinism*, edited by T. F. Glick. Austin: University of Texas Press.

Burton, James. 1986. Robert FitzRoy and the early history of the Meteorological Office. *British Journal for the History of Science* 19: 147–76.

Busk, George and Falconer, Hugh. 1865. On the fossil contents of the Genista Cave, Gibraltar. [Read 8 March 1865.] *Quarterly Journal of the Geological Society of London* 21: 364–70.

[Butler, Samuel.] 1865. *The evidence for the resurrection of Jesus Christ as given by the four evangelists, critically examined.* London: [the author].

[——.] 1873. *The fair haven. A work in defence of the miraculous element in our Lord's ministry upon earth, both as against rationalistic impugners and certain Orthodox defenders. By the late John Pickard Owen, edited by William Bickersteth Owen, with a memoir of the author.* 2d edition. London: Trübner.

——. 1923. *A first year in Canterbury settlement and other early essays.* Vol. 1 of *The Shrewsbury edition of the works of Samuel Butler*, edited by Henry Festing Jones and Augustus Theodore Bartholomew. London: Jonathan Cape. New York: E. P. Dutton & Company.

Bynum, William F. 1984. Charles Lyell's *Antiquity of man* and its critics. *Journal of the History of Biology* 17: 153–87.

Byrne, John Francis. 1964. The *Reader*: a review of literature, science and the arts, 1863–1867. PhD thesis. Northwestern University.

Calendar: *A calendar of the correspondence of Charles Darwin, 1821–1882. With supplement.*

Edited by Frederick Burkhardt *et al.* 2d edition. Cambridge: Cambridge University Press. 1994.

Cambridge University calendar: *The Cambridge University calendar.* Cambridge: W. Page [and others]. 1796–1950.

Camerini, Jane R. 1993. Evolution, biogeography, and maps: an early history of Wallace's line. *Isis* 84: 700–27.

Campbell, George, trans. 1789. *The four Gospels, translated from the Greek. With preliminary dissertations, and notes critical and explanatory.* 2 vols. London: A. Strahan.

Campbell, George, *et al.*, trans. 1818. *The New Testament translated from the original Greek.* London: John Lepard.

Campbell, George Douglas. 1860. Opening address, 1860–1 session. [Read 3 December 1860.] *Proceedings of the Royal Society of Edinburgh* 4 (1857–62): 350–77.

[——.] 1862. [Review of *Orchids* and other works.] *Edinburgh Review* 116: 378–97.

——. 1864. Opening address, 1864–5 session. [Read 5 December 1864.] *Proceedings of the Royal Society of Edinburgh* 5 (1862–6): 264–92.

——. 1865. The reign of law. *Good Words* (1865): 52–8, 126–33, 227–32, 269–74.

Campbell, Ina Erskine, ed. 1906. *George Douglas, eighth duke of Argyll . . . (1823–1900): autobiography and memoirs.* 2 vols. London: John Murray.

Campbell, John Francis. 1865a. *Frost and fire: natural engines, tool marks and chips: with sketches taken at home and abroad by a traveller.* Edinburgh: Edmonston and Douglas.

——. 1865b. *A short American tramp in the fall of 1864. By the editor of 'Life in Normandy.'* Edinburgh: Edmonston and Douglas.

[Carlyle, Thomas.] 1838. *Sartor resartus: the life and opinions of Herr Teufelsdröckh.* London: Saunders and Otley.

——. 1841. *On heroes, hero-worship, & the heroic in history.* London: James Fraser.

Caroe, Gwendolen Mary. 1985. *The Royal Institution: an informal history.* With a final chapter by Alban Caroe. London: John Murray.

Carter, Henry John. 1857. Transformation of the vegetable protoplasm into Actinophrys. *Annals and Magazine of Natural History* 2d ser. 19: 259–63.

——. 1858. On fecundation in *Eudorina elegans* and *Cryptoglena. Annals and Magazine of Natural History* 3d ser. 2: 237–53.

Caspary, Johann Xaver Robert. 1865. *Sur les hybrides obtenus par la greffe.* N.p.: n.p. [Reprinted from *Bulletin du Congrès International de Botanique et d'Horticulture réuni à Amsterdam* (1865): 65–80.]

Castelnau, Francis de. 1850–7. *Expédition dans les parties centrales de l'Amérique du Sud . . . exécutée par ordre du gouvernement français pendant les années 1843 à 1847, sous la direction de F. de Castelnau.* 7 parts. Paris: P. Bertrand, Libraire-éditeur.

Catalogue général de la librairie française: *Catalogue général de la librairie française pendant 25 ans (1840–1865).* Edited by Otto Lorenz. 4 vols. Paris: O. Lorenz. 1868.

Catalogue of the officers and graduates of Yale University. New Haven, Conn.: Tuttle, Morehouse and Taylor. 1905.

Catholic encyclopedia: *The Catholic encyclopedia, an international work of reference on the constitution, doctrine, discipline, and history of the Catholic Church.* Edited by C. G. Herbermann, *et al.* 16 vols. and supplement. New York: Robert Appleton. 1907–22.

CDEL: *A critical dictionary of English literature, and British and American authors, living and deceased, from the earliest accounts to the middle of the nineteenth century . . . with forty indexes of subjects*. By S. Austin Allibone. 3 vols. London: Trübner. Philadelphia: Childs & Peterson; J. B. Lippincott. 1859–71. *A supplement to Allibone's critical dictionary of English literature and British and American authors. Containing over thirty-seven thousand articles (authors), and enumerating over ninety-three thousand titles*. By John Foster Kirk. 2 vols. Philadelphia and London: J. B. Lippincott. 1891.

Challinor, John. 1978. *A dictionary of geology*. 5th edition. Cardiff: University of Wales Press.

Chamberlain, Joshua Lawrence, ed. 1898–1900. *Universities and their sons; history, influence and characteristics of American universities, with biographical sketches and portraits of alumni and recipients of honorary degrees*. 5 vols. Boston, Mass.: R. Herndon.

Chambers: *Chambers 20th century dictionary*. New edition. Edited by E. M. Kirkpatrick. Edinburgh: Chambers. 1983.

[Chambers, Robert.] 1844. *Vestiges of the natural history of creation*. London: John Churchill.

Chambers biographical dictionary. Edited by Melanie Parry. New York: Larousse Kingfisher Chambers. 1997.

Chapman, John. 1863a. A new method of treating disease by controlling the circulation of the blood in different parts of the body. *Medical Times and Gazette* (1863), pt 2: 60–2.

———. 1863b. *Functional diseases of women: cases illustrative of a new method of treating them through the agency of the nervous system by means of cold and heat*. London: Trübner and Co.

———. 1864. *Functional diseases of the stomach. Part 1. Sea-sickness: its nature and treatment*. London: Trübner & Co.

Chapuis, Félicien. 1865. *Le pigeon voyageur belge*. Verviers: Imprimerie de Ch. Vinche.

Churchill, Frederick B. 1979. Sex and the single organism: biological theories of sexuality in mid-nineteenth century. *Studies in History of Biology* 3: 139–77.

Clark, Ronald W. 1968. *The Huxleys*. London: Heinemann.

Clergy list: *The clergy list . . . containing an alphabetical list of the clergy*. London: C. Cox [and others]. 1841–89.

'Climbing plants': On the movements and habits of climbing plants. By Charles Darwin. [Read 2 February 1865.] *Journal of the Linnean Society of London* (*Botany*) 9 (1867): 1–118.

Climbing plants: *On the movements and habits of climbing plants*. By Charles Darwin. London: Longman, Green, Longman, Roberts & Green; Williams & Norgate. 1865.

Climbing plants 2d ed.: *The movements and habits of climbing plants*. 2d edition. By Charles Darwin. London: John Murray. 1875.

Clokie, Hermia Newman. 1964. *An account of the herbaria of the department of botany in the University of Oxford*. Oxford: Oxford University Press.

Cobbold, Thomas Spencer. 1864. *Entozoa: an introduction to the study of helminthology, with reference, more particularly, to the internal parasites of man*. London: Groombridge & Sons.

Cockburn, Henry. 1852. *Life of Lord Jeffrey, with a selection from his correspondence.* 2 vols. Edinburgh: A. and C. Black.

Cohn, Ferdinand Julius. 1860. Ueber contractile Gewebe im Pflanzenreiche. [Read 1 November 1860.] *Abhandlungen der Schlesischen Gesellschaft für vaterländische Cultur. Abtheilung für Naturwissenschaften und Medicin* 1 (1861): 1–48

——. 1865. Sur la culture des algues marines. *Bulletin du Congrès International de Botanique et d'Horticulture réuni à Amsterdam* (1865): 116–29.

Coleman, William. 1964. *Georges Cuvier, zoologist: a study in the history of evolutionary theory.* Cambridge, Mass.: Harvard University Press.

Colenso, John William. 1862–79. *The Pentateuch and Book of Joshua critically examined.* 5 vols. London: Longman, Green, Longman, Roberts & Green.

Collected papers: *The collected papers of Charles Darwin.* Edited by Paul H. Barrett. 2 vols. Chicago and London: University of Chicago Press. 1977.

Colloms, Brenda. 1975. *Charles Kingsley: the lion of Eversley.* London: Constable. New York: Barnes & Noble.

Colp, Ralph, Jr. 1977. *To be an invalid: the illness of Charles Darwin.* Chicago and London: University of Chicago Press.

——. 1978. Charles Darwin: slavery and the American Civil War. *Harvard Library Bulletin* 26: 471–89.

——. 1998. *To be an invalid*, redux. *Journal of the History of Biology* 31: 211–40.

Columbia gazetteer of the world: *The Columbia gazetteer of the world.* Edited by Saul B. Cohen. 3 vols. New York: Columbia University Press. 1998.

Commercial directory for Shropshire: *Pigot & Co.'s national commercial directory; 1828–9; comprising a directory and classification of the merchants, bankers, professional gentlemen, manufacturers and traders, in all the cities, towns, sea-ports, and principal villages in the following counties, viz. Cheshire Derbyshire Nottinghamshire Shropshire.* London and Manchester: J. Pigot & Co. 1828–9.

Complete peerage: *The complete peerage of England, Scotland, Ireland, Great Britain, and the United Kingdom, extant, extinct, or dormant.* By George Edward Cokayne. Revised edition. Edited by Vicary Gibbs, *et al.* 12 vols. London: St Catherine Press. 1910–59.

Comte, Auguste. 1851–5. *Système de politique positive, ou traité de sociologie, instituant la religion de l'humanité.* 4 vols. Paris: L. Mathias.

Copland, James. 1866. *A dictionary of practical medicine.* London: Longmans, Green, and Co.

Coral reefs: *The structure and distribution of coral reefs. Being the first part of the geology of the voyage of the* Beagle, *under the command of Capt. FitzRoy RN, during the years 1832 to 1836.* By Charles Darwin. London: Smith, Elder & Co. 1842.

Correns, C. 1916. Friedrich Hildebrand. *Berichte der deutschen botanischen Gesellschaft* 34 (pt 2): 28–49.

Correspondence: *The correspondence of Charles Darwin.* Edited by Frederick Burkhardt *et al.* 12 vols. to date. Cambridge: Cambridge University Press. 1985–.

Corsi, Pietro. 1983. 'Lamarckiens' et 'Darwiniens' à Turin (1812–1894). In *De Darwin au Darwinisme: science et idéologie*, edited by Yvette Conry. Paris: J. Vrin.

Corsi, Pietro and Weindling, Paul J. 1985. Darwinism in Germany, France and Italy. In *The Darwinian heritage*, edited by David Kohn. Princeton, N.J.: Princeton University Press in association with Nova Pacifica.

County families: *The county families of the United Kingdom; or, royal manual of the titled & untitled aristocracy of Great Britain & Ireland.* By Edward Walford. London: Robert Hardwicke; Chatto & Windus. 1860–93. *Walford's county families of the United Kingdom or royal manual of the titled and untitled aristocracy of England, Wales, Scotland, and Ireland.* London: Chatto & Windus; Spottiswoode & Co. 1894–1920.

[Cowell, F. R.] 1975. *The Athenaeum Club and social life in London, 1824–1974.* London: Heinemann.

Cox, George W. 1888. *The life of John William Colenso, D.D., Bishop of Natal.* 2 vols. London: W. Ridgway.

Crockford's clerical directory: *The clerical directory, a biographical and statistical book of reference for facts relating to the clergy and the Church. Crockford's clerical directory etc.* London: John Crockford [and others]. 1858–1900.

Crook, David Paul. 1975. *Diplomacy during the American Civil War.* New York: John Wiley and Sons.

Cross and self fertilisation: *The effects of cross and self fertilisation in the vegetable kingdom.* By Charles Darwin. London: John Murray. 1876.

Crüger, Hermann. 1864. A few notes on the fecundation of orchids and their morphology. [Read 3 March 1864.] *Journal of the Linnean Society (Botany)* 8 (1865): 127–35.

Curle, Richard. 1954. *The Ray Society: a bibliographical history.* London: Ray Society.

Curvey, Mary F. and Johnson, Robert E. 1969. A bibliography of Sir John Richardson (1787–1865)—printed books. *Journal of the Society for the Bibliography of Natural History* 5 (1968–71): 202–17.

Cuvier, Georges. 1805. *Leçons d'anatomie comparée.* 5 vols. Paris: Crochard; Fantin.

——. 1812. *Recherches sur les ossemens fossiles de quadrupèdes, où l'on rétablit les caractères de plusieurs espèces d'animaux que les révolutions du globe paroissent avoir détruites.* 4 vols. Paris: Deterville.

——. 1817. *Le règne animal distribué d'après son organisation, pour servir de base a l'histoire naturelle des animaux et d'introduction a l'anatomie comparée.* 4 vols. Paris: Deterville.

——. 1840. *Cuvier's animal kingdom . . . Mammalia, birds, and reptiles, by E. Blyth.– The fishes and radiata, by R. Mudie.– The molluscous animals, by G. Johnston.– The articulated animals by J. O. Westwood.* London: Wm. S. Orr and Co.

DAB: *Dictionary of American biography.* Under the auspices of the American Council of Learned Societies. 20 vols., index, and 10 supplements. New York: Charles Scribner's Sons; Simon & Shuster Macmillan. London: Oxford University Press; Humphrey Milford. 1928–95.

Dallas, William Sweetland, trans. 1869. *Facts and arguments for Darwin.* By Fritz Müller. London: John Murray.

Dana, James Dwight. 1852. *Crustacea.* 2 pts. Vol. 13 of *United States Exploring Expedition. During the years 1838, 1839, 1840, 1841, 1842. Under the command of Charles Wilkes, U.S.N.* Philadelphia: C. Sherman.

Dana, James Dwight. 1853a. *On the classification and geographical distribution of Crustacea: from the report on Crustacea of the United States Exploring Expedition, under Captain Charles Wilkes, U.S.N., during the years 1838–1842*. Philadelphia: C. Sherman.

——. 1853b. On an isothermal oceanic chart, illustrating the geographical distribution of marine animals. *American Journal of Science and Arts* 2d ser. 16: 153–67, 314–27.

——. 1854–5. On the geographical distribution of Crustacea. *American Journal of Science and Arts* 2d ser. 18 (1854): 314–26; 19 (1855): 6–15; 20 (1855): 168–78, 349–61.

——. 1855. On American geological history: address before the American Association for the Advancement of Science. [Read August 1855.] *American Journal of Science and Arts* 2d ser. 22 (1856): 305–34.

——. 1856. On the plan of development in the geological history of North America. *American Journal of Science and Arts* 2d ser. 22: 335–49.

Darwin, Charles and Wallace, Alfred Russel. 1858. On the tendency of species to form varieties; and on the perpetuation of varieties and species by natural means of selection . . . Communicated by Sir Charles Lyell . . . and J. D. Hooker. [Read 1 July 1858.] *Journal of the Proceedings of the Linnean Society* (*Zoology*) 3 (1859): 45–62. [*Collected papers* 2: 3–19.]

Darwin, Erasmus. 1794–6. *Zoonomia; or, the laws of organic life*. 2 vols. London: J. Johnson.

——. 1800. *Phytologia, or the philosophy of agriculture and gardening. With the theory of draining morasses and with an improved construction of the drill plough*. London: J. Johnson.

Darwin, Francis. 1914. William Erasmus Darwin. *Christ's College Magazine* 29: 16–23.

——. 1920. *Springtime and other essays*. London: John Murray.

Darwin pedigree: *Pedigree of the family of Darwin*. Compiled by H. Farnham Burke. N.p.: privately printed. 1888. [Reprinted in facsimile in *Darwin pedigrees*, by Richard Broke Freeman. London: printed for the author. 1984.]

Daubeny, Charles Giles Bridle. 1860a. *Remarks on the final causes of the sexuality of plants, with particular reference to Mr Darwin's work on the origin of species*. Oxford: printed by James Wright, printer to the University.

——. 1860b. Remarks on the final causes of the sexuality of plants, with particular reference to Mr Darwin's work 'On the origin of species by natural selection'. *Report of the thirtieth meeting of the British Association for the Advancement of Science held at Oxford*, Transactions of the sections, pp. 109–10.

Daunton, Martin J. 1985. *Royal Mail: the Post Office since 1840*. London and Dover, N.H.: Athlone Press.

Davidson, Edward H. and Scheick, William J. 1994. *Paine, scripture and authority.* The age of reason *as religious and political idea.* Bethlehem, Pa.: Lehigh University Press. London and Toronto: Associated University Presses.

Davies, Gordon Leslie. 1969. *The earth in decay. A history of British geomorphology 1578–1878*. London: Macdonald Technical and Scientific.

Davis, William C. 1991. *Jefferson Davis. The man and his hour.* New York: Harper Collins Publishers.

Davy, Humphry. 1813. *Elements of agricultural chemistry: in a course of lectures for the Board of Agriculture.* London and Edinburgh: Longman, Hurst, Rees, Orme, and Brown.

Davy, John. 1855. Some observations on the ova of the salmon, in relation to the distribution of species; in a letter addressed to Charles Darwin, Esq., M.A., V.P.R.S. &c. [Read 26 April 1855.] *Philosophical Transactions of the Royal Society of London* 146: 21–9.

——. 1856. On the vitality of the ova of the Salmonidæ of different ages; in a letter addressed to Charles Darwin, Esq., M.A., V.P.R.S. &c. [Read 7 February 1856.] *Proceedings of the Royal Society of London* 8 (1856–7): 27–33.

DBE: *Deutsche biographische Enzyklopädie.* Edited by Walter Killy *et al.* 12 vols. in 14. Munich: K. G. Saur. 1995–2000.

DBF: *Dictionnaire de biographie Française.* Under the direction of J. Balteau *et al.* 17 vols. and part of vol. 18 (A–Latreille) to date. Paris: Librairie Letouzey & Ané. 1933–.

DBI: *Dizionario biografico degli Italiani.* Edited by Alberto M. Ghisalberti *et al.* 57 vols. (A–Gonzaga) to date. Rome: Istituto della Enciclopedia Italiana. 1960–.

DBL: *Dansk biografisk leksikon.* Founded by Carl Frederick Bricka, edited by Povl Engelstoft and Svend Dahl. 26 vols. and supplement. Copenhagen: J. H. Schultz. 1933–44. 3d edition. Edited by Sv. Cedergreen Bech. 16 vols. Copenhagen: Gyldendal. 1979–84.

De Beer, Gavin, ed. 1958. *Charles Darwin and Alfred Russel Wallace. Evolution by natural selection.* Cambridge: Cambridge University Press. Reprint edition. New York and London: Johnson Reprint Corporation. 1971.

——, ed. 1959. Darwin's journal. *Bulletin of the British Museum (Natural History). Historical Series* 2: 3–21.

Decaisne, Joseph and Naudin, Charles Victor. 1862–71. *Manuel de l'amateur des jardins, traité générale d'horticulture.* 4 vols. Paris: Didot frères, fils et cie.

Dempster, W. J. 1996. *Natural selection and Patrick Matthew. Evolutionary concepts in the nineteenth century.* Durham: Pentland Press.

Denney, Robert E. 1992. *The civil war years: a day-by-day chronicle of the life of a nation.* New York: Sterling Publishing.

Denny, Henry. 1842. *Monographia Anoplurorum Britanniæ; or an essay on the British species of parasitic insects belonging to the order of Anoplura of Leach.* London: Henry G. Bohn

——. 1844. Report of the progress of the investigation of exotic *Anoplura*. *Report of the fourteenth meeting of the British Association for the Advancement of Science held at York*, p. 392.

Descent: *The descent of man, and selection in relation to sex.* By Charles Darwin. 2 vols. London: John Murray. 1870–1.

Desmond, Adrian. 1982. *Archetypes and ancestors: palaeontology in Victorian London, 1850–1875.* London: Blond & Briggs.

——. 1994–7. *Huxley.* 2 vols. London: Michael Joseph.

Desmond, Adrian and Moore, James. 1991. *Darwin.* London: Michael Joseph.

Desmond, Ray. 1992. *The European discovery of the Indian flora.* Oxford: Oxford University Press [in association with the] Royal Botanic Gardens.

Desmond, Ray. 1994. *Dictionary of British and Irish botanists and horticulturists including plant collectors, flower painters and garden designers*. New edition, revised with the assistance of Christine Ellwood. London: Taylor & Francis and the Natural History Museum. Bristol, Pa.: Taylor & Francis.

——. 1995. *Kew: the history of the Royal Botanic Gardens*. London: Harvill Press with the Royal Botanic Gardens, Kew.

——. 1999. *Sir Joseph Dalton Hooker, traveller and plant collector*. Woodbridge, Suffolk: Antique Collectors' Club with the Royal Botanic Gardens, Kew.

Desmond, Ray and Hepper, F. Nigel. 1993. *A century of Kew plantsmen. A celebration of the Kew Guild*. Richmond: Kew Guild.

Diccionario de historia de Venezuela. Under the direction of Manuel Pérez Vila. 3 vols. Caracas: Fundación Polar. 1988.

Dictionary of British sculptors: *Dictionary of British sculptors 1660–1851*. By Rupert Gunnis. New revised edition. London: The Abbey Library. 1968.

Dictionnaire biographique des Genevois et des Vaudois. By Albert de Montet. 2 vols. Lausanne: George Bridel Editeur. 1877–8.

Dictionnaire historique & biographique de la Suisse. Under the direction of Marcel Godet *et al.* 7 vols. and supplement. Neuchâtel: Administration du Dictionnaire Historique & Biographique de la Suisse. 1921–34.

Dictionnaire universel des contemporains: *Dictionnaire universel des contemporains contenant toutes les personnes notables de la France et des pays étrangers . . . Ouvrage rédigé et continuellement tenu à jour avec le concours d'écrivains et des savants de tous les pays*. Edited by Louis Gustave Vapereau. Paris: Libraire Hachette. 1858. 3d edition, 1865. 4th edition, 1870. 5th edition, 1880. 6th edition, 1893.

Di Gregorio, Mario A. 1984. *T. H. Huxley's place in natural science*. New Haven and London: Yale University Press.

'Dimorphic condition in *Primula*': On the two forms, or dimorphic condition, in the species of *Primula*, and on their remarkable sexual relations. By Charles Darwin. [Read 21 November 1861.] *Journal of the Proceedings of the Linnean Society* (*Botany*) 6 (1862): 77–96. [*Collected papers* 2: 45–63.]

Disraeli, Benjamin. 1864. *Church policy: a speech delivered . . . at a meeting of the Oxford Diocesan Society for the augmentation of small livings*. London: Rivingtons.

Dixon, Edmund Saul. 1851. *The dovecote and the aviary: being sketches of the natural history of pigeons and other domestic birds in a captive state, with hints for their management*. London: John Murray.

DNB: *Dictionary of national biography*. Edited by Leslie Stephen and Sidney Lee. 63 vols. and 2 supplements (6 vols.). London: Smith, Elder & Co. 1885–1912. *Dictionary of national biography 1912–90*. Edited by H. W. C. Davis *et al.* 9 vols. London: Oxford University Press. 1927–96.

DNZB: *A dictionary of New Zealand biography*. Edited by G. H. Scholefield. 2 vols. Wellington, New Zealand: Department of Internal Affairs. 1940. *The dictionary of New Zealand biography*. Edited by W. H. Oliver *et al.* 5 vols. Auckland and Wellington, New Zealand: Department of Internal Affairs [and others]. 1990–2000.

Dobson, Jessie. 1954. *William Clift.* London: William Heinemann medical books.

Dodd, Victor. 1983. *Beemasters of the past.* Hebden Bridge, Yorkshire: Northern Bee Books.

Dod's parliamentary companion: *The parliamentary pocket companion . . . compiled from official documents, and from the personal communications of members of both houses. Dod's parliamentary companion.* London: Whittaker, Treacher, & Arnot; Whittaker & Co. 1833–1914.

Doody, Margaret Anne. 1974. *A natural passion; a study of the novels of Samuel Richardson.* Oxford: Clarendon Press.

Dowson, John. 1861. *Erasmus Darwin: philosopher, poet, and physician. A lecture to the Literary and Philosophical Society of Whitby.* London: H. K. Lewis.

DSAB: *Dictionary of South African biography.* Edited by W. J. de Kock *et al.* 4 vols. Pretoria and Cape Town: Nasionale Boekhandel Beperk [and others]. 1968–81.

DSB: *Dictionary of scientific biography.* Edited by Charles Coulston Gillispie and Frederic L. Holmes. 18 vols. including index and supplements. New York: Charles Scribner's Sons. 1970–90.

Duncan, Peter Martin. 1865. A description of the echinodermata from the strata on the south-eastern coast of Arabia, and at Bagh on the Nerbudda, in the collection of the Geological Society. [Read 8 March 1865.] *Quarterly Journal of the Geological Society of London* 21: 349–63.

Dupree, Anderson Hunter. 1959. *Asa Gray, 1810–1888.* Cambridge, Mass.: Belknap Press of Harvard University.

Dutrochet, René Joachim Henri. 1843. Des mouvements révolutifs spontanés qui s'observent chez les végétaux. *Comptes rendus hebdomadaire des séances de l'Académie des Sciences* 17: 989–1008.

——. 1844. Recherches sur la volubilité des tiges de certains végétaux et sur la cause de ce phénomène. *Comptes rendus hebdomadaire des séances de l'Académie des Sciences* 19: 295–303.

Duval-Jouve, Joseph. 1863. Sur la floraison et la fructification du *Leersia Oryzoides. Bulletin de la Société Botanique de France* 10: 194–7.

Earl, George Windsor. 1971. *The eastern seas.* Singapore: Oxford University Press.

EB: *The Encyclopædia Britannica. A dictionary of arts, sciences, literature and general information.* 11th edition 29 vols. Cambridge: Cambridge University Press. 1910–11.

Eights, James. 1833–52. Description of a new crustaceous animal (*Brongniartia trilobitoides*) found on the shores of the South Shetland Islands, with remarks on their natural history. *Transactions of the Albany Institute* 2: 53–69

'Elevation and subsidence in the Pacific and Indian Oceans': On certain areas of elevation and subsidence in the Pacific and Indian Oceans, as deduced from the study of coral formations. [Read 31 May 1837.] By Charles Darwin. *Proceedings of the Geological Society of London* 2 (1838): 552–4. [*Collected papers* 1: 46–9.]

Elie de Beaumont, Jean-Baptiste-Armand-Louis-Léonce. 1831. Researches on some of the revolutions which have taken place on the surface of the globe; presenting various examples of the coincidence between the elevation of beds in certain systems of mountains, and the sudden changes which have produced the lines of demarcation observable in certain stages of the sedimentary deposits. *Philosophical Magazine, or Annals of Chemistry [etc.]* n.s. 10: 241–64.

Eliot, George. 1859. *Adam Bede.* 3 vols. Edinburgh: William Blackwood.

——. 1860. *The mill on the Floss.* 3 vols. Edinburgh: William Blackwood and Sons.

——. 1861. *Silas Marner: the weaver of Raveloe.* Edinburgh: William Blackwood and Sons.

Elliott, Clark A. 1979. *Biographical dictionary of American science: the seventeenth through the nineteenth centuries.* Westport, Conn., and London: Greenwood Press.

Ellis, Ieuan. 1980. *Seven against Christ. A study of 'Essays and reviews'.* Leiden, Netherlands: E. J. Brill.

Emma Darwin (1904): *Emma Darwin, wife of Charles Darwin. A century of family letters.* Edited by Henrietta Litchfield. 2 vols. Cambridge: privately printed by Cambridge University Press. 1904.

Emma Darwin (1915): *Emma Darwin: a century of family letters, 1792–1896.* Edited by Henrietta Litchfield. 2 vols. London: John Murray. 1915.

Enchiridion symbolorum: *Enchiridion symbolorum definitionum et declarationum de rebus fidei et morum.* 33d edition. Edited by Henry Denzinger. Rome: Herder. 1965.

Endlicher, Stephan Ladislaus. 1836–42. *Genera plantarum secundum ordines naturales disposita.* With 4 supplements; in 2 vols. Vienna: Friedrich Beck.

Engels, Eve-Marie, ed. 1995. *Die Rezeption von Evolutionstheorien im neunzehnten Jahrhundert.* Suhrkamp Taschenbuch Wissenschaft, Nr. 1229. Frankfurt am Main: Suhrkamp.

England, Richard. 2001. Natural selection, teleology, and the logos: from Darwin to the Oxford neo-Darwinists, 1859–1909. *Osiris* 2d ser. 16: 270–87.

English catalogue of books: *The English catalogue of books. Comprising the contents of the 'London' and the 'British' catalogues, and the principal works published in the United States of America and continental Europe.* Compiled by Sampson Low. 3 vols. London: Sampson Low, Marston, Low, & Searle. 1864–82.

Erasmus Darwin. By Ernst Krause. Translated from the German by W. S. Dallas, with a preliminary notice by Charles Darwin. London: John Murray. 1879.

Erfurt-Congress: *Zweiter Congress deutscher Gärtner, Botaniker und Gartenfreunde und allgemeine deutsche Ausstellung von Gemüsen und landwirthschaftlichen Produkten, Obst, Pflanzen, Blumen, Gartengeräthschaften etc. in Erfurt im September 1865.* Erfurt: A. Stenger. [1865].

Essays and reviews. London: John W. Parker. 1860.

Eudes-Deslongchamps, Jacques Armand. 1842. Note sur les cochons à pendeloques, variété ou monstruosité du cochon domestique qui parait n'avoir point été remarquée par les naturalistes. *Mémoires de la Société Linnéenne de Normandie* 7: 41–4.

Evans, George Eyre. 1897. *Vestiges of Protestant dissent: being lists of ministers, sacramental plate, registers, antiquities, and other matters pertaining to most of the churches (and a few others) included in the National Conference of Unitarian, Liberal Christian, Free Christian, congregations.* Liverpool: F. & E. Gibbons.

Expression: *The expression of the emotions in man and animals.* By Charles Darwin. London: John Murray. 1872.
Falconer, Hugh. 1857. Description of two species of the fossil mammalian genus *Plagiaulax* from Purbeck. [Read 11 March 1857.] *Quarterly Journal of the Geological Society of London* 13: 261–82.
——. 1868. *Palæontological memoirs and notes of the late Hugh Falconer . . . with a biographical sketch of the author.* 2 vols. Compiled and edited by Charles Murchison. London: Robert Hardwicke.
Farley, John. 1982. *Gametes & spores: ideas about sexual reproduction, 1750–1914.* Baltimore and London: The Johns Hopkins University Press.
Farrar, Frederic William. 1864a. Traditions, real and fictitious. [Read 23 February 1864.] *Transactions of the Ethnological Society of London* 3 (1865): 298–307.
——. 1864b. On the universality of belief in God, and in a future state. [Read 5 April 1864.] *Anthropological Review* 2: ccxvii–ccxix.
——. 1864c. On hybridity. [Read 5 April 1864.] *Anthropological Review* 2: ccxix–ccxxv.
——. 1865. *Chapters on language.* London: Longmans, Green, and Co.
Faulkner, Thomas. 1829. *An historical and topographical description of Chelsea, and its environs; interspersed with biographical anecdotes of illustrious and eminent persons who have resided in Chelsea during the three preceding centuries.* 2 vols. Chelsea: T. Faulkner [and others].
Ferguson, George. 1854. *Ferguson's illustrated series of rare and prize poultry, including comprehensive essays upon all classes of domestic fowl.* London: n.p.
'Fertilization of orchids': Notes on the fertilization of orchids. By Charles Darwin. *Annals and Magazine of Natural History* 4th ser. 4 (1869): 141–59. [*Collected papers* 2: 138–56.]
FitzRoy, Robert. 1863. *The weather book. A manual of practical meteorology.* London: Longman, Green, Longman, Roberts and Green.
Forbes, David. 1860. On the geology of Bolivia and southern Peru. [Read 21 November 1860.] *Quarterly Journal of the Geological Society of London* 17 (1861): 7–62.
Forbes, Edward. 1845. On the distribution of endemic plants, more especially those of the British Islands, considered with regard to geological changes. *Report of the 15th meeting of the British Association for the Advancement of Science held at Cambridge*, Transactions of the sections, pp. 67–8.
——. 1846. On the connexion between the distribution of the existing fauna and flora of the British Isles, and the geological changes which have affected their area, especially during the epoch of the Northern Drift. *Memoirs of the Geological Survey of Great Britain, and of the Museum of Economic Geology in London* 1: 336–432.
Forchhammer, G. *et al.* 1851–5. *Underslgelser i geologisk–antiquarisk Retning.* Occasional papers of the Kongelige Danske Videnskabernes Selskabs Oversigter for Aarene.
Foreign Office list: *The Foreign Office list.* London: Harrison & Sons. 1852–1965.
Forms of flowers: *The different forms of flowers in plants of the same species.* By Charles Darwin. London: John Murray. 1877.

Fossil Cirripedia (1851): *A monograph on the fossil Lepadidæ, or, pedunculated cirripedes of Great Britain.* By Charles Darwin. London: Palaeontographical Society. 1851.

Fossil Cirripedia (1854): *A monograph of the fossil Balanidæ and Verrucidæ of Great Britain.* By Charles Darwin. London: Palaeontographical Society. 1854.

Fossil Mammalia: Pt 1 of *The zoology of the voyage of H.M.S. Beagle . . . during the years 1831 to 1836.* By Richard Owen. Edited and superintended by Charles Darwin. London: Smith, Elder and Co. 1839–43.

Foster, Joseph. 1874. *Pedigrees of the county families of Yorkshire.* 3 vols. London: W. Wilfred Head.

Foster, Michael and Lankester, Edwin Ray, eds. 1898–1903. *The scientific memoirs of Thomas Henry Huxley.* 4 vols. and supplement. London: Macmillan. New York: D. Appleton.

Foundations: *The foundations of the Origin of Species. Two essays written in 1842 and 1844 by Charles Darwin.* Edited by Francis Darwin. Cambridge: Cambridge University Press. 1909. [Reprint edition. New York: Kraus Reprint Co. 1969. Also reprinted in De Beer ed. 1958.]

France, Peter, ed. 1995. *The new Oxford companion to literature in French.* Oxford: Clarendon Press.

Franklin, John. 1823. *Narrative of a journey to the shores of the Polar Sea, in the years 1819, 20, 21, and 22.* London: John Murray.

——. 1828. *Narrative of a second expedition to the shores of the Polar Sea, in the years 1825, 1826, and 1827.* London: John Murray.

Fraser, H. Malcolm. 1958. *History of beekeeping in Britain.* London: Bee Research Association.

Freeman, Richard Broke. 1977. *The works of Charles Darwin: an annotated bibliographical handlist.* 2d edition. Folkestone, Kent: William Dawson & Sons. Hamden, Conn.: Archon Books, Shoe String Press.

——. 1978. *Charles Darwin: a companion.* Folkestone, Kent: William Dawson & Sons. Hamden, Conn.: Archon Books, Shoe String Press.

Gage, Andrew Thomas and Stearn, William Thomas. 1988. *A bicentenary history of the Linnean Society of London.* London: Academic Press.

Gaimard, Joseph Paul, ed. [1843–8]. *Voyages de la Commission scientifique du Nord en Scandinavie, en Laponie, au Spitzberg et aux Faröe, pendant les années 1838, 1839 et 1840, sur la corvette La Recherche, commandée par M. Fabvre.* Astronomie et hydrographie; Météorology (3 vols.); Magnetisme terrestre (3 vols.); Aurores boréales; Géologie, minéralogie et métallurgie, by Eugene Robert; Géologie, minéralogie, métallurgie et chimie, by J. Durocher; Géographie physique, géographie botanique, botanique et physiologie (2 vols.); Relation du voyage; Littérature Scandinave; Histoire de la Scandavie; 7 atlases. Paris: Arthus Bertrand.

Garrod, Alfred Baring. 1863. *The nature and treatment of gout and rheumatic gout.* 2d edition. London: Walton and Maberly.

Gärtner, Karl Friedrich von. 1844. *Versuche und Beobachtungen über die Befruchtungsorgane der vollkommeneren Gewächse und über die natürliche und künstliche Befruchtung durch den eigenen Pollen*. Pt 1 of *Beiträge zur Kenntniss der Befruchtung der vollkommeneren Gewächse*. Stuttgart: E. Schweizerbart.

Gärtner, Karl Friedrich von. 1849. *Versuche und Beobachtungen über die Bastarderzeugung im Pflanzenreich. Mit Hinweisung auf die ähnlichen Erscheinungen im Thierreiche, ganz umgearbeitete und sehr vermehrte Ausgabe von der Königlich holländischen Akademie der Wissenschaften*. Stuttgart: E. Schweizerbart.

Gaskell, Philip. 1972. *A new introduction to bibliography*. Oxford: Clarendon Press.

Gates, Barbara T. 1997. Revisioning Darwin with sympathy: Arabella Buckley. In *Natural eloquence: women reinscribe science*, edited by Barbara T. Gates and Ann B. Shteir. Madison, Wis., and London: University of Wisconsin Press.

Gayon, Jean. 1998. *Darwinism's struggle for survival: heredity and the hypothesis of natural selection*. Translated by Matthew Cobb. Cambridge: Cambridge University Press.

Geikie, Archibald. 1863. *On the phenomena of the glacial drift of Scotland*. Glasgow: John Gray.

——. 1865. *The scenery of Scotland viewed in connexion with its physical geology*. London and Cambridge: Macmillan & Co.

——. 1895. *Memoir of Sir Andrew Crombie Ramsay*. London and New York: Macmillan.

Geison, Gerald L. 1969. Darwin and heredity: the evolution of his hypothesis of pangenesis. *Journal of the History of Medicine* 24: 375–411.

General index to the Journal of the Linnean Society: *General index to the first twenty volumes of the Journal (Botany), and the botanical portion of the Proceedings, November 1838 to June 1886, of the Linnean Society*. London: Linnean Society of London. 1888.

Geoffroy Saint-Hilaire, Isidore. 1836. *Histoire générale et particulière des anomalies de l'organisation chez l'homme et les animaux, ouvrage comprenant des recherches sur les caractères, la classification, l'influence physiologique et pathologique; les rapports généraux, les lois et les causes des monstruosités, des variétés des vices de conformation, ou traité de tératologie*. 3 vols. Paris: J.-B. Baillière.

Geology of the 'Beagle': *Geological observations on coral reefs, volcanic islands, and on South America*. By Charles Darwin. London: Smith, Elder & Co. 1851.

Gerstaecker, Carl Edouard Adolph. 1863a. Bericht über die wissenschaftlichen Leistungen im Gebiete der Entomologie während des Jahres 1862. *Archiv für Naturgeschichte* 29: 315–18.

——. 1863b. Arthropoden. In *Handbuch der Zoologie*, by Carl Edouard Adolph Gerstaecker and Julius Victor Carus, vol. 2. Leipzig: Wilhelm Engelmann.

Gilbert, Pamela. 1977. *A compendium of the biographical literature on deceased entomologists*. London: British Museum (Natural History).

Glaister, Geoffrey Ashall. 1996. *Encyclopedia of the book*. 2d edition. London: Oak Knoll Press and the British Library.

Goebel, K. von. 1926. *Wilhelm Hofmeister. The work and life of a nineteenth century botanist*. With a biographical supplement by Constanze Ganzenmüller. Translated by H. M. Bower. Botanical editor, F. O. Bower. London: Ray Society.

Grand Larousse de la langue française. 7 vols. Paris: Librairie Larousse. 1971–8.

Gray, Asa. 1856. *Manual of the botany of the northern United States: second edition; including Virginia, Kentucky, and all east of the Mississippi: arranged according to the natural system.* New York: George P. Putnam.

——. 1858. Note on the coiling of tendrils of plants. [Read 12 August 1858.] *Proceedings of the American Academy of Arts and Sciences* 4 (1857–60): 98–9.

——. 1858–9. Diagnostic characters of new species of phænogamous plants, collected in Japan by Charles Wright, botanist of the US North Pacific Exploring Expedition . . . With observations upon the relations of the Japanese flora to that of North America, and of other parts of the northern temperate zone. [Read 14 December 1858 and 11 January 1859.] *Memoirs of the American Academy of Arts and Sciences* n.s. 6 (1857–9): 377–452.

[——.] 1860. Darwin on the origin of species. *Atlantic Monthly* 6: 109–16, 229–39. Darwin and his reviewers. *Atlantic Monthly* 6: 406–25.

——. 1861. *Natural selection not inconsistent with natural theology. A free examination of Darwin's treatise on the origin of species, and of its American reviewers.* London: Trübner. Boston, Mass.: Ticknor & Fields.

——. 1863. Variation and mimetic analogy in Lepidoptera. *American Journal of Science and Arts* 2d ser. 36: 285–90.

——. 1865a. Dioico-dimorphism in the primrose family. [Review of 'Observations on the functions and structure of the reproductive organs of the *Primulaceæ*', by John Scott (*Journal of the Linnean Society London* (*Botany*) 8 (1865): 78–126).] *American Journal of Science and Arts* 2d ser. 39: 101–4.

——. 1865b. Observations upon dimorphic flowers. [Review of 'Ueber dimorphic Blüten', by Hugo von Mohl (*Botanisches Zeitung* (1863): 309–15, 321–8).] *American Journal of Science and Arts* 2d ser. 39: 104–6.

——. 1865–6. On the movements and habits of climbing plants; by Charles Darwin. *American Journal of Science and Arts* 2d ser. 40: 273–82, 41: 125–30.

Gray, George Robert. 1841. *A list of the genera of birds, with their synonyma and an indication of the typical species of each genus.* 2d edition. London: Richard and John E. Taylor.

Gray, Jane Loring, ed. 1893. *Letters of Asa Gray.* 2 vols. London: Macmillan and Co.

Grayson, Donald K. 1983. *The establishment of human antiquity.* New York: Academic Press.

——. 1985. The first three editions of Charles Lyell's *The geological evidences of the antiquity of man. Archives of Natural History* 13: 105–21.

[Greene, Joseph Reay.] 1865. The Linnean Society's *Transactions. Natural History Review* 5: 189–202.

Grisebach, August Heinrich Rudolf. 1859–64. *Flora of the British West Indian Islands.* London: Lovell Reeve & Co.

——. 1864. Die geographische Verbreitung der Pflanzen Westindiens. [Read 3 December 1864.] *Abhandlungen der Königlichen Gesellschaft der Wissenschaften zu Göttingen, Mathematisch–Physikalische Klasse* 12 (1864–6): 3–80.

Grote, George. 1846–56. *A history of Greece.* 12 vols. London: John Murray.

Grote, George. 1980. *The new Grove dictionary of music and musicians*. Edited by Stanley Sadie. 20 vols. London: Macmillan.

GSE: *Great Soviet encyclopedia*. A. M. Prokhorov, editor-in-chief. Translation of 3d edition. New York: Macmillan. London: Collier Macmillan. 1970.

Guenée, Achille. 1852. *Noctuélites*. 3 vols. Paris: Librairie Encyclopédique de Roret.

Guépin, Jean-Baptiste Pierre. 1830–54. *Flore de Maine-et-Loire*. 3d edition with 2 supplements. Paris: G. Baillière.

——. 1853. Note sur l'hybridation des orchidées. *Annales de la Société Linnéenne d'Angers* (1853): 100–14.

Guillard, Achille. 1857. Note sur les vrilles des cucurbitacés. [Read 25 February 1857.] *Bulletin de la Société Botanique de France* 4: 142–5.

Guizot, François Pierre Guillaume. 1864. *Méditations sur l'essence de la religion chrétienne*. Paris: Michel Lévy Frères.

Gulliver, George, ed. 1846. *The works of William Hewson, F.R.S.* London: Sydenham Society.

Gunn, Mary and Codd, L. E. 1981. *Botanical exploration of Southern Africa*. Cape Town: A. A. Balkema.

Gurney, John Henry. 1864. A list of birds collected in Damara Land by Mr. C. J. Andersson. [Read 12 January 1864.] *Proceedings of the Zoological Society of London* (1864): 1–8.

——. 1865. On a raptoral bird transmitted by Mr. Andersson from Damara Land. [Read 14 November 1865.] *Transactions of the Zoological Society of London* 6 (1869): 117–18.

Guy, Jeff. 1983. *The heretic: a study of the life of John William Colenso 1814–1883*. Pietermaritzberg, South Africa: Natal Press. Johannesburg, South Africa: Ravan Press.

Haast, Heinrich Ferdinand von. 1948. *The life and times of Sir Julius von Haast, explorer, geologist, museum builder*. Wellington, New Zealand: privately published.

Haast, John Francis Julius von. 1861. *Report of a topographical and geological exploration of the western districts of the Nelson Province, New Zealand*. Nelson, New Zealand: Nelson provincial government.

——. 1862a. [*Address delivered by Mr Haast at the dinner of the Philosophical Institute of Canterbury*.] Canterbury, New Zealand: Philosophical Institute of Canterbury.

——. 1862b. Notes on the geology of the province of Canterbury, NZ. *New Zealand Government Gazette. Province of Canterbury*, 9 (24 October 1862): 121–31.

——. 1867. Notes on the geology of the province of Canterbury, N.Z., principally in reference to the deposits of the glacial epoch at the western base of the Southern Alps. *Quarterly Journal of the Geological Society of London* 23: 342–52.

——. 1879. *Geology of the provinces of Canterbury and Westland, New Zealand: a report comprising the results of official explorations*. Christchurch, New Zealand: The 'Times' office.

Haeckel, Ernst. 1864. Beschreibung neuer craspedoter Medusen aus dem Golfe von Nizza. *Jenaische Zeitschrift für Medicin und Naturwissenschaft* 1: 325–42.

Haeckel, Ernst. 1865a. *Die Familie der Rüsselquallen (Geryonida.)* Vol. I of *Beitrage zür Naturgeschichter der Hydromedusen.* Leipzig: Wilhelm Engelmann.

——. 1865b. Über eine neue Form des Generationswechsels bei den Medusen und über die Verwandtschaft der Geryoniden und Aeginiden. *Monatsberichte der Königlichen Preußs. Akademie der Wissenschaften zu Berlin* (1865): 85–94.

——. 1865c. Über fossile Medusen. *Zeitschrift für wissenschaftliche Zoologie* 15: 504–14.

——. 1865d. Über den Sarcodekörper der Rhizopoden. *Zeitschrift für wissenschaftliche Zoologie* 15: 342–70.

——. 1866. *Generelle Morphologie der Organismen. Allgemeine Grundzüge der organischen Formen-Wissenschaft, mechanisch begründet durch die von Charles Darwin reformirte Descendenz-Theorie.* 2 vols. Berlin: Georg Reimer.

Haight, Gordon S. 1940. *George Eliot & John Chapman: with Chapman's diaries.* New Haven: Yale University Press.

Haldeman, Samuel Steman. 184[2–]5. *A monograph of the freshwater univalve mollusca of the United States, including notices of species in other parts of North America.* Philadelphia: Conchological Section of the Academy of Natural Science [and others].

——. 1843–4. Enumeration of the recent freshwater Mollusca which are common to North America and Europe; with observations on species and their distribution. *Boston Journal of Natural History* 4: 468–84.

Hall, A. Rupert. 1969. *The Cambridge Philosophical Society: a history, 1819–1969.* Cambridge: Cambridge Philosophical Society.

Hall, Marie Boas. 1984. *All scientists now: the Royal Society in the nineteenth century.* Cambridge: Cambridge University Press.

Hall, N. John. 1991. *Trollope. A biography.* Oxford: Clarendon Press.

Hall, Philip B. 1987. Robert Swinhoe (1836–1877), FRS, FZS, FRGS: a Victorian naturalist in Treaty Port China. *Geographical Journal* 153: 37–47.

Hallam, Henry. 1818. *View of the state of Europe during the Middle Ages.* London: John Murray.

——. 1848. *Supplemental notes to the view of the state of Europe during the Middle Ages.* London: John Murray.

Harris, Miriam Coles. 1860. *Rutledge.* New York: Derby & Jackson.

Harris, P. R. 1998. *A history of the British Museum library, 1753–1973.* London: The British Library.

Harris, Thaddeus William. 1841. *A report on the insects of Massachusetts, injurious to vegetation.* Cambridge, Mass.: Folsom, Wells and Thurston.

Harrow School register: *The Harrow School register, 1800–1911.* 3d edition. Edited by M. G. Dauglish and P. K. Stephenson. London: Longmans, Green, and Co. 1911.

Hart's army list: *The new annual army list . . . with an index.* Compiled by H. G. Hart. London: John Murray. 1840–1900.

Harvey, Joy. 1997. *'Almost a man of genius': Clémence Royer, feminism, and nineteenth-century science.* New Brunswick, N.J., and London: Rutgers University Press.

Harvey, William Henry. 1838. *The genera of South African plants, arranged according to the natural system.* Cape Town, South Africa: A. S. Robertson.

Harvey, William Henry. 1859–63. *Thesaurus Capensis: or, illustrations of the South African flora, being figures and brief descriptions of South African plants selected from the Dublin University Herbarium.* 2 vols. in 1. Dublin: Hodges, Smith & Co. London: John van Voorst.

Harvey, William Henry and Sonder, Otto Wilhelm. 1859–65. *Flora Capensis: being a scientific description of the plants of the Cape Colony, Caffraria, & Port Natal.* 3 vols. Dublin: Hodges, Smith & Co.

HBLS: *Historisch-Biographisches Lexikon der Schweiz.* Under the auspices of the Allgemeine Geschichtsforschende Gesellschaft der Schweiz. Edited by Heinrich Türler *et al.* 7 vols. and supplement. Neuchâtel, Switzerland: Administration des Historisch-Biographischen Lexikons der Schweiz. 1921–34.

Hearne, Samuel. 1795. *A journey from Prince of Wales's Fort, in Hudson's Bay, to the Northern Ocean . . . in the years 1769, 1770, 1771, & 1772.* London: A. Strahan and T. Cadell.

Heer, Oswald. 1855. Ueber die fossilen Pflanzen von St. Jorge in Madeira. [Read 5 November 1855.] *Neue Denkschriften der allgemeinen Schweizerischen Gesellschaft für die gesammten Naturwissenschaften* n.s. 5 (1857): paper 2.

——. 1855–9. *Flora tertiaria Helvetiae. Die tertiäre Flora der Schweiz.* 3 vols. Winterthur, Switzerland: J. Wurster.

——. 1860. *Untersuchungen über das Klima und die Vegetationsverhältnisse des Tertiärlandes.* Winterthur, Switzerland: Anstalt von Wurster.

——. 1861. *Recherches sur le climat et la végétation du pays tertiaire.* Translated from the German by Charles Théophile Gaudin. Winterthur, Switzerland: Jean Wurster & Comp.

——. 1864. *Discours prononcé à l'ouverture de la 48e* session de la Société Helvétique des Sciences Naturelles. N.p.: n.p. [Reprinted from *Bibliothèque Universelle et Revue Suisse (Archives des sciences physiques et naturelles)* n.s. 21: 335–69.]

Henshaw, Samuel. 1878. *Entomological writing of John L. Leconte.* Cambridge, Mass.: George Dimmock.

Henslow, George. 1865. Note on the structure of *Medicago sativa*, as apparently affording facilities for the intercrossing of distinct flowers. [Read 16 November 1865.] *Journal of the Linnean Society (Botany)* 9 (1867): 327–9.

——. 1866. Note on the structure of *Indigofera*, as apparently offering facilities for the intercrossing of distinct flowers. [Read 19 April 1866.] *Journal of the Linnean Society (Botany)* 9 (1867): 355–8.

Henslow, John Stevens. 1837. Description of two new species of *Opuntia*; with remarks on the structure of the fruit of Rhipsalis. *Magazine of Zoology and Botany* 1: 466–9.

——. 1838. Florula Keelingensis. An account of the native plants of the Keeling Islands. *Annals of Natural History* 1: 337–47.

——. 1858. *Illustrations to be employed in practical lessons on botany; adapted to beginners of all classes.* London: Chapman & Hall.

Herschel, John Frederick William, ed. 1849. *A manual of scientific enquiry; prepared for the use of Her Majesty's Navy: and adapted for travellers in general.* London: John Murray.

——, ed. 1851. *Admiralty manual of scientific enquiry.* 2d edition. London: John Murray.

Herschel calendar: *A calendar of the correspondence of Sir John Herschel.* Edited by Michael J. Crowe. Cambridge: Cambridge University Press. 1998.

Hildebrand, Friedrich Hermann Gustav. 1861. Die Verbreitung der Coniferen in der Jetztzeit und in den früheren geologischen Perioden. *Verhandlungen des naturhistorischen Vereines der preussischen Rheinlande und Westphalens* 18: 199–384.

——. 1865. Experimente zur Dichogamie und zum Dimorphismus. *Botanische Zeitung* 23: 1–6, 13–15.

Historical register of the University of Cambridge: *The historical register of the University of Cambridge, being a supplement to the* Calendar *with a record of university offices, honours, and distinctions to the year 1910.* Edited by J. R. Tanner. Cambridge: Cambridge University Press. 1917.

Historical register of Yale University: *Historical register of Yale University 1701–1937.* New Haven, Conn.: Yale University. 1939.

History of the king's works: *The history of the king's works.* Edited by H. M. Colvin. 6 vols. London: HMSO. 1963–82.

Hodge, M. J. S. 1985. Darwin as a lifelong generation theorist. In *The Darwinian heritage*, edited by David Kohn. Princeton, N.J.: Princeton University Press in association with Nova Pacifica.

Hodgson, Brian Houghton. 1847. On various genera of the ruminants. *Journal of the Asiatic Society of Bengal* 16: 685–711.

Hoff, Karl Ernst Adolf von. 1822–41. *Geschichte der durch Überlieferung nachgeweisen natürlichen Veränderungen der Erdoberfläche.* 5 vols. Gotha: Justus Perthes.

Hofmeister, Wilhelm Friedrich Benedict, ed. 1865–77. *Handbuch der physiologischen Botanik.* 4 vols. Leipzig: Wilhelm Engelmann.

——. 1867. *Die Lehre von der Pflanzenzelle.* Leipzig: Wilhelm Engelmann.

——. 1868. *Allgemeine Morphologie der Gewächse.* Leipzig: Wilhelm Engelmann.

Holland, Henry. 1855. *Medical notes and reflections.* 3d edition. London: Longman, Brown, Green, and Longman.

[——.] 1859. Life and organisation. *Edinburgh Review* 109: 227–63.

Hooker, Joseph Dalton. 1844–7. *Flora Antarctica.* 1 vol. and 1 vol. of plates. Pt 1 of *The botany of the Antarctic voyage of HM discovery ships* Erebus *and* Terror *in the years 1839–1843, under the command of Captain Sir James Clark Ross.* London: Reeve Brothers.

——. 1853. *Introductory essay to the flora of New Zealand.* London: Lovell Reeve.

——. 1853–5. *Flora Novæ-Zelandiæ.* 2 vols. Pt 2 of *The botany of the Antarctic voyage of HM discovery ships* Erebus *and* Terror, *in the years 1839–1843, under the command of Captain Sir James Clark Ross.* London: Lovell Reeve.

——. 1854. *Himalayan journals; or, notes of a naturalist in Bengal, the Sikkim and Nepal Himalayas, the Khasia Mountains, &c.* 2 vols. London: John Murray.

Hooker, Joseph Dalton. 1859. *On the flora of Australia, its origin, affinities, and distribution; being an introductory essay to the flora of Tasmania.* London: Lovell Reeve.

——. 1860. Outlines of the distribution of Arctic plants. [Read 21 June 1860.] *Transactions of the Linnean Society of London* 23 (1862): 251–348.

——. 1863. On *Welwitschia*, a new genus of Gnetaceae. [Read 18 December 1862 and 16 January 1863.] *Transactions of the Linnean Society of London* 24 (1864): 1–48.

——. 1864–7. *Handbook of the New Zealand flora: a systematic description of the native plants of New Zealand and the Chatham, Kermadec's, Lord Auckland's, Campbell's, and MacQuarrie's Islands.* 2 vols. London: Lovell Reeve & Co.

——. 1865. Moraines of the Tees Valley. *Reader* 6: 71.

——. 1867. Lecture on insular floras. [Abstract of a lecture before the British Association for the Advancement of Science in 1866.] *Journal of Botany* 5: 23–31; *Gardeners' Chronicle and Agricultural Gazette* (1867): 6–7, 27, 50–1, 75–6.

——. 1872–97. *Flora of British India.* 7 vols. London: Reeve.

——. 1881. On geographical distribution. Address [of the president of section 'E']. *Report of the fifty-first meeting of the British Association for the Advancement of Science, held at York*, Transactions of the sections, pp. 727–38.

Hooker, Joseph Dalton and Thomson, Thomas. 1855. *Flora Indica: being a systematic account of the plants of British India, together with observations on the structure and affinities of their natural orders and genera.* London: W. Pamplin.

Hope, Frederick William. 1837. Descriptions of some species of Carabidae, collected by Charles Darwin, Esq., in his late voyage. [Read 1 May 1837.] *Transactions of the Entomological Society of London* 2 (1837–40): 128–31.

Horsfield, Thomas. 1851. *A catalogue of the Mammalia in the museum of the Hon. East-India Company.* London: W. H. Allen & Co.

[Houghton, William.] 1865. Gleanings from the natural history of the tropics. *Quarterly Review* 118: 166–93.

Howarth, Osbert John Radcliffe. 1931. *The British Association. A retrospect 1831–1931.* Centenary (2d) edition. London: British Association.

Huet, Marie-Hélène. 1993. *Monstrous imagination.* Cambridge, Mass., and London: Harvard University Press.

Hull, David L. 1973. *Darwin and his critics: the reception of Darwin's theory of evolution by the scientific community.* Cambridge, Mass.: Harvard University Press.

Humboldt, Alexander von. 1814–29. *Personal narrative of travels to the equinoctial regions of the New Continent, during the years 1799–1804, by Alexander de Humboldt and Aimé Bonpland . . . translated into English by Helen Maria Williams.* 7 vols. London: Longman, Hurst, Rees, Orme, & Brown.

Hunter, Alexander. 1777. *Georgical essays.* York: J. Dodsley [and others]. London: T. Durham. Edinburgh: W. Creech.

Huntley, Michele A. *et al.* 1972. A bibliography of Sir John Richardson (1787–1865)—articles in learned journals. *Journal of the Society for the Bibliography of Natural History* 6 (1971–4): 98–117.

Hutchinson, Horace Gordon. 1914. *Life of Sir John Lubbock, Lord Avebury.* 2 vols.

London: Macmillan.

Huxley, Leonard, ed. 1900. *Life and letters of Thomas Henry Huxley.* 2 vols. London: Macmillan.

——, ed. 1918. *Life and letters of Sir Joseph Dalton Hooker, OM, GCSI.* Based on materials collected and arranged by Lady Hooker. 2 vols. London: John Murray.

Huxley, Thomas Henry. 1856. On natural history, as knowledge, discipline, and power. *Notices of the Proceedings at the Meetings of the Members of the Royal Institution of Great Britain* 2 (1854–58): 187–95. [Reprinted in M. Foster and Lankester eds. 1898–1903, 1: 305–14.]

——. 1858. On the agamic reproduction and morphology of *Aphis*. *Transactions of the Linnean Society of London* 22 (1859): 193–220, 221–36. [Reprinted in M. Foster and Lankester, eds. 1898–1903, 2: 26–80.]

——. 1860a. On species and races, and their origin. *Notices of the Proceedings at the Meetings of the Members of the Royal Institution of Great Britain* 3 (1858–62): 195–200. [Reprinted in M. Foster and Lankester eds. 1898–1903, 2: 388–94.]

[——.] 1860b. Darwin on the origin of species. *Westminster Review* n.s. 17: 541–70.

——. 1862. *On our knowledge of the causes of the phenomena of organic nature.* 6 pts. London: Robert Hardwicke.

——. 1863a. *Evidence as to man's place in nature.* London: Williams & Norgate.

——. 1863b. *On our knowledge of the causes of the phenomena of organic nature. Being six lectures to working men, delivered at the Museum of Practical Geology.* London: Robert Hardwicke.

——. 1864a. *Lectures on the elements of comparative anatomy. On the classification of animals and on the vertebrate skull.* London: John Churchill & Sons.

[——.] 1864b. Science and 'Church Policy'. *Reader* 4: 821.

[——.] 1864c. Criticisms on 'The Origin of Species'. *Natural History Review* 4: 566–80.

——. 1865. On the methods and results of ethnology. *Fortnightly Review* 1: 257–77.

——. 1866. *Lessons in elementary physiology.* London: Macmillan.

Huxley, Thomas Henry and Etheridge, Robert. 1865. *A catalogue of the collection of fossils in the Museum of Practical Geology, with an explanatory introduction.* London: Her Majesty's Stationery Office.

IBF: *Index biographique français.* Edited by Helen and Barry Dwyer. 4 vols. London: K. G. Saur. 1993.

IBN: *Index bio-bibliographicus notorum hominum.* Pt C: *Corpus alphabeticum sectio generalis.* Edited by Jean-Pierre Lobies *et al.* 116 vols. (A–Hyzrle z Choduv). Osnabrück, Germany: Biblio Verlag. 1974–.

'Illegitimate offspring of dimorphic and trimorphic plants': On the character and hybrid-like nature of the offspring from the illegitimate unions of dimorphic and trimorphic plants. By Charles Darwin. [Read 20 February 1868.] *Journal of the Linnean Society of London (Botany)* 10 (1869): 393–437.

Index Kewensis: *Index Kewensis: plantarum phanerogamarum, nomina et synonyma omnium generum et specierum . . . nomine recepto auctore patria unicuique plantae subjectis.* 4 vols.

and 20 supplements. Compiled by Benjamin Daydon Jackson, *et al.* Oxford: Clarendon Press; Royal Botanic Gardens, Kew. 1893–1996.

India list: *The East-India register . . . from the official returns received at the East-India House. The India Office list . . . Compiled from official records by direction of the secretary of state for India in council.* London: Wm. H. Allen [and others]. 1806–1937.

Insectivorous plants. By Charles Darwin. London: John Murray. 1875.

Jackson, Benjamin Daydon. 1906. *George Bentham.* London: J. M. Dent. New York: E. P. Dutton.

——. 1917. [Obituary notice of Daniel Oliver.] *Proceedings of the Linnean Society of London* 129: 53–7.

——. 1928. *A glossary of botanic terms with their derivation and accent.* 4th edition. London: Gerald Duckworth & Co.

Jamieson, Thomas Francis. 1863. On the parallel roads of Glen Roy, and their place in the history of the glacial period. [Read 21 January 1863.] *Quarterly Journal of the Geological Society of London* 19: 235–59.

——. 1865. On the history of the last geological changes in Scotland. [Read 11 January 1865.] *Quarterly Journal of the Geological Society* 21 (1865): 161–203.

Jenkins, Brian. 1974–80. *Britain & the war for the Union.* 2 vols. Montreal, Quebec, and London, Ontario: McGill-Queen's University Press.

Jensen, J. Vernon. 1977. 'The most intimate and trusted friend I have': a note on Ellen Busk, young T. H. Huxley's confidante. *Historical studies* 17 (1977): 315–22.

Johnson, Robert. 1976. *Sir John Richardson: arctic explorer, natural historian, naturalist.* London: Taylor & Francis.

Johnson, Samuel. 1755. *A dictionary of the English language: in which the words are deduced from their originals, and illustrated in their different significations by examples from the best writers.* 2 vols. London: J. & P. Knapton [and others].

Johnson, William S. 1990. *Nineteenth-century photography: an annotated bibliography, 1839–79.* Boston: G. K. Hall & Co.

Jones, Hugh Percy, ed. 1900. *A new dictionary of foreign phrases and classical quotations . . . with English translations or equivalents.* London: Charles William Deacan.

Jones, William, trans. 1869. *Institutes of Hindu Law; or, the ordinances of Menu according to the gloss of Cullúca, comprising the Indian system of duties, religious and civil.* Collated with the Sanscrit text by G. C. Haughton. 3d edition, with preface and index by S. G. Grady. London: Wm. H. Allen & Co.

Jordan, Alexis. 1864. *Diagnoses d'espèces nouvelles ou méconnues, pour servir de matériaux à une flore réformée de la France, et des contrées voisines.* Paris: F. Savy.

Journal and remarks: *Journal and remarks. 1832–1836.* By Charles Darwin. Vol. 3 of *Narrative of the surveying voyages of His Majesty's ships Adventure and Beagle between the years 1826 and 1836, describing their examination of the southern shores of South America, and the Beagle's circumnavigation of the globe.* London: Henry Colburn. 1839. [Separately published as *Journal of researches.*]

Journal of researches: *Journal of researches into the geology and natural history of the various countries visited by HMS Beagle, under the command of Captain FitzRoy, RN, from 1832 to*

1836. By Charles Darwin. London: Henry Colburn. 1839.

Journal of researches 2d ed.: *Journal of researches into the natural history and geology of the countries visited during the voyage of HMS Beagle round the world, under the command of Capt. FitzRoy RN.* 2d edition, corrected, with additions. By Charles Darwin. London: John Murray. 1845.

Jukes, Joseph Beete. 1862a. On the mode of formation of some of the river-valleys in the south of Ireland. *Quarterly Journal of the Geological Society of London* 18: 378–403.

——. 1862b. Address to the geological section. *Report of the thirty-second meeting of the British Association for the Advancement of Science; held at Cambridge*, Transactions of the sections, pp. 54–65.

Junker, Thomas. 1989. *Darwinismus und Botanik. Rezeption, Kritik und theoretische Alternativen im Deutschland des 19. Jahrhunderts.* Stuttgart: Deutscher Apotheker Verlag.

Karsten, Gustave Karl Wilhelm Hermann. 1861. On the sexual life of plants, and parthenogenesis. *Annals and Magazine of Natural History* 3d ser. 8: 81–9, 200–9.

Kaye, John William. 1864–76. *A history of the Sepoy War in India 1857–1858.* 3 vols. London: W. H. Allen & Co.

Kay-Shuttleworth, James Phillips. 1860. *Scarsdale; or, life on the Lancashire and Yorkshire border, thirty years ago.* 3 vols. London: Smith, Elder & Co.

Keilhau, Baltazar Mathias. 1838–40. Theory of granite, and the other massive rocks; together with that of crystalline slate; proposed in lectures in geology, in the University of Christiania in Norway, in the year 1836. *Edinburgh New Philosophical Journal* 24 (1838): 387–403; 25 (1838): 80–101, 263–72; 28 (1840): 366–71.

——. 1844. On the mode of formation of crystalline limestone . . . with preliminary observations on the present state of geology, and on the methods of investigation pursued in that science. *Edinburgh New Philosophical Journal* 36 (1844): 341–62; 37 (1844): 143–76.

Keller, Ferdinand. 1856–66. Die keltischen Pfahlbauten in den Schweizerseen. *Mittheilungen der antiquarischen Gesellschaft in Zürich* 9 (1853–6): 65–100; 12 (1857–8): 111–56; 13 (1858–63): i–x; 14 (1858–63): 1–34, 129–88; 15 (1863–66): 245–321.

——. 1866. *The lake dwellings of Switzerland and other parts of Europe.* Translated and arranged by John Edward Lee. London: Longmans, Green, and Co.

Keynes, Margaret. 1943. *Leonard Darwin, 1850–1943.* Cambridge: privately printed at Cambridge University Press.

Keynes, Randal. 2001. *Annie's box. Charles Darwin, his daughter and human evolution.* London: Fourth Estate.

Keynes, Richard Darwin, ed. 2000. *Charles Darwin's zoology notes & specimen lists from H.M.S. Beagle.* Cambridge: Cambridge University Press.

King-Hele, Desmond. 1999. *Erasmus Darwin. A life of unequalled achievement.* London: Giles de la Mare Publishers.

Kingsley, Charles. 1855. *Glaucus; or, the wonders of the shore.* 2d edition. Cambridge.

——. 1858. My winter garden. *Fraser's Magazine* 57: 408–25.

——. 1890. How to study natural history. In *Scientific lectures and essays.* New edition. London: Macmillan and Co.

Kingsley, Frances, ed. 1877. *Charles Kingsley: his letters and memories of his life.* 2 vols. London: Henry S. King & Co.

——, ed. 1883. *Charles Kingsley: his letters and memories of his life.* London: Kegan Paul, Trench & Co.

Knoll, Elizabeth. 1986. The sciences of language and the evolution of mind: Max Müller's quarrel with Darwinism. *Journal of the History of the Behavioral Sciences* 22: 3–22.

Kohn, David. 1980. Theories to work by: rejected theories, reproduction, and Darwin's path to natural selection. *Studies in History of Biology* 4: 67–170.

Kölliker, Rudolf Albert von. 1864a. Über die Darwin'sche Schöpfungstheorie. *Zeitschrift für wissenschaftliche Zoologie* 14: 74–86.

——. 1864b. *Über die Darwin'sche Schöpfungstheorie.* Leipzig: Wilhelm Engelmann.

——. 1864c. On Darwin's theory of the origin of species. *Reader* 4: 199–200, 234–5.

Krohn, August David. 1859. Beobachtungen über den Cementapparat und die weiblichen Zeugungsorgane einiger Cirripedien. *Archiv für Naturgeschichte* 25 (pt 1): 355–64.

——. 1861. Beobachtungen über den Bau und die Fortpflanzung der Eleutheria Quatref. *Archiv für Naturgeschichte* 27: 157–70.

——. 1862. Observations on the structure and reproduction of *Eleutheria*, Quatref. *Annals and Magazine of Natural History* 3d ser. 9: 1–9.

Lacordaire, Jean Théodore. 1854–75. *Histoire naturelle des insectes. Genera des Coléoptères, ou exposé méthodique et critique de tous les genres proposés jusqu'ici dans cet ordre d'insectes.* 11 vols. in 12. Paris: Librairie encyclopédique de Roret.

Lamarck, Jean Baptiste Pierre Antoine de Monet de. 1809. *Philosophie zoologique; ou exposition des considérations relatives à l'histoire naturelle des animaux; à la diversité de leur organisation . . . et les autres l'intelligence de ceux qui en sont doués.* 2 vols. Paris: Dentu; the author.

——. 1815–22. *Histoire naturelle des animaux sans vertèbres, présentant les caractères généraux et particuliers de ces animaux, leur distribution, leurs classes, leurs familles, leurs genres, et la citation des principales espèces qui s'y rapportent; précédée d'une introduction offrant la détermination des caractères essentiels de l'animal, sa distinction du végétal et des autres corps naturels, enfin, l'exposition des principes fondamentaux de la zoologie.* 7 vols. Paris: Verdière [and others].

Larousse du XXe siècle. Edited by Paul Augé. 6 vols. Paris: Librairie Larousse. 1928–33.

Laugel, Antoine Auguste. 1860. Nouvelle théorie d'histoire naturelle. L'origine des espèces. *Revue des Deux Mondes* 26: 644–71.

——. 1863. La guerre civile aux Etats-Unis, 1861–1863. Le gouvernement fédéral, les armées et les partis. *Revue des Deux Mondes* 47: 872–97.

——. 1864. Les Etats-Unis pendant la guerre. I. L'élection présidentielle de 1864. *Revue des Deux Mondes* 54.

——. 1865. Les Etats-Unis pendant la guerre. II. De l'Atlantique au Mississippi; l'Américain de l'ouest. *Revue des Deux Mondes* 56: 874–910.

——. 1866. *Les Etats-Unis pendant la guerre (1861–1865).* Paris: Germer Baillière.

Leach, John. 1987. *The book of Buxton: spa of blue waters.* Buckingham: Barracuda.

Lecky, William Edward Hartpole. 1865. *History of the rise and influence of the spirit of rationalism in Europe.* 2 vols. London: Longman, Green, Longman, Roberts, & Green.

Le Conte, John Lawrence. 1853. Descriptions of twenty new species of Coleoptera inhabiting the United States. *Proceedings of the Academy of Natural Sciences of Philadelphia* 6 (1852–3): 226–35.

——. 1859. The Coleoptera of Kansas and Eastern New Mexico. *Smithsonian Contributions to Knowledge* 11 (1859): Article 6.

——. 1861–2. Classification of the Coleoptera of North America. *Smithsonian Miscellaneous Collections* 3: Article 3.

Lecoq, Henri. 1845. *De la fécondation naturelle et artificielle des végétaux et de l'hybridation, considérée dans ses rapports avec l'horticulture, l'agriculture et la sylviculture . . . Contenant les moyens pratiques d'opérer l'hybridation et de créer facilement des variétés nouvelles.* Paris: Audot.

Le Fanu, Joseph Sheridan. 1864. *Uncle Silas: a tale of Bartram-Haugh.* London: Richard Bentley.

Leftwich, A. W. 1973. *A dictionary of zoology.* 3d edition. London: Constable.

Leighton, William Allport. 1866. The fecundation of *Lupinus polyphyllus. Journal of Botany* 4: 36–8.

Léon, Isidore. 1858. Recherches nouvelles sur la cause du mouvement spiral des tiges volubiles. *Bulletin de la Société Botanique de France* 5: 351–6, 610–14, 624–9, 679–85.

Leopold, Joan. 1980. *Culture in comparative and evolutionary perspective: E. B. Tylor and the making of* Primitive culture. Berlin: Dietrich Reimer Verlag.

Leslie, Charles Robert and Taylor, Tom. 1865. *The life and times of Sir Joshua Reynolds: with notices of some of his contemporaries.* Commenced by C. R. Leslie and continued and concluded by T. Taylor. 2 vols. London: John Murray.

Lesquereux, Leo. 1859–63. On some questions concerning the coal formations of North America. *American Journal of Science and Arts* 2d ser. 28 (1859): 21–37; 30 (1860): 63–74, 367–84; 32 (1861): 15–25, 193–205; 33 (1862): 206–16; 35 (1863): 375–86.

Lewes, George Henry. 1853. The development hypothesis of the 'Vestiges'. *Leader* (1853): 784–5, 812–14, 832–4, 883–4.

Lewis, Charlton T. and Short, Charles. 1969. *A Latin dictionary, founded on Andrews' edition of Freund's Latin dictionary.* Revised, enlarged, and in great part rewritten. Oxford: Clarendon Press.

Lewis, George Cornewall. 1862. *Historical survey of the astronomy of the ancients.* London: Parker, Son, and Bourn.

Liebig, Justus. 1840. *Organic chemistry in its applications to agriculture and physiology.* Edited by Lyon Playfair. London: Taylor and Walton.

Lilljeborg, Wilhelm. 1859. Liriope och Peltogaster H. Rathke. *Öfversigt af Kongl. Vetenskaps-Akademiens Förhandlingar* 16: 213–17.

Lilljeborg, Wilhelm. 1860. On the genera *Peltogaster* and *Liriope* of Rathke. *Annals and Magazine of Natural History* 3d. ser. 6: 162–73, 260–7.

Lindley, John. 1853. *The vegetable kingdom; or, the structure, classification, and uses of plants, illustrated upon the natural system.* 3d edition with corrections and additional genera. London: Bradbury & Evans.

List of the Geological Society of London. London: [Geological Society]. 1864–1934.

List of the Linnean Society of London. London: [Linnean Society of London]. 1805–1939.

Living Cirripedia (1851): *A monograph of the sub-class Cirripedia, with figures of all the species. The Lepadidæ; or, pedunculated cirripedes.* By Charles Darwin. London: Ray Society. 1851.

Living Cirripedia (1854): *A monograph of the sub-class Cirripedia, with figures of all the species. The Balanidæ (or sessile cirripedes); the Verrucidæ, etc.* By Charles Darwin. London: Ray Society. 1854.

LL: *The life and letters of Charles Darwin, including an autobiographical chapter.* Edited by Francis Darwin. 3 vols. London: John Murray. 1887–8.

Loiseleur Deslongchamps, Auguste Louis Armand, trans. 1833. *Manava–Dharma–Sastra: Lois de Manou, comprenant les institutions religieuses et civiles des Indiens.* Paris: Imp. de Crapelet.

Loudon, John Claudius. 1841. *An encyclopædia of plants.* London: Longman, Orme, Brown, Green, and Longmans.

Low, David. 1845. *On the domesticated animals of the British Islands.* London: Longman, Brown, Green & Longmans.

Lubbock, John. 1857. An account of the two methods of reproduction in *Daphnia*, and of the structure of the ephippium. *Philosophical Transactions of the Royal Society of London* 147: 79–100.

——. 1861. The kjökkenmöddings: recent geologico-archæological researches in Denmark. *Natural History Review* n.s. 1: 489–504.

——. 1862a. On the ancient lake habitations of Switzerland. *Natural History Review* n.s. 2: 26–52.

——. 1862b. On the evidence of the antiquity of man, afforded by the physical structure of the Somme Valley. *Natural History Review* n.s. 2: 244–69.

——. 1863a. A visit to the ancient shell-mounds of Scotland. *Natural History Review* n.s. 3: 415–22.

[——.] 1863b. [Review of Charles Lyell's *Antiquity of man.*] *Natural History Review* n.s. 3: 211–19.

——. 1864. Cave-men. *Natural History Review* n.s. 4: 407–28.

——. 1865. *Pre-historic times, as illustrated by ancient remains, and the manners and customs of modern savages.* London and Edinburgh: Williams & Norgate.

——. 1869. *Pre-historic times, as illustrated by ancient remains, and the manners and customs of modern savages.* 2d edition. London and Edinburgh: Williams & Norgate.

Lucas, Prosper. 1847–50. *Traité philosophique et physiologique de l'hérédité naturelle dans les états de santé et de maladie du système nerveux avec l'application méthodique des lois de la procréation au traitement général des affections . . .*. 2 vols. Paris: J. B. Baillière.

Lurie, Edward. 1960. *Louis Agassiz: a life in science.* Chicago: University of Chicago Press.

Lyell, Charles. 1830–3. *Principles of geology, being an attempt to explain the former changes of the earth's surface, by reference to causes now in operation.* 3 vols. London: John Murray.

——. 1853. *Principles of geology; or, the modern changes of the earth and its inhabitants considered as illustrative of geology.* 9th edition, entirely revised. London: John Murray.

——. 1855. *A manual of elementary geology: or, the ancient changes of the earth and its inhabitants as illustrated by geological monuments.* 5th edition London: John Murray.

——. 1859. On the occurrence of works of human art in post-Pliocene deposits. *Report of the 29th meeting of the British Association for the Advancement of Science held at Aberdeen*, Transactions of the sections, pp. 93–5.

——. 1863a. *The geological evidences of the antiquity of man with remarks on theories of the origin of species by variation.* London: John Murray.

——. 1863b. *The geological evidences of the antiquity of man with remarks on theories of the origin of species by variation.* 2d edition, revised. London: John Murray.

——. 1863c. *The geological evidences of the antiquity of man with remarks on theories of the origin of species by variation.* 3d edition, revised. London: John Murray.

——. 1863d. *The geological evidences of the antiquity of man, with remarks on theories of the origin of species by variation.* 2d American edition, from the latest London edition. Philadelphia: J. W. Childs.

——. 1864. Presidential address. *Report of the thirty-fourth meeting of the British Association for the Advancement of Science; held at Bath*, pp. lx–lxxv.

——. 1865. *Elements of geology; or, the ancient changes of the earth and its inhabitants as illustrated by geological monuments.* 6th edition London: John Murray.

——. 1867–8. *Principles of geology; or, the modern changes of the earth and its inhabitants considered as illustrative of geology.* 10th edition 2 vols. London: John Murray.

Lyell, Katherine Murray, ed. 1881. *Life, letters and journals of Sir Charles Lyell, Bart.* 2 vols. London: John Murray.

McCracken, Donal P. 1997. *Gardens of empire: botanical institutions of the Victorian British empire.* London and Washington: Leicester University Press.

Macgillivray, William. 1837–52. *A history of British birds, indigenous and migratory.* 5 vols. London: Scott, Webster, and Geary; William S. Orr and Co.

MacGregor, Arthur, ed. 1994. *Sir Hans Sloane: collector, scientist, antiquary, founding father of the British Museum.* London: British Museum Press in association with Alistair McAlpine for the Trustees of the British Museum.

MacLeod, Roy M. 1971a. Of medals and men: a reward system in Victorian science, 1826–1914. *Notes and Records of the Royal Society* 26: 81–108.

——. 1971b. The Royal Society and the government grant: notes on the administration of scientific research 1849–1914. *Historical Journal* 14: 323–58.

McOuat, Gordon R. 1996. Species, rules and meaning: the politics of language and the ends of definitions in nineteenth century natural history. *Studies in the History and Philosophy of Science* 27: 473–519.

McPherson, James M. 1988. *Battle cry of freedom. The Civil War era.* New York and Oxford: Oxford University Press.

'Males and complemental males of certain cirripedes': On the males and complemental males of certain cirripedes, and on rudimentary structures. By Charles Darwin. *Nature* 8 (1873): 431–2. [*Collected papers* 2: 177–82.]

Mammalia: Pt 2 of *The zoology of the voyage of HMS Beagle.* By George Robert Waterhouse. Edited and superintended by Charles Darwin. London: Smith, Elder and Co. 1838–9.

Manton, Jo. 1965. *Elizabeth Garrett Anderson.* London: Methuen. [Reprint edition 1987.]

Marchant, James, ed. 1916. *Alfred Russel Wallace: letters and reminiscences.* 2 vols. London: Cassell and Co.

Marginalia: *Charles Darwin's marginalia.* Edited by Mario A. Di Gregorio with the assistance of Nicholas W. Gill. Vol. 1. New York and London: Garland Publishing. 1990.

Markham, Clements R. 1880. *Peruvian bark. A popular account of the introduction of chinchona cultivation into British India.* London: John Murray.

Martin, Gerald P. R. and Uschmann, Georg. 1969. *Friedrich Rolle 1827–1887, ein Vorkämpfer neuen biologischen Denkens in Deutschland.* Leipzig: Johann Ambrosius Barth.

Martin, William Charles Linnaeus. [1849a.] *The sheep.* In vol. 2 of *The farmer's library. Animal economy.* London: Charles Knight.

——. [1849b.] *Poultry.* In vol. 2 of *The farmer's library. Animal economy.* London: Charles Knight.

Masters, Maxwell Tylden. 1855. An abnormal stem of a species of *Dipsacus.* [Read 6 March 1855.] *Proceedings of the Linnean Society of London* 2 (1848–55): 369–71.

——. 1860. On the relation between the abnormal and normal formations in plants. [Read 16 March 1860.] *Notices of the Proceedings at the Meetings of the Members of the Royal Institution of Great Britain* 3 (1858–62): 223–7.

——. 1861. On prolification in flowers, and especially on that form termed median prolification. [Read 17 January 1861.] *Transactions of the Linnean Society of London* 23 (1860–2): 359–70.

——. 1862a. Vegetable morphology: its history and present condition. *British and Foreign MedicoChirurgical Review* 29: 202–18.

——. 1862b. On prolification in flowers, and especially on that kind termed axillary prolification. [Read 20 February 1862.] *Transactions of the Linnean Society of London* 23 (1860–2): 481–93.

——. 1863a. Observations on the morphology and anatomy of the genus *Restio*, Linn., together with an enumeration of the South African species. [Read 16 April 1863.] *Journal of the Linnean Society (Botany)* 8 (1865): 211–55.

——. 1863b. Note on certain influences regulating the forms of leaves, &c. *Report of the 33d meeting of the British Association for the Advancement of Science, held at Newcastle-upon-Tyne*, Transactions of the sections, p. 97.

Masters, Maxwell Tylden. 1863c. On the existence of two forms of peloria. *Natural History Review* n.s. 3: 258–62.

——. 1864. On a peloria and semidouble flower of *Ophrys aranifera*, Huds. [Read 16 June 1864.] *Journal of the Linnean Society* (*Botany*) 8 (1865): 207–11.

——. 1869. *Vegetable teratology, an account of the principal deviations from the usual construction of plants.* London: Ray Society.

Matthew, Patrick. 1831. *On naval timber and arboriculture; with critical notes on authors who have recently treated the subject of planting.* London: Longman, Rees, Orme, Brown, & Green. Edinburgh: Adam Black.

——. 1864. *Schleswig-Holstein, etc.* London: Spottiswoode & Co.

Maw, George. 1865a. On some deposits of chert, white sand, and white clay in the neighbourhood of Llandudno, North Wales. *Geological Magazine* 2: 200–4.

——. 1865b. On an extensive distribution of white sands and clays in North Wales antecedent to the boulder clay drift. *Report of the thirty-fifth meeting of the British Association for the Advancement of Science; held at Birmingham in September 1865*, Transactions of the sections, p. 68.

Max Müller, Friedrich. 1861. *Lectures on the science of language delivered at the Royal Institution of Great Britain in April, May, and June, 1861.* London: Longman, Green, Longman & Roberts.

——. 1864. *Lectures on the science of language.* 2d series. London: Longmans.

Mayr, Ernst. 1942. *Systematics and the origin of species from the viewpoint of a zoologist.* New York: Columbia University Press.

MDCB: *The Macmillan dictionary of Canadian biography.* Edited by W. Stewart Wallace. 4th edition. Revised by W. A. McKay. Toronto: Macmillan of Canada. 1978.

Medical directory: *The London medical directory ... every physician, surgeon, and general practitioner resident in London.* London: C. Mitchell. 1845. *The London and provincial medical directory.* London: John Churchill. 1848–60. *The London & provincial medical directory, inclusive of the medical directory for Scotland, and the medical directory for Ireland, and general medical register.* London: John Churchill. 1861–9. *The medical directory ... including the London and provincial medical directory, the medical directory for Scotland, the medical directory for Ireland.* London: J. & A. Churchill. 1870–1905.

Medical who's who: *The medical who's who.* London: London & Counties Press Association [and others]. 1913–27.

Meinert, Frederik Vilhelm August. 1864. Miastor Metraloas: yderligere Oplysning om den af Prof. Nic. Wagner nyligt beskrevne Insektlarve, som formerer sig ved Spiredannelse. *Naturhistorisk Tidsskrift* 3d ser. 3: 37–43.

Mellersh, Harold Edward Leslie. 1968. *FitzRoy of the Beagle.* London: Rupert Hart-Davis.

Men and women of the time: *The men of the time in 1852 or sketches of living notables. Men and women of the time: a dictionary.* 15th edition. By Victor G. Plarr. London: David Bogue [and others]. 1852–99.

Men-at-the-bar: *Men-at-the-bar: a biographical hand-list of the members of the various inns of court, including Her Majesty's judges, etc.* By Joseph Foster. London: Reeves & Turner. 1885.

Messerli, Jonathan. 1972. *Horace Mann: a biography.* New York: Alfred A. Knopf.

Metcalfe, Richard. 1906. *The rise and progress of hydropathy in England and Scotland.* London: Simpkin, Marshall, Hamilton, Kent & Co.

Meteyard, Eliza. 1865–6. *The life of Josiah Wedgwood from his private correspondence and family papers ... with an introductory sketch of the art of pottery in England.* 2 vols. London: Hurst & Blackett.

——. 1871. *A group of Englishmen (1795 to 1815), being records of the younger Wedgwoods and their friends, embracing the history of the discovery of photography.* London: Longmans, Green, and Co.

——. 1970. *The life of Josiah Wedgwood from his private correspondence and family papers ... with an introductory sketch of the art of pottery in England.* Reprint edition, with an introduction by R. W. Lightbown. 2 vols. London: Cornmarket Press.

Michel, Francisque Xavier. 1861. *Du passé et de l'avenir des haras: recherches sur le commerce, les dénominations et la production des chevaux, principalement en France, avant 1789.* Paris: Librairie de Michel Lévy frères; London and Edinburgh: Williams and Norgate.

Mill, Hugh Robert. 1930. *The record of the Royal Geographical Society 1830–1930.* London: Royal Geographical Society.

Milne, David. 1841–4. Notices of earthquake-shocks felt in Great Britain, and especially in Scotland, with inferences suggested by these notices as to the causes of such shocks. *Edinburgh New Philosophical Journal* 31 (1841): 92–122, 259–309; 32 (1842): 106–27, 362–78; 33 (1842): 372–88; 34 (1843): 85–106; 35 (1843): 137–59; 36 (1843–4): 72–86, 362–76.

——. 1847. On the parallel roads of Lochaber, with remarks on the change of relative levels of sea and land in Scotland, and on the detrital deposits in that country. [Read 1 March and 5 April 1847.] *Transactions of the Royal Society of Edinburgh* 16 (1849): 395–418. [Also printed in *Edinburgh New Philosophical Journal* 43 (1847): 339–64.]

Milne-Edwards, Henri. 1834–40. *Histoire naturelle des Crustacés, comprenant l'anatomie, la physiologie et la classification de ces animaux.* 3 vols. Paris: Librairie encyclopédique de Roret.

Milne Home, Grace. 1891. *Biographical sketch of David Milne Home LL.D, F.R.S.E., F.G.S., etc.* Edinburgh: David Douglas.

Minor, W. C. 1865. Further remarks on larve budding. *American Journal of Science and Arts* 2d ser. 39: 362–3.

Modern English biography: *Modern English biography, containing many thousand concise memoirs of persons who have died since the year 1850.* By Frederick Boase. 3 vols. and supplement (3 vols.). Truro, Cornwall: printed for the author. 1892–1921.

Moggridge, John Traherne. 1864. Observations on some orchids of the south of France. [Read 3 November 1864.] *Journal of the Linnean Society* (*Botany*) 8

(1865): 256–8.
Moggridge, John Traherne. 1865–8. *Contributions to the flora of Mentone.* 4 pts. London: Lovell Reeve and Co.
——. 1869. *Ueber Ophrys insectifera L. (part.)* Dresden: E. Blochmann & Sohn. [*Verhandlungen der Kaiserlichen Leopoldino-Carolinischen deutschen Akademie der Naturforscher* 35 (1870): (3d paper) 1–16.]
Mohl, Hugo von. 1863. Einige Beobachtungen über dimorphe Blüthen. *Botanische Zeitung* 21: 309–15, 321–8.
Moir, Martin. 1988. *A general guide to the India Office records.* London: The British Library.
Möller, Alfred, ed. 1915–21. *Fritz Müller. Werke, Briefe und Leben.* 3 vols in 5. Jena: Gustav Fischer.
Montgomery, William M. 1988. Germany. In *The comparative reception of Darwinism*, with a new preface, edited by Thomas F. Glick. Chicago and London: University of Chicago Press.
Moore, David and More, Alexander Goodman. 1866. *Contributions towards a Cybele Hibernica.* Dublin: Hodges, Smith & Co.
Moore, James Richard. 1977. On the education of Darwin's sons: the correspondence between Charles Darwin and the Reverend G. V. Reed, 1857–1864. *Notes and Records of the Royal Society* 32 (1977–8): 51–70.
——. 1979. *The post-Darwinian controversies: a study of the Protestant struggle to come to terms with Darwin in Great Britain and America, 1870–1900.* Cambridge: Cambridge University Press.
——. 1985. Darwin of Down: the evolutionist as squarson-naturalist. In *The Darwinian heritage*, edited by David Kohn. Princeton, N.J.: Princeton University Press in association with Nova Pacifica.
Moquin-Tandon, Horace Bénédict Alfred. 1841. *Eléments de tératologie végétale, ou, histoire abrégée des anomalies de l'organisation dans les végétaux.* Paris: P.-J. Loss.
Morlot, Charles Adolphe. 1859. Etudes géologico-archéologiques en Danemark et en Suisse. [Read January 1859.] *Bulletin des séances. Société Vaudoise des Sciences Naturelles* 6: 263–328.
——. 1861a. General views on archæology. *Annual report of the Board of Regents of the Smithsonian Institution showing the operations, expenditures, and condition of the institution for the year 1860* 15 (1861): 284–343. Translated by Philip Harry, Esq., for the Smithsonian Institution.
——. 1861b. *General views on archæology.* Translated for the Smithsonian Institution by Philip Harry, Esq. Washington: n.p.
Morris, Paul J. 1997. Louis Agassiz's arguments against Darwinism in his additions to the French translation of the *Essay on classification. Journal of the History of Biology* 30: 121–34.
Mowbray, William. 1822. [Letter to the secretary of the Horticultural Society of London, 30 November 1822.] *Transactions of the Horticultural Society of London* 5 (1824): 274–5.

Müller, Fritz. 1859. Polypen und Quallen von Santa Catharina. *Archiv für Naturgeschichte* 25: 310–25.

——. 1862a. Die Rhizocephalen, eine neue Gruppe schmarotzender Kruster. *Archiv für Naturgeschichte* 28: 1–9.

——. 1862b. On the *Rhizocephala*, a new group of parasitic Crustacea. Translated by William S. Dallas. *Annals and Magazine of Natural History* 3d. ser. 10: 44–50.

——. 1864. *Für Darwin.* Leipzig: Wilhelm Engelmann.

——. 1865a. The Darwinian hypothesis supported by observations on Crustacea. *Annals and Magazine of Natural History* 3d ser. 15 (1865): 410–16.

——. 1865b. Notes on some of the climbing-plants near Desterro, in south Brazil. By Herr Fritz Müller, in a letter to C. Darwin. *Journal of the Linnean Society* (*Botany*) 9 (1866): 344–9.

——. 1865c. Ueber Darwinella aurea, einen Schwamm mit sternförmigen Hornnadeln. *Archiv für mikroskopische Anatomie* 1: 344–53.

——. 1866. Ueber das Holz einiger um Desterro wachsenden Kletterpflanzen. *Botanische Zeitung* 24: 57–60, 65–9.

——. 1867. Ueber Balanus armatus und einen Bastard dieser Art und des Balanus improvisus var. assimilis Darw. *Archiv für Naturgeschichte* 33: 329–56.

——. 1868a. On *Balanus armatus*, and a hybrid between this species and *Balanus improvisus*, var. *assimilis*, Darw. Translated by W. S. Dallas. *Annals and Magazine of Natural History* 4th ser. 1: 393–412.

——. 1868b. Notizen über die Geschlechtsverhältnisse brasilianischer Pflanzen. *Botanische Zeitung* 26: 113–16.

——. 1870. Die Bewegung des Blüthenstieles von Alisma. *Jenaische Zeitschrift für Medizin und Naturwissenschaft* 5: 133–7.

Munro, Robertson. 1868. On reproduction and cross-fertilisation of Passifloras. [Read 11 June 1868.] *Transactions of the Botanical Society of Edinburgh* 9: 299–402.

Murchison, Roderick Impey. 1864a. Presidential address. [Read 23 May 1864.] *Proceedings of the Royal Geographical Society* 8 (1863–4): 170–253.

——. 1864b. *On the relative powers of glaciers and floating icebergs in modifying the surface of the earth.* London: W. Clowes and Sons.

Murray, Andrew. 1860. On the pediculi infesting the different races of man. [Read 17 December 1860.] *Transactions of the Royal Society of Edinburgh* 22 (1857–61): 567–78.

Nachtrag: *Nachtrag zum Programm für den zweiten Congress deutscher Gärtner, Botaniker und Gartenfreunde und für die allgemeine deutsche Ausstellung von Gemüsen und landwirthschaftlichen Produkten, Obst, Pflanzen, Blumen, Gartengeräthschaften u. s. w. in Erfurt im September 1865.* Erfurt: A. Stenger. [1865.]

Narrative: *Narrative of the surveying voyages of His Majesty's ships Adventure and Beagle, between the years 1826 and 1836.* [Edited by Robert FitzRoy.] 3 vols. and appendix. London: Henry Colburn. 1839.

Nathusius, Hermann von. 1864. *Vorstudien für Geschichte und Zucht der Hausthiere zunächst am Schweineschädel.* Berlin: Wiegandt und Hempel.

Natural selection: *Charles Darwin's Natural selection: being the second part of his big species book written from 1856 to 1858*. Edited by R. C. Stauffer. Cambridge: Cambridge University Press. 1975.

Naudin, Charles Victor. 1853. *Melastomacearum, quæ in Musæo Parisiensi continentur monographicæ descriptionis et secundum affinitates distributionis tentamen*. Paris: V. Masson.

——. 1855. Organographie végétale. Observations relatives à la nature des vrilles et à la structure de la fleur chez les cucurbitacées. *Annales des Sciences Naturelles* (*Botanique*) 4th ser. 4: 5–19.

——. 1856. Observations relatives à la formation des graines sans le secours du pollen. *Comptes Rendus Hebdomadaires des Séances de l'Académie des Sciences* 43: 538–42.

——. 1863. Nouvelles recherches sur l'hybridité dans les végétaux. *Annales des Sciences Naturelles* (*Botanique*) 4th ser. 19: 180–203.

——. 1865a. Nouvelles recherches sur l'hybridité dans les végétaux. Mémoire présenté a l'Académie des Sciences par M. Ch. Naudin en Décembre 1861 et couronné dans la séance du 29 Décembre 1862. *Nouvelles Archives du Muséum d'Histoire Naturelle* 1: 25–176.

——. 1865b. Note on hybridism in vegetables. *Natural History Review* 5: 567–73.

Navy list: *The navy list*. London: John Murray; Her Majesty's Stationery Office. 1815–1900.

NBL: *Norsk biografisk leksikon*. Edited by Edvard Bull *et al.* 19 vols. Oslo: H. Aschehoug & Co. 1923–83.

NBU: *Nouvelle biographie universelle depuis les temps les plus regulés jusqu'à nos jours, avec les renseignements bibliographiques et l'indication des sources a consulter*. Edited by Jean Chrétien Ferdinand Hoefer. 46 vols. in 23. Paris: Firmin Didot Frères. 1852–66.

NDB: *Neue deutsche Biographie*. Under the auspices of the Historical Commission of the Bavarian Academy of Sciences. 20 vols. (A–Püterich) to date. Berlin: Duncker & Humblot. 1953–.

Neighbour, Alfred. 1865. *The apiary; or, bees, bee-hives, and bee culture: being a familiar account of the habits of bees, and the most improved methods of management, with full directions, adapted for the cottager, farmer, or scientific apiarian*. London: Kent & Co.; Geo. Neighbour & Sons.

Newman, William. 1993. Darwin and cirripedology. *History of Carcinology. Crustacean Issues* 8: 349–434.

New school room map of South Africa: *A new school room map of South Africa containing Cape Colony, Natal, South African Republic, Orange Free State . . . and other territories*. New and revised edition. Cape Town: J. C. Juta & Co. 1888.

Newspaper press directory: *The newspaper press directory and advertiser's guide. The newspaper press directory . . . A directory of the class papers and periodicals*. London: C. Michell. 1856–1900.

Newton, Alfred. 1860. On some hybrid ducks. *Proceedings of the Zoological Society of London* 28: 336–9.

Newton Alfred. 1863. On an illustration of the manner in which birds may occasionally aid in the dispersion of seeds. [Read 21 April 1863.] *Proceedings of the Zoological Society of London* (1863): 127–9.

——. 1864. On the irruption of Pallas's sand-grouse (*Syrrhaptes paradoxus*) in 1863. *Ibis* 6: 185–222.

——. 1865a. On some recently discovered bones of the largest known species of Dodo (Didus nazarenus, Bartlett). [Read 14 February 1865.] *Proceedings of the Zoological Society of London* (1865): 199–201.

——. 1865b. On a remarkable discovery of Didine bones in Rodriguez. [Read 28 November 1865.] *Proceedings of the Zoological Society of London* (1865): 715–18.

——. 1866. Note on an interesting discovery of Didine bones in Mauritius. *Ibis* n.s. 2: 128.

Newton, Edward. 1865. On a remarkable discovery of bones of *Didus* in the Island of Rodriguez. *Report of the thirty-fifth meeting of the British Association for the Advancement of Science, held at Birmingham*, Transactions of the sections, p. 92.

North, John S. 1997. *The Waterloo directory of English newspapers and periodicals, 1800–1900*. 10 vols. Waterloo, Ontario: North Waterloo Academic Press.

Notebooks: *Charles Darwin's notebooks, 1836–1844. Geology, transmutation of species, metaphysical enquiries.* Transcribed and edited by Paul H. Barrett *et al.* Cambridge: Cambridge University Press for the British Museum (Natural History). 1987.

Nowak, Ronald M. 1999. *Walker's mammals of the world.* 6th edition. 2 vols. Baltimore and London: The Johns Hopkins University Press.

NUC: *The national union catalog. Pre-1956 imprints.* 685 vols. and supplement (69 vols.). London and Chicago: Mansell. 1968–81.

Nyhart, Lynn K. 1995. *Biology takes form. Animal morphology and the German universities, 1800–1900.* Chicago and London: University of Chicago Press.

O'Brien, Charles F. 1970. *Eozoön Canadense*: 'the dawn animal of Canada'. *Isis* 61: 206–23.

O'Byrne, William R. 1849. *A naval biographical dictionary: comprising the life and services of every living officer in Her Majesty's Navy, from the rank of admiral of the fleet to that of lieutenant, inclusive.* London: John Murray.

OED: *The Oxford English dictionary. Being a corrected re-issue with an introduction, supplement and bibliography of a new English dictionary.* Edited by James A. H. Murray *et al.* 12 vols. and supplement. Oxford: Clarendon Press. 1970. *A supplement to the Oxford English dictionary.* 4 vols. Edited by R. W. Burchfield. Oxford: Clarendon Press. 1972–86. *The Oxford English dictionary.* 2d edition. 20 vols. Prepared by J. A. Simpson and E. S. C. Weiner. Oxford: Clarendon Press. 1989. *Oxford English dictionary additional series.* 3 vols. Edited by John Simpson *et al.* Oxford: Clarendon Press. 1993–7.

Olby, Robert. 1963. Charles Darwin's manuscript of pangenesis. *British Journal for the History of Science* 1: 251–63.

——. 1985. *Origins of Mendelism.* 2d edition. Chicago and London: University of Chicago Press.

Olby, Robert. 1993. Constitutional and hereditary disorders. In *Companion encyclopedia of the history of medicine*, edited by William F. Bynum and Roy Porter. 2 vols. New York: Routledge.

Oldham, J. Basil. 1952. *A history of Shrewsbury School 1552–1952*. Oxford: Basil Blackwell.

Oldroyd, David Roger. 1980. *Darwinian impacts: an introduction to the Darwinian revolution*. Milton Keynes: Open University Press.

O'Leary, Patrick. 1989. *Sir James Mackintosh: the Whig Cicero*. Aberdeen: Aberdeen University Press.

Oliver, Daniel. 1862. The Atlantis hypothesis in its botanical aspect. *Natural History Review* n.s. 2: 149–70.

[——.] 1864. Dimorphic flowers. *Natural History Review* n.s. 4: 243–8.

——. 1868–77. *Flora of tropical Africa*. 3 vols. London: L. Reeve and Co.

——. 1881. *Lessons in elementary botany. The part of systematic botany based upon material left in manuscript by the late Professor Henslow*. New edition. London: Macmillan and Co.

Oppenheim, Janet. 1985. *The other world: spiritualism and psychic research in England, 1850–1914*. Cambridge: Cambridge University Press.

Orbigny, Alcide Charles Victor Dessalines d'. 1835–47. *Voyage dans l'Amérique Méridionale (le Brésil, la République orientale de l'Uruguay, la République Argentine, la Patagonie, la République du Chili, la République de Bolivia, la République du Pérou), exécuté pendant les années 1826 . . . 1833*. 6 vols. in 7 and 4 atlases. Paris and Strasbourg: Pitois-Levrault et Cie, P. Bertrand.

Orchids: *On the various contrivances by which British and foreign orchids are fertilised by insects, and on the good effects of intercrossing*. By Charles Darwin. London: John Murray. 1862.

Orchids 2d ed.: *On the various contrivances by which British and foreign orchids are fertilised by insects, and the good effects of intercrossing*. By Charles Darwin. Revised. London: John Murray. 1877.

Origin: *On the origin of species by means of natural selection, or the preservation of favoured races in the struggle for life*. By Charles Darwin. London: John Murray. 1859.

Origin 2d ed.: *On the origin of species by means of natural selection, or the preservation of favoured races in the struggle for life*. By Charles Darwin. London: John Murray. 1860.

Origin 3d ed.: *On the origin of species by means of natural selection, or the preservation of favoured races in the struggle for life*. By Charles Darwin. With additions and corrections. London: John Murray. 1861.

Origin 4th ed.: *On the origin of species by means of natural selection, or the preservation of favoured races in the struggle for life*. By Charles Darwin. With additions and corrections. London: John Murray. 1866.

Origin 5th ed.: *On the origin of species by means of natural selection, or the preservation of favoured races in the struggle for life*. By Charles Darwin. With additions and corrections. London: John Murray. 1869.

Origin 6th ed.: *The origin of species by means of natural selection, or the preservation of favoured races in the struggle for life.* By Charles Darwin. With additions and corrections. London: John Murray. 1872.

Origin US ed.: *On the origin of species by means of natural selection, or the preservation of favoured races in the struggle for life.* By Charles Darwin. A new edition, revised and augmented by the author. New York: D. Appleton. 1860.

Ornithological notes: Darwin's ornithological notes. Edited by Nora Barlow. *Bulletin of the British Museum (Natural History). Historical Series* 2 (1959–63): 203–78.

Ospovat, Dov. 1977. Lyell's theory of climate. *Journal of the History of Biology* 10: 317–39.

——. 1981. *The development of Darwin's theory. Natural history, natural theology, and natural selection, 1838–1859.* Cambridge: Cambridge University Press.

Osten-Sacken, Charles Robert. 1864. Ueber den wahrscheinlicher Dimorphismus der Cynipiden-Weibchen, *Entomologische Zeitung* 25: 409–13.

Owen, Alex. 1989. *The darkened room: women, power and spiritualism in late Victorian England.* London: Virago Press.

Owen, Richard. 1838. Description of some fossil remains of *Chæropotamus*, *Palæotherium*, *Anoplotherium*, and *Dichobunes*, from the Eocene formation, Isle of Wight. [Read 7 November 1838.] *Transactions of the Geological Society of London* 2d ser. 6 (1841–2): 41–5.

——. 1849. *On parthenogenesis, or the successive production of procreating individuals from a single ovum.* London: John van Voorst.

——. 1850. On the development and homologies of the molar teeth of the wart-hogs (*Phacochœrus*), with illustrations of a system of notation for the teeth in the class Mammalia. [Read 7 February 1850.] *Philosophical Transactions of the Royal Society of London* 1850 (pt 1): 481–98.

——. 1852. Teeth. In vol. 4 of *The cyclopædia of anatomy and physiology*, edited by Richard B. Todd. 4 vols. and suppl. London: Sherwood, Gilbert, and Piper; Longman, Brown, Green, and Longman. 1836–59.

——. 1857. On the characters, principles of division, and primary groups of the class Mammalia. [Read 17 February and 21 April 1857.] *Journal of the Proceedings of the Linnean Society (Zoology)* 2 (1858): 1–37.

[——.] 1860. [Review of *Origin* & other works.] *Edinburgh Review* 111: 487–532.

——. 1862. On the characters of the aye-aye, as a test of the Lamarckian and Darwinian hypothesis of the transmutation and origin of species. *Report of the 32d meeting of the British Association for the Advancement of Science, held at Cambridge*, Transactions of the sections, pp. 114–16.

——. 1866. On the osteology of the Dodo (Didus ineptus *Linn.*). [Read 9 January 1866.] *Transactions of the Zoological Society of London* 6 (1869): 49–85.

Owen, Richard Startin. 1894. *The life of Richard Owen . . . With the scientific portions revised by C. Davies Sherborn; also an essay on Owen's position in anatomical science by the Right Hon. T. H. Huxley, F.R.S.* 2 vols. London: John Murray.

Oxford classical dictionary. 3d edition. Edited by Simon Hornblower and Antony Spawforth. Oxford: Oxford University Press. 1996.

Paez, Ramon. 1863. *Wild scenes in South America, or life in the llanos of Venezuela.* 2d edition. New York: Charles Scribner.
Paget, James. 1853. *Lectures on surgical pathology delivered at the Royal College of Surgeons of England.* 2 vols. London: Longman, Brown, Green, and Longmans.
——. 1859. On the chronometry of life. [Read 8 April 1859.] *Proceedings of the Royal Institution of Great Britain* 3 (1858–62): 117–24.
Paine, Thomas. 1794. *The age of reason: being an investigation of true and fabulous theology.* Paris: printed by Barrois. London: sold by D. I. Eaton.
——. 1795. *The age of reason. Part the second: being an investigation of true and of fabulous theology.* London: printed for H. D. Symonds.
Pakula, Hannah. 1996. *An uncommon woman. The Empress Frederick: daughter of Queen Victoria, wife of the crown prince of Prussia, mother of Kaiser Wilhelm.* London: Weidenfeld & Nicolson.
Palgrave, William Gifford. 1865. *Narrative of a year's journey through central and eastern Arabia (1862–63).* 2 vols. London and Cambridge: Macmillan and Co.
Pallas, Pyotr Simon. 1780. Mémoire sur la variation des animaux; première partie. *Acta Academiæ Scientiarum Imperialis Petropolitanæ* 4, pt 2: 69–102.
Pancaldi, Giuliano. 1991. *Darwin in Italy. Science across cultural frontiers.* Translated by Ruey Brodine Morelli. Updated and expanded edition. Bloomington and Indianapolis, Ind.: Indiana University Press.
Paradis, James G. 1978. *T. H. Huxley: man's place in nature.* Lincoln and London: University of Nebraska Press.
'Parallel roads of Glen Roy': Observations on the parallel roads of Glen Roy, and of other parts of Lochaber in Scotland, with an attempt to prove that they are of marine origin. By Charles Darwin. [Read 7 February 1839.] *Philosophical Transactions of the Royal Society of London* (1839), pt 1: 39–81. [*Collected papers* 1: 89–137.]
Parmenter, John C. 1981. *District post cancellations 1840–1857.* Edited by Peter Forrestier Smith *et al.* London: London Postal History Group.
Parry, William Edward. 1824. *Journal of a second voyage for the discovery of a north-west passage from the Atlantic to the Pacific; performed in the years 1821–22–23, in His Majesty's ships Fury and Hecla, under the orders of Captain William Edward Parry, R.N., F.R.S., and commander of the expedition.* London: John Murray.
——. 1826. *Journal of a third voyage for the discovery of a north-west passage from the Atlantic to the Pacific; performed in the years 1824–25, in His Majesty's ships Hecla and Fury, under the orders of Captain William Edward Parry, R.N., F.R.S., and commander of the expedition.* London: John Murray.
Payer, Jean Baptiste. 1845. Note sur les vrilles des Cucurbitacées. *Annales des Sciences Naturelles. Botanique* 3d ser. 3: 163–4.
Peckham, Morse, ed. 1959. *The origin of species by Charles Darwin. A variorum text.* Philadelphia: University of Pennsylvania Press.
Pemberton, W. Baring. 1954. *Lord Palmerston.* London: Batchworth Press.
Pennell, Henry Cholmondeley. 1863. *The angler-naturalist: a popular history of British*

fresh-water fish, with a plain explanation of the rudiments of ichthyology. London: John Van Voorst.

Pennell, Henry Cholmondeley. 1865. *The book of the pike; a practical treatise on the various methods of jack fishing.* London: Robert Hardwicke.

Peters, James L., *et al.* 1931–87. *Check-list of birds of the world.* 16 vols. Cambridge, Mass.: Harvard University Press and Museum of Comparative Zoology.

Petteys, Chris. 1985. *Dictionary of women artists. An international dictionary of women artists born before 1900.* Boston, Mass.: G. K. Hall & Co.

Phillips, John. 1865. Address of the president. *Report of the thirty-fifth meeting of the British Association for the Advancement of Science held at Birmingham*, pp. li–lxvii.

Physicians: *The roll of the Royal College of Physicians of London.* By William Munk. 2d edition, revised and enlarged. 3 vols. London: Royal College of Physicians. 1878. *Lives of the fellows of the Royal College of Physicians of London.* Compiled by G. H. Brown *et al.* 5 vols. London: Royal College of Physicians. Oxford and Washington, D.C.: IRL Press. 1955–89.

Planchon, François Gustave. 1864a. *Etude des tufs de Montpellier au point de vue géologique et paléontologique.* Montpellier: Boehm et fils. Paris: F. Savy.

——. 1864b. *Des modifications de la flore de Montpellier depuis le seiziéme siécle jusqu'à nos jours.* Montpellier: Boehm et fils. Paris: F. Savy.

Plant, Marjorie. 1965. *The English book trade. An economic history of the making and sale of books.* 2d edition. London: George Allen & Unwin.

Plarr, Victor Gustave. 1930. *Plarr's lives of the fellows of the Royal College of Surgeons of England.* Revised by Sir D'Arcy Power. 2 vols. London: Simpkin Marshall.

Porter, Duncan M. 1985. The *Beagle* collector and his collections. In *The Darwinian heritage*, edited by David Kohn. Princeton, N.J.: Princeton University Press in association with Nova Pacifica.

——. 1986. Charles Darwin's vascular plant specimens from the voyage of HMS *Beagle. Botanical Journal of the Linnean Society* 93: 1–172.

——. 1987. Darwin's notes on *Beagle* plants. *Bulletin of the British Museum (Natural History), Historical Series* 14: 145–233.

——. 1993. On the road to the *Origin* with Darwin, Hooker, and Gray. *Journal of the History of Biology* 26: 1–38.

Porter, Roy and Rousseau, G. S. 1998. *Gout: the patrician malady.* London and New Haven, Conn.: Yale University Press.

Post Office directory of Cambridgeshire, Derbyshire, Nottinghamshire, Leicestershire, & Rutlandshire. London: Kelly & Co. 1864.

Post Office directory of the county of Durham: *The Post Office directory of the county of Durham, and the principal towns and adjacent places in Northumberland, Cumberland & Westmorland. The Post Office directory of Durham and Northumberland.* London: Kelly and Co. 1873–.

Post Office directory of Gloucestershire, with Bath, Bristol, Herefordshire, and Shropshire. London: Kelly & Co. 1856–79.

Post Office directory of Northumberland: *Post Office directory of Westmoreland, Cumberland, Northumberland, and Durham.* London: Kelly and Co. 1858.

Post Office directory of the six home counties: *Post Office directory of the six home counties, viz., Essex, Herts, Kent, Middlesex, Surrey and Sussex*. London: W. Kelly & Co. 1845–78.

Post Office Edinburgh directory: *Post-office annual directory and calendar. Post-office Edinburgh and Leith directory*. Edinburgh: Ballantyne & Hughes [and others]. 1845–1908.

Post Office London directory: *Post-office annual directory. . . . A list of the principal merchants, traders of eminence, &c. in the cities of London and Westminster, the borough of Southwark, and parts adjacent . . . general and special information relating to the Post Office. Post Office London directory*. London: His Majesty's Postmaster-General [and others]. 1802–1967.

Post Office London suburban directory: *The Post Office London suburban directory. Kelly's London suburban directory*. London: Kelly & Co. 1860–1903.

Powell, Baden. 1855. *Essays on the spirit of the inductive philosophy, the unity of worlds, and the philosophy of creation*. London: Longman, Brown, Green, and Longman.

Poynter, F. N. L. 1950. John Chapman (1821–1894): publisher, physician, and medical reformer. *Journal of the History of Medicine and Allied Sciences* 5: 1–22.

Praeger, Robert Lloyd. 1949. *Some Irish naturalists: a biographical note-book*. Dundalk, Co. Louth: W. Tempest, Dundalgan Press.

Prichard, Iltudus Thomas. 1864. *How to manage it: a novel*. 3 vols. London: R. Bentley.

Pritchard, Michael. 1994. *A directory of London photographers 1841–1908*. Watford, Hertfordshire: PhotoResearch.

Raby, Peter. 2001. *Alfred Russel Wallace: a life*. London: Chatto & Windus.

Radick, Gregory. 2000. Morgan's canon, Garner's phonograph, and the evolutionary origins of language and reason. *British Journal for the History of Science* 33: 3–23.

Radlkofer, Ludwig. 1857. On true parthenogenesis in plants. *Annals and Magazine of Natural History* 2d ser. 20: 204–10.

Ramsay, Andrew Crombie. 1862. On the glacial origin of certain lakes in Switzerland, the Black Forest, Great Britain, Sweden, North America, and elsewhere. [Read 5 March 1862.] *Quarterly Journal of the Geological Society of London* 18: 185–204.

——. 1864. *The physical geology and geography of Great Britain. Six lectures to working men delivered in the Royal School of Mines in 1863*. 2d edition. London: Edward Stanford.

——. 1865. *Sir Charles Lyell and the glacial theory of lake-basins*. London: n.p. [Extracted from *Philosophical Magazine* 29 (1865): 285–98.]

Record of the Royal Society of London: *The record of the Royal Society of London for the promotion of natural knowledge*. 4th edition. London: Royal Society. 1940.

Redfield, William C. 1831. Remarks on the prevailing storms of the Atlantic coast of the the North American states. *American Journal of Science and Arts* 20: 17–51.

——. 1834. Observations on the hurricanes and storms of the West Indies and the coast of the U. States. *American Journal of Science and Arts* 25: 114–121.

Reeve, Lovell Augustus and Walford, Edward, eds. 1863–7. *Portraits of men of eminence in literature, science, and art, with biographical memoirs*. 6 vols. London: Lovell Reeve and Alfred Wm. Bennett.

Registers of Wadham College, Oxford: *The registers of Wadham College, Oxford. (Part II) From 1719 to 1871*. Edited by Robert Barlow Gardiner. London: George Bell. 1895.

Rehbock, Philip. 1975. Huxley, Haeckel, and the oceanographers: the case of *Bathybius haeckelii. Isis* 66: 504–33.

Reichenbach, Heinrich Gottlieb. 1851. *Icones florae Germanicae et Helveticae.* Vols. 13 and 14 (2 vols in 1). Leipzig: Friedrich Hofmeister.

Reptiles: *Reptiles.* Pt 5 of *The zoology of the voyage of HMS Beagle.* By Thomas Bell. Edited and superintended by Charles Darwin. London: Smith, Elder & Co. 1843.

Repton School register: *Repton School register 1557–1910.* Edited for the O. R. Society by the widow of G. S. Messiter. Repton, Derbyshire: A. J. Lawrence. 1910.

'Review of Bates on mimetic butterflies': [Review of 'Contributions to an insect fauna of the Amazon Valley', by Henry Walter Bates.] [By Charles Darwin.] *Natural History Review* n.s. 3 (1863): 219–24. [*Collected papers* 2: 87–92.]

Rheinberger, Hans-Jörg. 1983. Naudin, Darwin, Nägeli: Bemerkungen zu den Vererbungsvorstellungen des 19. Jahrhunderts. *Medizinhistorisches Journal* 18: 198–212.

Richards, Evelleen. 1994. A political anatomy of monsters, hopeful and otherwise. Teratogeny, transcendentalism, and evolutionary theorizing. *Isis* 85: 377–411.

Richards, Robert J. 2001. The linguistic creation of man: Charles Darwin, August Schleicher, Ernst Haeckel, and the missing link in nineteenth-century theory. In *Experimenting in tongues: studies in science and language*, edited by Matthias Doerres. Stanford: Stanford University Press.

Richardson, Edmund William. 1916. *A veteran naturalist; being the life and work of W. B. Tegetmeier.* London: Witherby & Co.

Richardson, John. 1829–37. *Fauna Boreali-Americana; or, the zoology of the northern parts of British America.* Assisted by William Swainson and William Kirby. 4 vols. London and Norwich: John Murray; Richard Bentley; J. Fletcher.

Richardson, Samuel. 1747–8. *Clarissa; or, the history of a young lady.* 7 vols. London: [the author].

Ringer, Sydney. 1869. *A handbook of therapeutics.* London: H. K. Lewis.

Ritvo, Harriet. 1992. Race, breed, and myths of origin: Chillingham cattle as ancient Britons. *Representations* 39: 1–22.

Rodgers, Andrew Denny III. 1968. *'Noble fellow': William Starling Sullivant.* London and New York: Hafner Publishing.

Roe, Shirley A. 1981. *Matter, life, and generation: eighteenth-century embryology and the Haller–Wolff debate.* Cambridge: Cambridge University Press.

Rolle, Friedrich. 1863. *Ch. Darwin's Lehre von der Entstehung der Arten im Pflanzen- und Thierreich in ihrer Anwendung auf die Schöpfungsgeschichte.* Frankfurt: J. C. Hermann.

——. 1866. *Der Mensch, seine Abstammung und Gesittung, im Lichte der Darwin'schen Lehre von der Art-entstehung und auf Grundlage der neuern geologischen Entdeckungen dargestellt.* Frankfurt: J. C. Hermann.

[Rorison, Gilbert.] 1861. *The three barriers: notes on Mr. Darwin's 'Origin of species'.* Edinburgh and London: Blackwood & Sons.

Rose, R. N. 1953. *The Field, 1853–1953.* London: Michael Joseph.

Ross, James Clark. 1847. *A voyage of discovery and research in the southern and Antarctic regions during the years 1839–43*. 2 vols. London: John Murray.

Ross, John. 1835. *Narrative of a second voyage in search of a North-west passage: and of a residence in the Arctic regions during the years 1829, 1830, 1831, 1832, 1833: including the reports of Commander, now Captain, James Clark Ross, R.N., F.R.S., F.L.S., &c. and the discovery of the Northern Magnetic Pole*. London: A. W. Webster.

Royal blue book: *The royal blue book; or fashionable directory and canvassing guide. Royal blue book, fashionable directory. Royal blue book, fashionable directory and parliamentary guide. Royal blue book. Court and parliamentary guide*. London: T. Gardiner and Son [and others]. 1822–1940.

Royal Society catalogue of scientific papers: *Catalogue of scientific papers (1800–1900). Compiled and published by the Royal Society of London*. 19 vols. and index (3 vols.). London: Royal Society of London. Cambridge: Cambridge University Press. 1867–1925.

Royer, Clémence Auguste, trans. 1862. *De l'origine des espèces ou des lois du progrès chez les êtres organisés. Par Ch. Darwin. Avec une préface et des notes du traducteur*. Translated from the 3d English edition. Paris: Guillaumin & Cie; Victor Masson.

——, trans. 1866. *De l'origine des espèces par sélection naturelle ou des lois de transformation des êtres organisés. Par Ch. Darwin. Augmentée d'après des notes de l'auteur*. 2d edition. Translated from the 3d English edition. Paris: Guillaumin & Cie; Victor Masson.

Royle, J. Forbes and Headland, Frederick W. 1865. *A manual of materia medica and therapeutics; including the preparations of the British Pharmacopœia, and many other approved medicines*. 4th edition. London: John Churchill & Sons.

Rudwick, Martin. 1974. Darwin and Glen Roy: a 'great failure' in scientific method? *Studies in History and Philosophy of Science* 5: 97–185.

Rugby School register: *Rugby School register*. Vol. 1, from April 1675 to October 1857. Revised and annotated for the Old Rugbeian Society by Godfrey A. Solly. Rugby: George Over. 1933.

Rupke, Nicolaas A. 1976. *Bathybius Haeckelii* and the psychology of scientific discovery: theory instead of observed data controlled the late 19th century 'discovery' of a primitive form of life. *Studies in History and Philosophy of Science* 7: 53–62.

——. 1994. *Richard Owen, Victorian naturalist*. New Haven, Conn., and London: Yale University Press.

Russell, Edward John. 1966. *A history of agricultural science in Great Britain. 1620–1954*. London: George Allen & Unwin Ltd.

Russell, W. H. 1865. *The Atlantic telegraph*. London: Day & Son.

Russell-Gebbett, Jean. 1977. *Henslow of Hitcham: botanist, educationalist and clergyman*. Lavenham, Suffolk: Terence Dalton.

Rütimeyer, Ludwig. 1861. *Die Fauna der Pfahlbauten in der Schweiz. Untersuchungen über die Geschichte der wilden und der Haus-Säugethiere von Mittel-Europa*. Basel, Switzerland: Bahnmaier's Buchhandlung (C. Detloff).

——. 1863. *Beiträge zur Kenntniss der fossilen Pferde und zu einer vergleichenden Odontographie der Hufthiere im Allgemeinen*. Basel, Switzerland: Schweighauserische Buchdruckerei.

——. 1865. *Beiträge zu einer palæontologischen Geschichte der Wiederkauer, zunächst an Linné's Genus Bos*. Basel, Switzerland: Schweighauserische Buchdruckerei.

Rütimeyer, Ludwig. 1866. Über Art und Raçe des zahmen europäischen Rindes. *Archiv für Anthropologie* 1: 217–50.

——. 1867a. Versuch einer natürlichen Geschichte des Rindes in seinen Beziehungen zu den Wiederkauern im Allgemeinen. Erste Abtheilung. *Neue Denkschriften der allgemeinen schweizerischen Gesellschaft für die gesammten Naturwissenschaften* 22: 1–103.

——. 1867b. Versuch einer natürlichen Geschichte des Rindes in seinen Beziehungen zu den Wiederkauern im Allgemeinen. Zweite Abtheilung. *Neue Denkschriften der allgemeinen schweizerischen Gesellschaft für die gesammten Naturwissenschaften* 22: 1–175.

Sabine, Edward. 1864. Anniversary address. [Read 30 November 1864.] *Proceedings of the Royal Society of London* 13: 497–517.

Sarjeant, William A. S. 1980–96. *Geologists and the history of geology: an international bibliography.* 10 vols. including supplements. London: Macmillan. Malabar, Fla.: Robert E. Krieger Publishing.

Sarton, George. 1942. Second preface to volume 34. Lesquereux (1806–89). *Isis* 34: 97–108.

Say, Thomas. 1823–4. Descriptions of the Coleopterous insects collected in the late expedition to the Rocky Mountains, performed by order of Mr. Calhoun, Secretary of War, under the command of Major Long. *Journal of the Academy of Natural Sciences of Philadelphia* 3 (1823–4): 139–216, 238–82, 298–331, 403–62; 4 (1824): 83–99.

SBL: *Svenskt biografiskt lexikon.* Edited by Bertil Boëthius *et al.* 29 vols. and 4 fascicles of vol. 30 (A–Rudbeck) to date. Stockholm: Albert Bonnier & P. A. Norstedt. 1918–.

Schaum, Hermann Rudolph. 1860. *Naturgeschichte der Insecten Deutschlands. Erste Abtheilung: Coleoptera.* Berlin: Nicolaische Verlagsbuchhandlung.

Scherren, Henry. 1905. *The Zoological Society of London. A sketch of its foundation and development and the story of its farm, museum, gardens, menagerie and library.* London: Cassell.

S[chlechtenda]l, [D. F. L.]. 1864. Personal-Nachrichten [Hermann Crüger]. *Botanische Zeitung* 22: 119–20.

Schleicher, August. 1863. *Die Darwinsche Theorie und die Sprachwissenschaft. Offenes Sendschreiben an Herrn Dr Ernst Häckel.* Weimar, Germany: Hermann Böhlau.

——. 1864. Die Darwin'sche Theorie und die Thier- und Pflanzenzucht. *Stöckhardt's Zeitschrift für deutsche Landwirthe* 15: 1–11.

——. 1869. *Darwinism tested by the science of language.* Translated from the German, with preface and additional notes, by Alex V. W. Bikkers. London: John Camdem Hotten.

Schofield, Robert E. 1963. *The Lunar Society of Birmingham. A social history of provincial science and industry in eighteenth century England.* Oxford: Clarendon Press of Oxford University.

Schönherr, Christoph Joseph. 1806–17. *Synonymia insectorum, oder Versuch einer Synonymie aller bisher bekannten Insecten; nach Fabricci Systema eleutheratorum geordnet.* 4 vols. Stockholm and Upsala: Heinr. A. Nordström [and others].

Schrempp, Gregory. 1983. The re-education of Friedrich Max Müller: intellectual appropriation and epistemological antinomy in mid-Victorian evolutionary thought. *Man* n.s. 18: 90–110.

Sclater, Philip Lutley. 1862. Exhibition of drawings representing the four known species of wild asses. *Proceedings of the Zoological Society of London* (1862): 163–4.

Scoresby, William. 1820. *An account of the Arctic regions, with a history and description of the northern whale-fishery.* 2 vols. Edinburgh: Archibald Constable and Co., Edinburgh, and Hurst, Robinson and Co., London.

Scott, John. 1864a. On the individual sterility and cross-impregnation of certain species of *Oncidium.* [Read 2 June 1864.] *Journal of the Linnean Society* (*Botany*) 8 (1865): 162–7.

——. 1864b. Observations on the functions and structure of the reproductive organs in the *Primulaceæ.* [Read 4 February 1864.] *Journal of the Linnean Society* (*Botany*) 8 (1865): 78–126.

——. 1864c. Notes on the sterility and hybridization of certain species of *Passiflora, Disemma,* and *Tacsonia.* [Read 16 June 1864.] *Journal of the Linnean Society* (*Botany*) 8 (1865): 197–206.

——. 1867. On the reproductive functional relations of several species and varieties of *Verbasca. Journal of the Asiatic Society of Bengal* 36 (pt 2): 145–74.

Secord, Anne. 2002. Botany on a plate: pleasure and the power of pictures in promoting early nineteenth-century scientific knowledge. *Isis* 93: 28–57.

Secord, James A. 1981. Nature's fancy: Charles Darwin and the breeding of pigeons. *Isis* 72: 163–86.

——. 1985. Darwin and the breeders: a social history. In *The Darwinian heritage,* edited by David Kohn. Princeton: Princeton University Press in association with Nova Pacifica.

——. 2000. *Victorian sensation: the extraordinary publication, reception, and secret authorship of* Vestiges of the natural history of creation. Chicago: University of Chicago Press.

Sedgwick, Adam. 1833. *Discourse on the studies of the University of Cambridge.* Cambridge and London: John W. Parker.

——. 1850. *A discourse on the studies of the University of Cambridge.* 5th edition. London: John W. Parker. Cambridge: John Deighton.

Seeley, H. G. 1901. *Dragons of the air.* London: Methuen & Co.

Seemann, Berthold Carl. 1852–7. *The botany of the voyage of HMS Herald . . . during the years 1845–51.* London: Lovell Reeve.

Seville, Catherine. 1999. *Literary copyright reform in early Victorian England: the framing of the 1842 Copyright Act.* Cambridge: Cambridge University Press.

Seward, Anna. 1804. *Memoirs of the life of Dr. Darwin.* London: J. Johnson.

Shaw, George. 1800–26. *General zoology, or a systematic natural history.* 14 vols., vols. 9–14 continued by J. F. Stephens. London: George Kearsley [and others].

Shorter, Clement King. 1913. *George Borrow and his circle.* London: Hodder and Stoughton.

Shrewsbury School register: *Shrewsbury School register. 1734–1908.* Edited by J. E. Auden. Oswestry: Woodall, Minshall, Thomas and Co., Caxton Press. 1909.

Shrosbee, Colin. 1988. *Public schools and private education: the Clarendon Commission, 1861–4, and the Public Schools Acts.* Manchester: Manchester University Press.

Siebold, Karl Theodor Ernst von. 1857. *On a true parthenogenesis in moths and bees; a contribution to the history of reproduction in animals.* Translated by William S. Dallas. London: John van Voorst.

——, trans. 1864. Weitere Erklärungen über die von Prof. Nic. Wagner beschriebene Insectenlarve, welche sich durch Sprossenbildung vermehrt. [Translation, with commentary, of Meinert 1864.] *Zeitschrift für wissenschaftliche Zoologie* 14: 394–9.

Sinclair, John. 1837. *Memoirs of the life and works of the late right honourable Sir John Sinclair, bart.* 2 vols. Edinburgh: William Blackwood and Sons. London: T. Cadell.

Sloan, Phillip R. 2001. 'The sense of sublimity': Darwin on nature and divinity. *Osiris* 2d ser. 16: 251–86.

Smith, Andrew. 1849. *Illustrations of the zoology of South Africa . . . collected . . . in the years 1834, 1835, and 1836.* 4 pts. London: Smith, Elder & Co.

Smith, Charles H., ed. 1991. *Alfred Russel Wallace: an anthology of his shorter writings.* Oxford: Oxford University Press.

Smith, Fabienne. 1990. Charles Darwin's ill health. *Journal of the History of Biology* 23: 443–59.

——. 1992. Charles Darwin's health problems: the allergy hypothesis. *Journal of the History of Biology* 25: 285–306.

Smith, John. 1839. Notice of a plant which produces perfect seeds without any apparent action of pollen. [Read 18 June 1839.] *Transactions of the Linnean Society of London* 18 (1841): 509–12.

Smith, Kenneth G. V. 1987. Darwin's insects: Charles Darwin's entomological notes. *Bulletin of the British Museum (Natural History). Historical series* 14: 1–143.

Smith, Ray F., *et al.* eds. 1973. *History of entomology.* Palo Alto: Annual Reviews.

SMK: *Svenska män och kvinnor biografisk uppslagsbok.* Edited by Nils Bohman *et al.* 8 vols. Stockholm: Albert Bonniers. 1942–55.

Sorby, Henry Clifton. 1856. On the microscopical structure of mica-schist. *Report of the 26th meeting of the British Association for the Advancement of Science held at Cheltenham*, Transactions of the sections, p. 78.

Sorensen, W. Conner. 1995. *Brethren of the net. American entomology, 1840–1880.* Tuscaloosa and London: University of Alabama Press.

South America: *Geological observations on South America. Being the third part of the geology of the voyage of the Beagle, under the command of Capt. FitzRoy RN, during the years 1832 to 1836.* By Charles Darwin. London: Smith, Elder & Co. 1846.

'Specific difference in *Primula*': On the specific difference between *Primula veris*, Brit.

Fl. (var. *officinalis* of Linn.), *P. vulgaris*, Brit. Fl. (*var. acaulis*, Linn.), and *P elatior*, Jacq.; and on the hybrid nature of the common oxlip. With supplementary remarks on naturally produced hybrids in the genus *Verbascum*. By Charles Darwin. [Read 19 March 1868.] *Journal of the Linnean Society* (*Botany*) 10 (1869): 437–54.

Spencer, Herbert. 1860–2. *First principles*. London: George Manwaring; Williams & Norgate.

——. 1864–7. *The principles of biology*. 2 vols. London: Williams & Norgate.

——. 1904. *An autobiography*. 2 vols. London: Williams and Norgate.

Sprengel, Christian Konrad. 1793. *Das entdeckte Geheimniss der Natur im Bau und in der Befruchtung der Blumen*. Berlin: Freidrich Vieweg.

Spruce, Richard. 1908. *Notes of a botanist on the Amazon & Andes, being records of travel . . . during the years 1849–1864*. Edited by Alfred Russel Wallace. 2 vols. London: Macmillan and Co.

Stace, Clive A. 1980. *Plant taxonomy and biosystematics*. London: Edward Arnold.

Stafford, Robert A. 1989. *Scientist of empire. Sir Roderick Murchison, scientific exploration and Victorian imperialism*. Cambridge: Cambridge University Press.

Stansbury, Howard. 1852. *An expedition to the valley of the Great Salt Lake of Utah: including a description of its geography, natural history, and minerals, and an analysis of its waters: with an authentic account of the Mormon settlement*. Philadelphia: Lippincott, Grambo & Co.

Statutes and rules for the British museum: *Statutes and rules for the British Museum and the British Museum* (*Natural History*). *Made by the Trustees in pursuance of the act of incorporation 26, Geo. II, cap. 22, para. xv*. London: Trustees of the British Museum. 1932.

Stearn, William T. 1956. Bentham and Hooker's *Genera plantarum*: its history and dates of publication. *Journal of the Society for the Bibliography of Natural History* 3 (1953–60): 127–32.

——. 1981. *The Natural History Museum at South Kensington: a history of the British Museum* (*Natural History*), *1753–1980*. London: Heinemann in association with the British Museum (Natural History).

Stebbing, E. P. 1922–6. *The forests of India*. 3 vols. London: John Lane, The Bodley Head.

Steenstrup, Johannes Japetus Smith and Lubbock, John. 1866. On the flint implements recently discovered at Pressigny-le-Grand. [Read 26 June 1866.] *Transactions of the Ethnological Society of London* n.s. 5 (1867): 221–7.

Stenhouse, John. 1990. Darwin's captain: F. W. Hutton and the nineteenth-century Darwinian debates. *Journal of the History of Biology* 23: 411–42.

Stenton, Michael. 1976. *Who's who of British members of Parliament, 1832–1885*. Brighton: Hanover Press.

Stephens, John Lloyd. 1837. *Incidents of travel in Egypt, Arabia, Petaea, and the Holy Land*. 2 vols. New York: Harper and brothers.

——. 1841. *Incidents of travel in Central America, Chiapas, and Yucatan*. 2 vols. New York: Harper and brothers.

Stephens, John Lloyd. 1843. *Incidents of travel in Yucatan.* 2 vols. London: John Murray.
Stevens, Peter F. 1994. *The development of biological systematics: Antoine-Laurent de Jussieu, nature, and the natural system.* New York: Columbia University Press.
Stewart, John. 1990. *Antarctica: an encyclopedia.* 2 vols. Jefferson, N.C., and London: McFarland & Co.
Stocking, George Ward, Jr. 1971. What's in a name? The origins of the Royal Anthropological Institute (1837–71). *Man* n.s. 6: 369–90.
——. 1987. *Victorian anthropology.* New York: The Free Press. London: Collier Macmillan.
Stoddart, D. R. 1976. Darwin, Lyell, and the geological significance of coral reefs. *British Journal for the History of Science* 9: 199–218.
Stokes, John Lort. 1846. *Discoveries in Australia.* 2 vols. London: T. & W. Boone.
Storrie, Margaret. 1997. *Islay. Biography of an island.* 2d edition. Islay: Oa Press.
Stresemann, Erwin. 1975. *Ornithology: from Aristotle to the present.* Translated by Hans J. and Cathleen Epstein. Edited by G. William Cottrell. With foreword and epilogue by Ernst Mayr. Cambridge, Mass., and London: Harvard University Press.
Strick, James. 2000. *Sparks of life: Darwinism and the Victorian debates over spontaneous generation.* Cambridge, Mass.: Harvard University Press.
Strickland, Hugh Edwin, *et al.* 1863. *Rules for zoological nomenclature . . . authorised by section D of British Association at Manchester, 1842.* Edinburgh: British Association for the Advancement of Science.
Sulivan, Henry Norton, ed. 1896. *Life and letters of the late Admiral Sir Bartholomew James Sulivan, KCB, 1810–1890.* London: John Murray.
Sullivan, Alvin, ed. 1984. *British literary magazines: the Victorian and Edwardian age, 1837–1913.* London and Westport, Conn.: Greenwood Press.
Swinhoe, Robert. 1864. Descriptions of four new species of Formosan birds; with further notes on the ornithology of the island. *Ibis* 6: 361–70.
Taxonomic literature: *Taxonomic literature. A selective guide to botanical publications and collections with dates, commentaries and types.* By Frans A. Stafleu and Richard S. Cowan. 2d edition. 7 vols. Utrecht: Bohn, Scheltema & Holkema. The Hague: W. Junk. 1976–88.
Taylor, Tom. 1869. *Our American cousin. A drama in three acts.* N.p.: privately printed.
Tegetmeier, William Bernhard. 1854. *Profitable poultry; their management in health and disease.* 2d edition. London: Darton and Co.
——. 1856. On the very remarkable peculiarities existing in the skulls of the feathercrested variety of the domestic fowl, now known as the Polish. *Proceedings of the Zoological Society of London* 24: 366–8.
——. 1867. *The poultry book: comprising the breeding and management of profitable and ornamental poultry, their qualities and characteristics; to which is added 'The standard of excellence in exhibition birds,' authorized by the Poultry Club.* London and New York: George Routledge & Sons.

Tennyson, Alfred. 1864. *Enoch Arden, etc.* London: Edward Moxon.

Thom's Irish almanac: *Thom's Irish almanac and official directory of the United Kingdom of Great Britain and Ireland.* Dublin: Alexander Thom. 1846–80.

Thomson, George Malcolm. 1884–5. Frederick Wollaston Hutton. *The New Zealand Journal of Science* 2: 301–6.

[Thomson, Thomas.] 1865. Species and subspecies. *Natural History Review* 5: 226–42.

'Three forms of *Lythrum salicaria*': On the sexual relations of the three forms of *Lythrum salicaria.* By Charles Darwin. [Read 16 June 1864.] *Journal of the Linnean Society* (*Botany*) 8 (1865): 169–96. [*Collected papers* 2: 106–31.]

Thwaites, George Henry Kendrick. 1858–64. *Enumeratio plantarum Zeylaniæ: an enumeration of Ceylon plants, with descriptions of the new and little-known genera and species, observations on their habits, uses, native names, etc.* Assisted in the identification of the species and synonymy by J. D. Hooker. 5 pts. London: William Paplin; Dulau & Co.

Times atlas: *'The Times' atlas of the world. Comprehensive edition.* 9th edition. London: Times Books. 1992.

Todd, Dennis. 1995. *Imagining monsters. Miscreations of the self in eighteenth-century England.* Chicago and London: University of Chicago Press.

Tolles, Winton. 1940. *Tom Taylor and the Victorian drama.* New York: Columbia University Press.

Topham, Jonathan Richard. 1993. 'An infinite variety of arguments': the *Bridgewater treatises* and British natural theology in the 1830s. PhD dissertation. University of Lancaster.

Tort, Patrick. 1996. *Dictionnaire du Darwinisme et de l'evolution.* 3 vols. Paris: Presses Universitaires de France.

Travers, Henry Hammersley. 1864. Notes on the Chatham Islands (lat. 44°30' S. long. 175° W.). [Read 3 November 1864.] *Journal of the Linnean Society* (*Botany*) 9 (1867): 135–44.

Treasures from India: *Treasures from India. The Clive collection at Powis Castle.* By Mildred Archer *et al.* London: Herbert in association with The National Trust. 1987.

Treviranus, Ludolph Christian. 1863a. Ueber Dichogamie nach C. C. Sprengel und Ch. Darwin. *Botanische Zeitung* 21: 1–7, 9–16.

——. 1863b. Nachträgliche Bemerkungen über die Befruchtung einiger Orchideen. *Botanische Zeitung* 21: 241–3.

Trimen, Roland. 1864. On the structure of *Bonatea speciosa*, Linn. sp., with reference to its fertilisation. [Read 1 December 1864.] *Journal of the Linnean Society* (*Botany*) 9 (1867): 156–60.

——. 1866. Notes on the butterflies of Mauritius. [Read 3 September 1866.] *Transactions of the Entomological Society of London* 3d ser. 5 (1865–7): 329–44.

Tristram, Henry Baker. 1864a. On a bone breccia with flints in Lebanon. *Report of the thirty-fourth meeting of the British Association for the Advancement of Science, held at Bath*, Transactions of the sections, pp. 72–3.

Tristram, Henry Baker. 1864b. On the sulphur and bitumen deposit at the south-west corner of the Dead Sea. *Report of the thirty-fourth meeting of the British Association for the Advancement of Science, held at Bath*, Transactions of the sections, p. 73.

——. 1864c. On the physical and political geography of the Jordan Valley and eastern Palestine. *Report of the thirty-fourth meeting of the British Association for the Advancement of Science, held at Bath*, Transactions of the sections, p. 148.

Trollope, Anthony. 1864–5. *Can you forgive her?* 2 vols. London: Chapman & Hall.

Tucker, Jennifer. 1996. Voyages of discovery on oceans of air: scientific observation and the image of science in an age of 'balloonacy'. In *Science in the field*, edited by Henrika Kuklick and Robert E. Kohler. (*Osiris* 2d ser. 11: 144–76.)

Tuke, Margaret J. 1939. *A history of Bedford College for Women, 1849–1937*. London: Oxford University Press.

Turnbull, Constance Mary. 1989. *A history of Singapore 1819–1988*. 2d edition. Oxford: Oxford University Press.

Turner, John M. 1995. *Victorian arena. The performers. A dictionary of British circus biography. Volume one.* Formby, England: Lingdales Press.

Turrill, W. B. 1953. *Joseph Dalton Hooker. Botanist, explorer, and administrator.* London: Thomas Nelson and Sons.

'Two forms in species of *Linum*': On the existence of two forms, and on their reciprocal sexual relation, in several species of the genus *Linum*. By Charles Darwin. [Read 5 February 1863.] *Journal of the Proceedings of the Linnean Society (Botany)* 7 (1864): 69–83. [*Collected papers* 2: 93–105.]

Two Shropshire ironmasters: *Two Shropshire ironmasters. Biographical extracts, letters and memoranda relating to Richard & William Reynolds.* Compiled by Shropshire County Library, n.d.

Twyman, Michael. 1998. *The British Library guide to printing: history and techniques.* London: British Library.

Tylor, Edward Burnett. 1865. *Researches into the early history of mankind.* London: John Murray.

——. 1870. *Researches into the early history of mankind.* 2d edition. London: John Murray.

[Tyndall, John.] 1864. Science and the spirits. *Reader* 4: 725–6.

Unger, Franz. 1860. *I. Die versunkene Insel Atlantis. II. Die physiologische Bedeutung der Pflanzencultur. Zwei Vorträge gehalten im Ständehause im Winter des Jahres 1860*. Vienna: Wilhelm Braumüller.

Uschmann, Georg. 1959. *Geschichte der Zoologie und der zoologischen Anstalten in Jena 1779–1919*. Jena: VEB Gustav Fischer.

——. 1984. *Ernst Haeckel. Biographie in Briefen.* Gütersloh: Prisma Verlag.

Van Riper, A. Bowdoin. 1993. *Men among the mammoths: Victorian science and the discovery of human prehistory.* Chicago and London: University of Chicago Press.

Variation: *The variation of animals and plants under domestication.* By Charles Darwin. 2 vols. London: John Murray. 1868.

Vaucher, Jean Pierre Etienne. 1841. *Histoire physiologique des plantes d'Europe ou exposition des phénomènes qu'elles présentent dans les diverses périodes de leur développement*. 4 vols. Paris: Marc Aurel Frères.

Verlot, Bernard. 1864. Mémoire sur la production et la fixation des variétés dans les plantes d'ornement. *Journal de la Société Impériale et Centrale d'Horticulture* 10: 243–56, 305–20, 375–84, 420–32, 468–80, 518–28, 571–6, 624–40.

——. 1865. *Sur la production et la fixation des variétés dans les plantes d'ornement*. Paris: J. B. Baillière.

Victoria county history of Shropshire: *The Victoria county history of Shropshire*. Edited by G. C. Baugh *et al.* 7 vols. to date (1–4, 8, 10, 11). London and Oxford. 1908–.

Vogt, Carl. 1864. *Lectures on man: his place in creation, and in the history of the earth*. Edited by James Hunt. London: Longman, Green, Longman, and Roberts.

Volcanic islands: *Geological observations on the volcanic islands, visited during the voyage of HMS Beagle, together with some brief notices on the geology of Australia and the Cape of Good Hope. Being the second part of the geology of the voyage of the Beagle, under the command of Capt. FitzRoy RN, during the years 1832 to 1836*. By Charles Darwin. London: Smith, Elder & Co. 1844.

'Volcanic phenomena and the formation of mountain chains': On the connection of certain volcanic phenomena in South America; and on the formation of mountain chains and volcanos, as the effect of the same power by which continents are elevated. By Charles Darwin. [Read 7 March 1838.] *Transactions of the Geological Society of London* 2d ser. 5 (1840): 601–31. [*Collected papers* 1: 53–86.]

Wagner, Nikolai Petrovich. 1862. *Samoproizvol'noe razmnozhenie gusenits u nasekomykh* [Spontaneous conception in insects (by caterpillars)]. Kazan: n.p.

——. 1863. Beitrag zur Lehre von der Fortpflanzung der Insectenlarven. *Zeitschrift für wissenschaftliche Zoologie* 13: 513–27.

——. 1865. Uber die viviparen Gallmückenlarven. *Zeitschrift für wissenschaftliche Zoologie* 15: 106–15.

Walford, Edward. 1868. *Representative men in literature, science and art*. London: A. W. Bennett.

Walker, Francis. 1838. Descriptions of some Chalcidites discovered by C. Darwin, Esq. *Entomological Magazine* 5: 469–77.

——. 1839. *Monographia Chalciditum*. London: Hyppolitus Bailliere.

——. 1842–3. Descriptions of Chalcidites discovered by C. Darwin, Esq. *Annals and Magazine of Natural History* 10 (1842): 113–17, 271–4; 11 (1843): 30–2, 115–17, 184–8; 12 (1843): 45–6.

Wallace, Alfred Russel. 1853. *A narrative of travels on the Amazon and Rio Negro, with an account of the native tribes, and observations on the climate, geology, and natural history of the Amazon valley*. London: Reeve.

——. 1856a. Some account of an infant 'Orang-utan.' *Annals and Magazine of Natural History* 2d ser. 17: 386–90.

——. 1856b. On the orang-utan or mias of Borneo. *Annals and Magazine of Natural History* 2d ser. 17: 471–6.

Wallace, Alfred Russel. 1856c. On the habits of the orang-utan of Borneo. *Annals and Magazine of Natural History* 2d ser. 18: 26–32.

——. 1862. List of birds from the Sula Islands (east of Celebes), with descriptions of the new species. [Read 13 January 1862.] *Proceedings of the Zoological Society of London* (1862): 333–46.

——. 1863a. List of birds collected in the Island of Bouru (one of the Moluccas), with descriptions of the new species. [Read 13 January 1863.] *Proceedings of the Zoological Society of London* (1863): 18–36.

——. 1863b. *On the physical geography of the Malay Archipelago.* London: W. Clowes and Sons. [Read 8 June 1863. Reprinted from *Journal of the Royal Geographical Society* 33: 217–34.]

——. 1863c. A list of birds inhabiting the Islands of Timor, Flores, and Lombock, with descriptions of the new species. [Read 24 November 1863.] *Proceedings of the Zoological Society of London* (1863): 480–97.

——. 1864a. The origin of human races and the antiquity of man deduced from the theory of 'natural selection'. [Read 1 March 1864.] *Journal of the Anthropological Society of London* 2: clvii–clxx.

——. 1864b. On the parrots of the Malayan region, with remarks on their habits, distribution, and affinities, and the descriptions of two new species. [Read 28 June 1864.] *Proceedings of the Zoological Society of London* (1864): 272–95.

——. 1864c. On the varieties of man in the Malay Archipelago. [Read 26 January 1864.] *Transactions of the Ethnological Society of London* 3: 196–215.

——. 1864d. On the phenomena of variation and geographical distribution as illustrated by the *Papilionidæ* of the Malayan region. [Read 17 March 1864.] *Transactions of the Linnean Society of London* 25 (1865–6): 1–71.

——. 1865. List of the land shells collected by Mr Wallace in the Malay Archipelago, with descriptions of the new species by Mr Henry Adams. [Read 25 April 1865.] *Proceedings of the Zoological Society of London* (1865): 405–16.

——. 1869. *The Malay Archipelago: the land of the orang-utan, and the bird of paradise. A narrative of travel, with studies of man and nature.* 2 vols. London: Macmillan and Co.

——. 1905. *My life: a record of events and opinions.* 2 vols. London: Chapman & Hall.

Wallace, Robert, ed. 1899. *A country schoolmaster: James Shaw.* Edinburgh: Oliver and Boyd.

Waller, Jessie Oriana. 1889. Mental and physical training of children. *Nineteenth Century* 26: 659–67.

Walsh, Benjamin Dann. 1863. Observations on certain North American Neuroptera, by H. Hagen, M.D., of Kœnigsberg, Prussia; translated from the original French MS., and published by permission of the author, with notes and descriptions of about twenty new species of North American Pseudoneuroptera. *Proceedings of the Entomological Society of Philadelphia* 2: 167–272.

——. 1864a. On dimorphism in the hymenopterous genus *Cynips*, with an appendix, containing hints for a new classification of Cynipidæ and a list of Cynipidæ, including descriptions of several new species, inhabiting the oak-galls of Illinois. *Proceedings of the Entomological Society of Philadelphia* 3: 443–500.

Walsh, Benjamin Dann. 1864b. On the insects, coleopterous, hymenopterous and dipterous, inhabiting the galls of certain species of willow. *Proceedings of the Entomological Society of Philadelphia* 3: 543–644.

——. 1864c. On certain entomological speculations of the New England school of naturalists. *Proceedings of the Entomological Society of Philadelphia* 3: 207–49.

——. 1864–5. On phytophagic varieties and phytophagic species. *Proceedings of the Entomological Society of Philadelphia* 3: 403–30 and 5: 194–216.

——. 1865. The new potato-bug, and its natural history. *Practical Entomologist* 1: 1–4.

——. 1866. On the insects, coleopterous, hymenopterous and dipterous, inhabiting the galls of certain species of willow.– Part 2d and last. *Proceedings of the Entomological Society of Philadelphia* 6: 223–88.

Walters, Stuart Max and Stow, E. Anne 2001. *Darwin's mentor. John Stevens Henslow 1796–1861*. Cambridge: Cambridge University Press.

Ward, Thomas Humphry. 1926. *History of the Athenaeum, 1824–1925*. London: printed for the Athenaeum Club.

Warner, Donald Frederick. 1960. *The idea of continental union: agitation for the annexation of Canada to the United States, 1849–1893*. Lexington, Ky.: Mississippi Valley Historical Association; University of Kentucky Press.

Waterhouse, George Robert. 1845. Descriptions of coleopterous insects collected by Charles Darwin, Esq., in the Galapagos Islands. *Annals and Magazine of Natural History* 16: 19–41.

Watson, Hewett Cottrell. 1845. On the theory of 'progressive development,' applied in explanation of the origin and transmutation of species. *Phytologist* 2: 108–13, 140–7, 161–8, 225–8.

Waugh, Francis Gledstanes. [1888.] *Members of the Athenæum Club 1824 to 1887*. N.p.: privately printed.

Weale, James Philip Mansel. 1865. Natural history in Natal. *Natural History Review* n.s. 5: 145–6.

——. 1867. Notes on the structure and fertilization of the genus *Bonatea*, with a special description of a species found at Bedford, South Africa. [Read 7 March 1867.] *Journal of the Linnean Society (Botany)* 10 (1869): 470–6.

——. 1870a. Notes on a species of *Disperis* found on the Kagaberg, South Africa. [Read 3 November 1870.] *Journal of the Linnean Society (Botany)* 13 (1873): 42–5.

——. 1870b. Notes on some species of *Habenaria* found in South Africa. [Read 3 November 1870.] *Journal of the Linnean Society (Botany)* 13 (1873): 47–8.

——. 1870c. Observations on the mode in which certain species of Asclepiadeæ are fertilized. [Read 3 November 1870.] *Journal of the Linnean Society (Botany)* 13 (1873): 48–58.

——. 1870d. Some observations on the fertilization of *Disa macrantha*. [Read 23 November 1870.] *Journal of the Linnean Society (Botany)* 13 (1873): 47–8.

——. 1891. *The truth about the Portuguese in Africa*. London: Swan Sonnenschein.

Wedgwood, Barbara and Wedgwood, Hensleigh. 1980. *The Wedgwood circle, 1730–1897: four generations of a family and their friends.* London: Studio Vista.

Wedgwood, Hensleigh. 1859–65. *A dictionary of English etymology.* 3 vols. in 4. London: Trübner & Co.

Wellesley index: *The Wellesley index to Victorian periodicals 1824–1900.* Edited by Walter E. Houghton *et al.* 5 vols. Toronto, Ontario: University of Toronto Press. London: Routledge & Kegan Paul.

Wells, Kentwood D. 1973. William Charles Wells and the races of man. *Isis* 64: 215–25.

Wells, Stanley and Taylor, Gary, eds. 1988. *William Shakespeare: the complete works.* Compact edition. Oxford: Clarendon Press.

Wells, William Charles. 1818. *Two essays: one upon single vision with two eyes; the other on dew. A letter to the Right Hon. Lloyd, Lord Kenyon and an account of a female of the white race of mankind, part of whose skin resembles that of a negro; with some observations on the causes of the differences in colour and form between the white and negro races of men.* London: Archibald Constable and Co. [and others].

Whitaker, Joseph. 1868. *Almanack for the year of our Lord 1869.* London: J. Whitaker.

White, Adam. 1841. Descriptions of new or little known Arachnida. *Annals and Magazine of Natural History* 7: 101–28.

White, Charles. 1799. *An account of the regular gradation in man, and in different animals and vegetables; and from the former to the latter.* London: C. Dilly.

Wichura, Max Ernst. 1865. *Die Bastardbefruchtung im Pflanzenreich erläutert an den Bastarden der Weiden.* Breslau: E. Morgenstern.

[Wilberforce, Samuel.] 1860. [Review of *Origin.*] *Quarterly Review* 108: 225–64.

Willis, Robert and Clark, John Willis. 1886. *The architectural history of the University of Cambridge, and of the colleges of Cambridge and Eton.* 4 vols. Cambridge: University Press.

Wilson, Dudley. 1993. *Signs and portents. Monstrous births from the Middle Ages to the Enlightenment.* London and New York: Routledge.

Wilson, Leonard Gilchrist, ed. 1970. *Sir Charles Lyell's scientific journals on the species question.* New Haven, Conn., and London: Yale University Press.

Wilson, Leonard Gilchrist. 1996. Brixham Cave and Sir Charles Lyell's . . . *the Antiquity of man*: the roots of Hugh Falconer's attack on Lyell. *Archives of Natural History* 23: 79–97.

——. 2002. A scientific libel: John Lubbock's attack upon Sir Charles Lyell. *Archives of Natural History* 29: 73–87.

Wimmer, Christian Friedrich Heinrich. 1853. Wildwachsende Bastardpflanzen, hauptsächlich in Schlesien beobachtet. In *Denkschrift zur Feier ihres fünfzigjährigen Bestehens herausgegeben von der Schlesischen Gesellschaft für vaterländische Kultur.* Breslau: Joseph Max & Komp.

Wingfield, William and Johnson, George William. 1853. *The poultry book; comprising the characteristics, management, breeding, and medical treatment of poultry; being the results of personal observation and the practice of the best breeders, including Captain W. W. Hornby,*

R N.; Edward Bond, Esq.; Thomas Sturgeon, Esq.; Charles Punchard, Esq.; Edward Hewitt, Esq.; and others. London: Wm. S. Orr and Co.

Wingfield, William and Johnson, George William. 1856–7. *The poultry book; comprising the characteristics, management, breeding, and medical treatment of poultry; being the results of personal observation and the practice of the best breeders, including Captain W. W. Hornby, R N.; Edward Bond, Esq.; Thomas Sturgeon, Esq.; Charles Punchard, Esq.; Edward Hewitt, Esq.; and others.* Another edition. Re-arranged and edited by W. B. Tegetmeier. London: Wm. S. Orr and Co.

Winsor, Mary Pickard. 1976. *Starfish, jellyfish and the order of life: issues in nineteenth-century science.* New Haven and London: Yale University Press.

——. 1979. Louis Agassiz and the species question. *Studies in History of Biology* 3: 89–117.

——. 1991. *Reading the shape of nature: comparative zoology at the Agassiz museum.* Chicago: University of Chicago Press.

Wollaston, Alexander Frederick Richmond. 1921. *Life of Alfred Newton, professor of comparative anatomy, Cambridge University, 1866–1907.* With a preface by Sir Archibald Geikie. London: John Murray.

Wollaston, Thomas Vernon. 1856. *On the variation of species with especial reference to the Insecta; followed by an inquiry into the nature of genera.* London: John van Voorst.

Wood, Ellen (Mrs Henry). 1864. *Trevlyn Hold; or, Squire Trevlyn's heir.* London: Tinsley Brothers.

Woodcock, George. 1969. *Henry Walter Bates: naturalist of the Amazons.* London: Faber & Faber.

Wright, Edward Perceval. 1864. On a new genus of *Teredininae. Transactions of the Linnean Society* 24: 451–4.

——. 1865. On a new genus of *Alcyonidae. Quarterly Journal of Microscopical Science* 2d ser. 5: 213–217.

Wright, Terence R. 1986. *The religion of humanity: the impact of Comtean positivism on Victorian Britain.* Cambridge: Cambridge University Press.

WWW: *Who was who: a companion to Who's who, containing the biographies of those who died during the period* [*1897–1995*]. 9 vols. and cumulated index (1897–1990). London: Adam & Charles Black. 1920–96.

Wyman, Jeffries. 1854. Cause of contractility in some vegetable tissues. [Read 8 November 1854.] *Proceedings of the American Academy of Arts and Sciences* 3 (1852–7): 167–8.

Yaldwyn, John and Hobbs, Juliet, eds. 1998. *My dear Hector. Letters from Joseph Dalton Hooker to James Hector, 1862–1893.* Museum of New Zealand Te Papa Tongarewa Technical Report 31.

Yarrell, William. 1836. *A history of British fishes.* 2 vols. London: John Van Voorst.

——. 1837–43. *A history of British birds.* 37 pts in 3 vols. London: John Van Voorst.

Yonge, Charlotte Mary. 1856. *Ben Sylvester's word.* London: J. and C. Mozley.

Youatt, William. 1834. *Cattle: their breeds, management, and diseases.* London: Baldwin and Cradock.

Zoology: *The zoology of the voyage of HMS Beagle, under the command of Captain FitzRoy RN, during the years 1832 to 1836*. Edited and superintended by Charles Darwin. 5 pts. London: Smith, Elder & Co. 1838–43.

NOTES ON MANUSCRIPT SOURCES

The majority of the manuscript sources cited in the footnotes to the letters are located either in the Darwin Archive, Cambridge University Library, or at Down House, Downe, Kent. Further details about the Darwin Archive are available in the *Handlist of Darwin papers at the University Library Cambridge* (Cambridge: Cambridge University Press, 1960) and the unpublished supplementary handlist available at the library; a new catalogue of the papers is currently being prepared. Further details about the manuscripts at Down House are available in Philip Titheradge, ed. *The Charles Darwin Memorial at Down House, Downe, Kent*, revised ed. ([Downe: Down House Museum], 1981) and from the curator (The Curator, Down House, Downe, Kent, BR6 7JT). In addition, there are a number of named sources that are commonly used in the footnotes: for each of these, the editors have provided brief descriptive notes.

CD's Account books (Down House MS). This series of seventeen account books begins on 12 February 1839, a fortnight after CD and Emma's marriage, and ends with CD's death. The books contain two sets of accounts. From the start, CD recorded his *cash account* according to a system of double-entry book-keeping. On each left-hand page he recorded credits (i.e., withdrawals from the bank, either in the form of cash paid to himself or cheques drawn for others), and on each right-hand page, he recorded debits (i.e., cash or cheques paid to others). CD also recorded details of his *banking account* from the start, but only noted them down in a single column at the bottom of the left-hand page of his cash account. In August 1848, however, he began a system of detailing his banking account according to double-entry book-keeping, in a separate chronological section at the back of each account book. On the left, he recorded credits to the account in the form of income (i.e., investments, rent, book sales, etc.). On the right, he recorded debits to the account (i.e., cash or cheque withdrawals).

CD's Classed account books (Down House MS). This series of four account books, covering the years 1839–81, runs parallel to CD's Account books. For each year, September – August (after 1867, January – December), CD divided his expenditure into different classified headings; in addition, he made a tally for the year of his income, expenditure, cash in hand, and money in the bank. From 1843, CD also compiled at the back of each book a separate account of the total expenditure under the various headings in each year, and from 1844 he added a full account of his income in each year, and of capital invested and 'paid' up. The classification of expenditure varied over time, but for 1864–5 the classes were: 'Miscellaneous', 'Personal', 'Governess & Girls School & Teachers',

'House', 'Boys & William', 'Coals & Firewood', 'Beer', 'Wine', 'Man-servants', 'Repairs, Furniture', 'Medical attendance', 'Stationery, newspapers postage', 'Books', 'Science', 'Gardens', 'Cow, Pigs, Field', 'Stable', 'Travelling Expences', 'Taxes Rates', 'Charities', and 'Gifts, Annual Subscriptions'.

CD's Experimental notebook (DAR 157a). This notebook contains notes on some, though by no means all of the experiments carried out between 13 November 1855 (with some back references) and 20 May 1868, although the majority of the notes date from before 1863. In many cases only the details of the experiment attempted are given, often with cross-references to results recorded in CD's portfolios of notes. The notebook also contains a number of letters to CD.

CD's Investment book (Down House MS). This book records under each of CD's investments the income received during the period 1846–81.

CD's 'Journal'. *See* Appendix II.

CD's Library catalogue (DAR 240). This manuscript catalogue of CD's scientific library was compiled by Thomas W. Newton, assistant librarian of the Museum of Practical Geology, London, in August 1875. Additions to the catalogue were subsequently made by Francis Darwin (who inherited most of his father's scientific library) and by H. W. Rutherford, who apparently used this catalogue as a basis for compiling his *Catalogue of the library of Charles Darwin now in the Botany School, Cambridge* (Cambridge: Cambridge University Press, 1908). However, there are items listed in this manuscript catalogue that do not appear in Rutherford's published catalogue, and which must have been dispersed subsequently to being listed.

Darwin's copy of the *Gardeners' Chronicle*: account of numbers and index to marks & memoranda (DAR 222). This is a bound manuscript index to CD's collection of the *Gardeners' Chronicle and Agricultural Gazette*, 1841–71 (which is at the Cory Library, Cambridge Botanic Garden). It comprises a list showing which numbers of the journal are missing from CD's copy, together with a 'List of the numbers of special interest to Darwin and kept by him in separate parcels'. The latter details the page-numbers and subjects of the articles that were of particular interest to (and annotated by) CD in the numbers that he kept separately. The handwriting of the list is unknown, but probably dates from the late nineteenth century.

Down Coal and Clothing Club account book (Down House MS). CD was for some years treasurer of this charitable organisation. The account book records subscriptions made by honorary subscribers between 1841–76; between 1848 and 1869 the entries are in CD's handwriting. For the years 1841–8 and 1868–76, there is also a statement of expenditures, though not in CD's handwriting.

Emma Darwin's diary (DAR 242). This comprises a series of small pocket diaries, in which Emma recorded details of the health of family members, trips made by herself, CD, and the children, school holidays, and visits to Down by others. The collection at CUL comprises diaries for the years 1824, 1833–4, 1839–45, and 1848–96.

H. E. Litchfield's autobiography (DAR 246). This unfinished autobiography, written in 1926 on forty-two loose leaves, and chiefly concerning Henrietta Emma Darwin's childhood, has never been published.

List of pamphlets (DAR 252.4). This is a catalogue of CD's pamphlet collection prepared by CD and Francis Darwin in 1878 (see the letter from Emma Darwin to Henrietta Emma Litchfield, [June 1878] (DAR 219.9: 175)). From about 1878 CD began to arrange the articles, papers, and reprints he received into a numbered collection. CD maintained this reprint collection until his death, when it was taken over by Francis Darwin. Francis continued the collection, adding new items, the numbers running consecutively from those of his father. Evidently, until this catalogue was prepared, CD used a working index similar to that of his 'List of reviews'. The catalogue is in two sections, a list of the quarto collection and one of the general collection. Both sections are alphabetically arranged with the entries pasted on sheets in a loose-leaf folder.

List of reviews (DAR 262.8: 9–18 (English Heritage MS: 88206151–60)). This manuscript, headed 'List Reviews of Origin of Sp & of C. Darwins Books', was CD's working index to his collection of the reviews of his own books. It corresponds approximately to the review collection in the Darwin Pamphlet Collection–CUL, but includes some items that were dispersed subsequently to being listed.

Reading notebooks. *See Correspondence* vol. 4, Appendix IV. These notebooks comprise sections entitled 'Books Read' and 'Books to be Read'. CD's entries in 'Books Read' often include a brief opinion of the work.

Scrapbook of reviews (DAR 226.1 and 226.2). Many of the reviews contained in these two volumes bear CD's annotations and thus were evidently collected by CD. However, the scrapbook seems to have been assembled by Francis Darwin: the tables of contents are in the handwriting of H. W. Rutherford, an assistant at Cambridge University Library who acted as a copyist for Francis on several occasions (see *ML*, 1: x, and Francis Darwin, ed. *The foundations of the Origin of Species. Two essays written in 1842 and 1844 by Charles Darwin* (Cambridge: Cambridge University Press, 1909)). In addition, the scrapbook is identified as Francis's in a note (DAR 226.1:132a) made in 1935 by Arthur Keith, whose appeal led to the purchase of Down House as a Darwin memorial (see Arthur Keith, *An autobiography* (London: Watts & Co., 1950)). DAR 226.1 bears the inscription 'Reviews of C. Darwin's works' on the spine, and contains, among others, reviews of *Origin* and *Orchids*; DAR 226.2 is inscribed: 'Reviews. Descent. Expression. Insect. Pl. Eras. D.'

W. E. Darwin's botanical notebook (DAR 117). This notebook contains observational and experimental notes on plants made by William, often in consultation with CD. The first observation bears the date 13 July 1862, and, although the date of the last observation is 26 June 1870, most of the notes were made between 1862 and 1864. The notebook originally contained letters from CD, but these were subsequently removed. William entered notes made from botanical textbooks in a separate notebook (DAR 234).

W. E. Darwin's botanical sketchbook (DAR 186: 43). This sketchbook, which contains entries dated 1862–72, was evidently begun in parallel to William's botanical notebook. It contains ink drawings of various parts of plants, and of sections, together with descriptions, which in some cases are very extensive.

INDEX

The dates of letters to and from Darwin's correspondents are listed in the Biographical register and index to correspondents and are not repeated here. Darwin's works are indexed under the short titles used throughout this volume and listed in the bibliography.

Tabl

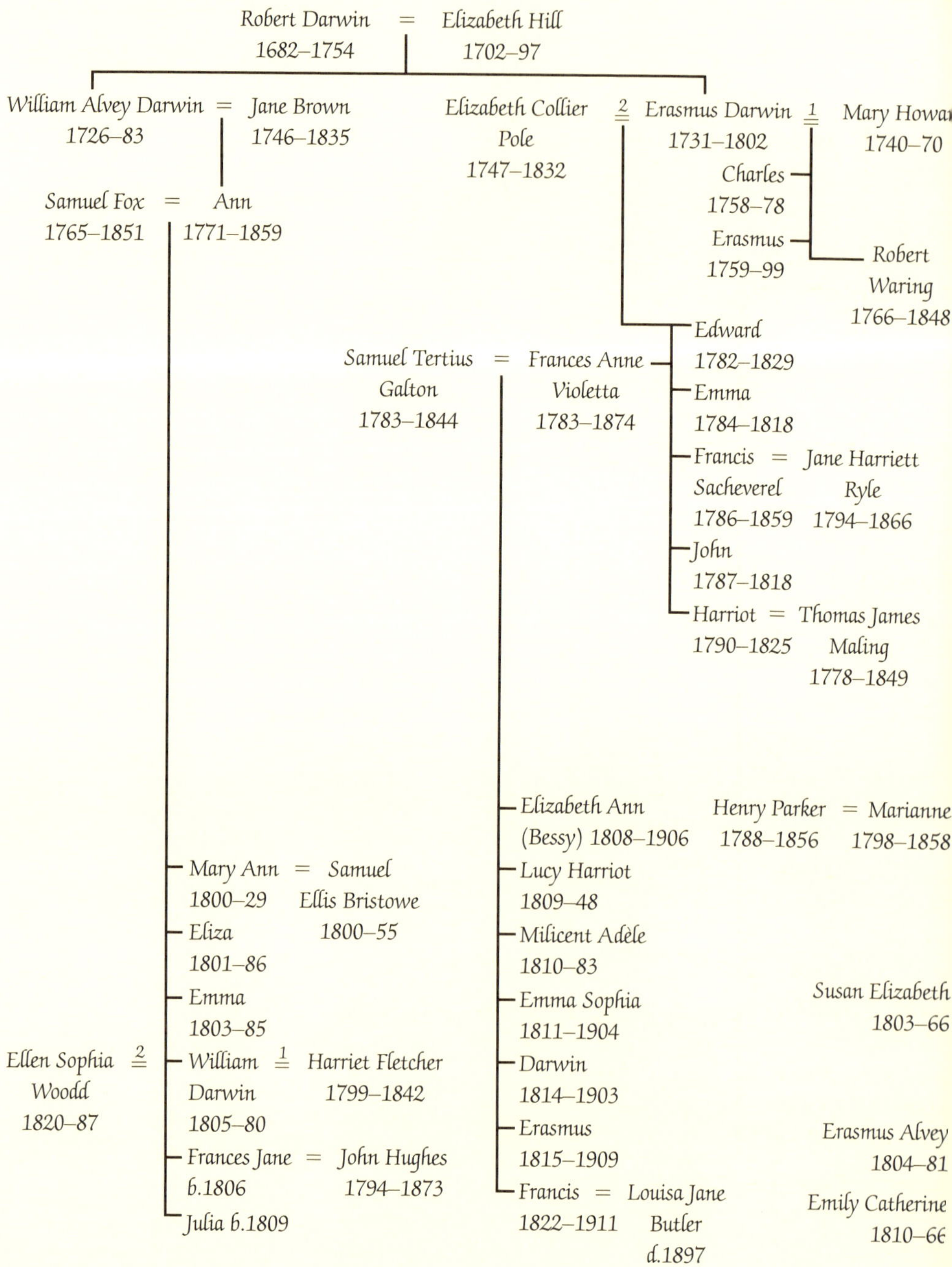
Robert Darwin = Elizabeth Hill
1682–1754
1702–97
William Alvey Darwin = Jane Brown
1726–83
1746–1835
Elizabeth Collier Pole 1747–1832
2
Erasmus Darwin 1731–1802
1
Mary Howa
1740–70
Samuel Fox = Ann
1765–1851
1771–1859
Charles 1758–78
Erasmus 1759–99
Robert Waring 1766–1848
Edward 1782–1829
Samuel Tertius Galton 1783–1844 = Frances Anne Violetta 1783–1874
Emma 1784–1818
Francis Sacheverel 1786–1859 = Jane Harriett Ryle 1794–1866
John 1787–1818
Harriot 1790–1825 = Thomas James Maling 1778–1849
Elizabeth Ann (Bessy) 1808–1906
Henry Parker 1788–1856 = Marianne 1798–1858
Mary Ann 1800–29 = Samuel Ellis Bristowe 1800–55
Lucy Harriot 1809–48
Eliza 1801–86
Milicent Adèle 1810–83
Emma 1803–85
Emma Sophia 1811–1904
Susan Elizabeth 1803–66
Ellen Sophia Woodd 1820–87
2
William Darwin 1805–80
1
Harriet Fletcher 1799–1842
Darwin 1814–1903
Erasmus 1815–1909
Erasmus Alvey 1804–81
Frances Jane b.1806 = John Hughes 1794–1873
Francis 1822–1911 = Louisa Jane Butler d.1897
Emily Catherine 1810–66
Julia b.1809